SERIES ENTOMOLOGICA

EDITOR

E. SCHIMITSCHEK
GÖTTINGEN

VOLUMEN 4

SPRINGER-SCIENCE+BUSINESS MEDIA, B.V. 1968

ECOLOGY AND BIOGEOGRAPHY OF HIGH ALTITUDE INSECTS

BY

Prof. Dr. M. S. MANI, M. A., D. Sc.,
Of the School of Entomology, St. John's College, Agra

SPRINGER-SCIENCE+BUSINESS MEDIA, B.V. 1968

ISBN 978-90-481-8511-5 ISBN 978-94-017-1339-9 (eBook)
DOI 10.1007/978-94-017-1339-9

Originally published by Dr. W. Junk N.V. Publishers, The Hague in 1968
Softcover reprint of the hardcover 1st edition 1968

Printed by Belinfante N.V.

Dedicated to my daughter
Prema
in affectionate token of her
discovery of the Coccinellid
Mass Assemblage on the Himalaya
a discovery that stimulated
interest in high altitude
entomology

CONTENTS

PREFACE

In my book *Introduction to High Altitude Entomology*, published in 1962, I summarized the results of eight years' studies, mainly on the Himalaya. I have since then had the opportunity of studying the collections of high altitude insects from the Alps, Carpathians, Caucasus, Urals, Alai-Pamirs, Tien Shan, Altai and other important mountains of the world in different museums and institutions in Europe. Through the courtesy and generosity of the Academy of Sciences of the Union of Soviet Socialist Republics, I was also able to personally collect insects and make valuable field observations on the Caucasus, the Alai-Pamirs, Ala-Tau and the Tien Shan mountains. Through comparative studies I have tried to synthesize the fundamental principles of high altitude entomology.

I have described here the distinctive characters of the high altitude environment, the ecological specializations of the high altitude insects, their ecological interrelations and the outstanding peculiarities of their biogeography. I have also presented here an outline of the high altitude entomology of the principal mountains of the world, with brief accounts of their orogeny, geology and vegetation. This book differs from all other contributions in the field in its comparative ecological approach and in the fact that the main emphasis is throughout on the evolution of the high altitude ecosystem as an integral part of the orogeny. High mountains are, in all parts of the world, important and independent centres of origin and differentiation of distinctive and highly specialized ecosystems and faunas. In addition to discussing the characteristic insects, I have also dealt with other Arthropods like Arachnida, Chilopoda and Diplopoda, the species of which are among the dominant members of diverse high altitude communities throughout the world. In these discussions, the problem of synonymies of species in such a diversified group of Arthropods has presented insurmountable difficulties; I have, therefore, largely followed widely used names of genera and species.

The book is divided into sixteen chapters. The first chapter discusses the concept of high altitude and defines the high altitude insects, on the basis of their origin and differentiation on mountains. The next three chapters deal with the ecology of high altitudes and high altitude insects. The fifth chapter surveys the major groups of high altitude insects and other related Arthropods in systematic order. The peculiarities of distribution of insects on mountains are summarized in the sixth chapter. The outstanding characters of the tropical mountains are discussed in the seventh and eight chapters. The ninth and the tenth chapters are devoted respectively to the Himalaya and the Pamirs-Tien Shan Systems. Then follows a chapter on the Alps and the Carpathians and the twelfth chapter gives the salient facts about boreo-alpine insects, in so far as they relate to the high altitude entomology of the European mountains. The other mountains of the Palaearctic Realm and the North American mountains are dealt with in the next two chapters. The fifteenth chapter is devoted to the subarctic and arctic mountains and the last chapter to the south-temperate and Antarctic mountains. The insect life of most of these mountains is only very imperfectly known and there are large gaps in our knowledge of the taxonomy, ecology and biogeography of even the better explored regions of the world. It is hoped, however, that this book will stimulate further research in an extremely interesting and little explored field of biology.

There is at the end of the book a bibliography of over 1100 titles, arranged alphabetically authorwise and numbered serially; the serial numbers are quoted in the text. The bibliography does not claim to be exhaustive, and includes only the works which are relevant to the main theme of the book. The book is illustrated by eighty figures, most of which are original and the sources of the others are suitably acknowledged.

I take this opportunity of expressing my heartfelt thanks to the authorities of the various institutions in India and Europe for facilities for my work. I am particularly indebted to the Institute of Zoology, Academy of Sciences, USSR, Leningrad, for facilities for work, donations of specimens and important literature and numerous other courtesies. Prof. OLGA L. KRYZHANOVSKY of the Institute of Zoology, Leningrad, most generously placed at my disposal his valuable collections of Coleoptera from the Alai-Pamirs. The celebrated Soviet botanist, Academician Prof. E. P. KOROVIN, Moscow, a few weeks before his death and despite his rapidly deteriorating health, discussed with me in considerable detail the salient features of the ecology of the Middle Asiatic mountains. I am greatly indebted to Prof. S. M. YABLAKOV-KHNZORYAN of the Institute of Zoology, Academy of Sciences, Armianskii SSR, Erivan, for valuable advice and discussions on the high altitude Coleoptera of the Armenian mountains. Dr. VLADIMIR ALEXANDROVICH TRJPITZIN of the Institute of Zoology, Leningrad, helped me greatly in collecting insects on the Caucasus. I am extremely thankful to Dr. IGOR BELOLIPOV, Botanic Garden, Academy of Sciences, Uzbek SSR, Tashkent, for identifications of the subalpine and alpine zone plants from the Pamirs, Ala-Tau and the Tien Shan and for his enthusiastic help in collecting many unique specimens of insects on the Ala-Tau and the Central Tien Shan mountains. The Director, Institute of Botany, Azerbaijan Academy of Sciences, Baku, donated a number of herbarium specimens of the alpine-zone plants from the Eastern Caucasus. My sincere thanks are due to my colleague Dr. K. K. TIWARI, Zoological Survey of India, Calcutta, for his willing help in collecting insects, particularly subalpine and alpine Lepidoptera, on the Trans-Ili Ala-Tau and other mountains of the Central Tien Shan. I am extremely grateful to Prof. Dr. E. SCHIMITSCHEK, the Editor of the Series Entomologica, for many valuable suggestions and for his personal interest in this book. Last and by no means the least, my thanks are also due to Mr. S. GHOSAL, Librarian, Zoological Survey of India, Calcutta, for numerous courtesies.

M. S. MANI

CHAPTER I

INTRODUCTION

The high altitude insects, in contra-distinction to the lowland forms, occur exclusively at high elevations on mountains. They are essentially inhabitants of high mountains. As, however, insects are found in great abundance and diversity, continuously from the sea-shore to the foot of mountains and thence upward, without a break, to almost within the shadow of some of the highest mountain summits in the world, it is not always easy to determine whether a given insect is a true high altitude species. The difficulty is in part also due to the vagueness of the expression "high altitude".

I. THE CONCEPT OF HIGH ALTITUDE

When we speak of high altitude, we assume elevation in relation to the surrounding lowland, which is of course variable, or the land that might theoretically exist at sea-level at the latitude of the mountain, which is unfortunately misleading. The socalled "mean sea-level" is itself more or less in the nature of a myth. As is well known, the sea-level varies considerably from place to place and from time to time (1056). Even the mean of observations extending over several years and the simultaneous means of the level vary appreciably from place to place. The British Ordnance Survey's Datum level at Newlyn in Cornwall, observed during the period 1915–1921 is ordinarily used in the United Kingdom. Other countries have naturally different means for the sea-level, observed over long periods of time. A variety of complex factors like the tides, ocean currents, glacial climatic changes, crustal movements and other unknown cosmic events underlie the changes in sea-level. The eustatic sea-level has, for example, undergone very significant changes, often amounting to several hundred metres, consequent on the unloading and melting of the inland ice-mass at the end of the Pleistocene. Recent evidence seems to indicate that the Antarctic ice-mass is increasing, so that the sea-level as a whole is falling all over the world. Even if we ignore these changes over such long periods as too small to be of practical importance, particularly in high altitude biology, we are nevertheless faced with another difficulty. It does not seem to have been satisfactorily determined so far as to how high a region must be above the general level of the surrounding lowland in order to be considered as high altitude. For example, some workers consider elevations of even 1500 m as high altitude, but others refer to mountain regions above an elevation of 2290 m as high altitude (648). As we shall see further on, altitude as such is really of very little biological significance, except as merely an indirect measure of certain complex ecological conditions and specializations in organisms. The concept of high altitude is also biologically meaningful only when considered in correlation to the presence of a relatively dense and moisture-laden atmosphere and the other complex conditions under which organisms have evolved on our planet.

2. MOUNTAINS AS INSECT HABITAT

Regardless of the limits that define high altitude, it must be emphasized here that the high altitude insects belong to the immense ecosystem in which the mountains are habitats. The terms mountains and hills are, however, only relative and cannot be precisely expressed in units of altitude. A hill is defined, for example, as a high mass of land, less than a mountain and a mountain is a high hill. Their use is often more or less arbitrary. Mt. Washington in the eastern United States of America is, for example, not so high as the Black Hill, but it is called, by general consent, a mountain. Again, the Rocky Mountains of America are nowhere so high as most of the "hills" in India. Aside from altitude, an essential feature that distinguishes a mountain from a plateau is the relatively limited width at the summit. Mountain masses do not generally stand alone, but form parts of irregular groups (for example, the Adirondack Mountains of the eastern United States) or of large regular belts, extending more or less unbroken over vast areas, often the entire length of a continent, and constituting ranges, systems and chains. A mountain range is a complex or a series of related and more or less continuous ridges. A mountain system is a group of mountain ranges of similar form, structure, alignment and origin. We have for example, the mighty Alpine System or the Himalayan System. The expression "mountain chain" is applied to any elongate mountain unit of several groups and systems, regardless of similarity and relationship. V. HUMBOLDT called a series of mountain chains, systems and ranges that form a more or less compact elevated area of vast extent in America as the cordillera.

As mountains are the habitats of the high altitude insects, it is perhaps useful to refer to certain fundamental, though well known, facts about the structure and uplift of mountains. Broadly speaking, mountains are of two types, viz. the tectonic or original mountains and the relict or subsequent mountains. The tectonic mountains are the result of piling up of material at the surface of the earth or of subterranean action leading to folding and rupturing of the earth's crust. The relict mountains are merely the residual portions of former highlands. They are relicts of more or less elevated tracts, which have been gradually reduced in extent and largely subdued by forces of decay. They are truly remnants of former plateaux.

The tectonic mountains are of two major groups: i. the accumulation mountains and ii. the deformation mountains. The accumulation mountains are further subdivided into the volcanic types and the epigene types. The volcanic types include the debris cones, the lava cones of viscous lava, lava cones of liquid lava, composite cones, geyser cones and air volcanoes or the pseudovolcanic mountains. The epigene types comprise the glacial hills and the aeolian hills. The deformation mountains are subdivided into i. the fold mountains, ii. the dislocation mountains and iii. the laccolith mountains.

Volcanoes consist of material ejected from and accumulated around an orifice in the earth's crust. Some of them are composed entirely of rocks, others of sheets of masses of lava and still others of debris in part and lava for the rest. A volcano is, therefore, essentially an opening in the earth's crust, through which heated material is brought from inside to the surface. Nearly all volcanoes assume the form of hills or mountains, often rising to great elevations. As is well known, volcanic mountains are of three general types, viz. the active, dormant and extinct. They are all originally built up by their own eruption, but exhibit at present wide differences in shape, structure and behaviour. The extinct volcanoes, like the Chilean Aconcagua and the Mexican Ixtaccihuatl, do not have either craters or lava cones. They are really peaks of volcanic origin, rather than actual volcanoes. Others like the Mt. Kilimanjaro (fig. 30) and the Chimborazo, though since long cold at their core, still retain all the external features of a typical volcano. Dormant volcanoes, found in the Cascade Range of the United States of America, are cones that have remained since long inactive but their internal mechanism makes them capable of further eruption. Active volcanoes differ greatly in size and other characteristics. The great Hawaiian craters, forming part of the high mountain system, are continuously active, but are seldom really destructive to human civilization. Lower isolated mountains like the Vesuvius and the Pelé (West Indies Island of Martinique) have remained quiescent for centuries, only to break out suddenly in violent eruptive activity. The most cataclysmic volcanic eruption in recent years was unquestionably that of Krakatoa in 1883. Strictly speaking, Krakatoa is not a mountain at all, so much as a volcanic island, which exploded and all but completely vanished into the sea. Paricutin, which appeared in 1943, is the most spectacular of recent eruptions and did not emanate in a mountain, but simply spewed forth from a flat Mexican corn-field. There are at present nearly four hundred active volcanoes on the earth. As is well known, they are mostly situated fairly close to the sea in regions, where the folding and faulting of the crust are still in active progress. To mention only a few, we have the Vesuvius, the Etna, Stromboli and Iceland in Europe; Erebus and Terror beneath the permanent ice and snow, with the former still in an active state, in the Antarctica; the Pacific shores beginning from the tip of South America, extending the whole length of the cordillera in the Andes of Chile, Peru, Bolivia, Ecuador, Guatemala, Mexico, northern California, Oregon, Washington, Alaska, the Aleutian Islands, Japan, Formosa, Philippines and the East Indies to New Zealand (151, 888, 1053, 1124).

Nearly all the great mountain ranges and mountain chains, of ecological and biogeographical interest to the entomologist, are really the fold mountains. As the name itself implies, they consist mainly of much folded and steeply inclined rock strata. They owe their origin to either one long lateral thrust or two or more repeated thrusts, separated by more or less wide intervals of time, so that the younger folds flank older folds. The relief of a fold mountain chain often coincides more or less closely with the disposition of its rocks. Long parallel ranges and intervening depressions correspond to the undulations or the foldings of the strata. The ridges are formed of anticlines or saddle-backs, with rooflike arrangement of the strata, inclined in opposite directions from a central axis plane. The intervening longitudinal valleys occupy synclines or trough-shaped structures, where the strata dip from opposite directions towards a central axis.

Some of the dislocation mountains also constitute chains and ranges, but they do not equal the great fold mountain systems like the Alps, the Himalaya or the Andes. They owe their origin to the fracturing of the crust, by unequal subsidence of the ground, along lines of vertical displacement. They represent essentially segments of the crust which have maintained their relative position, while the neighbouring tracts have broken away from them and have subsided. The best known dislocation mountain ranges are perhaps the parallel ranges of the Great Basin of North America. We have here a remarkable plateau, surrounded on all sides by lofty mountains, extending north and south for almost 1300 kilometres and stretching east-west between the Sierra Nevada of California and the Wahsatch Mountains for about 800 kilometres. The long parallel ranges, rising to 900-1500 m above their bases, are bounded by normal down-throw faults. The Vosges Mountains in Europe are also true dislocation mountains. The laccolith mountains are the result of subterranean igneous action, the crust being bulged upwards, owing to the pressure of concealed mass of molten rock from below. The Henry Mountains of Utah are typical laccolith mountains.

3. MOUNTAIN INSECTS

High mountains are found on all the continents. Every type of mountain, not excluding even some of the active volcanoes (483, 747, 797), supports a rich and more or less diversified insect life (211, 296, 298, 300, 518, 730, 731, 788, 1021, 1023). Nevertheless, not all the insects that occur on any mountain can be considered as high altitude species. Indeed many typically lowland species of insects and numerous other related Arthropoda often occur at remarkably high elevations on nearly all the mountains in the world. A great many of the insects one often comes across on mountains are either accidental or are regular but nevertheless temporary visitors from the surrounding lowlands and do not breed on the mountains. Incredibly large numbers of insects are lifted from the lowland by warm updraft air currents to high altitudes, to be chilled dead, blown passively and eventually cast on high mountain slopes (730). There are others that are, however, true residents and habitually breed on the mountains and are not as a rule met with on the lowlands. These are the true mountain autochthonous species—the species that have evolved on the mountains. The differentiation and evolution of mountain autochthonous species have conceivably taken place more or less both within the forest that covers the mountain slopes and also at much higher elevations, above the limits of the forest, indeed on barren summits. Sufficiently marked and characteristic ecological differences and specializations, closely correlated with numerous physiological, biogeographical and evolutionary peculiarities, appear among the mountain autochthonous insects at the upper limits of the forest on mountains and serve to distinguish them from those of the montane forest and of the lowland biota. The mountain autochthonous insects that habitually exist at elevations above these limits are the true hypsobiont or high altitude insects.

4. THE FOREST-LINE ON MOUNTAINS: THE THRESHOLD OF HIGH ALTITUDE BIOTA

The upper altitudinal limit of the closed and continuous forest or the socalled forest-line on mountains constitutes a relatively sharply defined boundary between two essentially different classes of ecosystems. The forest-line is ecologically a more or less wide transitional altitudinal zone at which the forest, with its characteristic fauna, gives place to the open vegetation of the elevated areas, with a more or less specialized fauna. The upper limit of the forest on mountains is often called treeline or timber-line, but it must be remembered that a distinctive zone of stunted shrubs lies between the edge of the forest and the typically open vegetation higher above. The forest-line may, therefore, be either a treeline as for example on the Rocky Mountains or a shrub-line as on the Presidential Range of the United States of America.

The upper altitudinal limits of the forest on mountains are subject to considerable variation and depend on whether the mountain is situated in the northern or the southern hemisphere, the latitude of the mountain, the nature of the general climate of the region in which the mountain range is situated, the direction or aspect of the mountain slope but particularly the north-south slope, the presence and arrangement of ridges and valleys, the general trendline direction of the mountain ranges and numerous other complex factors (96, 97, 227, 433, 436, 634-636, 779, 959, 960). The local shifting of the altitude of the forest-line higher or lower is also governed by the atmospheric temperature, atmospheric humidity, the extent and duration of the winter snow-cover on the mountain, the prevailing winds, soil temperature and moisture, the gradient of the mountain slope and finally by the massiveness of the mountain range. The forest-line altitude is, for example, higher on massive mountains than in the case of isolated mountains. A close parallelism exists on the Alps and some other mountains of the northern hemisphere between the timber-line altitude and the mean midday atmospheric temperature of the vegetative season, from May to September. In most parts of the world, the timber-line coincides with the altitude of 10-12° C isotherm of the mean midday temperatures during May-September. KÖPPEN (634) found, for example, that the 10° C isotherm and the forest-line coincide both at high altitudes on mountains and at high north latitudes on lowlands.

The timber-line is by no means permanent on any mountain, but oscillates within wide limits with the passage of time. The trees gain sometimes and the forest ascends higher, but they often also lose ground so that the timber-line comes to lie at lower elevations. Complex climatic and other environmental conditions, some of which seem to be more or less indirectly correlated with sunspot cycles, govern these long-term periodical oscillations and non-periodical changes in the timber-line. As mountain building is still active in many parts of the world, even a relatively small increase in the elevation of the mountain shifts the timber-line. Profound changes in and sometimes even permanent recessions of the timber-line on mountains are brought about by attacks of insects in mass outbreaks on the trees at the timber-line. NUORTEVA (796) has, for example, reported a permanent recession of the timber-line on the arctic hill Ailigas (623 m above mean sea-level) at Karigasniema in North Finland. The forest on the slopes of this hill are the subapine *Empetrum-Lichenes-Pleurozium* or the *Empetrum-Myrtillus* formations. The scrub-like *Betula pubescens* EHR. (= *Betula tortuosa* LDB.) on the timber-line suffered complete defoliation as a result of mass outbreaks of the Geometrid *Oporinia autumnata* (BKH.) during 1927, followed also by secondary damage by the stem-borer Coleoptera *Hylecoetes dermestoides* LINN. and *Agrilus viridis* LINN. These outbreaks are particularly most intense at the timber-line, because of the absence of predatory ants and the parasitic Ichneumonids in this zone. Recovery of the scrub vegetation over extensively devastated areas may need no less than one hundred and fifty years, in the course of which favourable climatic conditions may not perhaps continue long, so that the forest-line has suffered a permanent recession in the region.

From the foregoing, it is evident that the forest-line on mountains must differ in different parts of the world and also on different mountains even on the same latitude and hemisphere. In the northern hemisphere, it is fundamentally different both in its altitudinal magnitude, species composition and other ecological characters from that on mountains of the southern hemisphere. On Mt. Elgon, situated almost on the equator, an elevation of 3400 m above mean sea-level is the upper limit of forest, but on the north Fennoscandian mountains it lies at elevations between 400 and 900 m. The forest-line on some of the important mountains of the world is shown in fig. 1. On the Alps it lies at elevations between 2000 and 2100 m on the

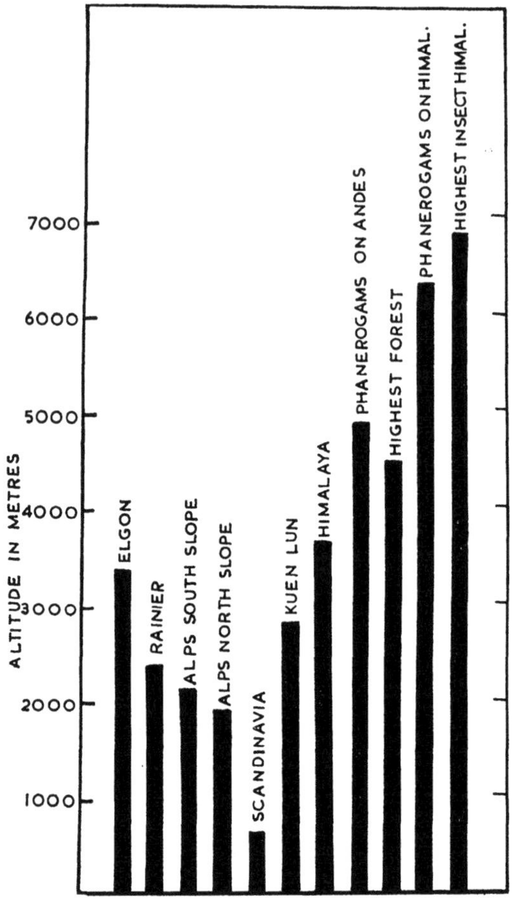

Fig. 1. The upper altitudinal limits of the closed forest on different mountains and the maximum altitudes reached by Phanerogams and insects.

south slope and between 1900 and 1800 m on the north slope. On the Himalaya, the forest-line is at an elevation of about 3620 m. On some parts of the Andes the forest reaches as high as 3690 m. On Mt. Marcy in North America the forest-line is really an irregular zone, about 1500 m above sea-level. In the Pamirs Region the treeline is at 2400 m in Dshailgan, but a zone of shrubs ascends to an elevation of 2800 m and a grassy meadow zone to 3400 m. On the Trans-Alai Mountains the meadow zone is at elevations between 3200 and 3400 m. The highest forests in the world ascend up to an elevation of 4600 m in Tibet. In the New World, at the same latitude, viz. 48° NL, the timber-line varies from 3500 m on the Rocky Mountains to 2000 m on Mt. Rainier. The timber-line on most mountains in the northern hemisphere really indicates the beginning of the socalled alpine vegetation, characterized by *Empetrum*, *Rhododendron lapponicum*, etc. Although the forest disappears at the socalled forest-line, isolated trees and a most remarkable group of Phanerogams grow up to much higher elevations, often above 6000 m above mean sea-level, on the Himalaya. Phanerogams grow at elevations of 5700 m in parts of Tibet, 3780 m on the Alps and at 4870 m on the Andes. On the Bolivian and Chilean Andes, trees (*Polylepis*) grow even at elevations of 4900–5000 m.

5. THE BIOME ABOVE THE FOREST

The biome immediately above the upper limits of forest on mountains is generally called the *alpine zone* in most parts of the world (235, 236, 237, 259, 1084). The alpine zone extends thus from the forest-line up to the permanent snowline, above which lies the socalled *arctic zone*. It is also the general practice to describe the region above the forest as the montane tundra, in contra-distinction to the taiga zone of the conifer belt near the upper limits of the forest on the mountains of the northern hemisphere. The alpine zone is usually further subdivided into a number of subzones, like the subalpine, the eualpine, the high alpine, the subnival, eunival and high nival zones. There is, however, no general uniformity in the names applied for the different subdivisions of the life zones above the forest and in their precise limits. The boundary lines between the various zones are naturally not sharply defined, but show considerable variations on different mountains, depending on a complex of local factors. Sometimes long narrow tongues of the high-alpine zone descend to even within the upper limits of the forest zone. High alpine enclaves may sometimes be found within the upper reaches of the forest, giving rise to peculiar local inversions of the life zone succession (96). Of the seven biotic zones recognized by GAMS (395) on the Central European mountains, four are above the timber-line: i. the nival zone from the highest peaks down to the level of abundant growth of vegetation mats, at elevations between 2900 and 3100 m on the north slopes of the Alps; ii. the subnival zone or the zone of vegetation mats up to the closed meadows, 2400 to 2600 m on the north slope and 2600 to 2750 m on the south slope; iii. the upper alpine zone or the grassy meadow zone, between the closed meadows and the shrub zone, at elevations of 2040 to 2100 m in some localities and 2100 to 2150 m in other localities and 2150 to 2230 m in still other localities; and iv. the lower alpine zone or the shrub zone, reaching as high as 2400 m and as low as 1900 m. The subalpine, the upper montane and the lower montane zones are below the forest-line. HOLDHAUS (526, 531) differentiates a high-alpine zone, extending from the lowest limit of snow patches in July to the upper limits of animal life, from the transitional or the intercalary zone down to the timber-line, corresponding roughly to the lower alpine zone of GAMS. FRANZ (369, 378) does not differentiate between the nival and the subnival zones, but combines the two into a single nival-subnival zone. Others apply the term arctic zone for the nival zone on mountains.

6. HYPSOBIONT INSECTS

The mountain autochthonous insects that exist above the forest-line are generally described as alpine insects. This practice overlooks, however, the existence of a most remarkable group of insects and other related terrestrial arthropods at elevations higher than the strictly alpine zone on mountains. It also ignores the fact that the life zones do not exactly correspond on different mountains. The life zones above the forest are not also strictly comparable on all mountains. The zone immediately above the forest on the Himalaya should really correspond altitudinally at least not with the lower alpine zone but to the eunival zone of the Alps and is therefore in no sense ecologically comparable with the Alps. The elevated areas above the forest in the Pamirs Region or on the Andes cannot also be correctly described as truly alpine areas. The term alpine zone is also quite misleading when applied to the Ethiopian and East African high mountains. Some workers have indeed used the expression Afro-alpine to describe the biotic zone on these mountains (387). Furthermore, we have also seen that the alpine zone may commence at elevations of 400 m or also above 3000 m. The alpine zone commencing at an elevation of 400 m can hardly be described ecologically as high altitude. Considered from the point of view of ecology, biogeography and evolution, high altitude is the region on mountains that is sufficiently elevated above the surrounding lowlands to be characterized by significant climatic differences, different flora and fauna. Pronounced ecological differences and characteristic high altitude specializations appear among the mountain autochthonous insects in the socalled alpine zone that commences at elevations above 2000-2500 m only. We prefer, therefore, the more general and also more comprehensive expression hypsobiont or high altitude insects to alpine insects. The high altitude insects may thus be described as an ecologically highly specialized, mountain autochthonous group existing exclusively in the biome above the forest, at elevations above 2000-2500 m. The upper limit of high altitude insect life extends, at least as at present known, up to elevations of 6800 m—it is perhaps quite conceivable that insects exist even much higher, perhaps on the highest mountain summit on the earth. Defined in this way, the expression high altitude insects embraces the insect life of the subalpine, the eualpine, high-alpine, subnival, eunival, the high nival and higher zones, to the very extreme high elevations at which insects possibly exist on the earth at present. Our criteria in determining a species as high altitude insect are not therefore taxonomic, but exclusively ecological and evolutionary.

We shall attempt to describe in the following chapters the outstanding and distinctive characters of the high altitude environment, the ecological and biogeographical specializations of high altitude insects, the high altitude insect types and the comparative high altitude ecology and entomology of the principal mountains of the world.

CHAPTER II

THE HIGH ALTITUDE ENVIRONMENT

This chapter describes the outstanding properties of high altitude environment, with particular reference to insect life. Considerable emphasis is laid on the fundamental differences between the high altitude and lowland environments. The high altitude environment is also differentiated from the arctic environment, with which it has been compared by earlier workers.

I. THE CONCEPT OF HIGH ALTITUDE ENVIRONMENT

The high altitude environment, compared to that of the lowlands, represents a highly specialized extreme. While in a sense it constitutes a wholly new type, the greatest majority of the environmental factors that influence the high altitude insect life operates also near the sea-level. Strictly speaking, there are exceedingly few new factors at high altitudes on mountains that are not already familiar to us in the montane forest or in the lowland prairie. Their magnitudes, relative importance mutual interactions, effects on insects and their reactions are, however, fundamentally different at high altitudes (730). The key to this difference is altitude. It is altitude that profoundly modifies, either directly or indirectly, every environmental factor, so as to give rise to the specialized environment of the high altitudes on mountains.

It must, however, be recognized that altitude as such is not an ecological factor. Altitude is merely a measure of the mean environmental complex that is distinctive from the prevalent environment of the lowland. The ecological importance of altitude rests, in the final analysis, on the characteristic conditions under which life exists at present on this planet, particularly two major fundamental factors, viz. the presence of a relatively dense atmosphere containing water vapour and the action of gravity on this atmosphere. Under the action of the earth's gravity, the molecules of the component gases of the air tend to be heavily concentrated close to the surface of the lowland, so that the atmosphere is densest at sea-level and becomes progressively less dense as the altitude above the surface of the earth increases. The complex ecological peculiarities that we ordinarily attribute to altitude are in reality the resultant of this action of the earth's gravity on the atmosphere. The high altitude environment is, therefore, essentially the environment of reduced atmosphere. On a waterless satelite like our moon without an atmosphere or also on a planet like Mars with only a relatively thin atmosphere, a high altitude environment, as distinct from the lowland environment, would not possibly exist. The high altitude conditions of the Earth would be prevalent in a rather highly exaggerated condition even at the bottom of the craters on the moon. The general conditions on the summits of the mountains on the moon are not also likely to be ecologically very different from those of the surrounding lowland. The atmospheric pressure on the general surface of Mars is, for example, only 83 millibars and it approximates to that of about 18000 m above mean sea-level or more than twice the altitude of Mt. Everest on the Earth. The Martian atmosphere is composed roughly of 98.5% nitrogen, 1.2% argon, 0.25% carbon dioxide and 0.1% oxygen. Even on the lowlands on the planet Mars extreme high altitude environment of the Earth-type should prevail at the bottom of such a thin atmosphere. On such a planet, it would be therefore quite meaningless to speak of high altitude environment as we understand on the earth, even if we assume that there are high mountains and that earth-type of organisms exist on Mars*.

2. REDUCED ATMOSPHERIC PRESSURE

As is well known, the immediate effect of increase of altitude is the fall in atmospheric pressure, as a result of the reduction in the density of the air. Atmospheric pressures above the forest-line on the Himalaya are, for example, about two-thirds that at sea-level. At elevations of about 6000 m above mean sea-level, the atmospheric pressure on most mountains is less than half that at sea-level. At higher elevations, the rarefaction of the air becomes even more rapidly pronounced. The fall in the atmospheric pressure with the increase in altitude is not, however, uniform at all elevations. Oxygen, about one-fifth of the air at sea-level, becomes increasingly deficient with the rarefaction of the air as the altitude increases. The oxygen tension of the atmosphere at the forest-line on the Himalaya is, for example, only 68% that of the air at sea-level and about 45% at elevations of 6000 m (*vide* table 1). Compared to nitrogen, oxygen is somewhat heavier (density of oxygen is 1.429 g/l, density of nitrogen is 1.251 g/l) and thus also tends to be rarefied more rapidly than nitrogen, as the elevation increases (fig. 2).

The mean atmospheric pressure at the same elevation differs within fairly wide limits on different mountains, depending on the latitude of the mountain and other complex sets of atmospheric conditions.

The effect of oxygen deficiency in the air at high elevations begins to be felt by human beings at pressures less than 350 mm Hg, by monkeys at 300 mm, pigeons at 350 mm, cats at 270 mm, dogs at 250 mm, rabbits at 200 mm, but only at 100 mm by frogs (928). The greatest altitude climbed by man without the use of artificial oxygen apparatus is 8540 m above mean sea-level by Norton on Mt. Everest. The highest altitude of permanent human habitations is 4800 m above mean sea-level in Tibet, but most shepherds on the Himalaya ordinarily climb to summer pastures at elevations of 5500 m. The mountain sheep, ibex and yak are found up to elevations of 5800 m on the Himalaya. These sink into relative insignificance when we recollect that a remarkable group of insects, mites and spiders habitually flourish at elevations of at least 6800 m on the Himalaya (fig. 3).

In the case of insects, however, the increasing oxygen deficiency of the air at high elevations does not seem to have any appreciable direct effect. Most insects are known

* Recent American Mariner IV (launched in November 1964) photograph project has shown that the atmosphere on Mars is actually thinner than has been believed so far. On the general surface, the pressure corresponds to what we find at altitudes of 28 000-30 000 m above mean sea-level on the earth—in other words three to three and a half times higher than the Mt. Everest. Owing to the weaker gravity, the molecules of the Martian atmosphere may have largely escaped into the interplanetary space.

TABLE I

Atmospheric pressure and oxygen tension at different elevations

Altitude in m	Atmospheric pressure in mm Hg	Percentage of oxygen tension to that at sea-level
Sea-level	760.00	100.00
304.8	733.00	96.40
1000.0	673.90	88.60
1524.0	634.40	83.21
2000.0	595.00	78.20
3000.0	519.40	68.30
4000.0	468.10	61.50
5000.0	398.00	52.40
6000.0	346.10	45.50
7000.0	280.50	36.88
9144.0	225.40	24.68

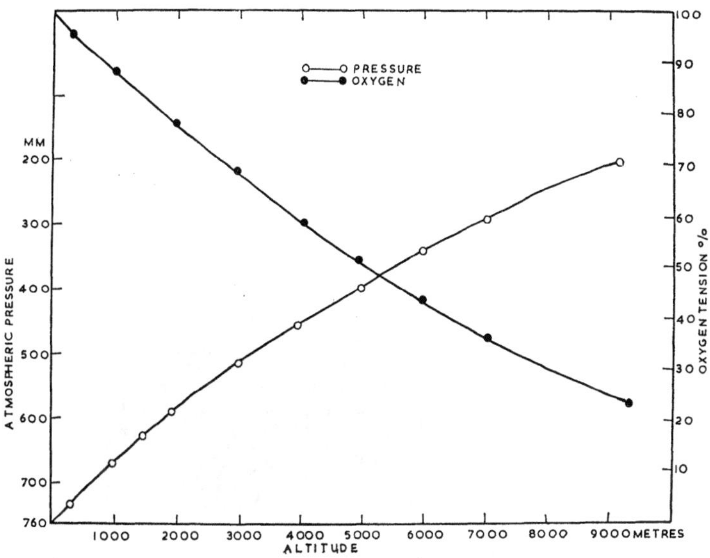

Fig. 2. The atmospheric pressure and the percentage of oxygen tension at different altitudes in the free atmosphere.

to be extraordinarily resistant to not only extremely low atmospheric pressures, but also to rapid alternations of low and high atmospheric pressures.

Recent laboratory investigations with diverse reptiles and mammals, especially with rats and dogs, inside reduced air-pressure chambers to simulate the conditions of high altitude approximating to 7620 m above mean sea-level, have shown that

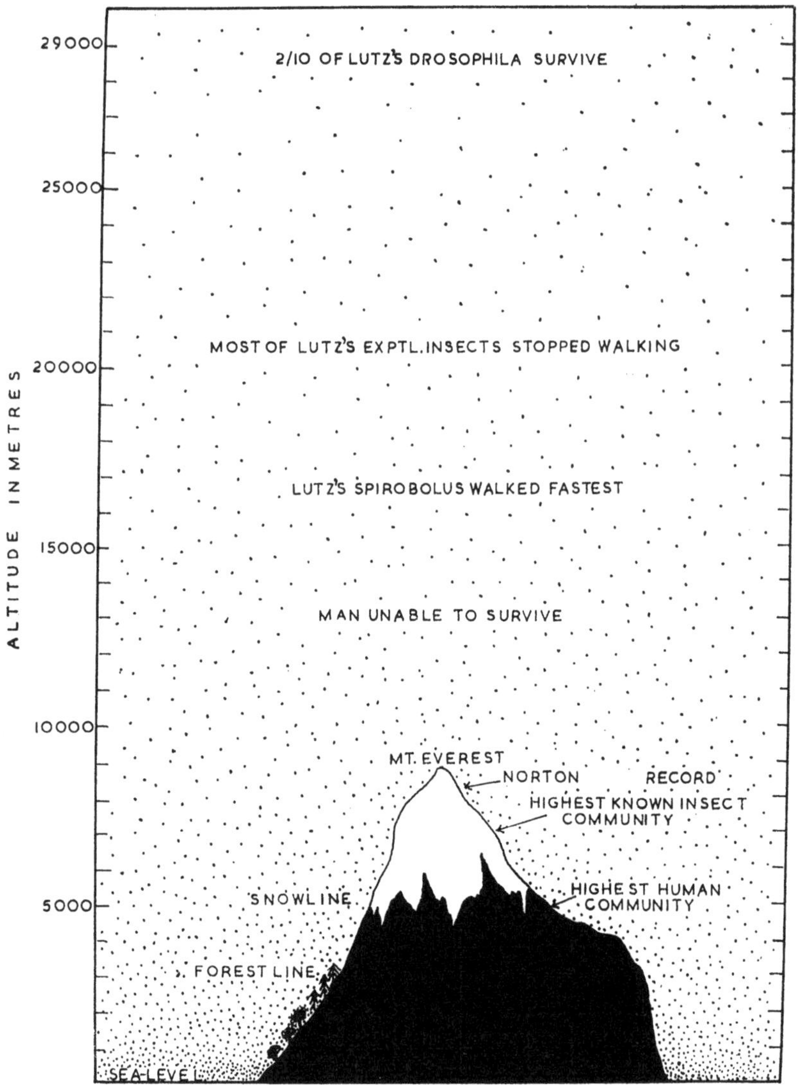

Fig. 3. Under the action of the Earth's gravitation, the molecules of the air tend to crowd near sea-level, so that the air is progressively rarefied as the elevation increases. Insects are able to withstand a high degree of hypoxia under experimental conditions and also seem to be capable of surviving at much higher elevations than the highest mountain summits on the earth.

these animals are less capable to tolerate very low atmospheric pressures and high degree of hypoxia and even anoxemia, than insects (40-43, 230, 512-514, 715, 716). Exposures to artificially induced hypoxia have given rise to numerous characteristic pathological conditions in these experimental animals. COLE (216), BACK & COTTON (72) and other workers have shown that insects are capable of withstanding an almost complete vacuum for several hours and even days. Their ultimate death under these conditions is really due not to the direct hypoxia or anoxemia, but to the exceedingly rapid evaporation of their body fluids (111, 462). LUTZ (715) exposed millipedes, *Drosophila*, beetles, butterflies and bees to low atmospheric pressures in the neighbourhood of 1/10000 mm Hg for 90 seconds without visible harm (fig. 3). This would approximate the conditions at an elevation of about 10 000 m above the general surface of the planet Mars or on a planet that has an atmosphere about one and a half times less dense than even on Mars! These experiments doubtless indicate that most insects might be able to survive low air pressures equivalent to altitudes much greater than normally might be experienced on the earth at present. While undoubtedly of great physiological and pathological significance, these experiments do not, however, reproduce exactly complex ecological conditions prevailing at high altitudes on mountains. The oxygen deficiency and the low barometric pressures are no doubt simulated, but the other environmental conditions, such as the high rate of evaporation, rapid insolation, radiation, high intensity of sunlight, intense ultra-violet radiation and atmospheric cold, are completely ignored. These experiments do not, therefore, provide any real clues to the actual conditions, under which the high altitude insects flourish on mountains. The ground effects under conditions of thin atmosphere are also completely lost sight of. Furthermore, such experiments merely serve to test the physiological limits of protoplasmic possibilities, of which there can be no serious doubt, but completely fail to yield any satisfactory and meaningful indications of the ecological properties of insect life at high altitudes on mountains (730). Such experimental investigations do, however, emphasize the fact that insects are not likely to be directly influenced, to any significant extent, by the reduced atmospheric pressure at high altitudes in the same way as Vertebrata are affected. The oxygen deficiency of the air at high elevations, while seriously impairing the normal metabolic and reproductive activities of various warm-blooded animals, is seemingly without a direct effect on the high altitude insects. As is well known, the generally low atmospheric temperatures at high elevations most probably tend to depress the general respiratory rate of organisms and the insect does not apparently suffer any serious ill effects from the low oxygen tension of the semi-rarefied air at an elevation of 6000 m above mean sea-level. The lesions commonly produced at high altitudes in the warm-blooded animals are not perhaps wholly due to the oxygen deficiency, but also in great part to the escape of bubbles of free nitrogen gas from the dissolved air into the blood stream, as a result of the abrupt lowering of the pressure. Since, however, the blood has no respiratory significance in insects, comparable to that of the Vertebrata, the tracheal respiration probably enables insects to withstand great and sudden changes of atmospheric pressures. Although the direct effects of the reduced atmospheric pressures at high altitudes are thus almost negligible on the high altitude insects, the indirect influence through the modifying action on diverse other environmental factors is very far-reaching and may be more or less readily perceived in nearly every peculiarity of the high altitude insect life. The dominance of the atmospheric cold and the atmospheric aridity may both unquestionably be observed as important ecological factors even at sea-level in different arid regions of the world, but it is their combination and co-ordinated interaction with the reduced atmospheric pressure that make the high altitude environment fundamentally different from that of the Arctic North or of the Antarctic landmass. In spite of the atmospheric cold, ice and snow, which are common to these areas, the environmental conditions for insect life at high altitude on mountains are essentially different from those prevailing in Spitzbergen, Greenland or the Arctic Alaska (fig. 8).

The modifying influence of reduced atmospheric pressure on the high altitude environment as a whole is no doubt readily apparent even to the casual observer, but it is not always easy to evaluate its effects on the individual factors. Some of the more important known effects of the reduced atmospheric pressure on the other environmental factors at high altitudes on mountains, in so far as insects are concerned, include the following:

1. Reduced atmospheric pressure results in high transparency of the air.

2. The high transparency of the air favours low atmospheric temperatures, so that the dominant high altitude ecological factor, viz. atmospheric cold is the result. Atmospheric cold retards evaporation from exposed surfaces and thus serves partly to counteract the harmful effects of high atmospheric aridity.

3. Reduced atmospheric pressure is also correlated with the reduced water vapour tension of the air and thus directly favours high atmospheric aridity. The atmospheric aridity progressively increases with the increase in elevation. This has the result of low precipitation.

4. Low atmospheric temperature leads to precipitation of the atmospheric moisture as snow and ice.

5. High transparency of the air and the low water vapour tension result in the greatly increased intensity of ultra-violet part of the sunrays and also favour high glare. The water vapour of the air absorbs solar radiation in the neighbourhood of 8000 Å or also higher. In the clear sky the amount of solar radiation absorbed is about 15% of the total incident radiation, but almost all the terrestrial radiation is absorbed, so as to give rise to the well-known greenhouse effect. The transpatent atmosphere at high altitudes is unable to do either of these (279). The atmosphere normally screens the radiation from the inter-planetary space, so that very little or none reaches the lowland areas. These radiations include cosmic rays from the interstellar space (85% protons), of 10^9 to 10^{18} eV energy (maximum), solar flares of protons of 10^9 to 10^{10} eV energy, low energy particles, etc. In the van Allen belt of the atmosphere, the inner parts are characterized by the absorption of protons of 10^8 eV energy and electrons of 10^6 eV and the outer part of 10^5 eV energy. The energetic primary particles have well-known biological effects, different from those of electromagnetic radiations. The Roentgen-equivalent of human beings or the dose in rads multiplied by RBE or the relative biological effectiveness (depending on complex conditions like the particle type, its energy, dose rate, size of the target, etc.) sufficient to cause nausea in man is 25 rads. Doses of 800 rads interfere seriously with the blood-forming function in human beings. Doses of 400-500 rads lead to death in 50% of the cases within two months. The risk of leukaemia is doubled with doses of 30-40 rads. Permanent sterility arises on exposures to 500 rads. There is a much larger frequency of such injurious elementary particles at higher elevations than

near the sea-level. The reader will find useful discussions on the general biological importance of radiations by DILL (263), DUGGAR (286) and GIESE (407).

6. The rarefied and the arid atmosphere allows a high rate of insolation and radiation.

7. Rapid insolation and radiation produce wide differences in the atmospheric and ground temperatures and also tend to favour great diurnal fluctuations in the surface temperatures of bodies exposed to the direct sunshine.

8. The rarefied and arid condition of the atmosphere accelerates the rate of evaporation from exposed surfaces and thus favours rapid desiccation of delicate and soft-bodied organisms.

9. On account of the atmospheric cold and aridity, the ecological importance of soil temperatures and soil moisture comes to be very greatly increased for insect life.

10. The ecological importance of snow-cover as a favourable factor for the high altitude insects is enormously increased for the same reason.

11. The great aridity of the atmosphere and the increased rate of evaporation at high altitudes tend to greatly magnify the ecological importance of atmospheric cold as a favourable factor for the high altitude insects.

It is necessary to emphasize that regardless of the diversity and complexity of the factors, the high altitude environment is a whole. The different individual environmental factors influence some specific character or aspect of the high altitude insect more than other factors and more directly than they influence other characters. In addition to this main factor, numerous other auxiliary factors influence the same character. Thus under the action of several inter-dependent and mutually interacting factors, the character becomes modified in a specific manner. The modification now triggers off a chain of other modifications. The ultimate controlling mechanism in this complexity of ecological factors, actions and the reactions of high altitude insects is, however, the reduced atmospheric density, itself the result of altitude.

In this ecological factor-complex, interacting with the insects at high altitudes on mountains, the atmospheric cold, atmospheric aridity and the micro-microclimate may perhaps be said to be dominant. Although these three dominant factors are, in the final analysis, but the secondary effects of the reduced atmospheric pressure, their influence on the high altitude insects is more direct than that of most other environmental factors. This is to be attributed to the peculiarities of the body organization of insects. Nearly every factor seems to influence all the aspects of insect life at high altitudes through these dominant factors rather than directly. It must also be pointed out that the dominant factors more or less modify the action of all other factors on the high altitude insects. As all these factors are, in addition, mutually interactive, the reactions of the high altitude insects to these dominant factors lead automatically to adjustments for nearly all the other factors also. The dominance of the three factors mentioned above is the result of altitude. While altitude alone has no remarkable direct effect on insects, it influences them by modifying the other environmental factors. The more important effects of altitude include the reduction of atmospheric pressure, atmospheric cold and aridity, high intensity of solar radiation and snow-cover. The sum-total of these effects lies outside the limits of optima observed in case of lowland insects, and in this sense the environmental conditions at high altitudes on mountains may be described as extreme. There is in Nature but a single indivisible and complex environmental factor, viz.

the high altitude environment, in which we cannot isolate any element from the other. The high altitude insects and the high altitude environment constitute integral parts of a single high altitude ecosystem. The high altitude insects and the high altitude environment are intertwined and interpenetrated so deeply with each other that it is difficult or even impossible to define a satisfactory dividing boundary between them—indeed such a boundary does not exist. The high altitude insects react not to any isolated individual factor as such, but to the whole complex—to the environment above the forest, at high elevations on mountains in toto.

3. ATMOSPHERIC TEMPERATURE

The relatively low density of the dust-free air, with scanty water vapour, and its high transparency at high elevations on mountains readily permit the sunrays to pass through, without the air itself being warmed up as appreciably as the denser and more or less humid air near the sea-level. When the sun is not shining, the air readily permits the rapid radiation of heat from the ground, again without itself undergoing a rise of temperature. The mean atmospheric temperatures are, therefore, lower than at sea-level. The low atmospheric temperature at high elevations on mountains is thus one of the direct effects of the reduced atmospheric pressure.

a. The Temperature Lapse Rate

With the fall in the atmospheric pressure, the mean atmospheric temperature also decreases as the elevation above the mean sea-level increases. In free air the rate

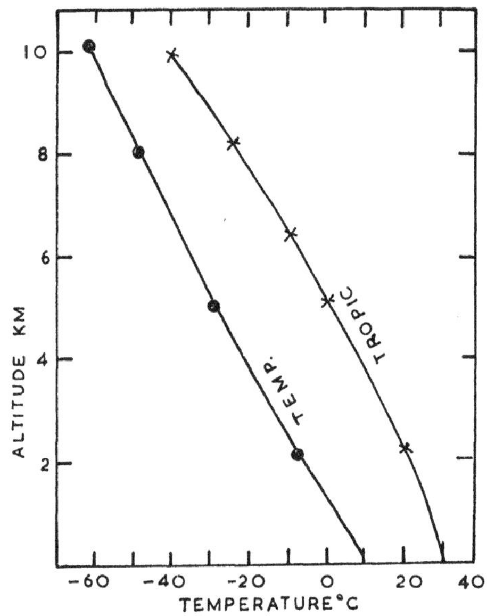

Fig. 4. The lapse rate of atmospheric temperature with increase of elevation above mean sea-level in free air in tropical and in temperate regions of the northern hemisphere.

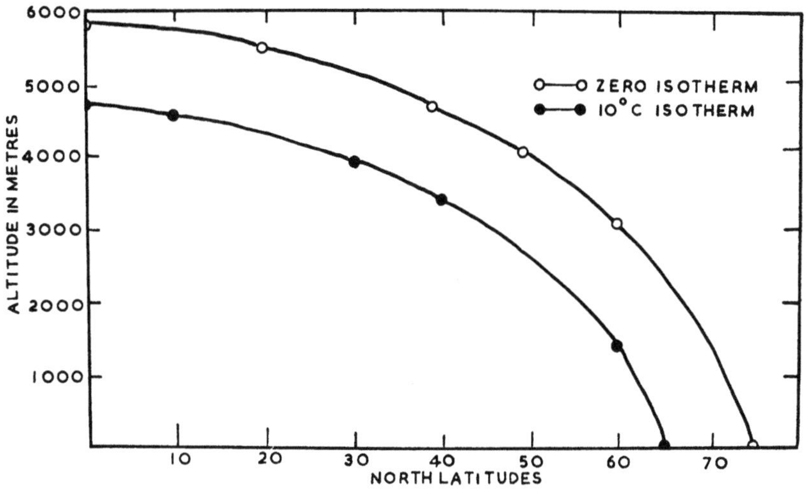

Fig. 5. The 0°C and the 10°C isotherms of the mean atmospheric temperatures and their relation to latitude and altitude in the northern hemisphere.

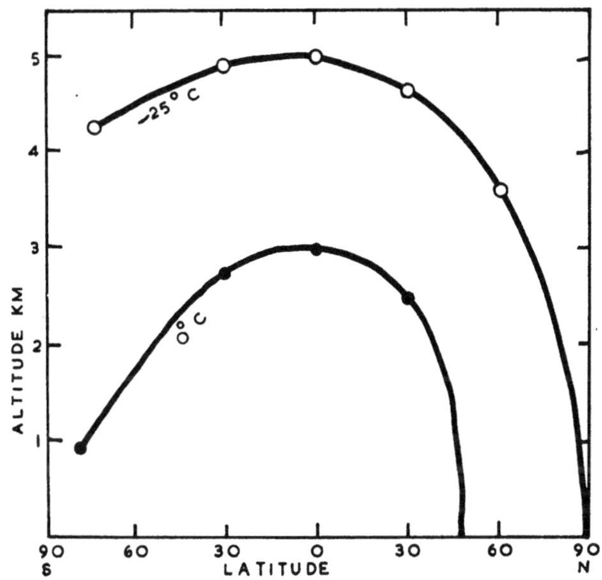

Fig. 6. The altitudinal differences of the zero °C and the -25°C isotherms of the mean atmospheric temperatures in the northern and southern hemispheres.

of fall of the mean atmospheric temperature or the socalled lapse rate of temperature amounts on an average to about 6.2° C for each increase of 1000 m in elevation. The lapse rate in free air in the tropical regions of the world differs from that in the temperate areas. The mean temperatures of free air at different altitudes in the tropics and in the temperate regions are plotted graphically in figs. 4, 5 and 6. The temperature lapse rate in free air is influenced by latitude; it is also different in the southern and northern hemispheres at the same latitude. We find, for example, in fig. 6 a pronounced difference in the zero and —25° C isotherms between the southern and northern hemispheres. The temperature lapse rate is also very profoundly influenced by other complex factors, particularly the presence of large continental masses and massive mountains. On mountains, the lapse rate is influenced by numerous conditions, so that the rate of fall of temperature with increase of elevation is not uniform on all mountains or even on the same mountain. On the Central Alps, for example, the fall of temperature with increase of elevation amounts to about 7° C for each increase of 1000 m in elevation. On the Caucasus the lapse rate is 6.6°, on Mt. Washington on the Presidential Range in the eastern United States of America it is 5.4°, on the Rawenzori Range in Equatorial East Africa it amounts to only 4° and on the tropical parts of the Andes to 5° C. The temperature lapse rate is fairly uniform on the Himalaya up to the upper limits of the forest, but becomes abruptly and irregularly rapid at higher elevations, especially above 4000 m. The temperature lapse rate is conspicuously different on the different ranges of the Himalaya and on the south and north slopes of the same range. As a rule, the lapse rate is smaller on massive mountains than on isolated mountains. It is also smaller on equatorial mountains than on those in the extreme northern latitudes.

Some examples of these differences may be given here. The 10° C isotherm is reached at an elevation of about 5000 m on equatorial mountains but even at sea-level at 65° NL. This isotherm lies at an elevation of 3400 m on Mt. Elgon in equatorial East Africa, at 2350 m on Mt. Rainier, between 1900 and 2100 m on the Alps, at 2775 m on the Kuen Lun Ranges, about 3625 m on the Himalaya and at 600 m on Fennoscandian mountains. The zero degree isotherm lies at different altitudes on different mountains, depending mainly on the latitude and continentality of the mountain, its massiveness, the slope aspect and other conditions. The freezing point is reached on Mt. Kilimanjaro (fig. 7) in equatorial East Africa at an elevation of 5500 m, on the Atlas Mountains at 3350 m, on the Alps and Pyrenees about 2590 m, on the Caucasus at about 3000 m, on the Nepal Himalaya at 4480 m, on the Kumaon Himalaya at 5180 m, on the Northwest Himalaya at 5790 m, in western Tibet at 6090 m, on the Karakoram Range at elevations between 5500 and 5650 m, on the Tien Shan System at 3350 m, on the Alai Mountains at 4260 m and on the Scandinavian mountains at 610 m.

The mean atmospheric temperature at the same elevation differs, therefore, within wide limits on different mountains. The mean values do not also provide any clue to the actual temperatures prevalent in many localities. At an elevation of about 4500 m on the Great Himalaya (*vide* Chapter IX), for example, the range of atmospheric temperature lies between —10.5° C and 9.5° C and at elevation of 5600 m it may range from —12.0° C to 7.0° C. On the Karakoram Ranges the mean atmospheric temperature is reported to be about —30° C at an elevation of 7300 m. On the summit of Mt. Everest the atmospheric temperature is —45° C.

The relation between the altitudinal lapse rate of temperature and the latitude of the mountain is illustrated, at least in part, by the well known HOPKIN's bioclimatic law (540, 541). According to this law, a mountain region, about 1525 m above mean sea-level, corresponds in its mean atmospheric temperature and general climatic conditions at sea-level to a latitude of 12° 30′ further north. In other words,

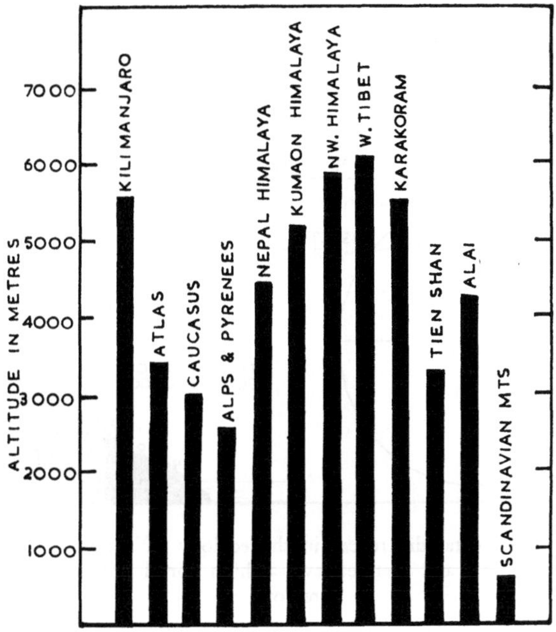

Fig. 7. The altitudes of the zero °C isotherms of the mean atmospheric temperature on some of the principal mountains of the world.

the lag of seasonal events is about four days for each 122 m elevation or 1° latitude in the northern hemisphere. A number of workers have been so greatly impressed by the apparent parallelism of temperature lapse rate at high altitude and latitude that they have attempted to equate the general climatic conditions at high elevations on mountains with those prevailing at high north latitudes and have often identified the altitudinal biotic zonation on mountains with that at sea-level as one proceeds from the equator to the north pole. While the mean atmospheric temperature conditions may perhaps be related to the latitude-altitude law, the environment at high altitudes on mountains is by no means identical or even comparable to that of the subarctic or the arctic regions at sea-level (fig. 8). The conditions are also fundamentally very different with regard to the mountains in the southern hemisphere. Although the atmospheric temperature conditions on the mountains of the southern hemisphere approach more or less those prevailing at high elevations of equatorial mountains, the high altitude environment is otherwise not identical with that of the Antarctic lowlands. The arctic and the antarctic cold environment is conditioned not by the reduced atmospheric pressure of the high altitude as on mountains, but entirely by the differences in the angle of insolation. The atmospheric cold of the high latitudes is the result of fewer sunrays falling on unit area on surface, but of the high altitude is the result of actually less air being warmed, although there are more sunrays per unit area. Although the mean atmospheric pressure is influenced by latitude to some extent, at higher latitudes the atmosphere is certainly not as thin as on the mountains and the climatic conditions are, therefore,

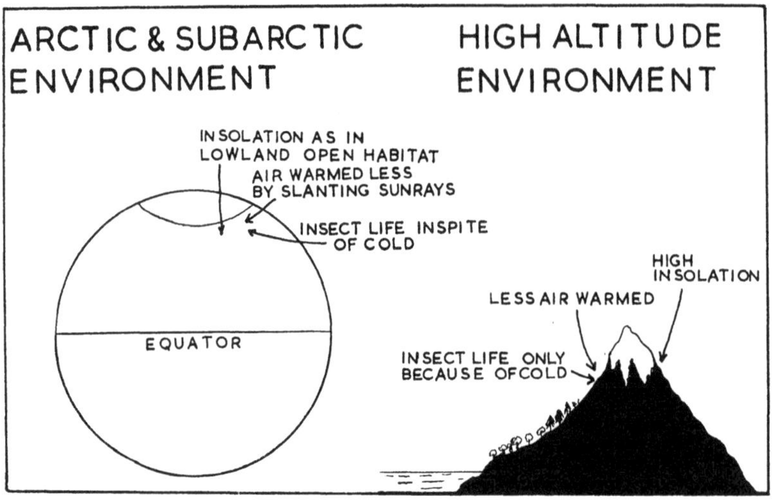

Fig. 8. The fundamental differences in the ecology of atmospheric cold at high altitudes on mountains and near sea-level at high boreal latitudes and in the arctic regions.

not ecologically comparable. The high altitude cold is conditioned by the same factors and is also more related to the cold of the interplanetary space than to the arctic cold. The intensity of insolation in the arctic is no more than that ordinarily observed in a lowland open habitat, but at high altitudes it tends to become progressively very intense. The subarctic and the arctic environments are fundamentally that of a lowland open habitat, under a pressure of one atmosphere, but the high altitude environment is an open habitat under a reduced atmosphere. Open habitats of normally one atmosphere pressure and permanently thinner atmosphere are evidently ecologically different in their properties, modifications and interrelations. All earlier workers have apparently overlooked these fundamental differences between the high altitude and the high boreal latitude environments.

b. Shade and Sun Temperatures

The difference between the mean atmospheric temperature in the shade and in the direct sunshine in the high altitude environment is far more pronounced than in the open habitat in lowland areas and this difference tends also to increase markedly with the increase in altitude (fig. 9). As may be expected, this peculiarity is closely correlated with the fact that insolation is far more intense and rapid than near sea-level, owing to the thinner atmosphere at high elevations on mountains. For example, at an elevation of about 3900 m on the southern slopes of the Northwest Himalaya, with the atmospheric temperature reading 4° C in the shade, the thermometer exposed to the direct sunrays registers 35° C. BÄBLER (74) observed that, with the temperature of free air reading —2.5° C at 1 p.m. in the shade during August at an elevation of 4560 m on the Swiss Alps, the atmospheric temperature in direct sunshine was 37° C. According to HANDSCHIN (471, 473), the difference between the

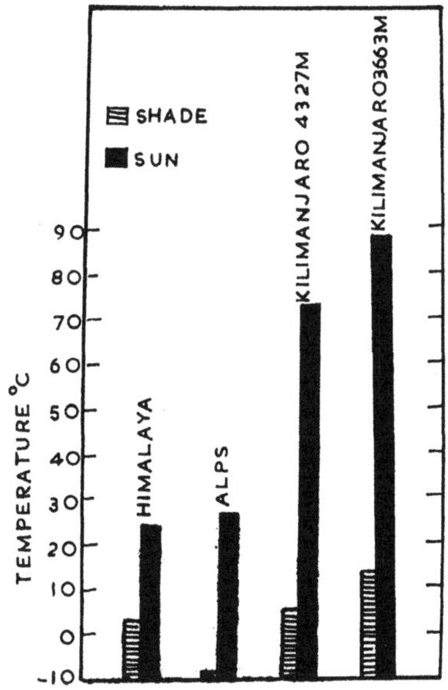

Fig. 9. Insolation is more intense at higher elevations than near sea-level, so that the difference between the mean atmospheric temperatures in the shade and in the sun is great on mountains. On Mt. Kilimanjaro insolation is very intense, both due to the vertical direction of the sunrays and to the high transparency of the air because of the high altitude of the mountain.

mean atmospheric temperature in the shade and in the direct sunshine on the Swiss Alps often amounts to 50° C. While, as explained above, the mean atmospheric temperature in the shade falls with increase in elevation on mountains, the direct-sun temperature tends to increase, slowly at first up to about 3500 m and rather more rapidly as the elevation increases above 4200 m. On high mountains in the northern hemisphere, the intensity of sunrays is specially greater during the winter months, when the atmosphere is also thinner, than during the summer. At high altitudes on mountains on the equator, the insolation is particularly intense, both due to the vertical direction of the sunrays and to the thinner atmosphere. On Mt. Kilimanjaro, the mean sun temperature is, for example, 73° C at an elevation of 4327 m, and 87.5° C at an elevation of 3663 m. The mean difference between the shade and sun temperatures at an elevation of 4150 m on Mt. Kilimanjaro is 43.2° C. With the thermometer reading 14° C in the shade, the sun temperature reads 87.5° C and with the shade temperature 6° C, the sun temperature amounts to 73° C. It must, therefore, be concluded that regardless of the low atmospheric temperatures, objects exposed to the direct sunshine become warmed up far more rapidly at higher elevations than within the forest and than in open habitats at sea-level. The ecological significance of this fact is enormous in high altitude entomology.

c. Diurnal Variations of Atmospheric Temperature

The amplitude of diurnal atmospheric temperature fluctuations tends to differ at different elevations and on different mountains, depending largely on the latitude of the mountains, its massiveness and on the continentality of the general climate of the surrounding country. The diurnal fluctuations of atmospheric temperatures are generally more pronounced on the equatorial and tropical mountains and to some extent also on the mountains in the southern hemisphere than on the mountains of the northern temperate latitudes. The amplitude of the night temperature minima is significantly narrower than that of daytime atmospheric temperature maxima, both on the Alps and on the Himalaya. On both these mountain systems, the diurnal fluctuations of the atmospheric temperatures are more pronounced on the south slopes than on the north.

The diurnal atmospheric temperature fluctuations are far more marked than the seasonal fluctuations on the equatorial and other tropical mountains and on the mountains of the south temperate and subantarctic regions (*vide* also chapters VIII and XVI).

d. Annual Variations in Atmospheric Temperature

The difference between the mean atmospheric winter and summer temperatures on the mountains in northern hemisphere is more pronounced than on equatorial, tropical and south temperate mountains. On the Himalaya, at elevations of 3500-4500 m, the mean of the annual temperature fluctuations is about 16° C. On the Alps the annual temperature amplitude is 15.6° C at an elevation of about 2000 m, 15° C at 2500 m, 14° C at 3000 m and 19.8° C at 3106 m. At an elevation of about 3535 m on Colorado mountains, it amounts to 17.2° C. On the extreme continental mountains like the Middle Asiatic Tien Shan Ranges, the difference between the mean winter and summer atmospheric temperatures is very pronounced; it averages, for example, to about 63° C at an elevation of 2780 m, 70.8° C at an elevation of 3000 m and 61.5° C at 3600 m. On the other hand, the annual temperature fluctuation does not exceed 5 - 6° C even at very high elevations, above 4000 m, on the equatorial, tropical and south temperate mountains.

e. Ecological Importance of Atmospheric Cold at High Elevations

The pronounced atmospheric cold at high elevations on mountains must, in the final analysis, be rated as perhaps the most important favourable environmental factor for the high altitude insects (137, 138, 263, 267, 730, 986). As already indicated, the low atmospheric temperatures tend to greatly counteract the injurious effects of the high atmospheric aridity, retard the excessively rapid evaporation, favour the precipitation of atmospheric moisture, lower the rate of oxygen consumption by depressing the respiratory activity of the high altitude insects, restrict the flight of insects, retard their growth and thus lead to diminution in the mean body size, restrict the vegetative period and generally magnify the ecological importance of microclimate. The reduction and atrophy of the wings of the high altitude insects and the development of a heavy body pigmentation are also partly to be attributed to this factor. It is also the atmospheric cold that enables the high altitude insects to tolerate and withstand effectively the severe desiccating effects of high atmospheric aridity. The ecological significance of atmospheric cold at high elevations on mountains is, therefore, fundamentally different from that observed in open habitats near sea-level in the high north latitudes and in the subarctic and arctic regions. We have thus the apparently paradoxical situation that while in the lowland open habitats of the subarctic and arctic areas the insects exist *inspite* of the great atmospheric cold, at high elevations on mountains they are able to exist *only because of* the atmospheric cold. This is perhaps the strongest argument against the attempts at equating the ecology of atmospheric cold at high altitudes on mountains with that in the arctic regions (fig. 10).

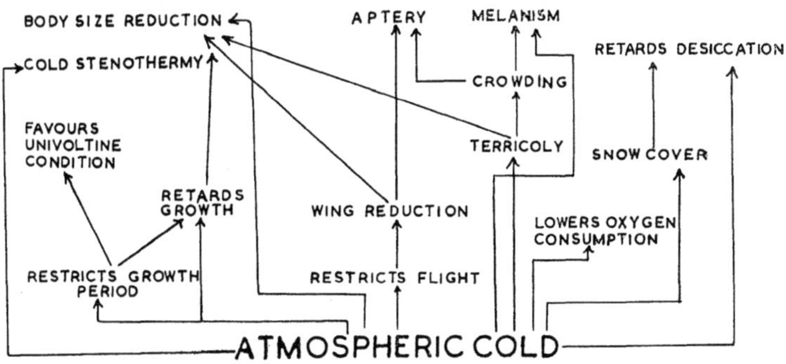

Fig. 10. The ecological importance of atmospheric cold for insect life at high altitudes on mountains.

4. ATMOSPHERIC PRECIPITATION

In the high altitude environment on mountains, next perhaps only to atmospheric cold, the most important ecological factor is atmospheric precipitation. Precipitation depends, among other complex factors, on the atmospheric humidity. As already indicated, the mean atmospheric relative humidity decreases and the mean evaporating power on exposure rises with the increase in altitude. On some mountains, however, there is an increase in the relative humidity and the mean annual precipitation as the altitude increases, up to a certain limit. The investigations of SCHUBERT (936) have demonstrated differences in the rates of evaporation within the forest and in the open habitat on high mountains. The atmospheric aridity at high elevations is more pronounced on massive and continental mountains than on isolated mountains or those situated near large seas. The north slopes of the mountains in the north temperate regions of the world, especially in the case of the ranges with east-west trends, are often more arid than the south slopes. At comparable elevations and latitudes, the mountains in the northern hemisphere are more arid than those of the southern hemisphere.

The pronounced high atmospheric aridity at high altitudes on mountains is determined partly by the distance of the mountain from oceans. The absolute humidity of the atmosphere and the total precipitation in any locality, in relation to the mean elevation above sea-level, become greater or smaller, according as the general climate is more continental or oceanic. GAMS (393) has attempted to measure the oceanity or continentality of the Alps in terms of the ratio of total precipitation in mm and the altitude in metres. He represents the total precipitation on the Swiss Alps in a coördinate system, with the elevation on the abscissa and the mean annual precipitation on the ordinates. The degree of oceanity is expressed in terms of the angle subtended by the curve on the X-axis and is obtained according to the formula $\tan \angle W \text{ of oceanity} = \frac{\text{Mean annual precipitation in mm}}{\text{Elevation in metres}}$. He found that the extremes for Switzerland are 18° $\tan \angle W = \frac{530 \text{ mm}}{1628 \text{ m}}$ at Grächen and 83° 55′ $\tan \angle W \frac{2070}{210}$. As the continentality increases, so does the atmospheric aridity. Compared with the Alps, the Northwest Himalaya, the Pamirs Region, the Tien Shan and other Middle Asiatic high mountains are far more continental and are, therefore, pronouncedly more arid at comparable elevations and aspect exposure. The close correlation between atmospheric temperature and precipitation is emphasized in evaluating the ecological effects at high altitudes by extending and applying EMBERGER'S and KÖPPEN'S formulae. EMBERGER (318) expresses the ecological effects of atmospheric temperature-precipitation by the formula

$$\frac{P}{2\left(\frac{M+m}{2}.(M-m)\right)} . 100,$$

where P = the annual rainfall in mm, M = the mean atmospheric temperature in C of the warmest month of the year and m = the mean temperature in C of the coldest month. KÖPPEN expresses the relation differently by the formula

$$R_{16} = R_t \frac{1}{1 + \frac{1}{40}(t+16)}$$

where R_{16} = the rainfall at 16° C, R_t = the rain fall at t° C. Ecologically, the atmospheric aridity is closely interlinked with the atmospheric cold, high rate of evaporation, rapid insolation and radiation, high light intensity, high wind velocities and other factors of the high altitude environment. The atmospheric aridity determines also the limits of the snowline on mountains. By favouring precipitation as snow, the atmospheric cold brings about an intensification of the atmospheric aridity. Atmospheric aridity raises the snowline, atmospheric cold retards evaporation, the rate of which is accelerated by the atmospheric aridity. The atmospheric aridity at high altitudes is the result of both low water vapour tension due to the less dense air and is also the result of atmospheric cold.

In a recent contribution, VACHON *et al.** have shown that pronounced habitat aridity seems to be closely correlated with a high resistance among scorpions to the action of gamma radiation under experimental conditions. The high atmospheric aridity and the great intensity of radiation in the high altitude environment might conceivably be similarly correlated with the high altitude insects, but we are at present largely ignorant of the precise conditions in this connection.

* VACHON, M., P. NIAUSSAT, J. H. ERRSOLE & C. GRENOT, 1964. *C.R. Acad. Sci. Paris*, 259: 3389-3391 (Nov. 9, 1964).

5. SNOW-COVER

Though considerable rain falls at moderate elevations on some of the equatorial mountains, like the Ruwenzori Range, precipitation at high elevations on most mountains is largely as snowfall. The principal factors governing snowfall are the atmospheric cold, atmospheric humidity and the continentality of the mountains. The mean annual snowfall increases with altitude above a certain limit, but again begins to diminish at higher elevations, because of the increasing atmospheric aridity. As we ascend above the snowline, we find the depth of the permanent snow tending to increase. We do not know, however, at what altitude in any given region the accumulation of snow becomes a minimum.

a. *The Winter Snow*

The bulk of the snow at high elevations falls during the winter, but snowfall is not infrequent during the summer also on many mountains. On the equatorial mountains of sufficient elevation, snowfall occurs perhaps every night. On high mountains in the northern hemisphere the winter snowfall starts often as early as September and may continue up to May of the next year. In contrast to the permanent snow, the winter snow-cover is only seasonal and with the coming of summer in May-June, the winter snow begins to melt. The mean thickness of the winter snow-cover at any locality varies greatly from year to year, so that only after prolonged observations, extending over a number of years, it is possible to get the thickness of the winter snow-cover of the locality. The depth and the duration of the snow-cover are not wholly meteorological peculiarities, but depend on the altitude, aspect of slope exposure, the mountain range, its trend-line direction, the exposure, gradient of the slope and diverse other complex conditions. During the summer, part of the winter snow melts and part of it also directly evaporates away. It is estimated, for example, that about 82% of the winter snow on the Himalaya evaporates into the atmosphere and only 18% of it contributes to the melt-water. The rate of melting of the winter snow-cover is usually indicated in terms of the altitudinal recession of the snow-edge in a period of ten days. The mean rate of melting of the winter snow on the Swiss Alps is about 100 metres in ten days (369). On the Northwest Himlaaya, MANI (730) reported the rate to vary from 200 m to 240 m in ten days on the south slope and the edge of the winter snow recedes to an elevation of about 5200 m by about the third week of August, when the summer is also well advanced (Fig. 11). The snow-cover persists longer on the northern slopes of the Alps and the Himalaya than on the southern slopes. It is not only the amount of the snowfall, but also the duration of the snow-cover that is ecologically very important at high altitudes on mountains.

GAMS (396) and STEINHAUSER (1008) have prepared diagrams for the winter snow-cover on the Alps. The recession and the advance of the winter snow-cover edge with the advancing of summer at different altitudes in the East Alps during the year 1890 are summarized in table 2.

The recession of the edge of the winter snow on the south slope is earlier by about 20 days than on the north slope of the Alps. The difference between the recession rates on the north and south slopes on the Northwest Himalaya lies between five to ten days. The rate of recession of the winter snow is influenced by a number

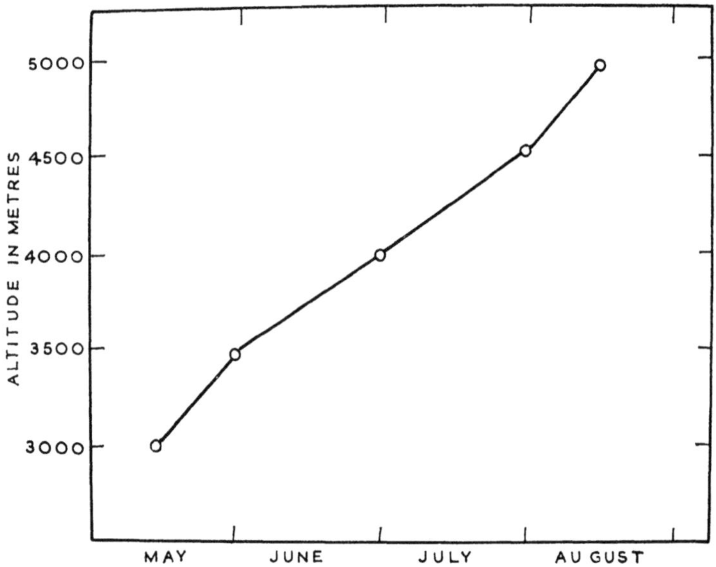

Fig. 11. The progress of the melting of the winter snow, with the advance of summer, on the Northwest Himalaya.

TABLE 2

Recession and advance of the winter snow on the East Alps

Month	Altitude of the snow-edge in m
March	1200
May	1800
July	2450
September	2300
October	1930
November	1250

of complex factors, especially the altitude, the aspect, the depth of the winter snow-cover, the mean atmospheric temperature during the summer, the duration of summer, exposure to the sun, the latitude of the mountain, frequency of avalanches and snow storms during the summer, etc.

b. *Snowline*

It must be evident from the foregoing pages that above the altitude of zero degree isotherm, the snow-cover is permanent. While the upper layers of snow remain practically unchanged, the lower layers are compressed, by pressure of the snow above, into a granular mass known as firn or neve. Neve is intermediate in character between the soft snow and the hard ice. Partly under pressure of the layers above and partly also due to the percolation of the melt-water during the summer insolation, the air in between the granules of neve is squeezed out and the lower layer thus becomes compacted into hard ice. This consolidation into neve and ice occurs more frequently with increasing altitude, particularly on mountains in the north temperate latitudes. As fresh snow continues to accumulate above the permanent snowline, enormous masses of snow and ice come down in great avalanches and thus often contribute to the store of melting winter snow during the summer. The glacier ice moves slowly down the valley until the glacier snout reaches sufficiently low to melt. The more or less wide transitional zone, above which the snow resists the heat of the summer so that the snow-cover is perpetual and below which the snow melts for a short period during the summer each year, is the socalled snowline. It represents the mean altitude above which the snow-cover is, therefore, permanent. It must be emphasized here that the snow often remains unmelted in sheltered patches in ravines even well below the permanent snowline, long after it has disapperead from the ridges and lower summits. There are likewise more or less extensive patches of snow-free areas of barren rock and permanently frozen soil above the permanent snowline. Even above the permanent snowline, snow and ice do not cover every summit, ridge or rock mass; the presence or absence of snow under these conditions depend on the peculiarities of steepness and prevalent local winds.

The snow-cover is permanent even at the sea-level in the high north and south latitudes. The relation between snowline altitude and the latitude in the northern hemisphere is summarized in fig. 12.

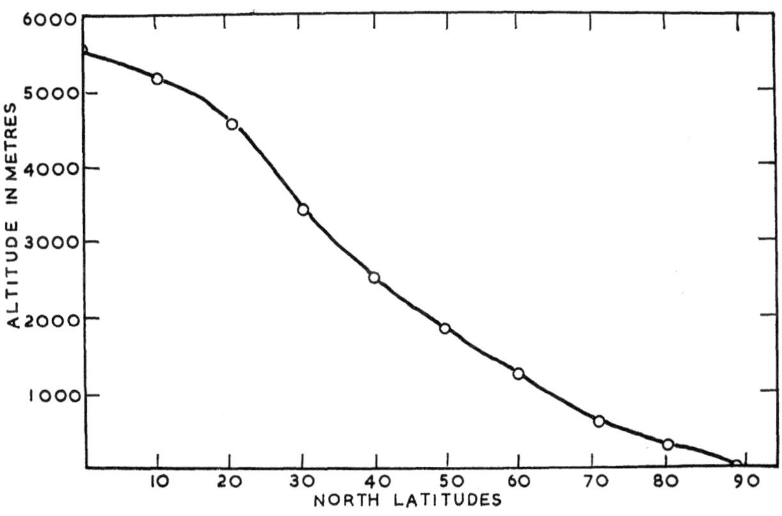

Fig. 12. The mean snowline at different latitudes in the northern hemisphere.

As may be expected, the snowline is influenced by the mean atmospheric temperature during the summer, the snowfall and to some extent by the prevailing wind. Scanty snowfall raises the snowline, which is usually somewhat higher on the southern slopes than on the northern slopes of the Alps and the Himalaya. The continentality of the mountain has also a profound influence on the snowline. The problem of the snowline is complicated by considerations other than exclusively meteorological ones. Snow accumulates on a flat-topped summit far more than on a precipitous and pointed peak. Mountain peaks like K_2 (Mt. Godwin Austen), Nanga Parbat, Rakaposhi, etc. on the Himalaya are, for example, too sharp to allow the snow to lie in any considerable quantity. As soon as the snow falls on these summits, it descends in the form of great avalanches and hardens to glacier ice at lower elevations.

The snowline differs within very wide limits on different mountains. On the south slope of the Himalaya, the snowline is about 900 m lower than on the north slope, the difference being mainly due to the fact that the moisture-laden winds from the Indian Ocean in the south shed their moisture on the south slope before crossing these mountains. Tibet and the Tien Shan Mountains are, therefore, extremely arid and the snowline on these mountains is therefore much higher than on the Himalaya or any other mountain at the same latitude in the northern hemisphere. The snowline altitudes on the Himalaya and on the Middle Asiatic mountains are summarized in table 3.

TABLE 3

Snowline on the Pamir-Himalayan Region

Mountain	Latitude	Slope	Snowline
Nepal Himalaya	28N	south	4480
Southeast Tibet	29N		3960
Kumaon Himalaya	30° 30′N	south	4720
Kumaon Himalaya	30° 30′N	north	5650
Punjab Himalaya	34N	south	5180
Punjab Himalaya	34N	north	5790
Zaskar Range	34N	south	6090
Zaskar Range	34N	north	5940
Ladak Range	34N	south	5790
Ladak Range	34N	north	5650
Kailas Range	31N	south	5940
Western Tibet	34N		6090
Karakoram	36N	south	5650
Karakoram	36N	north	5500
Tien Shan	42N	north	3350
Alai	40N		4260

The snowline on the northern Chilean Andes is 6090 m and in Patagonia is 760 m. The highest snowline is perhaps not over 6100 m on the earth at present.

c. Ecological Importance of Snow-cover

The snow-cover, particularly the winter snow-cover, is of great ecological importance for the high altitude insects; it must indeed be rated as perhaps the most favourable environmental condition that makes insect life possible on the arid elevated regions, baked mercilessly by the unfiltered sunrays. As early as 1800 WOEIKOFF (1120) recognized the great ecological importance of winter snow-cover for organisms and since then its general ecological importance has been increasingly appreciated (356, 357, 730, 932, 1003, 1006, 1052). While the winter snow-cover near sea-level, especially in the high north latitudes, is important for the life of insects and other poikilotherm organisms, its ecological importance increases very greatly under the conditions of thinner atmosphere at high altitudes on mountains. Snow-cover is important for insects at high latitudes and absolutely indispenable for the high altitude insects (fig. 13). It is snow-cover that assures optimal conditions for nearly all the high altitude insects on mountains and the life of every species and individual is inexorably bound up with it. Though the snow-cover is influenced by the atmospheric temperature, the latter is also influenced in its turn by the snow-cover. With the coming of summer, the mean atmospheric temperature does not, for example, rise as rapidly as one would expect, because enormous quantities of the sun's heat are absorbed as latent heat by the melting snow. The snow-cover exerts a most remarkable ameliorating influence on the extremes of atmospheric temperature fluctuations. The mean atmospheric temperature on snow-fields and on glaciers does not exhibit such pronounced secular or diurnal variations as in snow-free areas. On and in the immediate vicinity of snow and ice, the temperature fluctuations are relatively small. At an elevation of 4500 m on the south slope of the Great Himalaya, the temperature of the air in the shade during late June registers a maximum of —2° C and a minimum of about —10° C. During the winter the mean air temperature on snow-fields is as low as —17° C at the same elevation. The mean atmospheric temperatures reveal an interesting gradient from the snow-edge (Table 4).

TABLE 4

Gradient of mean atmospheric temperature from the snow-edge on the south slope of the Himalaya

Distance from the snow-edge in m	Max. temp.	Min. temp.
10	5.0	−10.0
100	7.0	− 1.0
200	12.0	+ 7.5

The snow-cover is also one of the important contributory factors for the great atmospheric aridity during the winter. In addition to contributing to the soil moisture and melt-water for the mountain lakes and streams, the snow-cover serves as an effective protective blanket for the high altitude insects during their hibernation and also prevents the freezing of the ground during the winter. The densest concentrations of high altitude insects are, therefore, always found in the immediate neighbourhood of the snow-fields on mountains (*vide* also Chapter III). As the intense insolation removes the soil moisture very rapidly at high elevations, if the melting of the snow does not replace the moisture thus lost, the growth of the high altitude vegetation and the breeding of the high altitude insects would both be impossible. It would be utterly meaningless to speak of high altitude insect life in total absence of snow-cover. The more important ecological relations of snow-cover for the high altitude insect life are shown diagrammatically in fig. 13.

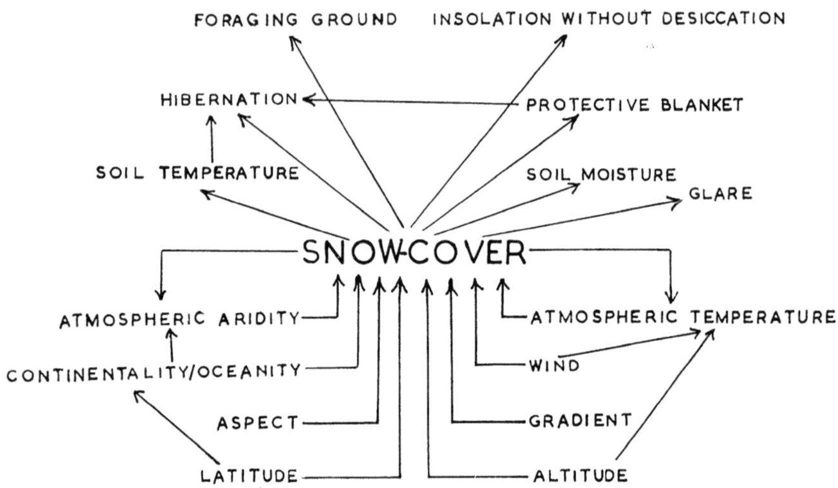

Fig. 13. The ecological importance of the winter snow-cover for high altitude insect life.

6. WIND

As is well known, wind velocity increases with altitude on most mountains and it would appear that the upper layers of the atmosphere, at elevations above 5000 m, are generally regions of strong winds. Strong and almost constant winds that often attain the force of storms are characteristic of the high elevations on the Himalaya, the Pamirs, Alps, Kilimanjaro, Tien Shan, Andes and other mountains. Even on the lesser mountains like the White Mountains of North America, wind is the most striking atmospheric feature. On the Pamirs Region violent winds spring up between 2 and 3 p.m. daily and become storms at 5-6 p.m. The wind velocities attain an intensity of 54-58 km/hr (*vide* Chapter X). The highest summit on Mt. Kilimanjaro is subject to a complex system of strong winds (*vide* Chapter VII). While the winds near sea-level are largely seasonal, the fierce winds at high altitudes on mountains are far more constant and appear to be part of the general turbulance of the upper atmosphere. The general ecological effects of wind are contrary to those of insolation and the net result is a considerable lowering of the mean atmospheric temperature. The temperature of all bodies exposed to strong winds at high altitudes is also considerably lowered. High wind velocities also accelerate the rate of evaporation from exposed surfaces and thus favour rapid desiccation of bodies under insolation. The high wind velocity should be considered as one of the major contributory factors accentuating the severe effects of atmospheric aridity at high altitudes. In addition to the various physical and mechanical effects of strong winds on other environmental factors at high altitudes, we also must not overlook the direct effects of constant and strong winds on the high altitude insects. Strong winds restrict or even totally eliminate the flight of insects, by compelling them to remain close to the ground. They also tend to bring down the body temperature of the high altitude insects and subject them to a rapid desiccating action. The ecological importance of wind at high altitudes on the Alps and other high mountains is discussed by a number of workers like KRÜGER (657), FRANZ (365), KRÜGER & DUSPIVA (658) and others.

7. LIGHT

As pointed out earlier in this chapter, the denser and moisture-laden layers of the atmosphere near the sea-level serve effectively to filter off a great part of the sunrays and thus act as a powerful protective blanket against the injurious parts of the sun's radiation. The relatively thin and dry air at high elevations on mountains offer, however, no such protection and therefore the intensity of the solar radiation increases appreciably with increase in altitude. The intensity of the direct sunlight in the high altitude biome above the forest-line on mountains is significantly greater than within the forest zone on the mountain or in any lowland open habitat. According to the estimates of ADAMS *et al.* (6), for example, it is known that, taking the mean total solar radiation received above the forest-line on Mt. Marcy as 100, the amount falling on the top of the fir-forest zone is 51, but only 4 within the fir-forest. Not only the direct sunrays are much more intense at high elevations than in lowland areas, but also the total light, including also the diffuse daylight and the light reflected from the surface of snow and ice, is very strong. There is also a conspicuously higher proportion of the shorter waves of sunlight, not only in the direct sunrays, but also in the reflected light. The general condition is that there is considerably more blue in the high altitude biome than in the lowland. The ecological importance of high intensity of light, especially the blue and the ultra-violet parts of the spectrum, is unquestionably great at high altitudes on mountains, but is at present largely obscure (286, 407, 516). It is known, however, that a high proportion of ultra-violet rays tends to strongly diminish the rate of oxygen consumption in aquatic organisms. Under the action of the ultra-violet rays, the proportion of oxygen dissolved in the water is lowered, but at the same time, the respiratory activity of the aquatic animals also seems to be greatly influenced by exposure to the action of ultra-violet radiation. The presence of a heavy body pigmentation in the high altitude insects is evidently closely correlated with the high intensity of light, especially the ultra-violet, against which the body pigmentation serves as a protective mechanism (730,733).

8. MICROCLIMATE

The difference between the general atmospheric condition, as indicated by ordinary meteorological observations, and the environmental conditions in the immediate surroundings of insects in their niches, of considerable ecological importance even in lowlands, tends to be very much more pronounced at high altitudes on mountains. Mainly owing to their size, characteristic properties of their bodywall and other peculiarities of their body organization, insects are exposed to conditions often wholly different from those which influence human beings and the larger animals (730, 988). The climate of the immediate surroundings of insects inside their special niches is in reality micro-microclimate or the climate of square and cubic centimetres and millimetres on or in the soil, under stones, on rock surfaces, inside rock crevices, in underground cavities, under snow, in ice crevasses, under vegetation mats and cushions of low-matted plants and in other similar situations. This greatly exaggerated importance of the microclimate is a direct consequence of the atmospheric cold and aridity at high altitudes on mountains. Other factors that underline this importance of microclimate include the ruggedness and irregularities of ground on the mountains, the more or less extensive alpine sculpturing of the mountains by glacier action, the presence of rock, snow and soil, the peculiarities of the trend-lines of the valleys and ridges, the crestlines, ridges, spurs, the alignment of the valleys and gorges, the location of high peaks, the disposition of large gerolls, boulders, erratics, perched boulders, avalanche debris, etc. The distance from the snow-edge, the altitude of the snowline, the amount and duration of the winter snowfall, high wind velocities, etc. also profoundly influence the microclimate in the high altitude environment. The effects of air viscosity, deep and shallow air currents depend largely on the peculiarities of the ground and largely modify the microclimatic conditions. The layer of air immediately above and in close contact with a rather rugged mountain ground is influenced much more by the ground irregularity than in the case of the air layer at higher levels above the ground. This layer of air does not also move as fast as the air at standard meteorological height above ground. Such slow-moving air does not, therefore, follow the general trend of the wind characteristic of the region, but often leaves behind pockets with wholly different climatic conditions. Wind shields near large stones and boulders, inside rock crevices, in spaces under large boulders, in ice crevasses, etc. have exposures to the local wind entirely different from the general average for the locality. Irregular islands of practically windless calms often thus exist in a vast wind-swept desert. As the increase in wind velocity is normally maximum on bare and exposed ground, vegetation cover of any kind, including even the low-matted dwarf plants or also the incrustation of lichen and cushions of moss, retards the extreme effects of strong winds and reduces the rate of evaporation. The difference between the macroclimate and the microclimate at high altitudes on mountains is principally the result of rapid and strong insolation and radiation and the thinner air, in complex interaction with diverse local conditions. The ecological importance of microclimatic conditions at high altitudes thus differs essentially from those prevailing near sea-level at higher latitudes in the north. Though the mean atmospheric temperatures are low even in summer at high altitudes on most mountains, rapid and strong insolation accounts for the relatively high temperatures in most microclimatic niches of the high altitude insects. Insolation becomes stronger and more rapid as the elevation increases, so that even above the permanent snowline relatively warm niches, suitable for insect life, are met with. Irrespective of the mean atmospheric conditions, the microclimatic conditions thus make possible niches of a wide variety of optimal conditions for a great many different types of insects, from the forest-line to elevations far above the permanent snowline.

a. Microclimatic Conditions on Rock

The effects of intense insolation and the rather wide range of microclimatic conditions may perhaps be most readily observed on barren rock even at elevations of 4000 m on most mountains. During the hours of bright sunshine and in the absence of clouds and strong winds, the surface of the exposed barren rock on the south slope of the Himalaya becomes warmed up rapidly on exposure to the sunrays, until the temperature of the rock surface rises to 30 - 35° C. At the same time, the atmospheric temperature in the shade may range from only 2° C to 7° C. On the shady side of the same boulder, perhaps only a few centimetres away, the temperature of the rock surface is no higher than that of the atmosphere. The temperature of the layer of air, about 5 cm above the warm rock surface, may

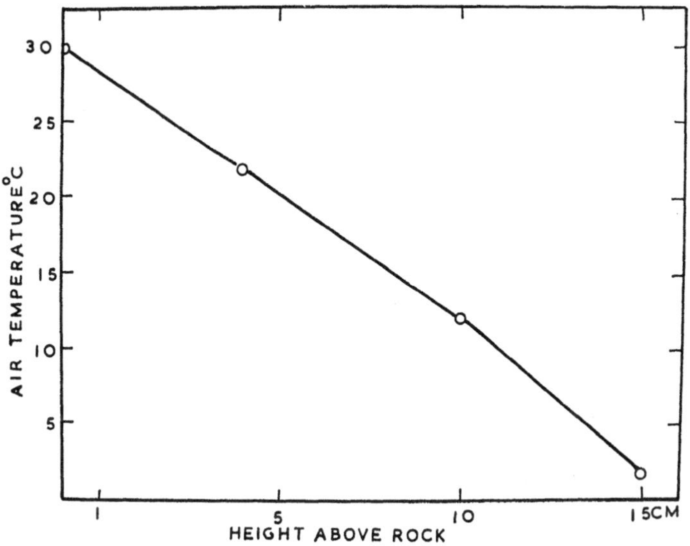

Fig. 14. The temperature gradient of the air above the rock surface, exposed to the action of insolation by direct sunrays, at an elevation of 3200 m, on the south slope of the Northwest Himalaya.

often be between 16° C and 25° C and at a height of about 10 cm between 8° C and 18° C (fig. 14). At greater heights above the warm rock surface, the ordinary atmospheric sun temperatures prevail. Nocturnal radiation from the rock surface is also extremely rapid, so that the temperature falls as low as the atmospheric minimum. The difference between the atmospheric and the rock surface temperatures is thus chiefly in the daily maxima and that also in direct sunshine only. The difference may amount to 10-30° C. SHREVE (966) reports this difference to be 31° C on the foothills of Mt. Catalina during August, when the atmospheric temperature maximum is 13° C. The rock surface is, therefore, characterized by rapid and great fluctuations of temperature, often as much as 37° C within the course of a few hours.

Consequent on the rapid warming, the humidity conditions of the layer of air, within about 5 cm above the rock surface, undergo rapid changes and the rate of evaporation becomes greatly accelerated. The saturation deficiency of the layer of air 5 cm above the warm rock surface (at a temperature of 29° C) is 10 mm at a temperature of 22° C at an elevation of 3200 m on the south slope of the Pir Panjal Range in the Northwest Himalaya. The saturation deficiency of the air at a height of 150 cm above the same rock surface amounted to 8.5 mm at a temperature of 14° C (the atmospheric temperature at standard meteorological height above ground at the time of the observation was 12° C). There is thus a sharp gradient of temperature-humidity conditions in the air in immediate contact with a rock surface exposed to the direct sunshine. The desiccating action of insolation and wind is thus significantly high on the rock surface.

While the conditions are so highly variable on barren rock surface, even a cover of lichen or moss tends to narrow down the range of these fluctuations. Inside the rock crevices and in the spaces in between sheltered rocks, the microclimatic conditions are, however, far more uniform.

b. *Microclimatic Conditions under Stones*

In sharp contrast to the rock surface, the temperature-humidity conditions of the air in spaces under boulders and large stones partly buried in the ground are more constant and optimal to a much larger number of species. For example, at an elevation of 4000 m on the south slope of the Northwest Himalaya, while the rock surface becomes warmed up to 35° C during the daytime insolation or also cooled to nearly —10° C during the nocturnal radiation, the temperature of the air in the hypolithic space fluctuates only between 5° and 7° C in the course of twenty-four hours during summer (early July) (fig. 15). At the same time the diurnal fluctuation of the atmospheric shade temperature near the stone may amount to 10-14° C. Sheltered from wind and from the intensive insolation, the rate of evaporation is also naturally low in the hypolithic space, so that the humidity conditions also remain fairly uniform and relatively high throughout. The relative humidity of the hypolithic air is often as high as near saturation and does not usually fall below 80-95% at the temperatures ranging from 18° C to 22° C. The average microclimatic conditions approximate to those prevailing in underground spaces and in the deep soil. In case of niches under stones still covered by the winter snow, the microclimatic conditions typical of winter at the elevation continue to prevail, even though summer conditions obtain in the open above, as long as the winter snow remains unmelted. The ecological importance of such hypolithic niches, in which the extreme atmospheric conditions are more or less greatly ameliorated and which provide refugial optima from the inhospitable conditions prevailing outside, increase as great elevations.

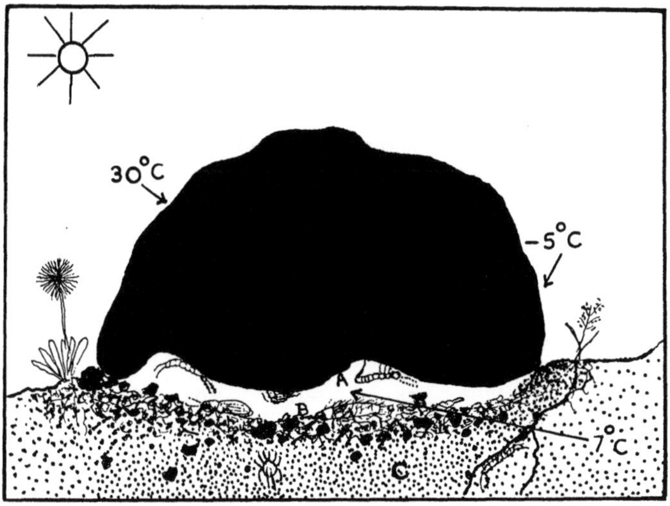

Fig. 15. The microclimatic conditions on rock and in the hypolithic spaces at high altitudes on the Northwest Himalaya. Diagrammatic representation of the ecological stratification of the hypolithic communities: the bottom surface of the rock, the soil surface in the hypolithic space and the soil under the stone.

c. Soil Temperature

The temperature conditions of the soil are of considerable ecological importance even for the lowland insects. Altitude enhances this ecological importance very greatly (596, 752). With the melting of the winter snow during summer, insolation results in the gradual warming up of the soil freshly exposed from the protective snow cover. The surface layers of the soil are more readily warmed by insolation and also radiate away the heat during the night more quickly than the deeper layers. The surface soil thus exhibits greater diurnal temperature fluctuations than the deeper layers. The diurnal temperature variations of the top soil are also greater than those of the atmosphere. It is also the top soil temperature maxima which have greater diurnal variations than the minima. The surface soil temperature maxima fluctuate more than the minima at high altitude on the Alps. BÄBLER (74) found, for example, that the top soil temperature reads 37° C at an elevation of 3647 m on the south slope of the Monte Rosa during August, when at the same time the atmospheric temperature was only —2° C. The difference between the atmospheric and soil temperatures at elevations of 1000, 1600 and 2200 m were 1.5° C, 2.4° C and 3.6° C respectively. The diurnal temperature fluctuations of the atmosphere and soil on some places of the Alps are summarized in table 5.

On the Himalaya, at elevations between 3500 and 4000 m on the south slope, the difference between the top soil maxima and the mean atmospheric temperature during the month of June is 13.5° C, but only 5.0° C in the case of the minima (730). The diurnal range of atmospheric temperatures at an elevation of about 3500 m on the south slope of the Himalaya is about 16.8° C, but at the same time the temper-

TABLE 5

Mean diurnal fluctuations of ground and atmospheric temperatures at different altitudes on the Alps

Altitude in m	Exposure	Gradient	Ground temperature Min.	Max.	Diff.	Atmospheric temperature
2692	W	20	−1.0	24.0	25.0	5.0
2730	N	6	1.5	14.0	15.5	2.0
2743	S	38	0.0	5.0	5.0	−2.0
2743	S	38	−0.5	4.5	5.0	1.0
2750	SW	2	14.0	15.5	1.0	12.0
2790	SSW	26	—	22.0	—	13.0
2800	SE	28	−1.0	7.0	8.0	4.5
2840	ESE	32	1.0	27.5	26.5	4.5
2847	W	4	−3.0	22.0	25.0	7.0
2893	peak	—	2.5	9.0	7.5	8.0
3000	NE	20	−1.0	24.0	25.0	4.5
3000	ENE	22	3.0	27.0	24.0	—
3000	S	24	1.0	33.0	32.0	9.0
3058	NE	44	−3.0	4.0	7.0	−1.0
3100	SE	42	−2.5	18.0	20.5	5.0
3190	SSW	48	−4.5	7.0	11.5	−2.0
3200	ridge	—	−2.0	37.0	39.0	−2.0
3237	SE	22	−3.5	22.0	25.5	1.0
3305	S	40	−4.5	15.5	20.0	3.0
3570	SSE	42	−4.5	16.0	20.5	−1.0
3647	S	46	−2.0	37.0	39.0	−2.0

ature of the top soil shows a diurnal range of 30° C. The mean atmospheric temperature and the mean maxima and minima of the top soil temperatures, at elevations between 3000 and 5200 m on the Himalaya are summarized in table 6.

On the Pamirs Region, with the atmospheric temperature in the shade reading 14-15°C, the ground temperatures average often 28°C. REINIG (878) has summarized (from OLUFSEN's records) the hourly temperatures of the atmosphere and soil and the intensity of insolation during July in the Alai Valley and in Jashil-kul. It is interesting to note that in correlation with the ecological peculiarities of the Pamirs Region, the surface soil temperature fluctuations during the day amount to about 35° C, while the atmospheric temperature fluctuation amounts only to 12° C. In the Jashil-kul and Alai areas, the fluctuations of atmospheric and surface soil temperatures during the day reveal more or less similar peculiarities (Table 7).

SCHRÖDER's observations (934) on the differences between the atmospheric and ground temperatures due to insolation on a bright sunny day, at about 2 p.m. during February 1912, at an elevation of about 4690 m on Mt. Kilimanjaro are extremely interesting. Insolation on sandy ground showed a maximum temperature of 39.6° C, but a moving thermometer at the same time registered only 7.2° C. The solar thermometer showed as much as 84.5° C. The atmospheric temperature at 7 p.m.

TABLE 6

Atmospheric and soil temperatures at elevations between 3000 and 5200 m on the south slope of the Himalaya during June 1954-1956

Altitude in m	Mean air temp.	Ground mean	Surface temp. Max.	Min.	Range	Soil temp. at depths between 15 and 20 cm
3000	9.5	30.0	33.0	1.0	34.0	19.0
3100	9.5	28.0	32.5	−2.0	34.5	17.2
3200	7.0	26.0	28.0	−2.0	30.0	15.5
3300	4.5	20.0	32.5	−4.0	36.5	13.0
3500	6.6	17.0	32.5	2.5	35.0	12.0
3600	3.0	10.4	22.2	−1.5	23.7	11.2
3700	3.2	9.5	22.0	−2.5	24.5	10.4
3800	3.0	8.2	21.0	−4.0	25.0	8.5
4000	1.5	4.5	14.0	−4.5	18.5	8.0
5200	−4.5	2.5	11.0	−5.6	14.6	6.5

sank to —3.4°C. The soil temperature maximum was 18.8° C and the minimum during a clear night —8.6° C. The magnitude of insolation differs within wide limits, depending upon the latitude, altitude, distribution of land and water, topographical peculiarities, etc. In the north temperate regions during the summer months, insolation results usually in plus-summation of the soil temperature, so that the mean soil temperature gradually rises as the summer advances. SHREVE (966), who studied the soil temperatures at elevations of 184 m, 1219 m, 1524 m, 1820 m and 2436 m on the foothills of Santa Catalina near Tucson, Arizona, reports a fall of the soil temperature with the increase in altitude. He also found that the rate of rise from the minimum to the maximum is more rapid than the corresponding fall from the maximum to the minimum.

The fluctuations of the surface soil temperatures at high altitudes are paralleled by those on the surface of a planet like Mars with a thinner atmosphere than the Earth. Fig. 16 shows, for example, the temperature conditions observed on the Himalaya and those on Mars observed during 1926-1943 in the Lowell Observatory. The similarity of the temperature fluctuations of the ground at high altitudes on the Himalaya and on Mars is extremely striking. The high intensity of insolation due to the thin atmosphere is the underlying factor both on the Himalaya and on Mars.

As may be expected, the temperature of the soil tends to be more uniform as the depth increases, so that extreme temperature fluctuations are restricted only to the surface. The temperatures at different depths of the soil, at different elevations on the south slope of the Himalaya, show, for example, that below a mean depth of about 15 cm, the soil temperatures are relatively uniform during the course of the day (Table 8).

TABLE 7

Diurnal fluctuations of atmospheric and soil temperatures in the Pamirs Region in summer

Locality and altitude in m	Hours of the day															
	6	7	8	9	10	11	12	13	14	15	16	17	18	19	20	21
Alai Valley 3145																
Atmospheric temperature	7	9	11	14	16	18	17	19	18	17	15	14	13	10	9	—
Soil temperature	10	15	22	28	36	45	42	42	38	34	28	23	19	13	11	—
Insolation	—	—	1.45	—	—	1.54	—	—	—	—	—	—	1.16	—	—	—
Jashil-kul 3982																
Atmospheric temperature	—	—	10.1	12.8	15.6	19.0	20.0	21.0	20.8	21.2	20.0	19.8	17.8	14.5	8.5	8.0
Soil temperature	10.4	11.2	21.8	30.8	25.0	23.5	23.8	22.0	21.1	19.0	17.0	15.0	13.1	12.8	—	—

TABLE 8

Soil temperatures in summer at different depths on the south slope of the Himalaya

Elevation in m	Mean atmosph. temperature	Top soil temperature Max.	Min.	Deep soil temperature 5 cm depth Max.	Min.	15 cm	20 cm
3200–3500	6.6	32.5	2.5	22.3	3.6	15	15
5000–5200	1.5	11.0	−5.6	11.0	2.5	7.2–3.0	6.5–2

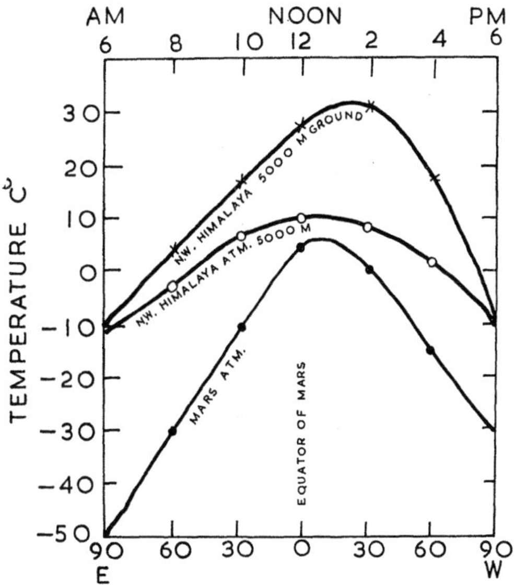

Fig. 16. Diurnal fluctuations of the mean atmospheric and surface soil temperatures at high elevations on the Himalaya and the atmospheric temperatures on the surface of the planet Mars. The marked similarity of the ground temperatures at high elevations on the Himalaya and the air temperatures on Mars is the result of the semi-rarefied atmosphere and intensity of insolation in direct sunshine.

Below a depth of about 15 cm there is, therefore, practically very little or no diurnal fluctuation in the soil temperature. The temperature of the deep soil thus tends to remain constant, irrespective of the fluctuations of the top soil temperature and of the atmospheric temperature. On the Swiss Alps also, the soil temperature at a depth of 10 cm was found by BÄBLER (74) to be 19° C at an elevation of 2790 m, while the atmospheric temperature was 13° C. The soil temperature fluctuations at different depths on the alpine meadow on the Central Tien Shan at an elevation of about 3000 m during August are summarized in table 9.

It is interesting to observe here also that as the depth of the soil increases, the diurnal fluctuations of the temperature become progressively smaller, approaching more uniform conditions.

TABLE 9

Soil temperature fluctuations during the day on an alpine meadow at an elevations of 3000 m on the Central Tien Shan during August

Depth of soil in cm	Hours of the day 14	16	18	20	22	Min. temp.	Range
Surface	43.0	36.4	24.2	3.2	2.0	−2.5	45.5
25	20.7	21.5	18.7	7.6	4.6	1.5	20.0
50	19.3	19.8	17.4	8.9	6.9	4.6	15.2
100	18.4	18.8	16.4	9.2	8.3	5.8	13.0
150	17.0	15.0	15.9	9.2	8.5	6.4	10.6

These remarks hold good only in the case of soil protected by snow-cover during the winter and also after the winter snow has started melting in the summer. With the increasing distance from the edge of the melting snow, there is usually a pronounced rise in the mean values of soil temperature maxima and minima. The mean soil temperature at a depth of 5 cm from the surface on the Himalaya increases more rapidly than at a depth of 20 cm. At a distance of about 10 m from the edge of the receding snow, the top soil temperature fluctuation amounts to nearly 16° C, but below a depth of 20 cm it amounts to only 1° C. At a distance of about 200 m from the snow-edge the diurnal temperature fluctuation in the top soil is 11° C, but only 2° C at depths below 20 cm. Soil temperatures at different depths and at different distances from the edge of the receding winter snow at an elevation of about 5000 m on the south slope of the Himalaya during the summer are summarized in table 10.

TABLE 10

Soil temperature gradient with increasing distance from the snow-edge on the Himalaya

Distance from snow-edge in m	Soil depth cm	Temperature Max.	Min.	Range
10	5	8.0	−8.0	16
100	5	11.0	4.0	15
200	5	16.0	5.0	11
10	20	3.0	2.0	1
100	20	7.0	6.0	2
200	20	10.0	8.0	2

We may summarize some of the observations on soil temperatures at high altitudes on mountains as follows: The mean soil temperatures tend to decrease with increase in altitude, but much more gradually than the atmospheric temperatures.

The soil temperature maxima show wider differences with differences in altitude than the minima. On mountains in the northern hemisphere, the southern slopes exhibit somewhat higher soil temperatures than the north slopes at corresponding elevations. The ecological importance of soil temperatures at high altitudes on mountains includes the influence on germination of seeds, growth of roots, depth of penetration of melt-water, the rate of evaporation or transpiration of the soil moisture, the optimal conditions for hibernation of the high altitude insects, etc. The soil temperature influences also the temperature of the layer of air immediately above the ground and in contact with it. The ecological importance of soil temperatures increases with the increase of altitude; its importance at very high altitudes is far greater than within the forest zone on mountains.

d. Soil Moisture

Next only to the soil temperature is the ecological importance of soil moisture at high altitudes. The soil moisture is great at the edge of the receding winter snow, near melt-water torrents, glacial lakes and ponds. The deep soil has also a higher moisture content than the surface soil. The soil at high altitudes, not covered by the winter snow or also only covered by scanty snow during the winter, is permanently frozen on some mountains. Permafrost has, for example, been described at high altitudes on Mt. Kilimanjaro and even on Mt. Washington. With the atmospheric temperature during the winter reaching to —13.7° C and the snow-cover not exceeding 27 cm, there is ground frost to nearly 0.75-1.3 m or even 2 m depth on Mt. Washington. Ground frost has also been reported in the black soil at an elevation of 1470 m (Madison Springs) during July, to a depth of 0.61 m and often extending down to 1.2 m depth. During August the ground frost was recorded at 0.9 m down to 1.8 m at an elevation of 1525 m. Permafrost depends on atmospheric temperature and snow-cover—low atmospheric temperatures and scanty snow of short duration favour permafrost at high altitudes. The ratio of the winter temperature in °C at high altitudes on mountains in the northern hemisphere to the mean thickness of the winter snow-cover measured in centimetres during January must be below 0.5 for permafrost to occur (933).

9. SLOPE ASPECT AND MASSIVENESS OF THE MOUNTAINS

On the mountain ranges like the Himalaya, the Caucasus, the Alps or the Pyrenees, which stretch nearly east-west in the northern hemisphere, especially north of the Tropic of Cancer, the environmental conditions and the biota at high altitudes are usually markedly different on the northern and southern slopes. The influence of altitude on the mean environmental conditions is profoundly modified by the differences of the north-south slope exposure. These differences are ultimately to be traced to the differences in the angle of incidence of the sunrays and therefore to the degree of insolation and often also to the differences in the gradient of the opposed slopes of the mountain range.

The differences of the mean atmospheric temperatures on the opposed slopes immediately above the treeline on a mountain may often amount to 10-15° C. There is, for example, on the Himalaya a mean altitudinal tilt of about 500-600 m from the south to the north in the isotherm of the mean atmospheric temperature during summer. The south-north tilting of the isotherm of the mean atmospheric temperature is subject to considerable variation on different mountains and even on the same mountain range. The differences generally tend to diminish appreciably and may even become obscured at very high elevation, especially above the permanent snowline. As already indicated, even the snowline differs on the north and south slopes of the same mountain range. The influence of the north-south aspect is perhaps more pronounced in the case of the soil temperatures than even of the atmospheric temperatures. The soil temperatures are, for instance, higher on the south slope than on the north slope at the same altitude on a mountain. The soil temperature maxima show, as a rule, greater differences than the minima on the opposed slopes. Above an elevation of about 4500 m on the Northwest Himalaya, for example, there is, however, very little difference in the soil temperature maxima of the north and south slopes. The soil temperature maxima of the north and south slopes at these elevations are nearly the same in most cases. The difference between the atmospheric temperature and the soil temperature tends to increase with altitude on the south slope of the Himalaya far more rapidly than on the north slope. The soil temperature maxima are usually reached before 13 hours during the summer on the north slope, but only sometimes later on the south slope of the Northwest Himalaya. The differences in the surface soil temperatures on the north and south slopes of the Northwest Himalaya are summarized in table II.

SHREVE (966) studied the influence of north-south exposures on the soil temperature at different altitudes on mountains. He found that the difference between the soil temperature on the north and south slopes is very great at higher latitudes. The north-south slope differences in the soil temperatures studied by SHREVE at an elevation of 2745 m on the Santa Catalina Mountains are shown in fig. 17. The north-south slope differences are not, however, very pronounced under arid conditions and are also often more or less obscured under extreme moist conditions of the soil, but are most marked under intermediate moisture conditions. The life zones on mountains also exhibit the characteristic north-south tilt, corresponding to that of the isotherm of the altitude of mean atmospheric temperatures. The vegetational character and the insect life of the north and south slopes are often fundamentally different, particularly to be observed in the case of the Himalaya.

TABLE II

Differences of the surface soil temperatures on the north and south slopes of the Pir Panjal Range of the Northwest Himalaya during May-June

Altitude in m	Max. temp. North	Max. temp. South	Min. temp. North	Min. temp. South	Mean temp. North	Mean temp. South
3000	12.0	20.0	8.0	15.0	10.0	16.4
4200	10.0	15.0	1.5	2.5	6.2	8.2

The atmospheric and to some extent even the ground temperatures are known to be appreciably higher at corresponding elevations and latitudes on massive mountains than on the lesser and isolated mountains that, like the Mt. Marcy or Mt. Kilimanjaro, rise abruptly from a peneplain. This difference is to be traced primarily to the effect of the ground temperature on the layer of air which is in immediate contact above. The temperature lapse rate with the increase of elevation is also generally smaller on massive mountains than on isolated ones. The diurnal variations of both the atmospheric and the ground temperatures are also conspicuously great on the massive mountain ranges.

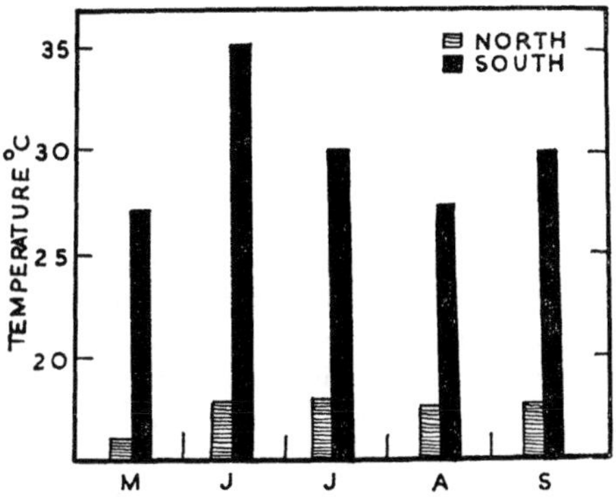

Fig. 17. Differences in the soil temperatures at an elevation of about 2745 m on the north and south slopes on the Santa Catalina Mountains, from May to September.

10. GRAVITY

As already indicated at the beginning of this chapter, all the complex environmental peculiarities of high altitudes on mountains are essentially part of the ecological importance of gravity, resulting from the stratification of the atmosphere under the action of gravity. The altitudinal zonation of life on mountains, the progressive impoverishment of species with the increase in altitude and all the other peculiarities associated with high mountain biota are results of the action of gravity on the atmosphere. Attention is being increasingly directed in recent years to the effects of abrupt changes of gravity on organisms, but the biological effects of constant gravity of different intensities have largely been neglected. High mountain habitats provide numerous opportunities for observations on this aspect. There is at any rate very little doubt about the importance of gravity as an environmental factor at high altitudes on mountains like the Himalaya. The enormous masses of rock that compose these mountain regions exert without doubt an appreciable force of attraction, as is readily evident from the deflections of the plumbline and pendulum experiments. Of the various biological effects of different constant gravities, the generally small size of the high altitude insects on massive mountains seems to be closely correlated, at least in part, with the increased gravitational effect. The possibility should not also be overlooked that even in the case of human beings, the short stature of the local inhabitants of the high Himalayan valleys is due to the effects of this gravity. It would seem that the difficulties of movement, the great effort required for even relatively small work and the rapid fatigue on even slight exertion at high elevations on the Himalaya may not all be due to simple oxygen deficiency, as has been generally assumed at present, but may be attributed, at least in part, to the gravitational differences between mountain and lowland regions. The skeletal muscles are normally adapted and adjusted for the gravity on the lowlands and do not, therefore, seem to respond immediately at high altitudes on massive mountains, until the necessary acclimatization has been achieved. There is of course a very urgent need for intense research on organisms exposed to different gravitational fields.

II. BIOTIC FACTORS

a. *Absence of Trees: Open Habitats*

The great majority of the biotic factors of the high altitude environment are ultimately to be traced to the absence of trees and to the restriction of the growth period of the scanty, characteristically low-matted dwarf vegetation to the all-too-short summer. Considered from this point of view, the high altitude environment deals with open habitats. In addition to the effects on the formation and character of the soil, the absence of trees exerts a far-reaching influence on nearly every aspect of the high altitude insect life by greatly accentuating the effects of reduced atmospheric pressure. The absence of trees, by increasing exposure, favours rapid and strong insolation and radiation, widens the amplitude of temperature-humidity changes of the soil, assists rapid transpiration and desiccation, eliminates wind shield, restricts the food resources, deprives the high altitude insects of a number of important microclimatic niches like the tree-bark, tree-trunk, foliage, etc. In a forest, the intensity of light on the forest-floor, covered by forest plants growing in the shade, amounts often to only about one-hundredth or even one-thousandth of that on the top of the forest (711). GEIGER (402, 403) has also given an interesting description of the differences introduced by the absence of trees in the intensity of solar radiation at high altitudes on mountains.

The principal biotic factors of importance to the high altitude insects include the nutritional conditions, the effects of crowding and isolation in the ecological niches. Most of the other biotic factors can be traced eventually to the numerous inter-specific relations at high altitudes.

b. *Nutritional Conditions*

The most striking peculiarity of the nutritional conditions of the high altitude insects is the severe restriction of the active feeding period to the short summer on most high mountains. On some of the mountains, especially in the northern hemisphere, the total period of active feeding is perhaps no more than a few sunny hours in the whole year. Even this is often irregular and frequently interrupted by unexpected unfavourable conditions. Many species are, therefore, often unable to continue feeding throughout the summer, when snow storms and avalanches may force them back to their winter quarters or perhaps refreezing after an early thaw may result also in high mortality. In nearly all cases, no feeding is also possible until all the winter snow has melted and the melt-water has revived the dormant high altitude plants to active germination and growth.

The abbreviation of the active feeding period becomes more and more pronounced with the increase in altitude and in the neighbourhood of the permanent snowline on high mountains like the Himalaya, it may sometimes be interrupted for one or even two years in succession. The mean total period of active feeding for most insects is about ten weeks in the altitudinal zone between 3000 and 4000 m on the south slope of the Northwest Himalaya. The period begins somewhat later and is also of much shorter duration on the north slope than on the south. Except in the case of the true soil forms, the feeding period is restricted to warm sunny hours even in the case of underground species.

The food resources at high altitudes are extremely irregular, relatively scanty and often greatly restricted to localized patches. Ecologically considered, there are two principal sources of food at high altitudes. The autochthonous sources comprise the material derived from the mountain—from the plants and animals that belong to the high altitude environment. The high altitude vegetation, like the algal slime in the streams, lichen, moss and the dwarf matted plants, represents the foundation of the food pyramid, on which the lives of all depend, directly or indirectly as the case may be. The second and perhaps astonishingly far more important source of food at high altitudes is the wind-blown organic matter from the surrounding lowlands.

The most abundant supplies of food, especially locally enriched organic matter, are generally at the snow-edge. A great variety of organic debris, such as decaying vegatable matter and dead high altitude animals, is to be had here. Vast hordes of high altitude insects are attracted by the enormous amount of decaying vegetable matter and countless dead and decaying insects and other animals, which come to be exposed from under the edge of the receding snow in summer. The melt-water soaks through and softens the organic matter and thus makes it readily available for the high altitude insects. It is also here that, activated by the melt-water and the heat of the summer sun, the seeds of the high altitude plants germinate and numerous plants are rapidly sprouting up. Debris feeders, carrion feeders, scavengers, plant feeders, and predators alike are assured of abundant food at the snow-edge.

Not only the snow-edge, but also the very surface of the snow, ice and glacier constitutes an important foraging ground for the high altitude insects. The snow-fields and the glacier surfaces are truly littered with a choice variety of refrigerated food particles for an astonishing group of high altitude insects on nearly all the mountains of the world. It sounds incredible that this refrigerated food supply is not really of autochthonous origin, but is largely derived from the surrounding lowlands. The great bulk of the food particles found in such situations is in reality air-lifted from lower elevations and often from far-off plains of the surrounding country. Countless millions of inorganic and organic dust particles, including the spores of fungi, pollen grains of diverse plants, seeds, insects, spiders, etc. are lifted up from the plains by air currents, blown by the upper-air winds, suddenly chilled and finally cast on the snow-fields on high mountains. On the extensive glaciers and snow-fields on high mountains one often comes across large numbers of insects that are really and typically inhabitants of the plains and of lower elevations, valleys, etc. and have been brought here by wind. These insects are as a rule found frozen and dead on the ice and snow on the mountains. There are numerous scattered references to such wind-blown organic derelicts on snow-fields on high mountains from different parts of the world. The ecological importance of such wind-blown derelicts at high altitudes on the Alps has been known since 1881 and considerable literature has accumulated in recent years (192, 194, 200, 274, 275, 303, 310, 326, 340, 366, 411, 452, 456, 476, 600, 601, 621, 730, 734, 751, 797, 987, 1108, 1114, 1115). MANI (730) has described the wind-blown derelicts on the snow-fields of the Himalaya as consisting of aphids, jassids, aleurodids, winged ants, gnats, midges, flies, fruitflies, beetles, butterflies, moths, grasshoppers, spiders, mites, etc., in addition to bits of leaves and seeds of a great many different kinds of lowland plants. SWAN (1022) has recently emphasized the great ecological importance of these wind-blown derelicts on the Himalaya and has indeed separated a distinctive aeolian zone at these extreme high altitudes, characterized by the fact that the food resources are predominantly wind-blown from the lowlands. ERHARD (322) and STAUDER (997) have described the lifting up of millions of butterflies *Pieris* SCHR. by warm air currents far above the high peaks on the Alps, to be frozen and thrown back on glaciers at night. There are records of great swarms of grasshoppers like *Melanoplus* STÅL. being carried by air currents from far-off areas in the plains and frozen and preserved in ice on glaciers on high mountains. Myriads of the grasshopper *Caloptenus spretus* UHLER swarm from the Pacific Coast slopes of the western highlands during July-August and in their eastward journey, they are arrested by the high mountains, chilled and cast dead upon snow-fields of the Rocky Mountains in North America. We have the well known Grasshopper Glacier of Montana in the western United States, in which are stratified layers of frozen remains of grasshoppers, accumulated in successive years (456). The accumulations of dead grasshoppers are estimated to be at least 600 years old. Among the wind-blown insects commonly found on the Alps are Coccinellidae, Ipidae, *Pieris brassicae* LINN., *Pieris rapae* LINN., *Vanessa urticae* LINN., *Plusia gamma* LINN., *Vespa rufa* LINN. on the Oberaar Glacier at an elevation of 2900 m and the Unteraar Glacier at an elevation of 2600 m, *Coccinella septempunctata* LINN. on glaciers at elevations of 2800-3000 m, *Halyzia sedicimguttata* LINN. on snow and glaciers at elevations of 2800-3170 m, *Gaurotes virginea* LINN. at 3000 m, *Macroglossa stellatarum* LINN. at elevations of 3000-3250 m, etc. (469). BOWDITCH (144) has also referred to the common occurrence of numerous wind-blown beetles on Mt. Washington. Recently WILSON (1114, 1115) has reported on wind-blown organic matter on snow-fields on high mountains in New Zealand as containing even material from sea, found deposited at elevations of 2440 m. In addition to the diverse insects, spiders, etc. from the plains, quantities of insects from the lower valleys are also regularly lifted by strong air currents, to be suddenly chilled and cast ultimately on the high mountain slopes. On all mountains, except perhaps those situated in the extreme high latitudes, during the warm months and in bright sunny weather, updraft air currents rise from low hills and from the valleys below in the forenoon, attain their maximum often exceeding an upward velocity of 40 m/sec. in the afternoon and declining about sunset. The updraft air currents lift various flying insects, balooning spiders, etc. from the valleys and carry them up to the mountain summits. The downward night winds that often blow from the mountain summits to the valleys do not, however, bring down the high altitude insects, because none of them are on their wings at night on the mountains. Even in the absence of strong winds, under ordinary atmospheric conditions, a most remarkable variety of particles like pollen grains, spores of fungi, bacteria, small seeds, Protozoa, insects, spiders, etc. are carried aloft from the plains. The intensive heating of the ground causes vertical currents, by which even heavier bodies are lifted upward. The air being calm near the ground, such convection currents are surprisingly strong and lift relatively heavy insects that happen to be on the wing to great altitudes, where they are suddenly chilled and killed. Unlike those near sea-level, the air currents at high altitudes are stronger and more constant in direction and thus transport the uplifted insects often over long distances, across vast plains. GLICK (411) found, for example, that aphids are frequently carried in this way over 1280 km and the gypsy-moth over 180 km at an altitude of 4267 m. JOHNSON & TAYLOR (595) have recently reported that about 92 000 flies of *Oscinella frit* (LINN.) were air-borne to an elevation of 1500 m in this manner. KISHKO (621) has found an abundance of bacteria and fungal spores; fungal spores are abundant up to elevations of over 3000 m; the bacteria predominate over fungi in the air-borne spore material up to elevations of 6000 m, above which the fungi gain dominance. The wind-blown debris accumulates not only on the glaciers and snow-fields, but also collects in rock crevices and other sheltered situations at high elevations on mountains.

Such frozen wind-blown organic material constitutes perhaps the most important source of food for the greatest majority of high altitude insects. Even VAN DYKE (1078) observed the frozen remains of flies, ants, etc., on the snow-fields of Mt. Rainier and Mt. Lyell in North America to attract enormous swarms of predators like *Bembidion* LATR., *Platynus* BON., etc. The recent observations of MANI (730) and of SWAN (1022) on the Himalaya emphasize the great importance of such wind-blown debris on high mountains and it is indeed evident that as the elevation increases, especially above the permanent snowline, the aeolian food source is perhaps the most important for the high altitude insects. SWAN has suggested that even green plants thrive on snow at great elevations on the Himalaya, on account of the nitrogenous material contained in the wind-blown organic debris on the snow.

c. *Crowding and Isolation*

Although the general effects of crowding on insects have long since been recognized (28, 334, 464, 621, 705, 720, 773, 965), its ecological importance seems to have been completely overlooked so far. MANI (730) recently drew attention to the possible influence of crowding and its correlation to a complex of high altitude insect life, their structure, habits and life-cycles. His observations on the Northwest Himalaya indicate that the frequency and intensity of crowding of the high altitude insects in their optimal microclimatic niches increase with elevation. As normal activities are possible for these insects only in these niches, a great many different species, often taxonomically wholly unrelated, are brought together, leading to pronounced overcrowding during the major part of their development, growth, hibernation and metamorphosis. In addition to very profoundly influencing the microclimatic conditions, crowding in the insect niches at high altitudes affects the rate of growth, the size, structure, colour, general metabolism, sex and other characters of the high altitude insects.

The heavy concentrations of individuals of high altitude insects in localized islets of widely separated and sparsely scattered microclimatically optimal niches have thus the remarkable effect of breaking up the populations of high altitude insects into numerous small groups, so that a high degree of isolation is characteristic of high altitudes in general. Every niche may perhaps be described as an ecosystem by itself, almost completely isolated from other similar ecosystems genetically. Isolation seems to be of far greater ecological significance at high altitudes than in lowland biota.

12. INTERDEPENDENCE OF THE LOWLAND AND HIGH ALTITUDE ENVIRONMENTS

We have presented in the foregoing pages the salient features of the high altitude environment, as distinct from the lowland environment. The emphasis on the fundamental differences between these two extreme environments must not, however, be interpreted to mean that these different environments are in any sense independent. Not only a high and massive mountain system, like for example the Himalaya, profoundly influences the climate and vegetation of the surrounding lowlands, but also at the same time the prevailing climatic and vegetational characters of the surrounding lowlands more or less greatly accentuate, ameliorate or otherwise modify the altitude-conditioned environment of the mountain system. It is well known, for example, that the climate and vegetation of the whole of India are very profoundly influenced by the existence of the high Himalaya in the north. The pattern of wind and rainfall distribution in India is determined by the Himalaya. The presence of the Himalaya is also one of the principal factors for the pronounced aridity and other vegetational peculiarities of the Middle Asiatic regions to the north of the Himalaya. The general high altitude environment on Mt. Kilimanjaro is largely the result of this mountain being situated on the Equator. The high altitude environment modifies the lowland environment and is in its turn also modified by the lowland environment. We observe, therefore, that the high altitude environment differs on different mountains, situated in different climatic regions of the earth. As is well known, broadly speaking, there are three principal climatic regions in the world at present, viz. the tropical belt, the north temperate and the cold belt and the south temperate and cold belt (fig. 18). The tropical regions are characterized by the absence of sharply defined temperature-conditioned seasons. The north temperate and the north cold regions of the world are large landmasses, with typically continental climate, characterized by sharp temperature-conditioned seasons. In the southern temperate and cold regions of the world, we find relatively narrow meridional land areas or numerous scattered islands, surrounded by extensive oceans, so that the general climate is more oceanic and nearly isothermal conditions prevail the whole year round. These fundamental differences in the general climatic conditions of the lowlands very profoundly modify the high altitude environment. The environment at high altitudes on the tropical mountains is predominantly conditioned by diurnal temperature fluctuations. The high altitude environment on the mountains of the north temperate and boreal cold regions of the earth is characteristically seasonal-thermic. The high altitude environment on the mountains of the southern hemisphere is typically isothermal (fig. 18). The influence of the local conditions and the modifications of the high altitude environment by the lowland environment are summarized in table 12.

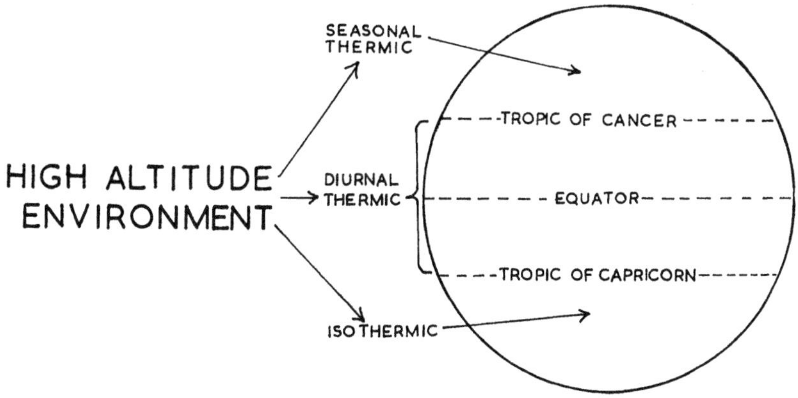

Fig. 18. The modification of the high altitude environment by the prevailing climatic conditions on the surrounding lowlands in different climatic zones of the earth. The high altitude environment in the northern latitudes is chiefly a seasonal-thermic one, with marked differences in the mean winter and summer atmospheric temperatures. In tropical lands, the high altitude environment is characterized by dominance of the diurnal range of atmospheric temperature over the seasonal, so that there is no sharp difference between the winter and summer, but we find nocturnal sub-zero temperatures every day at sufficiently high altitudes. In the southern latitudes, with vast areas of oceans, the high altitude environment is typically isothermic.

TABLE 12

The influence of the local climatic conditions on the high altitude environment on mountains situated in different climatic zones of the earth

Mountains in the north latitudes	Mountains in the equatorial/tropical areas	Mountains in the south latitudes
Climate: continental: seasonal thermic; seasonal temperature fluctuations greater than the diurnal and ecologically also dominant.	Climate: diurnal thermic: tropical.	Climate: Oceanic; isothermic
Atmospheric cold due to the slanting sunrays and thin air.	Atmospheric cold due only to thin air.	Atmospheric cold due slanting sunrays and thin air.
Insolation intense due only to the thin atmosphere.	Insolation intense due to i. vertical sunrays and ii. the thin atmosphere.	Insolation due only to the thin atmosphere.
Winter snow-cover pronounced.	No superabundant winter snow, but nocturnal frost, often though slight, every day in the year.	Neither winter nor nocturnal frost great.
Hibernation in winter.	No winter hibernation, but nocturnal dormancy every day.	–

On ecological, historical-geographical, evolutionary and other considerations, we shall, in subsequent chapters, attempt to describe the characteristic high altitude environment on the principal mountains of the world, their high altitude ecology, insect life and biogeography, under these major climatic zones of the earth. Brief accounts of the vegetational peculiarities of the mountains will also form part of the high altitude ecology of the mountains.

CHAPTER III

ECOLOGICAL SPECIALIZATIONS OF HIGH ALTITUDE INSECTS

This chapter presents a broad outline of the outstanding ecological specializations of the high altitude insects, with particular emphasis on the fundamental differences between the high altitude and lowland insects.

1. THE CONCEPT OF HIGH ALTITUDE SPECIALIZATION

The high altitude insects are an ecologically highly specialized group. Their specialization is total and represents the interaction of the high altitude insects with the complex high altitude environment in its totality. It would indeed be difficult to point out any single peculiarity of the high altitude insects as an exclusively high altitude specialization. We find, for example, nearly every one of these peculiarities even in the lowland forms. The pronounced melanism of the high altitude insects is by no means confined to high altitudes, but as is well known, melanism is common among the insects of the forest and the plains. Wing atrophy, so characteristic of the high altitude insects, is likewise equally widespread at sea-level. Prolonged hibernation under the protective snow-cover is not again confined to high elevations only. The question thus naturally arises: in what way are then the high altitude insects specialized? The high degree of specialization of the high altitude insects lies in the specific combination, at one place and time, of all these and a number of other complex characters—a unique combination associated with an equally unique combination of environmental conditions not met with elsewhere. It is the complex combination of melanism, wing atrophy and diverse other peculiarities of the body organization, development and habits, in intimate interaction with the high altitude environment in its totality, that really constitutes the ecological specializations of the high altitude insects. It would, therefore, be meaningless to speak of specialization with reference to any single isolated feature. The whole body organization of the high altitude insects and the sum-total of their habits and life-histories, taken together in their totality in the context of the high altitude environment, constitute the high altitude specialization.

The intricate complexity of the problems does not, however, obscure the major trends of specializations. The ecological specializations of the high altitude insects are in the main directed towards the utilization of some of the extreme environmental factors above the limits of the forest, so as to counteract the unfavourable effects of these and other extreme conditions. Paradoxically enough, some of the chain-effects of the reduced atmospheric pressure are taken advantage of by the high altitude insects in such a manner that the unfavourable effects of the other environmental factors are largely minimized or even counteracted. Some of the extreme factors have so modified the habits and the associated structures of the high altitude insects that this modification itself enables them to flourish under the whole set of environmental conditions. This specific utilization of the environmental factors is unknown among the insects of the forest and prairie near the sea-level. Although the specializations for life at high altitude on mountains thus involve the whole body organization and all the habits, it would be wholly outside the scope of the present volume to attempt an analysis of all of them. We shall, therefore, confine ourselves in the following pages to a broad general outline of only the more important or obvious characters and habits.

2. HIGH ALTITUDE MELANISM

Pronounced dark pigmentation of the body is one of the most striking characters of nearly all high altitude insects. Besides black, dark brown or reddish-brown, other colours commonly met with are red, orange and deeper tones of yellow than usually observed on the plains. Not only is the general ground colour on the whole darker in tone than in the lowland forms, but even the spots, bands and other markings on the body and wings tend to enlarge and often fuse together. The predominant colours at high elevations appear on the whole to belong to the group complimentary to blue or those that absorb or reduce the high intensity of the blue prevailing at high elevations above the forest on mountains. With the possible exceptions of a few true deep-soil forms, pale coloured or colourless species may be said to be absent at high altitudes. It is also common observation that the same species, which often occurs both below and above the forest-line on a mountain, is lighter coloured at lower elevation and within the forest zone, but conspicuously darker above the forest-line and at higher elevations. SCOTT (950, 951) observed, for example, more intense melanistic colours at higher elevations than at lower on East African mountains. *Thaumatomyia secunda* RICH. from elevations of 2000-2900 m on Mt. Elgon is, for example, paler than from elevations above 3000 m and 3000-4260 m on the High Semyen in Abyssinia. There is also definite increase in the frequency of darker coloured species with increase in altitude. The high altitude butterflies *Parnassius* LATR., which are common on most Old World high mountains, at least in the northern hemisphere, have, for example, wing-markings relatively lighter coloured at elevations of 2700-3000 m, but almost exclusively and conspicuously darker ones in tone at elevations above 3500 m on the Himalaya. Another high altitude butterfly, *Colias* FABR. from the Himalaya, occurring at elevations like 4000 m, have much deeper yellow or orange coloured wings than specimens found about the treeline. The high altitude Carabid *Bembidion* LATR. is pale brown, bright metallic-green or blue at lower elevations, but tends to be much darker brown or also mostly black at higher elevations. MANI (730) has described a remarkable massing of enormous numbers of darkly pigmented Collembola like *Isotoma* (BOURL.) BÖRN., *Proisotoma* BÖRN., *Hypogastrura* BOURL., etc. on snowfields on the Himalaya imparting a characteristic sooty-black appearance to the whole snowfield, conspicuous even from considerable distance. He has also pointed out that on the Himalaya, melanism is pronounced not only among insects that often frequent the open ground and snow surface, thus becoming exposed, but also among the others that ordinarily occur in relatively sheltered situations under stones and are not ordinarily exposed.

The high altitude melanism has been known for some time, at least in the case of the insects from the Alps. As nearly as 1836, HEER (493) showed, for example, that the high-alpine zone insects from the Alps tend to be darker and darker as the altitude increases. Since then, the problem has been discussed from various points of view and there is considerable diversity of opinion regarding the significance of the various underlying factors in the case of melanism in insects in general and of the high altitude melanism in particular (278, 348, 404, 458, 629, 703, 768, 842, 911, 994, 1042, 1096). A useful summary of the more important papers on the subject will be found in PROCHNOW (844, 845).

Recent studies on the Himalaya seem to demonstrate the fact that, at least in so far as the high altitude insect melanism is concerned, there is a significant correlation with the special high altitude environmental conditions, but particularly the atmospheric cold, the snow-cover, the high intensity of ultra-violet radiation, high light intensity, the excessive blue of sunlight and skylight, the intense insolation, the generally high environmental (niche) humidity, and the pronounced crowding during development and hibernation (fig. 19). The ecological importance of melanism at

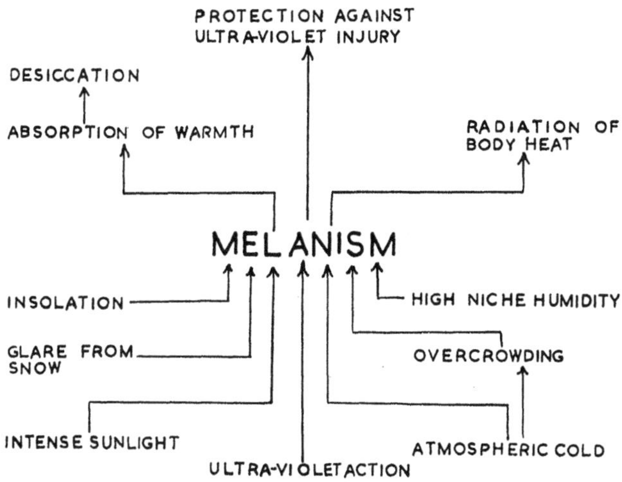

Fig. 19. The high altitude melanism and its ecological correlations.

high altitudes is closely bound up with the fact that dense pigments serve as effective protection against the injurious effects of the intense ultra-violet, the action of which is one of the contributory factors in the production of these pigments. ERHARD (322) and others (730) have suggested that the high altitude melanism has a protective value from injury due to the ultra-violet rays. Heavy pigmentation of the integument is the result of reaction to excessively intense solar radiation. Having risen under the action of this radiation, dense pigmentation serves to protect the deeper and delicate tissues from injury by the radiation. In addition to this main function, a number of other accessory functions may also be observed. It is also readily apparent that melanism has a significant relation to the intense insolation and its importance in absorbing warmth in direct sunshine has been recognized by a number of workers. HANDSCHIN (468, 471, 473) has, for example, drawn particular attention to this function. SCHRÖDER (934) has given an excellent account of the thermo-insulatory action of the dark body pigmentation of insects, with special reference to the alpine-zone species from Mt. Kilimanjaro. He has also given a general discussion of the views concerning this property, put forward by PROCHNOW in 1903, in the light of his own observations on Lepidoptera at elevations of 4690 m on Mt. Kilimanjaro. HANDSCHIN (471) observed the nival-zone Collembola from the Alps to absorb, with the help of their dense body pigments, enough warmth during the day in direct sunshine to survive the freezing during the night. It would seem, therefore, that but for the densely pigmented condition, the high altitude insects are unable to absorb warmth from the direct sunshine, without at the same time suffering serious injury. It must also be recollected that dark pigments not only facilitate insolation but also favour radiation of the warmth equally readily, depending upon the tone of colour of the background. As on darker background of rock, the body radiates away the warmth, the melanistic high altitude insects seek the snow surface for absorbing warmth from direct sunshine. Reference may also be made to a paper by RÜCKER (905) on the correlation between body pigmentation and the heat economy of the body.

3. REDUCTION AND LOSS OF WINGS

Reduction of wings and total apterous condition are frequent in nearly every group of insects even near sea-level and it would seem that many different and unrelated groups are at the present time evolving towards secondary apterous condition for one reason or the other. Wing atrophy is often associated with very diverse conditions, like sex dimorphism or with sedentary modes of life, degenerate modes of feeding, myrmecophily, termitophily, parasitism, etc. The proportion of short-winged and wingless species in the biota of the forest and prairie is on the whole comparatively small. At high elevations on mountains, there is, however, a significant dominance of species with reduced or vestigial wings or without any trace of wings in all orders and in nearly all communities. Even most of the winged species rarely, if ever, take to wings at high elevations above the forest on mountains.

The general preponderance of flightless, brachypterous and apterous insects on mountains has long been known from different parts of the world. BÄBLER (74), JANETSCHEK (567), SCHÖNMANN (932) and others have, for example, reported the great abundance of these types on the Alps. SCOTT (950, 951) and UVAROV (1070) similarly found the apterous condition to be common among the high altitude insects, at elevations between 3000 and 4000 m on the East African mountains. DARLINGTON (232, 233) has reviewed the problem of the apterous condition among the Carabidae on mountains and in islands. Carabidae, which constitute an important group of high altitude insects on all the well known high mountain ranges in the world, are remarkable in many ways. In temperate regions on the plains, the wingless condition is more frequent among the Carabidae than is the case in the tropical regions. According to VAN DYKE (1079), Carabidae are gradually losing their wings near sea-level in the arctic latitudes and on high mountains even within the tropical-temperate regions. He has also drawn attention to the important fact that Carabidae, which are endemic to the cold and wind-blown elevated areas of Middle Asia, show a general tendency for wing reduction and total loss of wings. MANI (730) showed that nearly 50% of the high altitude insects from the Northwest Himalaya are apterous. The frequency of the apterous condition increases to nearly 60% at elevations above 4000 m. The great majority of the remaining species, which have more or less well developed wings, rarely, however, fly on the Himalaya. In some of the typically winged orders like Coleoptera, for example, the percentage of the apterous and flightless species is much higher than the general average for the biota of the forest in the plains of India. The apterous species at high altitudes on the Himalaya are about twenty-five times more abundant than within the forest zone. The frequency of the apterous condition is distinctly much higher near the permanent snowline than in the zone immediately above the treeline. The apterous condition is also somewhat more pronounced in the biota of the north slope of the Himalaya than on the south slope. The mean frequencies of the flightless conditions and of the apterous species in different orders of the high altitude insects from the Northwest Himalaya are summarized in table 13.

TABLE 13

The frequency of apterous and flightless conditions among the high altitude insects of the Northwest Himalaya

Order or family	Percentage of apterous species	Percentage of flightless species
Orthoptera	70	25
Dermaptera	100	—
Heteroptera	56	?
Carabidae	90	10
Staphylinidae*	—	100
Tenebrionidae	100	—
Chrysomelidae	86	12
Curculionidae	75	25
Hymenoptera	40	10

* Although a great many species of Staphylinidae are apterous on the Himalaya, we have at present no means of estimating the exact proportion of the apterous and flightless species.

Apterous species are common among the following genera on the Northwest Himalaya: Orthoptera: *Dicranophyma* UVAROV, *Gomphomastax* BRUN., *Conophyma* ZUB., *Bryodema* FIEB. with reduced elytra and wings especially in the female, *Podisma* LATR. (flightless). Dermaptera: *Anechura* SCUDDER. Heteroptera: *Phimodera* HUTCH., *Dolmacoris* HUTCH., *Chlamydatus* CURTIS, *Tibetocoris* HUTCH. Coleoptera: *Bembidion* LATR., *Bradytus* STEPH., *Carabus* LINN., *Chaetobroscus* SEM., *Cymindis* LATR., *Nebria* LATR., *Trechus* CLAIRV., *Pterostichus* BON., *Notiophilus* DUM., *Patrobus* DEJ. (Carabidae); *Ascelosodis* REDTB., *Bioramix* BATES, *Blaps* FABR., *Cyphogenia* SOL., *Syachis* BATES (Tenebrionidae); *Longitarsus* LATR., *Merista* CHAPUIS, *Chaetocnema* STEPH. (Chrysomelidae); *Anchlaenomus*, *Lagenolobus* FAUST, *Leptomias* FAUST (Curculionidae).

Nearly 85% of the Pterygota on Mt. Kilimanjaro are brachypterous and are quite unable to fly. The Tipulidae *Tipula subaptera* FREEMAN (fig. 32) from Mt. Kilimanjaro, *Limnophila (Dactylolabis) rhicnoptiloides* ALEX. from the Alaskan mountains and *Limnophila (Dactylolabis) wodzickii* (NOW.), the larvae of which breed in gigantic cliffs over which water drips at elevations of 1800-2450 m on the West Carpathians, are brachypterous. HACKMAN (457) has recently given an extremely interesting account of the apterous and subapterous high altitude Diptera. Many endemic genera and mountain autochthonous species of other often widely distributed genera are apterous in both the sexes on the East African mountains. *Mesaptilotus* RICH. from the Ruwenzori Range, *Ocellipsis* RICH. from Mt. Elgon, *Binorbitalia* RICH. from the Aberdare-Kenya Mountains and from the Ethiopian Highlands, *Lobeliomyia* RICH., *Pismira* RICH., *Oribatomyia* RICH., *Atuligera* RICH., etc. are some of the important examples. Other high altitude apterous or subapterous Diptera from the East African mountains include *Limonia oreositropha* SPEISER from Mt. Kilimanjaro, *Limonia butoniana* ALEX., *Fur fugitivus* JONES, *Obstinocephala tali* JONES and *Wiedemannia reducta* JONES (Empididae) from the Ruwenzori Range, the Phoridae *Aptinandria effeminata* SCHMITZ from Mt. Kenya, the Sepsidae *Apterosepsis basilewskyi* RICH., from Mt. Meru, the endemic Chloropid *Alembus* RICH., etc. We have from the Alps records of the Borborid*Aptilotus paradoxus* MIK. without wings and halteres in both the sexes. Many species of *Tipula* LINN. and *Chionea* DALM. and *Alfredia acrobata* BEZZI are brachypterous in both the sexes. While in most Tipulidae from the Alps, the females are generally apterous (for example *Tipula sexspinosa* STROBL., *T. hemapteranda* BEZZI, *T. cisalpina* RIEDEL, *T. franzi* MANNH. *T. riedeliana* MANNH. and *T. sacci* MANNH.) and occur so far as is known only on the southern Alps (Tauern, Koralps and the Dolomites) and the southwest Alps (Monte Rosa, Aletsch, Pontresina and Sondrio), there are no apterous female tipulids on the Tyrolean Alps. LINDNER (697) found the apterous females of *T. sexspinosa* STROBL., known only as males on the Koralps. On the Cordilleran mountains of North America, wing reduction is met with in *Pedicia degenerata* ALEX., *P. subaptera* ALEX., *P. aspidoptera* COQ., etc. In *Limonia geyserensis* ALEX. the female is brachypterous but the male is normally winged. The Phoridae *Puliciphora glacialis* MALL. is apterous in the female and the Borborid *Pteremis unica* SPUL. is brachypterous in both the sexes and *Americaptilotus* spp. are apterous in both the sexes. The Tipulid *Gynoplistia* WESTW. from New Zealand mountains is remarkable for its wing reduction in the female. HEMMINGSEN (505) has discussed some aspects of wing reduction among Tipulidae. Wing reduction and associated reduction in palpi among Sciaridae are discussed by LENGERSDORF (687).

The wing atrophy and the flightless condition in high altitude insects are without doubt correlated with the high altitude environment and represent, therefore, part of the high altitude specializations (fig. 20). Indeed the rôle of the prevailing climatic conditions of the habitat in the reduction and loss of wings has been stressed by a number of workers (232, 233, 1079). The researches of DEWITZ (257, 258) demonstrate a close correlation between wing atrophy at high altitudes and the oxygen deficiency of the atmosphere and the atmospheric cold. The influence of atmospheric temperature on the flight of insects even at sea-level is readily observed in the case of the common cockroach. According to the observations of REHN (868), both sexes of *Periplaneta* spp. appear to be capable of flying normally in tropical and subtropical countries, in the north temperate climates the females appear to be on the whole incapable of flight, though of course they are provided with normally developed

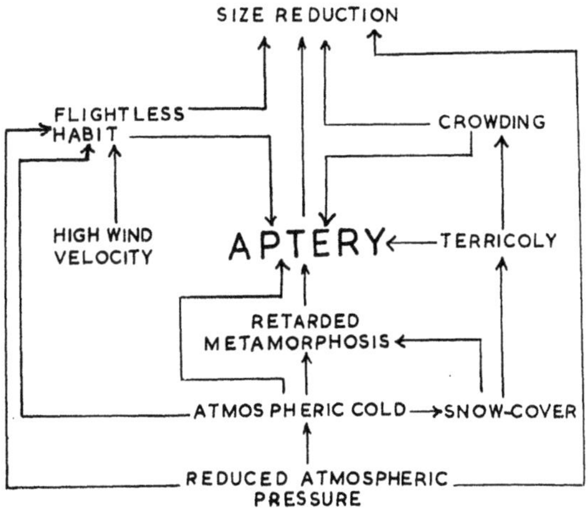

Fig. 20. The ecological correlations of the apterous conditions among high altitude insects.

wings. KRAMER (655) refers to MALLIS as having thrown the females of *Periplaneta americana* LINN. experimentally into air at a temperature of 27.7° C and found them to merely flutter the wings, but quite unable to gain either height above ground or to fly. KRAMER also found that most cockroaches fail to fly at temperatures below 21° C. DARLINGTON (233) has also referred to the possible influence of the prevailing atmospheric temperature in the flight of insects and on the loss of wings in Carabidae. While he admits that low atmospheric temperatures may retard or inhibit the normal development of wings, hinder active and normal movements of the flight muscles, thereby restricting or totally eliminating effective flight and thus lead to progressive wing atrophy, he fails to fully recognize the close correlation between the atmospheric temperature and wing atrophy among the high altitude insects. Low atmospheric pressures would also appear to make flight mechanically difficult, perhaps in much the same way as high wind velocities. DIGBY (262) concludes from field and experimental evidence that even at sea-level the effects of strong winds on the loss of wings are obviously great. ILLIES (563) traces the loss of wings in the Andiperlinae to the high intensity of the prevailing winds on the high mountain regions where the group is endemic. Limitation of the habitat on mountains is also considered by DARLINGTON (233) as one of the possible contributory factors. The experimental evidence summarized by KISIMOTO (622) seems to demonstrate the influence of crowding during development and metamorphosis on wing atrophy. It must, however, be emphasized that the loss of wings among the high altitude insects is by no means a simple phenomenon, governed by a few isolated or even predominantly important factors, as is only too often taken for granted. The absence of flight, vestigial wings, the reduction and total loss of wings among such a large majority of insects at high altitudes on nearly all the high mountains of the world are without doubt part of their ecological specializations, closely correlated with the high altitude environment in its totality.

4. SOME OTHER STRUCTURAL PECULIARITIES

In addition to the heavy body pigmentation and the wing atrophy, reduction in the mean body size is one of the most striking characters of high altitude insects. While certainly large or medium-sized insects are not altogether absent on mountains, there is, however, a very pronounced tendency for reduction in the mean size of the body in most high altitude groups of insects with increase in elevation. Among Coleoptera, relatively large forms like *Calosoma* WEBER do not generally ascend to elevations much above the treeline even on the Himalaya (where nearly every type of insect occurs at much higher elevations than on most other mountains). In any case no large species of this genus occur at elevations above 3500 m on the Himalaya. The smaller species occur, however, at the highest elevations at which Coleoptera have so far been found on the Himalaya. The great bulk of the species of *Bembidion* LATR. from the high meadows at elevations of 3000-4000 m on the Northwest Himalaya are, on an average, about 5-8 mm long, but the species occurring at elevations of 5000 m measure between 2-3 and 3-4 mm only (fig. 21). A number of species that occur within the upper reaches of the forest zone, at elevations of 2500-3000 m, average to lengths of 8-13 mm. Reduction in the mean body size, with the increase in altitude, is observed in several other altitude genera like *Amara* BON., *Nebria* LATR., *Trechus* CLAIRV., etc. on the Himalaya. As is well known, the Trechini as a whole are among the smallest Carabidae, but the proportion of Trechini in the total Carabidae near sea-level is nowhere significantly as high as at high altitudes. Both as we ascend higher on mountains and as we proceed to higher

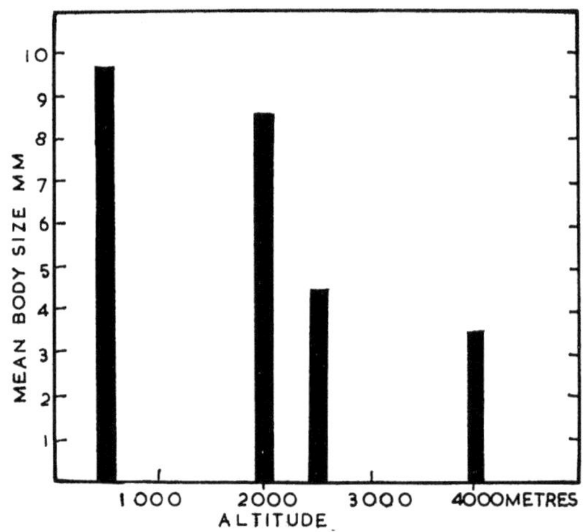

Fig. 21. The mean body size of the high altitude species of *Bembidion* LATR. at different altitude zones on the Northwest Himalaya. A significant reduction in the mean body size may be observed with increase in elevation.

north latitudes, the Trechini become progressively more abundant, until both at extreme high altitudes and north latitudes they are perhaps the only Carabidae that one comes across. The studies of THIELE & KIRCHNER (1034) on the relation between the mean body size reduction and elevation among Carabidae are extremely interesting. In *Pterostichus* BON., for example, the low-mountain forms are markedly smaller than the typical lowland forms. The high-mountain populations are still smaller in size than the low-mountain populations. In *Agonum* BON. the low-mountain forms do not differ very conspicuously from the lowland forms, but the high-mountain populations are remarkable for their pronounced reduction in size. RENSCH (881) has also referred to the mean smaller size of the mountain races of *Carabus* LINN. in comparison to the lowland forms.

Reference may also be made to the interesting observations of REINIG (874,877, 878) on the significant reduction in the mean body size of Tenebrionidae with increase in elevation. According to him, there is at the same time a progressive general flattening and increase in the width of the body, associated with increase in the convexity of the elytra. The elytral margin becomes also acute and the lateral carina of the pronotal shield becomes fainter or is also totally absent at higher elevations. There is often an over-proportional growth in the length of the mucro by about 160%. An increase of 50% in the mean length of the body from 22 mm to 34 mm is associated with a corresponding lengthening of the mucro from 1.6 to 4.2 mm (160%). The arching up and the flattening of the elytra of the steppecole species from the high Pamirs are correlated with the differences in the prevailing atmospheric temperatures. The space within the mucro is believed to serve as a reservoir in these species. The flatter species appear to have longer mucro than those with relatively convex elytra. In some strongly convex and apterous *Pimelia* spp. air is known to be stored up in the space between the dorsum of the abdomen and the elytra and this air is believed to help in protecting the body against the undue warming up by strong insolation. For example, the intense insolation in the Alai Valley of the Pamirs Region has the effect of mean high ground temperatures of 45-50° C and the nocturnal radiation results in the fall of ground temperature to —6 or even —12° C. The air stored up in the space between the dorsum of the abdomen and the elytra tends to mollify the extreme effects of both these factors. In *Prosodes* (*Uroprosodes*) *costifera* KR. the range in body length of the female lies between 22.0 mm and 26.6 mm, with a mean of 22.852 ± 0.228 mm in the lowland populations from Buchara area. In the females from the Alai Valley the range is only 19.5 to 24.2 mm, with a mean length of 21.484 ± 0.060 mm. The length of the mucro in females from populations of the lowland area of Buchara ranges from 1.0 to 1.5 mm, but in those from the Alai Valley the maximum is only 1 mm. The length of the elytra in the females ranges in the Buchara populations from 1.70 to 1.90 and in the Alai Valley populations from 1.61 to 1.79 mm, with respective means 1.80 and 1.68 mm. The means of the index of the elytral length to thorax width for the females from the Buchara lowlands and from the Alai Valley amount respectively to 0.55 and 0.61. *Prosodes alaiensis* KR. from the Buchara lowlands has a body length ranging from 22 to 26.6 mm in the female and 20-26 mm in the male, but in the eastern parts of the Alai Valley 19.5-24.2 in the female and 19-23 mm in the male, at an elevation of 3270 m. In the case of *Blaps caraboides* ALL. also similar conditions are described, though the distance separating the two populations is hardly 50 km. The mean body length is 18.9 mm, the range is 16.1-20.1 mm at an elevation of 3000 m and 20.5 mm with a range of 19.8-21.1 mm at an elevation of 1890 m. The range of body length in *Anatolica paphia* REITT. is 9.5-12.4 mm at an elevation of 3270 m (with a mean of 10.693 ± 0.067) and 11.0-13.5 mm (with a mean of 11.860 ± 0.147) at an elevation of 2800 m in the Pamirs Region. In the variable species *Platyscelis* (*Platyscelis*) *margellanica* KR., widely distributed in the Pamirs, the range of body length in the males and females at different elevations is summarized in table 14.

TABLE 14

Body length of *Platyscelis (Platyscelis) margellanica* KR. at different elevations on the Pamirs

Elevation in m	Female		Male	
	Range	Mean	Range	Mean
1800-2200	12.0-13.7	12.9	11.6-13.5	12.6
			10.9-12.0	
3100	11.0-12.0	11.5		11.5
3200-3400	10.1-12.0	11.1	9.5-13.0	9.9

It is also interesting to observe that there is not only a progressive reduction in the mean body size with increase in elevation in most orders of high altitude insects, but the smaller and more minute-sized orders like Diptera and Collembola gain increasing dominance at higher elevations and replace almost entirely every other type above the permanent snowline. There is, therefore, a very significant correlation between reduction in the mean body size of the high altitude insects and the elevation. Size reduction must be described as part of the high altitude specializations.

The body size reduction is apparently influenced by a number of complex environmental factors, with which it is also closely correlated. The reduction and loss of wings and the absence of flight even among the winged forms have the direct effect of reduction in the general body size. It is known, for example, that at sea-level in every group in which wing reduction and wing atrophy have occurred, the alate forms are larger than the subapterous and apterous forms. The atrophy of wings may, therefore, directly favour the reduction in the general size of the body as a whole. Atmospheric cold, the prolonged snow-cover, the predominant tendency for endogeous mode of life and the prolonged hibernation also tend to retard metamorphosis and cause in consequence a reduction in the size of the body. Crowding must also be considered as an important underlying factor in this connection. The small size of the body enables the high altitude insects to effectively reach and take advantage of the sheltered microclimatic niches under stones, in rock crevices and other similar situations. It is interesting in this connection to recollect that even among the high altitude plants the small size, due to the abbreviation of the internodes, must not be interpreted as indicative of young age in the plants, but as a general reduction in size, in close correlation to the high altitude environment.

As the elevation increases, we may observe an increasing preponderance of species with dense body clothing of bristles, setae, scales and waxy coatings, especially in the case of the insects that frequent the open snow surface or visit the rock surface during the hours of bright sunshine. LINDNER (697) has pointed out that a dense clothing of bristles forms a part of the high altitude adaptations among the high-alpine zone Anthomyiidae (Diptera) from the Alps. He ascribes to it the functions of not only absorbing the warmth from the sunshine but also of insulating the body warmth. A dense clothing of bristles would obviously be useful in retarding the rate of evaporation of moisture from the exposed body. The same peculiarity has also been observed at high altitudes among the Anthomyiidae from the Northwest Himalaya (730).

LINDNER (*op. cit.*) has described a characteristic forward bulging of the face in the high altitude Anthomyiidae from the Alps. According to him, the bulk of the nival-zone flies are readily recognized by their characteristic protruding frons. This bulging of the frons in the Anthomyiidae and other Muscoid flies has, according to him, an important physiological significance. These flies not only feed but also absorb the warmth of the sunshine and the warmth that is being radiated away by rock surface that had previously been exposed to insolation. The flies rest on the warm rock surface during cloudy intervals in order to absorb the warmth that is being radiated away. The forward bulging of the frons is believed to increase the effective heat-absorbing surface of the body and thus also to aid in searching for warm patches of rock and in testing the radiating power of the rock surface. Identical conditions have also been observed among the Anthomyiidae at elevations of 4000 m on the Northwest Himalaya (730). The shiny and the highly reflecting surfaces of the dorsum of the high altitude Coleoptera like *Orinodromus* KOLBE and *Peryphus* DE JEAN., the conspicuously arched elytra in *Peryphus* DE JEAN., from Mt. Kilimanjaro, enclosing air spaces within are also considered as some of the important high altitude ecological specializations.

5. COLD STENOTHERMY

The greatest bulk of the high altitude insects are characterized by their pronounced cold stenothermy. They have a relatively narrow range of optimal temperatures, usually in the neighbourhood of the freezing-point. They are ecologically and physiologically bound to low temperatures and may, therefore, be appropriately described as cryobiont types. All their metabolic activities are normally adjusted to relatively low temperatures. There are also some typically cryophile forms, existing side by side with the strictly cryobiont species, and these occur often not only under conditions of low atmospheric temperatures near glaciers but also in relatively warmer conditions in special niches. We may observe indeed nearly every transition from the extreme cryobiont type to the more or less thermophile one, in correlation with the wide range of optimal conditions of temperatures in the different environmental and microclimatic niches. Regardless of these differences, cold stenothermy is a distinctive character of the high altitude insects; it is a part of the high altitude specializations. On the basis of their cold stenothermy, we may recognize the following broad categories among the high altitude insects on most mountains: i. insects which are normally at subzero atmospheric temperatures, often as low as —10° C; ii. insects which are active at temperatures between 0° C and 5° C; iii. insects which are active only during the hours of bright sunshine and which are adjusted to atmospheric temperatures ranging from 5° C to 10° C and iv. insects that are habitually subjected to more uniform but low temperatures in the underground niches, in melt-water torrents, etc. The low temperature valence, which is a natural corollary of the pronounced cold stenothermy, is characteristic of the high altitude insects, whether they are from equatorial high mountains or from the mountains of the north temperate zones of the earth. The temperature valence becomes indeed progressively narrower as the altitude increases and is also more pronounced on mountains in the middle of large continental masses than in oceanic areas, as for example in the southern hemisphere, at comparable altitudes. While able to tolerate low temperatures in the neighbourhood of the freezing-point and often even much below this limit, most high altitude insects appear to be unable to withstand even a slight rise in the temperature above the upper limits of the temperature optimum, characteristic of the species in a given habitat. Most species which are adjusted to fairly uniform niche temperatures, such as, for example, the larvae of Ephemerida, Plecoptera, many aquatic Diptera like Blepharoceridae, Deuterophlebiidae, many Collembola, etc., developing normally under temperatures ranging from —1.5 to 5° C during the summer, are killed by a few minutes' exposure even to the warmth of the human hand (730). The unpigmented, hygrophile, blind land Isopod *Mesoniscus alpicola* (HELLER) from the Alps behaves also in a similar manner (384). CHAPMAN (200) and MARCHAND (739) have also reported that the snow Tipulid *Chionea* DALM. is likewise extremely sensitive to the warmth of the human hand.

A great deal of experimental work has been carried out in recent years on the cold stenothermy of insects (302, 351, 450, 654, 707, 708, 718, 818, 823, 907, 914, 943, 1083). Diverse insects have been experimentally cooled down to temperatures of —10, —20, and —26° C without apparent injury. The insects that normally hibernate during the winter and those that are actually undergoing hibernation are known to exhibit far greater tolerance to low temperatures than others. Some of them withstand, for example, exposures to a temperature of —50° C. A number of insects which are undergoing diapause have also been experimentally cooled down to —81° C without apparent harm.

The cold stenothermy of the high altitude insects differs, however, from the cold hardiness of the lowland species, observed under such experimental conditions, in several fundamental respects. The high altitude insects are characterized by their capacity not merely to survive, but also to continue their normal development and growth, undergo metamorphosis, to feed and carry on all their normal activities at temperatures, at which the lowland insects ordinarily suffer reversible cold stupor. They are further characterized by their capacity to utilize the prevailing atmospheric cold itself, in an effective manner, for counteracting the extreme injurious effects of the reduced atmospheric pressure at high elevations. The temperatures that are suboptimal or perhaps even lethal to the insects of the lowlands, forest and prairie are, on the other hand, completely optimal to the high altitude insects. Many high altitude insects are normally active at temperatures of —1.5° C and 1.7° C on the Himalaya. *Colias hecla glacialis* MCLACH. is active on its wings on the Klutlan Glacier at elevations of 2885 m, during the hours of bright sunshine, at temperatures of 9-13° C. The lowest temperature at which a high altitude insect has been found to be active is —10° C. At this temperature at night during June 1953, it was observed that Collembola like *Hypogastrura* BOURL. and *Proisotoma* BÖRN. crawl about actively on the surface of the Sonapani Glacier on the Northwest Himalaya, at an elevation of 5000 m. It may also be recalled that the cavernicolous Silphid *Astagobius angustatus* has been reported to be able to carry on normal activities in ice-grottoes, at temperatures ranging from —17 to 10° C. Among the high altitude insects, Collembola stand perhaps foremost in their cold stenothermy (570, 571, 572, 730,

1003, 1006). The recent observations of PRYOR (847) in the Antarctic seem to confirm these facts. Collembola have been observed there to withstand exposures to temperatures as low as —50° C. *Isotoma klovstadti* CARPENTER is, for example, not harmed by the low atmospheric temperature, but it is unable to withstand desiccation in sunny weather. As higher atmospheric temperatures favour rapid desiccation, the high altitude insects are inexorably bound to the low temperatures and their cold stenothermy is, therefore, of significance in preventing death by rapid loss of their body moisture due to high atmospheric aridity. The survival of the high altitude insects on mountains is possible only because of the atmospheric cold and the cold stenothermy is the specialization that makes the utilization of the atmospheric cold in this unique manner.

The cold stenothermy of the high altitude insects varies within certain limits, from species to species and even in the same species, each succeeding stage in the life-cycle shows a progressive diminution of the cold stenothermy. An insect that hatches from the over-wintered egg or reawakens from a hibernating stage during early summer, develops normally at increasingly higher temperatures, with the advance of summer. There is, therefore, a gradual fall in the cold stenothermy as the summer advances and the life-cycle progresses from the dormant to the active stage. Maximum abundance of the larvae of *Deuterophlebia mirabilis* EDW. is found, for example, in water with a mean temperature of 5.7° C, but for the pupae the mean water temperature is 7.5° C. The cold stenothermy of the high altitude insects is closely correlated to the winter snow-cover on mountains. The snow-cover acts as a most effective protective blanket for the ground against the great atmospheric cold during the winter and thus also sets a limit to the cold stenothermy of the insects hibernating underground. It must not also be overlooked that the cold stenothermy of the high altitude insects is the result of the chain action of atmospheric aridity. It has, for example, been shown experimentally that partial dehydration actually increases the cold hardiness of insects even near sea-level (823). SALT (914) showed, for example, that severe desiccation produces cold hardiness in species that are normally not cold hardy. While not a linear function of the prevailing atmospheric temperatures at high altitudes on mountains, the cold hardiness of the high altitude insects is undoubtedly the outcome of the prolonged exposures to cold.

6. HYGROPHILY AND TERRICOLY

Although the high altitude environment is largely characterized by more or less pronounced atmospheric aridity on most mountains, the greatest majority of the high altitude insects are typically hygrophiles or also hygrobionts. They are everywhere remarkably concentrated in humid, damp or moist localities. As snow-cover, especially the winter snow-cover, is the ultimate source of all moisture and water at high elevations, all terrestrial species of high altitude insects and other Arthropoda are heavily concentrated near the edge of the winter snow, the ice margin, the glacier snout, the edge of the melt-water ponds and torrents, glacial lakes and springs. Even the strictly true soil types that are not directly dependent on the atmospheric moisture, are found only in such localities as are covered by snow during the winter. As may be recollected, the snow-cover during the winter prevents the permafreeze of the soil and also ensures adequate soil moisture during the succeeding summer. It is not only the soft-bodied and delicate forms like Collembola, which are of course highly hygrophiles even at sea-level, but also other groups like Coleoptera, Dermaptera, Diptera, etc. (many of which are not necessarily hygrophiles in lowland) which are exclusively either hygrophiles or also true hygrobionts in the high altitude biota. The dominant hypsobiont genera like *Bembidion* LATR., *Nebria* LATR., *Trechus* CLAIRV., *Atheta* THOMS., the high altitude Dermaptera, Lepidoptera and others are to be found almost exclusively at the snow-edge. Remarkably enough, no hibernating insect may also be found in a locality that does not receive the winter snow-cover.

It is obvious that only in moist situations near the snow-edge or near the melt-water collections the high altitude insects can expose themselves in the open and absorb the warmth from the direct sunshine and still not suffer from the risk of rapid desiccation to death. The high altitude insects are therefore absolutely bound to the snow in a most intricate manner, mainly because of the extreme atmospheric aridity at higher elevations. With the increase in altitude on the Himalaya, the Alps and most other comparable high mountains in the world, there is a significant increase in the proportion of hygrophile and hygrobiont types of insects. It must, however, be pointed out that excessive soil moisture does not seem to favour the colonization by rich terricole and soil insect life on mountains. This is observed particularly on some of the East African mountains, where marshy conditions are generally associated with an extremely poor insect life (*vide* Chapter VII).

The high altitude insects are predominantly geobionts on most high mountains or many are also geophiles. Even the characteristically planticole species that generally occur in vegetation mats and vegetation cushions or among the moss and lichen incrustations show a very pronounced tendency to remain on the underside of the vegetation, close to the ground. Most species actually also spend the greatest part of their time on the ground or remain even underground and come to the plant above ground only during feeding time. It is also remarkable that only the adults of a relatively small number of the high altitude insects occur on the open surface of the ground; the rest are found exclusively under stones, inside crevices in rock, in the soil, under snow and other similar concealed situations on the ground. In the lowland biota, the proportion of true geophiles and geobionts is mostly small, except perhaps in hot deserts and the geophile and geobiont groups are not also dominant. At high altitudes on mountains, however, it is these groups that are predominant. Collembola, Dermaptera, Heteroptera, Carabidae, Staphylinidae, Histeridae, Tenebrionidae, Chrysomelidae, Curculionidae, Hymenoptera, Lepidoptera and Diptera are nearly wholly either geophiles or geobionts on mountains. The caterpillars of the high altitude Lepidoptera remain in the ground or under stones and the eggs are deposited not on the larval food plants as is usual with these groups in the lowland environment, but under stones. ZELLER (1132) has described this habit among the high-alpine zone Lepidoptera from the Alps. There is a progressive increase in the dominance of these forms with the increase in elevation, until at extreme elevations, especially above the snowline, these are the only types of insects that one finds.

The close correlation between hygrophily and terricoly may be seen in the fact that the insects can be truly geobiont and endogeous types only in localities where the soil is not frozen permanently, owing to the protective action of the winter snow-cover. The terricoly, hygrophily and the snow-cover are, therefore, all most intricately and closely interlinked and the habits of the insects are integral parts of the high altitude specializations.

7. OTHER HABITS

In all the high altitude insects active life is severely restricted to the summer, at least on the mountains in the northern hemisphere. The greater part of the year is passed in hibernation under snow-cover. With the coming of summer and the melting of the winter snow, the insects revive to a period of active development and growth, movements, feeding, oviposition, etc. On the equatorial mountains, however, all activities cease everyday nearly at or soon after sunset, to be revived again after the sunrise next morning. There is, therefore, no pronounced seasonal rhythm in the activities of the high altitude insects on the tropical and equatorial mountains. On these mountains and on the mountains of the south temperate regions of the world, the activities of the high altitude insects are typically characterized by a pronounced dominance of the diurnal rhythm over the seasonal.

We have already shown that the fluctuations of the atmospheric conditions influence also the microclimatic conditions to different degrees in different insect niches. There is actually wide selection of conditions of temperature and humidity for the diverse activities of different insects. Insects that belong to niches with uniform microclimatic conditions continue to be more or less normally active throughout the twenty-four hours of the day during the summer, if an occasional slowing down due to abnormal conditions is ignored. The activities of species in other niches are, however, largely governed by the temperature fluctuations. This applies particularly to the activities of insects above ground, like foraging, hunting, crawling, flight, mating, etc. The adults of most typically endogeous species come to the surface of the ground for brief intervals at some time or the other, depending on the temperature of the layer of air within 15 cm immediately above the surface of the ground. Variations of wind velocities and the intensity of sunlight also greatly modify these normal activities. Both strong winds and high intensity of light act as powerful deterrents in many cases and force the insects to remain underground. The insects that are capable of flight remain obstinately close to the ground and take to wings only in the absence of wind and also under definite conditions of atmospheric temperature.

Although some species of Dermaptera like *Anechura* SCUDDER, some Carabidae, Tenebrionidae, Thysanura and Collembola, a few Diptera and most Trichoptera remain concealed in their niches or shelters the whole day long and come out in the open only late in the evening, diurnal habits are the rule among the high altitude insects. Nocturnal species may sometimes be found about the treeline on some mountains, but as the elevation increases, they are completely replaced by strictly diurnal forms. Even the predominantly nocturnal or crepuscular groups like moths are either sparse at high altitudes or they have become completely diurnal (730, 967).

The appearance of even the apterous and the flightless insects on the open surface of the ground occurs only after the ground has been warmed by insolation, depending once again on wind and other atmospheric conditions being otherwise favourable. Like the winged insects, these also absorb the warmth that is radiated away from the ground or from the direct sunshine during the daytime. We have already referred to the peculiar habits of Diptera, especially the high altitude Anthomyiidae from the Himalaya and the Alps, resting on the warm rock surface and absorbing the warmth during a cloudy interval. Collembola, grasshoppers, bugs, beetles, butterflies and flies thus absorb the warmth in the morning hours before becoming active. During the hours of intensive sunshine of midday, most planticole species disappear from the surface view. These peculiarities depend largely on the moisture requirements of the terricole species. In humid weather, the top layer of the soil is usually highly saturated and moist, so that the terricole species generally occur on the upper layers of the soil. In case of prolonged drought or scanty snowfall, however, the upper layer of the soil is dry and the terricole species migrate with the disappearing moisture and go deeper into the more humid rock crevices and other similar sheltered situations. The diurnal wanderings of the high altitude terricole species from the upper to the deeper layers occur in fine weather during the summer on all important mountains like the Himalaya and the Alps. During the morning, about 9-10 hours, the upper layers of the soil have perhaps the richest terricole insect life. As the sun begins to shine and as the upper layers become further warmed and start to lose the moisture, the terricole forms now leave the upper layers and go deeper. They return to the surface layers about 16-17 hours in the evening. These diurnal movements are generally less pronounced at the snow-edge. During cloudy weather the terricole species remain completely concealed. The diurnal movements are not observed in the terricole species within the forest zone, so that these movements must be considered as part of the high altitude specialization.

The observations on the general activities of the high altitude insects of the Himalaya, at elevations between 3500 and 4000 m during June are of particular interest. In the vast majority of these insects, surface activities are initiated only after the ground has been sufficiently warmed by the action of insolation (fig. 22). On

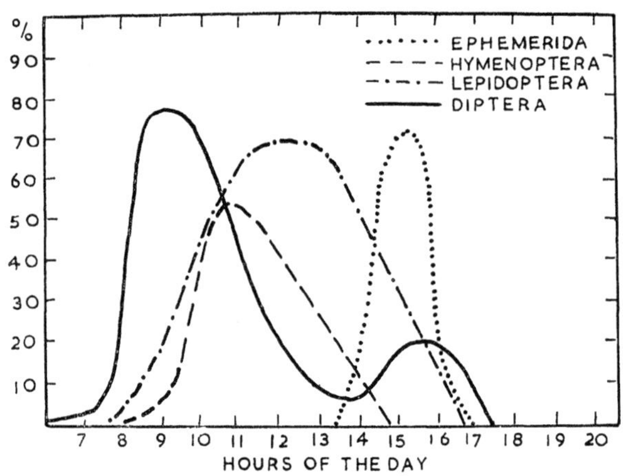

Fig. 22. The diel periodicity on the above-ground activities of adults of the high altitude Ephemerida, Hymenoptera, Lepidoptera and Diptera, at elevations between 3500 and 4000 m, on the Northwest Himalaya, measured in terms of the percentage frequency of the adults observed on wings from sunrise to sunset in June.

bright sunny days, with little or no wind, the surface activities, particularly on the wing, begin with many species about two hours after sunrise, by which time the intense insolation causes an appreciable rise of the ground temperature and also a rise of the temperature of the layer of air immediately above the warm ground, often to about 16° C. As the morning advances and with the further rise in temperature of the air above the ground to nearly 20° C or more, most activities of insects begin to diminish to a considerable degree, until a minimum is reached near about 14 hours. With the fall of temperature in the later afternoon, there is a partial resumption of activity, but never equal to that of the forenoon. There is an abrupt cessation of all surface activities at sunset. Many Diptera like Syrphidae, Anthomyiidae, Sarcophagidae, Tachinidae, terrestrial Chironomidae, Simuliidae, some Tipulidae, Hymenoptera, etc. show the peak of their above-ground activities between 9 and 11 hours and then disappear for the rest of the day to their nocturnal shelters under stones, underground or in rock crevices. Very few species active during the forenoon reappear in the afternoon. The observations of SWAN (1021) on Mt. Makalu in the Nepal Himalaya (fig. 23) show the close correlation of insect activity and temperature changes during the day.

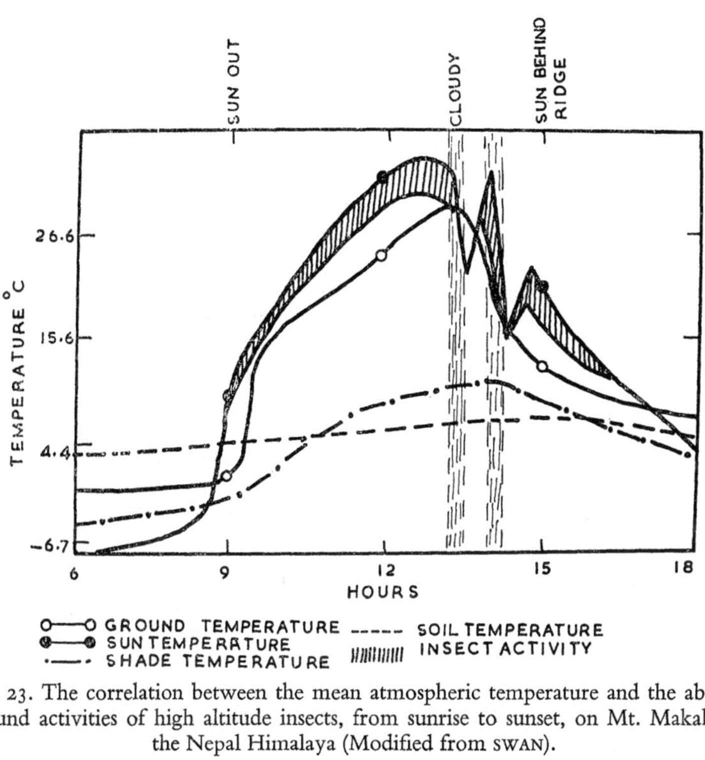

Fig. 23. The correlation between the mean atmospheric temperature and the above-ground activities of high altitude insects, from sunrise to sunset, on Mt. Makalu in the Nepal Himalaya (Modified from SWAN).

KEISER (612) found similar conditions regarding the times of appearance of the high-alpine zone Diptera on the Swiss National Park, with special reference to the temperature and other atmospheric conditions. The high-alpine zone Diptera from the Alps are only very slightly active during the first hour of sunshine, when the air temperature is between 3.5° C and 16.5° C, but begin to show high activity after the second hour, when the air temperature above the ground has reached between 22° C and 25° C. (fig. 24). Further rise in the air temperature to 30° C was

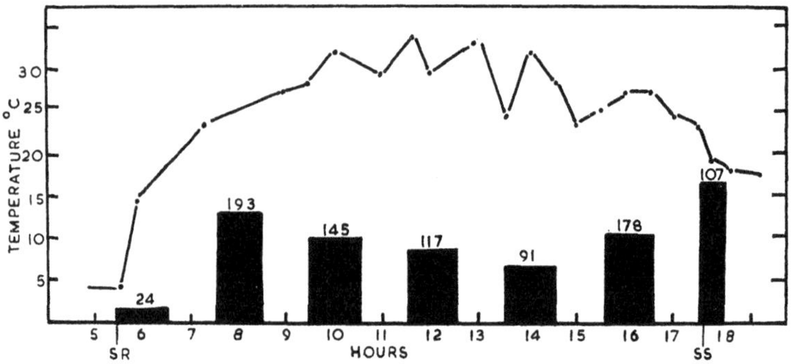

Fig. 24. The correlation between temperature and above-ground activities of insects at an elevation of 2050 m on the Grimmels Alps, in July 1945. The graph shows the air temperature and the histograms the insect activity from sunrise (SR) to sunset (SS)(modified from KEISER).

followed by minimum activity. Maximum aerial activity occurs between 8 and 11 hours. Continuous and strong winds cause a sharp break in the temperature-humidity conditions and are associated with cessation of all activities of insects on the surface of the ground. It is not simply the sunrise that is important, because sunrise behind clouds does not bring out the insects. Different groups of high altitude insects, belonging to different environmental niches, naturally behave differently in this connection. The period of activity of the high altitude Lepidoptera is perhaps longer than that of Diptera and extends from about 9 hours to nearly 16 or even 17 hours, although the peak hour of their flight is between 11 and 14 hours. A somewhat similar behaviour is reported by SWAN (1020) in the case of the high altitude insects from Mt. Orizaba in Mexico.

Reference should also be made to the interesting observations of REINIG (873, 875, 878) on the activities of Bombidae from the Pamirs Region. The flight seasons of *Bombus* LATR. on the Pamirs reveal remarkable specializations. The mature females fly from the end of June to the end of August, the workers appear from the end of July to September, the drones from early August to September and the young females about the middle of August. In each case, the total period of flight is shorter on the Pamirs than on the Alps, where on an average the bumblebees remain on wings from May to the end of October. On the Alai the period is reduced to one month. It is not only the length of the summer, but also the possible flight period during the day is an important factor that governs the flight of these bees on the high mountains and on the high steppes. In Novaya Zemlya, *Bombus hyperboreus* SCHÖNH., *B. balteatus* DAHLB., *B. kirbiellus* CURTIS and *B. lapponicus* FABR. are on their wings at atmospheric temperatures between 4.5° and 10° C and *B. melanurus griseofasciatus* REIN. and *B. separandus* VOGT fly on the Pamirs practically under similar conditions of atmospheric temperatures. Even with a light snowfall at an elevation of 4000 m, *B. separandus* VOGT disappears from flight. The most favourable hours for gathering pollen by the bumblebees in the elevated regions appear to be 10–12 hours and the stronger insolation after 12 hours seems to be unfavourable. The cold wind that regularly starts at about 2–30 p.m. on the Pamirs, naturally puts an end to flight not only of *Bombus* spp. but of all other insects also. The atmospheric temperature at a height of 1.5 m above the ground rises to 12–13° C at midday, but in the afternoon sinks rapidly with the starting of the wind, at an elevation of 4000 m. As the wind velocities reach 5 or more metres per second at about 17 hours, the atmospheric temperature is only a little above the freezing-point and sinks below this limit soon after sunset*.

The specializations in the food habits of the high altitude insects are largely correlated with the severe restriction and irregularity of the feeding periods, the peculiarities of the vegetation and other conditions. Although the sparse and specialized vegetation, growing up to the highest elevations of existence of insect life, constitutes the ultimate foundation on which the entire high altitude insect life rests, it is remarkable that only a relatively small proportion of the species are phytophagous at higher elevations. No more than 3% of the total high altitude insects of the Northwest Himalaya are, for example, phytophagous. The rest of the species are either feeders on diverse organic debris or they are carrion feeders, scavengers or active predators. In addition to a great preponderance of the carrion feeding or the active predacious types like Carabidae, Staphylinidae or Diptera, there is also a corresponding scarcity of the strictly phytophagous groups like Orthoptera, Chrysomelidae, Curculionidae, etc. Though some phytophagous insects depend on the algae, moss, lichen and Phanerogams that grow at elevations above 6000 m on the Himalaya, the phytophagous habit becomes progressively scarce as the elevation increases, until there is almost total dominance of the other feeding habits. Phytophagous species are fairly common, but by no means dominant even on alpine meadows at lower elevations, with relatively abundant soil and lush green grass and other matted plants. It is here that we find grasshoppers, caterpillars of high altitude butterflies, some Heteroptera and a few Coleoptera that feed on the roots or leaves of plants. The aquatic forms like the larvae of Ephemerida feed on the algal slime on the submerged stones. Some Staphylinidae and Collembola are known to feed on moss and the caterpillars of one or two typically high altitude butterflies feed on lichens. Wind-blown pollen grains and the spores of fungi constitute the food of numerous species of Collembola. Immediately above the forest-line, almost half the species of high altitude insects are ordinarily phytophagous. At elevations above 4000 m and particularly near the permanent snowline on the Himalaya, hardly 10% of the species are phytophagous. In marked contrast, there is a great abundance of debris feeders, carrion feeders and predators. Though no strict altitudinal zonation of the food habits exists, the increasing dominance of the habit of debris-feeding, carrion-

* The recent observations of SYCHEVSKAYA & SHAIDUROV (SYCHEVSKAYA, V. I. & V. S. SHAIDUROV, 1965. O temperature tela nekotoriikh synanthropniikh mukh na vostachnom Pamire. *Zoologicheskii Zhurnal, Akad. Nauk USSR*, 44: 779–783), at an elevation of 3860 m above mean sea-level on the East Pamir, show that the body temperature of the flies *Phormia terrae-novae* R.D. and *Calliphora uralensis* VILL. rise to 32–37° C, thus exceeding the ambient temperature by 11–22° C, under the insolation at a rate of 1660 cal/cc/min. The maximal activity of these flies coincides with the hours of maximal insolation.

feeding and predation, in sharp contrast to the conditions that prevail in the lowland biota, is without doubt part of the high altitude specialization.

Regardless of the nature of food, an important peculiarity of the food habits of the high altitude insects is correlated to the profound influence of the winter snow-cover. In the first place, as already pointed out, feeding is possible only after the winter snow has melted and the melt-water has irrigated the ground and has thus initiated the germination and growth of vegetation. As long as the snow persists, the warmth from the summer insolation disappears completely as latent heat of ice and snow and it is therefore only after the melting of the snow is complete that insolation can reawaken the hibernating insects to the active life of the summer. Secondly, nearly all the high altitude insects depend to an astonishing degree upon the snow itself for an abundant and varied source of food.

No matter whether an insect is phytophagous or feeds on pollen grains and spores, seeks after organic debris in various stages of decay, or is a carrion feeder or even an active predator, all species have developed the unique habit of utilizing the open surface of the snow and glaciers as the principal foraging ground. On the Himalaya, for example, the snow surface is indeed almost exclusive feeding ground at elevations of 5000 m and above*. Vast numbers of Collembola, Dermaptera, Carabidae, Staphylinidae, Hymenoptera, Brachycera-Diptera and even certain Lycaenidae come to the snowfields for feeding on the wind-blown derelicts, frozen and well preserved in or on snow and ice. There is indeed an almost inexhaustible source of such accumulations inside even a moderately large snowfield or glacier. Truly immense quantities of pollen grains of many different plants, spores of fungi and of Protozoa, seeds of a great variety of plants, flies, gnats, midges, aphids, winged ants, jassids, grasshoppers, all sorts of moths and butterflies and nearly every conceivable group of winged and apterous insects, spiders, etc. brought by the upper air currents, are frozen and entombed in the snow and glacier-ice. Surprisingly large numbers of Collembola feed on the pollen grains of juniper and other high altitude plants, scattered on the snow surface. Diverse other high altitude insects gorge themselves with the soft morsels of food, refrigerated in the snow. Numerous beetles like *Nebria* LATR., *Bembidion* LATR. and *Atheta* THOMS. may be seen in fine weather to dart from their shelters on the snow-field to bite off morsels of food from the dead bodies strewn on it, either to devour the morsels on the spot or to carry them away to their shelters to be eaten at leisure. A number of Diptera-Brachycera like Asilidae, Sarcophagidae, etc. swarm on the derelicts on snow, sucking the juices from the softened dead bodies. Other species of *Bembidion* LATR., *Nebria* LATR. and other

*The snow biocoenose of algae and other micro-organisms should not also be overlooked in this connection. The well known red-snow supports, for example, Volvocale-type of *Chlamydomonas nivalis* WILLE and *Chlamydomonas ohioensis* KERR., the Chlorococcale-type of *Scotiella nivalis california* KOL, *Trochiscia americana* KOL, etc. The green-snow derives its colour from the immense numbers of *Chlamydomonas yellowstonensis* KOL, with *Scotiella polyptera* FRITSCH, *Mycacanthococcus antarctica* WILLE, *Pleurococcus vulgaris cohaerens* WITTR., *Romeria elegans nivicola* KOL, *Sorochloria aggregata cryophila* KOL, *Scotiella nivalis* (CHOD.) FRITSCH, *Raphidonema tatrae yellowstonensis* KOL, etc. The yellow-snow, at the snow-edge, is due to the presence of iron bacteria *Siderocarpus coronata* RED., and *Leptothrix ochracea* KG. The reader will find an interesting account of these and other cryobiocoenoses on the Rocky Mountains in a recent contribution of KOL (KOL, E. 1964. Cryobiological research in the Rocky Mountains. *Arch. Hydrobiol.*, 60: 278-285).

Carabidae and Staphylinidae like *Aleochara* GRAV., spiders, etc. utilize the snow-field as an ideal hunting ground, to capture the large numbers of other high altitude insects that come there for foraging. The peregrinations of such hunters and predators often extend to long distances on the snow-fields and glacier surfaces. There are numerous scattered references in the literature to the habits of hunting on snow-fields on the Alps, on Fennoscandian mountains and on the high mountains of North America (600, 1078).

The high altitude insects not only utilize the open surface of the winter snow and the glacier-ice, but a large number of them and indeed the greatest majority of them on most high mountains, congregate at the edge of the receding winter snow and at the edge of the glacier and glacier snout and feed on various dead organisms, entombed in the snow and ice, as they come to be exposed with the melting of the snow during the summer. Collembola are of course the first to appear at the snow-edge and to swarm around in millions on every bit of dead and softened animals, as it falls free from inside the snow. Many species of Carabidae, including often some *Bembidion* LATR. and *Nebria* LATR., nearly all Staphylinidae, Tenebrionidae and diverse other carnivorous and debris-feeding Diptera thus come to the snow-edge for foraging not only on the food that is being uncovered from within and from under the winter snow-cover, but also to capture the other insects that have come to feed on the entombed material. We find at the snow-edge every class of feeder. The material that comes to be uncovered at the snow-edge or glacier snout is sometimes decades or even hundreds of years old and represents stratified layers of all sorts of dead insects, accumulated in the past. As the melting progresses at the snout and feeds the torrential streams, the melt-water carries with it the entombed debris and provides rich food for the larvae of Trichoptera and some aquatic Diptera of these streams. The ability to utilize the snow and glacier as a storehouse of food is one of the most important high altitude specializations among insects.

8. HIBERNATION AND LIFE-CYCLE

Prolonged hibernation under snow-cover and correspondingly short period of rapid development during the brief and fleeting summer are characteristic specializations of the high altitude insects of most mountains, particularly in the northern hemisphere. On the equatorial mountains and on the south temperate zone mountains, even at extreme elevations, the characteristic diurnal climate of the former and the nearly isothermal climate of the latter are closely correlated with the absence of a definite single cold-hardy hibernating stage in the life-cycle. In such cases, all the developmental stages of the high altitude insects have nearly equal capacity for remaining dormant during the night, with its daily frost even during the summer. In other cases, however, hibernation is always restricted to the winter and all active development and metamorphosis to the summer. While some high altitude insects like Orthoptera hibernate as eggs, others hibernate as last-stage larvae or as pupae and sometimes even as adults. Hibernation takes place always in ground that receives adequate snowfall during the winter. As pointed out earlier, the soil under snow-cover is not frozen during the winter, although the atmospheric temperature may drop to —45° C. The presence of a heavy snow-cover thus ensures optimal conditions for the hibernation of high altitude insects. It is not merely the successful initiation and completion and break of hibernation, but also the entire course of active development during the succeeding summer is very profoundly influenced by the extent and depth of the winter snowfall. Regardless of other conditions, the velocity of development of insects during the summer at high altitudes, the nature and abundance of the resulting individuals, the number of generations, the chances of any given species appearing as adults in a locality and indeed the whole course of the life-cycle of the high altitude insects depend on the snow-cover in the previous winter. Scanty snowfall in winter results in suboptimal conditions during the succeeding summer for nearly all species.

It must be observed that the beginning of summer does not at once mark the end of hibernation. Hibernation continues as long as the winter snow-cover persists and is only broken after the snow has melted and the melt-water has soaked through into the insect shelters. As mentioned above, as long as the snow-cover lasts, insolation does not bring about any rise in the soil temperature, so that it is only after the snow-cover has disappeared that the soil temperature rises and acts as a favourable factor for the active phase of development of the high altitude insects. The melt-water also seems to act in some unexplained manner as a sort of chemical or physiological trigger that breaks the dormant spell of the insects, which by now are physiologically in a state of readiness to return to active life. As soon as the snow-cover disappears, we find, therefore, every species rushing through its development, growth and metamorphosis. The release from under the winter snow-cover is thus followed by an extremely remarkable sudden acceleration of metabolism, development, growth and metamorphosis. The acceleration of development is closely correlated with the earlier exposure to sub-zero temperatures during the winter and the relatively warmer conditions during the summer. The stimulating effects of such alternating conditions of temperatures on insects are well known even at sea-level (137, 205, 399, 1083).

The velocity of development increases markedly with the advance of summer, so that the later stages in the life-cycle of a species succeed each other more quickly than the earlier ones. Regardless of the species and most other environmental conditions, there is a conspicuous gradient of developmental velocity and insect activity from the snow-edge. The life-cycle marches with maximum speed in all species in the immediate vicinity of the snow-edge and diminishes with increasing distance. On the ground moistened and enriched by the melt-water immediately below the edge of the receding winter snow, the hibernating insects burst forth into active life, the life-cycles march forward with a sudden bound, rapidly attain their peaks and slow down before the snow-edge has receded very high (fig. 25). In a narrow belt of the ground, freshly exposed from under the snow-cover, there spring forth, within a few days, dense carpets of low-matted vegetation, insects emerge, larvae feed, undergo metamorphosis and adults emerge. In the rapidly drying belt of ground further below, at a greater distance from the snow-edge, and uncovered earlier from under the snow, the activities and the abundance of the insects too, are already diminishing; most species have indeed moved up with the snow-edge. Still further away from the snow-edge, in the ground exposed in spring and early summer, the activities are nearly ended and most of the insects are already in the stage that is destined to hibernate under snow during the next winter.

Although the high altitude environment favours univoltine species, there are also some bivoltine forms. Species which need more than one year to complete a

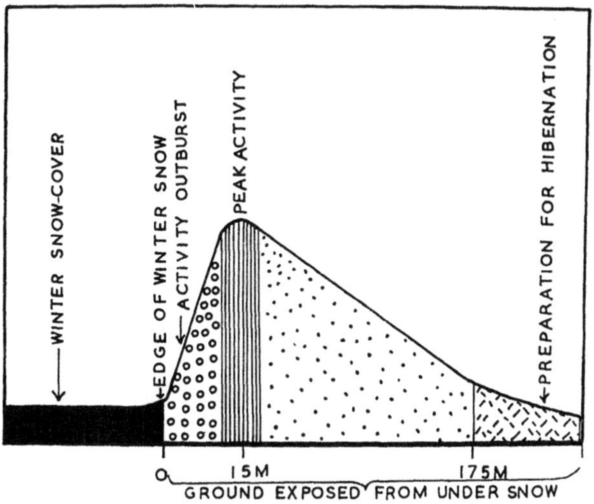

Fig. 25. The gradient of developmental velocities and intensities of total activity of the high altitude insects, from the snow-edge, on the south slope of the Northwest Himalaya.

single generation also appear to be favoured at high altitudes. The reduction in the number of generations with the increase in elevation is often obvious. Species like *Papilio machaon asiatica* (MÉN.) from the Northwest Himalaya are double-brooded at elevations of about 3000 m, but are only single-brooded at about 4300 m above mean sea-level.

The life-cycles are nearly always short and completed within one year and in such cases the hibernation takes place in the same stage year after year. The adults appear approximately about the same time during the spring-summer. In other cases, especially in Ephemerida, Plecoptera and many Carabidae, the life-cycles are rather prolonged over perhaps two or even three years. The larval stage is, therefore, prolonged and hibernation takes place in this stage in one year and perhaps in the pupal stage in the next year and the adults appear in spring during the third year. Small-sized larvae are not always, therefore, indicative of a new or a young generation; some of them may actually be one or two years old. Owing to the differences in the hibernating stage and other conditions, the adults of different species emerge at different times during spring-summer. The Collembola are perhaps among the first to appear even before the winter is properly over. The Dermaptera, most Heteroptera and many Coleoptera and Lepidoptera appear as adults in early summer. In the case of many Diptera and Lepidoptera the adults appear only in mid summer. Among those species in which the adults appear in late summer are Orthoptera, most Heteroptera, some Diptera and some Coleoptera.

C rtain peculiarities in the sex ratio of the high altitude insects seem to be correlate with the high altitude environment on the Himalaya. In many multivoltine Diptera a remarkable shifting of the sex ratio has been observed in successive generations. The spring or the early summer generations are mostly characterized by a marked preponderance of females, but in the mid summer the sexes largely tend to become approximately equally abundant. In some cases there is often a slight rise in the proportion of females in the last generation in late summer or early autumn (fig. 26). In the case of some univoltine species that extend from the treeline to an elevation of about 4500 m on the Northwest Himalaya, the sexes are approximately equally abundant near the treeline, but show a tendency for preponderance of females at an elevation of about 4000 m. The observations of REINIG (878) on the Pamirs

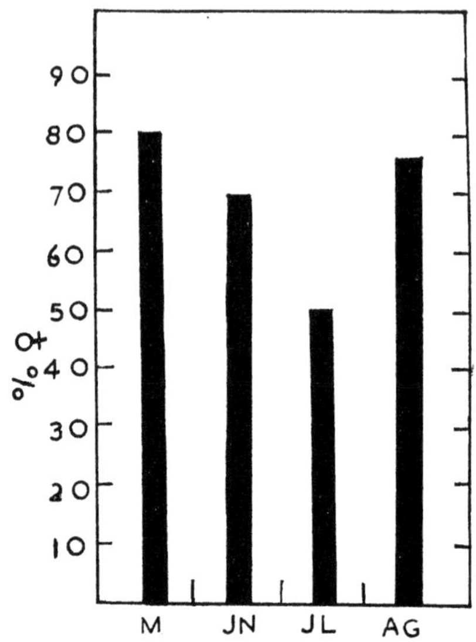

Fig. 26. The fluctuations of sex-ratio in the high altitude Diptera on the Northwest Himalaya, from spring to late summer.

Region are of particular interest in this connection. He found a conspicuous sparseness of the workers of *Bombus* spp., with increase in elevation from 3800 to 4050 m. Similar conditions have also been described in the case of arctic *B. hyperboreus* SCHÖNH., *B. balteatus* DAHLB. and the subarctic *B. lucorum* (LINN.) from Lappland.

CHAPTER IV

ECOLOGICAL INTERRELATIONS AT HIGH ALTITUDES AND HIGH ALTITUDE INSECT COMMUNITIES

While it is not possible here to deal with all the complex ecological interrelations and the communities of high altitude insects, we may, however, draw particular attention to the more important interrelations and the major types of communities and emphasize the fundamental differences between the high altitude and lowland biota.

Regardless of the spatial, geographical and genetic isolations, ecologically considered, every species of the high altitude biota is bound up either directly or indirectly with every other species of insects, other animals and plants. The ecological interrelations of the species are essentially ecological interdependence of the species and are intraspecific or also interspecific. The latter forms the basis for the community concept. Most of the interrelations may be eventually traced back to the food requirements of the species, but owing to the special environmental conditions prevailing on high mountains, the requirements for shelter from the extreme conditions very profoundly modify most basic patterns of ecological interrelations, with which we are generally familiar in the plains. Irrespective of the nature of the interrelation and the number and taxonomic affinities of the species involved, the characteristic environmental conditions above the forest-line on mountains and the high ecological specializations of the high altitude insects constitute the ultimate basis for the ecological interrelations of the high altitude biota. The complex threads that bind these insects with each other and with other organisms in a great variety of ways are primarily their urge to seek optimal conditions in the microclimatic niches. The ecological interrelations are, therefore, determined in the high altitude biota by a wholly different set of factors and have also a fundamentally different significance from those prevailing within the forest and the prairie. We do not find single and isolated relations or even a simple succession of one relation after another, but a simultaneous operation of many different types of interrelations to produce a complex pattern of ecological interdependence. Just as the member species of a community are linked together by complex interrelations, the different communities are also ecologically interdependent. The ecological interrelations are in no case rigidly fixed, but shift in their respective importance and also undergo various other changes from time to time, depending upon elevation and the nature and magnitude of the changes in the ecosystems. The ecological interrelations considered here are i. the interrelations of the high altitude, the montane forest and the lowland biota and ii. the interrelations within the high altitude biota.

I. INTERRELATIONS OF THE HIGH ALTITUDE, THE MONTANE FOREST AND THE LOWLAND BIOTA

Despite the great fundamental differences in the environmental conditions, the ecological specializations, composition, organization, spatial and genetic isolations, a remarkably complex ecological interrelationship links together the high altitude biota, the montane forest biota and the lowland biota of the region in which the mountain is situated. Precisely as in the case of the member species of a high altitude insect community and among the different high altitude communities, an extraordinarily high degree of ecological interdependence may be observed between the high altitude mountain-biota as a whole and the biota of the mountain forest-zone or of the surrounding lowlands. Of the various complex threads in this inter-relation, we may indicate only two of the more outstanding ones here.

Reference has already been made in the foregoing chapter to the passive transport of inorganic and organic dusts by air currents from the forest and from the lowlands and to the ecological importance of the wind-blown derelicts at high altitudes. A considerable amount of organic debris from the high altitudes on mountains is also carried down by the melt-water torrents to the plains, thus providing for a constant circulation of matter between the elevated regions above the forest-line and the surrounding plains. We may now refer to another remarkable inter-relation between the high altitude and the lowland biota. Some insects of the forest-zone on the mountain slopes and often even from the surrounding lowlands exhibit an enigmatic behaviour of actively seeking the summits of high mountains, above the forest-line and sometimes even above the permanent snowline, regularly at certain times of the year. The habit of summit-seeking has been observed in many different groups of winged insects like Formicoidea, Coccinellidae, etc. from different parts of the world. Although a number of possible explanations regarding this unique behaviour have been put forward by different workers, we are still largely ignorant about the underlying factors and precise significance of these movements to the species (202, 203, 204, 274, 545, 595, 600, 843, 997, 1078). The most important fact to be remembered in this connection is that, unlike in the case of the wind-blown derelicts on snow-fields, the summit-haunting insects are not dead and frozen and are perhaps also not all passively transported by wind throughout. Though favourable air currents may without doubt assist the active flight of the summit-seeking insects, they are not, however, completely dependent on air currents also; the summit-seeking seems to be a deliberate and active movement on the part of the insect. Enormous swarms of the adults of various winged insects from the valleys and from the forest-zone on the mountains and also from the far-off plains, rise up, even where there seems to be no updraft air current, and have indeed been observed to move against the prevailing wind at lower elevations, in the direction of high mountain summits. In swarming on the summits of high mountains, the insects seem to show a more or less decided and so far unexplained preference for some mountain peak or ridge over numerous others, often quite closeby and apparently equally or perhaps even more readily accessible. The summit-seeking insects are essentially not habitual residents of the mountain summits and mountain slopes, but are only more or less regular visitors from the biota at lower elevations. On arriving at the summits of the mountains, they assemble in enormous numbers under some suitable shelter. When thus assembled *en masse* on the mountain summits, they neither feed nor breed. *Colias euxanthe* FELD., found on snow-fields on the Andes, is reported, for example, to actively rise to the high elevations from the lower zones.

Although the summit-seeking and the mass assemblages of diverse insects on high mountain tops have been known since long, the behaviour of certain Coccinellidae is remarkable in many respects. During autumn great swarms of Coccinellidae like *Coccinella septempunctata* LINN., *Adonia variegata* GOEZE, *Hippodamia* spp., etc. move up the mountains from the lower forests and valleys. They often cover great distances and ascend to elevations of 4500 m on the Himalaya, crossing some high passes, before finally assembling in enormous numbers under some sheltered stone ledge, boulder or snow. Some of these mass assemblages contain from 5000 to 9000 individuals per square metre. Covered by heavy snowfall during the winter, they remain protected often under 10-29 m thick snow-cover throughout the winter. With the melting of the winter snow in summer, many of the dormant individuals revive and move down back to the valleys below for breeding. Enormous numbers of beetles in the mass assemblage, however, perish and the dead beetles can usually be found in all stages of disintegration during the height of summer on the Himalaya. As successive swarms assemble in the same area year after year, we often find considerable accumulations of the dead and frozen beetles from the swarms of the earlier years. *Coccinella septempunctata* LINN. is not in reality a member of the high altitude insect community of the Himalaya, but mass assemblages have been found up to elevations of 4200 m (730). MANI (730) has given an interesting account of one such mass assemblage of *Coccinella septempunctata* LINN., in association with *Hippodamia* spp., on a sheltered snow-field at an elevation of about 4260 m, near the Lakha Pass on the south slope of the Dhauladhar Range of the Northwest Himalaya, in the middle of May 1953. This assemblage had about two million individuals and is perhaps the largest and the highest mass assemblage of any Coccinellid so far described from the world.

Mass assemblages of Coccinellidae near the summits of high mountains have been described from other parts of the world also, particularly Europe, Middle Asia and North America (482, 545, 549). DOBZHANSKY (268) has reviewed the earlier records and has also given an account of his observations on the mass assemblage of *Hippodamia* MULS. in autumn in the valleys and on mountain summits. According to him, Coccinellidae have two generations in southern USSR and it is only the individuals of the second generation which assemble *en masse*. He found mixed mass assemblages of *Hippodamia heydeni* WES., *Semiadalia 11-notata* SCH., *Brumus octosignatus* GEBL., *Coccinella sinuatomarginata* FALL., *Adonia variegata* GOEZE and *Platynaspis luteorubra* GOEZE on the mountains of Ferghana in Turkestan (Uzbek SSR), at an elevation of about 1500 m. Within an area of about 0.9 m^2, there were over 6000 individuals of these beetles. Most of them were motionless and if disturbed, they stirred a little and settled down again. None of them attempted to fly. He also found assemblages of Coccinellidae on the Sjuru-tjube Mountains. The mass assemblages were always found under stones and well protected from winds on projecting and prominent peaks, summits, ridges, etc. In one locality he found over 30 000 individuals in an area of a single square metre. Gigantic mass assemblages of *Hippodamia heydeni* WES. were found inside rock crevices at an elevation of 3200 m on the Kugar Pass near Aubeck-tau Peak (3500 m above mean sea-level). There were also mass assemblages of *Hippodamia rickersi* WES. The North American *H. convergens* GUER. assembles under stones covered by snow within the forest-zone on mountains, but the Middle Asiatic *Hippodamia* spp. ascend not only above the forest-line but also far above the permanent snowline in order to assemble *en masse* and be well protected from winds when over-wintering under snow-cover. In a recent paper, EDWARDS (300) has drawn attention to a number of interesting problems connected with the behaviour of mass assemblages of insects on high mountains. He found large accumulations of *Hippodamia oregonensis* CR., *H. 5-signata* BBY. and *Coccinella nivicola monticola* MULS. on snow of the Pinnacle Peak, on the south side of Mt. Rainier in North America. He has also described the mass assemblages on the Allan Mountains in Montana. In some localities, he found large accumulations of the dead beetles, belonging to the mass assemblages of earlier years. The number of individuals in any of these mass assemblages is usually between 30 000 and 70 000.

The underlying factors and the biological and ecological significance of the habit to the summit-seeking species are completely unknown at present, but there seems to be little doubt about the importance of the habit from the point of view of the ecological interrelations at high altitudes. Although the summit-seeking Coccinellidae and other insects in the mass assemblages do not themselves feed when on the mountains, they nevertheless provide an abundant and readily accessible source of rich and concentrated food during the summer for the high altitude insects. Even insectivorous birds and the mountain bears regularly feed on the dead and living insects in these accumulations. CHAPMAN *et al.* (204) have described the remarkable habit of the grizzly-bear feeding on the ladybird beetles and on the army cutworm moth *Agrotis* OL. in the mass assemblages on the summits of high mountains in Montana. The accumulations of the dead ladybird beetles provide the basis for the colonization of necrophagous Collembola, Staphylinidae, Tenebrionidae, Diptera, Acarina, etc. It is, therefore, interesting to note that, in addition to the air-lift of refrigerated food particles from the lowlands, the high altitude or the hypsobiont insects depend also on the summit-seeking or hypsophile insects of the lowland biota for their food.

2. INTERRELATIONS WITHIN THE HIGH ALTITUDE BIOTA

The ecological interrelations within the high altitude biota are of three major categories, viz. i. the interrelations of the insects and the plants of the high altitude environment, ii. interspecific interrelations of the high altitude insects and iii. the interrelations of insects and other animals of the high altitude environment.

The primary basis of the interrelations of high altitude plants and insects is, in the final analysis, trophic as is the case even within the forest-zone on the mountains or in the plains. Depending upon this relation, the high altitude insects are of two ecological types, viz. the lichen-moss feeders and the Phanerogam feeders. There are then the types that are only indirectly dependent on the high altitude vegetation for their food.

Pioneer vegetation like algae, lichen and moss on the rock surface among snow and ice, often far above the permanent snowline, provides optimal conditions for characteristically pioneer insect types like Collembola and Diptera. Even the caterpillars of some remarkable Lepidoptera thrive on the sparse lichen incrustations on rock at very high elevations on the Alps and the Himalaya. The pioneer association insects attract, in their turn, predatory and parasitic Acarina and are also soon followed by a great variety of other types, including large numbers of carnivorous and predaceous species and detritus feeders. These interrelations are dominant at the high elevations on nearly all the high mountains, on which observations have so far been made. In addition to this direct trophic basis, the incrustations of lichen and the cushions of moss rapidly corrode the rock and thus favour the formation of sufficient layers of soil, on which the colonization of Phanerogams and diverse high altitude insects, especially Coleoptera, becomes possible. The appearance of high altitude Phanerogams naturally favours the provision of optimal conditions for a much larger number and variety of high altitude insects and eventually leads to ecological stability of the communities. As is well known, the ecological interrelations of insects and lichen-moss vegetation are extremely unimportant even within the forest-zone on mountains and more so on the plains.

Among the Phanerogam-feeders we may differentiate the polyphagous from the oligophagous and monophagous types, but the great majority of the species are dependent on Dicotyledons like Compositae, Leguminosae, Rosaceae, Cruciferae or they are also largely dependent on Gramineae. The Gramineae are perhaps the most dominant Phanerogams at higher elevations and the insects, the larvae of which feed on diverse grasses, are, therefore, abundant over an extensive range of altitudes on nearly all the mountains. Polyphagous species are common in the transitional zone immediately above the forest, but become progressively scarce as the elevation increases. Oligophagous types soon replace the polyphagous ones in the alpine and in other parallel zones on different mountains. At extreme elevations, especially above the permanent snowline, in the subnival and eunival zones and at higher levels as high as the phytophagous species continue to occur, there is a complete predominance of monophagous types. It may also be remarked here that the leaf-feeding types are common at lower elevations, but it is the root-feeding geobiont types that really predominate at higher elevations. At the base of the pyramid stand the phytophagous types near sea-level, but at high altitudes this place is really occupied by the detritus-feeders, so that the direct dependence of the high altitude insects on the higher plants for food is only slight. Even this comparatively light dependence diminishes with increase in altitude.

We may then consider the mutualistic relations of insects and plants involved in the cross-pollination of flowers by insects (630). In the forest and in the lowland environment, bees, especially the honeybees, represent perhaps the dominant flower pollinators and other insects like Lepidoptera and Diptera are on the whole of only secondary importance. Both within the forest-zone on mountains and in the plains, we find a great abundance of more or less strongly scented and white-coloured flowers that often open late in the evening or also at night and are cross-pollinated by diverse crepuscular or night-flying insects, particularly moths.

The situation is, however, fundamentally different at high altitudes above the forest on mountains. We have already emphasized the fact that an extremely small proportion of the high altitude insects is active on wings, only during the hours of bright sunshine and only when there is a lull in the fierce wind. Insects do not fly at night even on the equatorial mountains. Even the species that belong to groups, which are typically crepuscular or nocturnal in the forest, are exclusively diurnal at high elevations. In close correlation with these facts, all plants at elevations above the forest-line on mountains bloom only during the daytime, particularly only in the hours of bright sunshine. The flowers of these high altitude plants are gorgeously coloured yellow, orange, red, pink, etc. and only rarely blue, white or violet and are mostly massed together in an extraordinary manner, so as to produce marvellous colour effects on the ground, conspicuous even from a distance. In an environment where all the insects are active only during the daytime, white and strongly scented night-flowering plants would be meaningless from the point of cross-pollination by insects.

As we shall see in the next chapter, the honeybees are totally absent at very high elevations above the forest-line and even the nectar-sucking or the predominantly pollen-gathering semi-social or solitary bees are also extremely scarce and are generally restricted to elevations immediately above the forest-line. The Bombidae are perhaps the only bees that may be found at high altitudes, but they are exclusively day-fliers and are guided, in pollen gathering, by the gorgeous colours of the flowers rather than by their scent in a wind-blown environment. Cross-pollination of flowers by bees and bee-pollinated flowers are, for example, practically absent even immediately above the forest-line on the Himalaya and are wholly unknown at elevations of over 4000 m. We find, in their place, a characteristic increase in the dominance of flowers cross-pollinated by Lepidoptera and Diptera. The principal cross-pollinators of the high altitude Phanerogams are Lepidoptera like *Papilio* LINN., *Argynnis* FABR., *Colias* FABR., *Parnassius* LATR., *Pieris* SCHR., etc., Brachycera-Diptera like Stratiomyiidae, Empididae, Dolichopodidae, Asilidae (males), Syrphidae, Anthomyiidae, Sarcophagidae, Tachinidae, etc. Furthermore, insect-pollinated flowers occupy a minor place in comparison to the flowers pollinated by wind at high altitudes. The reader will also find the observations on the cross-pollination of flowers by insects on the Alps by MÜLLER (785) interesting in this connection.

The interspecific interdependence of insects and other animals seems to be far more intense at high altitudes than in the plains or even within the forest-zone on mountains. This is obviously correlated with the relatively extreme nature of the abiotic environment at high altitudes. In marked contrast to the forest and the lowlands, the ecological interrelations with animals, including other species of insects, in the high altitude biota, are characterized by the very pronounced dominance of predatism and parasitism. There is in nearly every group of the high altitude insects a conspicuous dominance of predators and parasites. In any forest or typically lowland community, the predators and parasites are by no means the dominant members, but this is likely to be so in the high altitude insect community. Predators are also more abundant than true entomophagous parasites. We may consider here, by way of example, a subnival-nival zone community from the Alps. The *Nebria atrata-Gnophos caelibarius* community is remarkable for the fact that nearly 60% of its character species and other associated species are either predators or they are parasites. Similarly about 50% of the members of the *Caeculus echinus-Chrysomela crassicornis norica* community of the subnival zone from the Alps (*vide infra*) are predator-parasites. In some cases, every member of a high altitude insect community may prove to be a parasite or a predator of some degree. We have, for example, at the base of the pyramid of numbers a more or less omnivorous Collembolan that tends to be as often predatory as it can, with a chain of other predators like Acarina, Dipterous larvae, the larvae and adults of Carabidae and Staphylinidae, scorpions, chelifers, centipedes and spiders. Such a community of species may be found apparently in more or less complete harmony (?) under the same stone. In such a community, it is often found that the Collembola depend for their food on the dead bodies of nearly every other insect or arthropod and they may also attack some of them alive if conditions are favourable. The Collembola are devoured by the Diptera larvae and masses of dead Collembola provide rich organic debris for large numbers of necrophagous Coleoptera, especially Staphylinidae. Acarina and Coleoptera feed on the larvae of Diptera, the latter breed on live spiders and the spiders suck the live flies dry! It would seem that almost every member of a high altitude community spends practically all its time devouring and predating on every other member species.

Within the forest and on the lowlands, the more common predatory insects belong to Odonata, Mantodea, Heteroptera, solitary and social wasps, many Neuroptera and some Coleoptera. The principal predatory species in the high altitude environment belong, however, to Dermaptera, Carabidae, Staphylinidae, Diptera, Araneida, Acarina, Scorpionida, Chelonethida and Chilopoda—predominantly terricole forms. The more important entomophagous parasites in the forest and lowland biota belong, as is well known, to the Ichneumonoidea, Chalcidoidea, Serphoidea, Bethyloidea and other parasitic Hymenoptera. These well known groups are, however, quite unimportant at high altitudes and are progressively replaced, with increase in elevation, by parasitic Diptera and Acarina. The absence of predatory and parasitic Hymenoptera even at the timberline on the arctic hills in Finnland has been reported by a number of observers (796). Acarina are indeed the foremost entomophagous parasites at extreme altitudes on the Himalaya and the Alps.

The predatism and parasitism by Acarina at high altitudes is associated with a most remarkable phenomenon. Just as the Collembola stand at the base of the pyramid of numbers in the organization of most high altitude insect communities, the predatory-parasite Acarina occupy the topmost position in the pyramid. There is indeed a most remarkable correlation between the fluctuations in the population abundance of Collembola and Acarina in the high altitude communities. The relative abundance of the predatory and parasitic Acarina in any high altitude community is inversely proportional to that of Collembola (375, 564, 730). The larger the numbers of Collembola, the smaller is the density of population of Acarina in the community. The dominance of Collembola is indicative of the relative immaturity of the community and the dominance of Acarina is characteristic of its maturity. The reciprocal fluctuations of the abundance of Collembola and Acarina represent an extremely important diagnostic character of most high altitude, particularly subnival and nival zone communities. JANETSCHEK (564) has, for example, described increase in the abundance of Acarina with fall in the abundance of Collembola on the Alps. The Collembola-Acarina fluctuations are indirect measures of the degree of soil maturity and the age of a biotope above the timberline on the Alps. Climax communities or communities with the conditions approaching the climax stage are indicated by the dominance of Acarina over Collembola.

The dominating urge to seek shelter from the extreme atmospheric conditions brings together, under the same snow-covered stone or other similar microclimatically optimal niche, taxonomically wholly unrelated prey, predator, parasite, etc., not merely during hibernation but also in summer. Under the same stone we may thus often find diverse Collembola, the larvae and adults of Coleoptera, Dermaptera, fly maggots, spiders, mites, scorpions, chelifers and centipedes. It is only under the stones that all of them find the absolutely minimal conditions, not merely for existence but also for successful hibernation during the winter and reawakening in the summer. We have here exactly the situation where several species are forced to live together under one shelter, because of the specialized habitat conditions, in a state of what may more appropriately be termed a sort of armed neutrality rather than peaceful coexistence!

3. HIGH ALTITUDE INSECT COMMUNITIES

The high altitude insect communities differ from most insect communities of the lowlands in a number of characters. In contrast to the lowland and forest communities, the high altitude insect communities are mostly composed of relatively fewer species, perhaps no more than three or four, but are characterized by a great abundance of individuals of these species. Owing to the generally extreme climatic and other conditions of the high altitude environment, the threads that bind the member species of a high altitude insect community are usually stronger and far more tangled than in the forest or the lowland community. The interdependence of the member species and the integration of the community as a whole are, therefore, far more pronounced than at sea-level. A high degree of interspecific integration, with the concomitant community independence and isolation, are generally peculiar characters of most high altitude insect communities. Stratification within the high altitude insect community involves phenomena, basically different from those obtaining in the forest and lowland communities. The size, composition, abundance, spacing, equilibrium and stability, the interrelations and succession of the insect communities differ on different mountains and at different altitudes on the same mountain. As the habitats vary with altitude, the ecosystems also differ. The community regulatory mechanisms, which promote the community stability, are not the same as in the plains and the relative importance of these mechanisms varies with altitude. The communities become increasingly sparse and smaller in size as the altitude increases. The spacing of the communities becomes more and more irregular and adjacent communities are very greatly separated. With the increase in altitude, there is also a conspicuous shift in the dominance of the different communities. The dominance of the vegetation-mats communities at elevations of about 3000 and 3500 m on the Northwest Himalaya gives place, for example, to the dominance of the hypolithic communities at elevations above 4200 m. The general community stability tends to diminish both above and below this elevation on the Himalaya. The maximum concentration of relatively stable communities of all types is generally found at an elevation of about 4200 m on the Northwest Himalaya. The pioneer communities are mostly aggregated immediately above the forest-line and also in the vicinity of the permanent snowline on the Himalaya.

The characteristic altitudinal stratification of the species in all groups of high altitude insects profoundly alters the qualitative composition of the high altitude communities. While the different communities constitute a more or less interrelated and interdependent mosaic at lower elevations, particularly within the forest zone, isolation, irregular heaping and absence of randomness become more and more pronounced above the forest-line. The high degree of community interdependence that is readily apparent immediately above the forest, diminishes markedly at higher elevations. The interspecific integration within the community is, however, relatively more complete only at higher elevations. On mountains with a decided east-west trend, especially in the northern hemisphere, the characters of the high altitude insect communities are more or less markedly influenced by the differences in the habitats on the south-north slopes. Such differences are readily observed on the Alps, Caucasus and the Himalaya, particularly at medium elevations. On the Nortwest Himalaya, for example, the differences due to the north-south slope exposures are particularly marked at elevations of 3000-3500 m, but tend to be more or less obscured at elevations above 5000 m. Not only the different altitudinal zones on a mountain have different characteristic communities, but even within each altitudinal belt we find different communities. These differences are governed largely by the local climatic and edaphic factors of different biotope, some of which are readily observed, but others are still largely obscure. By comparing the behaviour of different high altitude insect and plant communities in the high-alpine-nival zones of the Alps and the Himalaya, we find differences in the synecological behaviour of plants and insects; these differences are due primarily to the fact that in some cases the plants and in other cases the insects are more sensitive to a given environmental factor. It must, however, be observed here that inspite of these differences, an intimate relation exists between the plant and insect communities at all elevations above the forest-line.

The high altitude insect communities may be recognized by their character or indicator species or also on the basis of altitudinal zonation (369, 378, 471). FRANZ (368) has, for example, recognized the *Nebria atrata-Gnophos caelibarius* community of the Alps on altitudinal zonation as one of the subnival-nival zone communities and the *Carabus concolor-Zygaena exulans* community as a high-alpine meadow community (*vide* Chapter XI). On the basis of their ecological specializations, the communities may be described as transitional, terricole, endogeous, aquatic, cryophile or cryobiont, thermophile, hygrobiont, phytophile, etc. As a community is, however, an assemblage of ecologically interrelated species, an interdependent and self-sustaining unit, related as a whole in some specific manner to a given habitat, the high altitude communities may also be grouped according to their ecological habitats. BÄBLER (74) recognizes, for example, three major groups of high altitude insect communities on the Alps on the basis of habitats, viz. i. the communities of the snow-ice-free area, ii. the communities of snow-fields and glaciers and iii. the aquatic communities. These groups are further subdivided again on the basis of habitat differences into smaller communities. HANDSCHIN (471, 472) has also utilized the habitat differences in classifying the Collembola communities of the Alps. He recognizes, for example, i. the communities on the surface of stagnant waters, ii. spring and stream communities, iii. hygropatric communities or the communities of damp and moist lichen-moss covered boulders and stones, iv. snow-edge communities, v. snow communities, vi. phytophile communities, vii. petrophile communities and viii. soil communities. On the Northwest Himalaya, the habitats, with special reference to the grouping of the high altitude insect communities, are i. the snow-edge, ii. the snow surface, iii. underground spaces and cavities, iv. spaces under stones and boulders deeply sunk in the ground, v. crevices and fissures in rock, vi. barren rock, vii. dry moss-lichen incrusted rock, viii. damp or moist moss covered rock, ix. nunataks rock, x. soil, xi. vegetation mats, moss cushions, etc., xii. meadow with abundant grass, xiii. the edges of melt-water ponds, lakes, steams, etc., xiv. lakes and ponds, xv. melt-water torrential streams and xvi. thermal and chemical springs. We may also add to these habitats the cave that represents an important high altitude insect habitat on a number of mountains. We may consider here the following communities: i. the rock communities, ii. the hypolithic communities, iii. soil communities, iv. cavernicole communities, v. snow communities and vi. the aquatic communities.

4. THE ROCK COMMUNITIES

As an insect habitat, the rock surface reveals at high altitudes many interesting features of extreme conditions and specializations. The rock includes not only the "Muttergestein", but also the large and small boulders, fixed and loose and movable stones, pebbles of older glacier moraines, erratics, avalanche schutt, etc. They comprise mostly granites, gneiss and other crystalline rocks, non-fossiliferous sedimentaries, sandstones, slates, shales and often also dolomites and numerous fossiliferous materials. As a habitat of the high altitude insects, the rock is available from above the upper limits of the forest to the highest mountain summits. Even on the high grassy meadows and the alps, the rock habitat is abundant. Above the permanent snowline, a considerable area of rock surface remains snow-free during the summer and is thus available for colonization by diverse species. The rock habitat includes not only the surface of the rock, but also the cracks, crevices and fissures in the rock, the spaces between loose stones and the spaces between the large boulders that may often form miniature caves. The rock may be dry and barren or it may also be covered with more or less thin incrustations of lichen or matting of moss and other sparse higher vegetation. The rock is not, however, always dry, but often damp or even moist from being wetted by constant sprays from cataracts, by the melt-water trickling from the snow or it may also have a thin layer of water constantly running over it. The projecting rock masses in vegetation terraces on the slopes are also important habitats.

The rock that completely lacks a vegetation cover of any kind, including even the dry lichen incrustations, represents no doubt an extreme habitat. The intersurface between the rock and the air is in this case sharply defined and the rock surface is exposed to the full force of sunlight, it is baked hot and dry during the hours of bright sunshine under the action of intense insolation, it is blown by wind and is also subject to rapid erosion by running water, moving snow and ice or by falling and rolling stones. Evaporation is strong and the diurnal and seasonal temperature changes are the highest when compared to other habitats. Extremes of temperatures, ranging from 35° to —7° C occur sometimes even within the space of an hour. Constantly exposed to the direct sunrays, the rock surface heats up rapidly, but a cloudy halfhour interval is adequate to bring down the temperature of the rock surface below that of even the atmospheric minimum in the shade. The layer of air immediately above the rock surface has also the lowest relative humidity and the rate of evaporation is, therefore, extremely high. There is maximum scarcity of food. Vegetation of some sort, even including the pioneer plants like lichen and moss, profoundly alters the conditions and provides suitable optima for the petrophile and petrobiont communities. The initial colonization, composition and stability of the high altitude insect communities on the rock-surface habitat depend, therefore, more on the degree of exposure than on the chemical nature of the underground (527).

The rock communities include mostly pioneer associations of Collembola-Thysanura-Acarina, representing perhaps the relatively unimportant and more or less thermophile elements at high altitudes. Stratification within the rock communities is perhaps not always very marked, but the altitudinal zonation is usually very conspicuous. Diel periodism is pronounced but also rather irregular, when compared with the conditions prevailing in soil or moss communities. The seasonal aspect of the rock communities is often revealed in the diminishing hygrophily and in the perceptible increase in the dominance of the carnivorous and predatory forms as the summer advances.

The initial colonization of the rock surface by the lichen-moss association paves the way for other pioneer communities of Collembola like *Isotoma* (BOURL.) BÖRN., *Isotomurus* BÖRN., *Orchesella* TEMPL., *Tomocerus* NIC. and sometimes exceptionally also *Entomobrya* ROND., to be succeeded by diverse predatory and parasitic Acarina, *Ctenolepisma* ESCH., *Machilinus* SILV. (Thysanura). These are followed by some terrestrial Chironomidae, some brachycerous Diptera and a few Staphylinidae. Although Lepidoptera as a rule belong to phytophile communities, some species are known to be typically members of the communities on rock, incrusted with the yellow, orange or red lichen on the Himalaya and on the Alps (730, 1132). The caterpillars of *Dasydia tenebraria* ESP. and *Gnophos caelibarius spurcaria* LATR. from the Alps are members of lichen communities on rock surface at elevations of 2780-2825 m and the caterpillars of *Psodos alticola* MÉN. also occur on lichen-incrusted rock at an elevation of 3237 m on the Alps.

The rock surface is also visited by the adults of Ephemerida, Plecoptera, Trichoptera, Lepido a and Diptera from other neighbouring communities during the early hours of ...ght sunshine or especially in a cloudy interval in the late forenoon after some hours of sunshine for absorbing the heat that is being radiated away. The rock surface is also one of the common hunting grounds for predatory Coleoptera, Diptera and spiders.

Some of the most interesting rock communities are associated with nunataks. The word nunatak is derived from the language of the Eskimo people and describes the rock mass that projects as an island from a vast desert of snow and ice. Such nunataks are exceedingly common above the permanent snowline. Due to peculiarities of gradient, local exposures to wind, aspect, etc., the winter snow usually evaporates away during the summer or only insignificant amounts of the winter snow contribute to the film of moisture on the surface of the rock necessary for a sufficiently rich growth of lichen, algae or moss and for colonization by Collembola, Acarina, Tipulidae, Chironomidae, Anthomyiidae, Syrphidae and other Diptera and some Lepidoptera, but particularly Nymphalidae and Papilionidae.

The food resources of the rock communities are derived from the sparse vegetation of algae, lichen, moss and other higher plants that grow in between the rock masses. Considerable amounts of wind-blown organic dusts also contribute to the food supply for the members of the rock communities at all elevations. Though many Collembola depend on these and also on the organic debris, they represent in their turn the basic animal food for a great variety of debris-feeding or predatory species. There is on the whole a large population of predatory forms in the rock communities. Although most rock communities are essentially pioneer associations, some of them apparently attain considerable stability and even closely approach climax conditions at higher elevations, especially above 5000 m on the Himalaya.

5. HYPOLITHIC COMMUNITIES

The hypolithic communities embrace the immense and much varied insect associations that belong to the characteristic habitat in spaces under stones. The hollow spaces under stones, as insect habitat, represent perhaps the most favoured among the diverse habitats above the forest-line on all high mountains in the world. We must include here all the underground cavities and narrow spaces under a variety of stones and boulders, more or less deeply buried in the ground. The spaces under loosely lying stones that often slide or roll down are not, however, permanent habitats of the high altitude insects. The hypolithic habitat is available in all types of localities, from the treeline to the highest elevations of existence of insect life. Some of these hypolithic spaces are natural, but others have often been more or less enlarged and otherwise modified by the different species that occur in them. As habitat of the high altitude insect communities, the hypolithic spaces are characterized by optimal conditions for perhaps the largest majority of species. The environmental conditions in the hypolithic spaces approach everywhere alsmost maximum stability. The temperature-humidity conditions of the air within these spaces fluctuate perhaps the minimum during the day and even the seasonal fluctuations are not on the whole abrupt and extreme. The rise of temperature of the air inside the understone spaces progresses gradually with the coming of summer and is correlated with the heat summation in the ground and general rise of the deep-soil temperature and is not, therefore, largely influenced by the prevailing conditions of the atmosphere outside or even by the temperature changes on the surface of the rock. More or less uniformly high relative humidity is always ensured and the absence of light and the shelter from wind, the minimum risk of desiccation, etc., are some of the prevailing environmental conditions in the hypolithic habitat. There is, however, no marked ecological stagnation, such as one finds typically in the cave. Unlike in the caves, the general ecological conditions change, but do so only very gradually and within relatively narrow limits. The factors which immediately govern the prevailing conditions in the hypolithic habitat are the size of the stone, the presence and the nature of the vegetation cover of the locality, the distance from the snow-edge or from the melt-water ponds, lakes, streams, etc., the gradient of the floor under the stone, the thickness and the duration of the winter snow-cover on the stone, etc. The environmental conditions under smaller stones are generally subject to greater fluctuations than under the larger stones. It is also interesting to observe that regardless of the size and other suitable factors, no insect communities may be associated with a given hypolithic space, unless the stone is within the immediate vicinity of a snow-field or at least a melt-water pond or stream.

The hypolithic community is perhaps the single largest community in the high altitude biota and is characterized by much larger numbers of member species than other communities associated with other habitats. Dermaptera, Carabidae, Staphylinidae, Histeridae, Tenebrionidae, Chrysomelidae, Curculionidae, Formicidae, nearly all Lepidoptera, numerous Diptera, many Collembola, Thysanura, Chelifers, Acarina, spiders, scorpions, centipedes and millipedes are common representatives of the hypolithic communities. Nearly all the geobiont and geophile species show a pronounced tendency to inhabit every available space under suitable stones. The character species belong most usually to genera like *Anechura* SCUDD. (Dermaptera); *Calosoma* WEBER, *Carabus* LINN., *Harpalus* LATR. ,*Bembidion* LATR., *Amara* BON., *Nebria* LATR., *Trechus* CLAIRV., *Broscus* PANZ., *Chaetobroscus* SEM., *Atheta* THOMS., *Aleochara* GRAV., *Geodromicus* REDT., *Merista* CHAPUIS, *Chaetocnema* STEPH., *Longitarsus* LATR., *Blosyrodes* JEKEL, *Anchlaenomus* WATERHOUSE, *Catopionus* SCHÖNH., *Otiorrhynchus* GERM., *Ascelosodis* REDT., *Blaps* FABR., *Syachis* BATES, *Laena* LATR., *Cyphogenia* SOLS.

(Coleoptera); diverse species of ants, *Bombus* LATR. (Hymenoptera); many Lepidoptera like *Parnassius* LATR. (caterpillars only); *Isotoma* (BOURL.) BÖRN. and several other Collembola and many Thysanura, spiders like *Lycosa* LATR., etc. Mass assemblages of the individuals of several species often exist side by side under the same stone. Nearly all the hypolithic communities are practically climax associations and stratification within them is, therefore, very well marked and the altitudinal zonation of the communities is also well pronounced. Each zone of altitude has a characteristic community composition and although the total number of the member species in such communities diminishes with the increase in altitude, at any given altitude the average number of species in the hypolithic communities is still much larger than most other communities. Though the size of the community shows a definite gradient from the snow-edge or with increasing distance from any considerable body of water, the dominance in numbers of the member species is never surpassed by other communities. The hypolithic communities at elevations above the permanent snowline contain many pioneer species, but on the whole the community stability is never obscured or disturbed. The diel periodism is minimum in the hypolithic communities and totally absent in some localities. Seasonal periodism is not very common and not also very well pronounced.

Every member species in the hypolithic community is typically a hygrobiont and has a very low temperature valence. The greatest bulk of the member species is either a debris feeder or carnivorous and predatory form. The interspecific dependence is most marked, but the communities themselves are perhaps the most isolated and also the most irregularly distributed.

The members of the hypolithic communities are termed hypolithion and are characterized by a conspicuous discontinuity of their distribution and numerous other peculiarities of behaviour such as avoidance of strong light, a high degree of hygrophily, lack of special fossorial structural adaptations and by their distinct aspect sequence. The discontinuity of their distribution corresponds to the discontinuous distribution of the stones themselves. To the biochores of centres of action belong the roots of plants in the soil below the hypolithic space, the decaying vegetable matter found under the stones, the excreta of various animals, etc. Stones about 100-400 cm^2 large and about 20 cm thick typically provide the most optimal conditions for the hypolithion, and stones larger or smaller than these usually have smaller numbers of species and individuals. SCHÖNBORN (931) has recently described a peculiar stratification of the hypolithion. The first layer is on the under surface of the stone itself, the second layer on the surface of the ground in the hypolithic space and the third layer belongs to the more or less decomposed vegetable debris that accumulates under the stone. All the three layers exhibit peculiarities in population density and in the specific composition and have also a definitive function within the hypolithic community. The under surface of the stone, depending on its unevenness, has a first layer up to 10 mm thick, but this may locally be larger also. The interspace between the local unevenness of the under surface of the stone represents the most optimal place for the member species. Where the layer is only 5 mm thick, relatively few species flourish and these are generally also minute in size. The limiting factors of this layer include the presence of vegetable matter, so that when the latter decays, considerable empty space is left above and between the underside of the stone and the stone does not generally sink further below in the ground. The mechanical activities of the animals under stones also often contribute to enlarge the range of the second and third layers; the Carabidae in particular take considerable part in this activity. The melt-water that percolates and runs under the stones must also serve to maintain the optimal conditions of moisture and humidity and at the same time the water brings about a gradual sintering and breaking down of the lower surface of the stone, so as to enlarge the general area of the first layer. The size of the interspace between the first and the second layers of the hypolithion depends largely on the nature of the ground and the vegetation of the locality. The third layer contains the detritus from the under surface and also from the living beings. This layer has usually a specific composition and structure and the thickness ranges about 30 mm. The ecological types of the hypolithic communities are lithophile in the first layer, but endogeous in the other layers. The character species of the first layer are predominantly zoophagous, of the second layer detritus feeders and of the third layer either phytophagous or also detritus feeders. The third layer may be taken as a sort of base of the pyramid within the hypölithic community. In the hypolithic community, we find an example of the extreme case of rational utilization of strictly limited space, with deficiency of food, coincidence barriers and a little of pressure of concurrence.

6. SOIL COMMUNITIES

The usually scanty soil, collected in irregular patches at favourable localities, constitutes the habitat of a surprisingly large number of associations of high altitude insects, mostly dominated by Collembola. The optimal conditions for different soil associations are primarily governed by i. the depth of the soil, ii. the nature of the soil, iii. the gradient of the ground, iv. aspect of the mountain slope, v. the temperature-humidity conditions of the soil, vi. the winter snow-cover and its duration and vii. the age or maturity of the soil (usually measured in decades). As already mentioned, the surface of the soil is characterized by greater fluctuations of conditions than the deeper soil. Different soil-species show slightly different reactions to the temperature fluctuations in the surface soil. The top soil temperature is usually an important factor and often even a decisive factor in the distribution, composition, organization and other conditions in the high altitude insect communities in any given soil. The depth of penetration in the soil is largely governed by the soil moisture conditions. The top soil absorbs and radiates away heat and also dries up more rapidly than the deeper layers. Relatively few species occur, therefore, in the surface layers of the soil; the deeper layers have, on the other hand, much larger numbers of species. Light is an important factor that governs the nature of the surface layer of the soil, but the deep-soil associations are only indirectly influenced by the action of light on the surface layer of the soil. Organic matter, including the roots of living plants and other dead material of both plants and animals, constitutes the principal food resources for the soil communities.

The character species of most soil communities belong to Collembola like *Frisea* D.T., *Achorutes* TEMPL., *Onychiurus* GERV., *Orchesella* TEMPL., etc., representing the deep-soil or the true soil species (the euedaphon). The mesedaphon Collembola *Lepidocyrtus* BOURL., *Entomobrya* ROND., *Tomocerus* NIC., *Isotoma* (BOURL.) BÖRN. are also rich in species. Nearly every Collembola-association in soil contains one or more species of bdellid and trombiculid Acarina also as character species. The depth to which the Collembola occur in the soil in the nival zones of the Alps and the Himalaya is dependent on the soil moisture.

Coleoptera are richly represented in soil communities in localities rich in organic matter. They are also associated with soils of greater maturity than immature soils. Coleoptera are succeeded in the ecological succession by Diptera, especially Brachycera. Other types include Isopoda and Diplopoda, in close association with the Coleoptera.

The soils of relatively small age are rich in pioneer communities, with a preponderance of Collembola, but with increasing maturity, the soil communities show a distinctive fall of the dominance of Collembola and a corresponding ascendency of the soil Acarina. The communities nearer climax conditions include numerous Coleoptera, Diptera and Acarina. According to JANETSCHEK (564), similar conditions of the fall of Collembola dominance with the increasing soil maturity are also met with on the Alps. The soil communities are on the whole characterized by their pronounced hygrophily, but their cold stenothermy is less than in the case of the members of the other communities. Their ecological valence is also narrow, but the stratification within the community is well marked. Altitudinal zonation of the soil communities is very pronounced. The frequency of climax communities in soil is greater than in the snow habitat. At elevations of about 5000 m, there are, however, more numerous pioneer communities even in the soil than at elevations of 4000 m on the Himalaya. The reader will find the contributions of DIEM (260), DOWDY (284) and FRANZ (370, 373, 374, 379) of considerable importance to our knowledge of soil communities at high altitudes.

7. CAVERNICOLE COMMUNITIES

The cavernicole communities of high altitudes are composed of troglobiont, troglophile and trogloxene elements. The troglobionts are the true cavernicolous species, found exclusively in the high altitude caves on mountains. The troglophile species find their normally optimal conditions within the environment of the cave, but often they occur also in the open. The trogloxene species are not always found in caves and cannot also always exist in the cave environment, but take advantage of this environment only occasionally. The majority of the troglobiont high altitude insects from the European mountains are Coleoptera, but some remarkable Orthoptera (Stenopelmatidae) and small numbers of Diptera have also been found.

The endemic cavernicole Orthoptera from Europe are species of *Dolichopoda* BOL. and *Troglophilus* KRAUSS. Although the species have well developed and also pigmented eyes, they are generally apterous. In *Dolichopoda* the legs and antennae are rather conspicuously long. The genus is widely distributed on the south European mountains and extends to Asia Minor and the Caucasus Region. *Troglophilus* occurs in Asia Minor, Crete, the Balkan and on the East Alps, northwards into the calcareous zone of the Wienerwald.

The cavernicole Carabidae, Dytiscidae, Staphylinidae, Silphidae, Pselaphidae and Curculionidae are mostly troglobionts like *Trechus* CLAIRV., *Spelaeodytes* MILL., blind species of *Pterostichus* BON. of the subgenera *Rambouskiella* KNIRSCH, *Speluncarinus* REITT. and *Troglorites* JEANNEL, the Pselaphidae *Troglomaurops* GGLB., *Machaerites spelaeus* MILL., *Bythinus* LEACH, etc. A number of species of Bathysciinae (Silphidae) are also character species of cavernicole communities. The microphthalmous Staphylinidae *Glyptomerus cavicola* MÜLL., *Atheta absoloni* RAMB., Colydiidae *Anommatus titanus* REITT., Curculionidae *Troglorrhynchus* SCHMIDT, *Absoloniella* FORMANEK, etc., sometimes found in caves on the European mountains, are not, however, troglobionts. Some of the blind Carabidae, Pselaphidae and Bathysciinae found within the cavernicole communities often occur outside also. All the strictly troglobiont Coleoptera are apterous, yellowish-brown or bright reddish-brown, with greatly reduced or atrophied eyes, which may also be totally absent. Their legs and antennae do not, however, differ very much from those of the open-terricole forms, but in some extreme troglobiont species they are rather conspicuously elongated. Two small Dytiscidae have also been found in cavernicole waters, viz. *Hydroporus (Siettitia) balsetensis* AB. and *Hydroporus (Siettitia) avenionensis* GUIGNOT. Both these species have reduced eyes and are uniformly pale yellowish-brown. Further accounts of cavernicolous Coleoptera may be found in ABSOLON (1, 3), AELLEN & STRINATI (8), BEZZI (122), BRITTON (155), BURESCH (172–174), CHAPPUIS (206), JEANNEL (574, 575, 586, 587), LELEUP (685), PEYERIMHOFF (832), VANDEL (1077) and WOLF (1123).

The troglobiont blind Coleoptera are abundant on the southern European mountains, but are absent in the north, and on the Central European mountains we find only some troglophile elements. *Parabathyscia wollastoni* JANS., not distinct from *P. muscorum* DIECK., from the Ligurian Apennines, and found in decaying vegetable matter on the ground, attains its maximum northward extension in northwest France and in south England. *Bathysciola fauveli* JEANNEL and *Trechus (Trichaphaenops) sollaudi* JEANNEL occur in caves of Faux-Monnayeaurs in Mouthiers. *Royerella villardi* BED. (Bathysciinae) occurs in the Grotte de Lejoux in the French Jura. *Trechus (Trichaphaenops) angulipennis* MEIXN. is reported from the caves of the Dachstein Massiv on the border of North Austria and Steiermark and is the only blind cavernicolous Coleoptera so far found on the North Limestone Alps; it does not occur on the Austrian Central Alps or even on the Swiss Alps. It is a fully isolated relict species. *Trechus (Duvaliopsis) pilosellus stobieckii* CSIKI is known from caves in parts of the Vysoké Tatry and occurs under stones in the subalpine zone of the Babia-Góra in the Beskids. The Bathysciinae are absent in the cavernicole communities of the Atlas Mountains, but the blind *Trechus jurjurae* PEYERIMH. and *Aphaenops iblis* PEYERIMH. occur in the caves of the Djurdjura Mountain. *Nebria (Spelaeonebria) nudicollis* PEYERIMH., with elongated legs and antennae and pigmented but small eyes, occurs in a deep cave on the same mountain. In the same cave were also found two blind Staphylinidae *Paraleptusa cavatica* PEYERIMH. and *Apteraphaenops longiceps* JEANNEL and the blind Curculionid *Troglorrhynchus mairei* PEYERIMH. Reference may be made to certain peculiarities in the behaviour of the cavernicolous Coleoptera. The troglophile *Troglorrhynchus anophthalmus* SCHMIDT is either a free terricole or also cavernicole in different localities. *Anophthalmus bielzi* SEIDL. is also either a free terricole or a cavernicole form on the Transylvanian Alps and both these species extend over the East Carpathians to the Vysoké Tatry.

8. SNOW COMMUNITIES

The snow communities are perhaps the most interesting of the high altitude insect communities. They embrace not only diverse pioneer associations, but also a number of semi-climax and near-climax communities. The snow habitat embraces the winter snow-edge, the surface of the winter snow, the so called red-snow or the snow mixed with mud and earthy matter including the debris brought down by avalanches, the glacier snout, the glacier margin, the glacier surface, the erratics on glaciers, the edge of the permanent snow, etc. In these diverse habitats the members of the snow communities are distinct from the casual visitors to the snow from other habitats.

Although varied in size, location and altitude, the snow, as a habitat for the high altitude insects, is characterized by the most remarkably uniform conditions of low temperatures of the atmosphere, intense glare, great abundance of organic matter and optimal conditions of humidity. Though the layer of air immediately above the snow surface is remarkably dry, the low temperature of the air and the evaporation from the exposed surface of the snow largely tend to minimize the risk of rapid desiccation of the delicate insects on the snow. Of the diverse habitats enumerated above, the snow-edge presents perhaps the most optimal conditions to the largest majority of the species. The associations of the winter snow-edge are unquestionably the most important among the high altitude insect communities. The members of the snow-edge communities are generally characterized by their capacity for tolerating the cold and the high atmospheric humidity. The surface soil is always cold, owing to the rapid evaporation of moisture and the consequent extraction of heat. The capillary force replenishes the moisture lost by evaporation from the edge of the melting winter snow. Low temperatures and high humidity are also characteristic of the edge of the permanent snow. We find, therefore, that it is only the strictly cold stenotherm species, which are members of the snow-edge communities. The character species belong mostly to Collembola like *Hypogastrura* BOURL., *Onychiurus* GERV., *Isotoma* (BOURL.) BÖRN., *Proisotoma* BÖRN., etc. The snow-edge communities on the ridges are generally different in composition and other characters from those of the intervening valleys.

The character species of the snow communities belong to Collembola like *Isotoma* (BOURL.) BÖRN., *Proisotoma* BÖRN., *Hypogastrura* BOURL. *Onychiurus* GERV., *Tetracanthella* SCHÖTT, *Isotomurus* BÖRN., *Lepidocyrtus* BOURL., *Orchesella* TEMPL., *Womerselya* SAL., etc. The character species of Coleoptera belong to *Bembidion* LATR., *Nebria* LATR., *Trechus* CLAIRV., *Atheta* THOMS., *Aleochara* GRAV., etc. Tipulidae, Chironomidae, Syrphidae, Anthomyiidae, Sarcophagidae, etc. are also commonly represented by a large number of species. The other important groups include Plecoptera, Ephemerida, Trichoptera and Lepidoptera, in addition to spiders, especially *Lycosa* LATR. and *Xysticus* C. KOCH and numerous Acarina. The organizational pyramid rests on some species of Collembola in most snow communities, but particularly the pollen and spore feeding Collembola stand at the very base of this pyramid. The snow communities are remarkable for the great abundance and frequency of the guest species, visitor species, etc., from nearly every other neighbouring communities.

The outstanding characters of the snow communities are the conspicuous variations in the species composition and the character species are on the whole minimal for the biotic province above the forest-line on mountains. The number of individuals is, however, relatively high in nearly all the component species. Although the true character species are relatively few in most snow communities, there are, however, large numbers of other associated species. Inspite of the constant changes in the snow communities, it is remarkable that the true character species retain their fidelity nearly throughout, so that the altitude zonation is not really due to them but to the other member species. Even this altitudinal zonation is not really so well marked in the snow communities as is ususally observed in the other high altitude insect communities. Both diel and seasonal periodisms are exceedingly well marked in the snow communities. The snow communities have also perhaps the maximum proportion of carnivorous and predatory forms. The snow communities occur from the forest-line to elevations far above the permanent snowline.

9. AQUATIC COMMUNITIES

Three principal groups of aquatic communities, depending on the diversity of the habitat, may be recognized, viz. i. the communities of stagnant water, ii. the communities of torrential streams and iii. the communities of the thermal and chemical springs. The communities of the subterranean waters are best considered under the cavernicole communities.

a. Communities of Stagnant Waters

The stagnant water habitat at high altitudes on mountain embraces the melt-water ponds, the glacial lakes, etc. Great deal of attention has been devoted to the study of various aspects of the limnology of the high altitude lakes and ponds on different mountains (57, 81, 94, 355, 460, 461, 829, 905, 1016, 1138). PENNAK (825) and RAWSON (860) have recently made valuable contributions to our knowledge of the high mountain lakes of North America. The most important works dealing with the high altitude lakes of the Himalayan System are by HUNTINGTON (552), HUTCHINSON (557) and LUNDQVIST (712).

Unlike most lowland lakes, the high altitude lakes on most mountains are glacial in their origin. Most of the glacial lakes at elevations above 5000 m remain frozen almost the whole year and the lakes at elevations between 5000 and 4000 m are only partly ice-free for a brief period during the summer. Useful information on the hydrology and other peculiarities of an extremely interesting high altitude lake in Middle Asia may also be found in BRODSKII (160, 161).

The temperature of the water in the glacial ponds and lakes at high altitudes is subject to greater fluctuations during the day in summer than that of the high altitude streams. In addition to altitude, the slope aspect and the latitude of the mountains, the size and the depth of the lake have a profound influence on the temperature of the water in the high altitude ponds and lakes. The high altitude lake water shows a conspicuous temperature stratification. The surface of most high altitude lakes on the Himalaya is about 2-5° C warmer than the deeper water during the hours of bright sunshine in summer. Exposure to wind naturally lowers this difference to about 0.5° C. With masses of floating ice on the water or also with a snow-bank, the difference between the surface and the bottom temperatures is not, however, very pronounced. The inflow and outflow are other important conditions that govern the water temperature of the high altitude lakes. The deeper lakes are on the whole more homothermous than the shallow ones, but some deep lakes also show remarkable temperature stratification on the Himalaya.

The greatest majority of the high altitude lakes are transparent to depths of nearly four metres. The amount of oxygen dissolved in the water in these lakes is very much less than in the high altitude torrential streams and is often only 70-80% of that required to saturate the water at the prevailing temperature of the water and at the prevailing atmospheric pressure. EDMONDS & HUTCHINSON (292) found, however, that the oxygen tension in some of the fresh-water lakes in Ladak is nearly comparable to that of lowland lakes, mainly because of the lower atmospheric and water temperatures at these altitudes.

In the absence of rooted plants, the water in the high altitude lakes never shows the oxygen supersaturation that we find in the case of the torrential streams. The proportion of dissolved oxygen falls somewhat with the advance of summer and the disappearance of floating ice from the surface of the water. There is often a distinctly perceptible oxygen gradient from the surface to the bottom and sometimes also from the shore to the middle of the lake. The shallower portions of the water near the shore, with abundant submerged stones covered by algal slime, are generally richer in dissolved oxygen than the water in the middle.

Owing to the low temperature of the water, most of the lakes at higher elevations on the Himalaya have abundant algal growth, but the zooplankton is surprisingly very poorly developed in many of them. In comparison with the freshwater lakes at lower elevations and within the forest-zone on mountains and on the plains, the high altitude lakes are remarkable for a much shorter open season, and their characteristically low temperature. They are entirely fed by melt-water from ice and snow. The high altitude lakes have also a less marked temperature zonation and stratification and have a higher transparency than most typical lowland lakes. Their oxygen content, though not very different from that of lowland lakes, is, however, below the saturation point.

The insect communities of the freshwater ponds and lakes at high altitudes on mountains are perhaps among the smallest. With the notable exception of Collembola, exceedingly few species are common in this habitat at all elevations. The surface-haunting Collembola in the high altitude freshwater lakes belong mostly to *Hypogastrura* BOURL., *Onychiurus* GERV., *Proisotoma* BÖRN., *Sminthurides aquaticus* BOURL., etc. Excluding the Collembola, the surface forms are relatively extremely scanty but the species that dive under the water surface are fairly common. The character species belong to Dytiscidae like *Agabus* LEACH, *Potamonectes* ZIMMERMANN and *Dytiscus* LINN. and Hydrophilidae like *Helophorus* FABR. There is a conspicuous absence of the surface-haunting Heteroptera in most of the high altitude lakes, but MANI (730) has recorded *Gerris sahlbergi* DIST. as occurring in small numbers in isolated patches of melt-water at an elevation of 3400 m on the south slope of the Northwest Himalaya. There are also some species of Chironomidae and *Ephydra* FALL. The last mentioned genus is, however, more typical of saline and alkaline lakes than of freshwater ones. The dominant higher insects of most high altitude lakes are, therefore, mostly Coleoptera, almost to the exclusion of most other groups. The Coleoptera of the glacial lakes of the Alps belong to species which are mostly widely distributed even on the lowlands of Central Europe. They are mostly carnivorous, but often also phytophagous, almost without exception winged and are never blind. Exclusively montane autochthonous stagnant water forms like *Helophorus brevitarsis* and *Hydroporus kraatzi* also occur on the Carpathians. Most high altitude stagnant water Coleoptera have typically retained their ability to swim, but the Hydrophilidae swim to a lesser extent than the Dytiscidae. In some of the freshwater ponds we also find the freshwater fairy-shrimps *Branchinecta* spp. (730). In addition to the true pond and lake communities, mention may also be made of the important and often large associations of Collembola-*Atheta* at the edges of the ponds and lakes on nearly all the important high mountains of the world.

b. Communities of Torrential Streams

The high altitude streams are mostly torrents and are also of glacial origin on most mountains. The bottom of nearly all the torrential streams is typically stony, with little sand or mud. The fall in the gradient is high in most streams and may amount to as much as 500-2000 m/km and the current velocities vary from 1 to 6 m/sec. The great majority of the high altitude torrents are generally very swollen and highly turbid during the afternoon in summer, and some of them indeed bring down considerable rock debris and mud and often also blocks of ice and snow. They are, therefore, conspicuously turbid in the afternoon, but clear for the rest of the time. GILBERT (408) has given an interesting account of the transportation of debris by running water.

The temperature of the water in the high altitude torrential streams is far more uniform than in the lakes and ponds, not only over several weeks during summer but also over a fairly wide range of altitude. Regardless of the altitude, aspect and perhaps also latitude of the mountain range, the mean temperature of the water during summer remains nearly constant in most high altitude torrents. The diurnal temperature fluctuations are thus extremely small at any locality and do not generally exceed 1.0-1.5° C even on the south-facing slopes. The marked uniformity of the water temperature is the result of high current velocities. The influence of current velocites on the water temperature of the torrents is best observed in the absence of a pronounced south-north difference on the Himalaya. The temperature of the water in the R. Issyk, which is a torrent within the Tien Shan mountains, observed by BRODSKII (160, 161), ranges from 11° C at elevations between 1360 and 1450 m, 10.6° C at elevations between 1450 and 1725 m, 8.6° C at elevations between 1725 and 2020 m and 0.6° C at elevations of about 2600 m. The mean atmospheric temperature during July is 24.5° C and the mean water temperature at the same time is 10.5° C; with the atmospheric mean at 4° C the mean water temperature is 0° C. The diurnal fluctuation of the atmospheric temperature in the R. Issyk Gorge was between 19 and 23° C, but that of the water in the river only between 10.8 and 10.9° C. On the Northwest Himalaya, the water temperature at an elevation of about 2700 m is 5.5° C, with the mean atmospheric temperature reading 16° C. The mean water temperature in the same torrent at an elevation of about 4500 m is only 4.5° C.

On account of the low temperature of the water in the torrents, the oxygen tension in the water is apparently not less than that in most lowland waters. On the other hand, while the oxygen tension of the atmosphere falls with the increase in altitude, that of the water in the torrential streams does not necessarily decrease. The oxygen tension in the water of the torrential streams is governed partly by the temperature of the water also. While the reduced atmospheric pressure lowers the solubility of oxygen in the water, the lowering of the temperature increases the solubility. The result is that often a much higher proportion of oxygen is found dissolved in the high altitude melt-water torrents than could be expected from the conditions of altitude alone. At an elevation of about 2750 m on the Himalaya, a relatively slow-flowing stream on the south slope, with a mean water temperature of 12° C, contains about 7 cc of oxygen per litre of water (fig. 27). The amount of oxygen required to saturate the water at this temperature is only 5 cc/l, so that the torrential stream is actually supersaturated with oxygen. At elevations between 3700 and 4000 m, the oxygen tension in torrential streams with a temperature of about 4.5° C is about 5 cc/l. This quantity is again considerable supersaturation for the atmospheric pressure prevailing at the elevation. According to the observations of DODDS & HISAW (270–273), the torrential streams at an elevation of about 2500 m on Colorado mountains are likewise supersaturated with oxygen by about 150%.

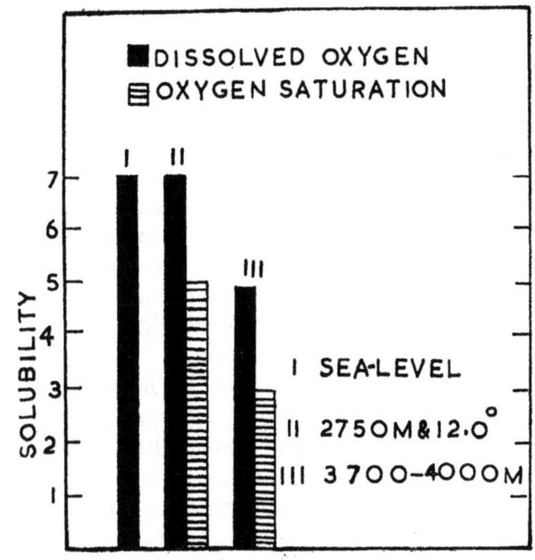

Fig. 27. Dissolved oxygen in lowland waters and in the high altitude melt-water streams on the Himalaya.

BRODSKII (160) similarly found the oxygen content of the torrents of the R. Issyk system on the Tien Shan mountains to be above the saturation value, being 9 cc/l at a temperature of 10° C and at an atmospheric pressure of 760 mm Hg. The most important ecological factor is the current itself. The high current velocity ensures efficient and rapid removal of the layer of water deficient in oxygen and increasing proportion of carbon dioxide from the immediate surrounding of an aquatic organism. In this immediate surrounding of the insect, the fall in oxygen tension is accompanied by an increase of the carbon dioxide tension. The rapid current serves, however, to maintain a uniform and regular supply of freshly dissolved oxygen for respiration. It also contributes to the maintenance of a uniform temperature of the surrounding water, thus permitting the occurrence of species over a wider range of altitudes than is otherwise possible. The current velocities are variable in different parts of the torrent and may range from 0 to 6 m/sec.

Considerable attention has been paid to the temperature of the water and the amount of dissolved oxygen in mountain torrents as decisive ecological factors.

Most workers have, however, overlooked the fact that the biota of the mountain torrents are governed not only by these factors, but also by a complex set of other factors, including the nature and character of the bank, the stream bottom, the width and the depth of the stream, the current velocity, the total water discharge per second, etc. BRODSKII (160) is perhaps the most important worker to realize the complexity of the ecological factors in the high altitude torrents.

Current velocities alone do not by any means give a satisfactory clue to the type of organisms, their structural and other adaptations in any given stream. Broadly speaking, there are three distinct ecological groups of communities in mountain torrents. The first group, typified by Blepharoceridae and Deuterophlebiidae, consists of species that are permanently exposed to the full force of the strong current. The second group, comprising most Ephemerida and some Plecoptera, is exposed only temporarily to the full intensity of the current, but for the most part remains under submerged stones and thus escapes from the extreme intensity of current velocities. To the third group belong the species that remain permanently under the submerged stones and are not, therefore, exposed to the full force of the current at all. We have accordingly the extreme rheobiont, the rheophile and the rheoxene types in these torrents. Most of the torrenticole insects are also characterized by their narrow cold stenothermy. BRODSKII (160) reports, for example, that cold stenotherm rheophiles occur in the littoral zone of the R. Issyk and the cold stenotherm rheobionts in the upper reaches of the torrent. Lithophile and bryophile species are also abundant in all the high altitude torrents, but strictly limnophile and open-water associations are of course totally absent from them.

Some of the outstanding features of the insect life in the mountain torrents have been discussed by a number of workers (160, 270-273, 546, 562, 1009). The most striking feature of the insect communities of the high altitude torrents is the enormous abundance of individuals of relatively few species, some of which appear again and again in many localities, often over a fairly wide range of altitude. The character species belong to Ephemerida, Plecoptera, Trichoptera, Deuterophlebiidae, Blepharoceridae, Simuliidae, Chironomidae, some Collembola and Hydracarina. Except perhaps in some cases of Collembola and Hydracarina, it is only the larval and other immature stages of the other groups that are found in the torrents. Aquatic torrenticole Coleoptera are relatively scarce and Heteroptera, so common in still-waters at sea-level, are totally absent in the high altitude torrents. There are on the Himalaya three ecological groups of associations of torrential communities: i. *Baetis* LEACH-*Nemoura* LATR. associations, with *Capnia* PICT., *Iron* EATON, *Ecdyurus* EATON, etc. ii. the Deuterophlebiidae-Blepharoceridae-Simuliidae associations and iii. the Trichoptera-Chironomidae associations.

The members of the torrenticole associations have taken advantage of the high current velocity in a most complex manner and indeed survive only because of the high velocity of flow of water and not inspite of it. The streamlining effects of the current on the body that behaves almost like a fluid are well known and have been so well discussed by THOMPSON (1037). Both in the still water and in the slow-flowing streams, the mean body size of the insect is larger than in the case of the torrential streams. The mean size of the body of the insect thus tends to diminish with the increase in the velocity of the water flow in the mountain torrents. The body also exhibits a conspicuous tendency to become flattened and depressed in the case of the species that spend most of their time under submerged stones. In the case of the species that are exposed to the full force of the current, the body is certainly not at all flattened, as is most often erroneously asserted, but the shape of the body is such that it offers the least resistance to the current. It must be observed that not only the larvae of torrential insects, but in reality everything submerged in a rapid-flowing current of water, including the very pebbles, are shaped exactly alike and are streamlined by the current iself. The shape of the body is directly moulded by the very current that ensures diverse other environmental optimal conditions. The total gill surface is relatively small in the case of torrenticole species. In stagnant waters, the respiration of enormous numbers of overcrowded larvae of Ephemerida, Plecoptera and other forms considerably lowers the oxygen tension and at the same time increases the carbon dioxide content of the water within the immediate surroundings of these larvae. Increased concentration of CO_2 stimulates the gill lamellar movements and thus at the same time serves to stir up the water, thereby dissipating the water enriched with CO_2 and eventually restoring a fresh supply of oxygen. In the torrential streams, however, such gill movements are totally absent and are in fact wholly unnecessary, since the current itself ensures prompt removal of CO_2 and renewal of oxygen. The gill lamellae of the torrenticole larvae are therefore characteristically small. The larvae of Plecoptera in the glacial torrents on the Himalaya have thus remarkably small gills and compared to the larvae of Ephemerida are also poorly equipped with gills. They are characterized by a lower tolerance for oxygen deficiency than the larvae of Ephemerida or Trichoptera. As may be expected, the organs of anchorage like hooks, suckers, discs, etc. are remarkably well developed. The species that withstand the strong current and remain exposed and anchored to stones and other rock surfaces under water have relatively rounded bodies or have also semi-flattened bodies. The Ephemerid *Baetis* LEACH has rounded larvae and *Iron* EATON has semi-flattened larvae. In the case of the species that avoid the current by remaining concealed under submerged stones and inside crevices, the body is relatively bulky and has also rather sprawling legs. It must be remembered that a flattened body is by no means always a torrential adaptation, but rather rounded bodies seem to be more common among the torrenticole species. The flattened *Ecdyurus* EATON does not, for example, occur exposed to the full force of the current and the flattening is here an adaptation that in effect enables the insect to get under stones and thus avoid the strong current. In the case of the Trichoptera, the larval cases rather than the larvae themselves exhibit conspicuous modifications for life in high current velocities. The size, shape and the material used in the construction of the case are largely governed by the current velocity and are therefore characteristic.

The rich algal slime on the submerged stones is the primary source of food for the species in the torrential stream communities. The larvae of Ephemerida that stand at the base of the pyramid of numbers in the torrenticole communities feed largely on the algal slime. The current brings an abundance of food in the shape of organic particles entombed in the winter snow and in the glacier ice and released when the ice and snow melt during the summer.

The principal specializations of the members of the torrenticole communities include, therefore, the ability not only to withstand the current but also to utilize the current itself for supplies of oxygen and food, in their organs of anchorage, in the absence of swift movements of locomotion, in the restriction of locomotion, in their small body size, reduced respiratory organs, streamlining of the body, habit of feeding from the current or by scraping the food from the stone surface, in the narrow temperature valency and in the low tolerance of oxygen deficiency.

c. Communities of Thermal and Chemical Springs

The high altitude thermal or hot springs are characterized by the temperature of the water being higher than the mean atmospheric temperature of the locality. In some of the more well known hot springs of the world, the water emerges almost at the boiling point, but cools off soon after leaving the spring (164, 547). We find, therefore, a series of biotopes with graded temperature differences. Most high altitude hot springs are also rich in dissolved mineral salts, hydrogen sulphide and carbon dioxide and are, therefore, also at the same time chemical springs. The spring water in such cases is at a higher osmotic pressure than the body fluids of insects. The salts generally met with in these springs are the chlorides of sodium, potassium, etc., borax and in the case of the alkaline springs, sodium carbonate also. Unlike the stagnant water and the torrential streams, the high altitude thermal and chemical springs are pronouncedly deficient in dissolved oxygen.

The insect communities in the thermal and mineral springs at high altitudes inhabit waters at temperatures of 30-51° C. On the Northwest Himalaya, the larvae of *Ephydra* FALL. occur in springs, with a temperature of 49.1° C at elevations of 4500-4750 m. In North America they occur in the hot springs at temperatures of 43° C. Although Dytiscidae, Hydrophilidae and Gyrinidae occur sometimes in the hot springs, they are on the whole sparse and relatively unimportant elements in most high altitude hot springs. Hydracarina like *Protziella hutchinsoni* LUNDB. occur commonly in many hot springs, at temperatures of 22° C on the Himalaya. Other character species of the high altitude hot springs belong to Chironomidae, Ceratopogonidae, Stratiomyiidae like *Stratiomyia* GEOFFR., *Odontomyia* MEIG., *Oxycera* MEIG. and *Nemotelus* GEOFFR., Syrphidae like *Eristalis* LATR., etc. *Helophorus* (*Atractelophorus*) *montanus* D'ORCH. occurs in sulphur springs on the Himalaya. Mention may also be made of remarkable associations of species at the edge of the hot springs. We have, for example, the associations of *Helophorus* (*Lihelophorus*) *ser* ZAIT. and *Bembidion hutchinsoni* ANDR. at the edge of the Khyam Hot Spring on the Northwest Himalaya.

CHAPTER V

SOME TYPICAL HIGH ALTITUDE INSECTS AND OTHER ARTHROPODA

While the insect life of the forest and the prairie at lower elevations is remarkable for its enormous diversity, exceedingly few orders have successfully colonized the biome above the upper limits of the forest on high mountains. The dominant high altitude insects belong to groups like Plecoptera, Coleoptera, Lepidoptera, Diptera and Collembola, which also occur at the highest elevations in the vicinity of snow and ice. Some other orders like the Ephemerida, Orthoptera, Dermaptera, Heteroptera, Hymenoptera, Trichoptera and Thysanura usually occupy a minor place among the insects at extreme high altitudes and most of these groups do not also occur much above the limits of the forest. Even in the typically mountain autochthonous groups, there is an abrupt fall in the general abundance of species at the upper limits of the forest. Some of these orders, relatively insignificant at the sea-level, become conspicuously dominant at high elevations. In all cases, the high altitude species are wholly different from those with which we are familiar in the plains. Broadly speaking, the high altitude insects belong to the following ecological groups: 1. Widely distributed eurytype hypsophiles, characteristic of the forest and prairie and abundant at lower elevations, but occurring regularly in the subalpine and sometimes even in the alpine zones on high mountains in different parts of the world. 2. Mountain autochthonous hypsobionts; these exist at the highest elevations and are as a rule never found within the forest zone. Some of them are generally characteristic of all or nearly all the principal mountains of the world. Others are, however, restricted to specific regions only. 3. Lowland and forest elements that occur accidentally above the upper limits of the forest on mountains, but do not breed at high elevations. This chapter aims at an outline of the taxonomy of the dominant hypsobiont types, found on the principal mountains of the world. Mention is also made of some unique forms which are exclusively characteristic of well known mountains.

A. INSECTS

ORDER I. EPHEMERIDA

Ephemerida apparently reach their maximum development in the north temperate areas and on the mountainous regions and yet exceedingly few species are at present known from above the upper limits of the forest on most mountains. The order flourishes, however, almost up to the permanent snowline on the Alps and the Himalaya.

In contrast to the species from the plains, the larvae of the high altitude Ephemerida are almost exclusively torrenticole rheobionts, truly immense numbers of which may be found clinging to the submerged stones in the melt-water streams. Only exceptionally do we find the larvae of a few species on the submerged stones near the edge of stagnant ponds and lakes, especially in windy and exposed localities, where the strong winds ensure constant movements of the shallow layers of the water and thus provide a sort of ecological substitute to current. As a rule, the larvae move but little, since even at the edge of the stream the current in most torrents is strong. The free tracheal-gill lamellae are generally relatively small and practically or also completely immovable. The optimal temperatures of the water for the larvae of most species on the high mountains of the world lie between 5° and 8° C. The larvae feed on the rich algal slime growing on the surface of the submerged stones and on the rocky bottom of the streams, but often they also turn into detritus feeders and occasionally also predators. Hibernation during the winter is most usually in the last larval stage. The entire life-cycle of the high altitude mayflies may be completed within a single summer, especially immediately above the forest-line, but more usually it requires two or even three years. The subimago may often be collected in great numbers during the summer and all on a sudden on a calm sunny afternoon, huge swarms of the adults emerge. The neighbouring snow-fields are littered next morning with immense numbers of the dead adult mayflies, to be devoured by vast hordes of carrion feeders.

The typical mountain autochthonous species belong to *Ameletus* EATON, *Rithrogena* EATON, *Ecdyurus* EATON, *Ecdyonurus* EATON, *Caenis* STEPH., *Baëtis* LEACH, *Cloëon* LEACH, *Palingenia* EATON, *Ephemera* LINN., *Ephemerella* WALSH, etc. *Iron* spp. are common in torrential streams on the mountains of Middle Asia, at elevations between 1300 and 2300 m, with the water temperature ranging from 11° to 7° C and at about an elevation of 1900-2000 m on the Alps. On the North American mountains, the genus occurs at relatively higher elevations of 2750-3400 m. *Ecdyonurus* EATON seems to be confined to lower elevations of 1300-1500 m on the Tien Shan System, but sometimes ascends to 1700-2130 m; it flourishes in water at temperatures of 9-7° C. *Ephemerella* WALSH is a typical rheobiont, characteristic of high mountain torrents at elevations of about 1300 m on the Middle Asiatic mountains and over 2900 m on the North American Cordilleran System; it flourishes well in water at a temperature of 11-10° C. *Baëtis* LEACH flourishes best in streams with stony bottom, usually at elevations above 2300 m, at temperatures ranging from 11° to 5° C, but on the North American Cordillera the genus ascends up to 3500 m and on the Himalaya to nearly 4000 m. The genus *Ororotsia* TRAV. is endemic on the high Himalaya. The reader will find useful accounts of mountain authochthonous Ephemerida in BERTRAND & VERRIER (121), BRODSKII (159), DODDS & HISAW (271-273), KAPUR & KRIPALANI (606), LESTAGE (690), TRAVER (1045) and UÉNO (1055).

ORDER 2. PLECOPTERA

The Plecoptera have a wide Holarctic distribution and are particularly abundant on mountains. Although relatively few species have actually been described so far from most mountains, the stoneflies constitute nevertheless an important group of high altitude insects. As with the Ephemerida, there is in this order also a remarkable quantitative abundance of the individuals of relatively few species at high altitudes. The larvae require cold and well aerated running water and are, therefore, wholly confined to the torrential streams on mountains. While the mayflies are common enough even near the sea-level, the stoneflies are almost absent on the plains, but come into dominance at higher elevations. They flourish almost up to the permanent snowline. The temperature of the water optimal for the Plecoptera lies between 4° and 8° C on the Himalaya and though the larvae of many species are normally active even at temperatures in the neighbourhood of 1.5° C, exposures to temperatures of 10–12° C usually prove rapidly lethal to most larvae. The stonefly adults emerge indeed in most parts of the world in the coldest part of the year. The filamentous gills on the neck, prosternum and coxae are poorly developed when compared to those of the larvae of the species that are exclusively characteristic of the upper reaches of the forest-zone on mountains. The larvae are mostly phytophagous and feed on the algal slime on the submerged stones on which they occur. In many species carnivorous habits have, however, been observed and the larvae predate on the larvae of other aquatic insects, particularly Chironomidae, Hydracarina and also on mayfly larvae. The larvae of some species are undoubtedly omnivorous. Hibernation during the winter takes places mostly in the second larval instar, but nearly every developmental stage has the capacity to pass into a dormant condition if unfavourable conditions make it necessary. Though some species complete a generation within a single year, especially immediately above the forest-line on the south slopes of the mountains of east-west trend like the Alps and the Himalaya, most species require two or even three years to complete a single generation. We do not, therefore, always come across the adults of all the species in any locality in the summer. The subimago apparently waits for favourable conditions, ready to emerge as adults for several days during the summer. The emergence of the adults takes place late in the afternoon or also early in the evening and seems always to synchronize with a lull in the wind. The adults are short-lived, but many of them seem to survive longer than the mayflies. Except perhaps in the case of *Rhabdiopteryx* KLAPÁLEK, the adult stoneflies do not feed. After mating, the adult females may be observed crawling into the edges of the streams for egg-laying. The emergence of the adults is followed by a most phenomenal increase of the dead stoneflies on snow-fields and on the grassy meadows and by a marked addition to the food supplies for numerous carrion feeders. The dead and dying stoneflies are largely devoured by swarms of Collembola, Carabidae, Staphylinidae, Diptera and by birds.

The commonest stoneflies above the limits of the forest on mountains belong to *Chloroperla* NEWM., *Capnia* PICT., *Nemoura* LATR., *Perlodes* BANKS, etc. On the high mountains of Middle Asia *Perlodinella* KLAPÁLEK takes the place of *Perlodes* BANKS. On these mountains we also find *Amphinemura* RIS. *Kyphopteryx* KLAPÁLEK and *Rhabdiopteryx* KLAPÁLEK flourish at elevations of 3300–3700 m and 5000 m respectively in the Mt. Everest area of the Himalaya. *Gripopteryx* PICT. is characteristic of South American mountains; the group is remarkable for the absence of true anastomosis of wing veins and closely approximated fore coxae. *Andiperla* AUB., *Megandiperla* ILLIES and *Apteryoperla* WISLEY are unique in several respects. The former two are apterous and lack ocelli and have semi-terrestrial larvae and are found in the southern Andes (560, 561, 562, 563). The terrestrial habits of the larvae are correlated with the extreme climatic conditions prevailing in the high Patagonia. *Apteryoperla* WISLEY is also apterous and occurs at an elevation of 2000 m on New Zealand Mountains. *Dictyopterygella recta* KEMP., *D. septentrionis* KLAP. and *Acrynopteryx dovrensis* MARTYN. from the East Alps are believed to be boreo-alpine forms. Recent phyletic taxonomic and biogeographical evidence seems to indicate that the Plecoptera came to be differentiated on the Antarctic area or at least on a southern land mass (427 a). Further reference to Plecoptera may be found in AUBERT (59), BERTRAND & AUBERT (120),

BRINCK (152, 153), CASTEL (191), DESPAX (252), FRISON (389), JEWETT (591-593), KAWAI (610), KIMMINS (614), SÁMAL (915, 916) and WISLEY (1117).

ORDER 3. ORTHOPTERA

Though essentially a thermophile group typically characteristic of the hot deserts of tropical and subtropical lowlands, it is remarkable that a great many different types of Orthoptera have become completely adapted for life in the cold deserts, both in the extreme north latitudes and at great elevations on high mountains in all parts of the world. The literature on mountain Orthoptera is much scattered, but the reader will find useful information in ADELUNG (7), ALEXANDER (19-23), ANDER (45), BEI-BIENKO (99), CAVALHO & SOUTHWOOD (193), CHOPARD (209-211), EBNER (291), FRUHSTORFER (390), GRABER (421), HARZ (479b), HEBARD (486-488), MIRAM (777), RAMME (854-857), REHN & GRANT (869), REHN & HEBARD (870), SCUDDER (953), SJÖSTEDT (981), STOLJAROV (1013), TARBINSKY (1028), UMNOV (1057), UVAROV (1056-1075) and WALKER (1093, 1094). Even in the plains many Orthoptera, especially the Acrididae, are preferably terricole forms and occur generally on loose and sandy soils. On the high mountains they are exclusively geophiles and occur on the grassy meadows and slopes. The Acridoidea are dominant but some very remarkable Tettigoniids are also found at high elevations. Widely distributed eurytypes, typical of the forest and prairie, often ascend regularly to truly remarkable elevations on high mountains in nearly all parts of the world. Most of the grasshoppers recorded so far from high elevations on the equatorial East African mountains are indeed widely distributed in the lowlands of East Africa and cannot therefore be strictly described as true mountain autochthonous species. They are fundamentally lowland elements that have secondarily become hypsophile.

The grasshoppers commonly met with on high mountains belong to *Acrydium* OLIV., *Bryodema* FIEB., *Chorthippus* FIEB., *Podisma* LATR., *Aeropus* GISTL., *Gomphocerus* THUNB., *Aeropedellus* HEB., *Xanthippus* SAUSS., *Melanoplus* STÅL, *Dociostaurus* FIEB., *Gomphomastax* BRUN., *Conophyma* ZUB., *Hypernephia* (Mt. Everest area of the Nepal Himalaya), *Dicranophyma* UVAROV (endemic on the Northwest Himalaya), *Pamiracris* RAMME (endemic on the Pamirs), *Sphingonotus* FIEB. (Middle Asiatic mountains and the Northwest Himalaya), *Chrysochraon* FISCHER, *Metrioptera* WESM., etc. On the Caucasus we find *Acryptera* SERV., *Nocarodes* FISCHER DE WALDHEIM (on the alpine meadows at elevations of 2700-2750 m), *Schizonotinus* RAMME, *Poecilimon* FISCHER, *Semenovites* TARB., *Decticus* SERV., *Polysarcus* FIEB. (at elevations of 1800-3000 m), *Stenobothrus* FISCHER, *Meconema* SERV., *Olinthoscelis* SERV., *Mecostethus* FIEB., etc. From the North American mountains we have *Nemotettix* HANCOCK, *Neotettix* HANCOCK (from the Appalachian Mountains), *Tetrix* LATR. (on the Cordilleran mountains from Alaska southwards to the Californian Sierra Nevada), *Paratettix* BOL., *Eumorsea* HEB., etc. also from the Cordilleran mountains. *Psychomastax* REHN & HEB. occurs from sea-level to an elevation of about 3350 m on the Sierra Nevada in California. *Hyphinomos* UVAROV is a tettigoniid, found at elevations of 4575-4880 m on the Northwest Himalaya and in Tibet. *Pezotettix* BRUN. is characteristic of the mountains of New Zealand. The typical Orthoptera from the southern Andes are *Trimerotropis* STÅL, *Dichroplus* STÅL and *Bufonacris* WALKER.

The highest elevation at which specifically identified Orthoptera seem to flourish and breed permanently is 4900 m on the Mt. Everest Massif. *Dasynema* UVAROV occurs in the same region at elevations of 4575-4880 m, but the nymphs of an unidentified Acridid are also reported from the same area at an elevation of 5490 m; this is at present the highest altitude record for the order in the world. *Parasphena* BOL. occurs under stones at elevations of 4880 m on the alpine desert zone on Mt. Kilimanjaro. *Acrodectes philophagus* REHN & HEB. occurs at an elevation of 4420 m on Mt. Whitney. The socalled mormon-cricket *Anabrus simplex* HALD. is known at an elevation of 3960 m on Colorado Mountains and the cave-cricket *Ceuthophilus alpinus* SCUD. is reported above the timberline on Mt. Lincoln in Colorado. The occurrence of certain Siberian forms, like *Gomphocerus* THUNB. and *Podisma* LATR., on high mountains is of considerable biogeographical interest. The whole of the alpine grasshopper elements on the Middle Asiatic mountains are fundamentally different from those of the European mountains. *Podisma* LATR. is replaced by the superficially similar *Conophyma* ZUB. and it is believed by UVAROV that these grasshoppers had already become widely distributed before the uplift of the mountains began, on which several characteristic species now occur. *Aeropedellus variegatus* FISCHER DE WALDHEIM and *Podisma frigida* BOH. are boreo-alpine species found on the Altai, the southern European mountains and in Alaska (*vide* Chapter XII). In addition to these socalled resident types, ALEXANDER (19, 22) has drawn attention to the interesting fact that a number of typical lowland species like *Melanoplus mexicanus mexicanus* (SAUSS.) are often accidental visitors at high elevations on mountains and do not really breed above the forest-line. We have, for example, an interesting record of *Locusta migratoria danica* (LINN.) at an elevation of 4270-4575 m on the north side of the Mt. Everest Massif on the Himalaya.

The true high altitude grasshoppers are mostly flightless or they have more or less reduced wings or they are also totally apterous. *Podisma* LATR., *Bryodema* FIEB., *Conophyma* ZUB., etc. are, for example, apterous at high elevations. SCOTT (950) has made the interesting observation that the grasshoppers found at elevations of 3200-3400 m on the Gughé Highlands of Abyssinia are also brachypterous and quite incapable of flight. The truly hypsobiont Orthoptera represent as a whole a small fraction of the high altitude insect life and constitute only about 3% of the total number of species of all orders even on the Himalaya (730). Nocturnal Orthoptera like Gryllidae are absent in the alpine zones of high mountains and all the mountain autochthonous species are completely diurnal at high elevations.

The greatest majority of the species occur on grassy meadows, but some are found even on the stony ground. Hibernation of the true hypsobiont Orthoptera takes place in the egg stage in soil, usually at depths of about 10-20 cm, under the winter snow-cover and usually lasts from September to the end of May or sometimes even early June, depending on diverse local conditions. After the winter snow has melted, the young nymphs hatch and remain active only during the hours of bright sunshine and mass assemblages of the nymphs may sometimes be observed on warm rock-surface, absorbing the warmth that is being radiated away. On becoming sufficiently warmed up, they scatter and begin feeding. The adults emerge after about the middle of August, but sometimes early in September. Most usually, however, by the beginning of September the eggs have in most cases been laid underground and the adults have already disappeared nearly completely.

ORDER 4. DERMAPTERA

Although other subfamilies also occur at high elevations, the Anechurinae are without doubt the dominant Dermaptera on most high mountains. The Anechurinae differ from the rest of the Dermaptera in their broad and transverse mesosternal and metasternal plates, dilated abdomen and in the curiously curved forceps. *Anechura* SCUD. (fig. 28) has characteristically short elytra, with parallel sides and the wings are greatly abbreviated. Other Dermaptera from high mountains are also more or less conspicuously apterous or at any rate flightless. *Forficula sjöstedti* BURR, found at elevations of 1970-3970 m on Mt. Kilimanjaro and Mt. Meru, is flightless.

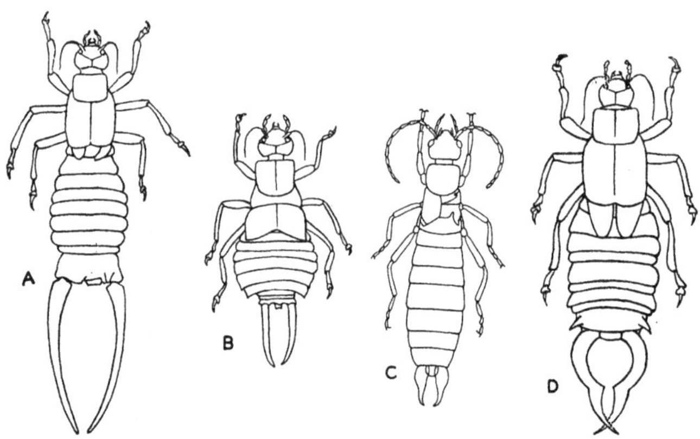

Fig. 28. High altitude Dermaptera. A. *Anechura himalayana* SANTOKH, B. *Himanechura lahaulensis* SANTOKH and D. *Anechura bipunctata* FAB. from the Himalaya; C. *Forficula triangulata*, micrapterous from Mt. Kilimanjaro.

All the hypsobiont forms are pronouncedly hygrophiles and without exception occur under stones, deeply sunk boulder, etc. in the immediate neighbourhood of streams and the edge of snow. Both as nymphs and as adults they are predominantly debris feeders or sometimes also predators; some species feed on the roots of plants. Cannibalism is not infrequent among them. Both the nymphs and the adults spend most of their time under stones and are generally associated with Carabidae. Some widely distributed lowland forms like *Forficula* LINN., *Pygidicrana* SERV., *Anisolabis* FIEB., *Gonolabis* BURR, *Apterygida* WESTW., *Chaetospinia* KARSCH, *Opisthospinia* KARSCH, etc. often occur at great elevations. *Forficula* LINN. ascends, for example, up to elevations of 4300 m on Mt. Kilimanjaro and Mt. Meru. *Forficula cavallii* BOR., found at an elevation of 4880 m on Mt. Kilimanjaro, represents at present the highest altitude record for Dermaptera in the world (517). *Forficula sjöstedti* is widely distributed in the lowlands of Kenya, Tanganyika, the former Belgian Congo and other areas and becomes increasingly abundant at higher elevations on Mt. Kilimanjaro and Mt. Meru. *Anechura* SCUD. is found in the subalpine and alpine zones of the Alps, Caucasus, Middle Asiatic mountains and on the Himalaya. The genus *Lithinus* BURR is endemic on the Spanish Sierra Nevada and occurs at elevations between 2400 and 2900 m.

ORDER 5. GRYLLOBLATTODEA

The Grylloblattids or the socalled alpine rock-crawlers occur at elevations of 1000-3000 m on the mountains of Japan and North America. *Grylloblatta* WALKER, apterous, occurs under stones often near snow, with microclimatic temperature of about 1.1° C, especially on moist northern slopes, in the subalpine and alpine zones. Reference may also be made to PLETSCH (838) and CHAPMAN (201) for further accounts on these insects.

ORDER 6. HETEROPTERA

The Heteroptera are minor elements in the high altitude ecosystem and do not usually exceed 2% of the total hypsobiont and hypsophile insects on most of the principal mountain systems of the world. Miridae, Lygaeidae, Pentatomidae and Saldidae are the families most frequently met with above the upper limits of the forest. Being essentially a pronounced thermophile group of insects, with the optimal temperatures lying above 10-12° C, exceedingly few species appear to have adapted for permanent colonization at high elevations. The shortening of the summer at high elevations in the temperate latitudes and the daily nocturnal frosts on the tropical mountains are perhaps the most decisive limiting factors for the order.

The commonest hypsobiont Heteroptera are *Nysius* DALL. and *Chlamydatus* CURTIS. The greatest bulk of the hypsobiont bugs are found on grassy meadows, but many species occur also in typically semi-arid localities. Both the nymphs and the adults of most species occur under stones or under the low-matted vegetation on the ground. While aquatic bugs are common enough among the lowland Heteroptera, they are relatively extremely sparse at high elevations. Even the typically aquatic groups of the lowlands have become adapted at high altitudes for semi-aquatic mode of life on or under moss and stone at the edge of streams. The life-cycle is completed in about 6-8 weeks and there is but a single generation. The hibernating adults become active in late spring or early summer and the later instars succeed each other more rapidly than the earlier one. In addition to the strictly mountain autochthonous hypsobiont bugs, certain interesting lowland types seem to have evolved relatively recently into secondary high altitude forms. Purely non-montane types like *Coriomeris* WESTW. (Coreidae) often occur at extreme elevations of 5000 m near the Mt. Everest region of the Himalaya. The dominant hypsobiont Miridae include *Chlamydatus* CURTIS, *Macroplax* FIEB. (at elevations of 2500-3350 m on the Spanish Sierra Nevada), *Stenodema* LAP. (Pamirs), *Dicyphus* FIEB. (Pamirs, Northwest Himalaya and the Spanish Sierra Nevada), *Myrmecophyes* FIEB. (Pamirs and the Tien Shan), *Ischnocoris* FIEB. (2600-3200 m on the Spanish Sierra Nevada), *Parahypsitylus* FIEB. (2500-2600 m on the Spanish Sierra Nevada), *Systellonotus* FIEB. (2600 m on the Alps and the Spanish Sierra Nevada), etc. To the Lygaeidae belong *Apterola* MULSANT, *Microplax* FIEB., *Geocoris* FALL., *Nysius* DALL., *Trapezonotus* FIEB., *Emblethis* FIEB. *Dolmacoris* HUTCHINSON (fig. 49), *Lamprodema* FIEB., etc. The Pentatomidae include *Dolycoris* STAL., *Phimodera* GERM. (fig. 49), *Eurydema* STAL., *Carbula* STAL., *Acanthosoma* CURTIS, etc. Some interesting Coreidae like *Stictopleurus* STÅL occur at very high elevations on the Himalaya. There are also a number of extremely interesting endemic mountain autochthonous genera like *Phimodera* GERM., *Dolmacoris* HUTCHINSON, *Tibetocoris* HUTCHINSON, etc. from the Northwest Himalaya. KIRIT-

SHENKO (617, 618) has published an interesting list of the endemic mountain genera from Middle Asia, of which *Mimula* JAK., *Pheocoris* JAK., *Mormidella* HORV., etc. are the more important. *Arctocorisa carinata* SAHLB. and *Glaenocorisa cavifrons* THOMS. are boreo-alpine elements that occur on the mountains of Central and southern Europe. Most species occur up to a maximum of 2500-3000 m, but some flourish even at an elevation of 4000 m and a few others exist at 5000 m. *Tibetocoris margaretae* HUTCHINSON occurs, for example, at elevations of 5000 m on the Northwest Himalaya. *Chiloxanthus* REUTER occurs at elevations of 4300 m on the Pamirs and on the Northwest Himalaya. It may be remarked that only terricole types ascend up to an elevation of 2600 m and above. Although the atmospheric temperatures may fluctuate over 20° C at elevations of 2000-2500 m, the ground temperature fluctuations are never more than 3-4° C that of the air in the localities in which the Heteroptera abound. For further accounts of high altitude and high boreal Heteroptera reference may be made to VAN DUZEE (289), HOFFMÄNNER (524), KIRITSHENKO 615-620), OSHANIN (804), REUTER (882) and WAGNER (1090).

ORDER 7. COLEOPTERA

Coleoptera are the most important mountain autochthonous hypsobiont insects in all parts of the world. On the immense Alpine-Himalayan System of Tertiary mountains they amount to almost half the total number of insects that have colonized the biome above the forest-line and they are also the dominant insects on the American Cordilleran System. Although about 20 families have so far been found at high elevations on different mountains, Carabidae, Dytiscidae, Staphylinidae, Histeridae, Tenebrionidae, some Chrysomelidae and Curculionidae are among the character forms of most mountain regions. Interesting examples of Hydrophilidae, Byrrhidae, Silphidae, Dryopidae, Coccinellidae and Scarabaeoidea are also often found at very high elevations. The percentage abundance of species of different families is summarized in table 15.

The greatest majority of the hypsobiont Coleoptera are true geophiles or geobionts that occur under stones, in cavities under deeply sunk boulders, in rock crevices and in the soil, in the vicinity of streams, ponds, lakes and the edge of melting snow. Debris feeders, carrion feeders and predators outdominate the other types and strictly phytophagous species are extremely scarce at higher elevations. The following ecological analysis of the hypsobiont Coleoptera is based on the records from nearly all the principal mountain ranges of the world (Table 16).

The greatest bulk of the mountain autochthonous hypsobiont species are confined exclusively to the zone above the forest-line and only a small number of them occur within the upper reaches of the coniferous-forest zone and extremely few species descend to the foot of the mountain. Exceedingly few species, which are widely distributed in the surrounding lowlands, occur above the forest zone on mountains. *Cychrus rostratus* LINN., *Byrrhus fasciatus* FORST., *Chrysomela marginata* LINN., etc. extend, for example, from relatively low elevations to nearly the permanent snowline. The typically lowland *Carabus violaceus* LINN., *C. convexus* FABR., *Cychrus rostratus* LINN. and *Chrysomela marginata* LINN. have developed distinctive local subspecies on high mountains. These local subspecies are characteristically smaller in size than the typical lowland races. This is, for example, readily observed in the case

TABLE 15

The percentage abundance of different families of Coleoptera at high altitudes

Family	Percentage of species in total Coleoptera above the forest-line
Cicindelidae	0.05
Carabidae	54.00
Paussidae	0.10
Amphizoidae	0.05
Dytiscidae	2.00
Silphidae	1.00
Staphylinidae	16.00
Histeridae	0.40
Hydrophilidae	2.00
Coccinellidae	1.50
Byrrhidae	1.00
Dryopidae	0.05
Elateridae	0.10
Meloidae	0.10
Tenebrionidae	10.50
Chrysomelidae	2.00
Curculionidae	7.00
Scarabaeoidea	3.00

TABLE 16

Percentage analysis of the ecological types of hypsobiont Coleoptera

Ecological groups	Percentage in total hypsobiont Coleoptera
Transitional	1.0
Endogeous (including hypolithion)	80.0
Phytophile	15.0
Terricole (on open ground)	0.5
Aquatic (stagnant water)	5.0
Aquatic (rheophile or rheobiont)	2.0
Xerophile or xeroxene	1.0
Cryophile or cryobiont	99.0
Thermophile	0.5
Hygrophile or hygrobiont	97.0
Phytophagous	10.0
Predators, carrion feeders, debris feeders etc.	85.0

of *Cychrus rostratus pygmaeus* CHAUD. from the Alps and the Carpathians, when compared to the typical lowland *C. r. rostratus* LINN. *Carabus violaceus neesi* HOPP from the high-alpine zone of the Tyrolean Alps and Kärnten is also smaller than the typical form from the lowlands. Unlike Orthoptera, the number of widely distributed lowland forms occurring at high elevations is, however, exceedingly small among the Coleoptera. Most species occur at elevations between 2500 and 4000 m and the highest altitude record for the Coleoptera is at present 5600 m. The total number of strictly mountain autochthonous hypsobiont Coleoptera so far described from the world is probably about 8000 species. As may be expected, the literature on mountain Coleoptera is very extensive and it is not possible to review even a small fraction of the papers, containing records and descriptions and notes on distribution and ecology of species from different parts of the world. The reader will, however, find valuable material for further work in ALLUAUD (30-35), AMANN & KNABL (44), ANDREWES (47-50), APFELBECK (53), ARROW (56), AUSTIN & LECONTE (65), BASILEWSKY (86-88), BATES, F. (89-90), BATES, H. W. (99-93), BLAIR (133-135), BREUNING 150), DARLINGTON (234), EDWARDS (295-301), ESPANOL (323), FAIRMAIRE (327-328), FRANZ (367, 372), FRIDÉN (386), GRIDELLI (430-432), HEER (492, 494), HÖLZEL (520), HOLDHAUS (525, 527, 532), HOLDHAUS & DEUBEL (536), JEANNEL (574-590), KRYZHANOVSKII (659-661), LINDBERG (693,694), LINDROTH (698-702), LUTSCHNIK (714), MANDL (723), MANI & SANTOKH (734), MATEU (745, 746), MEDVEDEV (755, 756), NETOLITSKY (789-791), DE PEYERIMHOFF (832, 833), PIC (835), REINIG (871, 874, 877), ROUDIER (903, 904), SCHÖNMANN (932), SCHUSTER (941), WÖRNDLE (1121) YABLAKOV-KHNZORYAN (1128) and others listed in the bibliography.

Family Carabidae

The Carabidae represent almost 55% of the total hypsobiont Coleoptera on the high mountains of the world. The family is also remarkable for flourishing at much higher elevations on mountains and in the high boreal latitudes at sea-level than any other Coleoptera. *Carabus* LINN. is, for example, represented within the Arctic Circle by many species like *C. polaris* POPPIUS in Asia, *C. truncaticollis* ESCH. and *C. vietinghoffi* AD. in North America. *Nebria* LATR. is also commonly found on snow and ice on the Alps, the Himalaya, in the Arctic Alaska, Spitzbergen, Iceland and in Greenland. The lowland forest-zone Carabidae are often arboreal, but the mountain autochthonous species are almost exclusively terricole and live under stones, in moss, etc. The tropical species often fly, but the mountain Carabidae and the Carabidae from the temperate-zone lowlands rarely if ever take to their wings. The greatest majority of the hypsobiont Carabidae are small or minute forms, with a compact body of a mean length of about 2 mm. They are either subapterous or completely apterous. Even the few species with more or less developed wings never fly on the high mountains and when thrown into the air, they merely spread their elytra and soon fall to the ground. Almost without exception, the hypsobiont species are inhabitants of moist or damp localities, moss and subterranean cavities under large boulders, edges of glacial streams, lakes and melting snow. Although most species are predatory, many carrion-feeding forms are also known. Nearly all the species visit the snow-fields either for capturing their prey or for feeding on the wind-blown dead organic matter.

The Bembidine tribe Anillini are among the smallest Coleoptera, hardly 1-2 mm long, completely flightless, with the compound eyes entirely atrophied or at any rate greatly reduced. The body is generally pale yellowish. They are subterranean and live under deeply sunk boulders or inside fissures in rock and damp soil. On most mountain ranges in the northern hemisphere, they seem to prefer the north slope to the south slope. Being endogeous, their antennae, palpi and legs are short, in sharp contrast to the cavernicolous species that have long appendages to serve as tactile organs in the prevailing darkness. The Anillini are believed by competent authorities to have differentiated on the Gondwana landmass. The Trechinae, which are among the dominant Carabidae at high elevations, are considered by JEANNEL as a separate family Trechidae. They include swift-running species, 3-8 mm long and usually occur at the edge of snow and are all flightless or also have their wings more or less reduced. Important contributions to our knowledge of the mountain autochthonous hypsobiont Trechini are those of SCHÖNMANN (932), JEANNEL (575, 577, 578, 579, 580, 581, 583, 585, 589, 590), FRANZ (367), SCHWEIGER (944), etc. The tribes commonly met with at high elevations, often at the snow-edge on different mountains, include Carabini, Nebrini, Notiophilini, Loricerini, Scaritini, Omophronini, Broscini, Bembidiini, Trechini, Pterostichini, Amarini, Harpalini, Licini, Abacetini, Chlaenini, Lebiini, Brachini and Dryptini. To the typical hypsobiont Carabidae belong *Cychrus* FABR., *Procerus* DEJ., *Carabus* LINN. with a number of subgenera like *Chrysotribax* REITT., *Platycarabus* MOR., *Pseudocechenus* MOR., etc., *Orinocarabus* KR., *Carpathophilus* REITT., *Leistus* FRÖL. with the subgenera *Oreobius* DAN. and *Leistidius* DAN., *Nebria* LATR. with the subgenera *Nebriola* DAN. and *Oreonebria* DAN., *Dyschirius* BON., *Bembidion* LATR. with numerous subgenera, *Scotodipnus* SCHAUM, *Anillus* DUV., *Trechus* CLAIRV., *Anophthalmus* STURM, *Licinus* LATR. with the subgenus *Neorescius* BED., *Zabrus* CLAIRV., *Amara* BON. with the subgenera *Leiromorpha* GGLB. and *Leirides* PUTZ., *Percus* BON., *Molops* BON., *Pterostichus* BON. with the subgenera *Tapinopterus* SCHAUM., *Pseudorthomus* CHAUD., *Orites* SCHAUM., *Cryobius* CHAUD., *Oreophilus* CHAUD., etc., *Agonum* BON., *Platynus* BON., *Metabletus* SCHM.-GOEB., *Cymindis* LATR., *Drypta* LATR., *Brachynus* WEB., *Calosoma* WEB., etc. *Oreonebria* DAN. occurs at the snow-edge on the Alps; FOCARILE (351) has recently given an account of the cryotropism in this genus. The subgenus *Spelaeonebria* PEYER. of *Nebria* BON. is remarkable for its minute eyes, elongated antennae and legs. It occurs in caves on the Atlas Mountains. *Cryobius* CHAUD. occurs on the mountains of Angar-Beringea and also in Alaska-Yukon Territory. *Calathus* BON. extends as far south as the Ethiopian Highlands, where it is represented by about 17 endemic species, but there are no species further south on the East African mountains. The blind *Scotodipnus* SCHAUM, *Kenyacus* ALLUAUD, *Plocamotrechus* JEANNEL, the apterous subgenus *Carabomorphus* KOLBE of *Calosoma* WEBER and *Orinophonus* ALLUAUD are characteristic of the tropical East African mountains. *Plocamotrechus* JEANNEL is absent from the East African lowlands and is confined to the mountains of the eastern and western branches of the Rift Valley System. *Chaetobroscus* SEM. is endemic to the Northwest Himalaya. *Pelmetellus* BATES occurs on the Andes and on the mountains of Arizona and New Mexico. The genus *Colpodes* M.L. is represented by numerous species on the Andes. Mention may also be made of the extremely interesting Carabid *Macranillus coecus* SHARP, with reduced and unpigmented eyes and lacking distinct facetting, from the Hawaiian Mountains. From the same mountains (elevation of 2750 m), are also known the apterous Proterhinid Carabids, 1.5 to 5.5 mm long, with weakly beak-like head in the male and pronouncedly beak-like head in the female. *Mecyclothorax* SHARP and

Thriscothorax SHARP are other Carabids recorded at elevations of 2700-3000 m on the Hawaiian Mountains. Many mountain Carabidae like the subgenus *Cechenus* FISCHER of *Carabus* LINN., *Pseudocechenus* MOR., *Platycarabus* MOR., *Platynus* BON., etc. have a depressed body. Some very unique Carabidae like *Merizodus* SOLIER, *Migadops* WATERH., *Pseudomigadops* JEANNEL, *Metius* CURT. and *Abropus* WATERH. are reported from the southern Andes.

Family Paussidae

Although a number of species of ants occur at fairly high elevations on nearly all the principal mountains of the world, the myrmecophilous Paussidae have so far been recorded only from the alpine zone of the Abyssinian Highlands. It is also interesting to observe that, though abundant in the alpine zone, Paussidae are absent in the subalpine zone in the region (950, 951).

Family Amphizoidae

This family is characteristic of cold streams of North America, parts of Tibet and the Northwest Himalaya (1080). The adult beetles do not swim and their legs are not also modified into the typically natatory appendages that we find among the Dytiscidae. There is at present a single genus *Amphizoa* LEC., the species of which live under submerged stones. The family is considered to be a relict group (295, 297).

Family Dytiscidae

Although widely distributed in the plains in most parts of the world, the family is fairly abundant in cold-water ponds amid ice and glaciers. The species seem to prefer torrential streams at high elevations and become sparse in stagnant waters with increase in elevation. The common mountain autochthonous hypsobiont species belong to the genera *Dytiscus* LINN., *Agabus* LEACH, *Rhantus* LAC., *Laccophilus* LEACH, *Coelambus* THOMS., *Gyrodytes* SEIDL., *Cybister* CURT., *Deronectes* SHARP and *Potamonectes* ZIMMERMANN. Some of the species are known to occur at extreme high elevations on the Himalaya.

Families Silphidae and Scydmaenidae

The hypsobiont mountain autochthonous species from the Alps and other Tertiary mountains include the blind *Bathyscia* SCHIÖDTE, *Pteroloma* GYLL. and *Leptoderus* SCHMIDT (Silphidae), *Neuraphes* THOMS., *Stenichnus* THOMS. and *Leptomastax* PIR. (Scydmaenidae). The specis occeur mostly under stones and are nearly always debris feeders or carrion feeders. Some species are predatory on Acarina. Among the Silphidae many blind cavernicolous species are also known. *Rybinskiella* REITT. occurs on the Carpathians, the Tien Shan and the Northwest Himalaya.

Families Staphylinidae and Pselaphidae

The Staphylinidae are a large family of elongate, slender geophile beetles, with greatly abbreviated and truncated elytra, leaving the 7-8th segments of abdomen exposed. Both larvae and adults feed on diverse decaying vegetable or animal matter and on fungi, but many are also predaceous. The Aleocharinae are mostly predatory forms. The bulk of the high altitude species are black or very dark reddish-brown and are also conspicuously densely clothed with hair-like setae. Nearly all the high altitude species are also apterous and in any case none of them ever takes to wings. All are pronounced hygrophiles and occur typically under stones, on moss and in decaying organic matter near the edge of the melting winter snow. A number of species frequent the snow-fields for foraging. Like Carabidae, the Staphylinidae also occur at extreme elevations, in the immediate vicinity of snow and ice. Staphylinidae are abundant on all principal mountains of the world and extend north up to the 70 parallel. The highest altitude record for Coleoptera in the world is at present held by the Staphylinid *Atheta (Dimetrota) hutchinsoni* CAM. that occurs at an elevation of 5600 m on the Himalaya. A number of species which are really typical of the upper reaches of the high coniferous forest zone regularly ascend to the subalpine and even to the alpine zones both on the Alps and on the Himalaya. The great bulk of species at high elevations, above the forest-line, belong to *Atheta* THOMS., *Aleochara* GR., *Geodromicus* RDT., *Lesteva* LATR., *Ocyusa* KRAUSS, *Oxypoda* MANNERH., *Philonthus* CURT., *Tachinus* GR., etc. Others belong to the genera *Anthobium* STEPH., *Anthophagus* GR., *Niphetodes* MILL., *Lathrobium* GR. with the subgenus *Typholinus* SHARP, *Staphylinus* LINN., *Quedius* STEPH., *Leptusa* KR., etc. The cavernicole *Paraleptusa* PEYERH. and *Apteraphaenops* JEANNEL from the Atlas may also be mentioned in this connection. *Amaurops* FAIRM., *Bythinus* LEACH, *Pselaphus* HBST., etc. are all apterous and represent the more important high altitude Pselaphidae. Descriptions of high altitude Staphylinidae may be found in ABSOLON (3), BERNHAUER (112, 113), FAUVEL (335), GRIDELLI (430), JARRIGE (573) and others listed in the bibliography.

Family Histeridae

Histeridae are small or medium-sized, mostly semi-flattened beetles, which are predatory both as larvae and as adults, but sometimes also feed on organic debris. While the greatest bulk of the species are typically thermophiles, some interesting forms are characteristic of the temperate regions. The family is represented by some remarkable species of *Hister* LINN. at elevations above 3000 m on the Northwest Himalaya (730).

Family Hydrophilidae

The Hydrophilidae are perhaps most abundant in temperate latitudes, but a number of interesting species occur even well within the Arctic Circle. The mountain autochthonous hypsobiont species are aquatic and breed most usually in melt-water torrents and only sometimes in glacial lakes at the edge of ice and snow. Some typically high altitude forms are, however, semi-aquatic and live on moss, under stones and other similar situations at the snow-edge. The most widely distributed mountain forms belong to *Helophorus* FABR. Diverse species of this genus occur in sulphur and hot springs at elevations of 5400 m on the Northwest Himalaya. *H. glacialis* VILLA. is a typical glacier-zone species from the Alps. The widely distributed lowland *Hydrous* DAHL sometimes occurs above the forest-line on the Himalaya. The sulphur spings, in which the Hydrophilidae occur on the Himalaya, have a temperature of 24-27° C and contain considerable amounts of dissolved hydrogen sulphide. Though most species are confined to fresh water, some occur even in distinctly brackish waters in the Tibetan Plateau (277).

Family Byrrhidae

The Byrrhidae are a widely distributed group that occurs also in the arctic latitudes. The larvae of the mountain autochthonous species feed mostly on moss, but some are also predators. On high mountains, the family is confined to the edges of streams and ponds and melting snow. *Simplocaria* STEPH. and *Byrrhus* LINN. are the common forms in the subalpine and alpine zones. *Carpathobyrrhulus* GGLB. is characteristic of the North and East Carpathians. *Simplocaria metallica* STURM is a boreo-alpine species (*vide* Chapter XII).

Family Dryopidae

The often characteristically flattened larvae and adults of Dryopidae may be found attached to submerged stones and plants in mountain torrents. The larvae have a curiously shaped lateral expansion on each segment, well developed antennae and legs and posterior gills. They are predominantly phytophagous. *Psephenus* HALD., with its characteristic trilobite-like larvae, occurs in Himalayan rivers and glacial streams. The larvae are anchored to the submerged stones with so much force that it is not possible to dislodge them without crushing their bodies. Other members belong to *Dryops* OLIV. *Helmis* LATR. occurs in Fennoscandian mountain streams. The species generally occur up to the subalpine zone and are abundant immediately above the forest-line.

Family Tenebrionidae

The Tenebrionidae are an important group of high altitude beetles, especially on the Tertiary mountains. The greatest bulk of the Tertiary mountain Tenebrionidae belong to endemic genera. Most species occur at elevations of 3000-3800 m, but some species ascend to much higher elevations. *Blaps* FABR. occurs, for example, at an elevation of 5000 m and *Ascelosodis* REDTB. at elevations of 4400-5600 m on the Himalaya. The highest altitude record for the family in the world is 5600 m. Most hypsobiont species are black or very dark brown and are almost all pronounced hygrophiles, concentrated near the edge of streams and snow-fields. On the Himalaya they are less conspicuous on the steeper southern slopes than on the more gentle northern slopes of the ranges. They occur under stones and feed on moss and diverse other debris. The high altitude species of the family have been studied intensively by a number of workers like BLAIR (133, 134, 135), GRIDELLI (431) and REINIG (871, 874, 877). Some of the more interesting and outstanding facts of the ecology and biogeography of the mountain autochthonous Tenebrionidae are described in the chapter on the Pamirs (*vide* Chapter X). Of the characteristic mountain autochthonous hypsobiont genera, mention may be made of *Itagonia* REITT., *Prosodes* ESCHTZ., *Blaps* FABR., *Platyscelis* LATR., *Bioramix* BATES, *Trichomyatis* REINIG, *Trichoplatyscelis* REINIG, *Myatis* BATES, *Ascelosodis* REDTB., *Scythis* SCHAUM, *Chianalus* BATES, *Cyphogenia* SOL., *Syachis* BATES, *Laena* LATR., *Asida* LATR., *Pimelia* FABR., *Heliopathes* MULS.,

Phylan STEPH., *Micrositus* MULS., *Opatrum* FABR., *Gnaptorina* REITT., *Epipedonota* SOL., *Nyctelia* LATR., *Scotobius* GERM., *Emmallodera* SOL., *Praocis* ESCHTZ., etc. Most of these are typical endemics on well known mountain systems and often on single massifs of a mountain range. *Phrynacolus* LAC. occurs in the alpine zone of the East African mountains. *Epipedonota* SOL. is dominant on the Andes, between 36° and 48° SL. The remarkable endemic genera *Nyctelia* LATR., *Scotobius* GERM., *Emmallodera* SOL., *Praocis* ESCHTZ., etc. are characteristic of the Patagonian steppes-zone of the southern Andes. *Parahelops* WAT. and *Hydromedion* WAT. are confined to the southern Andes. There are a number of endemic high altitude Tenebrionidae on the Spanish Sierra Nevada also (*vide* Chapter XIII). *Oxycara* SOL. and *Trichopedius* BEDEL have endemic species on the mountains of Cape Verde Islands. *Przewalskia* SEM. is endemic to the Tibetan Plateau.

Family Chrysomelidae and Anthribidae

This predominantly phytophagous group of beetles occurs at relatively high elevations on some of the principal mountains of the world. In all cases, the species are apterous and occur under stones. The bulk of the mountain autochthonous species occur immediately above the forest-line, but a number of them are found at elevations above 3900 m. *Apaksha himalayensis* MAULIK, an endemic apterous species, occurs at an elevation of 3500 m and *Chaetocnema alticola* MAULIK at an elevation of 4575 m on the Himalaya. There are two other interesting endemic hypsobiont Chrysomelidae from the Himalaya, viz. *Swargia nila* MAULIK at an elevation of 4880 m and *Shaira krishna* MAULIK at elevations of 4200–4575 m. The highest altitude record for the family in the world at present is 4880 m in the Mt. Everest area (*Swargia nila* MAULIK). Mention must also be made of the records of apterous Galerucids from the upper levels of the forest-zone on Mt. Kilimanjaro. *Sjöstedtinia* WSE. is an endemic genus, found at elevations between 3000 and 3500 m on Mt. Kilimanjaro. *Longitarsus* LATR. occurs in soil communities at an elevation of 4270 m on the same mountain. Other hypsobiont Chrysomelidae include species of *Chrysomela* LINN., *Orina* MOTSCH., *Gynandrophthalma* LAC. (3400 m on the Trans-Alai Mountains), *Entomoscelis* CHEVR. and *Gastroidea* HOPE at 3100 m from the Pamirs Region, *Galeruca* GEOFFR. at 4050 m from the Pamirs, *Leptosonyx* WSE. from the Pamirs, Siberia, Caucasus and the Himalaya, *Haltica* GEOFFR., *Merista* CHAP., *Plagiodera* REDT. at 3200–4000 m on the Himalaya, *Podagrica* FOUDR. and *Phytodecta* KIRBY on the Himalaya. *Chrysomela crassicornis* HELLIES and *Phytodecta affinis* GYLL. are boreo-alpine species (*vide* Chapter XII).

Family Curculionidae and Anthribidae

The mountain autochthonous curculionids belong mostly to the genera *Otiorrhynchus* GERM., *Blosyrodes* JEKEL, *Catapionus* SCHÖNH., *Polydrosus* GERM., *Thylacites* GERM. *Bathynoderes* SCHÖNH., *Lepyrus* GERM., *Dorytomus* STEPH., *Pissodes* GERM., *Orestia* GERM., *Phaenotherion* FRIV., *Dichotrachelus* STIERL., *Plinthus* GERM., *Neoplinthus* BED., *Liosoma* STEPH., *Leptomias* FAUST, *Scepticus* ROELOFFS, etc. Some of the species occur at very high elevations. *Otiorrhynchus* GERM. occurs, for example, at elevations of 3000–4200 m on the Pamirs, *Heteronyx* GUÉRIN-MEN. and *Lagenolobus* FAUST at elevations of 3500–4500 m on the Northwest Himalaya. A number of endemic forms are characteristic of the Pamirs, the Himalaya and of the Andes. *Scepticus* ROELOFFS occurs at 3200-4600 m on the Turkestan Mountains and also on the Altai Range. *Helicorrhynchus* OLLIFF, related to *Otiorrhynchus* GERM., is endemic to the Andes and occurs at elevations of 4757 m. The highest elevations at which the family has so far been observed in the world is 4976 m; at this elevation *Macropscoelorum* OLLIFF occurs on the Andes. *Amathynetes* OLLIFF is another endemic form from the Andes that occurs at elevations between 3600 and 4800 m. *Cylindrorhinus* GUÉRIN-MEN., *Adioristus* WAT. and *Scotoeborus* SCHÖNH. are dominant on the Andes between the 36th and 48th south parallel. The southern Andes are remarkable for a number of brachypterous and apterous endemic forms like *Listroderes* SCHÖNH. with the subgenera *Antarctobius* FAIRM., *Puranius* GERM., *Falklandius* ENDERL., *Falklandiellus* ENDERL., etc. Almost without exception, the high altitude species of Curculionidae live under stones on grassy meadows, near streams and close to the snow-edge. Nearly all the species are either flightless or are completely apterous. Some like *Troglorrhynchus* SCHMIDT are blind cavernicolous forms. A number of species of *Otiorrhynchus* GERM. and *Barynotus squamosus* GERM. are boreo-alpine forms.

Other Families

The other families of Coleoptera occupy only a minor place above the forest-line on mountains. Anthicidae have been recorded from stones at elevations of 4000-5200 m on the Eastern Himalaya. Meloid beetles occur on the Himalaya and the Pamirs at elevations of 3000-4600 m. Reference has already been made to the mass assemblages of Coccinellidae on the Himalaya, Pamirs and other high mountains. The Telephoridae *Podabrus* WESTW. and *Absidia* MULS., found on the European mountains, are interesting because of their boreo-alpine distribution. Elateridae like *Lacon* CAST., *Cardiophorus* ESCH., *Agriotes* ESCH. have been recorded from the Pamirs and *Corymbites* LATR. is a boreo-alpine element. A number of interesting Scarabaeoidea have been observed at high elevations on the Himalaya and some other mountain ranges. *Copris* GEOFFR., *Aphodius* ILLIGER and *Onthophagus* LATR. are some of the important Scarabaeids from the Himalaya and the Pamirs and species of these genera are often found at elevations of about 3900 m. *Taurocerastes* PHIL. occurs on the southern Andes.

ORDER 8. HYMENOPTERA

Although the high altitude Hymenoptera seem to have been inadequately investigated, it is nevertheless evident that these insects are but poorly represented above the timberline on most mountains. The dominant high altitude Hymenoptera are Bombidae and Formicidae, but several other families have also been sporadically found on different mountains. Sawflies have been described from the alpine zones of the Scottish Mountains and also from the Swiss Alps. BENSON (103-107) found Nematinae, closely related to the arctic forms, on the alpine zone of the Swiss Alps. He collected a number of interesting species on snow patches, at an elevation of 2500 m near the Alpid du Zaté. According to him, the alpine meadows are free from ants, but are on the other hand rich in Tenthredinidae. *Dolerus aeneus* KLUG, one of the commonest West-European species, was observed by BENSON (104,

105) at an elevation of 2500 m. *Nematus* JUR., *Amauronematus* KONOW, *Pristiphora* LATR., *Arge* SCHR., *Aglaostigma* KIRBY, *Paracharactus* MCGILL., etc. are some of the Tenthredinids found by him up to elevations of 2750 m. Although some Ichneumonid parasites of spiders and of Carabidae are sparsely found in the subalpine zone, the dominant entomophagous parasitic insects above the forest-line on high mountains are not, however, Hymenoptera, but Tachinidae (Diptera).

We have the following records from the Pamirs Region: *Ammophila* KIRBY, *Tachysphex* KOHL, *Crabro* FABR. and *Pompilus* FABR. at elevations of about 3500-4200 m. These are believed to be essentially the Turkestan steppe elements that have secondarily colonized the high elevations and are not therefore in the true sense of the term mountain autochthonous, though found at such high elevations. The Apoidea from the Pamirs Region include *Prosopis* FABR. (3700-4200 m), *Colletes* LATR. (4200 m), *Andrena* FABR. (3500-3600 m), *Nomada* SCOP. (4100-4200 m), *Xylocopa* LATR. (3000 m), *Melecta* LATR. (3400-3600 m), *Osmia* PANZ. (3500-4500 m), *Chalicodoma* LEPEL (4300 m), *Megachile* LATR. (3000-3900 m), *Anthidium* FABR. (3000-4000 m) *Coelioxys* LATR. (3800-4200 m) and *Anthophora* LATR. (3500-4300 m). The typical mountain autochthonous hypsobiont Bombidae have been extensively studied by VOGT (1084), SKORIKOV (984) and REINIG (873). Some of the hypsobiont species have a typical circumpolar distribution and are also peculiar to the Tertiary mountains. *Sibiricobombus* VOGT, *Lapidariobombus* VOGT, *Subterraneobombus* VOGT, *Mendacibombus* SKOR., *Mucidobombus* SKOR., *Bombus* LATR., *Alpinobombus* SKOR., etc. are the principal mountain genera on the great Alpine-Himalayan Systems of Tertiary mountains. Above the forest-line, the Formicidae are found exclusively under stones. The common species belong to *Formica* LINN., *Cataglyphis* FÖRST., *Camponotus* MAYR, *Myrmica* LATR., etc. Most of the species occur at elevations between 1500 and 3000 m, but *Formica* (*Serviformica*) spp. have been found up to an elevation of 4100 m on the Himalaya. *F.* (*S.*) *picea* NYL., found at an elevation of 4800 m on the Himalaya, represents at present the highest altitude record for ants in the world.

ORDER 9. TRICHOPTERA

The greatest bulk of the mountain autochthonous Trichoptera are confined to the altitude zone between 2500 and 4000 m, but some species of *Pseudohalesus* MARTYNOV and *Pseudostenophylax* MARTYNOV flourish at elevations between 4600 and 4800 m on the Himalaya. The highest altitude at which Trichoptera seem to exist is 5000 m on the Himalaya. *Apatania* KOL., *Dolophilodes* ULMER, *Dinarthrum* MCLACH., *Agapetus* CURT., *Brachycentrus* CURT., *Rhyacophila* PICT., *Oligoplectrodes* MARTYNOV, etc. are some of the more important hypsobiont types. The larvae of the high altitude Trichoptera are found in the melt-water torrents, rivers and springs and they remain anchored to the submerged stones near the margin. Rarely, some species seem to prefer the quieter waters in these streams. The cases of the larvae, inhabiting the torrents, are as a rule constructed of relatively heavy material, such as large-sized stones and small pebbles and are on the whole compact and rigid. They are characteristically flat beneath and conspicuously arched or also convex above in cross-section. The adults emerge from mid summer to late summer on most mountains in the northern hemisphere. Interesting accounts of the mountain Trichoptera may be found in BRODSKII (160), ROSS (897, 898) and SCHMID (925).

ORDER 10. LEPIDOPTERA

The typical mountain autochthonous hypsobiont Lepidoptera belong to Papilionidae, Parnassiidae, Nymphalidae, Satyridae, Pieridae, Lycaenidae, Arctiidae, Geometridae and Noctuoidea. Most high altitude species occur typically at elevations between 2500 and 4000 m, but some interesting forms have been found even at 6000 m. A number of species habitually breed at elevations of 5000 m and do not occur below 4000 m. It is only exceptionally that we find widely distributed lowland forms, like *Vanessa urticae* LINN. and *Papilio machaon* LINN., at relatively high elevations on various mountains. Localized subspecies and races are common on most high mountains. The caterpillars of the high altitude species feed in most cases on the subalpine and alpine-zone vegetation. *Erebia* DALM. feeds on grasses, but others like *Scoparia* HAW. (Pyraloidea), Arctiidae, *Callimorpha* LATR., etc. feed on lichens. *Agrotis* OL. is generally polyphagous. In nearly all cases, the caterpillars remain under stones most of the time or inside rock crevices, and do not appear on the food plant as is their habit near sea-level. The caterpillars have the remarkable habit of remaining quiescent, without feeding and without pupating, for varying periods in the absence of adequate snowfall. The caterpillars of most species are remarkably concentrated near the snow-edge and are in many cases strictly confined to the snow-edge. Almost without exception, all the high altitude species of Lepidoptera are diurnal, both as larvae and as adults, including even the moths like Arctiidae, Geometridae and Noctuoidea. Crepuscular and nocturnal forms are practically unknown at higher elevations. The commonest Lepidoptera at high elevations are *Parnassius* LATR., *Melitaea* FABR., *Argynnis* FABR., *Erebia* DALM., *Pieris* SCHR., *Colias* FABR., *Arctia* SCHR., *Larentia* TR., *Eupithecia* CUR., *Gnophos* TR., *Dasydia* GU., *Agrotis* OL., *Psodos* TR., *Bryophila* FR., *Hadena* SCHR., etc. Important contributions to our knowledge of the high altitude Lepidoptera include those of ALPHERAKY (36-39), BARTEL (85), CARADAJA (184), CARNELUTTI (187), EHRLICH (304), ELLER (308), EVANS (325), FAWCETT (336), FORSTER (359), FORSTER & ROSEN (361), GROSS (438-440), HOFFMANN (523), HORMUZAKI (543), HOVANITZ (544), KITSCHELT (625), KRZYWICKI (662), MELL (759), PETERSEN (831), POBOLNY & MOUCHA (839), PRÜFFER (846), REAKIRT (861), REBEL (862), REBEL & ZERNY (863), RILEY (886, 887), RONDOU (894, 895), ROSEN (896), ROTHSCHILD (900, 901), SHELJUZHKO (961), SPEYER (989), STAUDINGER (998, 999), TABUCHI (1026), WARREN (1103), WNUKOWSKY (1119) and WOJTUSIAK & NIESIOLOWSKI (1122).

Family Parnassiidae

The family Parnassiidae, characteristic of the high mountains of Europe, Asia and North America and of the high north latitudes in these lands, is remarkable for the high elevations at which the species are found. The family is recognized by a number of distinctive characters. Two genera *Hypermnestra* MÉN. and *Parnassius* LATR. are represented by a number of unique endemic species and subspecies on different mountains. The latter genus is distinguished by a characteristic horny anal pouch, extruded after copulation in the female. The wings are semi-diaphanous. The larvae are gregarious, black coloured and conspicuously narrowed at both ends. The majority of the species occur at elevations of 3000 to 4000 m, but some very interesting forms are found even at an elevation of 6000 m on the Himalaya. There are several valuable contributions on the taxonomy and biogeography of the Parnassiidae and the more important works include those of AUSTANT (61-63), AVINOFF (67-70), BANG-HAAS (82, 83), BRYK (166-169), BRYK & EISNER (170), ELWES (312), FAWCETT (336), OBERTHÜR (798), SKINNER (982) and STAUDINGER & BANG-HAAS (1000).

P. apollo LINN. occurs in Spitzbergen and at an elevation of about 2000 m on the mountains of southern and Central Europe*. It is considered by some workers to be a glacial relict, but others do not agree with this view (235, 1101). *P. phoebus* FABR. is a boreo-alpine species that occurs on the Alps, the northern Urals, the Altai Mountains, Tarbagatai, Sayanskii Range, the eastern Tien Shan Mountains, Mongolia, Alaska, British Columbia and the North American Sierra Nevada. HERING (508) has discussed the problem of subspecies of *Parnassius* LATR. on a morphological basis (*vide* also Chapter IX).

Families Nymphalidae and Satyridae

The Nymphalidae-Satyridae complex are characteristic of high altitudes on most Tertiary mountain systems, on which the species often occur at remarkably high elevations. Except perhaps the East African equatorial mountains, most other high mountains have distinctive endemic species of the complex. The common mountain autochthonous forms belong to *Precis* HBN., *Vanessa* FABR., *Erebia* DALM., *Melanargia* MEIG., *Epinephele* HBN., *Satyrus* LATR., *Coenonympha* HBN., *Oeneis* HBN., *Argynnis* FABR., etc. *Argyrophenga* DOUBLEDAY, *Heteronympha* WALLENGREN, *Xenica* WESTW., *Erebiola* FERED. and *Percnodaimon* BUTLER are peculiar to the high mountains of New Zealand (1000-1900 m above mean sea-level). As recently shown by GROSS (439, 440), the high altitude Satyridae-Nymphalidae complex seems to have become differentiated from the early Tertiary primary and the late Tertiary secondary centres. The two major Tertiary centres of differentiation are Fennoscandia and the Angaran area. The Nymphalidae appear to have had their focus of dispersal in the Angaran area and the Satyridae in Fennoscandia. The West Palaearctic ancestral forms of the Satyridae were already present in Fennoscandia during the early Tertiary and the ancestors of the others were present in the East Palaearctic. The West Palaearctic forms are believed to have disappeared during the Pleistocene. Faunal exchanges during the late Tertiary and Quaternary were perhaps more intensive in the north than in the south. Almost 50% of the present day Satyridae on the Turkestan mountains are autochthonous, about 25% have had their origin in Asiatic Tertiary area and 70% in the West Palaearctic Tertiary mountains. The Himalaya functioned as an important dispersal centre throughout the Pleistocene; about 80% of the present day Satyridae are autochthonous and 75% are of Asiatic origin and therefore autochthonous since the Tertiary. Only 16% are of West Palaearctic origin. The Turkestan mountains and the Himalaya have been the scenes of differentiation of mountain autochthonous species of *Precis* HBN., *Vanessa* FABR., *Erebia* DALM., *Melanargia* MEIG., *Epinephele* HBN., *Satyrus* LATR., *Coenonympha* HBN., *Oeneis* HBN. and *Argynnis* FABR. (fig. 44).

* PLANEIX (PLANEIX, P. 1965. *Alexanor* 4: 73-80) makes interesting observations on the altitudinal distribution of this species.

Family Pieridae

The Pieridae occur up to the snowline on most mountains. *Aporia* HBN., *Pieris* SCHR., *Euchloë* HBN. and *Colias* FABR. are typical members at high altitudes. *Colias* FABR. is restricted to the temperate and arctic regions and the whole of Eurasia and North America (excluding most of India). A small number of species occur in the extreme north Africa and Abyssinia and on the East African mountains. The species on the South American Andes belong to a distinct group, different from the species from the Central American Cordilleran System and the rest of the world. The South American mountain forms occur at elevations of 2440 m and descend to the sea-level in Argentina-Chile and in Tierra del Fuego. There is considerable parallelism in the dark and sober coloured species from the arctic north of North America and Eurasia and the species from South America and southern South America. As recently discussed by HOVANITZ (544), the small greenish-yellow species of the Ecuadorian and Peruvian alpine-zones is reproduced phenotypically in the alpine meadows of the North American Sierra Nevada, the Rocky Mountains, the Alps and the Himalaya. The basic genetic stock of each of these groups seems to have been different. The females of *C. lesbia* and *C. dimera* DBLD-HEWN. are whiter at higher elevations than at lower on the Andes and this peculiarity is also met with in the Himalayan species of *Colias*. *Baltia* MOORE is common on the Himalaya and on the Middle Asiatic mountains. On the Himalaya, none of the species appear to descend below an elevation of 3900 m and some species occur at elevations of about 5000 m. The species of *Colias* from the Himalaya and from the Andes are among the smallest in the genus. *Pontia* FABR. occurs as high as 4300 m.

ORDER II. DIPTERA

The Diptera are among the dominant hypsobiont insects on nearly all the high mountains in the world and also occur at the highest elevations. They are abundant even at the highest altitude at which evidence of permanent existence of insect life has been found; they are, for example, quite common at elevations between 6000 and 6300 m on the Himalaya. The vast majority of the high altitude Diptera breed in organic debris and are severely restricted to the snow-edge and to the immediate vicinity of melt-water. A large number of them, especially at higher altitudes, are carnivorous and active predators. We find a great predominance of Stratiomyiidae, Empididae, Rhagionidae, Asilidae, Dolichopodidae, Calliphoridae, Anthomyiidae and Tachinidae. Among the flower visiting insects on the high-alpine grassy meadows, the Diptera stand next only to the Lepidoptera. Not only the Syrphidae, but the adults of most other families also regularly visit flowers. Diptera become active after the winter hibernation much earlier than most other groups and often more than one generation is completed within the short summer. The literature on the mountain autochthonous Diptera is very extensive, but the more important papers include those of ALEXANDER (13-18), AUSTEN (64), BECKER (98), BERTRAND (114, 115), BEZZI (122, 123), BISCHOFF (129-131), EDWARDS (293, 294), FRANZ (377), KOMÁRCK (644, 645), LINDNER (697), etc. HACKMAN (457) has recently given an interesting account of the brachypterous and apterous high altitude Diptera.

Nematocera

Family Tipulidae

The Tipulidae are often represented by a number of remarkable endemic genera and species above the forest-line on most high mountains. A number of subapterous and even completely apterous species are also known on nearly all the high mountains. Some of the species occur at extreme high elevations. The family has been recorded from within 900 km of the North Pole. Most of the subarctic and hypsobiont species of Tipulids are exclusively hygrophiles and breed at the snow-edge or close to the melt-water. The typical members belong to *Tipula* LINN., *Cylindrotoma* MACQ., *Limonia* MEIG., *Pedicia* LATR., *Gonomyia* MEIG., *Cladura* OST.-SACK., *Molophilus* CURT., *Pachyrhina* MACQ., *Dicranota* ZETT. and *Chionea* DALM. (13-18). The great bulk of the hypsobiont Tipulidae occur at elevations between 1200 and 3000 m, but a number of species of *Tipula* LINN. occur on the Himalaya up to elevations of 5200 m

Family Bibionidae

The Bibionidae are robust flies, with stout, short legs and antennae and eyes contiguous in the male. They are mostly black, yellow or orange and have a heavy arched torax. The larvae feed in decaying vegetable matter at the edge of the melting snow or springs and torrents, under stones at the snow-edge and other similar situations. The family is widely distributed on most high mountains and extends also right into the Arctic Circle. *Bibio* GEOFFR. and *Plecia* WIED. contain all the typical mountain autochthonous species.

Family Blepharoceridae

The Blepharoceridae are extremely abundant in the melt-water torrents on the Himalaya, the Caucasus, the Tien Shan and other high mountains. The aquatic larvae cling to the submerged stones in clear and well aerated fresh water in torrential streams. They are curiously shaped, with a large head-segment, consisting of the head fused together with the three thoracic segments and the first five abdominal segments, followed by six conspicuously constricted body segments. Ventrally the body segments are provided with well developed and powerful circular suctorial pads. The composite head segment is provided with well developed mandibles and long antennae. The pupae are oval, convex or nearly hemispherical and permanently anchored to the submerged stones by their suctorial discs. Both the larvae and the pupae occur only in that part of the stream, where the velocity of the current is maximum. The adult females appear to be predacious on Chironomid flies. Considerable literature exists on the habits and life-histories of Blepharoceridae (98, 129-131, 161, 522, 643, 645, 696, 738, 1039). The typical high altitude species belong to the genera *Blepharocera* MACQ., *Bibiocephala* OSTEN-SAC., etc. The species occur up to elevations of 3900 m. *Tianschanella* BRODSKII is endemic to the Tien Shan Range. On the Colorado Mountains, the family ascends to an elevation of nearly 3660 m.

Family Simuliidae

The Simuliidae are widely distributed all over the world, but are perhaps best developed in the temperate and subarctic regions. They are abundant in Alaska, northern Canada, Greenland and Iceland (18). A number of species of *Simulium* LATR. occur at elevations of about 3600 m on the Colorado Mountains and at 3900 m on the Himalaya. *Prosimulium* ROUB. and *Nevermannia* END. are other typical mountain forms. The larvae of the mountain autochthonous Simuliids feed on the algal slime on submerged stones in torrential streams and the pupae also occur under identical situations. The adults emerge under water and rise quickly to the surface, though the current carries them to some distance downstream.

Family Deuterophlebiidae

The Deuterophlebiidae are well known as mountain-midges and are distributed on the Northwest Himalaya, the Tien Shan, the Altai, North Korea, Formosa, Japan, Canada, Wyoming, Colorado, Utah, Oregon, California and the Central Chilean Andes (696, 889, 890, 1016a). The family was erected by EDWARDS (293) in 1922 for *Deuterophlebia mirabilis* EDW., which he described from the adult found by MITCHELL, floating on the surface of the water in the Gangabal Lake, at an elevation of about 3000 m, below the Harramukh Glacier (5000 m above mean sea-level) on the Northwest Himalaya. In 1924, PULIKOVSKY (849, 850) recorded the species from the Altai Mountains and also described the characteristic larvae and pupae and their remarkable habits. In 1926, BRODSKII & BRODSKII (157) reported the family from different parts of Middle Asia. MUTTKOWSKI (787) reported finding the family in different parts of North America in 1927. Since then, different species have been described from Japan by KITAKAMI (623, 624) and KAWAMURA (611), from Japan and Formosa by YIE (1129), from Canada by SHEWALL (963) and from the U.S.A. by STONE (1015), PENNAK (827) and WIRTH (1116). PENNAK (826-828) has summarized the available information on the family and tabulated the species known to him. KENNEDY (613) has described the larva, pupa and adult of *D. nielsoni* KEN. from eastern California and has also given a detailed account of the habitat, habits and life-history of the species. He has also included in his paper a synoptic key to the species from the world. The Deuterophlebiidae provide an interesting example of distribution, along high mountains, of Middle Asiatic elements to the South American Andes.

Deuterophlebiidae are extraordinarily delicate flies, with truly immense wings, clothed with fine pubescence, but with all the veins nearly obliterated. They have a most curious superficial resemblance to certain small Ephemerids, but the wing is creased fanlike in transverse folds. The antennae have only six segments, but are much longer than the body, especially in the male. The most striking character of the adult is the complete absence of ocelli and mouthparts. The larvae and pupae inhabit clear cold-water torrents, with a mean temperature of 4.5° C, but often they also occur in torrents at a temperature of 1.5° C. In Colorado, they are reported to occur in streams at a temperature of 18° C. The body is very conspicuously segmented and bears seven paired, stout, lateral prolegs, armed with circlets of peculiarly shaped hooks that serve as powerful organs of anchorage. The antennae and the tip of abdomen bear blood-gills. The pupa is somewhat oval in outline, highly convex above and flattened below and very much like the pupa of Blepharoceridae. The pupa is also anchored to the submerged stones. The larvae feed by scraping the rich algal slime and diatomaceous growth on the submerged stones. There are four larval instars. The winter is passed in hibernation in the first larval stage. The over-wintered larva becomes active in early spring and completes its development by the end of June or sometimes early July. A second generation is then completed during July and there is sometimes a third generation also during August. On the Himalaya, the family occurs at least up to an elevation of 4000 m and passes through only two generations, viz. an early and a late summer generation. The early summer generation seems to consist predominantly of females and the mid summer and late summer generations have a larger proportion of males on the Himalaya (730).

Family Chironomidae

Although widely distributed, the family seems to attain its maximum development in the arctic and subarctic regions (18) and is also abundant on most high mountains. The larvae feed mostly in torrents and glacial lakes (1035, 1036), but a number of high altitude forms like, for example, *Euphaenocladius alpicola* GTGH. from the Alps, are typically semi-terrestrial or even completely terrestrial and thus feed in damp or wet moss and under stones. The typical mountain autochthonous species belong to the genera *Brillia* KIEFF., *Metriocnemus* V. D. WULP, *Orthocladius* V. D. WULP, *Chironomus* MEIG., *Clunio* HAL., *Palpomyia* MEIG., *Anatopomyia* MALLOCH, *Psilotanypus* KIEFF., *Eutanypus* COQ., *Cricotopus* V. D. WULP, etc. *Orthocladius* V. D. WULP is also known to occur in Siberia, Lappland, Spitzbergen and Greenland. *Diamesa* WALTL. occurs on the Colville Mountains of Alaska. Most of the mountain autochthonous species occur up to elevations of 3900 m.

Family Psychodidae

The Psychodidae are represented only extremely sparsely on high mountains. On the Himalya, *Pericoma* WALK. has been found up to an elevation of 3700-3800 m, near glacier snouts.

Family Culicidae

The typically alpine-zone mosquitoes *Aëdes pullatus* COQ., *A. cataphylla* DYAR and *Theobaldia niveitaeniata* (THEOB.) occur on mountains, in tundra and muskeg regions and should be considered as essentially subarctic or arctic-alpine or boreo-alpine elements. Some of these species occur further north than the boreal timberline and also at elevations above the forest-line on high mountains. On the Rocky Mountains of North America, mosquitoes have been recorded at elevations of 3960 m. *Aëdes pullatus* COQ. is considered by competent authorities to be a relict of the Würm Glaciation in Norway. *A. pullatus* and *Theobaldia* BLANCH. occur at elevations of 3600-3700 m on the Northwest Himalaya.

Family Sciariidae

Some Sciariidae have also been reported from high altitudes. *Sciara* MEIG. occurs at elevations of 2900-3000 m on the Alps; from the same region are also known *Neosciara* PETT., *Caenosciara* LENGERSDORF and the brachypterous *Orinosciara* LENGERSDORF at elevations of 2400-2700 m. Most of these high altitude species breed in soil

Brachycera-Orthorrhapha

The high altitude Stratiomyiidae breed in organic debris, under stones, in humid localities. *Stratiomyia* GEOFFR. and *Odontomyia* MEIG. occur at elevations of 3500 m on the Himalaya. The Rhagionidae (= Leptidae) are common at high elevations as pronounced hygrophiles, under stones covered by wet moss near streams and snow-edge. Both the larvae and the adults of Rhagionidae are active predators on diverse insects. The common high altitude species belong to *Rhagio* FABR., *Hilarimorpha* SCHINER. *Ptiolina* ZETTERSTEDT, *Symphoromya* FELD., etc. The family has been found up to elevations of 3300-3600 m on the Northwest Himalaya. The Tabanids *Sziladynus* ENDERLEIN and *Tylostypia* ENDERLEIN have been recorded at elevations of 3400-3700 m on the Pamirs. The Thereviid *Reinigellum* ENDERLEIN is endemic on the Pamirs and has been recorded at elevations of 4400 m. The larvae are detritus-feeders and occur in soil, but the adult flies are predacious. The Asilidae are generally restricted to the snow-edge on high mountains and the adults are predacious on adult mayflies and on Chironomids. *Cyrtopogon* LÖW and *Stenopogon* LÖW are typical with high altitude Asilids that are endemic on the Pamirs and the Northwest Himalaya. Bombyliidae are represented commonly by species of *Anthrax* SCOP., *Bombylius* LINN., *Systoechus* LÖW and *Argyromoeba* SCHINER at elevations of 4100-4200 m on the Pamirs. The larvae of Empididae feed in organic debris under damp moss near streams and the adults are predacious on Acarina and Collembola. *Rhamphomyia* MEIG., *Tachypeza* MEIG. and *Clinocera* MEIG. may be mentioned as the commonest high altitude genera. A number of species of Dolichopodidae are found at the snow-edge and in the transitional belt immediately above the treeline both on the Alps and on the Himalaya. The adults are predacious on Chironomidae. *Argyra* MACQ., *Rhaphium* MEIG., *Neurigona* ROND., *Dolichopus* LATR. *Eucoryphus* MIK., etc. are some of the more common forms at high altitudes. On the Alps, the family has been recorded up to an elevation of 3000 m and on the Himalaya specimens have been taken even at an elevation of 4000 m (730).

Brachycera-Cyclorrhapha

Aschiza

Syrphidae are the most important family of this subgroup at high altitudes. A number of species are characteristic of grassy meadows and snow-edge at relatively high elevations, ranging from 3000 to nearly 4500 m. The genera represented at high altitudes include *Melanostoma* SCHINER, *Platycheirus* LEP., *Eriozona* SCHINER, *Mallota* MEIG., *Temnostoma* LEP. et SERV., *Milesia* LATR., *Criorhina* MEIG., *Chrysotoxum* MEIG., *Arctophila* SCHINER, *Syrphus* FABR., *Helophilus* MEIG., *Volucella* GEOFFR., etc. The adults of nearly all the high altitude species visit the alpine-zone flowers (730).

Hypocera

The high altitude Phoridae are generally cavernicoles and some of them have been recorded at elevations of 3200 m. *Megaselia* ROND. and *Metopina* MACQ. at elevations of 2900-3200 m are peculiar to the Spanish Sierra Nevada. Other high altitude species belong to *Trupheonura* MALLOCH and *Paraspiniphora* MALLOCH.

Schizophora

The Cyclorrhaphous Schizophora are abundant even at elevations of 5500 m on the Himalaya. The Acalypteratae are represented by Ochthiophilidae, Ulidiidae, Piophilidae, Helomyzidae, Ephydridae, Sciomyzidae, Sepsidae (= Borboridae) and Chloropidae. Mention must also be made of the apterous *Sphaerocera* LATR., *Titanocera* DUN., the apterous *Binorbitalia* RICHARDS, *Sepsis* FALL. at elevations of 4100-4200 m on the Pamirs, the Ortalidae *Melieria* ROBINEAU and *Tenea* MALLOCH at elevations of 4400 m also on the Pamirs. The Ephydridae are hygrophile flies, associated with swamps, marshes, shores of lakes, ponds and snow-edge. Many high altitude species breed in saline and alkaline waters, at temperatures ranging from 43° to 49° C on the Northwest Himalaya, at elevations of 4500-4800 m. BECKER (98), CRESSON (224) and DAHL (228) have given interesting accounts of the habits and distribution of the Ephydridae from the mountains of Middle Asia and the Northwest Himalaya.

The Muscoid flies are extremely abundant at higher elevations and include the dominant Anthomyiidae, Sarcophagidae and Tachinidae. The Cordyluridae (=Scatophagidae) are represented by *Scatophaga* MEIG. at elevations of 4200 m in the Pamirs Region. The adult flies are common flower-visitors on the high-alpine grassy meadows and they also frequent the snow-fields (730). The Anthomyiidae have been found at extreme high elevations. SWAN (1021) has, for example, recorded the interesting find of Anthomyiid flies at elevations of 6100-6200 m in the Mt. Everest area of the Himalaya. LINDNER (697) has described the interesting habits of high altitude Anthomyiidae from the Alps. Among the commonest high altitude muscoid flies, the following should be particularly mentioned: *Chortophila* MACQ., *Limnophora* R.D., *Hydrophoria* R.D., *Anthomyia* MEIG., *Phaonia* ROB., *Trichopticus* ROND., *Hydrotae* R.D., *Bithoracochaeta* STEIN., *Hoplogaster* ROND., *Pollenia* R.D., *Calliphora* R.D., *Brachicoma* R.D., etc.

ORDER 12. THYSANURA

The Thysanura are often abundant on barren rock, inside rock crevices and under stones on the high-alpine grassy meadows. Great swarms of them have been found, in parts of the Northwest Himalaya, to come out and feed on vegetable debris during the early parts of the warm summer nights. Most species are common at elevations of 2500-3600 m. SWAN (1021) observed dark coloured *Machilinus* SILV. to be active at elevations of 5335-5800 m on Mt. Makalu in the Nepal Himalaya. These glacier-fleas absorb the warmth of the sun during the hours of bright sunshine or from the warm rock surface in cloudy intervals, but dart off into cracks and fissures in rock when disturbed. *Ctenolepisma* ESCH. has also been found at elevations of 3600 m on the Himalaya (730). Lepismatids have been recorded at high elevations on the Alps and other high mountains, including also the equatorial East African mountains.

ORDER 13. COLLEMBOLA

The Collembola are *par excellence* the dominant insects of the highest elevations, at which the existence of an insect seems at all possible. Collembola flourish on the

Himalaya at elevations of 6300 m, in localities where there seems hardly a trace of any other kind of insect (730). The most outstanding character of the high altitude Collembola is the truly stupendous number of individuals of numerous species, incredibly large in proportion to what we find in other orders. At whatever elevation we may explore, the individuals of a number of species are sure to be found in enormous masses. In such enormous swarms do they often occur on the high mountain snows, that vast stretches of the snow-fields at very high altitude appear sooty-black even from a distance (676, 730, 812). While the Collembola are fairly common in a variety of situations even near sea-level, their position in the lowland insect life as a whole is insignificant, but at the upper limits of the forest on high mountains they become rather abruptly dominant. The forest-line must indeed be described as a veritable threshold for the world of Collembola—such is indeed the life zone above the forest on mountains. Not only do they exist at much higher elevations on mountains than perhaps every other insect, but, judging from the extensive literature (9, 71, 138, 352, 469, 564, 919, 920, 983, 1109), Collembola are also remarkable for living in very large numbers on ice masses in Greenland and Antarctica. It is also interesting to recall in this connection the fact that these insects are abundant on snow during the winter even near sea-level in different temperate and cold countries of the world.

Their small size enables them to flourish in humid, semi-arid, cold or warm conditions on the plains and at high altitudes they are almost exclusively hygrobiont and cold stenotherm forms. The temperature-humidity conditions seem to be the most decisive limiting factor for these insects. The extreme sensitiveness of the high altitude Collembola to temperatures even slightly higher than their normal micro-environmental temperature result in rapid and fatal desiccation. The dense covering of scales, setae, waxy material and the heavy body pigmentation are some of the more striking adaptative characters at high elevations. A number of the high altitude Collembola are characteristically highly sensitive to light. *Hypogastrura* BOURL. and *Tomocerus* NIC. do not generally seem to require much light and thus prefer dark shelters. Others like *Onychiurus* GERV., *Folsomia* WILL., *Proisotoma* BÖRN. and *Isotoma* (BOURL.) BÖRN. are also highly sensitive to light. Remarkably enough, most high altitude forms that have furca, if at all, rarely spring; the furca-bearing Arthroploeona very rarely spring, but the Symphyploeona do so occasionally.

Ecologically, the high altitude Collembola belong to the following types: i. Inhabitants of the surface of stagnant waters, such as the glacial lakes and melt-water ponds, mostly sparse at higher elevations. Typical examples include species of *Hypogastrura* BOURL., *Onychiurus* GERV., *Proisotoma* BÖRN., *Isotoma* (BOURL.) BÖRN. and *Sminthurides* BÖRN. ii. Inhabitants of springs and glacial torrents, found predominantly under submerged stones along the stream edge. They also occur on stones, covered by moss and algae and wetted constantly by water sprays and dripping water from streams. The species belong mostly to *Onychiurus* GERV., *Proisotoma* BÖRN., *Isotoma* (BOURL.) BÖRN., *Isotomurus* BÖRN. and *Orchesella* TEMPL. iii. Inhabitants of damp lichen and moss on sheltered rock surfaces belong to species of *Hypogastrura* BOURL., *Isotomurus* BÖRN. and *Orchesella* TEMPL. iv. Inhabitants of snow and ice-edges are extremely cold stenotherm and hygrobiont and relatively scarce during the daytime but common at night. They usually occur under stones, inside cracks and crevices in rock, in ice crevasses and other similar situations at the snow-edge. Gigantic swarms of *Hypogastrura* BOURL., *Onychiurus* GERV., *Proisotoma* BÖRN.,

Isotomurus BÖRN and *Lepidocyrtus* BOURL. occur at the snow-edge. v. Inhabitants of glacier moraines are similar to the snow and ice-edge forms and include species of *Lepidocyrtus* BOURL., *Isotomurus* BÖRN. and *Orchesella* TEMPL. vi. Inhabitants of snow and ice, above the permanent snowline, at elevations of 6000 m include *Isotoma* (BOURL.) BÖRN., *Hypogastrura* BOURL. and *Proisotoma* BÖRN. vii. Inhabitants of vegetation mats are mostly *Entomobyra* ROND., *Bourletiella* BANKS and *Sminthurides* BÖRN. viii. Inhabitants of decaying vegetable matter belong mostly to *Onychiurus* GERV. ix. Myrmecophile *Cyphoderus* (NIC.) TULLB., *Entomobrya* ROND., *Lepidocyrtus* BOURL. and *Onychiurus* GERV. are mostly restricted to the zone immediately above the forest-line, especially to the grassy meadows. x. Petrophile and semi-xeroxene species of *Entomobyra* ROND. and *Orchesella* TEMPL. may be sparsely found on the rock surface. Although considerable numbers of species are fungicole in the upper reaches of the forest, true fungicole Collembola do not seem to be important elements in the subalpine and alpine zones.

The ecological types of the Collembola from the Alps and their characters have been described in detail by GISIN (409). He recognizes a series, in which the powers of movement and springing, the intensity of the body pigmentation, the number of ommata, the lengths of the furca, antennae and the body pubescence and the general body size increase, as we pass from the typical soil-biont to the surface forms.

The true geobiont high altitude Collembola lack ommata or have only reduced ones and the body pigment is also confined to the reduced ommata, their furca is greatly reduced, the antennae, legs and the general body size are short. Being typically blind, they have other complex sense organs. The typical members belong to *Folsomia* WILL. and *Onychiurus* GERV. The mesophiles are found on the soil surface and have relatively better developed body pigment, moderately long antennae, dense clavate setae or forked hairs on the tibiotarsus. The phytophile forms have 8-8 ommata and long antennae. Other important contributions to our knowledge of the high altitude Collembola include ABSOLON (1, 2), BAIJAL (76-78), CARL (185, 186), DELAMARE-DEBOUTTEVILLE (244), DENIS (248), GISIN (409), HANDSCHIN (467-473), LATZEL (677-678), SCHUBERT (937), STACH (990-993), STEINBÖCK (1003-1008) and YOSHII (1130, 1131).

B. OTHER ARTHROPODA

An account of the high altitude insect life would perhaps be incomplete without some reference to at least the more important Arthropoda, other than insects, particularly the Crustacea, Arachnida, Chilopoda and Diplopoda that are nearly always ecologically closely associated with the high altitude insects. Some of them often represent dominant members of diverse high altitude insect communities. Like the typical high altitude insects, these arthropods are also predominantly cryobiont, hygrobiont and geobiont forms that occur under stones, in the immediate vicinity of snow and ice. They also exhibit the essential ecological specializations, which are so very characteristic of the high altitude insects. A great proportion of them are active predators, but many are also carrion feeders and still others are parasitic on the high altitude insects. Some remarkable aquatic species of these arthropods are also known from relatively high elevations.

Crustacea

The greatest bulk of the high altitude Crustacea are aquatic and occur in the glacial lakes at very high elevations on the Alps, the Himalaya and the Andes. The Branchipod *Artemia salina* (LINN.) occurs, for example, at elevations of 3600-5000 m in glacial lakes on the Himalaya and on other Middle Asiatic high mountains. *Branchinecta* spp. are particularly abundant in the melt-water ponds at elevations of 3600 m, and sometimes even at 5220 m on the Northwest Himalaya and the Pamirs Region (730). *Gammarus pulex* (LINN.) is a widely distributed amphipod that is found in various high altitude glacial lakes on the Himalaya, at elevations ranging from 3700 to nearly 5335 m. A number of other interesting high altitude Crustacea have been described by BOND (140), DEES (239), FREUDE (384), UÉNO (1054) and others (*vide* Bibliography). *Mesoniscus* spp. and *Protracheoniscus* spp. represent some of the more common hypsobiont Isopoda from the Alps and the Northwest Himalaya. Endogeous Isopoda have been collected at elevations of 4800 m on the Northwest Himalaya.

ARACHNIDA

Scorpionida

Although otherwise a typically thermophile lowland group, scorpions are often found under stones, in association with diverse hypsobiont insects like *Bembidion* LATR., *Nebria* LATR., *Atheta* THOMS., Acarina and Collembola, on the high-alpine grassy meadows and other situations on the Alps and the Himalaya. *Buthus* LEACH occurs, for example, at elevations of 4000 m on the Himalaya. *Chaerilus* SIMON is common at elevations of 3600-3900 m on the Northwest Himalaya. The highest elevations at which a scorpion has so far been found in the world is 4300 m; at this elevation, *Scorpiops rohtangensis* MANI is found on the Northwest Himalaya, under snow-covered stones.

Chelonethida

The pseudoscorpions are common under stones, under moss and also under various organic debris on the grassy meadows. They are most usually found in close association with species of Acarina, Collembola, Dermaptera, Salticid spiders and sometimes also Carabidae. *Neobisium (Neobisium) jugorum* (KOCH) is found under stones in the high-alpine zone of the Alps and *N. (N.) nivale* (BEIER) occurs on snow-fields on the Spanish Sierra Nevada. *Microcreagris kaznakovi* (REDIK.) occurs on the Pamirs and on the Northwest Himalaya, at elevations of about 4300 m. *Centrochthonus kozlovi* (REDIK.), widely distributed in Middle Asia and in Tibet, occurs at elevations of 3600 m on the Himalaya. Other hypsobiont pseudoscorpions belong to the genera *Microbisium* CHAMBERLIN, *Microcreagris* BALZAN, *Atemnus* BERLESE, etc. *Tetanotemnus montanus* BEIER occurs on Mt. Kilimanjaro. The phenological peculiarities of some high altitude pseudoscorpions have been recently described by BEIER (102).

Araneida

Spiders flourish at remarkably extreme elevations, often far above the permanent snowline, amidst ice, glacier and snow, on nearly all the important high mountains of the world. The dominant groups are, as may be expected, typically terricole families like Thomisidae, Gnaphosidae, Lycosidae and Salticidae, but some Argiopidae, Agelenidae, Drassidae, Clubionidae and Linyphiidae are also found at elevations of 3000-5200 m. The greatest majority of the hypsobiont species belong, however, to genera that are common enough and also generally widely distributed even on the lowlands. It is, therefore, sometimes difficult to consider any single genus as typically and exclusively hypsobiont. The species are, however, often endemic and naturally exclusively mountain autochthonous. The greatest majority of the species occur typically under stones, inside rock crevices and micro-caves. Most species are hygrophiles and are concentrated near melt-water torrents and glacial lakes, but some more or less semi-xerophiles have also been reported to occur on barren rock surfaces. Most spiders at high altitudes dart on the surface of the glacier or on snow-fields during the hours of bright sunshine, in order to capture their prey. Some of them, particularly the Lycosidae and Salticidae, are typical of the aeolian zone on the Himalaya (730, 1029). Spiders have been recorded at elevations of 4500 m on the Alps, at 4250 m on Mt. Kilimanjaro and even at an elevation of 6700 m on the Nepal Himalaya. Many hypsobiont species of spiders belong to genera that are known from the subarctic and arctic regions (320). Descriptions of high altitude spiders may be found in AUSSERER (60), CAPORIACCO (182), DENIS (246, 247), GLENNIE (410), HOLM (539), DE LESSERT (689), SCHENKEL (922) and others listed in the bibliography.

Among the large numbers of species recorded from different high mountains, the following are worthy of special mention: Drassidae: *Drassodes* WESTERLING at elevations of 2900-3000 m on the Alps and on the Spanish Sierra Nevada; *Haplodrassodes* CHAMBERLIN at an elevation of 2400 m on the Spanish Sierra Nevada. Lycosidae: *Pardosa* C. KOCH at elevations of 2500-3200 m on the Alps and on the Spanish Sierra Nevada, and at an elevation of 4000 m on the Himalaya; *Lycosa* LATR. at an elevation of 2200 m on the Spanish Sierra Nevada, but often even above 6000 m on the Himalaya*. Gnaphosidae: *Gnaphosa* LATR. at elevations of 2900-3000 m on the Alps and on the Spanish Sierra Nevada, and up to 4000 m on the Himalaya. Thomisidae: *Xysticus* C. KOCH and *Thanatus* C. KOCH at elevations of 2400-3200 m on European mountains and at an elevation of 4000 m on the Himalaya. Linyphiidae: *Diplocephalus* BERTKAU, *Macrargus* DAHL, *Microneta* MENGE, *Micryphantes* C. KOCH, *Stylotector* SIMON, *Cornicularia* MENGE, *Hilaira* SIMON, *Lephthyphantes* MENGE and *Erigone* AUDOI at different elevations between 2500 and 3100 m on the Alps. Salticidae: *Heliophanus* C. KOCH, *Aelurillus* SIMON and *Euophrys* KOCH. Clubionidae: *Clubiona* LATR. at an elevation of 4250 m on Mt. Kilimanjaro and *Liocranum* L. KOCH at an elevation of 3460 m on the Spanish Sierra Nevada. Theridiidae: *Theridion*

* In reporting on the collections of the Cho-oyu Expedition, TIKADER [TIKADER, B. K. 1964. *Rec. Indian Mus.*, 59(3): 257-267] has recorded *Tegenaria lunakensis* TIK., at 5180 m in the Nepal Himalaya, in addition to other species like *Xysticus roonwali* TIK., *Lycosa tatensis* TIK., *L. nigrotibialis* SIMON, a widely distributed species from India and Burma, on the Kumbu Glacier at 5334 m.

WALCKENAER at elevations between 2400 and 3400 m on the Spanish Sierra Nevada. Agelenidae: *Hahnia* C. KOCH. Argiopidae: *Nephila* LEACH and *Zilla* C. KOCH.

Acarina

The Acarina are the most dominant Arachnida at high altitudes and are abundant both as species and as individuals. Like the spiders, the mites also flourish at extreme limits of life, at elevations over 6300 m on the Himalaya and at 5000 m on the Peruvian Andes. Most high altitude genera of mites are also found in the high north latitudes, in the subarctic, arctic and also in the antarctic regions (228a, 229, 320, 427a, 570, 571, 572). The great bulk of the hypsobiont mites are geobiont, hygrophiles that occur typically under stones, on moss, in soil, on snow, etc. The majority of the high altitude mites are also parasitic on the high altitude insects; indeed nearly every species of the high altitude insect is subject to the attack of one or more species of parasitic mites. Descriptions of high altitude mites may be found in DADAY (226), DEES (239), EVANS (324), LUNDBLAD (710), PSCHORN-WALCHER (848), SCHWOERBEL (949), WALTER (1098) and WILLMANN (1113).

The typical high altitude species belong to *Caeculus* DUFOUR, *Trombidium* FABR. *Rhyncholophus* DUGÈS, *Erythraeus* BERLESE, *Rhagidia* THORELL, *Tetranychopsis* CANESTRINI, *Bryobia* KOCH, *Damaeus* KOCH, *Ceratoppia* BERLESE, *Hypoaspis* CANESTRINI, *Cyrtolaelapis* BERLESE, *Pergamasus* BERLESE, *Eugamasus* BERLESE, *Bdella* LATR., *Cyta* V. HEYDEN, etc. Some, like *Africoribates* EVANS from Mt. Kilimanjaro, are endemic mountain autochthonous genera.

A large number of Hydracarina are also known in the glacial lakes, springs and in torrential streams, at extreme high elevations, on different high mountains. *Halacarus* GOSSE, *Neocalonyx* WALTER and *Hygrobates* KOCH occur at elevations of 5000 m on the Peruvian Andes. LUNDBLAD (710) has given an interesting account of the high altitude Hydracarina from the Himalaya. *Eylais hamata* KOENIKA occurs at an elevation of 4340 m on the Northwest Himalaya and is also known at an elevation of 2450 m from the Alps. *Protziella hutchinsoni* LUNDB. occurs in springs at elevations of 4200 m on the Northwest Himalaya, with a water temperature of 7-8° C. *Calonyx montanus* LUNDB. and *Calonyx flagellatum* LUNDB. occur on stones in torrential streams at an elevation of about 3000 m on the Northwest Himalaya, with the water temperature at 19° C. *Parathyas primitiva* LUNDB. and *Kashmirothyas hutchinsoni* LUNDB. are other species recorded from the Northwest Himalaya. On Norwegian mountains, about half a dozen species have been recorded at elevations of 1200 m. On Mt. Elgon in equatorial East Africa, four species occur at elevations between 2400 and 3200 m, two species each at elevations between 3200 and 3300 m and 3300-4200 m. It is important to observe that *Eylais hamata* KOENIKE, mentioned above as having been found at an elevation of 4340 m on the Himalaya, is also known to occur at sea-level. It is indeed not a typical mountain species, but essentially a widely distributed form that seems to have secondarily become hypsophile. *Protziella* LUNDB. and *Kashmirothyas* LUNDB. are endemic to the Northwest Himalaya. Some extremely interesting subterranean water mites like *Stygomomonia* SZALAY, *Lethaxenia* VIETS, *Atractides* THORELL, *Porohalacarus* THORELL, *Torrenticola*, PIERSIG, *Stygothrombium* VIETS, etc. are also known at high elevations on the Alps (949).

CHILOPODA-DIPLOPODA-SYMPHYLA COMPLEX

A number of species of Geophilidae, Scolopendromorpha, Lithobiomorpha, Oniscomorpha, Juloidea (especially *Leptoiulus* spp.) occur at relatively lower elevations of 1200-1800 m on most mountains. *L. (L.) alemannicus alemannicus* VERH. is an endemic form from the Alps that occurs under stones, at elevations of 1300-2800 m. *L. (L.) simplex simplex* VERH. is an alpine-zone to high-alpine zone form in the East Alps. *Hypsoiulus alpivagus* VERH. (Juloidea) occurs at the snow-edge, at elevations of 2450-2800 m on the Swiss Alps. *Taueriulus aspidiorum* VERH. is known so far only from the Hohe Tauern and the Radstädter Tauern on the Alps, at elevations of 2450 m. *Leptophyllum nanum* LATZ occurs at an elevation of 2250 m on the Tyrolean Alps. The Chilopoda are generally associated with Coleoptera under snow-covered stones and are predators on Trichoptera, Plecoptera and other insects. Scolopendridae, Scutigeridae and Geophilidae are the principal types at high elevations on the Alps and the Himalaya. *Lithobius* LEACH occurs commonly at elevations of 3500-5200 m on the Himalaya. The Chilopod *Cryophilus alaskanus* CHAMBERLIN has been recorded from under stones on the high-tundra zone on Alaskan mountains. Other high altitude Diplopoda belong to *Julus* LINN., *Ceratosoma* VERHOEFF, *Trimeropheron* ROTHENBÜHLER, *Trimerophorella* VERHOEFF, *Polydesmus* BOISD., *Orotrechosoma* VERHOEFF, *Leptoiulus* VERHOEFF, etc. The Symphyla *Scutigerella* RYDER occurs at an elevation of 2700-2800 m on the Alps and at about an elevation of 3000 m on the Himalaya. Descriptions of high altitude species of the complex may be found in the works of ATTEMS (58), BIGLER (124, 125, 126), EMERTON (320), LIGNAU (692) and VERHOEFF (1081, 1082).

CHAPTER VI

DISTRIBUTION OF HIGH ALTITUDE INSECTS

The peculiarities of distribution of high altitude insects, with special reference to the Alps and other principal European mountains, have been discussed by a number of workers like BÄBLER (74), FRANZ (368, 369, 370, 378), HOLDHAUS (526, 527, 530, 531, 535), JANETSCHEK (565, 566, 567), LINDROTH (698-702) and others mentioned in the bibliography. MANI & SANTOKH (733, 735) have dealt with some of the important aspects of the distribution of high altitude insects from the Himalaya and MANI (730) has recently given an outline of the fundamental patterns of distribution of insects at high altitudes. The outstanding peculiarities of the distribution of high altitude insects are closely correlated with the high altitude environment, the high ecological specializations of the insects, the geographical position of the mountain, its orogeny and tectonics, the distribution of the high altitude plants of the mountain, the Pleistocene glaciations and other complex factors. We shall consider here in broad outline some of the major factors governing the distribution of high altitude insects, the outstanding peculiarities of their vertical and horizontal distributions, the faunal and biogeographical characters and the evolution of high altitude insects in general.

I. FACTORS GOVERNING THE DISTRIBUTION OF HIGH ALTITUDE INSECTS

The factors governing the distribution of high altitude insects differ within wide limits on different mountains and in different parts of the world, depending partly on the ecological conditions prevailing at present and partly on the past distribution. Compared, however, with the distribution of insects in the plains, the limiting factors in the distribution at high altitudes are strikingly characteristic.

To begin with, it must be observed that the extreme ecological specialization of the high altitude insects severely restricts their means of dispersal. The predominantly flightless and apterous conditions, the very pronounced geophily and geobiont habits and the restriction of all active development and movements to the brief summer on most high mountains are some of the major factors that retard rapid and wide dispersal of species. The means of dispersal of the high altitude insects are, therefore, mostly passive in the majority of cases. Even in Diptera and Lepidoptera*, which have largely retained their powers of flight even on the highest summits, active dispersal is by no means more rapid and extensive than among other high altitude insects. The localization of the food plant of the larvae or of the host insects, besides numerous other ecological conditions, restricts the active dispersal of these species also. The fact that most of the high altitude insects are bound to the snow-edge acts against a rather wide dispersal, even if active flight over relatively extensive area is assumed.

* Some species of Lepidoptera migrate over high mountains.

In the plains of the tropics and the temperate regions of the world, even if not within the subarctic and arctic areas, a great majority of Coleoptera are winged and only about 5% of these species are apterous and incapable of flight. The high altitude Coleoptera, on the other hand, are characterized by their very pronounced wing atrophy and it has been estimated that over 70% of all the high altitude Coleoptera known so far from the world are apterous. Many of them are also microphthalmous or even totally blind and never come on the surface of the ground. The migration of the blind and apterous terricole species is, therefore, largely governed by factors of their habitat, such as the structure of the soil, the soil temperature and moisture.

The high altitude Trechini (Carabidae) are, for example, remarkable for being small, compact and apterous, the last character being closely correlated with their occurrence under stones in the vicinity of snow-fields. Outside the snow-fields, they are found, as mentioned already, only in humid situations. Where the ground is dry, the species occur deeper in the soil. *Trechus glacialis* HEER can, for example, only be collected from spaces under deeply buried boulders after continued fair weather on the East Alps. The species that occur under such conditions show greater reduction of their compound eyes than in the shallow layer forms and often, as in the case of true cavernicolous species, they are also totally blind. To summarize, the Trechini that constitute such an important group of high altitude Coleoptera on most high mountains, possess little or no migration capacity and are, therefore, well suited to illustrate the biogeographical problems at high altitudes. In the case of the greatest majority of high altitude insects, dispersal by active migratory movements must be considered as quite insignificant. The dispersal of the high altitude insects is, therefore, almost exclusively by passive means. As we shall see, most high altitude insect faunas have evolved *in situ* on the mountain and the influence of immigration is almost negligible.

The discontinuity of distribution of mountains, the unevenness of the ground on mountains and the general ruggedness of the terrain are other effective barriers to dispersal and migration. Finally, the species that exist on high mountains are, like the species on oceanic islands, isolated effectively by the barrier of the intervening lower elevations. Even in the case of the high altitude species that sometimes descend to the foot of the mountain, the presence of loose soil and the absence of rock in the plains is a barrier in distribution. The other ecological conditions that serve as effective barriers are the prevailing environmental conditions at high altitudes. The distribution of vegetation, its composition and other characters of the flora of a mountain region govern the distribution of the high altitude insects, so that it is also necessary to take into consideration the general vegetational character of a mountain in discussing the biogeographical peculiarities.

In the greatest majority of the high altitude insects, the passive means of dispersal include the relatively slow movements of glaciers and erratics and the rolling down of stones in torrents. As may be readily understood, altitude by itself is no barrier to dispersal, as indeed altitude is not even an ecological factor. The complex environmental effects caused by high altitudes profoundly influence the distribution. For a species characteristic of a definite zone of altitude, the absence of isohypse sets a limit to distribution. It is also remarkable that some of the topographic and physical barriers to dispersal have at the same time routed the dispersal of immense faunas and faunal elements at high altitudes.

The limiting factors comprise, therefore, the ecological specializations of the insects themselves, the climatic factors at high altitude, the topographic and physiographic barriers, the vegetational conditions and finally also the edaphic factors. The ecological importance of the edaphic factors in the distribution of high altitude insects is seen not only in the case of the typically terricole and endogeous forms, but also in the case of the planticole and aquatic types. The wide variations of the geological conditions ensure a diversified edaphic factor. It may also be recalled in this connection that most high mountains, with rich and well differentiated insect life, are typically Tertiary mountains. The rocks, which on weathering produce soil rich in nutrient minerals and of high water retention capacity, contain as a rule a much larger and a more diversified montane fauna than those which on weathering contain little minerals and have also a low capacity for retention of water. Faunistically rich soils are generally derived from limestones, basic eruptive rock, quartz-deficient sandstones and conglomerates, calcareous marl or slate, most basic crystalline shales, etc. Dolomites, quarzite, quartz-rich sandstones and conglomerates, highly acidic eruptive rocks and crystalline shales and poorly weathering shales give rise to faunistically poor soils. Rocks which have undergone more or less prolonged weathering and are therefore rich in nutritive minerals support a much larger number of forms than unweathered rocks. Even the gradient of the mountain slope has an important influence on the distribution of high altitude insects. Localities not excessively strongly sloping appear to be most favourable for an abundant and diversified high altitude insect life. The steeper slopes of most high mountains have a poorer fauna than the gentler slope, since the former is likely to become drier more rapidly and also retains much less of the winter snow-cover.

2. VERTICAL DISTRIBUTION

The vertical distribution of high altitude insects is characterized by the progressive and often rapid fall in the abundance of species with increase in elevation, the extreme high elevations of permanent existence and the variable width of the altitudinal range of distribution, the dominance at certain zones of altitudes and the more or less pronounced altitudinal stratification of the species. Although these distributional peculiarities are closely correlated with, and strictly speaking, also inseparable from the peculiarities of horizontal distribution, we shall, for the sake of convenience, consider the vertical distribution separately at first.

Taking first the fall in abundance of species with increase in elevation, it is interesting to observe that while this is noticeable even within the forest-zone, there is a pronounced abruptness in the fall at the forest-line. From above the forest-line, the impoverishment of the species becomes so rapid that at extreme limits of a group on a mountain, there are perhaps no more than one or two species. We find, for example, that as we ascend from elevations between 3000 and 3500 m to the zone between 3500 and 4200 m on the Northwest Himalaya, the number of species of mountain autochthonous Carabidae is reduced to nearly one-eighth. At elevations above 4200 m, it is less than one-tenth that of the abundance immediately above the forest-line. The hypsometric analysis of the Coleoptera from Mt. Kilimanjaro is equally interesting and is summarized in table 17.

TABLE 17

Hypsometric analysis of the high altitude Coleoptera from Mt. Kilimanjaro

Altitude zone in m	Percentage of species
2600–3000	37.0
3000–3500	4.6
3500–4250	67.0
4250 and over	16.0

At elevations above 4250 m, the abundance of species is only one-seventh that of the zone below 3500–4250 m and one-eleventh that of the zone of 2600–4250 m. Only a single species extends over a wide range of altitudes, from 2600 m to 4250 m. According to the observations of PRÜFFER (846), the fall in the abundance of species of Lepidoptera on the Polish part of the Tatra Mountains with the increase in elevation is very striking. His data are summarized in table 18:

TABLE 18

Hypsometric analysis of the high altitude Lepidoptera from the Polish part of the Tatra Mountains

Altitude zone in m	Percentage of species
900–1545	74.0
1545–1789	16.0
1960–2250	8.0
2250–2663	2.0

It is not only in the case of the strictly terrestrial species, but also in the case of the high altitude aquatic forms that we may observe a significant fall in the abundance with increase in elevation. The hypsometric analysis of the Hydracarina on the Himalaya is of particular interest in this connection, Table 19.

Extremely similar conditions have also been described in the case of the vertical distribution of Hydracarina on Mt. Elgon in the equatorial East Africa (*vide* Chapter VII).

The rate of fall in the abundance of species with increase in elevation differs on different mountains and even on the same mountain it tends to become higher as the elevation increases. A number of complex local factors have a profound modifying influence on the rate of fall in abundance. As a rule, under otherwise more or less similar conditions, the rate of fall in the abundance of species is less pronounced on massive mountains than on isolated ones. Pronounced and abrupt differences appear in localities where deep gorges push the upper limits of the forest higher up and result in an unusually low level of unmelted winter snow patch during the summer.

TABLE 19

Hypsometric analysis of the Hydracarina from the Northwest Himalaya and from the Alps

Altitude zones in m	Percentage of species
Northwest Himalaya	
1500–1600	30.0
1600–2800	30.0
2800–3000	20.0
3000–4100	10.0
4100–4300	10.0
Alps	
1800–2000	37.0
2000–2200	28.0
2200–2400	20.0
2400–2600	13.0
2600–2800	2.0

The rate of fall in the species abundance differs also on the north and south slopes of the mountains with a general east-west trend, especially in the northern hemisphere. Owing to these differences and to the different rates of fall in abundance in different groups of insects, we observe an interesting increase in the dominance of different orders and families like Carabidae and Staphylinidae at higher elevations.

In the fall in the absolute abundance of species with increase in elevation we may observe not only a disappearance of the species typical of lower elevations, but also a partial or even a total replacement by other species that do not occur at lower elevations. There is thus a more or less pronounced, though sometimes overlapping altitudinal stratification of species. This stratification deals with a qualitative difference in the altitudinal distribution, in addition to the quantitative differences mentioned earlier. As we ascend higher and higher above the limits of the forest on any mountain, we find a most interesting succession of species; some species replace numerous others, which disappear at successively higher elevations.

On the Himalaya, for example, the species complex at elevations between 2500 and 3000 m is different from that between 3000 and 3500 m and this latter is again essentially different, both qualitatively and quantitatively, from that characteristic of elvations between 4000 and 4500 m. Although the mean altitudinal range is largely specific to a group, and even in the same species it is subject to considerable variation, exceedingly few species occur over a wide range of elevations on either side of the forest-line. On the Himalaya, for example, *Bembidion bracculatum* BATES, *B. fuscicrus* MOTSCH. and *B. irregulare* NET. are perhaps the only Coleoptera that occur at elevations from nearly 1800 to 4500 m. There is on most high mountains unmistakable evidence of intrusion of transitional elements from the upper limits of the forest-zone into the biome immediately above the forest-line, but only exceptionally do we find a similar intrusion of the typically high altitude elements into the upper reaches of the forest-zone. On the Northwest Himalaya, for example, the Coleoptera of the transitional zone immediately above the forest-line show distinct evidence of such instrusive elements from below, reaching sometimes even up to an elevation of 3500 m. Over 75% of these species disappear altogether at elevations above 3500 m and are replaced by others that are, strictly speaking, typical high altitude elements, characteristic of the zone reaching up to an elevation of 4000 m, and are never found at elevations below 3000 m. Nearly one half of the species that are characteristic of elevations between 3500 and 4000 m do not occur above 4000 m and in their place we find others that are also never found at lower elevations. The following hypsometric qualitative analysis of high altitude species of *Bembidion* LATR. from the Northwest Himalaya is summarized from MANI & SANTOKH (733):

I. Species characteristic of altitude zone between 1800 and 3700 m: *B. algidum* ANDR., *B. betegara* ANDR., *B. exaratum* ANDR., *B. himalayanum* ANDR., *B. orinum* ANDR. and *B. satanas* ANDR.
II. Species characteristic of the altitude zone between 1800 and 4200 m: *B. bracculatum* BATES, *B. fuscicrus* MOTSCH. and *B. irregulare* NET.
III. Species characteristic of the altitude zone between 2400 and 3700 m: *B. aquilum* ANDR., *B. gagates* ANDR. and *B. livens* ANDR.
IV. Species characteristic of the altitude zone between 2400 and 4300 m: *B. hasurada* ANDR. and *B. luntaka* ANDR.
V. Species extending from an elevation of 2400 to 4800 m: *B. pluto* ANDR.
VI. Species characteristic of the altitude zone between 3000 and 3700 m: *B. ajmonis* NET., *B. beesoni* ANDR., *B. ladakense* ANDR., *B. ladas* ANDR., *B. leve* ANDR. and *B. pamiricola* LUTCH.
VII. Species characteristic of the altitude zone between 3000 and 4300 m: *B. bucephalum* BATES.
VIII. Species characteristic of the altitude zone between 3700 and 4300 m: *B. nivicola* ANDR. and *B. petrimagni* NET.
IX. Species characteristic of the altitude zone between 4300 and 5000 m: *B. hutchinsoni* ANDR.

The altitudinal overlapping of different species, which we observe in a simple hypsometric analysis, is in reality only an illusion and must not, therefore, be interpreted as a real ecological or a spatial overlap. Indeed there is rarely an overlapping of the distributional range, mainly because most true high altitude species, at least those that are autochthonous on the Tertiary mountains, are severely restricted to greatly isolated summits. In absence of accurate data regarding the horizontal distribution of the species, our interpretation of vertical distribution is thus bound to be misleading.

While the true hypsobiont species never descend below the forest-line, their upper altitudinal limits of permanent existence differ in different groups and on different mountains. Even the early contributions on the vertical distribution of insects by SCHLAGINTWEIT (924), TRENTINAGLIA-FELVENBERG (1046) and others recognized these differences. The mean for the Himalaya is about an elevation of 5120 m and is much higher than the mean for all other mountains in the world. The absolute maximum altitude at which insects have so far been found to be permanent residents and to breed is about 6800 m on the Himalaya. The highest altitude record for different groups of high altitude insects are summarized below (Table 20).

The highest altitudes are reached naturally by the strictly endogeous and predatory types, especially the Collembola, Acarina, spiders, Coleoptera and Diptera rather than by those that belong to the communities of the open surface. The adult Lepidoptera that are found at elevations far above 6000 m breed mostly about this altitude. The maximum altitudes of permanent existence of a group is higher on the Tertiary

TABLE 20

The highest altitude records for different groups of hypsobiont insects and other arthropods, in m

Order/Family	Himalaya	Andes	Kilimanjaro	Other mountains
Ephemerida	5200			2000 New Zealand
Plecoptera	5030			
Orthoptera				
Tettigoniidae	4900	4900		4420 Mt. Whitney
Acrididae	5490			
Dermaptera	4500		4880	
Heteroptera	5335			
Coleoptera	5600			
Carabidae	5300	4820		
Staphylinidae	5600			
Hydrophilidae	5400			
Tenebrionidae	5600			
Chrysomelidae	4600		4270	
Curculionidae		4976		
Formicidae	4800			
Trichoptera	5000			
Lepidoptera	6000			
Diptera	6300			
Tipulidae		5600		
Thysanura	5800			
Collembola	6300			
Scorpions	4300			
Spiders	6700		4250	4500 Alps
Acarina	6100			

mountains than on older mountains and also on the more massive mountains than on isolated eminences. It is also interesting to observe that much higher elevations are attained by a group in glaciated areas than in the unglaciated. The north temperate mountains support insect life at much higher elevations than the mountains at comparable latitudes in the south temperate regions of the world.

As may be readily observed, in most cases the mean of the maximum altitude of different groups of arthropods lies about the permanent snowline; but this certainly is not the absolute upper limit of existence of insects. The treeline no doubt marks a sharp lower boundary for the true hypsobiont insects, but diverse high altitude species flourish at much higher elevations than the permanent snowline. Permanent snow and ice are in no sense to be taken to rule out the existence of insect life; insects at these extreme elevations may be sparse and rather restricted in kind, but not altogether absent.

Mention may be made here of the observations of MANI (730) on an interesting peculiarity of the vertical distribution of high altitude insects on the Northwest

Himalaya. The highest altitudes at which the species of different orders exist do not necessarily indicate their dominance at these extreme limits. The altitudinal zone of relative dominance is in most cases at about 4200 m above mean sea-level. This

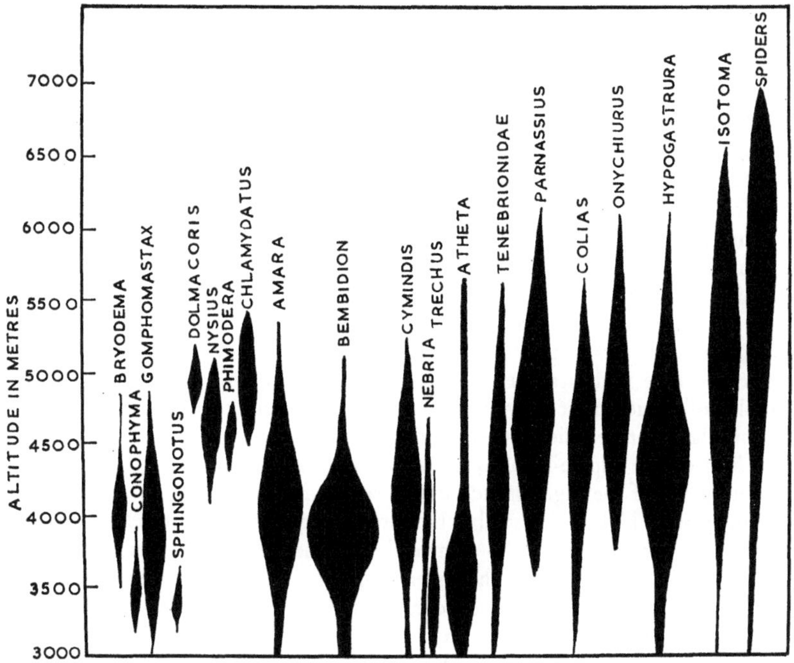

Fig. 29. The altitudes of dominance and the highest altitudes of permanent existence of different high altitude insects on the Himalaya

altitude is indeed a measure of the maximum frequency of occurrence of the largest numbers of species of a group and represents the zone in which the communities containing the member species are also dominant. Communities with species of *Bembidion* LATR. as the dominant members are dominant, for example, at an elevation of about 3900 m. It is at this elevation that we may observe a dominance of the exclusively hypsobiont species of the genus. Although somewhat larger numbers of species of this genus may be found at lower elevations, it is due not to the presence of true hypsobiont species, but to the occurrence of considerable transitional elements from the upper reaches of the high forest-zone. The mean altitudes of dominance of different high altitude insects on the Northwest Himalaya are summarized in fig. 29.

3. HORIZONTAL DISTRIBUTION

The horizontal distribution of high altitude insects is characterized by the concentration of relatively few species in greatly isolated, discontinuous and localized patches. Most high-alpine and nival zone species on all the principal mountains are typically restricted to extremely small areas, frequently around single summits or ridges. They occur in greatly isolated, irregularly spaced, unequal and severely localized pockets. These pockets are generally small and restricted to islets of optimal conditions, separated by extremely wide stretches of unfavourable conditions. The species may not necessarily always be found in every locality, where apparently optimal conditions obtain and where we should ordinarily expect the species. The horizontal distribution of the typically high altitude species is, therefore, characteristically patchy, non-random and is thus of the contagious type, with a more or less pronounced tendency for aggregation. The non-randomness is evidently the direct consequence of the habitat irregularity and the spacing of populations is also, therefore, irregular. The patches tend to be greatly localized in well defined areas, determined by characteristic topographical peculiarities. Both localization and discontinuity favour isolation, but at the same time the topographic and physiographic barriers seem to have also routed the distribution of many species. A number of species are very significantly and exclusively localized together in the same area, resulting in their pronounced concentration in small, isolated pockets in distinctive localities. There is thus a very characteristic clustering together of a distinctive complex of species, most of which are not found outside the area of their concentration. They are true indicator species of the locality.

At least on the Alpine-Himalayan Systems of Tertiary mountains, the localized concentrations of the high altitude insects reveal a most significant tendency for massing around high peaks, so that the intervening areas of lower elevations are mostly sparsely populated or also completely lack the species. A massing around on high peaks is not only true of the endemic species, but also of other species to a considerable extent. The massed concentrations of the high altitude species are also conspicuously localized in areas which were more or less heavily glaciated during the Pleistocene. These glaciated areas have indeed typically indicator species both on the Alps and on the Himalaya. There is also a pronounced tendency for massing of the species in the immediate vicinity of the present-day glaciers.

On most high mountains, two broad categories of distributional patterns may be recognized, viz. forms widely distributed in different mountain regions and others that are restricted to only particular mountain systems and are in a sense indicator species of the mountains. Some typically high altitude genera are of nearly cosmopolitan distribution, but others are restricted to the mountain; we have accordingly widely distributed genera and endemic genera, with intermediate transitional types. The genera like *Trechus* CLAIRV., *Bembidion* LATR., *Pselaphus* HBST. are cosmopolitan mountain types, but numerous species of these genera are found even on the lowlands in many parts of the world, especially in the north temperate regions. The Holarctic genus *Carabus* LINN. extends southwards to the north Mexican mountains and has even transgressed into the Oriental Realm and contains some lowland and forest species. The genus *Erebia* DALM. has also some lowland species. The genera with restricted distribution are *Agelaea* GEN. on the mountains of Sardinia, *Synuchidius* APFB. on the mountains of Albania, *Anchomenidus* HEYD. on the Austrian Alps, *Coryphiodes* BERNH. on the Rodna Mts., *Scotoplectus* REITT. on the Kraiste and Croatian mountains, *Trogaster* SHARP on the mountains of Corsica and the northern Apennines, *Ablepton* FRIV. on the Banat Mountains, *Carpathobyrrhulus* GGLB. on the north and east Carpathians, *Analota* BRUNN. from Schlern to Arlberg on the Alps, the

Middle Apennines and the Jura Mountains, etc. The Himalaya and the Pamirs Region have numerous severely restricted types in Orthoptera, Heteroptera and Tenebrionidae (*vide* Chapters IX and X). A number of hypsobiont genera have wide but discontinuous distribution. *Broscosoma* ROSH. (Carabidae) is known, for example, from Mt. Lessini, Mt. Baldo, Cima Tombea, Mt. Cadria, the Sikkim Himalaya, north Sze-tschuan (3050–3960 m above mean sea-level) and Japan (2133–2438 m above mean sea-level). The Staphylinid *Trigonurus* MULS. occurs on the Maritime Alps, the Basses Alps, Caucasus and on the North American Cordillera. The Silphid *Rybinskiella* REITT. is represented by one species on Czornahora in the East Carpathians (1700 m above mean sea-level), two species on the Northwest Himalaya and one species each on the Tien Shan and Daurien. The Dermestid *Montandonia* JAQ. is represented on the south Carpathians, southern East Carpathians and on the Altai Mountains.

Exceedingly few species of hypsobiont Coleoptera are common to more than one mountain system. *Carabus fabricii* PANZ. occurs, for example, in the subalpine and high-alpine zones of the Alps and the Carpathians, *C. irregularis* FABR. on the mountains of Central Germany, the Jura, the Alps, the Bohemian Mass, the Carpathians, Krain, Croatia and on the mountains of Bosnia, *Byrrhus glabratus* HEER occurs on nearly all the mountains of Central Europe, the Pyrenees and in the northern parts of the Balkan Peninsular Mountains and *Aphodius mixtus* VILL. occurs on the Carpathians, the Alps, northern Apennines, the Pyrenees, Mont-Dore in Puy-de-Dome, etc. The greatest majority of the species are, however, extremely intensively localized. *Nebria schusteri* GGLB. is, for example, restricted to the high-alpine zone of the Koralps, *Carabus olympiae* SELLA to the Valle Sessera in the Piemont Alps, *C. cychroides* BAUDI to the northern Cottian Alps, *Trechus rudolphi* GGLB. and *Tr. regularis* PUTZ to the Koralps, *Tr. montis-rosae* JEANNEL to Monte Rosa, etc.

4. THE COMPONENT ELEMENTS OF THE HIGH ALTITUDE FAUNAS, THEIR ORIGIN AND EVOLUTION

The component elements, origin and evolution of the high altitude insect fauna form an integral part of the history of uplift and other changes associated with the mountain system and are therefore largely characteristic of the mountain. The evolution of the flora of the mountain system and the Pleistocene glaciations, particularly in the region of the mountain, also very profoundly influence not only the composition but also the evolution of the hypsobiont insect fauna of nearly all the important mountains. Comparative biogeographical studies on the characters of the principal mountains of the world show that the fundamental elements of the present-day hypsobiont insect fauna belong to the following major groups: 1. Fauna which has had its origin within the area of the uplift of the mountain and simultaneous evolution *in situ* of the autochthonous endemics, *pari passu* with the uplift of the mountain; 2. Fauna that represents colonizations from outside the area of uplift of the mountain and 3. relicts. The absolute quantitative dominance and the relative importance of the individual groups on any given mountain system depend on the peculiarities of the orogeny and tectonics of the system. The Tertiary fold-mountain systems are everywhere characterized by a significant dominance of the autochthonous endemics over the other groups. The hypsobiont insect fauna of isolated volcanic mountains like, for example, Mt. Kilimanjaro, is predominantly composed of the second group. The older mountain systems, like the Altai Mountains, are readily recognized by absence of a significant predominance of any group in particular. The relicts are everywhere sparse and are perhaps localized on Tertiary mountains in certain parts of the world.

Autochthonous endemic elements are extremely important ecologically and biogeographically. Their origin and evolution form part of the complex history of the orogeny of the mountain system more than in the case of the other groups. The uplift of any mountain is a gradual process, in the course of which the ground inhabited by the original fauna of essentially lowland character is subjected to a more or less complex series of thrusts, folds, crumplings, etc., culminating eventually in the rise of the area to a high elevation, above the surrounding country. As the area is being raised, the flora and the fauna are at the same time literally uplifted *en masse* from lowland conditions to higher elevations. The organisms that belong to the area behave like a plastic mass and are successively modified and remoulded in close correlation with every phase of the uplift of the mountain. The entire flora and fauna evolve successively into the low montane, the high montane, the subalpine, alpine, high-alpine, subnival, eunival, high-nival and other hypsobiont types. There is therefore neither a migration of the original flora and fauna away from the area, because of the changes involved in the uplift of the mountain, nor are they driven to extinction, as has often been assumed. On the other hand, the entire biota is reshaped and their modifications and evolution synchronize closely with the process of uplift of the mountain itself.

The dominant component elements of the hypsobiont insect life of the Northwest Himalaya arose, for example, in this manner, from a northern lowland ancestral stock that flourished in the area that was being uplifted during the early and middle Tertiary. The vast region of the Northwest Himalaya and the Pamirs is indeed remarkable for the great preponderance of true autochthonous endemic genera and species in nearly every group of insect (730). The present-day hypsobiont insects of this region became differentiated during the Pliocene phase of the Himalayan orogeny. SKORIKOV (984) and MANI (730) have independently recognized the region of Tertiary mountains, within the Palaearctic Realm, as an important and immense centre of differentiation of many characteristic faunas of high altitude insects. The principal component elements of the high altitude fauna of the East Himalaya have had a wholly different history from those of the Northwest Himalaya (730), but here also we may observe unmistakable evidence of a comparable *in situ* origin and evolution from a Himalayan-Chinese and an Indo-Malayan ancestral stock. Even at very high altitudes on the Himalaya, we find distinct evidence of this differentiation of the local lowland stock into the typically high altitude forms.

The second major group of hypsobiont insects represent colonizations from outside the actual area of uplift of the mountain system. As may be expected, this colonization seems to have proceeded mostly from the surrounding lowland region. There have, in addition, been progressive colonizations from distant boreal and even arctic elements, as the uplift of the mountain system progressed. The lowland elements have in both these cases gradually pushed upwards to higher elevations, with each successive phase of uplift of the mountain. They have often simultaneously become more or less greatly modified and have also given rise to a variable number of neoendemic species and localized subspecies in many parts of the world. In some cases the principal phase of colonization seems to have been subsequent to the major orogenic movement of the mountain system, even in Post-Pleistocene times.

As distinct from the mountains of the Pamirs Region, the Pamirs proper have a most remarkable high-steppe element in their high altitude insect life that cannot be strictly described as alpine type in any sense of the term. The Middle Asiatic steppe forms constitute an important component element of the Pamirs fauna. Most of these steppe forms have become secondarily adapted for life at high elevations, some of which are much higher than the highest summit of the Alps, but are nevertheless closely related to or are even completely identical with those of the lowland steppes of the Buchara area in Turkestan. The fundamental component elements of the high alpine-desert insect life on Mt. Kilimanjaro have also evolved from forms that have had their ancestral stock on the surrounding Masai steppes. Here again a great many species, occurring even at very high elevations, are identical with those living in the surrounding steppe country. The higher elevations of Mt. Kilimanjaro were evidently colonized by insects only after the cessation of the volcanic activity, in other words, after the mountain had already been completely uplifted. Some typically South African elements have spread northwards and have colonized the high altitudes of equatorial East African mountains. A number of boreal forms have likewise also reached their maximum southward limits on these mountains. As we shall see in the next chapter, the South African and the boreal elements do not, however, meet on the same mountain. Extensive colonizations of the high altitudes by boreal elements are observed on the equatorial, tropical and south temperate Andes. The southward spread of the boreal elements has been possible only because of the high elevation and also because of the continuous meridional trend of the Cordillera. Some of the arctic elements extend to the mountains of Colorado, at elevations of 3500-3900 m (744a; *vide* also Chapters VIII and XVI).

Unlike the autochthonous endemics and the elements representing colonizations subsequent to the uplift of the mountain, the relicts constitute rather older elements on the mountains. The relicts are generally remarkable for their pronounced discontinuous distribution and they are also on the whole extremely sparsely found both on the older and on the Tertiary mountain systems. Some of the relict species of the Altai Mountains are found in widely separated areas like Labrador (*vide* Chapter XIII). The relicts of the Tertiary mountains are either Tertiary relicts or they may be also Pleistocene relicts (boreo-alpine species) that have survived the glaciations of the Pleistocene times on nunataks on the high mountains (*vide* Chapter XII). While a considerable number of relicts are common on the Alps, the Himalayan System is strikingly poor in relict species; they may almost be said to be totally absent on the Himalaya. These peculiarities are apparently closely correlated with the important facts that Europe is biogeographically merely an appendage of Asia and that the Pleistocene glaciations have been far more extensive and intense in Europe than in Asia and lastly also the fact that the Himalaya as a whole is richer in autochthonous endemics than perhaps any other mountain system in the world. The conditions on the Himalaya favour a very high rate of speciation and the high altitude fauna of the Himalaya is at present most intensely evolving, so that the persistence of relicts on the Himalaya seems unlikely.

The fundamental component elements of the high altitude insect faunas of most high mountains in the world appear to have become practically fully differentiated and also more or less stabilized by the Pliocene times (665, 667, 730). The Pleistocene glaciations brought about an unprecedented intensification of isolation on nearly all the high mountains, at least in the northern hemisphere, so that we find at the present time a very pronounced tendency for the formation of a number of characteristic local races and subspecies, some of which are indeed restricted wholly to single peaks. The evolution in the past seems to have been directed towards ecological specializations, in the course of which relatively few species evolved. At the present time, however, the general evolutionary trend in the high altitude faunas everywhere seems to be directed to an enrichment of species.

CHAPTER VII

THE MOUNTAINS OF EQUATORIAL EAST AFRICA

The mountains of equatorial East Africa are of exceptional geographical, geological, ecological and biogeographical interest. An attempt is made in this chapter to give a brief outline of some of the outstanding peculiarities of the ecology and biogeography of the high altitude insects of tropical East Africa. Emphasis is laid on the fundamental differences between these mountains and other well known mountains like the Alps and the Himalaya. Attention is also directed to the evident relation between the high altitude insect life and the Rift Valley System of Africa.

1. THE EAST AFRICAN MOUNTAINS AND THE RIFT VALLEY SYSTEM

The occurrence of high altitude insects in the equatorial East Africa is rendered possible by the existence of high mountains, covered with permanent snow. The following are such elevated regions: three massifs in Abyssinia, the two peaks of Mt. Kilimanjaro (6010 m above mean sea-level) and Mt. Meru (4730 m); Mt. Kenya (5600 m) and the mountainous area of Leikipia (4270 m), Mt. Elgon (4230 m) and the Ruwenzori Range (5500 m). These elevated areas are closely related to the well known Rift Valley Fracture System of East Africa, stretching nearly 6400 km from the Jordan Valley to almost south of Zambesi, formed by the sinking down of narrow blocks of the earth's crust between high steep walls. The Rift Valley contains more than thirty lakes. The rifts consist of two gigantic troughs, 32 to 80 km wide and nearly parallel to one another, with an interval of about 6° Longitude in between, cutting northward through the continent, from the Lake Nyasa. The Western Rift or the so-called Tanganyika Trough lies between the 29° and the 30° meridian. The Eastern Rift or the Rift Valley proper stretches from the north to the south, across Kenya and Tanganyika and contains the Lakes Rudolf, Baringo and Naivasha. The Western Rift contains the Lakes Albert, Edward, Kivu and Tanganyika. The Lake Tanganyika surface lies at an elevation of 760 m above mean sea-level and its depth is about 1555 m. Almost all the larger lakes of East Africa lie in the Rift Valley System; the only exceptions are the Lakes Victoria and Tana. Both the rifts include thus a nearly continuous chain of lakes, mountains, volcanic cones and craters. The Ruwenzori Range in the western series of mountains is an ancient structure and is thus an exception to their generally volcanic origin. The western series include otherwise the Birunga (Virunga) Volcanoes, some of which are active, Karisimbi (4507 m) and Nyamuragira (3050 m). To the eastern series belong the mountains of Central and Southern Ethiopia. The extinct cone of Zuquala and the Massif Chillalo stand on opposite sides of the Galla Depression, on the western edge of which, further south, rise the isolated Massif of Mt. Damota and the great block of Gughé Highlands. Further south are the Mt. Kenya and the Aberdare Mountains (including Mt. Kinangop, 3900 m above mean sea-level) on the east side of the Rift Valley; Mt. Elgon and Cherangani (3380 m) on the west side; Mt. Kilimanjaro, with the associated Mt. Meru, to the southwest and the

Chyulu Hills to the northeast; these mountains are all related to the Great Eastern Trough. An excellent account of the Rift Valley System and the general geology of the region may be found in GREGORY (423). Permanent snow and glaciers exist on the higher massive mountains, especially on Mt. Kilimanjaro, Mt. Kenya and on the Ruwenzori Range. On these three mountains, rising to elevations over 5000 m, the lowest limit of eternal snow is at an elevation of about 4800 m. The snowline is thus much higher than, for example, on the Alps or the Pyrenees.

Broadly speaking, the East African mountains have on their slopes a zone of montane forest, which, with the gradual Post-Pleistocene warming and desiccation, has evidently given place to dry bush and thorn-scrub formations on the intervening plateau. This montane-forest belt, extending from an elevation of about 1980 to 3050 m, comprises three sub-zones: i. The lower montane forest or the ombrophile forest, dominated by conifers, especially *Podocarpus*, and extending to an elevation of about 2380 m; some *Helichrysum* and the giant *Lobelia gibberoa* occur as low as this zone. ii. The bamboo sub-zone, from an elevation of 2380 m to nearly 3050 m, is an almost pure botanical formation. iii. The highest forest or the *étages des forêts supérieures*, dominated by *Hagenia*, from an elevation of 3050 m up to the timberline. The second and the third sub-zones lie above the limits of maximum atmospheric precipitation on the mountains. Above the forest-zone lies the heath-zone, in which *Erica arborea* predominates. The heath-zone extends from an elevation of about 3050 m to nearly 3780 m above mean sea-level. This zone may also take the form of grassy expanses, with sparse growth of heath and other plants, and is characterized by giant *Lobelia* and arborescent *Senecio* and *Helichrysum*. High above the heath-zone is the high-montane zone, consisting of the *Alchemilla* sub-zone, in which Phanerogams are still to be met with. This zone is characterized by *Alchemilla*, giant *Lobelia* and arborescent *Senecio*. Beginning generally at an elevation of about 3780 m, this zone extends to nearly 4200 m on Mt. Kenya, a little over 4270 m on the Ruwenzori Range and even higher on Mt. Kilimanjaro. A second sub-zone is also sometimes recognized, viz. the moss-lichen sub-zone, extending up to the altitude at which all vegetation ceases (584). The altitudinal limits of the zones and the sub-zones vary naturally within wide limits on the different mountains and on the different slopes of the same mountain. Mt. Meru has, for example, luxuriant rain-forest and bamboo sub-zones on the south slope, but there is no bamboo and the forest is also much reduced on the north slope. *Protea kilimandscharica* forms open or dense scrub on the northeast slope of Mt. Kenya, at elevations between 2750 and 3660 m, particularly where there is little bamboo. The presence of *Protea* may probably be due to the low rainfall on this slope. The giant *Lobelia* also grows here. Some of the other interesting peculiarities of the biotic zonation on the equatorial and tropical high mountains are dealt with in the next chapter. See also COE (215a).

2. MT. KILIMANJARO

Mt. Kilimanjaro (Tanganyika, 3° 55′ SL), the highest summit in Africa, is an extinct volcano, the mighty conical mass of which rises to over 5000 m above the level of a vast arid steppe that is hardly 900 m above mean sea-level. Along with Mt. Meru, 40 km to the west and Mt. Kenya 320 km to the north, Mt. Kilimanjaro lies on a line of volcanoes on the east-west fault. The major axis of the mountain, about 83 km long at its base, is also east-west; the base is about 58 km from the north to the south. Its snow-capped summit is about 5963 m above mean sea-level. The northern and southern slopes rise in a great sweep from the surrounding steppe to its summit, so that the mountain appears as a single volcanic cone when seen from the east. Its two principal peaks are, however, seen separately from the north or the south (fig. 30). The Kibo Peak on the west is the biggest and also the younger truncated volcanic cone and is flanked by broad shoulders. It is covered by permanent ice, nearly 60 m thick, with the glaciers descending to elevations of about 4200 m above mean sea-level on the western slope and to 4875 m on the northwest slope. At the end of the eastern shoulder, the so-called Saddle, is situated the lesser Mawenzi Peak (4320 m above mean sea-level) and this peak is separated from the Kibo Peak by a straight distance of about 8 km. The western shoulder or the Shira Plateau ends in the remains of another ancient, but smaller crater, the Shira Crater. The Saddle, about 36 sq. km in area, is nearly level for about 6.5 km from the base of the Kibo Peak and the scree slopes of the lesser Mawanzi; its general elevation is never less than 4270 m above mean sea-level. The northern and the larger part of the Saddle, sloping gently northeastwards, is covered with sand and stones. On the south, it is marked off by a series of low hills, glacial troughs and water-worn valleys. Its much broken surface, with numerous boulders, provides excellent shelter from wind, so that the vegetation flourishes on the Saddle to relatively high elevations. The Shira Plateau slopes gradually from the Kibo Peak to the Shira Ridge, for about a distance of 10 km. The Plateau measures about 8 km north-south and has an area of about 62 sq. km. This represents the largest

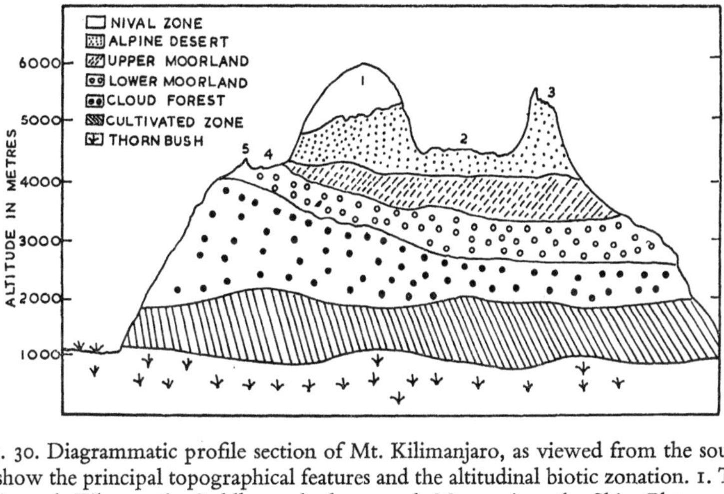

Fig. 30. Diagrammatic profile section of Mt. Kilimanjaro, as viewed from the south, to show the principal topographical features and the altitudinal biotic zonation. 1. The main peak Kibo, 2. the Saddle, 3. the lesser peak Mawenzi, 4. the Shira Plateau and 5. the Shira Ridge (Crater).

and relatively flat area on the mountain. On the west and on the south, the remains of the Shira Crater form its boundary, but on the east the relatively level ground passes gently into the slope of the Kibo Peak and the northern flank of the mountains. On the west is the Shira Ridge that rises 245 m above its surface, as a series of hills, like the Shira Needle, the Shira Dome and the East Shira Hill on the south. The Kibo cone rises to the east and shelters the Shira Plateau from the southeast and northeast monsoon winds. The morning shadow of the Kibo Peak retards also the too rapid warming up of the plateau. The Platz Peak and the Lent Valley are conspicuous topographical features on the Shira Plateau. The Platz Peak is a dome-shaped hill, about 210 m above the ground in the southwest corner and probably represents the eruption cone of the Shira Volcano. The Lent Valley in the east is of glacial origin and has high lateral moraines. A large number of melt-water streams flow in the eastern part of the Shira Plateau. An extensive bog on the southeast receives streams from the Platz Peak. The rock, exposed on the Shira Plateau, is chiefly trachydolerite of Shira origin on the west and rhombporphyry of Kibo origin on the east, showing evidence of frost and arid weathering under desert conditions. The rocks of Kibo and Shira were laid down as lava and have hard and evenly weathered surfaces. Numerous boulders of this rock are scattered on the Shira Plateau. There are also small patches of well developed soil in the area of the volcanic sand.

i. Ecological Conditions on Mt. Kilimanjaro

The outstanding features of the ecology and biogeography of the high altitude insects of Mt. Kilimanjaro are to be attributed primarily to its isolated eminence in an arid steppe, its location near the equator, its relative proximity to the Indian Ocean and its relation to the Rift Valley System and lastly its volcanic history.

Mt. Kilimanjaro presents a wide range of climatic and ecological conditions, from the equatorial arid plain at its base (900 m above mean sea-level) to the perpetual snow on the summit, with a relatively well defined altitudinal zonation of life on its slopes. The ecology of this mountain has been studied by a number of workers and the reader will find an excellent recent review in SALT (913). From the ecological point of view, Mt. Kilimanjaro is a tropical mountain and is characterized by an essentially diurnal-thermic climate.

KLUTE'S (628) observations give us some idea of the general atmospheric temperatures at an elevation of about 4000 m on the mountain.

TABLE 21

Mean diurnal fluctuations of atmospheric temperatures at an elevation of 4000 m on Mt. Kilimanjaro

Time of the day	Temperature ° C
7 hrs	0.1
14 hrs	5.6
21 hrs	0.5

The mean minimum atmospheric temperature was found to be —1.8° C, the absolute minimum temperature was —4.0° C and the mean diurnal temperature range 10° C. The mean atmospheric temperature on the Shira Plateau was 2° C and the minimum —2.4° at 4 a.m. KLUTE's screened thermograph, placed at an elevation of 4160 m, recorded a minimum of 1.8° C, but the unscreened thermograph recorded —7.3° C at an elevation of 4340 m. The high intensity of insolation on Mt. Kilimanjaro is due both to the thinner atmosphere because of the high altitude and also to the vertical direction of the sunrays characteristic of the equator. MEYER (771) measured the sun-temperature at an elevation of about 4327 m on the mountain to be 73° C and at an elevation of 3643 m to be 87.5° C. The rapid radiation and the presence of large masses of ice have the result of low atmospheric temperatures on the Upper Kilimanjaro, even near a heated surface. There is, therefore, considerable difference between the sun and shade temperatures (628). For example, at an elevation of 4150 m the sun temperature was found by KLUTE to be 15.5° C and at the same time the surface temperature on a stone measured 39.5° C, while the shade temperature of the atmosphere was only 8.3° C. With the atmospheric shade temperature reading 14° C, he found the sun temperature to be 87.5° C and with the atmospheric temperature 6° C, the sun temperature was 73° C, thus giving differences of 73.5° and 67° C respectively. Because the loss of heat from exposed surface is rapid and the air temperature is low, the surface temperatures at higher elevations fall very quickly, as soon as the sun is obscured by a cloud. There is thus not only a very pronounced difference between the sun and shade temperatures, but we also find here a rapid and rather abrupt change from conditions of high temperatures to intense cold. The violence of these temperature changes is striking in the ecology of Mt. Kilimanjaro.

The prevalent directional winds on Mt. Kilimanjaro are i. the southeast monsoon, ii. an anti-trade wind usually above an elevation of 3865 m, iii. the northeast monsoon and iv. a local southwest wind. None of these winds are ecologically important on the Shira Plateau, because the enormous dome of the Kibo Peak, rising 2133 m above the general level of the plateau on the east, protects the region effectively from the two principal winds and the Shira Ridge and the Shira Hills on the south shield large areas of the Plateau from the southwest wind. In addition to these directional winds, there is an updraft of warm air from the Masai Steppe during the daytime and a downdraft of cold air from the ice-cap on the Kibo Peak during the night.

The mean annual precipitation on Mt. Kilimanjaro increases at first with the increase in elevation, up to nearly 2440 m above mean sea-level, above which there is, however, a marked fall in total precipitation, so that the higher elevations of the mountain are relatively arid. Maximum precipitation at higher elevation is found on the southwest. The bulk of the rainfall, between 174 mm and 1016 mm, occurs respectively during April-June and November-December. The rising air currents from below carry clouds from the forest-zone over the moorland during November, so that the Shira Plateau is shrouded in dense clouds on the south after 10 a.m. Later on during the day, especially in the afternoon, damp mists are frequent on at least the southern half of the Plateau.

Unlike in the case of the mountains of the north temperate and cold regions, the high altitude climate of the equatorial and tropical mountains is characterized by greater ecological importance of diurnal fluctuations of atmospheric temperature. There is indeed a marked cycle of high and low temperatures within twenty-four hours of the day on Mt. Kilimanjaro. Both day and night are approximately of the same duration almost throughout the year. The atmospheric temperature falls below the freezing point every night on the moorland on Mt. Kilimanjaro and is always below the freezing point on the alpine desert. The fall of atmospheric temperature is very rapid after sunset every day, so that there is night frost for nine to ten hours through the year. There is, however, a period of about two hours in the morning, before the surface temperature rises too high for the insects to move about and again one or two hours in the evening, before the ground becomes too cold for them to come out after sunset. Clouds naturally modify and greatly influence this cycle and generally serve to greatly ameliorate the severity of the conditions. Similar conditions prevail on Mt. Kenya also. This diurnal climate of the Upper Kilimanjaro has a very profound influence on the habits of the high altitude insects. In the absence of clouds, the surface temperatures during the daytime, even at high elevations, correspond to those of the lowland hot desert-like conditions. The insects do not, however, become nocturnal, as in the lowland hot deserts, because during the night both the atmospheric and the ground temperatures fall below the freezing point, so that they cannot remain exposed on the surface. We find, therefore, that the high altitude insects are completely endogeous on Mt. Kilimanjaro. Underground, they are not exposed to the extremes of the ground temperature changes.

ii. Biotic Zones on Mt. Kilimanjaro

The belt of forest that is characteristic of the equatorial East African mountains in general, lies at elevations between 1800 and 2600 m on the east slope and up to 3200 m on the west slope of Mt. Kilimanjaro (fig. 30). Near the upper limits of the forest and often also above these limits, *Erica arborea* LINN. may be found, so that ecologically the timberline on Mt. Kilimanjaro is not strictly comparable to the treeline on the European mountains. According to SJÖSTEDT (979, 980), a number of species of Phasmida occur at elevations above the forest-line on Mt. Kilimanjaro, a situation that is unknown either on the Alps or on the Himalaya. Above the upper limits of the forest, the zone of alpine-prairie extends up to an elevation of about 4200 m on Mt. Kilimanjaro. Above this limit lies the zone of high-alpine steppe (zone du désert alpin) and reaches up to the snowline. The soil above the forest zone on Mt. Kilimanjaro is not marshy or excessively moist and is everywhere clothed with dense growth of dwarf grass, so that a greater variety of terricole insects find much wider optimal conditions than, for example, on the other equatorial East African mountains like Mt. Kenya and the Ruwenzori Range.

The socalled alpine-zone on Mt. Kilimanjaro commences at elevations between 2600 and 3000 m and is recognized by the presence of plants like *Helichrysum* and the arborescent *Senecio johnstoni*. This zone extends higher to nearly the permanent snowline, to elevations between 4800 and 5200 m above mean sea-level. ALLUAUD (30) recognizes the following life zones on Mt. Kilimanjaro: i. the "zone inférieure" (brousse épineuse), extending up to an elevation of about 1000 m; ii. the "zone des cultures", extending to an elevation of about 2000 m; iii the forest zone extends from an elevation of 2000 to 2600-3000 m; iv. the prairie zone extends to an elevation of about 4000 m; v. the alpine zone or "désert alpin", from an elevation of about 4200 m to the permanent snowline (4800-5200 m); vi. the glacier zone from an elevation of about 5000 m to the summit of the mountain. In his recent discussion on the ecology of Mt. Kilimanjaro, SALT (913) recognizes nearly the same biotic zones. According to him, above the thorn-bush and grass of the surrounding steppes, there is a cultivated zone of coffee, banana and other garden crops, reaching to an elevation of about 1830 m. A belt of cloud-forest (Nebelwald or fog forest) flourishes at elevations between 1820 and 2745 m on the east, northwest and southwest and somewhat narrower on the northeast and north. The moorland zone, above the forest, extends from an elevation of about 2745 m to 4270 m. SALT subdivides this zone into a lower moorland zone, from the forest margin to an elevation of 3500 m and characterized by a variety of high and low shrubs, including *Erica*, *Philippia*, *Protea*, *Adenocarpus*, *Hypericum* and *Artemisia*, with the giant groundsel *Senecio kilimanjara*, diverse grasses, sedges and herbs. In the upper moorland zone, extending from an elevation of 3500 to 4270 m, the flora is less varied and is dominated by *Philippia jaegeri*, *Euryops dacrydioides*, *Helichrysum* spp. and the giant *Senecio cottonii*, grasses, sedges and some herbs. The vegetation becomes increasingly more open above an elevation of 3965 m and at about 4270 m, the moorland zone is succeeded by the alpine-desert zone. The lower reaches of the alpine desert zone are characterized by sparse growth of grass-tufts and some plant cushions, especially under the shelter of overhanging stones. Still higher up, the vegetation becomes increasingly sparse. The alpine-desert zone may be taken to extend the limits of Phanerogams to an elevation of 4878 m. Angiosperms are extremely sparse here, except perhaps in sheltered places. Only moss and lichens persist among the rocks near the permanent snow.

The vegetation types of the Shira Plateau comprise the following: 1. The Bog Formation: i. Sedges and grass communities, ii. *Alchemilla cinerea* community and iii. the Bog-heath community. 2. Heath Formation: i. *Helichrysum* community, ii. *Philippia* community, iii. *Euryops* community, iv. Valley and gorge community, v. *Senecio cottonii* community and vi. *Erica arborea* community.

The sedges and grass communities are composed mainly of *Carex monostycha* A. RIC. and *Deschampsia* sp., with low bushes of *Alchemilla argyrophylla*. The *Alchemilla cinerea* ENGL. community includes also *A. argyrophylla* OLIV. and *Lobelia deckenii* HEMSL. The bog ecotone is invaded by *Alchemilla cinerea* ENGL., *A. argyrophylla* OLIV., *Euryops dacrydioides* OLIV., *Helichrysum hoehnelii* SCHWEINF. and *H. newii* OLIV. & HIERM. The heath formation is dominated by Ericaceae, Compositae and Rosaceae. The *Helichrysum*-community is naturally dominated by *H. hoehnelii* SCHWEINF. and *H. newii* OLIV. & HIERM.

The vegetation of the moorland zone includes *Helichrysum* spp., at elevations of 4270 m, some clumps of *Euryops dacrydioides* and rosettes of *Trachydium abyssinicum* (Umbelliferae). The vegetation of the alpine desert includes *Helichrysum newii* OLIV. & HIERM., *H. fruticosum compactum* VATKE and *Senecio volkensii* HOFFM. and the grasses *Koeleria cristata* PERS. *supina* ENGL. and *Pentaschistis minor* BALL. & HUBB. The mosses include *Grimmia ovalis* HEDW. (LINDB.), *Hypnum cupressiforme* HEDW., *Tortula cavallii* NEGRI and *Webera afrocruda* (C. MÜLL.) PAR. Even as high as 4880 m above mean sea-level on the Kibo Peak, clumps of grass flourish in abundance.

iii. Early Studies on Insects from Kilimanjaro

Mt. Kilimanjaro was discovered in 1848 by JOHANNES REBMANN of the Church Mission Society, but it was not till 1884 that JOHNSTON collected insects from the region for the first time. His collections were described by WATERHOUSE (1105) and GODMAN (412). The next important attempts at collecting insects were by KERSTEN of the Expedition of VON DECKEN, VOLKENS, KERTSCHNER and HANS MEYER (770, 771), though mainly from the forest-zone on the mountain. The three monumental volumes of SJÖSTEDT's Kilimanjaro-Meru Expedition of 1905-1906 contain much valuable information on the general geography, physiography, geology, vegetation and fauna of Mt. Kilimanjaro (979, 980). SJÖSTEDT collected insects and spiders from within the forest-zone and higher up to elevations of nearly the snowline. The most important contributions to our knowledge of the insect life of the mountain and its general ecology are, however, those of ALLUAUD (30-35), JEANNEL (575, 579, 582) and SALT (913). ALLUAUD himself collected insects largely at elevations between 2800 and 3200 m on Kilema, on the north slope, during 1903-1904; from the southwest slope on Kiboscho, at the snout of glacier on the Kibo Peak (4000-4800 m), in February 1904 and again he made extensive collections during 1908-1909, and in association with JEANNEL during 1911-1912. The Coleoptera, especially the Carabidae, from Mt. Kilimanjaro were described by KOLBE (637), FAIRMAIRE (328), ALLUAUD (30), ALLUAUD & JEANNEL (35), JEANNEL (575, 579, 583), BASILEWSKY (86) and PIC (835). FAUVEL (335) and TOTTENHAM (1041) have described the Staphylinidae, BRYANT (165) the Halticinae and MARSHALL (740) the Curculionidae collected by the different Kilimanjaro Expeditions. Except for the short notes of GODMAN (412) and BUTLER (177), we know very little of the Lepidoptera from the higher elevations of the mountain. The Orthoptera of Mt. Kilimanjaro were described by SJÖSTEDT (979). Dermaptera were studied by HINCKS (517), Hymenoptera (Pompilidae) by ARNOLD (55) and Diptera by VAN EMDEN (319). The Collembola from the mountain were described by DELAMARE-DEBOUTTEVILLE (244). EVANS (324) has described the Acari and BEIER (100) the Pseudoscorpions from the mountain. Important contributions to our knowledge of the spiders form Mt. Kilimanjaro are by DE LESSERT (689), DENIS (246) and HOLM (539).

iv. Ecological Characteristics of the High Altitude Insects from Kilimanjaro

The ecological studies so far undertaken on Mt. Kilimanjaro seem to have been mostly confined to the lower elevations and also in particular to the easily accessible southeast slope. Ecological explorations at higher elevations, near the permanent snow, may, however, be said to have been almost completely neglected. The most complete and recent accounts of the ecology of the mountain are perhaps those of ENGLER (321), COTTON (222) and SALT (913). COTTON was indeed the first to discover the giant *Senecio* at elevations above 3660 m. The fauna of the mountain is unfortunately much less known than the flora and SJÖSTEDT's collections and observations deal largely with the steppes and the cultivated and the forest-zones on the mountain and only extremely few records refer really to the true high-alpine levels. In the 58 fascicles of ALLUAUD and JEANNEL's Expedition of 1912, we find that only three species are recorded from elevations above 3200 m.

Even a casual observer will not fail to note a certain characteristic similarity to ordinary desert conditions in the insect life of the upper Kilimanjaro, but there are certain fundamental differences. On the upper Kilimanjaro, due to the daily nocturnal frost, the insects are unable to avoid the extreme high daytime temperatures of the ground by becoming nocturnal. They are, therefore, cryptozoic-terricole and endogeous throughout. There is a dominance of forms with shiny and highly reflecting surfaces, such as, for example, *Orinodromus* KOLBE and *Peryphus* KOLBE, with pronouncedly convex elytra so as to enclose an air space above the dorsum of abdomen as in the Curculionid *Oreonyscius* MSHLL. or special coloration for selective absorption, as for example, in the case of *Parasphena* BOLIVAR and *Orinodromus albomaculatus* BASILEWSKY. Most species remain thus concealed underground or under large stones and even when occasionally found above ground (as in the case of Tipulidae and Trypetidae), they never attempt to fly. The whole of the insect life of the upper moorland of Mt. Kilimanjaro is described by SALT (912) to be "sedentary, unwilling to fly or to move over the ground". These peculiarities are even more pronounced on the alpine-desert zone than on the moorland. Nearly 85% of the typical Pterygota from the higher elevations of Mt. Kilimanjaro have either reduced wings or are wholly unable to fly.

A most remarkable peculiarity of the insect life of the higher elevations on the mountain is that there is no single cold-hardy, true hibernating stage in the life-cycle of any species, as we find in the case of the insects from the Alps and the Himalaya, but all the developmental stages are exposed to the full range of climatic changes during the twenty-four hours of the day all the year round. Strictly speaking, there is no winter hibernation, as there is also no true winter on the mountain. Each stage exhibits, on the other hand, the ability of passing into a dormant condition during the nocturnal winter and of re-awakening to active life during the daytime summer every day. Unless an insect can complete both ecdysis and metamorphosis within a single period of daytime warmth, it must have the capacity for withstanding low temperatures at night, even in the course of its moulting. It is this daily alternation of winter and summer conditions that serves as the most important and decisive limiting ecological factor for the high altitude insects on Mt. Kilimanjaro.

The highest elevations at which ALLUAUD and JEANNEL collected insects on Mt. Kilimanjaro are between 2700 and 2800 m, so that their observations do not provide information about the ecological conditions above the upper moorland zone. In his 1908-1909 Expedition, ALLUAUD did not collect at elevations above 3200 m. In the first expedition, he collected at elevations between 2800 and 3200 m on the southeast slope, above Kilema. He also went up the south slope, above Kiboscho and camped at an elevation of 4200 m in February 1904. The collections made on this occasion include *Phrynacolus* (= *Amiantus*) *ater* (WATERH.), *Orinodromus glacialis* KOLBE and *Tachyusa pratensis* FVL., from an elevation of 3500 m. The type specimen of *Dismegistus funebris* AMYOT & SERV. was collected by JOHNSTON at elevations between 3050 and 4270 m. *Orinodromus deckeni* (= *nigripennis*) GERST. is recorded by VOLKENS at elevations between 3000 and 4000 m and both the species have since then been found at lower elevations also. According to KOLBE (in SJÖSTEDT, 979), the Scarabaeid *Trox montanus* KOLBE was taken at elevations between 4000 and 4400 m, but according to other authors, it was found only at elevations between 2500 and 3000 m. Some species found in the upper moorland zone of Mt. Kilimanjaro are also known to be more or less widely distributed in Africa and thus represent comparatively recent colonizations from lower elevations. Other species occur also on the other East

African mountains, but are absent on the steppe country between the mountains This is, for example, the case with *Parafolsomia africana* SALMON on the Ruwenzori Range; *Forficula sjoestedti* BURR on Mt. Meru, Mufumbiro Volcano and Mt. Kenya; *Spilostethus montislunae* BERG. on the Ruwenzori Range; *Cupido aequatorialis* SHARP on Mt. Kenya, Mt. Elgon, Mt. Birunga and the Ruwenzori Range; *Ceutorrhynchus bipunctatus* HUST. from Mt. Elgon; *Coenosia heterocnemis* EMD. from Mt. Elgon, Aberdare and the Ruwenzori Range; *Caricea tetriguttata* EMD. from Mt. Elgon; *Hahnia rouleti* LESS. from Mt. Meru, Mt. Kenya and the Ruwenzori Range. Some of the species are peculiar to Mt. Kilimanjaro and are thus true endemics. The common examples of such endemics are *Parasphena pulchripes* (GERST.) (fig. 31A), *Orinodromus glacialis* KOLBE, *Orinodromus albomaculatus* BASIL., *Afrotarus kilimanus* KOLBE, *Agonum johnstoni* ALL., *Hystrichopus alticola* ALL., *Omotaphus mixtus kilimanus* ALL., *Peryphus sjöstedti* ALL. (fig. 31D) and *Plocamotrechus kilimanus* JEANNEL (fig. 31C).

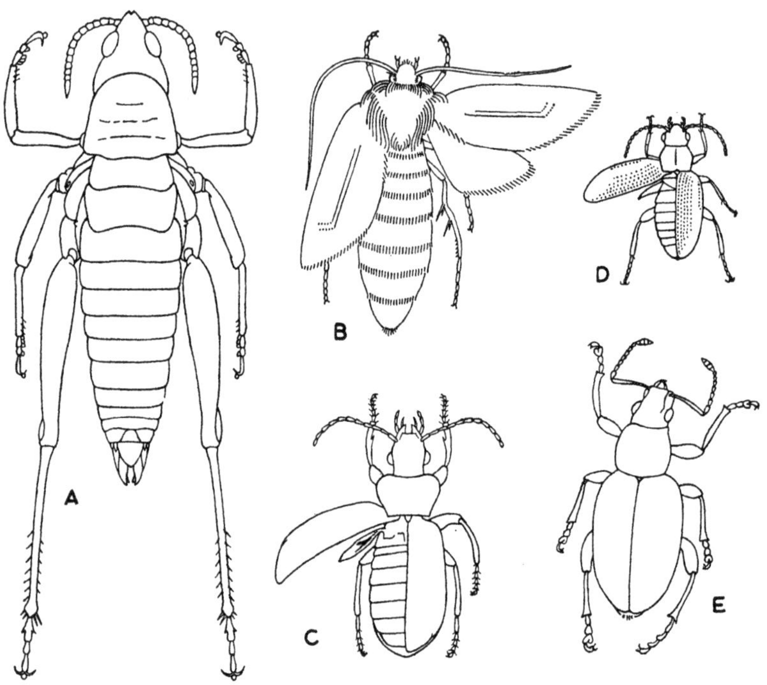

Fig. 31. Some typical high altitude insects from Mt. Kilimajaro. A. *Parasphena pulchripes* GERST. (apterous), B. *Saltia acrophylax* (brachypterous), C. *Plocamotrechus kilimanus* JEANNEL, D. *Peryphus sjöstedti* ALLUAUD, E. *Parasystatiella agrestis* MSHLL. (Modified from SALT).

Although at present our knowledge of the insect life at higher elevations on the mountain is extremely fragmentary, it is nevertheless evident that there is a progressive decrease in the number of species inhabiting the successive zones of the cloud-forest, the lower moorland, the upper moorland and the alpine-desert. Out of a total of about 600 species recorded from within the forest-zone, about 50% are known on the lower moorland zone, 25% on the upper moorland zone and only 7% from the alpine desert.

v. Distributional Characters of the High Altitude Insects from Mt. Kilimanjaro

Although widely distributed mountain genera like *Bembidion* LATR., *Agonum* BON., *Atheta* THOMS., etc. are found on Mt. Kilimanjaro, it may be observed that there is really no affinity between the high altitude Coleoptera of Mt. Kilimanjaro and of the European mountains. The high altitude Carabidae on Mt. Kilimanjaro are particularly remarkable for the minute, apterous subgenus *Orinodromus* KOLBE of *Calosoma* WEBER, related to *Carabophanus* KOLBE from the Abyssinian Highlands. The species are mostly non-metallic and dark reddish-brown or also black. Among the other high altitude Carabidae, mention must be made of *Bembidion kilimanum* ALL. (2600-3000 m), *Tachys ascendens* ALL. (2600-3000 m), *Plocamotrechus kilimanus* JEANNEL (2800 m) (fig. 31C), *Zagrochilus bedeli* ALL. (2800 m), *Orinophomus kilimanus* ALL. (2800 m), *Cymindis kilimana* KOLBE (2600-3000 m), *Hystrichopus alticola* ALL. (2600-2800 m), *Metabletus kilimanus* ALL. (2800-3000 m) and *Zuphium ascendens* ALL. (2800 m) (Fig. 34B). The Dytiscidae *Hydroporus (Nebrioporus) kilimandjarensis* RÉG. and *Agabus dytiscoides* are reported at elevations between 3000 and 3500 m. The principal Staphylinidae so far recorded include *Homalium algidum* FAUV., *Staphylinus dispersus* FAUV., *Philonthus altivagans* FAUV., *Atheta praticola* FAUV. and *Tachyusa pratensis* FAUV. The Tenebrionid *Phrynacolus ater* (WATERH.), occurring at elevations of 3000 and 4000 m, is also known from Mt. Kilimanjaro. SJÖSTEDT collected several species of Chrysomelidae above the timberline and also came across many interesting apterous Galerucinae even within the forest-zone on the mountain. The genus *Sjöstedtinia* WSE. is endemic to the mountain and is at present known only by a single species *Sjöstedtinia montivaga* WSE., from elevations between 3000 and 3500 m; specimens of this species were collected among the dry flowers of *Lobelia deckenii*. A number of Curculionidae like *Parasystates minor* AURIV., *Hypsomias lobeliae* AURIV., *Oreoscotus minor* AURIV., *Oreoscotus lobeliae* AURIV. and *Cossonus lobeliae* AURIV. are also recorded from the mountain; the genus *Parasystates* AURIV. is represented by a number of species within the rain-forest zone of both Mt. Kilimanjaro and Mt. Meru. Some species of Gyrinidae, Nitidulidae, Cryptophagidae, Cantharidae, Cleridae, Coccinellidae and Scarabaeidae have also been reported from above the forest-zone on Mt. Kilimanjaro.

Mt. Kilimanjaro seems to be significantly poor in typical mountain autochthonous Lepidoptera. Most of the species found at relatively high elevations on the mountain occur indeed also at much lower elevations, often also on other equatorial East African mountains and on the surrounding steppe country. *Cupido aequatorialis* SHARP, found on Mt. Kilimanjaro up to an elevation of about 4000 m, is also widely distributed on the other equatorial mountains of East Africa, such as Mt. Elgon at elevations of 2590 m, the Ruwenzori Range, Mt. Kenya and Mt. Birunga and is perhaps one of the very few typical mountain species in the region. The other species like *Vanessa cardui* LINN., *Colias electo* LINN., *Papilio demodocus* ESP., etc., usually found above the timberline on Mt. Kilimanjaro, occur also commonly at much lower elevations on the plains of East Africa. Satyridae, so very characteristic of the mountain autochthonous fauna of other areas in the northern hemisphere, seem to be completely absent on Mt. Kilimanjaro. SJÖSTEDT collected *Phryganopsis elongata* AURIV. (Lithosiidae); *Borolia eripygoides* AURIV. (Noctuidae); *Oreometra vittata* AURIV., *Hypsometra ericinellae* AURIV., *Hydrelia costata* AURIV., *Triphosa corticearia* AURIV., *Onychia (Eubolia) nictitaria* H.-SCH., *Larentia sjöstedti* AURIV. (Geometridae); *Gorgopsis alticola* AURIV. (Hepialidae); etc., some of which seem to be particularly confined to elevations above the forest-zone on the mountain.

A few Acrididae, like *Heteropternis couloniana* SAUSS., *Parasphena pulchripes* GERST. (fig. 31A) and *Chrysochraon kilimandjaricus* SJÖST., are known exclusively from above the timberline, at elevations between 3000 and 3500 m on Mt. Kilimanjaro. The discovery of *Chrysochraon* FISCH. on Mt. Kilimanjaro is of particular interest, for according to SJÖSTEDT, *Chr. kilimandjaricus* SJÖST. is closely related to the European *Chr. dispar* HEYER. The apterous Dermaptera *Forficula sjöstedti* BURR, found at elevations between 3000 and 4000 m, is common under stones above the forest, but the species has also been collected from within the forest zone, at elevations of 2000 and 3000 m; it is also found at elevations of 4000-4300 m on Mt. Meru.

Collembola occur at very high elevations on Mt. Kilimanjaro. SJÖSTEDT found, for example, *Mesira annulicornis* BÖRNER at an elevation of about 3800 m. They are most abundant on the lichen-incrusted rock at the edge of permanent snow, at an elevation of 5500 m. Recently SALT has also recorded *Hypogastrura* BOURL. at high elevations on the mountain.

vi. Vertical Distribution of Insects on Mt. Kilimanjaro

SALT (913) has recently listed the species of high altitude insects, associated with different vegetation communities in the different biotic zones on Mt. Kilimanjaro. The greatest bulk of the Invertebrates studied by him were found at elevations between 3400 and 4575 m. Nearly 62% of the species from the alpine-desert zone are endemic to Mt. Kilimanjaro.

The following synopsis of the vertical distribution of high altitude insects on Mt. Kilimanjaro is summarized from SALT (op. cit.):

I. Species characteristic of the bog-formation at an elevation of about 3680 m

Diptera: *Tipula subaptera* FREEMAN (fig. 32), flightless; *Pseudosmittia salti* FREEMAN, a true bog-chironomid, with the larvae feeding in small accumulations of water; *Coenosia heterocnemis melanomeros* EMD., *Hylemya salti* EMD. (muscoid flies). Arachnida: *Trochosa urbana* CAMB. and *Sphaerolophus salti* EVANS.

II. Species characteristic of the heath-formation

1. *Helichrysum*-community

Coleoptera: *Mnionomus bicolor* BRUCE (Cryptophagidae), *Sjöstedtinia* sp. prox. *montivaga* WSE. (Galerucidae), *Longitarsus salti* BRYANT. Diptera: *Paroxyna shiraensis* MUNRO, *Urophora pantomelas* BEZZI (Trypetidae); *Coenosia heterocnemis* EMD., *C. h. melanomeros* EMD., *Helina posterodorsalis* EMD., *Carica tetriguttata* EMD. and *Wagneria rufitibia abbreviata* MESN. (Tachinidae). Arachnida: *Dictyna montana* TULLG.

2. *Philippia*-community

Heteroptera: *Orthotylus* spp. (Miridae). Coleoptera: *Ceutorrhynchus bipunctatus* HUST., known also from Mt. Elgon at an elevation of 3500 m. Arachnida: *Dictyna montana* TULLG.

3. *Euryops*-community

Coleoptera: *Longitarsus salti* BRYANT and *Ceutorrhynchus bipunctatus* HUST. Diptera: *Paroxyna salina* MUNRO, *P. saltorina* MUNRO (Trypetidae); *Coenosia heterocnemis melanomeros* EMD., *Helina posterodorsalis* EMD. and *Wagneria rufitibia abbreviata* MESN. Collembola: *Lepidocyrtinus armillatus pallidus* DEL. Arachnida: *Dictyna montana* TULLG.

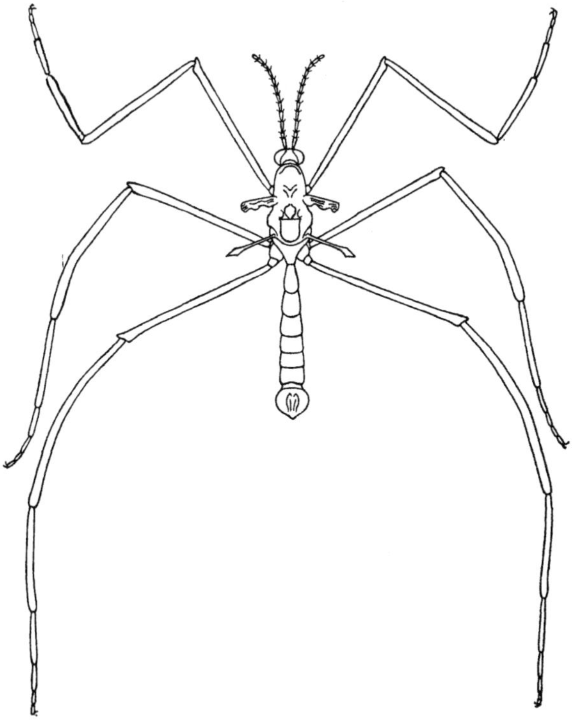

Fig. 32. *Tipula subaptera* FREEMAN (After SALT).

4. Moss community

Coleoptera: *Troposipalia africana* TOTT. (Staphylinidae) and *Phrynacolus ater* (WATERH.) (Tenebrionidae). Heteroptera: *Nysius salti* USINGER. Collembola: *Lepidocyrtinus armillata pallida* DEL. Arachnida: *Clubiona abbajensis kibonotensis* LESS., *Xysticus* sp., *Hahnia rouleti* LESS., *Pardosa karagonis* STRAND, *Oribata geniculata* LINN., *Bdella piggotti* EVANS, *Erythraeus nasalis* EVANS and *Anystis baccarum* LINN.

5. Hypolithic community

Orthoptera: *Parasphena pulchripes* GERST. Dermaptera: *Forficula sjöstedti* BURR. Coleoptera: *Afrotarus kilimanus* KOLBE, *Agonum johnstoni* ALL., *Hystrichopus alticola* ALL., *Omotaphus mixtus kilimanus* ALL., *Peryphus sjoestedti* ALL., *Leptacinus miltoni* TOTT., *Phrynacolus ater* (WATERH.), *Myrcina* sp., *Heterocossonus lobeliae* AURIV. Arachnida: *Pardosa karagonis* STRAND, *Lamyctes africana* (PORAT) and *Erythraeus nasalis* EVANS.

6. Soil community

(i.) At elevation of 4270 m. Coleoptera: *Longitarsus salti* BRYANT, *Parasystatiella* sp. Collembola: *Hypogastrura* sp., *Mesaphorura salti* DEL., *Parafolsomia africana* SALMON, *Lepidocyrtinus armillata pallida* DEL. Acarina: *Africoribates ornatus* EVANS, *Microtegeus undulatus* BERL., *Scheloribates laevigatus* KOCH, *Protoribates shiraensis* EVANS.

(ii.) At elevation of 4420 m. Coleoptera: *Plocamotrechus kilimanus* JEANNEL, *Oreosecus porculus* MSHLL. Collembola: *Lepidocyrtinus armillatus pallidus* DEL. Arachnida: *Clubiona abbajensis kibotensis* LESS., *Thanatus* sp., *Hahnia rouleti* LESS., *Pardosa karagonis nivicola* LESS.

(iii.) At elevation of 4575 m. Collembola: *Friesea africana* DEL. *Parafolsomia africana* SALMON. Arachnida: *Scheloribates laevigatus* KOCH, *Protoribates shiraensis* EVANS and *Africoribates ornatus* EVANS.

III. Species characteristic of the moorland-zone

Orinodromus albomaculatus BASILEWSKY on rock; *Tipula subaptera* FREEMAN, with greatly reduced wings and flightless in both the sexes, crawl about in hundreds on cushions of *Trachydium abyssinicum* (Umbelliferae).

IV. The Alpine-desert communities

1. Hypolithic community

Dermaptera: *Forficula triangulata* HINCKS; Heteroptera: *Spilostethus montislunae* BERG. and *Oreonysius rugosus* USINGER; Coleoptera: *Peryphus sjoestedti* ALL., *Oreoscotus salti* MSHLL., *Parasystatiella agrestis* MSHLL.; Lepidoptera: *Saltia acrophylax* TAMS, *Gorgopsis salti* TAMS; Chilopoda: *Lamyctes africana* (PORAT); Arachnida: *Allothrombium barbuligerum* ANDRE.

2. Soil community

Collembola: *Parafolsomia africana* SALMON; Arachnida: *Africoribates ornatus* EVANS, *Scutovertex africanus* EVANS, *Scheloribates laevigatus* KOCH, *Protoribates shiraensis* EVANS. SALT records the following species at an elevation of 4880 m: *Lepidocyrtinus armillatus pallidus* DEL., *Parasphena pulchripes* GERST., *Forficula cavallii* BOR. (the highest elevation for Dermaptera on Mt. Kilimanjaro); *Peryphus sjoestedti* ALL., *Parasystatiella arctipes* MSHLL. and *Hahnia* sp.

The following is a list of species collected by SALT at elevations of 3500-4250 m on the upper moorland zone of Mt. Kilimanjaro:

Orthoptera: *Parasphena pulchripes* GERST. (fig. 31A). Dermaptera: *Forficula sjöstedti* BURR. Corrodentia: *Mesopsocus montinus* ENDERLEIN. Ephemerida: *Baetis* sp. Heteroptera: *Lamprostenarus* sp. prox. *sjöstedti* POP., *Orthotylus* sp., *Nysius salti* USINGER, *Eupteryx salti* RIBAUT, *Naöasius acutipennis* RIBAUT. Coleoptera: *Afrotarus kilimanus* KOLBE, *Agonum johnstoni* ALL., *Hystrichopus alticola* ALL., *Omotaphus mixtus kilimanus* ALL., *Orinodromus albomaculatus* BASILEWSKY, *Peryphus sjöstedti* ALL. (fig. 31D), *Leptacinus miltoni* TOTT., *Philonthus swynnertoni* TOTT., *Troposipalia africana* TOTT., *Aposericoderus minor* PAUL., *Mnionomus bicolor* BRUCE, *Sjöstedtinia* sp. prox. *montivaga* WSE., *Longitarsus salti* BRYANT, *Phrynocolus ater* (WATERH.), *Biophinina obscura* PIC., *Ceutorrhynchus bipunctatus* HUST. and *Heterocnemis lobelliae* AURIV. Lepidoptera: *Gorgopsis salti* TAMS, *Eupithecia salti* FLETCHER, *Hypsometra ericinellae* AURIV., *Oreometra vittata* AURIV. Diptera: *Tipula subaptera* FREEMAN (fig. 32), *Pseudosmittia salti* FREEMAN, *Paroxyna salina* MUNRO, *P. saltorina* MUNRO, *P. shiraensis* MUNRO, *Urophora pantomelas* BEZZI, *Thereva* sp., *Coenosia heterocnemis melanomeros* EMD., *Helina posterodorsalis* EMD., *Hylemya salti* EMD., *Wagneria rufitibia abbreviata* MESN. Collembola: *Brachystomella parvula* (SCHÄFFER), *Friesea africana* DEL., *Hypogastrura* spp., *Mesaphorura iowensis* (MILLS), *M. kilimanjarica* DEL., *M. salti* DEL., *Pronura kilimanjarica* DEL., *Pseudachorutes* sp., *Ballistura excavata* FOLSOM, *Lepidocyrtinus armillatus pallidus* DEL. and *Sminthurinus salti* DEL. Chilopoda: *Lamyctes africana* (PORAT). Chelonethida: *Titanatemnus montanus* BEIER. Araneida: *Dictyna montana* TULL., *Drassodes kibonotensis* TULL., *Clubiona abbajensis* STRAND, *Xysticus fagei* LESS., *Heliophanus crudeni* LESS., *Pardosa karagonis* (STRAND), *Trochosa urbana* CAMB., *Hahnia rouleti* LESS., *Nesticus ambiguus* DENIS. Acarina: *Bdella piggotti* EVANS, *Anystis baccarum* (LINN.), *Balaustium angustum* EVANS, *B. bisetis* EVANS, *Erythraeus areolata* TGDH., *E. nasalis* EVANS, *Sphaerolophus salti* EVANS, *Oribata geniculata* (LINN.), *Oribatula setosa* EVANS.

The following is a list of the species recorded by SALT at an elevation of 4250 m on the alpine-desert zone of Mt. Kilimanjaro:

Orthoptera: *Parasphena pulchripes* GERST. (fig. 31A). Dermaptera: *Forficula cavalli* BOR., *F. triangulata* HINCKS. Heteroptera: *Oreonysius rugosus* USINGER, *Spilostethus montislunae* BERG., *Pterasthenia shiraensis* STOYRAN. Coleoptera: *Peryphus sjöstedti* ALL. (fig. 31D), *Plocamotrechus kilimanus* JEANNEL (fig. 31C), *Longitarsus salti* BRYANT, *Oreoscotus salti* MSHLL., *Oreosecus porculus* MSHLL., *Parasystatiella agrestis* MSHLL. (fig. 31E), *P. arctipes* MSHLL. Collembola: *Friesea africana* DEL., *Hypogastrura* sp., *Mesaphorura salti* DEL., *Parafolsomia africana* SALMON, *Lepidocyrtinus armillata pallida* DEL. Chilopoda: *Lamyctes africana* (PORAT). Araneida: *Clubiona abbajensis kibonotensis* LESS., *Thanatus* sp., *Pardosa karagonis nivicola* LESS and *Hahnia rouleti* LESS. Acarina: *Hypoaspis praesternalis* WILLM., *Allothrombium barbuligerum* ANDRE, *Oppia africanus* EVANS, *O. nasalis* EVANS, *Microtegeus undulatus* BERL., *Scutovertex africanus* EVANS, *Scheloribates laevigatus* KOCH, *Africoribates ornatus* EVANS, *Protoribates shiraensis* EVANS.

3. MT. MERU, MT. KENYA, MT. ABERDARE AND MT. ELGON

Mt. Meru, Mt. Kenya, Mt. Aberdare and Mt. Elgon are all volcanic formations of Tertiary times and are perhaps best considered at one place here.

Mt. Meru (near 2° 30′ SL and 36° 20′ EL), about 50 km to the west of Mt. Kilimanjaro, close to the Kenya-Tanganyika border, in Tanganyika, rises to an elevation of 4558 m above mean sea-level.

Mt. Kenya and Mt. Aberdare both rise from a vast elevated arid plain, with a mean altitude of 1600-1800 m, in the interior of East Africa. Mt. Kenya (0° 28′ SL) (elevation 5194 m above mean sea-level) is a somewhat irregular cone, with a shattered top. It is the second highest mountain in Africa and is in reality the ruined stump of an ancient volcano, which when in eruption must have naturally been hundreds of metres higher than at present and was perhaps loftier than even Mt. Kilimanjaro. It is deeply scoured by gullies, especially on the east and south slopes. Unlike in the case of Mt. Kilimanjaro, the base of Mt. Kenya is not inhabited and there is no cultivated zone also on this mountain. The forest-zone commences at elevations of about 2300-2400 m and ascends, somewhat higher than on Mt. Kilimanjaro, to nearly an elevation of 3300 m. The forest-zone consists of volcanic ash and tuff, with some lava flows and the rocks are better exposed in the alpine zone than at lower elevations. The central peak is of solidified rock in the throat of the volcano and is consolidated as a plug of hard rock. The alpine prairie on Mt. Kenya is characterized by taller grass than on the alpine zone of Mt. Kilimanjaro and is also far more humid. Above an elevation of 4500 m there are snow-fields and glaciers. Interesting accounts of the general geography and physiography of Mt. Kenya may be found in DUTTON (288), and GREGORY (424, 425) has given some account of the geology and glacial history of the mountain.

Mt. Aberdare is really a mountain ridge, extending nearly north-south between the equator and the 1° SL (somewhat east of 36° EL) in Kenya and about 100 km to the southwest of Mt. Kenya. It is not as high as Mt. Kenya; its highest points, Sattima and Kinangop, are only 3963 and 3960 m in elevation respectively. Mt. Aberdare overlooks Lake Naivasha on the west. The Peak Sattima is about 100 km to the northwest of Nairobi. The mountain is not at present capped by snow.

Mt. Elgon (1° NL in Kenya-Uganda), also known as Massawa, is an extinct volcano that forms a vast isolated mass, over 64 km in diameter at the base. Its slopes are precipitous on the north, west and south, but gentle on the east. The southern cliffs are remarkable for the presence of numerous caves that have apparently risen under the action of water. The mountain has the rim of an old crater, about

13 km in diameter. The highest point of the rim is 4321 m above mean sea-level. Steep spurs, separated by narrow ravines, provide for a large number of streams on the northwest and west, ending in the R.Nile through Lake Choga and on the south and southeast to Lake Victoria Nyanza and on the northeast to Lake Rudolf. The outer slopes of the mountain are covered by forests, especially bamboo forest, on the south and west, which also receive much heavier rainfall than the other slopes.

Traces of three separate glacial periods, with interglacial times, are found on the East African equatorial mountains and on parts of the Abyssinian Highlands. The last glacial period is also generally supposed to have produced the maximum glaciation in the area. Ancient moraines occur in the Nithi Valley on the east side of Mt. Kenya and glacial moraines occur also here and there and in other river valleys up to an elevation of 3000 m. The three glacial periods on the East African mountains paralleled the glacial times in Europe. There is at present no ice on the highest peak Sattima of Mt. Aberdare, but during the Pleistocene small glaciers flowed down to an elevation of 3700 m.

The top of Mt. Elgon was covered by a glacier cap, nearly 72 sq. km in size. Glaciation did not perhaps extend very low down on Kilimanjaro, but to elevations between 3600 and 3800 m only. The lowest limit of glaciers during the Pleistocene was perhaps 3550 m above mean sea-level. The glaciers are dwindling on Mt. Kenya, Mt. Kilimanjaro and on the Ruwenzori Range at present. Some glaciers have indeed completely disappeared on Mt. Kenya within the last 35 years.

FRIES & FRIES (387) have recently discussed some of the salient features of the general ecology and phytogeography of Mt. Kenya and Mt. Aberdare. They have given an interesting account of the altitudinal zonation of vegetation, the general characters of the vegetation, the principal plant formations and communities of the basal plateau and of the montane rain-forest zone, the marginal belt of the forest, the bamboo zone, the *Hagenia-Hypericum* zone, the alpine and the nival zones. They have also drawn attention to certain important aspects of the historical development of the flora of Mt. Kenya. See also COE (215a).

The altitudinal zonation of life on both Mt. Kenya and Mt. Aberdare agrees in its main features. The basal plateau is dominated by the steppe-vegetation, especially *Themeda*-steppe, widely spread all over Africa. There are also gully forests along the larger mountain streams. Beginning from the basal plateau, the montane rain-forest zone extends upward to an elevation of nearly 2380 m. Above this limit lies the sharply defined bamboo zone, reaching above to an elevation of 3200 m.The lower and higher parts of the bamboo zone differ more on the western slope of Mt. Kenya and on the eastern slope of Mt. Aberdare than on the other slopes of these mountains. The lower and the upper bamboo zones are separated from each other at approximately the boundary altitude of 2810 m. There is then a narrow belt of about 100-150 m of *Hagenia-Hypericum* zone. The alpine zone follows above this belt, from elevations between 3225 and 3300 m. The upper limit of Phanerogams is at about 4500 m on Mt. Kenya, above which lies the nival zone. There are no alpine and nival zones on the lower Mt. Aberdare. We can also clearly recognize on Mt. Kenya a lower-alpine zone as distinct from an upper-alpine zone, with the boundary altitude between them at about 3860 m. Relatively less rain falls on the north slope of Mt. Kenya. Erosion gullies, wide above and narrow lower down, often locally depress the zonal lines. In the upper-alpine zone the general ecological conditions are predominantly influenced by the prevailing climatic factors, but in the lower-alpine zone it is the edaphic factors that exert a significant influence on the ecological conditions.

The vegetation of the alpine zone on Mt. Kenya is remarkable for *Lobelia* and *Senecio*, at elevations of 3600 m. The lower limit of the alpine zone, towards the *Hagenia-Hypericum* belt, is generally sharply defined. Stray individuals of *Hypericum* grow at higher elevations, especially underneath steep cliffs. The alpine zone extends from the upper limit of the *Hagenia*-belt to approximately 4500 m, where it passes, without a sharp boundary line, into the nival zone. The presence or the absence of giant *Senecio* forms the only basis for the subdivision of the alpine zone into the lower and upper-alpine sub-zones on this mountain. The tree-like *S. keniodendron* is absent in the lower-alpine zone, where, however, the trunkless *S. brassica* is prominent; its uppermost limit is 3860 m. The upper-alpine zone is characterized by the dominance of the tree-like *Senecio*. A rosette-forming trunkless species occurs thus at lower elevations, but the arborescent species of the same genus grows at higher elevations. The last *Senecio* trees grow at about 4455 m above mean sea-level. Sparse in the lowermost part of the alpine zone, *S. brassica* becomes dominant at elevations of 3420 m, in a steppe composed mainly of *Festuca pilgeri*. The cabbage-like leaf rosettes, almost without stalks, are remarkable for their white and woolly, exposed undersides of the leaves. *Senecio keniodendron* appears at altitudes at which *S. brassica* disappears and its sparsely branched trunk grows up to heights of 6 m (or it may also be unbranched at higher elevations) and covered by leaf-fragments. Isolated plants grow at the very highest level of the alpine zone. *S. battiscombei* is tree-like and is more branched than the other species and is peculiar to Mt. Kenya, in the lower-alpine zone, on the slopes and below steep cliffs, at an elevation of 3600 m. There are also two other dominant species, viz. *Lobelia telekii*, followed by *L. keniensis* about 100 m higher up. The former species grows up to the nival zone (4450 m), but the latter species disappears at about 4085 m above mean sea-level. The characteristic Ericaceae include *Erica arborea*, *Philippia keniensis* and *Ph. excelsa* on ridges in the lower-alpine zone. The *Senecio brassica—Festuca* association occupies the greater part of the lower-alpine zone (the regio alpina inferior), but is absent in localities of pronounced solifluction and bare rocky ground. *Senecio brassica* grows luxuriantly on grassy steppe, mainly of metre-high tussocks of *Festuca pilgeri gemina*. Other species include *Agrostis trachyphylla*, *Lobelia keniensis*, *L. telekii*, *Carex monostachya*, the yellow-flowered *Haplocarpha rueppellii* with its leaf-rosettes pressed close to the ground, *Swertia leucantha* and *Sw. mearnsii* and the mosses *Polytrichum keniae* and *Campylopus stramineus*. The characteristic vegetation on the steep cliffs in the eroded ridges in the higher reaches of the lower-alpine zone includes *Senecio battiscombei* below and *S. keniodendron* at higher levels, Ericaceae bush, *Arabis cuneifolia*, *Sedum ruwenzoriense*, *Asplenium uhligii*, *Polypodium rigescens*, *Antholyza speciosa*, *Helichrysum stuhlmannii keniense* and *Cineraria kilimandscharica*. Lichens, especially the yellow *Usnea*, Gyrophorae and Parmeliae, are abundant on rock. The commonest mosses include *Andreaea cucullata*, *Bartramia afro-ithophylla*, *Brentelia subgnathophalea*, *Frimmea ovata*, *Rhacocarpus humboldtii* and *Rhacomitrium durum*.

Solifluction occurs as low as the lower-alpine zone, but is much more frequent higher up and such localities are characterized by the presence of *Cerastium caespitosum*, *C. keniense*, *Sagina* sp., *Oreophytum falcatum depauperatum* f. *leiophyllum*, *Tachydium abyssinicum*, *Senecio keniophytum decumbens* and *S. k. glabrior*. A number of grasses and other plants grow on wind-eroded solifluction localities in the upper parts of the upper-alpine zone, close to the upper limit, viz. 4480 m, in February. The more important of these plants are *Agrostis volkensii deminuta, Anthoxanthum nivale, Festuca pilgeri, Cerastium caespitosum, C. keniense, Sagina* sp., *Arabis cuneifolia meruensis, Oreophytum falcatum* var. *depauperatum* f. *leiophyllum, Alchemilla keniensis subuniloba, Swertia subnivalis, Myosotis keniensis, Calamintha kilimandschari, Galium simense hypsephilum, Valeriana keniensis, Lobelia telekii, Cardus platyphyllus, Helichrysum brownii, H. fruticosum compactum, Senecio keniodendron, S. keniophytum candidolanatus* and *Senecio platzii*.

The fundamental ecological difference between Mt. Kenya and the other East African mountains is more readily observed at higher altitudes than in the lower montane rain-forest zone. Only about 8% of the species of this zone are endemic on Mt. Kenya, 11% in the high-montane bamboo zone and *Hagenia* zone and about 53% in the alpine zone. The alpine flora and the terricole insect fauna of Mt. Kenya are isolated in the highest degree from those of the other East African mountains. The alpine zones of Mt. Kenya and of Kilimanjaro are separated from each other by nearly a gap of 300 km of hot arid steppes. Only 18 of the purely alpine species of Mt. Kenya occur on Mt. Kilimanjaro also. The two mountains, as close as Mt. Kenya and Mt. Aberdare, with their alpine zones only about 68 km apart, exhibit great differences in their alpine-zone flora. Only 14 species of the purely alpine-flora of Mt. Kenya occur, for example, on the alpine zone of Mt. Aberdare.

It is extremely difficult to determine the precise boundary altitudes of the different biotic zones on Mt. Aberdare, especially at higher elevations; there is also no nival zone on this mountain. The valley phenomenon is conspicuous in the partly depressed plateau, causing a curious zone inversion, so that the naturally altitude-conditioned "alpine zone" occurs often above the *Hagenia-Hypericum* zone—the alpine high plateau edaphically conditioned by the valley phenomena at the bottom of the depression (BJÖKMANN's regio alpina descensa on the North Scandinavian mountains). To the alpine zone belong all the vegetation, lacking bamboo or *Hagenia-Hyericum* association, i.e., the open treeless steppes, interrupted by thickets and tree-like *Senecio*. Two sub-zones may be recognized by these characters, but with vague boundaries of about 3500-3600 m in the valleys, at the lower levels of which *Senecio* trees no longer form stands. The limit is slightly higher on the slopes of the valleys. Giant *Senecio* are prominent on the alpine zone of Mt. Aberdare: *S. brassicaeformis* parallels *S. brassica* of Mt. Kenya. Lacking a trunk, its leaf rosettes are, however, open and the underside of the leaf lacks also the white woolly hairs. It grows as low as in the *Hagenia-Hypericum* belt and also extends to very high altitudes, nearly to the peak Sattima. *Senecio aberdaricus* is tree-like, nearly 7 m high, with the trunk more slender than that of *S. keniodendron* of Mt. Kenya and is dichotically branched at the top. It forms small forests underneath cliff precipices and occasionally also on the very top of Sattima.

The bearded giant *Lobelia telekii* occurs also on Mt. Aberdare. *L. sattimae* is a parallel to *L. keniensis*. None of the species play physiognomically the same rôle as on Mt. Kenya. *L. sattimae* grows up to the upper-alpine zone and *L. telekii* grows on the summit of Sattima.

As on Mt. Kenya, there are on Mt. Aberdare also thickets of the bushlike *Erica princeana aberdarica* up to an elevation of 3600 m and *Philippia excelsa*, 1.5 m high in the alpine zone, but 4-5 m high in the lower *Hagenia-Hypericum* belt. *Stoebe kilimandscharica* (Compositae) is physiognomically conspicuous, with its grey hairiness; *Cliffortia aequatorialis* (Rosaceae) grows to 4-5 m height and represents the northernmost limit of the genus in Africa. It is found sparingly also on Mt. Kenya, but many species occur in South Africa. The principal alpine-zone association on Mt. Aberdare is *Koeleria convoluta-Alchemilla cyclophylla*, characteristic of dry or slightly moist localities. It reaches on Sattima right up to the summit (387).

The most important contributions on the insect life of Mt. Kenya are by ALLUAUD & JEANNEL (34, 35). TOTTENHAM (1041) described an interesting collection of Staphylinidae from the mountain. HANCOCK & SOUNDRY (466) have described the fauna of Mt. Elgon. SALT (913) has also recently attempted to compare the insect fauna of Mt. Kilimanjaro with that of Mt. Kenya, Mt. Meru, Mt. Aberdare and Mt. Elgon.

Among the extremely interesting subterranean Coleoptera at the edge of the forest-zone on Mt. Kenya, mention may be made of the blind Carabid *Scotodipnus jeanneli* ALLUAUD at elevations between 2600 and 2800 m, the Pselaphid *Jeannelia microphthalmus* RAFFR., remarkable for its very minute and unpigmented eyes and blind Staphylinidae and microphthalmous Curculionidae. *Bembidion* (*Acropezum*) *mackinderi* ALLUAUD (at elevations of 3800-4400 m), *Plocamotrechus kenyensis* JEANNEL (4000-4500 m), *Kenyacus acrobius* ALLUAUD (3500-3700 m), *Kenyacus hypsibius* ALLUAUD (2800-3500 m), *Peryphus mackinderi* ALLUAUD, *Plocamotrechus unisetosus* JEANNEL, *Parasystatiella rugulipennis* MSHLL., *Oreoscotus fulvitarsis* HUST. and *Coenosia heterocneme* END., *Scarites hutchinsoni* ALLUAUD (2500 m), *Sc. kenyensis* ALLUAUD 2900 m) in the upper reaches of the bamboo forest and *Bembidion* (*Hypsipezum*) *kenyense* ALLUAUD in open areas within the upper reaches of the *Podophyllum*-bamboo forest are the other interesting records of high altitude insects from Mt. Kenya. It may be remarked that the genus *Plocamotrechus* JEANNEL is also known from South Africa and as already mentioned, it is represented on Mt. Kilimanjaro, Mt. Elgon and on the Ruwenzori Range and thus appears to be a typical mountain genus that has spread to the south temperate areas in Africa. JEANNEL records a second species *Pl. elgonicus* ALLUAUD on Mt. Elgon. *Trechus sjöstedti* ALLUAUD has also been found at higher elevations on Mt. Meru. It is remarkable that while the apterous *Calosoma* WEBER is absent on Mt. Kenya, the apterous *C.* (*Carabomorphus*) *catenatum* ROESCHKE extends on Mt. Aberdare from the edge of the forest to the alpine zone (3000-3100 m). Mention may also be made of *Tachys gilvus* SCHAUM. (3600 m), *Trechus kenyensis* JEANNEL (3000-4000 m), *Tr. k. atripes* JEANNEL (4500 m), *Orinophonus hypsinomus* ALLUAUD (2900 m) on Mt. Kenya, *Agonum kenyense* ALLUAUD (2800 m), *A. oribates* ALLUAUD (3000 m) and *Metabletus orinodromus* ALLUAUD (2500 m).

The following is a list of the species at different altitude levels on Mt. Kenya:

Species at elevations of 2500-2900 m: *Scarites hutchinsoni* ALLUAUD, *Sc. kenyensis* ALLUAUD, *Bembidion* (*Hypsipezum*) *kenyense* ALLUAUD, *Scotodipnus jeanneli* ALLUAUD, *Orinophonus hypsinomus* ALLUAUD, *Agonum kenyense* ALLUAUD, *A. oribates* ALLUAUD, *Metabletus orinodromus* ALLUAUD.

Species at elevations of 2900-3500 m: *Kenyacus hypsibius* ALLUAUD.

Species at elevations of 3500-3750 m: *K. acrobius* ALLUAUD.

Species at elevations of 3600-4200 m: *Tachys gilvus* SCHAUM., *Trechus kenyensis* JEANNEL and *Bembidion* (*Acropezum*) *mackinderi* ALLUAUD.

Species at elevations of 4000-4500 m: *Trechus kenyensis atripes* JEANNEL and *Plocamotrechus kenyensis* JEANNEL.

ALLUAUD & JEANNEL (*op. cit.*) report from Mt. Aberdare the interesting *Carabomor-*

phus catenatus ROESCHKE at elevations of 2650-3000 m; there are two subspecies, viz. *C. c. bastinelleri* ROESCHKE and *C. c. kolbei* ROESCHKE; *Scarites hypsipus* ALLUAUD at an elevation of 2650 m on the southwest slopes at the base of Kinangop; *Sc. abarderensis* ALLUAUD at an elevation of 3000 m in open localities in the upper reaches of the bamboo forest and on the alpine-prairie at the base of Kinangop; *Bembidion* (*Omotaphus*) *mixtum altipeta* ALLUAUD at an elevation of 2750 m in the upper reaches of the forest and the alpine-prairie; *Tachys ascendens* ALLUAUD at an elevation of 2700 m on the alpine-prairie; *Orinophonus hypsinomus* ALLUAUD at 3000 m on the alpine-prairie at the base of Kinangop; *Harpalus gregoryi* ALLUAUD at 2460 m on the west slope; *Agonum oribates* ALLUAUD at 3000 m and *A. kinangopum* ALLUAUD at 3000 m on the alpine-prairie. *Orinophonus hypsinomus* ALLUAUD occurs both on Mt. Aberdare at an elevation of 3000 m and on Mt. Kenya at an elevation of 2900 m and *Agonum oribates* ALLUAUD also occurs at an elevation of 3000 m both on Mt. Kenya and Mt. Aberdare; *Tachys ascendens* ALLUAUD occurs at an elevation of 2700 m both on Mt. Aberdare and on Mt. Kenya. There is the interesting record of the Staphylinid *Paederus alticola* FAUVEL at elevations of 3000-3500 m and of the Carabid *Hystrichopus meruensis* ALLUAUD at 4300 m on Mt. Meru. The Lepidoptera *Cupido aequatorialis* SHARP occurs on Mt. Elgon at an elevation of 2590 m, on Mt. Kenya and on Mt. Kilimanjaro at elevations of 3000-4000 m and on Mt. Birunga at 2900 m. *Forficula sjöstedti* BURR occurs at elevations of 4000-4300 m on Mt. Meru. The intensity of speciation seems to be high among the present-day high altitude Coleoptera from Mt. Elgon. According to the observations of SCOTT (950), more than 20 species and subspecies have been found on only one-tenth area of its crater (the circumference of which is only 16 km).

4. THE RUWENZORI RANGE

The Ruwenzori, commonly called the Mountains-of-the-moon, is also known under the names Ukonju, Bugombowa, Avuruku, Avirika, Ruwenzuru-ru, Ruwenjura, etc. meaning the King-of-clouds or the Rain-maker and Ru-nssoro, Kokora, Kiriba or the Mountain-of-mountains and Ansororo or snow. The range is situated less than half a degree north of the Equator in Uganda (about 30° EL), with a general direction north-northeast to south-southwest, in the shape of an enormous G. It is about 110 km long and 50 km wide in its broadest place. The Ruwenzori rises from the so-called Albertine Depression that lies about 180-200 m below the mean level of the surrounding land and contains the basins of the lakes Albert and Albert-Edward. This Depression is a part of the Western Rift. The Ruwenzori Range is indeed nearly completely surrounded by the Albertine Depression. It is without doubt the largest group of snow-covered high mountain range of Africa, situated in the middle of the continent and extending roughly in the direction of the principal axis of the landmass.

Of the six principal groups of high peaks, with glaciers and separated from each other by snow-free cols, the easternmost is Mt. Gessi and the westernmost is Mt. Emin. The latter joins Mt. Speke at the point where the range turns westward and rises to the highest peak group, viz. Mt. Stanley, and extends around in an eastward curve to Mt. Baker. To the south of this is Mt. Thomson. The Mt. Stanley group includes the highest twin peaks Margarita (5125 m) and Alexandra (5090 m), with other lesser peaks like Elena (4990 m) and Savoia (4980 m). The most important river basin to the east of the chain is that of the Bujuku Valley, surrounded by five mountain groups and fed by the glaciers from Mt. Stanley, Mt. Speke and Mt. Gessi. The upper Mobuku Valley receives the melt-waters from the Baker Glacier and other lesser glaciers on the east side of Mt. Luigi di Savoia. The western slopes are very much steeper (gradient 22°) than the eastern slopes (gradient 4°) and the distance from the foot of the mountain to the watershed is considerably less. At the south end, two narrow spurs project from the main mass into the Lake Edward Plain and reach almost to the edge of the lake. In the north the lower slopes gradually pass into the high country (1525 m above mean sea-level) below and continues along the east of the Lake Albert. The main ridge is, however, rather narrow at this point and extends into the Semliki Valley, nearly to the Lake Albert. Mt. Karisimbi (4507 m), in Ruanda near Lake Kivu (close to 1°30'SL and west of 30°EL), forms a part of the Albert National Park and should be considered to represent the southernmost extreme of the Ruwenzori Range.

It may be remembered that while both Mt. Kilimanjaro and Mt. Kenya are isolated high massifs, which were originally active volcanoes, the Ruwenzori is, on the other hand, a chain of mountains, consisting of primitive rocks. Three causes are generally believed to underlie the uplift of the Ruwenzori mountains: geo-tectonic, stratigraphic and lithographic. The main features are the upheaval *en bloc* of a whole portion of the Archaean rocks of Central Africa, with a main slope for the rise from the west to the east, related in the main to the gigantic western fracture, vertical displacements, etc., the ellipsoidal or anticlinal elevation, with the general direction from the north to the south and the strata more or less conspicuously uplifted in the Ruwenzori group. A central group of rocks like amphibolites, diorite, diabasis, amphibolic gneiss, etc. have resisted the weathering, but the gneiss rocks and mica schists of the lower zone are far less resistant to weathering.

i. General Ecological Conditions and Biotic Zonation on the Ruwenzori Range

The mean atmospheric temperatures at elevations of about 1830 m are never very high on the Ruwenzori mountains, but the temperatures do not also fall very low even at the summit. At an elevation of about 1980 m the mean atmospheric temperature is 23.35° C for the maximum and 14.9° C for the minimum. Above the permanent snowline, the maximum atmospheric temperature is 6.1° C and the minimum is —3.3° C. The following mean temperatures were recorded by the Italian Ruwenzori Expedition (241) at different altitudes during the summer:

TABLE 22

Mean atmospheric shade temperatures at different altitudes on the Ruwenzori Range

Elevation in m	Mean temperature ° C
1916	16.6
2910	0.6
3506	9.7
3795	6.6
3918	10.7
3933	6.2
4032	3.6
4045	3.2
4071	6.2
4117	8.7
4160	0.6
4188	4.0
4219	2.7
4243	4.2
4310	4.0
4326	3.2
4322	0.6
4347	4.0
4349	2.7
4390	3.7
4444	4.0
4465	1.0
4475	3.0
4516	3.6
4582	0.7
4770	3.6
4858	0.7

The mean summer temperature in the shade at elevations between 2000 and 4000 m is 8.4° C and at elevations above 4000 m it is 3.2° C. The temperature lapse rate is 4° C per 1000 m of increase of altitude.

The atmosphere around the Ruwenzori mountains is always misty and the mountains are always covered by dense clouds during the daytime at elevations above 2750 m. The rainfall must be described as indeed excessive; it generally rains from the middle of February to May or even June; June to August are mostly fine and fairly dry, though even during this period there may be some rains; from September to the middle of December the rainfall is rather heavy, these being perhaps the wettest months on the mountain. The dry season lasts from the end of December to February. There is a marked contrast between the climates of the southern and the central portions of the Ruwenzori Range; in the south there is usually much less rainfall than in the central parts. In proportion to the length and the massiveness of the mountains, the extent of snow on the Ruwenzori Range is really small and practically all the permanent snow is at present contained within an area hardly

16 km long and 13 km wide. The permanent snowline on the east slope is 4220 m. Being more protected from the sunrays than the east slope, the snowline on the west slope is also lower.

The existence of vast glaciers on the Ruwenzori Range during the Pleistocene is proved by a number of observations. Large glaciers filled the valleys of Mahoma, Mobuku and Bujuku, met below Nakitawa and flowed down to beyond Bihunga. The present-day glaciers are all small and are also receding; they are indeed only the insignificant relics of earlier glaciers that filled the valleys. The Mobuku Glacier ends at present at an elevation of 4170 m, but probably it extended earlier down to 1980 m in the valley. The Moore and the Semper Glaciers are the longest at present; the former descends as low as 4175 m and the latter as low as 4270 m. The largest present-day glaciers are found on Mt. Stanley, Mt. Speke and Mt. Baker and on the eastern slope of the Gessi group of peaks. The snowline is estimated to lie between 4510 and 4480 m. At an elevation of about 4270 m the rain always turns into snowfall on the mountains.

The life zones on the Ruwenzori Range, from the basal Albertine Depression to the summit, are as follows: the grass zone at elevations below 1980 m; the forest zone at elevations between 1980 and 2580 m; the bamboo zone above this reaches up to an elevation of about 3050 m; the tree-heath and moss zone above, reaches to an elevation of 3800 m; the *Senecio-Lobelia* zone lies between 3800 and 4267 m; above this lies the barren rock and snow zone. The bamboo zone extends between 2580 and 3050 m on the east slope and between 2133 and 2745 m on the west slope. The tree-heath-moss zone is succeeded by the *Senecio-Lobelia* zone, with *L. deckenii* up to elevations of 3962 m, *L. stuhlmanni*, appearing at an elevation of 3050 m, is replaced by *L. wollastoni* at an elevation of 3657 m.

Because of the prevailing higher humidity on the Ruwenzori Range and also because of the close proximity to the forests of Congo, the temperate forest is far more extensive than on Mt. Kenya and reaches up to nearly 4000 m above mean sea-level, at which elevation we find *Erica arborea*. The alpine-flora of *Lobelia, Senecio* and *Helichrysum* is more luxuriantly developed on the Ruwenzori Range than on Mt. Kenya and Mt. Kilimanjaro. The general ecological conditions are not, however, very favourable for great abundance of high altitude terricole species of insects at elevations above 3000 m on the Ruwenzori mountains. Although there are no dominant Palaearctic forms, there are nevertheless certain boreal genera of both plants and insects on these mountains. Butterflies, moths and Diptera have been collected on snow-fields at elevations of 4876 m. On barren rock above the snowline (4220 m) worms, lichen and moss have also been collected.

ii. High Altitude Insects from the Ruwenzori Range

The reader will find interesting data on the Ruwenzori mountains in FUCH's Report of the Lake Rudolf Rift Valley (391), GREGORY's Rift Valley and Geology of Africa (423), GRUNNE's book on the Ruwenzori (448), in MEYER (770) and REIST (880). The most important contributions to our knowledge of the high altitude insect ecology and the general conditions of life on the Ruwenzori Range are DE FILIPPI's (241) account of the Expedition of the Duke of Abruzzi and OGILVIE-GRANT's 1909/1910 Report of the British Museum Ruwenzori Expedition (800). The Coleoptera collected by the Italian Ruwenzori Expedition were described by CAMERANO

(180). In his interesting book *The Ruwenzori*, DE FILIPPI (241) has, in addition to describing the general geography, geology, climate and vegetation of the region, also given a brief account of the zoological studies by the Italian Expedition (pp. 394-397). His list of species collected by the Expedition unfortunately lacks exact data regarding the locality or altitude where the insects were found. The following are the more important records from his book:

Dermaptera: *Pygidicrana livida* BORELLI, *Anisolabis compressa* BORELLI, *Gonolabis picea* BORELLI, *Spongiophora aloysii sabandiae* BORELLI, *Chaetospina ugandana* BORELLI, *Opisthocosmia roccatii* BORELLI, *Apterygida cognii* BORELLI and *A. cavallii* BORELLI. Coleoptera: *Hydaticus rochei* CAM., *Cillaeus cavallii* CAM., *C. cagnii* CAM., *Hydrophilus sloanei* CAM., *Lixus roccatii* CAM., *Sipalus aloysii sabandiae* CAM., *Eumelosomus aloysii sabandiae* PONGELLA. The myriapods include species of the genera *Cryptops* LEACH, *Scutigerella* RYDER, *Phaeodesmus* COOK, *Habrodesmus* COOK, *Archispirostreptus* SILV. and *Odontopyge* BRANDT.

The British Museum Ruwenzori Expedition (1135) was led by R. B. WOOSNAM and collected over 100 Arachnida, 1015 Coleoptera, 33 Orthoptera, 25 Neuroptera, 47 Hymenoptera, 1372 Lepidoptera, 130 Hemiptera, 23 Homoptera and 55 Diptera. The Coleoptera were described by ARROW (56), GAHAN, WATERHOUSE, MARSHALL and CAMERON (1135), and particularly the Carabidae by JEANNEL (583), the Staphylinidae by TOTTENHAM (1041), Halticinae by BRYANT (165), Hemiptera by DISTANT (266), Lepidoptera by HAMPSON (465) and HERON (510) and the Diptera by AUSTEN (64) and JONES (598).

The following Coleoptera are recorded by the British Museum Expedition:

Carabidae: *Eudema erichsoni* HOPE (1830-3960 m), distributed from Sierra Leone to Nyasaland; a species close to *Platynus* (*Amchomenus*) *striatitarse* PER. at elevations between 1830 and 3960 m; *Harpalus* sp. (1850-3960 m); *Lebia* sp. (1830-3960 m); *Trechus alluaudi* JEANNEL, *Tropicoritus ruwenzorii* ALLUAUD, *Selenoritus ptolemaei* ALLUAUD (2700-2800 m); Histeridae: *Hister monitor* LEW. (1830-3960 m); Tenebrionidae: *Opatrum* sp. (1850-3960 m); Chrysomelidae: *Colasposoma consimile* GAHAN (1850-3960 m), *Diacontha passeti* ALLARD (1850-3960 m); Curculionidae: *Blosyrodes haroldi* HARTM. (1850-3960 m), *Bl. seminitidus* MSHLL. (1850-3960 m), *Amphitmetus leggei* MSHLL. and *Apotmetus montanus* MSHLL. (1850-3960 m).

Heteroptera comprise the following species, taken at elevations of 1850-3960 m: Scutelleridae: *Cryptacrus comes* (FABR.), also widely distributed in the plains of tropical and subtropical Africa; Pentatomidae: *Elachtheus lutulentus* (STÅL) known also from south and southeast Africa, *Awemba typica* DIST., *Aspongopus nigroviolaceus* (PAL.) and *Aspongopus alternatus* DIST.; Coreidae: *Holopterna wollastoni* DIST. Mention may also be made of the interesting Mirid *Meconema* FIEBER from tropical mountains of India and Africa. The University of Rhodesia–Nyasaland Mount Ruwenzori Expedition collected *Meconema ruwenzoriese* GH. at an elevation of 3932 m (405A). Most of the species listed so far occur at lower elevations also and are not, therefore, typically high altitude mountain autochthonous species.

The bulk of the Rhopalocera collected by the British Museum Expedition (510) from elevations of 1850-3960 m are also found on the plains, and are not, therefore, strictly speaking mountain autochthonous high altitude forms. *Amauris grogani* SHARPE, *Danaus chrysippus* (LINN.), *Gnophodes minchini magniplaga* HERON, *Melanitis leda* (LINN.), *Mycalesis dentata* SHARPE, *Yphthima itonia* HEW. and *Y. simplicia* BUTL., *Acraea sotikensis* SHARPE, *A. encedon* (LINN.), *A. disjuncta* SMITH, *A. amicitiae* HERON,

Argynnis excelsior BUTL., *Hypolimnas anthedon* DOUBLEDAY, *H. misippus* (LINN.), *Precis pelarga* (FABR.) (3960–4875 m), *Pr. tugela pyriformis* BUTL., *Salamis parhassus aethiops* (PALISOT DE BEAUV.), *Pseudargynnis hegemone nyassae* (GOD.), *Neptis agatha* (CRAM.), *Neptidopsis ophionae velleda* (MAB.), *Harpendyreus reginaldi* HERON, *Uranothauma nubifer* (TRIM.), *U. delatorum* HERON, *Mylothris clarissa* BUTL., *M. jacksoni* SHARPE, *Catopsilia florella* (FABR.), *Chioneigia leggei* HERON, etc. are some of the more important records. The highest altitude record for Lepidoptera on the Ruwenzori mountains is perhaps 4875 m. The moth *Episilia rhodopea* HAMPS. occurs at an elevation of 3850 m and *Larentia heteromorpha* HAMPS. at elevations of 1830–3850 m.

Among the Diptera recorded by the British Museum Expedition, mention may be made of the interesting Syrphid *Senaspis elliotii* AUSTEN and the Bombylid *Bombylius* sp. at elevations of 1830–3960 m.

5. THE ETHIOPIAN HIGHLANDS

The Ethiopian mountains differ conspicuously from other East African mountains in a number of respects. The mean elevation of the Ethiopian highlands is great, but the altitudes of isolated mountains are conspicuously less than those of Mt. Kilimanjaro, Mt. Meru, Mt. Elgon, Mt. Kenya and the Ruwenzori Mountains. The Semyen Massif that rises to an elevation of 4575 m (Ras Dashan) is the highest mountain in the Ethiopian highlands; Mt. Buahit rises to an elevation of 4267 m, Mt. Batu exceeds 4267 m and a number of other mountains are only 3657–3960 m above mean sea-level.

The Ethiopian massifs are likewise related to the Rift Valley Fracture System. They are associated with the trough extending northeast to the southwest diagonally across Ethiopia, from the point of meeting of the troughs of the Red Sea and of the Gulf of Aden to the northern end of the East Africa Rift at the Lake Rudolf. This Ethiopian Depression, containing the chain of the Galla Lakes, is generally known as the Galla Depression. The northeastern end of the Galla Depression is determined by two series of fractures, one of which is directed from the north to the south and forms the eastern escarpment of the high plateau of northern Ethiopia; the other from the east-northeast to west-southwest forms the Harar and North Somaliland Escarpment. The fractures lie parallel and the Depression between them encloses the furrow, in which the R. Awash (Hawash) descends from the highlands to flow in the Lake Abbe in the Danakil Desert. South of Awash, the trough contains a series of hollows, with the chain of Galla Lakes, from Lake Stefanie. From here it is continued to the northern end of Lake Rudolf, to unite with the meridional East African Rift. Many volcanic formations of the Aden Series are associated with the Galla Depression and have mainly Pliocene and Pleistocene basalts and trachytes, distinct from the earlier Trap series of dolerites and basalts of the early Tertiary or even Cretaceous times and forming great tablelands of the northern and western Ethiopia.

Although there are no active volcanoes at present in Central Ethiopia, those of the later series seem to be scarcely quiescent, particularly the eastern volcano Dubbi, south of Edd on the Danakil Coast (which erupted in 1961). Dofan in the lower Awash Valley has active fumaroles. Hot springs occur in the lower parts of Addis Ababa, near the Lakes Abaya, Dilla, etc. The Ethiopian series of volcanic structures is continued to the northeast of Harras and other volcanic formations of the foothills and highlands of southwest Arabia. The Ethiopian series unites in the south with that of the Rift Valley, in which the Teleki Volcano, at the southern end of Lake Rudolf, has been active during recent years.

Evidence of extensive glaciations in the area during the Pleistocene has recently been found. Almost 440 sq. km area of the Semyen Mountains were covered by glaciers. Even during the last glacial time, as much as 60 sq. km were covered by ice in the area. Evidence of former glaciers has also been found on Mt. Chillalo. It was formerly believed that due to the shifting of prevalent winds, the mountains of Ethiopia became desiccated in the Pleistocene, thus giving rise to cold dry conditions. These cold arid conditions were assumed to have caused the disappearance of flora and fauna more than the ice during the Pleistocene period. OMER-COOPER suggested in 1930 that the wide extent of the highlands would really favour condensation and, therefore, total destruction of flora and fauna due to aridity was unlikely to have really occurred. On the other hand, extensive fall of volcanic dust accompanying the last phase of plutonic activity in the area was perhaps the most decisive factor in the disappearance of the freshwater fauna. The curious absence of Crustacea in the freshwater lakes in the region seems to lend support to this view. Whatever the ultimate cause, the disappearance of the flora and fauna of the ancient highland seems to have happened not very long ago. The animals and plants that flourish at present at elevations up to 2130 m are essentially recent extensions of the savanna or the thorn-scrub elements.

i. Biotic Zones on the Abyssinian Highlands

Although the life zones on the southern Ethiopian highlands are broadly related to those of the other East African mountains, the general sequence of the altitudinal zones on the Ethiopian Highlands presents certain interesting peculiarities. As pointed out already, the mean elevation of the plateau is great, but the heights of the isolated massifs rising above it are less than those of the other East African mountains. Consequently, coniferous forests flourish in some localities somewhat above the general level of the plateau. We have, for example, *Juniperus procera* and *Podocarpus* of the Jem-Jem forest west of Addis Ababa, at elevations between 2500 and 2750 m and *Podocarpus* in the Magado forest in south Sidamo. Based on agriculture, we have in the Abyssinian region the lowest zone of the *quolla* (= hot land) from sea-level to an elevation of 1500-1800 m, *waina daga* (= mountain forest) from the upper limits of the quolla to about 2400-2500 m above mean sea-level and the zone of the higher *daga* (= hill) up to 3500 m above mean sea-level. While there is a pronounced wealth of giant lobelias and arborescent senecios on the other East African mountains, only two species of the former, viz. *Lobelia rhynchopetala* and *L. gibbosa* are found on the Ethiopian highlands and there is no arborescent *Senecio*. Mt. Kilimanjaro, Mt. Kenya and Mt. Elgon and the Ruwenzori Range, although situated near the Equator, are covered by permanent snow and have a characteristic biotic zone above the clouds. This zone is absent on the Ethiopian mountains; no mountain in Ethiopia attains indeed the snow level.

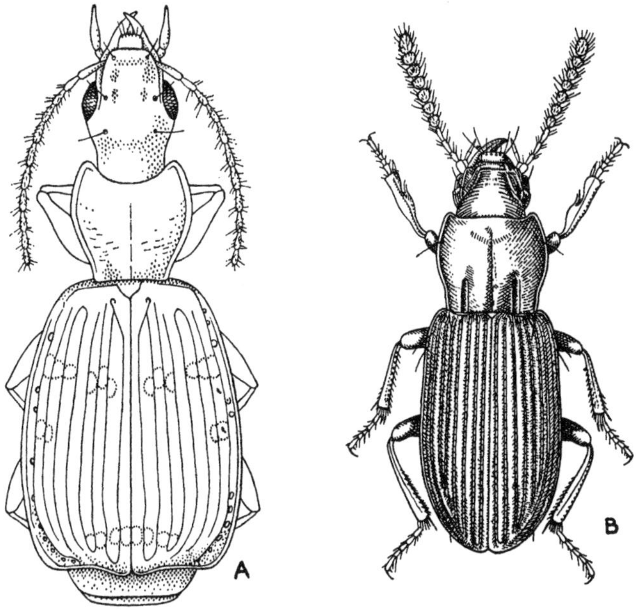

Fig. 33. High altitude Coleoptera from the Equatorial East African mountains. A. *Selenovites ptolemiei* ALLUAUD from the Ruwenzori Range (2700-2800 m), B. the blind and apterous *Caecocaelus* SCOTTI from the Abyssinian Highlands (2700 m). (After SCOTT).

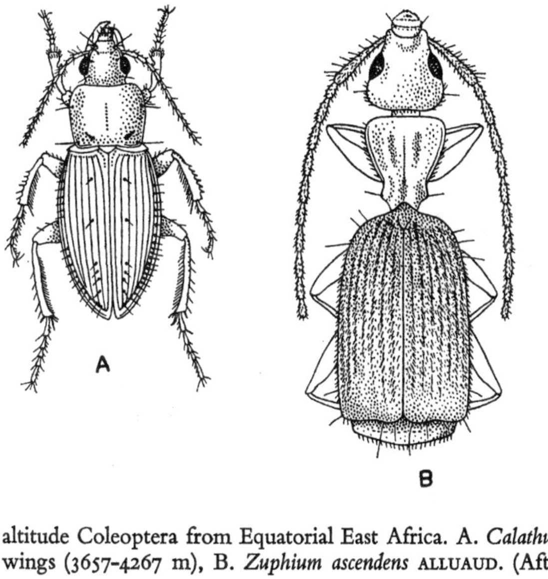

Fig. 34. High altitude Coleoptera from Equatorial East Africa. A. *Calathus ras* with atrophied wings (3657-4267 m), B. *Zuphium ascendens* ALLUAUD. (After SCOTT).

ii. Characteristics of the High Altitude Insect Life of the Ethiopian Highlands

The Ethiopian highlands have been explored by a number of entomologists. RAFFRAY (852, 853) was perhaps one of the first to collect insects from Mt. Abuna-Yosuf. A number of insects were collected by Marchese SAVERIO PATRIZI in 1940 from an elevation of 3250 m on Mt. Termaber, situated about 70 km from Addis Ababa on the road to Dessie. RAFFRAY (852, 853), BASILEWSKY (87, 88), JEANNEL (575, 578, 579, 580, 581, 584, 585, 589), CARPENTER (188), DYTE (290), MARSHALL (741), SCOTT (950, 951) and UVAROV (1070, 1072, 1075) have made valuable contributions to our knowledge of the high altitude insects from the region. The Carabidae from the Abyssinian mountains are described by BASILEWSKY (87, 88), JEANNEL (581, 585), RAFFRAY (852); Pselaphidae by DYTE (290); Curculionidae by HUSTACHE (554) and MARSHALL (741); Diptera by RICHARDS (884, 885); Orthoptera by UVAROV (1070) and Rhopalocera by CARPENTER (188). Some of the genera occurring at elevations between 3500 and 4000 m on the Ethiopian mountains are also known from the alpine zones of some of the south European mountains. On the Lasta Mountain Range in Abyssinia, at an elevation of about 3500 m, RAFFRAY *(op. cit.)* recorded many species of *Calosoma* WEBER, like *C. caraboides* RAFFR., which have distinct *Carabus*-like facies. A species of *Cymindis* LATR. is closely related to *C. humeralis*, two species of *Harpalus* LATR. are related to the European *H. goudoti* and *H. litigiosus*, *Amara* spp. analogous to *A. trivialis* from Europe, three species of *Calathus* BON., one of which seems to be related to the Californian (?) *C. ruficollis* and also two species each of *Trechus* CLAIRV. and *Bembidion* LATR. Two species of Dytiscidae, related to *Agabus cephalotes* from Corsica, are found under submerged stones in icy-cold torrents. The apterous subgenus *Carabophanus* KOLBE of *Calosoma* WEBER, found above the timberline on the Abyssinian mountains, is not, however, represented in Europe. Among the Staphylinidae, the species of *Ocyphus* STEPH. from the Abyssinian mountains are related to the European *O. fulvipennis*. A species of *Deleaster* ERICHS., occurring both in Europe and North America, is reported from the Ethiopian Highlands. Two Abyssinian high altitude species of *Otiorrhynchus* GERM. are related to *O. perdix* OLIV. from the Tyrolian Alps and another to *O. noui* from the Pyrenees. The Scarabaeid genera not known from Europe, but found above the forest zone on the Abyssinian mountains include *Simogonius* HAR., *Schizonycha* DEJ., *Centrantyx* FAIRM., etc. Endemic Paussidae are particularly abundant at higher elevations on the Abyssinian highlands, but the family is wholly absent in the sub-alpine zone. Except *Trechus* CLAIRV., *Calathus* BON. and *Amara* BON., the affinities of most other Carabid genera are, however, entirely African. The boreal forms that have extended their range to the Ethiopian Highlands become sparse southwards. There are, for example, over 17 endemic species of *Calathus* BON. on the Ethiopian Highlands, but only one further south. Fifteen endemic species of *Trechus* CLAIRV. (s. str.) are found on the Ethiopian Highlands and a number of species have also evolved on Mt. Elgon and still further south we have *T. sjoestedti* ALLUAUD on Mt. Meru. *T. gughéensis* JEANNEL is the southernmost species from Mt. Elgon to be so far found on the Ethiopian Highlands. The ancestral stock of *Trechus* CLAIRV. in Africa, which also extended southwards, seems to have developed primarily in the moist forests at lower elevations and the existing montane species began as colonizations from such types. The tropical African species of *Plocamotrechus* JEANNEL seems to have radiated northward from a southern dispersal centre during the Miocene and the species are at present abundant at all elevations in South Africa. While many relict species occur on Mt. Kilimanjaro, Mt. Aberdare, Mt. Kenya and particularly on the Ruwenzori Range, none is known at present on the Ethiopian Highlands. Both *Trechus* CLAIRV. and *Plocamotrechus* JEANNEL are absent in the tropical African lowlands and they are confined exclusively to the mountains of the western and eastern branches of the Rift Valley System. *Omotaphus simienensis* BASILEWSKY, found at elevations of 3050-4267 m, is related to forms from the Aberdare Mountain and from Mt. Elgon and *Liagonum scotti* BASILEWSKY is related to species from Mt. Kilimanjaro. All the species of *Trechus* CLAIRV. from the Abyssinian mountains, including *T. sublaevis* JEANNEL and *T. bipartibus* JEANNEL from Mt. Abuna-Yosuf described by RAFFRAY *(op. cit.)*, and four species from Mt. Chillalo and *T. gughéensis* JEANNEL from the Gughé Highlands, are completely apterous. Even *T. aethiopicus* JEANNEL from relatively lower elevations of 2800 m is apterous. Many species of *Calathus* BON. are likewise apterous. Some of the species are indeed known only from particular massifs. The four species of *Trechus* CLAIRV., recorded from Mt. Chillalo, do not, for example, occur elsewhere, but the other species occurring at lower elevations are more or less widely distributed. The vicarious species on the different East African mountains are shown in table 23.

TABLE 23

A conspectus of vicarious species on the different Equatorial East African mountains

Family	Genus	Kilimanjaro	Kenya/Aberdare	Ruwenzori
Carabidae	Peryphus	sjoestedi	mackinderi	ruwenzoricus
	Plocamotrechus	kilimanus	unisetosus	ruwenzoricus
Curculionidae	Parasystatiella	agrestis	rugulipennis	—
	Oreoscotus	salti	fulvitarsis	—
Lepidoptera	Salta	acrophylax	—	edwardsi
Diptera	Coenosia	melanomeros	heterocnemis	brunnigena

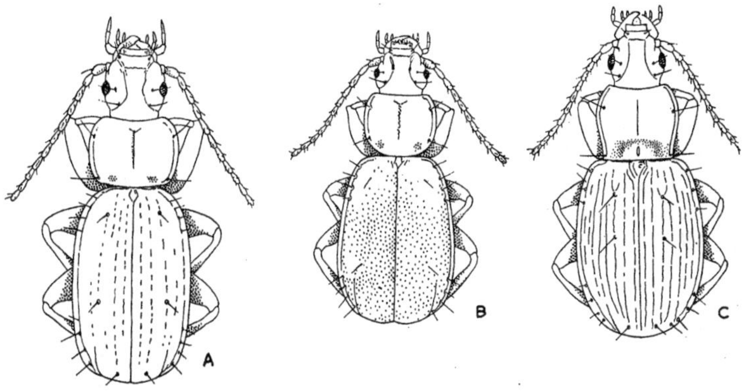

Fig. 35. Apterous high altitude *Trechus* CLAIRV. from Equatorial East Africa. A. *Trechus degiensis*, B. *Trechus pilosipennis*, C. *Trechus siminensis*

It is interesting to observe that the proportion of endemic species is on the whole greater at higher elevations than at the lower on all the East African mountains which we have discussed so far. Among the high altitude insects of the East African mountains, we find the following faunal component elements: i. tropico-alpine endemics like *Orinodromus* KOLBE, *Carabophanus* KOLBE (an analogue of the Neotropical-alpine *Blaptosoma* GÉH. and *Carabomimus* KOLBE from the high mountains of Mexico); ii. boreal temperate elements like *Bembidion* LATR. (including the subgenus *Testediolum* GANGLB.), *Amara* BON., *Cymindis* LATR. and *Otiorrhynchus* GERM.; iii. austral temperate elements like *Hystrichopus* BOH. and *Hipporrhinus* SCHÖNH.

In an interesting discussion on the biogeographical peculiarities of the high altitudes of the Gughé Highlands in southern Ethiopia, SCOTT (950) has recently suggested that the high mountain insects of this region are either descendents of stocks far more widely distributed in former times, when the prevailing climate of the surrounding lower elevations was cooler and more moist than at present or they are later evolved species, closely related to those of the surrounding lower levels, but which have developed montane characters. The species restricted at the present time to the highest altitudes appear to consist partly of the descendents of the ancestral forms, widely distributed in former periods. These views are largely in agreement with those of JEANNEL (585). The Trechinae and the Anillini from the Ethiopian Highlands constitute evident links between the Palaearctic fauna and the fauna of the East African high mountains. The presence of Trechinae on the Ethiopian Highlands also points to the mountains having been populated both from the north and from the south, the descendents of which interlock at the present time on the East African equatorial high mountains, but do not at the same time mingle on the same massifs. Most species seem to have evolved in close association with the high mountain plants on separate massifs by isolation. It seems, therefore, that these are relicts of an ancient Tertiary fauna, for the association of species appears to have been rather long-standing and the insects and plants have become specialized together. The high montane species may have also evolved directly during the Post-Pleistocene times from a parent stock that inhabited the surrounding plateau. LÖNNBERG (704) has shown that during the early Tertiary, including perhaps Miocene times, Africa was covered by a vast evergreen forest. The fauna was also endemic, but evidently with many types common to South Asia. The climate seems to have changed to arid conditions during the Pliocene and the fauna also became the restricted Asiatic steppe type, developed in East Asia. Pleistocene times saw the revival of the forests, but there were repeated wet and arid periods.

As mentioned above, the Trechinae restricted to the high elevations in Ethiopia link the Palaearctic fauna with that of the East African mountains. The Trechinae from the western series of high mountains are *Plocamotrechus* spp., a genus of southern origin. All the others belong to *Trechus* CLAIRV. of northern origin. According to JEANNEL (584), the chain of mountains to the west of Lake Rudolf and extending between the southwestern highlands of Ethiopia and Mt. Elgon served as migration routes southwards to the mountains, associated with the eastern trough of the Rift Valley System. Though *Trechus* CLAIRV. occurs on Mt. Elgon and on Mt. Meru, none of the species seems to have succeeded in colonizing the Ruwenzori Range or other mountains of the western series, separated from the eastern series by lower-level lands of Uganda and Sudan. The absence of *Trechus* CLAIRV. on Mt. Kenya, Mt. Kilimanjaro and on Mt. Aberdare (all of which have, however, *Plocamotrechus* JEANNEL) cannot also be readily explained in the present state of our knowledge. Not only the Trechinae of undoubted northern affinities are absent on the western series mountains, but even those of the southern origin (*Plocamotrechus* JEANNEL) are exceedingly few on the Ruwenzori Range. This poverty of species is usually explained, as pointed out above, on the basis of the excessive soil moisture on the Ruwenzori mountains. The fact that the four species of *Plocamotrechus* JEANNEL so far found on the Ruwenzori Range are represented by local races in different parts of the massifs seems to indicate that speciation is in active progress.

We may conclude, therefore, that Trechinae of northern and southern affinities interdigitate on the mountains of equatorial Africa, but do not mingle on the same massif. The descendents of species inhabiting relatively lower elevations in the region during the early Tertiary presumably took refuge on the high mountains during the Pliocene. Isolated on separate massifs, they have apparently survived the alternating wet and dry periods of the Pleistocene and the increasing aridity of the Post-Pleistocene. Localized new forms have evolved, isolated by wide stretches of tropical plateau.

The ancestral stock of the African *Trechus* spp. is thus believed to have extended southwards and the tropical African *Plocamotrechus* JEANNEL similarly extended northwards from a southern centre of differentiation and dispersal, where its ancestral forms arose during the Tertiary. The genus is at present rich in species in South Africa and it probably spread during the Miocene and many relict species occur on Mt. Kilimanjaro. No species of *Plocamotrechus* JEANNEL is known from the Ethiopian Highlands, but relict species occur on Mt. Aberdare, Mt. Kenya and the Ruwenzori mountains. Both *Trechus* CLAIRV. and *Plocamotrechus* JEANNEL do not occur in tropical African lowlands and are confined to the mountains.

The other descendent species, the ancestors of which spread northwards to the East African mountains, seem to have been restricted to the higher parts of isolated massifs by subsequent climatic changes. The northern ancestral stock arose on the primitive landmass of Angara (the part of Asia north of the Lake Baikal, between the Lena and Yenisei rivers; *vide* chapter XIII), and their descendents, spreading southwards and likewise isolated by subsequent climatic changes, have given rise to many forms, also confined to single massifs, but not on the same mountains that have been colonized by species of southern origin.

According to JEANNEL (582), the parental stock of *Plocamotrechus* JEANNEL differentiated at the beginning of the Tertiary on the vast mountainous landmass of South Africa, already folded at the end of the Palaeozoic Era and broken by fractures dividing the ancient continent of Gondwanaland, when the Atlantic Ocean and other seas originated as rifts. The Crozet and other Sub-Antarctic islands are held to be remnants of the Gondwanaland. The flightless *Amblystogenium* END., now confined to the Crozet Archipelago, and the parental stocks of the closely related South African *Plocamotrechus* JEANNEL were differentiated presumably not earlier than the Miocene, previous to which the climatic conditions were apparently unfavourable for Carabidae on the Crozet Islands. The great fractures of the Rift Valley System are also believed to have first opened during the Miocene. The separation and isolation of the descendent species of *Plocamotrechus* JEANNEL and *Trechus* LAIRV. on different East African massifs depended, therefore, on the formation of the Rift System.

Among the Anillini from the Ethiopian Highlands linking with the other East African mountain forms, mention may be made of the blind species of the anophthalmous genus *Microdipnus* JEANNEL, so far known only from Mt. Kenya. Another species has only vestiges of eyes and constitutes a monotypic genus *Cryptorites* JEANNEL and is related to *Microdipnus* JEANNEL, but with certain exceptional features.

Although several species of Lepidoptera* have been collected at elevations of about 3350 m, the highest elevation at which they have been found on the Ethiopian Highlands is 4267 m. Some typically boreal forms, like *Colias electo* LINN., *Pontia daplidice aethiops* J. & V., *Lycaena phaeas pseudophaeas* LUCAS, *Pararge maderakal* GUÉRIN, generally occur at elevations of about 3350 m. *Vanessa cardui* LINN. occurs at about the same elevation and *Argynnis hyperbius* R. & J. occurs on the southern Ethiopian mountains. Among the other Rhopalocera with distinctive boreal affinities, mention may be made of *Cacyreus palemon ghimirra* PALL. (Lycaenidae), with complex markings on the underside of the wings and widely distributed on the Ethiopian Highlands. The African Nymphaliid *Antanartia* ROTHSCHILD & JORDAN is sometimes superficially like *Vanessa atalanta* LISM.

Ants do not seem to occur above elevations of 2400-2750 m on the Abyssinian Highlands. Only rarely some species, like *Engramma ilgi* (FOREL) and *Acantholepis capensis* MAYR, are found at elevations of about 3200 m on the Semyen Massif.

The high altitude Diptera from the Ethiopian Highands include the interesting wingless Sphaerocerid (= Borboridae) fly *Binorbitalia triseta* RICH., related to two other species from the Gughé Highlands and Mt. Aberdare and Mt. Kenya, found at elevations of 3650 m and above. The Chloropid *Thaumatomyia* spp. are abundant. Other species are *Sphaerocera longipes* RICH. at an elevation of 4120 m southeast of Acruasie (Mai Datcha) and *Leptocera* spp. at elevations of 3350-4120 m.

SCOTT (951) has referred to the pronounced increase in melanism in the fly *Thaumatomyia secunda* RICH. with increase in elevation. The specimens collected at elevations of 3050-4260 m on the Ethiopian Highlands are darker than those taken on Mt. Elgon at elevations of 2133-2890 m; the specimens at elevations of 4290 m from the High Semyen are the darkest. Some Anthomyiidae like *Limnophora* R.D. and *Atherigona* ROND. were also found at high elevations.

Orthoptera were not met with at the highest elevations explored by SCOTT (950, 951). The maximum elevation for grasshoppers is 3050 m on Mt. Chillalo, 3250 m on Mt. Tola (Gughé Highlands) and 3350 m on the High Semyen. The high altitude grasshoppers from the region are generally brachypterous and have apparently evolved from the forms that inhabited the lower elevations. UVAROV (1075) does not find any trace of the Palaearctic elements among the Orthoptera of the region. Some apparently Mediterranean forms are really African, indicating a former northward extension of the fauna associated with dry acacia thorn-scrub savana. There is also a species of the Eremian element, widespread in the Palaearctic deserts, but penetrating the deep valleys intersecting the Abyssinian Highlands and reaching up to an elevation of 1670 m. There is also one montane Dermaptera, which is distinct from the lowland element. The Dermaptera from Mt. Kilimanjaro and the Ruwenzori Range have reduced wings or are apterous and each species is confined to single massifs.

In discussing the effects of isolation on separate massifs on the evolution of congeneric but distinct species, SCOTT (*op. cit.*) has drawn particular attention to the close association of the species of insects with certain high altitude plants, indicating remarkably parallel or perhaps also joint evolution in plants and insects, especially

* CARPENTER (CARPENTER, G. D. H. 1935. *Trans. R. ent. Soc. London* 83: 313-448) has made an interesting faunistic study of Abyssinian butterflies.

the assumption of giant statures in both. Species associated with *Lobelia* and arborescent *Senecio* on the summit of Mt. Chillalo include four Staphylinidae and two Cryptophagidae under the bark of the dead stem of *Lobelia rhynchopetala* and a species of Scolytidae found in fetid drying inflorescence of *L. gibberosa* on Zuquala. The high mountain species of Dermaptera and Orthoptera seem to have evolved directly from the parental stock that flourished on the surrounding plateau.

CHAPTER VIII

OTHER TROPICAL MOUNTAINS

The outline of the outstanding characters of the high altitude ecology of the mountains of the tropical regions, contained in this chapter, serves to illustrate the influence of local conditions on the high altitude environment and emphasize, at the same time, the fundamental differences from the high altitude ecology of the north and south temperate regions of the world.

In addition to the mountains of equatorial East Africa and of Ethiopia, dealt with in the foregoing chapter, several other tropical mountains attain subalpine, alpine or even nival zone elevations. In tropical West Africa itself there is the well known active volcano, Mt. Cameroon (near 4° NL and 9° EL), with two high peaks, the Great Cameroon (4075 m) and the Little Cameroon (1774 m), covered by forest nearly to the top (814). Mt. Cameroon erupted in 1909 and again in 1922. The mountain chains of Java, Sumatra, Borneo, New Guinea and Celebes are other interesting tropical mountains. In Java the Mt. Semeru (from the Sanskrit *maha-meru*; 8° SL and a little to the east of 112° EL) rises to an elevation of 3676 m. Mt. Kerintji (1° SL and a little to the east of 100° EL), reaching to an elevation of 3799 m is in Sumatra. Mt. Kinabalu* (6° SL and 117° EL), about 50 km east from the South China Sea coast, rises to an elevation of 4139 m in North Borneo and Mt. Rantekombola in Celebes (120° EL and 4° SL) is nearly 4338 m above mean sea-level. Most of these mountains are still largely unexplored entomologically and some of them, the insect life of which is known at least in part, present interesting problems of biogeography and faunal differentiation. The insect life is generally abundant and also greatly diversified within the forest-zone of these mountains. Mt. Mauna Kea (19° 35′ NL and about 4250 m above mean sea level) and Mt. Mauna Loa (19° 30′ NL and about 4168 m above mean sea-level) in Hawaii are of considerable interest from the point of their highly specialized insect life, its origin and faunal affinities. The remarkable Carabid *Macranillus oceus* SHARP, with greatly reduced compound eyes that lack also pigmentation and distinct facetting, occurs on the high plateau of the Kauai Island (Hawaii). Other high altitude Carabidae from the Hawaiian mountains like *Mecyclothorax pusillis* SHARP, *M. micans* BLACKB. and *Thriscothorax apicalis* SHARP are found at elevations of 2740-3050 m on Maui (Haleakala). The interesting family Proterhinidae of apterous beetles, that measure 1.5 to 5.5 mm long and have weakly beaklike head in the male and pronouncedly beaklike head in the female, is characteristic of elevations of about 2750 m on the Hawaiian mountains. Some remarkable insects have also been described from the volcanic mountains of Cape Verde Islands (15° NL). The easternmost island lies in the Atlantic Ocean about 500 km from the African mainland and the westernmost is about 200 km from this island. Interesting mountain autochthonous endemic species of Tenebrionidae like *Oxycara* SOL. and *Trichopedius* BEDEL

* CORNER (CORNER, E. J. H. 1964. *Proc. R. Soc. London*, (B) (982) 161: 1-91) has given an interesting account of the plant geography of the mountain flora of Mt. Kina Balu, the geological relations of this mountain to the neighbouring regions, the mountain soil, bird life and the zoogeographical characters.

and the Curculionid *Dinas* WOLL. occur in colonies under stones, both in the cloud-forest zone and at higher elevations on these mountains. The mountain of each island in the Cape Verde group has its own characteristic endemic species. There is also an older apterous terricole endemic element on the mountains of the Cape Verde Islands (695). While these records of high altitude and other mountain autochthonous insects are without doubt of considerable interest, two regions viz. the mountains of New Guinea and the Andes, deserve particular attention.

I. NEW GUINEA MOUNTAINS

The New Guinea mountains, situated in a region of exceptional biogeographical importance, attain impressive proportions and elevations. The Central Highlands of New Guinea constitute a sort of high mountainous backbone, nearly 2400 km long, extending from end to end of the island and reaching to elevations of nearly 5000 m above mean sea-level, thus ranking as one of the great mountain systems of the world. Though continuous, the backbone mountain is not a single chain, but comprises a series of complex ranges, separated by broad upland valleys and forming the water-parting divide between north-south flowing drainage systems of the island.

The principal topographic features include the Star Mountain, the Hindenburg, Behrmann, Mueller, Kubor, Schrader, Bismarck, Owen Stanley Ranges, etc., all rising to elevations of 3050 m above mean sea-level and sometimes even higher. The foothills of lesser elevations are best developed along the southern flank in the Western Papua, but are absent in the Bismarck Range, which rises rather steeply from the Markham-Ramu Valley to an elevation of 4694 m above mean sea-level in Mt. Wilhelm.

The Border Mountains (900 m above mean sea-level) of northern New Guinea constitute the eastern extremity of a highland that extends eastward from the former Dutch New Guinea (Irian Barat) into Australian New Guinea. It is linked to the Central Highlands by rather rugged and plateau-like low hills between Sepik and Sober. The Northern Ranges constitute a series of mountains, from the Indonesian New Guinea border south-eastward to Huon Peninsula. From the west to east are the Bewani, Torricelli, Prince Alexander, Adelbert and Finnisterre Mountains. The peaks rise to elevations between 1220 and 1370 m in Bewai and Torricelli Mountains. The Finnisterre and Saruwaged Mountains are high ranges, reaching to elevations of 3050-3915 m and rising rather abruptly from the sea in the north and east and the Ramu and Markham Rivers in the southwest. The Cyclops Serra Mountains are short ranges, with the Cyclops Mountain in the Indonesian area, the Oenake Mountains and the Serra Hills. The Cyclops Mountain rises to an elevation of 2119 m. In the Indonesian New Guinea we have the well known Sneeuw Gebergte (Pegunungan Saldju), which includes the Nassau and the Oranje Ranges. Mt. Carstensz (Puntjak Sukarno) on the Pegunungan Saldju rises to an elevation of 4998 m; the range has a west-east trend (about 4° SL and 136° EL). On the same range we have other well known peaks like Mt. Trikora (Mt. Wilhelmina) (4749 m), about 100 km to the east of Puntjak Sukarno. Mt. Mandala (Mt. Juliana) (5° SL and just east of 140° EL) has an elevation of about 4695 m.

The high mountain ranges of New Guinea, extending nearly east-west along the principal axis of the island, were elevated in part from under shallow seas during the Miocene and Pliocene times. The sea-level recessions of the Pleistocene led to the southern parts coming into direct contact with the northeastern parts of Australia (611A, 1056). Parts of eastern New Guinea and part of the north area, especially the Cyclops Mountains near Sukarnopura (Hollandia), are, however, of much greater age (207), and are believed to have been in existence since Pre-Cambrian times. The Cyclops Mountains, together with Manus and Solomon Islands, may perhaps represent the remnants of an ancient Melanesian Continent. Accurate dating of the coral limestones found at high elevations and other evidence like erosion patterns, changes in the river courses, the antecedent river systems, etc. indicate late uplift in certain areas, especially in the region of the Pegunungan Saldju (Sneeuwgebergte). The cliff on the south side of the Pegunungan Saldju is over 3000 m high and is perhaps the highest in the world. The cliffs seem to have originated from the snapping of rocky mass, in relieving crustal stress, combined with differential movements on the two sides of the fault. The old rock structures are much tilted and folded. The crustal movements appear to have taken place during the Pliocene. The high ranges, like the Bismarck Range and the Owen Stanley Range in the east, are of great age. The Pre-Cambrian schists and gneiss have been raised and lowered at different periods. Their subsidence was accompanied by deposit of limestone beds, now found high on the mountains and completely covering the old metamorphic rocks in the west. Volcanism and lava flows seem to have continued throughout the Tertiary. During the Pleistocene, glaciers descended to elevations of 3000 m or perhaps even less.

Our knowledge of the high altitude insects of the New Guinea mountains is based largely on the collections made by the well known American-Dutch Archbold Expedition to New Guinea during 1938-1939. The collections come from a variety of habitats, ranging from tropical swamps to alpine meadows above the forest, up to elevations of nearly 4570 m, on Mt. Trikora (Mt. Wilhelmina) and on the Pegunungan Saldju.

Despite the massiveness and length, the New Guinea mountains lack a true alpine insect community (1043, 1044). Diverse plants naturally grow, however, up to an elevation of 5000 m, but there are no conifer forests at higher elevations—the conifer-like Auracarian, Podocarp, Cupresses, etc., being confined to the lower elevations (73, 95). These peculiarities are interpreted by some workers as suggestive of the fact that the New Guinea region was connected with other areas at low elevations, which acted as effective barriers against the colonization of high altitude elements from other regions. The insects found at the present time at high elevations on the New Guinea mountains are recent derivatives of lowland forms that have become secondarily adapted to the cooler climatic conditions of the mountains. The Rhopalocera are rather rare or even absent at the highest elevations and the few species that are found at high elevations on the mountains are indeed identical with those occurring at lower elevations. DARLINGTON (234) has drawn particular attention to the pronounced neo-endemism among the Carabidae of these mountains. GRESSITT (427) reports the predominantly Papuan and common lowland longicorn beetle *Tmesisternus* LATR. at an elevation of 3600 m on Mt. Wilhelm. *Brontispa* SHARP (Hispinae: Chrysomelidae), found at an elevation of 2000 m in the east and western area, is represented by several species near sea -level also.The typical lowland butterfly *Troides* HÜBN. is also recorded by GRESSITT *(op. cit.)* at an elevation of 2000 m. Significant differences may be observed among the high altitude insects from the

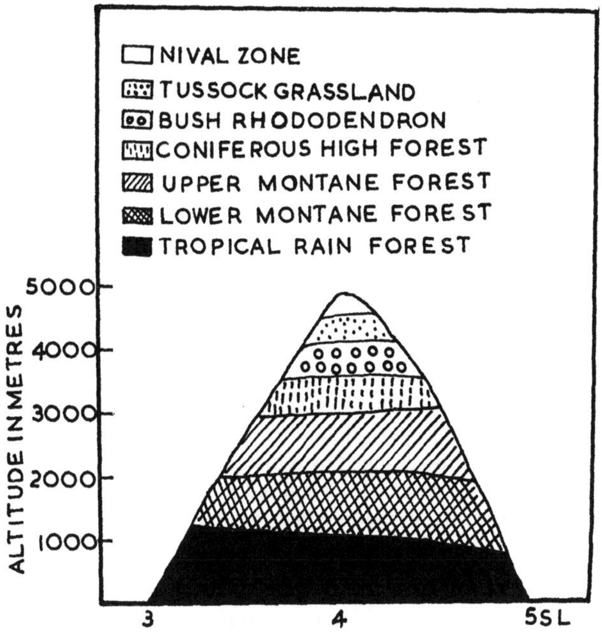

Fig. 36. Diagrammatic profile section of the Pegunungan Saldju (Sneeuwgebergte), in the area of Puntjak Sukarno (Mt. Carstenz), to show the altitudinal biotic zonation.

eastern and western mountain regions, separated by a stretch of nearly 1000 km on the island. According to GRESSITT (426, 427), broad-nosed Curculionidae are dominant on the northeast highlands, but Cryptorhynchinae predominate in the Paniai Peku (Wisselmeren) area.

While phytogeographers like GRIESBACH, ENGLER and WARBURG rightly believed the flora of New Guinea to be of Asiatic origin, zoologists have generally assumed the Papuan fauna to be of Australian origin (611a). TOXOPEUS (1044) has, however, recently shown, from his studies on the Rhopalocera collected by the Archbold Expedition, that there are, on these mountains, many typically Asiatic genera of butterflies, with some of the species related to those of Malaya and the rest are endemic. The Australian element is, on the other hand, exceedingly small and is also characteristically restricted to the extreme eastern parts of the island. More or less similar conclusions seem to have been independently reached by GUPTA (455) with regard to the general Asiatic affinities of the Papuan fauna.

There is high endemism of both genera and species, with numerous peculiarities, such as, for example, the green and golden *Troides* HÜBN., *Delias* HÜBN., *Tenaris* HÜBN., *Miletus* HÜBN., the luminous-blue *Thysanotis* HÜBN., etc. The great abundance of species and subspecies on the mountains of the island must be attributed to locally intense evolution. The Asiatic species are largely confined to the so-called Manukwari (Vogelkop = Bird's Head) region in the west. As in the case of all oceanic islands, the species endemism is generally high, but there is a relatively conspicuous poverty of species of butterflies in the alpine zone of the New Guinea mountains. Many Heterocera were, however, collected by the Archbold Expedition on the shores of the Lake Habbema, at an elevation of 3350 m. Only four species of butterflies appear to be characteristic of this lake area, one of which is a Satyrid, not found so far at lower elevations. No butterflies were also found above an elevation of 3960 m. This striking poverty of the subalpine and alpine insects is explained partly by the fact that the Pegunungan Saldju Mountains are of comparatively very recent uplift. New Guinea was submerged perhaps even up to the Miocene times and the Pegunungan Saldju attained alpine-zone elevations only during the Pleistocene. Mt. Trikora (Wilhelmina) is snow-capped at present, but was presumably unglaciated during the Pleistocene. It seems to have indeed reached its present altitude of 4785 m only during the last phase of the Pleistocene. The antecedent river system, such as the R. Balem, also strongly supports the view that these mountains are of comparatively very recent uplift. The mountains are in fact still believed to be actively rising. The species, which inhabited the high mountains that did not, however, rise to the alpine-zone elevations, had opportunities of spreading to the newly emerged land below. The mountain species, changed by prolonged isolation, had already advanced from the subspecies to the species level and did not mix with the offspring of the originally uniform elements, so that we find today a number of species duplices in the region.

2. THE ANDES

The Andes are part of the western highlands of South America. The western highlands extend from Panama to Cape Horn, a distance of about 7000 km. The Andes in Colombia are well defined ranges of mountains with deep valleys, converging near Pasto, north of Ecuador. South of Ecuador, the Andes have again three distinct ranges across Peru and converge to Cerro de Pasco in Peru. South of this there are two ranges, one on either side of the Bolivian Plateau. These two ranges unite at Aconcagua (7034 m above mean sea-level; 32° 31′ SL in Argentina), south of which there is a single range up to the Cape Horn.

Strictly speaking, the Andes are not the only tropical mountains in South America. Mt. Roraima of the Sierra Pacaraima on the border of Venezuela and British Guiana (somewhat to the north of 5° SL and the west of 60° EL), rises to an elevation of 2810 m. In the jungle-clad uplands of Brazil, there is a little known mountain range called the Serra do Roncador or the Snoring-Mountains (Land-of-Oz). The Brazilian Highlands have the peak Itatiaia (Agulhas Negras, 2820 m), just north of the Tropic of Capricorn (about 45° WL), the highest elevation in the eastern part of South America. Some interesting insects, like the Coleoptera *Charagmophorus lineatus* WATERH. and *Cryptocephalus quelchi* WATERH., are reported from an elevation of 2590 m on Mt. Roraima (1106, 1107). None of these mountains, however, compare with the Andes in elevation, massiveness and in their wealth of high altitude insects. The tropical parts of the Andes are also of considerably greater ecological and biogeographical interest than these mountains.

The Andes constitute the mountain system of the entire length of South America, narrowest in the south and broadest in the Central and Bolivian Andes and in the north. The Andes are beyond comparison the longest unbroken mountain range in the world, with its northernmost spurs from the Caribbean Sea on the coast of Colombia and Venezuela, stretching to the Cape Horn in the Tierra del Fuego. In Tierra del Fuego the general trendline of the ranges is nearly east-west, but on the mainland of South America the trendline is north-south. The Andes are also among the highest mountain ranges, second only to the Himalaya and other Middle Asiatic mountains. The summit of Aconcagua towers 7034 m above the nearby Pacific Ocean. Scattered through Peru and along the Argentine-Chilian border are scores of other high peaks that exceed 6100 m in elevation. Extending from the tropics, through the sub-tropics and the south temperate regions, the Andes enter into the Antarctic area. The ecological conditions on the Andes thus vary very considerably in the formation of the terrain, vegetation and climate. Permanent snow is found only at an elevation of 6090 m on the Atacama region in northern Chile, but in the cold and stormy Patagonia rain and snowfalls occur on an average for 300 days in the year and the snowline dips to a mere 760 m above mean sea-level. In Colombia, Venezuela and Ecuador the eastern flanks have rain-forests up to great elevations. In the Central Altiplano of northern Chile and Bolivia in times of drought thousands of square kilometres of peaks and plateaux lack vegetation, often including even a blade of grass.

The Andes are remarkable in many respects and consist largely of two parallel chains of mountains, with their crests separated by broad plateau-like valleys, 35-100 km wide and 2450-3950 m above mean sea-level. Near the northern termination, there are three diverging chains instead of two and near the southern end there is only one. Numerous cross swells of mountain knobs connect the range, thus separating the high plateau into a number of distinct basins, the broadest and the highest of which is the Altiplano Plateau of Bolivia. The elevation of the Andes increases from the Isthmus of Panama southwards to Bolivia, where the crests reach to 4875 m above mean sea-level. The width of the system is 300-550 km. The slopes are abrupt and deeply cut by transverse valleys and there are no longitudinal valleys on the western slope and very few on the eastern slope. The summit is not really a narrow ridge, but usually plateau-like and several kilometres broad, with numerous volcanic peaks.

The Cordillera of Patagonia are a series of massifs, separated by longitudinal valleys. The South Chilean Andes, northward from the 39th south parallel to the 30th parallel, consists of double crestlines. The valley in between is in fact a high-level plateau that lies at elevations of 3050-3960 m. The Puna de Atacama is the wide region north of 30° SL. The mean elevation of the Puna Basin lies between 3300 and 3960 m above mean sea-level and the peaks in the interior rise 300-1500 m above the general level of the Puna. The western border of the Puna is the Maritime Cordillera or the Cordillera de los Andes (high volcanoes). In the Central Andes the northern limit of the Puna de Atacama is a transverse ridge of high snow-capped peaks and ridges at about 23° SL. The high plateau continues northwards as the Altiplano of Bolivia, about 800 km long and about 160 km wide, rising to an elevation of nearly 3700 m at Lake Titicaca and Lake Poopo.

North of Aconcagua, the chain of the Andes extends along the Argentina-Chilean border, towards Bolivia and Peru. In this section of nearly 1300 km length stand a great array of high peaks, rising to elevations of 6720 m above mean sea-level, such as Mercedario, Ojos del Salado, Dos Corros and Llullaillaco, which are mostly pyramids of weathered debris of volcanic origin like Aconcagua. Inspite of the great altitudes, there are no snows, glaciers and streams on most of these high peaks. Beyond Atacama, the range crosses into Bolivia, where it is known as the Cordillera Occidental, with Sajama (6536 m above mean sea-level) as the highest peak. North and east of the Cordillera Occidental lies the Bolivian plateau of Altiplano, which may perhaps be properly described as the core of the Andes. Between 15° and 20° SL, the ranges attain the greatest breadth and represent one of the highest inhabited places on the earth. The Cordillera Oriental of southeastern Peru has truly alpine-zone areas, covered with snow.

Beyond the southern Cordillera, about the latitude of Lima, there is a great conglomeration of high mountain ranges known as the *Nudos* (knot). The Cordillera Blanca, the Raura Huagaruancha Range and the Cordillera de Huayhuash are important topographic features of the region. The Cordillera Blanca is the largest and Mt. Huascaran (6766 m above mean sea-level) the highest peak. The Huagaruancha Massif has a peak of the same name rising to an elevation of 5741 m. Yerupaja or the Carnicero or the Butcher Peak (6635 m) is in the Huayhuash Cordillera. The northern half of the Andes of Ecuador is characterized by well known volcanoes like Chimborazo (6272 m), Canhuairazo (5124 m), Quilatos (4790 m), Illiniza (5340 m), Pichincha (4791 m), Catacachi (4966 m), Cimbal (4830 m), Chiles (4764 m), Altar (5400 m), Tunguragua (5085 m), Cotopaxi (5943 m), Antisana (5660 m) and Cayambe (5810 m).

The Cordillera of Colombia have three regions of the Andes, viz. the Western Cordillera, the Central Cordillera or the Cordillera del Quindio and the Eastern Cordillera or the Cordillera de Bogota. The Central Cordillera is the highest and the shortest. In the extreme north of Colombia, close to the Caribbean Sea, we find the Sierra Nevada de Santa Marta.

The snow-capped Aconcagua Peak, visible from the Pacific Ocean at a distance of 160 km, is of volcanic origin, but is not itself a volcano. It has most appropriately been described as an "intolerably monotonous slag pile". South of this is the glacier-clad bulk of Tupungato (6797 m). The Mount Sarmiento on Tierra del Fuego, though not so high, is most spectacular for its icy grandeur.

The Andes are divided into three natural divisions: i. The Montana or the eastern slopes of the Amazon forests, ii. the Sierra or the temperate region of the slopes and of the secondary valleys of the plateau and iii. the Puna or the cold region of the Cordillera, at elevations of 3050–4575 m. The part of the Andes that extends north from Bolivia to Colombia is within the tropical belt.

A part of the cordilleran mountains of Mexico lies also within the tropical belt. The Western and Eastern Ranges of the Sierra Madre Mountains have plateaux in between them.

As in the case of the equatorial East African mountains, the diurnal atmospheric temperature fluctuations are far more pronounced and ecologically important than the annual and seasonal temperature fluctuations on the tropical parts of the Andes. The mean monthly variation of the atmospheric temperature in Quito, at an elevation of 2850 m above mean sea-level, on the Andes is only 0.8° C. The following table 24 summarizes the temperature conditions at different altitudes within the tropical parts of the Andes:

TABLE 24
Atmospheric temperatures on the tropical Andes

Elevation in m	Absol. Max. ° C	Absol. Min. ° C	Diurnal fluctuation Max.	Min.
3600	17.3	−1.5	14.9	2.2
3950	14.0	+1.5	10.2	3.6
4450	12.7	−1.8	9.2	1.1
4720	3.5	−2.0	4.3	0.6

The diurnal atmospheric temperature fluctuation within the paramos-zone in the Quito area, at an elevation of about 4720 m, near the permanent snowline, ranges from 4.3° C to 0.6° C. The temperature fluctuation in a period of four months lies between 3.5° C and —2.0° C only. The climate of Mexico City, at an elevation of 2280 m, is almost comparable with that of Arequipa in South Peru, at an elevation of 2450 m, and particularly with that of Cochabamba in Bolivia. The annual fluctuation of the atmospheric temperature is 6.4° C in Mexico City and 6.0° C in Cochabamba.

3. SOME ECOLOGICAL PECULIARITIES OF TROPICAL MOUNTAINS

A knowledge of the outstanding peculiarities of the high altitude ecology of the tropical mountains is necessary for a proper understanding of the characters of the insect life of the Andes. As already indicated in the foregoing chapter, the fundamental difference in the ecology of the high altitudes of the mountains of temperate and cold latitudes of the northern hemisphere on the one hand and of the temperate and cold zones on the mountains of the tropical regions on the other hand may be traced eventually to the significant differences in the periodical fluctuations of atmospheric temperatures. In the higher northern latitudes, it is the sharp contrast between the cold winter and the warm summer that is a dominant ecological factor that profoundly modifies the environmental conditions at high altitudes, but the tropical mountains are characterized by the absence of such distinct temperature-conditioned seasons. The tropical regions are, on the other hand, characterized by great diurnal atmospheric temperature fluctuations; the difference between the day and night temperatures is very great on the mountains of the tropical regions, far more than on the mountains in the north temperate regions of the earth. The diurnal fluctuations are also often greater than the annual. The mountains of the equatorial zone have thus a pure diurnal-temperature climate, but the mountains of the northern hemisphere, particularly at higher north latitudes, have a seasonal-temperature climate. On the fringe of the tropics, the diurnal-temperature climate and the seasonal-temperature climate are nearly in equilibrium. It is not this line of equilibrium, but only the equatorial limit of frost above mean sea-level that is employed in separating the two. The diurnal-temperature-fluctuation climate of the tropical mountains has a number of far-reaching effects on diverse other ecological factors at high altitudes. The foremost consequence of such a climate is nocturnal snowfall every day, all the year round, on sufficiently high tropical mountains. There is, however, no distinct superabundant winter snowfall, in sharp contrast to the summer thaw. There is, on the other hand, a sort of virtual winter every night and summer during the succeeding daytime sunshine, irrespective of the season. Depending upon the latitude, the massiveness of the mountain, local conditions of insolation and nocturnal radiation and other complex of environmental factors at a given elevation, the absolute lower altitude limit of daily nocturnal frost varies within fairly wide limits on the different mountains of the tropical regions. In the tropical parts of the Andes, in Quito (figs. 37, 38), under permanent humid conditions of high elevation

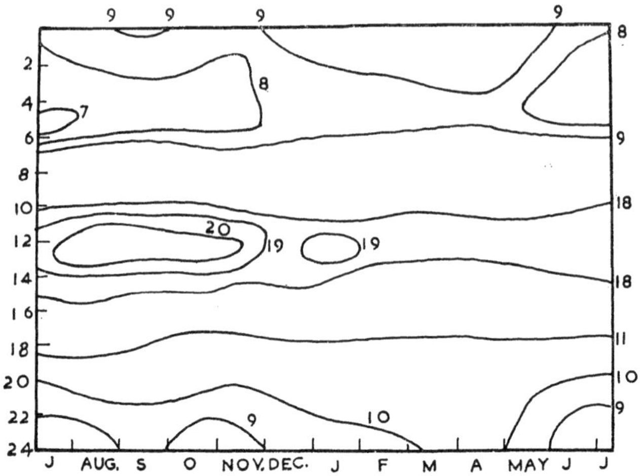

Fig. 37. Isotherms of a typical equatorial high altitude environment, marked by the absence of thermic-conditioned annual seasons (Modified from TROLL).

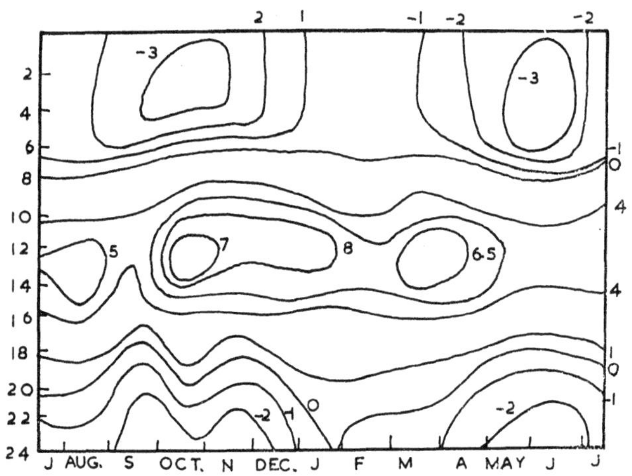

Fig. 38. Isotherms in a typical subnival environment, at an elevation of 4760 m, marked by nocturnal frost and strong daytime insolation throughout the year (Modified from TROLL).

of 2850 m above mean sea-level, directly on the equator, snowfall has been recorded only once in the course of several years and the lowest limit of snowfall is at an elevation of 3000 m in the area. In Arequipa in the semi-arid southwest Peru in the extreme tropical belt, at an elevation of 2360 m, snowfall is similarly exceptional, but is more frequent than in Quito. The mean snow limit here is somewhat below 3000 m. In the arid climate of the Puna-Andes (*vide infra*), due to the strong fluctuations of diurnal temperatures above an elevation of 4000 m, there is nightly snowfall nearly every day and at elevations of 4700–5200 m the snowfall is every night all the year round and is accompanied by a pronounced freezing of the ground during the night and considerable warming of the soil during the daytime. We have already referred to similar night frosts on Mt. Kilimanjaro. TROLL (1049) has recently represented graphically the vertical distribution of the number of frost-days, ice-days, frost-change days and frost-free days on the west slopes of the Andes, the El Misti Volcano in South Peru, from October 1893 to December 1895, El Misti Montblanc Station from January 1894 to December 1895, Alto de los Huesos from March 1894 to December 1895, Arequipa from January 1892 to December 1895 (fig. 39). Although the tropical mountains do not have a specially heavy and characteristic winter snowfall, the snow falls nevertheless even below the permanent snowline during the winter (of the northern hemisphere) on these mountains. The snow-cover thus produced is either of short duration and evaporates away or melts during the daytime insolation or it persists at the most for a few weeks only and that also not because of particularly favourable conditions of low atmospheric temperatures, but only because of local shelter from insolation due to cloud or also through the rainy season snowfall (the rainy season often coincides with summer on tropical mountains). The tongues of glaciers thus lie naked and directly exposed to the daytime insolation the whole year, while the glaciers in northern latitudes remain for most part of the year under the protective cover of the winter snow.

Significant edaphic differences also exist between the tropical mountains and the boreal temperate-zone mountains. The soil that is not covered by glacier-ice, but receives snowfall between the permanent snowline and the forest-line, both in the north latitudes on lowland and at high altitudes on mountains, is generally known, on pedological and geomorphological basis, as the periglacial or subnival zone. The periglacial zone is characterized by certain peculiarities of the mechanical effects of frost on the soil. The freezing of the soil in this belt gives rise to the well known solifluction on the surface of the ground. Solifluction is a special form of creep, under the action of gravity, on mountains, on which the ground is permanently frozen to considerable depths. The upper layers of the soil undergo thawing during the summer, but the deeper layer continues to remain frozen. The surplus water is thus unable to drain deeper downward into the ground. The upper thawed layer of the soil, often perhaps only a few centimetres, but sometimes even two or more metres thick, thus becomes saturated with the melt-water. This layer then behaves like a viscous liquid and flows sluggishly downwards under its own weight; this phenomenon is known as solifluction. This periglacial phenomenon in Subarctic and Arctic areas ranges from the polar winter, with its permafreeze of even the deep soil, to the summer thawing of the surface and the consequent seeping of the melt-water down into the soil to solifluction. On the Andes and on the high mountains of equatorial East Africa, the periglacial phenomena are, however, diurnal occurrences, with the characteristic nocturnal frost and the formation of the comb-ice on the

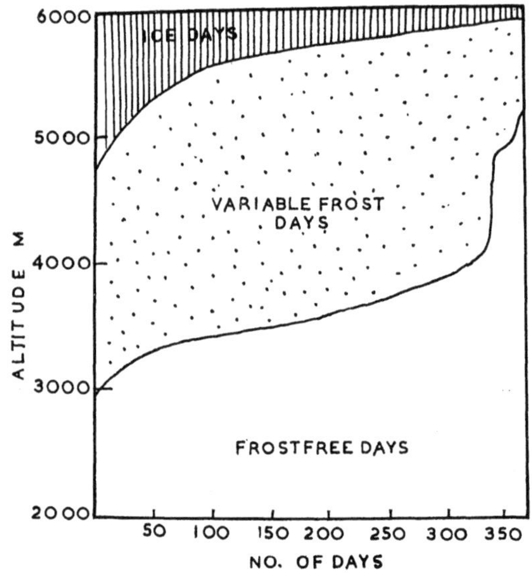

Fig. 39. Vertical distribution of frost-days, ice-days and days of frost change on the south Peruvian Andes. (Modified from TROLL).

surface of the ground. The comb-ice cover, however, melts away under the action of the forenoon insolation. The soil movements of solifluction or the comb-ice solifluction concern, therefore, only a superficial layer of a few centimetres of the soil. Yet as the ground frost and solifluction are of regular daily occurrence, they act as important unfavourable factors for the growth of vegetation and prevent large-scale colonizations by terricole and geobiont insects. It seems, therefore, that high tropical mountains are in reality characterized by a poorer and much less diversified high altitude insect life than the north temperate zone mountains.

The evidence of the decisive influence of the winter snow-cover on standing trees and on the dwarf plants near the upper limits of the forest on the boreal temperate latitude mountains, like the Alps or the Himalaya, is very striking. The stunted and dwarf plants, under the cover of the winter snow, are effectively protected from injury by the extreme winter conditions of high desiccation, intense frost, severe mechanical injury through winds, etc. Isolated trees that happen to project above the winter snow-cover exhibit a very characteristic weather-form, viz. injury due to frost and desiccation, directly above the surface of the snow. On the tropical mountains we do not, however, find this situation. The trees near the upper limits of the closed forest on the high tropical mountains have a broad, dome-shaped or umbrella-shaped crown, with dense peripheral foliage on the upper side (the socalled evergreen "Kugelschirmbäume"). The closed humid mountain forests and the high forests of the tropical areas are, therefore, relatively rich in evergreen types, with abundant foliage. Correlated with the prevailing high humidity, due to the updraft of moisture-laden winds from below, we find a most luxuriant growth of epiphytic moss, lichens, ferns and other flowering plants, giving rise to the typical "Nebelwald" or cloud forest ("Mooswald" or "Elfinwald"). The cloud forest passes above rather abruptly into the high grasslands or we may find a gradual dwarfing of the trees and partly also a more or less pronounced impoverishment of species and transition into evergreen bushes. This type of forest is typical of the high mountains of the Andes and Central America, equatorial East Africa, Indonesia and New Guinea.

Mention must also be made of another rather striking contrast between the tropical and boreal temperate zone mountains. On the temperate and high boreal latitude mountains, the trees and the forest typically cover the ridges rather than up the valleys. This is readily explained, when it is recollected that on these mountains, the cold air is strongly stagnant in the valleys, giving rise to a total local climatic inversion. Furthermore, in regions of abundant snowfall, there are frequent avalanches in the valleys and thus the snow tends to accumulate more and also last longer in the valley, so that the conditions are unfavourable for regeneration of stock. The vegetative period is also thus greatly abbreviated in the valleys more than on the ridges between them. The situation on the tropical mountains is, however, fundamentally different. The high cloud forests of the tropical mountains develop more in the valleys than on the ridges between them, though naturally there may be a few isolated patches of tree clumps in sheltered localities even on the ridges. As may be expected, there is no pronounced seasonal rhythm on the high forests of the tropical mountains, particularly in their upper levels. Flowering continues almost unbroken the whole year round. The ultimate altitudinal upper limits of the forests on the tropical mountains are influenced by the humidity-temperature factor, the gradual fall in temperature with increase in elevation, the definitive heat-summation at higher levels, but there is neither a winter-cold season nor a winter snow-cover to play a predominantly decisive rôle, more than any other factor.

4. ALTITUDINAL ZONATION OF LIFE ON THE ANDES

The following biotic zones may be recognized on the South American tropical Andes: i. tierra caliente, ii. tierra templada, iii. tierra fria and iv. tierra helada. The boundaries of these zones represent transitions, rather than sharply defined lines and depend upon diverse local differences of the massiveness of the elevation, the climatic character at different elevations, etc. In Mexico, the tierra caliente, from the sea-level to an elevation of about 900 m, is the hot zone. The annual mean atmospheric temperature in this zone is 48.3° C. The tierra templada extends from 900 to 1830 m above mean sea-level and has a mean atmospheric temperature of 24° C. The tierra fria extends from 1830 to 2750 m and has a mean atmospheric temperature of 18° C. These limits naturally differ further south. On mountains of medium elevation in the tropical permanent humid region, the lower montane forest or the *medio yungas* (v. HUMBOLDT's Region der Fieberrinde) or the zone of coffee-cinchona cultivation, receives no snowfall. The upper levels of the forest, especially in the valleys, are typically characteristic for their treeferns that descend sometimes right down to the sea-level in Australia, Formosa, South Brazil and other places at the outer fringe of the tropics. In most humid-tropical mountain forests, there is a zone of bamboo, especially in Africa, immediately below the grass-zone. The treefern-bamboo belt passes, in the high and cloudy forests (Cejawald), to the uppermost limits of the closed and dense forest zone that is enveloped in dense clouds for several hours, both during the day and during the night, due to the heavy condensation of the atmospheric moisture. It is remarkable that in all parts of the world, the Ericaceae occupy a dominant position in the high cloud forest zone, often taking the place of trees in the forest and higher up in the form of shrubs (Strauchparamo). v. HUMBOLDT has also referred to the Andes-rose zone, where other Ericaceae like *Vaccinium*, *Pernettya*, *Gaultheria*, *Gaylussacia*, *Cavendishia*, *Thibaudia*, *Andromeda*, etc. also occur on the Andes. In East Africa we find above all *Erica arborea*, as well as *Philippia*; in tropical Asia and in New Guinea *Rhododendron*, *Vaccinium*, *Gaultheria* and *Thibaudia*; and in Mexico we find *Arbutus* and *Andromeda*. Other affinities also may be observed with the woody plants, adapted for these conditions in the intermediate zone between the high forest and the high grassland. We have, for example, the bushlike or arborescent *Lobelia*, *Siphocampylos*, etc. from the Cordillera, East Africa, South India, Malaya, etc. or the evergreen woody *Hypericum* sp.

The Altiplano of Bolivia has a mean elevation of 3657 m, between the great Andean chains; many of the highest peaks in the western hemisphere are indeed situated in Bolivia. The altitudinal biotic zonation on the Bolivian Andes may be summarized as follows (359):

I. The Yungas region
   1. The Lower Yungas. Montaña. Tropical zone. The Chaco to the southeast. Elevation 150-1000 to 1700 m.
   2. The Middle Yungas. Subtropical zone. Elevation 1000-2500 m, sometimes reaching to 2800 m.
   3. The Upper Yungas. Cloud forest (Nebelwald), Ceja de Montaña. Elevation 2500 to 3400 m, sometimes 3500 m.

II. The Altiplano and the Paramo
   1. The Altiplano (West Cordillera). Elevation 3000-3500 m.
   2. Altacordillera (The High Cordillera).
   3. The Paramo. Elevation 3400-4500 m and also higher.

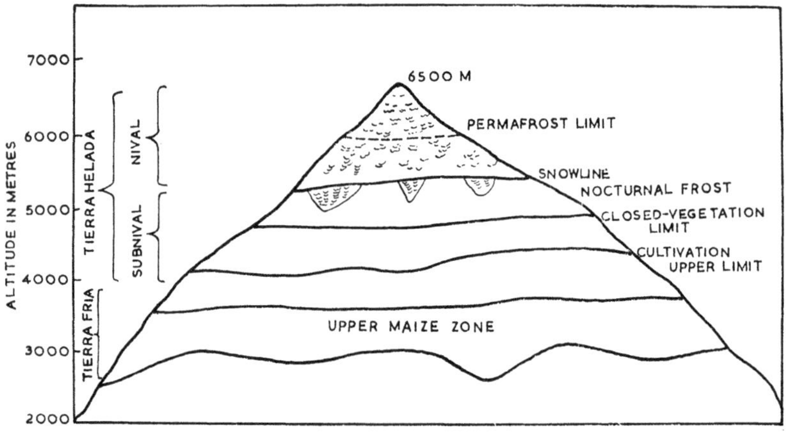

Fig. 40. Climatic altitudinal zonation of the high Andes of south Peru and north Bolivia or the socalled Puna-Andes. (Modified from TROLL).

The biotic zonation on the tropical parts of the Andes is greatly complicated by the orographic characters of the mountains, such as, for example, the rain-slope, the insolation-slope, diurnal wind equilibrium, valley winds and slope winds. The following synopsis of the biotic zonation on the Andes is modified from TROLL (1047, 1049) (figs. 40, 41, Table 25).

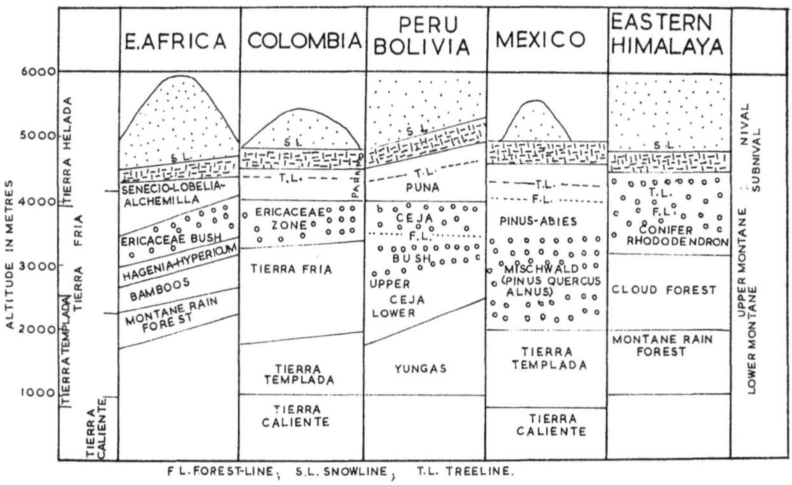

Fig. 41. Altitudinal biotic zonation on the principal equatorial mountains. (Modified from TROLL).

The atmospheric temperatures in the upper paramos are near the freezing point the whole year round, so that there is really no true frost-free season and the temperature does not also rise much above the freezing point. The plants grow, therefore, rather slowly and though often actually several years old, remain characteristically short and stunted, with short internodes. The dwarf condition is, therefore, no indication that the plants are young. The permanent humid paramos occupy the major area of the Andes and only on the west Peruvian Andes do we find variable humid-montane conditions under what is known as the *puna*. The higher zones of the rough High Andes are called *puna brava*. Depending on the conditions prevailing on the east slopes, with constant rains and on the coastward-sloping west slope of the Andes, we come across fewer rainy days and longer dry days, from the northwest to the southeast, in a succession that constitutes respectively the humid puna, the arid puna, thorn puna, succulent puna and the desert puna. With increasing aridity of the puna, the temperature conditions of the puna also differ progressively from those obtaining in the paramos. The annual temperature fluctuation increases by about 9-11° C, but owing to the slight vegetation cover, the diurnal fluctuation is also rather high. In place of the interruption of growth and development of plants and insects during the arid season, we find an interruption of growth due to the strong night frost as the dominant ecological character during all the seasons. At elevations of 4700 m, there is night frost every day during the year, but during the daytime the ground is strongly heated up. There is greater dominance of polster and

TABLE 25

Biotic zonation on the tropical Andes

Tierra helada	Paramo	Humid (grass) puna	Arid puna	Thorn-succulent puna	Desert or salt-puna	
Tierra fria	Tropical high and cloud-forest and high bush	Tropical humid sierra high bush	Tropical arid sierra high bush	Tropical thorn-succulent sierra high bush	Desert sierra Tropical high semi-desert	Tropical high desert
Tierra templada	Tropical montane forest	Tropical montane humid savanna	Tropical montane arid savanna	Tropical montane thorn-succulent trees	Desert valley Tropical montane semi-desert	Tropical montane desert
Tierra caliente	Tropical ever-green lowland rain-forest & semi-evergreen transition forest	Tropical humid savanna (forest & grassland)	Tropical arid savanna (forest & grassland)	Tropical thorn-succulent savanna (forest & grass-land)	Tropical desert savanna (semi-desert)	Tropical desert
Number of humid months in the year	12 11 10	9 8 7	6 5	4 3 2	1	0

rosette plants on the puna than on the paramos. Even the cacti tend here to grow into polster plants. The moors on the puna Andes are also pronouncedly hard-polster moors (the *Distichia*-moors). The coriaceous and scaly-leaved tola shrubs of *Baccharis*, *Lepidophyllum*, *Loricaria*, etc. are characteristically aromatic and resinous. On the arid-puna zone from the southwest Peru to the northwest Argentina giant clusters of *Lepidophyllum quadrangulare* and *Festuca orthophylla* on loose ground constitute the so-called tola zone. The ground water in certain river beds in the Puna de Atacama becomes frozen in the night, but melts during the day to form surface water. The ever-humid altitude zone of the tropics, in the region of the sub-nival zone, is known as the paramo in the Latin American countries. The term paramo is now generally applied in a comprehensive sense for the peculiar climate, vegetation, soil conditions and the general landscape and also for the ecologically similar situations and very similar biological conditions and vegetational forms on other tropical mountains, such as those of the African and Australasian regions. The paramos are dense stands of hard, bushy grass, with which are mixed the characteristic tree types like *Espeletia* on the Andes, *Senecio* and *Lobelia* on the East African mountains, *Anaphalis* in Indonesia; rosette-plants with hairy leaves and short woolly inflorescence like *Lobelia* in East Africa, *Lupinus* on the Andes; shrubs with small hard leaves like *Befaria*, *Gaultheria* and *Hypericum*; scaly and roll-leaved shrubs like *Loricaria*; woolly-leaved shrubs and herbs like *Helichrysum*; polster plants like *Azorella*, *Distichia*, *Plantago*, *Acichne* and *Oreobolus*; rosettes flattened close to the ground and with stout taproots as in the case of *Werneria*, *Nototriche*, etc.

5. SOME PECULIARITIES OF THE INSECT LIFE AT HIGH ALTITUDES ON THE ANDES OF THE TROPICS

The high altitude insects of the Andes are generally dealt with by biogeographers as a part of the Chilean Subdivision of the Neotropical Realm, but these insects are fundamentally different from the strictly Chilean faunal elements in a number of ecological and distributional peculiarities. Among the high altitude insects of the Andes, Coleoptera and Lepidoptera are better known than most other groups. As early as 1891, WHYMPER (1111) made extensive collections of Coleoptera, which were described by SHARP, BATES, GORHAM, OLLIFF, JACOBY and others. The high altitude Coleoptera of Mt. Chimborazo are remarkable for the presence of a number of typically boreal genera like *Pterostichus* BON., *Trechus* CLAIRV., *Bembidion* LATR., *Silis* LATR., *Erirrhinus* SCHÖNH., etc. *Pelmatellus* BATES is a Neotropical genus from the Andes, represented by one transgressive species in Arizona and New Mexico in North America. It is remarkable that neither *Calosoma* WEBER nor high altitude Chrysomelidae seem to have been found so far on Mt. Chimborazo. The genus *Colpodes* ML. is widely distributed in the warmer parts of the Andes, but the genus is absent in Europe and most parts of the Nearctic Realm. The genus, found up to the permanent snowline on Mt. Chimborazo, has developed ecological characters similar to those of *Nebria* LATR. or *Platynus* BON. According to WHYMPER, *Colpodes megacephalus* BATES occurs at an elevation of 4775 m on the summit of Guagua Pichincha, but in other localities it occurs at elevations of 3660-4270 m. The other species of *Colpodes* M.L. recorded from elevations above 4575 m are *C. pustulosus* BATES, *C. rotundiceps* BATES, *C. pichinchae* BATES, *C. orthomus* CHAUD., *C. fusipalpis* BATES,

C. oreas BATES (at an elevation of 4820 m), *C. diopsis* BATES and *C. steno* BATES (ascending up to an elevation of 4725 m). *C. alticola* BATES descends to an elevation of 2745 m. *Pterostichus* (*Agraphoderes*) *antisanae* BATES occurs at elevations of 3600-4050 m, *Pt.* (*Ag.*) *liodes* BATES between 2745 m and 4050 m, and *Pt.* (*Ag.*) *integer* BATES between 3565 and 4050 m. *Pelmatellus andium* BATES occurs at elevations of 3350-4114 m. The apterous *Bembidion fulvotinctum* BATES is found at elevations between 3660 and 4575 m. *B. andinum* BATES occurs at an elevation of 4820 m. *Silis chimborazona* GORH. (4575 m) and *Astylus bisexguttatus* GORH. (2745-4114 m) are the typical high altitude Cantharidae from the Andes. The Coprinini *Uroxys elongatus* HAR. is reported from Cotocachi at elevations of 3660-3962 m, the Melolonthid *Clavipalpus antisanae* BATES at 4050 m and the Meloid *Pseudomeloe sexguttatus* SHARP at 2740-4114 m. Some of these species were found under stones, cemented together by ice. Among the high altitude Curculionidae from the Andes there is a dominance of typically American forms, with some interesting endemic genera. The genus *Helicorrhynchus* OLLIFF, related to the Holarctic *Otiorrhynchus* GERM., is endemic on Mt. Chimborazo. *Helicorrhynchus vulsus* OLLIFF was collected at elevations of 4810-4880 m. *Amathynetes* OLLIFF, another endemic genus related to *Listroderes* SCHÖNH., is represented by *Amathynetes simulans* OLLIFF at an elevation of 4575 m on the same mountain. The other interesting Curculionidae from the tropical Andes are *Naupactus parvicollis* OLLIFF (4575-4810 m), *Macrops coelorum* OLLIFF (4050 m), *Erirrhinus glaber* OLLIFF (4575 m), *Hilipus longicollis* OLLIFF (4050 m) and *Erirrhinoides distinctus* OLLIFF (4810 m). It may be observed that the great majority of the high altitude Coleoptera from the Andes are typically small-sized, rather dull coloured and apterous forms.

Very few of the Andean Lepidoptera seem to have been taken near the permanent snowline. *Pieris xanthodice* LUC. is widely distributed at elevations of 2750-4575 m on the Andes, *Colias alticola* GODM. (3960-4876 m) sometimes occurs above the permanent snowline. *Lycaena koa* DRUC. (3050-4270 m) ranges from Ecuador to Peru and Bolivia and an undetermined species of *Pedaliodes* BUTL. has been recorded at elevations of 3050-4270 m. Most of the typically high altitude butterflies from the Andes are small-sized and are generally strongly reminiscent of the high altitude forms from the Palaearctic Region. The reader will find interesting data on the Andean Lepidoptera in PAGENSTECHER (809). A number of genera known from the Alps and the high mountains of Middle Asia are found at high elevations on the Andes and constitute evident links with characteristic South American genera. In addition to *Pieris* SCHR. and *Colias* FABR., already mentioned, we have also some genera which are peculiar to the Andes, like *Phulia* HERRICH-SCHÄFFER and *Trifucula* ZELL., but with their close relatives among the Middle Asiatic Pieridae. In place of the boreal *Erebia* DALM. and *Oeneis* HÜBN., we find on the Andes *Pseudomaniola* ROEBER, *Lymanopoda* WESTW., *Pedaliodes* BUTL., *Steroma* WESTW., etc. FORSTER (359, 360) has given an interesting account of the biogeographical peculiarities of the high altitude Lepidoptera, occurring at elevations above 3000-3500 m on the Bolivian Andes. The Lepidoptera of the yungas-zone (the rain-forest of the east slope of the Cordillera, up to an elevation of 3500 m) present many interesting features and are perhaps the richest in species in South America. The yungas-element extends from the tropical rain-forest belt (500-3500 m) of Santa Cruz de la Sierra on the Andes to Colombia, but the abundance of species falls off perceptibly from the north to the south. The differences in the species composition are far more pronounced at different altitudes than between the north and the southern parts of the tropical Andes.

Papilio isidorus DOUBL. is a typical yungas species in the lower forest belt. *P. wårscewiczii* HPFFR., *Catasticha* spp. and *Dismorphia* are typical of the upper areas. These species are relatively young elements that have apparently become adapted secondarily to the changed conditions on the tropical mountains, uplifted during Tertiary times. The Lepidoptera of the highlands and high mountains, especially of the treeless zones at elevations above 3000-3500 m, contain insignificant South American elements, but are composed of Holarctic elements like *Colias* FABR. and *Phulia* HERRICH-SCHÄFFER related to the Middle Asiatic *Baltia* MOORE. According to FORSTER (*op. cit.*), there has apparently been more than one wave of migration of the boreal elements to the high altitudes on the tropical mountains of the Bolivian Andes, the last of which occurred even as late as the Post-Pleistocene. The Pleistocene glaciations are also believed to have played an important rôle and may have favoured isolation in refugial areas and thus leading to subsequent intense speciation. The faunal movements were perhaps not as extensive as they have been in the Old World, due to the Pleistocene conditions in South America. The occurrence of such genera like *Phulia* HERRICH-SCHÄFFER and *Piericolias* STAUD. on the Bolivian Andes and on the Pamirs Region evidently indicate Pre-Pleistocene differentiation of these genera. The numerous closely related species of these genera from the Bolivian Andes represent, however, Post-Pleistocene origin. The High Andes species of *Colias* FABR. are not yet found in constant geographic forms and are still very closely related to the Holarctic species. It seems, therefore, that *Colias* FABR. has perhaps arrived on the Andes since the Pleistocene. The great similarity of the High Andes species to those from the Pamirs and other Middle Asiatic high mountains must either be due to true faunal affinity or must be attributed to convergent evolution of related groups under comparable conditions of ecology. Very few species, adapted to the prevailing extreme conditions, flourish at elevations between 4600 and 5000 m on the East Cordillera. Among these are *Piericolias andina* STGR., *Phulia paranympha* STGR. and *Argynnis inca* STGR. Whilst there is considerable wind-blown material from the Altiplano and the lower levels of the Yungas, *Colias euxanthe* FELD., found on the snow-fields, seems to actively rise to these elevations from below. The Altiplano, the highland between the mountains (4000-4600 m above mean sea-level), is rich in species. There are characteristic species of the Bolivian and Peruvian Altiplano and there are others that belong to Chile, such as, for example, *Tatochila* spp. The typical South American forms may be found in the deep valleys like La Paz Valley in the Altiplano. The Saturniid *Automeris stuarti* ROTHSCH. ascends, for example, up the river valleys. The Lepidoptera from above the forest zone of the East Cordillera, especially on the grassy meadows at elevations of 3500-4500 m, are very little known at present. The Upper Songotal seems to be faunistically related to Northwest Argentina, the mountains of Salta and Jujuy south of Tucuman area. The High Andes *Argynnis* FABR. and *Phulia* HERRICH-SCHÄFFER occur at high elevations in this area.

A number of interesting Tipulidae (Diptera) are known from the Bolivian Andes (17). Of the species occurring at elevations above 3000 m, almost 23% ascend to 4000-5600 m. It is only exceptionally that we find the same species occurring over a wide range of altitudes, from lower elevations to very high altitudes. This is, for example, the case with *Tipula (Eumicrotipula) glossophora* ALEXANDER that occurs at elevations of 2200-5000 m, from the Yungas de Arepucho to Illimani West slope. *Limonia (Dicranomyia) perretrata* ALEXANDER occurs at elevations of 2800-5000 m. The greatest bulk of the species described by ALEXANDER (17) were, however, taken at elevations of 3300-3700 m. The highest altitude record for Tipulidae in the world is at present 5600 m; at this elevation on the Bolivian Andes is recorded *Limonia* (*Dicranomyia*) *perexcelsior* ALEXANDER. About 30% of the high altitude species from the Andes belong to the subfamily Tipulinae, 60% to the Limoniinae and 10% to the Eriopterini. *L.* (*D.*) *excelsior* ALEXANDER is described as frequenting snow-fields and as resting on the stones projecting above the surface of the snow. *L.* (*D.*) *brevispina* OSTEN-SACKEN, one of the commonest North American tipulids, was also collected at an elevation of 3150 m in Cochabamba on the Bolivian Andes. *Tipula* (*Eumicrotipula*) *phalangioides* ALEXANDER, known from Ecuador and the Sikkim Himalaya, is also reported from the Andes. The following species are recorded at elevations above 4000 m: *Holorusia* (*Holorusia*) *neurotrichia* ALEXANDER, *Tipula* (*Pectinotipula*) *titicacae* ALEXANDER, *T.* (*Eumicrotipula*) *fortior fortior* ALEXANDER, *Limonia* (*Dicranomyia*) *altandina* ALEXANDER (3600-5000 m), *L.* (*D.*) *andicola* ALEXANDER (2900-5000 m), *L.* (*D*). *diversigladia* ALEXANDER also found on the Peruvian Andes, *L.* (*D.*) *hirsutissima* ALEXANDER, *Tipula* (*Bellardina*) *hypsistos* ALEXANDER (5181 m), *Limonia* (*Dicranomyia*) *puoensis* ALEXANDER (3810-5000 m), *L.* (*D.*) *regifica* ALEXANDER (4500-5000 m), *L.* (*D.*) *schindleri* ALEXANDER (4500-5000 m), *L.* (*D.*) *smythiana curtispina* ALEXANDER (4500-5000 m), *L.* (*D.*) *titicacana* ALEXANDER at 4500-5000 m on the Bolivian Andes and at 3810 m on the Peruvian Andes, *L.* (*Geranomyia*) *forsteriana* ALEXANDER (4500-5000 m). Most species were collected during January-February, but some were also collected during August-October. *Tipula* (*Eumicrotipula*) *fortior fortior* ALEXANDER was, for example, collected at elevations of 3800-4000 m in October and at 3150 m in February. A number of species were also found in April. *Tipula* (*Eumicrotipula*) *glossophora* ALEXANDER was, for example, taken at elevations of 4500-5000 m during April, but at elevations of 2200-2500 m in September. *Limonia* (*Neolimonia*) *diva* SCHINER, found at an elevation of 3150 m at Cochabamba on the Bolivian Andes, is also previously known to be widely distributed in tropical America (Cuba, Puerto Rico to Brazil).

Mention may also be made here of the extremely interesting records of high altitude Hydracarina from the fresh-water lakes on the Peruvian Andes (1097). The material was collected in 1915 by E. GODET, along with plankton, at an elevation of 5140 m in lakes on the Peruvian Andes and in the Lavandera Lake. Of the various species recorded by WALTER, *Halacarus* (*Halacarus*) *processifer* WALTER (Halacaridae); *Neocalonyx godeti* WALTER (genus described as new) (Limnocharidae); *Limnesia unguiculata* WALTER, *Hygrobates placophorus* WALTER, *Frontipoda ciliata* WALTER and *Arrhenurus hirsutipalpis* WALTER (Hygrobatidae) are important.

6. BOREAL FAUNAL ELEMENTS ON TROPICAL MOUNTAINS

As we have shown above, the range of many typical boreal and even subarctic and arctic insects extends quite distinctly into the mountains of the tropical zone, in some cases even south of the Equator. It is well known that even the high mountain flora of the Andes has, for example, considerable boreal elements. V. HUMBOLDT has described, for example, that in Mexico "Boden, Klima und Pflanzenformen, ja die ganze Ansicht des Landes einen Charakter annehmen, welcher der gemässigten Zone angehörigen scheint". The high mountains of the tropical and Central Mexico have, for example, broad-leaved forests and mixed forest of *Quercus* and Conifers at elevations of 2500-3400 m, above which there are only Conifers up to the forest-line, at an elevation of 4000 m. The boreal elements are relatively high. There are, for example, 39 species of *Pinus* and 200 species of *Quercus*. There are besides many other typically boreal deciduous trees like *Acer*, *Fraxinus*, *Tilia*, *Carpinus*, *Celtis*, *Crataegus*, etc. and evergreens like *Mahonia*, *Arbutus*, *Evonymus*, etc. An interesting group of Tertiary relicts like *Liquidambar*, *Liriodendron*, *Magnolia*, *Toxodium*, *Carya*, *Nyssa*, *Ostrya* and *Platanus* characteristic of the eastern U.S.A. has its discontinuous distribution on the Mexican and Central American high mountains. These boreal elements meet here the northern outposts of the pronouncedly austral and antarctic-tropical montane elements like *Podocarpus*, *Weinmannia*, *Fuchsia*, *Oreopanax*, *Gaultheria*, *Pernettya*, *Gunnera* and *Acaena* that have pushed their way on to the northern limits of the tropics on the Central American mountains by way of the Andes.

Comparative ecological evidence, from floristic and faunistic studies, indicates that the high altitude climate of Central American and Mexian mountains is essentially a tropical climate that actually corresponds to the latitudes. We must conclude, therefore, that the boreal plants and insects at high elevations on these mountains have spread here from northern areas and have subsequently become secondarily ecologically adjusted to the conditions of tropical high altitudes. The high altitude regions of Mexico must, therefore, be considered as typically tropical high mountains, populated by boreal elements under essentially tropical mountain conditions.

The differences in the ecology of the high mountains of the northern and the southern margins of the tropics of the New World cannot be explained entirely on the basis of the conditions prevailing at present. The peculiarities have evidently their roots in the past geological and developmental histories and in the possibilities of formation of new species locally. In spreading southward on the high mountains with meridional trendline, the boreal elements have also shown evidence of having undergone intense speciation. Speciation seems to have been most intense during the Pleistocene and perhaps even during the Post-Pleistocene, periods which are generally associated with profound and rapid changes of climate, rapid colonizations and changes in vegetation on the Tertiary mountains everywhere. The colonization of the high tropical mountains of Central and South America by so many boreal genera, with the evolution of a number of endemic species locally on these tropical mountains, are without doubt related closely to the Pleistocene and Post-Pleistocene conditions. It is remarkable that in this colonization of the tropical high mountains of South America by the boreal elements, the young fold mountain belts that stretch across the tropical zone have been the principal migration routes. The migration into and across the tropics and the formation of new species have occurred on the young mountains and seem to have been favoured by the uplift of the land, by volcanic and climatic changes during the Pleistocene.

The south temperate Andes are considered in Chapter XVI.

CHAPTER IX

THE HIMALAYA

The Himalaya* forms an integral part of the High Asia and is *par excellence* the mountain of India. The mountains of India fall under two broad groups, viz. the peninsular and the extra-peninsular (fig. 42). The peninsular group of mountains embrace the Western Ghats, the Eastern Ghats, the Aravalli Ranges, the Vindhya and the Satpura Ranges. The Himalaya belongs to the extra-peninsular group.

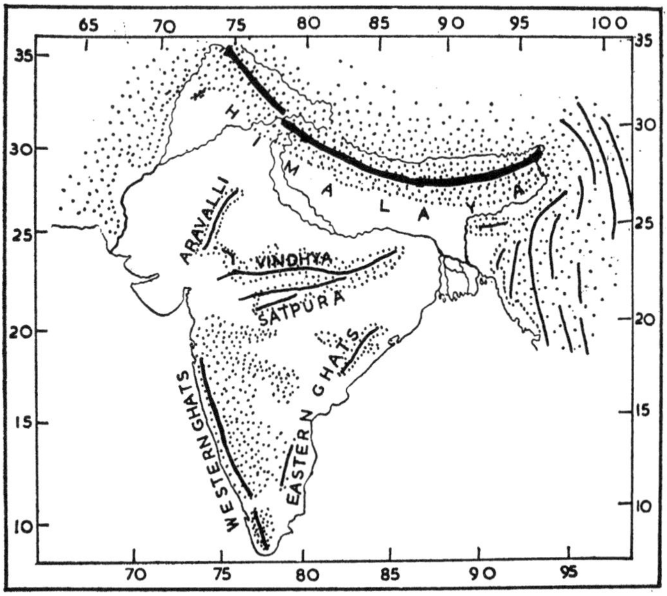

Fig. 42. Sketch-map of the Indian Region, showing the principal elevated areas (shaded) and the mountain ranges.

The Western Ghats are an unbroken chain of meridional ranges (mean elevation 1000 m) that flank the west coast of India. They are formed largely of Deccan lavas that poured forth over an area of nearly 512 000 sq. km about the beginning of the Eocene, when the uplift of the Himalaya was perhaps initiated. The Eastern Ghats comprise much broken ranges. In contrast to the Western Ghats, a relatively broad coastal plain fringes the hills of the Eastern Ghats. The Aravalli Ranges are three-fold mountains, elevated in the early days of the Earth's history and stretching northeast for about 700 km in Rajasthan. The mean elevation of the Aravalli moun-

* Not Himalayas, as only too often, but nevertheless quite erroneously termed.

tains varies from 300 to 1000 m; the highest peak, Mt. Abu (1772 m), is at the southwest extremity. The Aravalli Hills are generally clothed with but scanty and characteristically xerophytic vegetation. Large intervening areas are covered by sandy deposits near Alwar, Jaipur and other parts. The isolated hills near Delhi represent the extreme northeast spur of the Aravalli Ranges. The Vindhya forms an uninterrupted range of hills (600-1375 m above mean sea-level) from the northern parts of the Madhya Pradesh, separating the Indo-Gangetic Plain from the Deccan Plateau. They are composed mainly of little-disturbed and mostly horizontal sandstone beds. The Satpura Mountains (meaning in Sanskrit seven folds) form a range of hills between the basins of the R. Tapti and the R. Narbada in central India. The high plateau of Amarakantak is the starting point of the Satpura and the range continues in a westerly direction to the west coast of India. The mean elevation is about 750 m above mean sea-level, but at Amarakantak the plateau is 1075 m above mean sea-level. The range is about 960 km long and the hills are composed mainly of the Deccan lavas and also partly of granites. The peninsular mountain mass has exerted a profound influence on the Himalayan uplift (*vide infra*).

I. THE HIMALAYAN SYSTEM

The Himalaya (from the Sanskrit *him* = snow, *alaya* = abode) was the name applied in ancient India to the Great Snowy Range of mountains, visible in the north from the Indo-Gangetic Plain. As now understood, the Himalaya embraces the complex system of nearly parallel ranges of Tertiary mountains, extending over

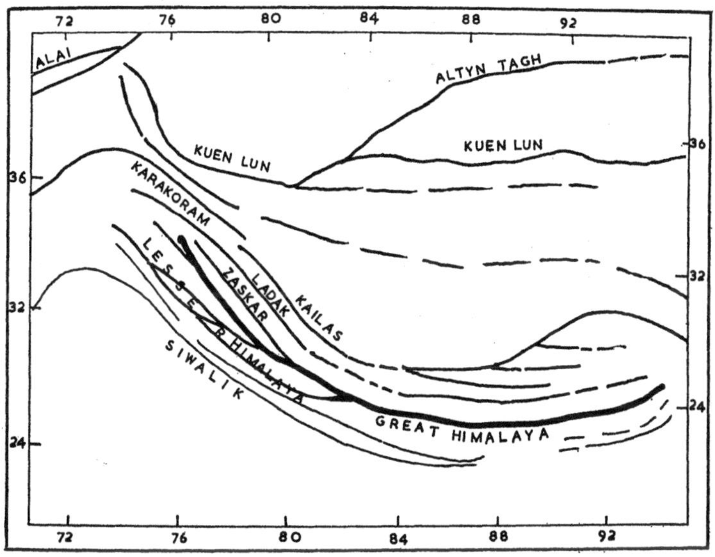

Fig. 43. Sketch-map of the Himalayan System, showing the crestlines of the principal mountain ranges.

3200 km, from North Burma in the east to nearly Afghanistan in the west (approximately between the east longitudes 72° and 91° and the north latitudes 27° and 36°) (fig. 43). The width of the Himalayan System is extremely variable and is in places only 80 km and in other places exceeds 300 km. Though the Himalaya appears to terminate in the east at the southward bend of the R. Brahmaputra and in the west similarly at the bend of the R. Indus, it is now generally admitted that the Himalayan System continues much further westward into the mountains of Afghanistan. Structurally and geologically, the Himalayan System is likewise continued into the meridional mountains of Burma. While the Himalayan System rises nearly abruptly from the plains of North India, it is continued in the north as a great series of folds, of which the Kuen Lun represents perhaps the northernmost. Between the Kuen Lun and the Himalaya proper lies the bleak high plateau of Tibet, at elevations between 4750 and 4880 m. The physical unity of Tibet with the Himalaya is evident in the general geological structure, in the parallelism of the Kailas Range, the Ladak Range, the Great and the Lesser Himalayan Ranges, all of which change the directions of their trendlines together. The relation of the Himalayan System to the mountain systems of Eurasia and Northeast Africa may be seen in fig. 44 and 45.

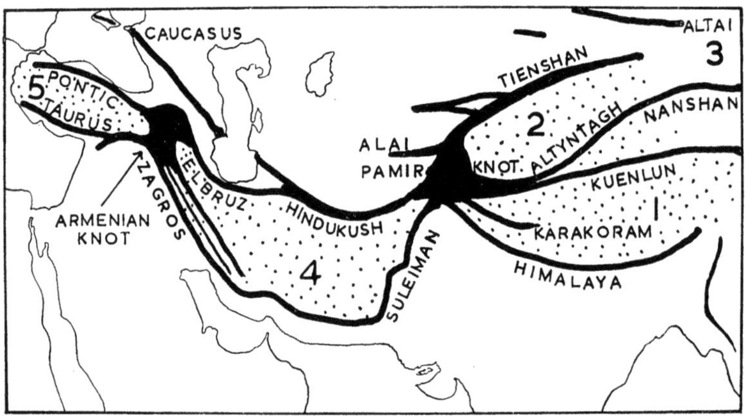

Fig. 44. Sketch-map of part of Asia showing the relation between the Himalayan System, the Caucasus and other Tertiary mountain systems to the west. The socalled Pamir-Knot and the Armenian-Knot may be noted. Between the parallel mountain folds lie the elevated plateaux: 1. the Tibetan Plateau, 2. Tarim Basin, 3. Mongolia, 4. Iran-Afghanistan and 5. Asia Minor.

The Himalaya is not a single continuous mountain chain, but a complex series of more or less parallel or often also converging ranges, intersected by enormous valleys and plateaux. Connected with the Himalaya are the other mountain ranges of High Asia, viz. the Hindu Kush, the Karakoram, the Kuen Lun, the Tien Shan, the Trans-Alai and the Alai Mountains, bound together in the socalled Pamir Mountain Knot (*vide* next chapter). In the general surface of Middle Asia, we observe two primary crustal elevations, separated by troughlike depressions, viz. the northern elevation or the Tien Shan Mountain chain and the southern elevation or the Tibetan

Plateau, with the intervening depression being the Tarim Basin (fig. 44). A second minor trough, south of Tibet is the alluvium-filled Indo-Gangetic Plain, separating that elevated region from the ancient Vindhya Mountains. These two elevations and their complementary depressions form the basis of the high mountains of Asia. This elevated area constitutes a major biogeographic subdivision of the Palaearctic Realm, viz. the Region of the Tertiary Mountains (730, 984).

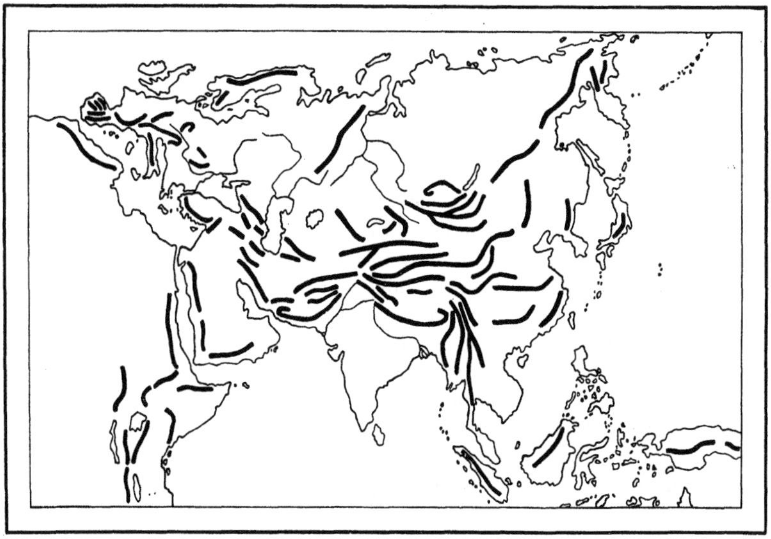

Fig. 45. Sketch-map of Eurasia and part of tropical East Africa, showing the relation of the principal mountain systems.

The mountainous area thus extends from the foot of the Siwalik Range in the south to the foot of the Kuen Lun in the north, but the principal plateau of Tibet lies between the Ladak Range and the Kuen Lun Range. It extends from the Karakoram Mountains to Kunsu and Szechuan Provinces of China. Southern Tibet consists of the troughs drained by the rivers Sutlej, Brahmaputra, Arun and others. Western Tibet has been compressed between the north, south and the central range-the Karakoram. Northern, northeastern and central Tibet forms a tableland, containing a large number of lakes, but southeastern Tibet is not really speaking a tableland but a rugged mountainous country that is cut by the rivers of China and Burma and is intersected by deep ravines. Tibet is wooded in its southeastern portion only. The alluvial deposits of the Tibet Plateau have been washed away by the rivers out of the greater number of troughs, south of the Ladak Range and north of the Kuen Lun Range. Some areas on the Indian side of the Ladak Range are, however, still filled with the original alluvium, of which the plains of Ngri Korsam and the Kashmir Valley, the Nepal Valley and the "duns" enclosed by the Siwaliks are the most important. The extensive plateau of Tsaidam lies north of the Kuen Lun. With its mean elevation of 4575 m above mean sea-level, the Tibet Plateau is connected at its northwest corner with the Pamirs (mean elevation 3660 m). The Pamirs are also connected by high mountains with the Tien Shan (mean elevation 3350 m). The interior of this enormous horse-shoe-shaped elevated mass is an inland desert basin, drained by the R. Tarim and its feeders. According to the well known explorer SVEN HEDIN, sand is continually accumulating in this basin, burying towns and will continue to accumulate until Lob Nor and the rivers of Tarim have been completely choked up. The lowest part is the lagoon of Lob Nor (elevation 670 m). There is no mountain range on the east, where its mouth is narrow, with the Gobi Desert beyond (mean elevation 600 m). On the north, west and south, there are disintegrating mountains.

During the middle of the Tertiary, the areas that are now Tibet and the Himalaya were covered by an extension of the Tethys Sea, in which deposition of immense sediments had continued for a vast period of time. The Tethys Sea separated Eurasia from the southern Gondwana landmass. Except the Altai of eastern Turkestan and the Aravalli Ranges of India, no other mountains existed in the area. The sedimentation, accumulated from the Palaeozoic Era, attained a thickness of perhaps 15200 m and was accompanied by slow sinking of the sea-bed, according to the principles of isostacy. During a period of crust movements, the floor of the sea began to rise gradually and was thrown into series of long, parallel, wave-like folds. The crests of the waves were eroded by rain and weather and the rising land became much broken and irregular. Drainage basins came to be carved out of the flanks of the folds and river systems, composed of the transverse valleys, were gradually formed. As the uplift progressed, the troughs of the folds emerged to form a series of longitudinal valleys, at right angles to the transverse valleys and parallel to the longitudinal axis of the crustal folds. A combination of concurrent uplift and erosion thus gave rise to the mountain systems of the Himalaya and Middle Asia. As denudation proceeded, deeper and deeper parts of the crust were exposed, but the forms of many folds may even now be traced and the trends of the longitudinal axis may clearly be followed over long distances. Folds were superimposed on folds, arches were overturned until almost horizontal and the whole region became greatly distorted and crumpled. The uplift of the mountainous region has thus been brought about by the horizontal pressure of the crust, acting in a meridional direction through long periods, right down to the present time. The wrinkling of the crust has taken diverse forms; the plateaux were wrinkled into ranges and the folded surfaces have been wrinkled and these wrinkled mountains have in turn been corrugated so as to form smaller folds. The intervening troughs were filled with debris from the mountains. In places where the stress exceeded the breaking strain of the crust, the rocks have fractured and have complicated the structural features of the mountains. Great parts of the crust have subsided and have moved horizontally. Considerable molten material has been forced up from below in places of weakness and fracture and has partly absorbed the original sediments also.

The uplift of the Himalayan System is believed by competent authorities to have taken place in four major impulses, respectively during the Eocene, the Middle Miocene, the Upper Pliocene and in the late Pleistocene. The pressure that led to the uplift of the Tethyan sediments originated mainly from the north and the peninsular mass produced an under-thrust towards the Middle Asiatic mass, so that the curvature of the Himalaya is convex on the Indian side.

The mountain ranges of the Himalaya are composed of consolidated rock, but the level portions of the elevated plateaux are troughs between the ranges, filled with loose debris of boulders, gravel, sand and mud washed down from the mountains and deposited in horizonatal layers. The mountain ranges vary in their width, in places bulging toward one another and in other places receding from one another, so that the intervening troughs and flat plains are either narrow or wide, as the case may be. The fossil remains of rhinoceros found at elevations of 5180 m in Tibet seem to suggest, according to GODWIN AUSTEN, that the Himalaya has risen up by about 2440-3050 m within comparatively recent times, since the Pleistocene. The action of wind contributed to the filling up of the troughs with loess and mud avalanches have also transported large boulders. The rivers have extensively cut back into the Tibet plateau, from all sides, by head-erosion and the high-level alluvial plains are, therefore, found only intact in the areas not so far reached by the rivers. On the north, south and the west, the great rivers are confined to a few troughs. On the east the rivers of China rise in the wide troughs of Central Tibet and have cut back comparatively easily into the soft alluvium.

2. THE RANGES OF THE HIMALAYA

The ranges of the Himalaya fall under two major groups: the Cis-Himalayan Ranges and the Trans-Himalayan Ranges. The Cis-Himalayan Ranges, to the south of the Great Himalaya, comprise i. the Lesser Himalaya, a series of nearly parallel ranges closely related to the Great Range and ii. the Siwalik that intervenes between the Lesser Himalayan Ranges and the Indo-Gangetic Plain. The Trans-Himalayan Ranges, north of the Great Himalaya, comprise the Zaskar, the Ladak and the Karakoram Ranges (fig. 46).

As commonly understood, the Great Himalaya Range refers to the portion between the bends of the R. Indus and the R. Brahmaputra, but it should be remembered that like the rivers Sutlej and Ganges, the rivers Indus and Brahmaputra also cut through gorges, which they have themselves carved. It is, however, not easy to trace the course of the Trans-Indus and Trans-Brahmaputra sections of the Himalayan ranges. The outer zone of mountains in the south is contiguous to the Indo-Gangetic Plain and contains the Siwalik Ranges and the valleys to the north of them. This region was elevated more recently than the Himalaya proper. The width of the Siwalik-zone varies from about 8 km to nearly 50 km, being narrow in areas where the Siwalik hills are jammed against the Lesser Himalayan Ranges and wide where open valleys intervene. To the north of the Siwalik-zone is a second zone of mountains, about 60-80 km wide, characterized by the longitudinal ranges that are generally parallel to the main Himalayan range in the north, in the Punjab and in Nepal. In the Kumaon Division, however, the topography is far more complex and the ranges do not appear to follow the distinct alignments of the maximum elevation, but to be scattered throughout the area. The mean elevation of the peaks of the mountains in this zone is 1830-3050 m. The third zone of mountains, about 16 km wide, is characterized by spurs projecting southwards from the Great Range. A few peaks here exceed 4575 m in elevation. The fourth zone, about 25 km wide, is occupied by the mighty line of snow-covered peaks, the mean elevation of which exceeds 6100 m. With the exception of some low ravines, the zone as a whole rises above the limits of permanent snowline. The fifth zone, about 33 km wide, contains the troughs of the rivers rising to the north of the Great

Himalaya. The mean elevation of the beds of these troughs is 4270 m and the mountains intersecting them rise to about 5790 m above mean sea-level. The highest snow-covered elevated areas are thus located about 150 km, north of the southern limits of the mountains. The Great Himalaya Range does not form a uniform curve, but follows different alignments. As the Range bends west to the northwest, it often bifurcates, gives rise to minor ranges on the convex side of the bend and at each bifurcation, the minor range tends at first to continue in the alignment of the Great Range, but gradually becomes parallel with it.

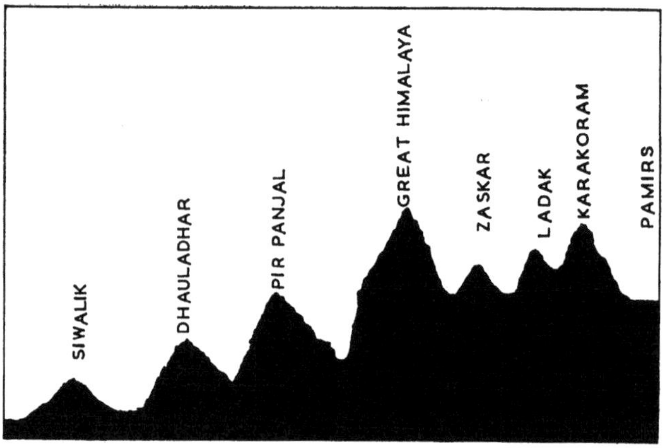

Fig. 46. Diagrammatic profile section of the Himalayan System in the area of the Northwest Himalaya, showing the principal ranges (not to scale).

The Siwalik Range separates the Himalaya proper from the Indo-Gangetic Plain and is in this respect the southern border-range of the Himalayan System. Though its upheaval was accompanied by movements of the Himalaya also and perhaps also by increase in the elevation of the main Himalayan range, the Siwalik is of more recent origin than the great mountains to the north. With the exception of a short distance of about 80 km, opposite the basins of the R. Teesta and the R. Raidak, the Siwalik Range lies, with remarkable uniformity, in front of the Himalaya, throughout its whole length, from the bend of the R. Brahmaputra to that of the Indus. There is a break at the passage of the R. Sutlej in the general alignment and the two lengths of the range appear to overlap, so that the range to the north of the R. Sutlej is not in direct prolongation of the one to the south. In some places, the Siwalik Range is pressed against the Lesser Himalayan Ranges and in other places it is separated from these by gaps of even 80 km, so as to enclose the characteristically shaped longitudinal valleys called duns, filled with deposits of rounded stones, gravel and sand, brought down from the Himalaya. The Siwalik Range is strongly developed opposite Dehra Dun, with a steep southern slope and a gentle northern slope.

The folding of the Siwalik strata is strongly suggestive of the view that the Himalaya must have advanced southwards. The Siwalik strata exceed in thickness 4575 m and consist entirely of freshwater deposits of the material brought down by the Himalayan rivers and upheaved in comparatively recent times. Unlike in the case of the main Himalaya, the striking features of the Siwalik are due to crustal deformation and not to river erosion.

The mountainous region, about 150 km wide, between the Great Himalaya and the Siwalik, constitutes an intricate system of the Lesser Himalayan Ranges. The contortions of the strata in this region show that the Lesser Himalayan Ranges have been compressed horizontally. These ranges are, however, the result of a series of crustal movements, with more complex history than in the case of the Siwalik Range. After their uplift, the Lesser Himalayan Ranges appear to have been forced to change their direction, so that the whole region has been subjected to successive compressions and the general wrinkling process seems to be still active. In parts of Nepal and in Kashmir, the outer ranges and the flat alluvial valleys north of the Lesser Himalayan Ranges are distinct.

The Lesser Himalayan Ranges comprise two principal groups: i. the ranges that branch off the Great Himalaya proper and ii. the ranges that form separate folds from the Great Himalaya. The branch range lies obliquely across the mountain area, but the separate folds follow a general curvilinear alignment parallel to the Great Range. The Great Himalaya bifurcates at points where there is a change in the alignment and each successive branch adopts the alignment forsaken by the trunk range. There are seven Lesser Himalayan Ranges, viz. the Nag Tibba Range, the Dhauladhar, the Pir Panjal, the North Kashmir, the Mahabharat, the Mussoorie and the Ratan Pir.

Of the Trans-Himalayan Ranges, the Zaskar branches off from the Great Himalaya near Nampa. The well known peak Mt. Kamet (7770 m) is situated on the Zaskar Range. The Ladak Range extends from Assam area to Baltistan, but its continuity north of the Great Himalaya is not distinct throughout. North of Assam, the Ladak Range is strongly developed and forms the principal water-parting between the Tibetan and Indian sections of the R. Brahmaputra. Westwards from Nyang Basin, for a distance of nearly 320 km, the Ladak Range is parallel to the Great Himalaya and the intervening trough is occupied by the R. Arun. North of its bifurcation from the Great Himalaya at Dhaulagiri, the mean elevation of the Ladak Range increases. South of Lake Mansarovar is the Gurla Mandhata Peak, west of which the continuation of the Ladak Ranges becomes obscured. It is generally believed that the Ladak Range has risen subsequent to the birth of the R. Indus.

Strictly speaking, the Karakoram and the Hindu Kush Ranges are different sections of the same crustal fold that stretches from the southeast to the northwest, curves round Hunza and Gilgit, passing north of Chitral and entering Afghanistan from the northeast to the southwest. The eastern part of the fold is known as the Karakoram and the western part is the Hindu Kush. East of the R. Shyok, there are no high peaks of elevations of 7315 m and west of the Tirich Mir group of peaks also the mean elevation does not reach so high. The rivers Shyok, Hunza, Gilgit and Kunar drain the trough north of the Karakoram. The R. Nubra rises on the Karakoram and the glacier at its source has cut a notch in the crestline of the mountain range. On a stretch of about 170 km of the Karakoram, there are several great peaks; the Aling Kangri Peak is generally considered to mark its extreme eastern limit. The mean elevation of the Karakoram Range in Tibet is between 5000 and

5500 m, in other words hardly 300-600 m above the general level of the plateau.

The Hindu Kush, about 800 km long from the Pamirs Region to the Koh-i-Baba Mountain west of Kabul, forms the water-parting divide between the R. Kabul and the R. Amu Darya (= R. Oxus). According to some authorities, the Hindu Kush is a branch from the Taghdumbash Pamir, where the latter connects with the Sarikol Range (*vide* next chapter). There are two parallel ranges; the southern range is a western extension of the Karakoram. The two ranges of the Hindu Kush conjoin, near the Tirich Mir Group of peaks, to continue as a single range. For a distance of about 160 km, it is a relatively flat-backed range, with small lakes and numerous passes at elevations of 3810-5330 m. For a distance of about 225 km from the Hunza-Gilgit water-parting to the Tirich Mir Peaks, the Hindu Kush does not rise to great elevations. The uplift of the Hindu Kush took place later than that of the Himalaya proper and was marked by overthrust of Cretaceous limestones on Recent shales. Like the Himalaya, the Hindu Kush is the facet of the northern plateau. Lake Gazkul, on the crestline of the Hindu Kush near the Taghdumbash Pamir, is believed to be a Pleistocene relic.

The Kailas Range is parallel to the Ladak Range, but about 80 km to the north of it. Mt. Kailas is the highest of a cluster of peaks, exceeding 6000 m in elevation, in the area of Lake Mansarovar. The Kailas Range appears to terminate in the Sajum Peak, but it actually continues further west and forms the water-parting between the R. Shyok and the R. Nubra. Further west, the Kailas Range is parallel to the Karakoram and the trough between the Kailas and Karakoram ranges is occupied by the Biafo, Hispar and Chogo Lugma Glaciers.

The Aghil, Kuen Lun and the Altyn Tagh Ranges are situated in Turkestan. The Kuen Lun Range extends east-west through the northern Tibet, between 76° and 83° EL and represents the northern border-range of the Tibetan plateau. West of the 83rd meridian, the Kuen Lun faces the Tarim desert and east of this meridian the Altyn Tagh, its branch range, forms the border. The Kashghar and the Sarikol Ranges are two parallel mountain chains that form the eastern flank of the Pamirs, with south-southeast to north-northwest trend. The Sarikol Range is separated from the Kashghar Range by a distance of about 60-65 km and it also forms the border between the Taghdumbash Pamir and the Little Pamir. The Pamirs-Tien Shan Complex is dealt with in the next chapter.

3. THE GEOGRAPHICAL AND BIOGEOGRAPHICAL DIVISIONS OF THE HIMALAYA

Geographically, the Himalaya is divided into i. the Eastern or the Assam Himalaya, ii. the Central or the Nepal Himalaya, iii. the Kumaon or the Western Himalaya and iv. the Northwest or the Punjab Himalaya (fig. 47).

The Assam Himalaya, approximately 720 km long, comprises the portion between the Namcha Barva Peak (7750 m above mean sea-level), east of which the R. Brahmaputra curves southward, and the R. Teesta in the west. There is very little of the Sub-Himalayan tract in this division, so that the Himalaya rises rather abruptly from the plain. In Sikkim, where the Assam Himalaya passes into the Nepal Himalaya, there is also a change of alignment with the ranges in the west extending from the east a little north of west. The lower and outer ranges also disappear here and paired

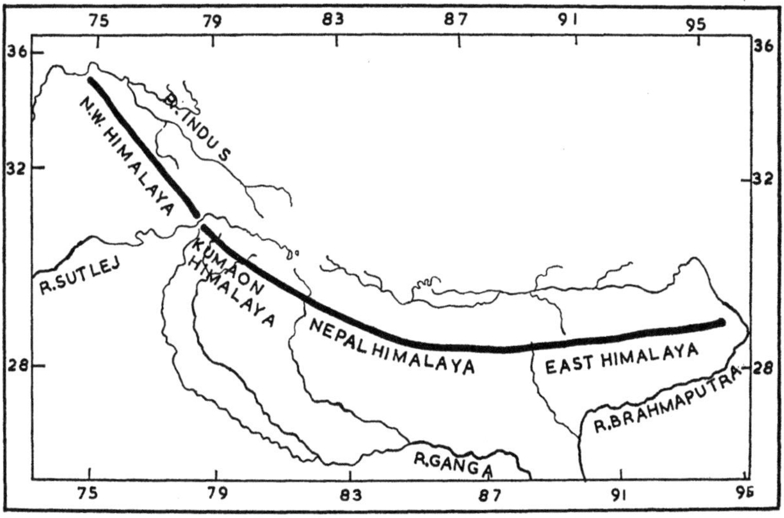

Fig. 47. Sketch-map of the Main Himalayan Range, showing the principal geographical subdivisions.

spurs, the Singalila Ridge from the Kinchinjunga and the Chola Ridge from Pauhurni, extend southwards. The Nepal Himalaya, between the R. Teesta in the east and the R. Kali in the west, is about 800 km long. A number of well known high peaks like Mt. Everest (8848 m above mean sea-level), Kinchinjunga (8579 m), Makalu (8470 m), Dhaulagiri (8245 m), Annapurna, Gosainthan (8010 m), etc. are situated in this division. The Great Himalaya bends and bifurcates near the Dhaulagiri Peak, west of which the mean elevation diminishes, so that none of the peaks rise to over 6700 m. Near the western end of the basin of the R. Karnali there are the Api-Nampa group of high peaks; there is another bifurcation at this place. The southern branch is the Great Himalaya, with Nanda Devi (7820 m) and the Badrinath Peak (7060 m), and the northerly branch, with Mt. Kamet (7770 m), is the Zaskar Range. The Kumaon Himalaya extends for about 320 km between the R. Kali and the great defile of the R. Sutlej. Naini Tal, Almora and Garhwal of the Uttar Pradesh State are within this division. Nanda Devi, Trisul, Mana, Badrinath, Kedarnath, Gurla Mandhata, Gangotri and Bandar Punch are some of the better known high peaks of the Kumaon Division. The Kumaon Himalaya is much corrugated. The Northwest Himalaya is the division west of the defile of the R. Sutlej, approximately 560 km long, extending to Mt. Nanga Parbat, west of which the R. Indus curves round southwestward.

Biogeographically, the defile of the R. Sutlej is a most important landmark (*vide infra*). To the east of it the Himalaya forms a zoogeographical unit that is fundamentally distinct from the Northwest Himalaya to the west of the defile. The Northwest Himalaya differs from the rest of the Himalaya in several essential features. At the defile of the R. Sutlej, there is an abrupt break in the general trendline of the mountain ranges. While to the east of the defile the ranges are mostly east-west, the trend of the ranges in the Northwest Himalaya is more southeast to the northwest. While the rest of the Himalaya lies between 27° and 29° NL, the Northwest Himalaya stretches from about 30° to nearly 36° NL and has thus a much greater width, with an extensive Sub-Himalayan tract. West of the defile of Sutlej there is an abrupt fall in the mean elevation of all the ranges. The peaks rise to elevations of 7600 m in the rest of the Himalaya and often above 8000 m; in the Northwest Himalaya very few peaks are above 6700 m. There are, however, numerous peaks with a mean elevation of 6000 m in the Northwest Himalaya. Two notable high peaks are the Nun Kun twin Peaks (7130 and 7086 m) and the great Nanga Parbat (8126 m). While many river gorges cut through the Himalaya east of the defile of Sutlej, no rivers pierce the main range in the Northwest Himalaya. The main water-parting between India and Tibet lies just north of the Great Himalaya in the divisions to the east of the Sutlej defile, but in the Northwest Himalaya the crestline of the main range actually coincides with the water-parting. North of the Northwest Himalaya is the Karakoram-Pamir mass. The succession of ranges from the south are the Siwalik, the Nag Tibba, the Dhauladhar, the Pir Panjal, the Great Himalaya, the Zaskar, the Ladak and the Karakoram. These ranges are essentially secondary undulations on the great broad elevated arch, the span of which reaches from the plains of the Punjab in the south to beyond the R. Indus in Tibet.

4. GEOLOGY AND GLACIOLOGY

Broadly speaking, three stratigraphic zones are generally recognized in the Himalaya: i. the Sub-Himalayan zone of sedimentaries, mostly of Tertiary origin; ii. the Himalayan zone and iii. the Tibetan zone north of the crestline of the Great Himalaya. The Sub-Himalayan zone is characterized by conglomerates, sandstones and clays of Tertiary age. The Upper Siwalik beds, immensely thick, are composed of loosely aggregated conglomerates and soft earthy deposits. Below these lie very considerable thicknesses of soft sandstones, resting on harder sandstones of the Nahan Stage. The Pleistocene high-level terraces of the Karewa beds in Kashmir and the ossiferous rocks of Ngri Korsam also belong to this zone. Two Series, the Sirmur Series and the Siwalik Series, are generally recognized. Between the Siwalik Series and the older deposits, there are the so-called reversed faults or the main boundary fault, in which the older rocks have been thrust up over the younger. The Siwalik Series do not overlap the boundary fault-line and are also never found among the mountain ranges further north. The total thickness of the Siwalik Series is 4875 m. The Sirmur Series are not generally observed east of the R. Jamuna. The Himalayan zone comprises the great bulk of the Lesser Himalaya and the Great Himalaya and is characterized by granites, gneisses, crystalline schists, slates, quartzites and unfossiliferous limestones. The central gneiss is intrusive. The axis of the Himalaya is thus composed of the oldest group of rocks. The crystalline axis is separated from the Sub-Himalayan zone by a belt of unfossiliferous sedimentaries. On the northern flank of the crystalline axis there are, however, fossiliferous sedimentaries, mostly of marine deposition. The Tibetan zone embraces the northern parts of the Himalaya and extends into Tibet also. We find here the most complete series of strata, from the earliest Cambrian to the Late Tertiary. The deposits are of immense thickness, often exceeding 6000 m of almost entirely marine sediments. These sedimentaries are in contact with the granite axis of the Great Himalaya in the south. For further details on the general geology and tectonics of the Himalayan System reference may be made to BURRARD & HAYDEN (175), WALKER & PASCOE (1095), WADIA (1086-1089), DE TERRA (253-256), HEIM (498-499) and HEIM & GANSSER (500).

There are hundreds of glaciers at the sources of streams and rivers, but most of them are small and generally unknown (175, 484, 485, 538). The larger and the better known glaciers are mostly relics of the older and more extensive ice-flows of the Pleistocene. During the Pleistocene, the Himalayan glaciers descended to localities of 1500 m elevation at the present time. As is well known, the Northwest Himalaya rose very considerably in the Late Pleistocene and in the Post-Pleistocene. Profound changes have also taken place in the region since the Pleistocene and the land may have been 300-900 m lower during the Pleistocene than at the present time. It is, therefore, obvious that during the Pleistocene, the Himalayan glaciers must have descended to levels much below those where we find evidence of the glaciers at present. Continuous ice-sheets are believed to have covered Ladak, Ngri Korsam and other areas. According to ANTEVS (51) and DE TERRA (253, 254), there are distinct evidences of at least four separate periods of glaciations during the Pleistocene on the Himalaya. The unloading of the mountains at the end of the Pleistocene was followed by a very pronounced increase in their elevation. As today, even during the Pleistocene general atmospheric aridity seems to have prevailed in the elevated areas of Middle Asia. The snowline depression in the Northwest Himalaya was about 800 m and on the south slope probably 1600 m. The total thickness of the ice-sheet that covered the Himalaya Middle Asia region is estimated at 150 m. The Pamir glaciers were 250 km long during the Pleistocene. There are at present over 1200 glaciers in the Pamirs Region. When compared with those of Europe and North America, the Pleistocene glaciations in the area of Middle Asia were on the whole small, mainly because of the enormous distances from the oceans and the great size of the continental mass and also because of the precipitation on the south slopes of the high mountains causing atmospheric aridity on the north.

The present-day Himalayan glaciers are mostly dirty-coloured, due to the enormous quantities of dust deposited by the upper air currents. No Himalayan glacier is at present longer than 32 km, but the Karakoram glaciers are the greatest, outside of the polar areas. Some of the Karakoram glaciers are 60 km long. Some glaciers in the Spiti area of the Northwest Himalaya descend to elevations of 3350 m, but the snout of the Hispar Glacier in the Northwest Himalaya is at an elevation of 3152 m. Some transverse glaciers like the Rakaposhi Glacier have their snouts at elevations of 2435 m. The glaciers on the south slope are generally shorter than on the north. Some of the Himalayan glaciers appear to be receding at the present time, but others like the Hanzanagar Glacier appear to be extending. Other glaciers on the Himalaya have, however, remained practically stationary.

The snowline, as commonly understood, would really convey no meaning when we speak of the conditions prevailing in the deep innermost Himalaya. Vast areas of rock are snowfree even at very high elevation, largely because of the extreme atmospheric aridity. Deep tongues of snow descend, however, to an elevation of 5485 m and large continuous slopes of the mountains remain free from snow even at an elevation of 7000 m on the north aspect of the ranges. The mean value of 5170 m usually given for the snowline on the Nepal Himalayas is also misleading. On the south slope of the Nepal Himalaya, snow is abundant during June-September even at elevations of 4575-4780 m, but in the inner Himalaya in the same area, the mean atmospheric precipitation drops off abruptly, so that the northern gradient of light snowfall raises the snowline. The term snowline is, therefore, strictly speaking, not very appropriate in the case of the Himalaya, where enormous tongues of snow descend to 5500 m and vast areas higher above are snowfree, sometimes even up to 6900 m above mean sea-level. The snowline is on the average about a thousand metres lower on the south slope of the Great Himalaya than on the north slope. In the Northwest Himalaya, the snowline is higher than in the rest of the Himalaya. As we proceed westward from the Assam Himalaya, there is a gradual increase in the general atmospheric aridity and consequently the snowline also rises. The following table summarizes the snowline altitudes in different sections of the Himalaya:

TABLE 26

Snowline on the Himalaya, in m

Division of the Himalaya	South slope	North slope	Mean
Nepal	4480	—	—
Kumaon	4725	5638	
Punjab	5180	5790	
Zaskar	6000	5945	
Ladak	5790	5640	
Tibet	—	—	6000
Karakoram	5640	5648-5485	

5. ECOLOGICAL CHARACTERS AND BIOTIC ZONES

The outstanding peculiarities of the high altitude ecology of the Himalaya may be traced to its enormous massiveness, the stupendously great elevations of the mountain ranges, their trendlines, their location in the middle of a vast continental landmass, their Tertiary orogeny and Pleistocene glaciations. The conditions commonly met with above the upper limits of the forest on other mountain ranges of the world can really give no adequate clue to those likely to prevail on the vastly amplified and much higher life zones above the forest on the Himalaya. The Himalaya is high and large enough not merely to influence and modify the climate and vegetation of the whole of India and to determine also the climate of Middle Asia, but also to make its own climate. It is difficult to compare the climate of the Himalaya with that of any other mountain. Although situated only a few degrees north of the torrid zone, owing to its enormous size and its unparalleled elevation, we find here a complete range from the tropical to the deep arctic conditions. Extending over 3200 km from the east to the west, it is obvious that the conditions must differ profoundly in the extreme eastern and western ends of the Himalaya. The general climatic conditions in the east are semi-oceanic, but become more and more continental as we proceed westward.

The mean annual 10° C isotherm lies at an elevation of 3960 m in the eastern Himalaya, but descends to 3050 m in the Northwest Himalaya. In the area of the Mt. Everest, the mean winter temperature, at an elevation of 5990 m, varies from —12° to —27° C and the atmospheric temperature at the summit of Mt. Everest is

—42° C. The mean monthly atmospheric temperatures and the mean precipitation at an elevation of about 3500 m on the Great Himalayan Range in the Northwest Himalaya are summarized in table 27.

The mean annual atmospheric temperature is 5° C and the total rainfall is 80 mm.

TABLE 27

Mean monthly temperatures and precipitation at an elevation of 3500 m on the south slope of the Great Himalaya in the Northwest Himalaya

Month	Mean temp. °C	Mean precipitation mm
January	−8.1	10.0
February	−6.1	7.5
March	−0.5	7.5
April	6.1	5.0
May	10.0	5.0
June	14.4	5.0
July	17.2	12.5
August	16.1	12.5
September	12.2	7.5
October	6.1	5.0
November	0.0	—
December	−5.5	5.0

South of the main crestline of the Great Himalaya, the lower hills are covered by broad-leaved wet forest (the lower monsoon forest) up to elevations between 900 and 1000 m. Above this zone are the middle and the upper montane evergreen forests (197, 607). At elevations between 2440 and 3050 m there is the broad-leaved sclerophyll or the *Quercus-Rhododendron* forest; in the Northwest Himalaya this zone reaches up to 1800 m. In the eastern divisions there is mixed forest of *Rhododendron* and conifers at slightly higher elevations. These give place above to the *Betula-Juniperus* zone, at elevations between 3000 and 3660 m in the western divisions and a shrub zone of *Abies spectabilis*, with thickets of *Rhododendron campanulatum*, succeeded by dwarf *Rh. anthopogon* and *Rh. setosum* at elevations between 4260 and 4575 m in the eastern divisions. This belt may often ascend to an elevation of 5180 m on the East Himalaya. In the west, the *Betula-Juniperus* zone marks the upper limits of the closed forest and gives place to the open vegetation of the alpine-zone type higher up. At much higher elevations, there are cushions of low plants and prostrate plants, with grasses and sedges. This zone may be met with even at an elevation of 4575 m and may extend higher up to 5790 m, particularly on the south slopes, where the snow meets the dwarf shrub vegetation. On the north slope the picture changes into a typical desert prostrate vegetation. At elevations above the zone of plant cushions, the vegetation is confined to the base of boulders and large rock pieces. Phanerogam cushions are found even far above 6000 m on the Nepal Himalaya. Cushions of *Stellaria decumbens* grow, for example, at elevations of 6140 m on the north slope of Mt. Makalu (1021), but there are reasons to believe that Phanerogams flourish at still higher elevations, perhaps even at 6300 m. Cultivation of potato

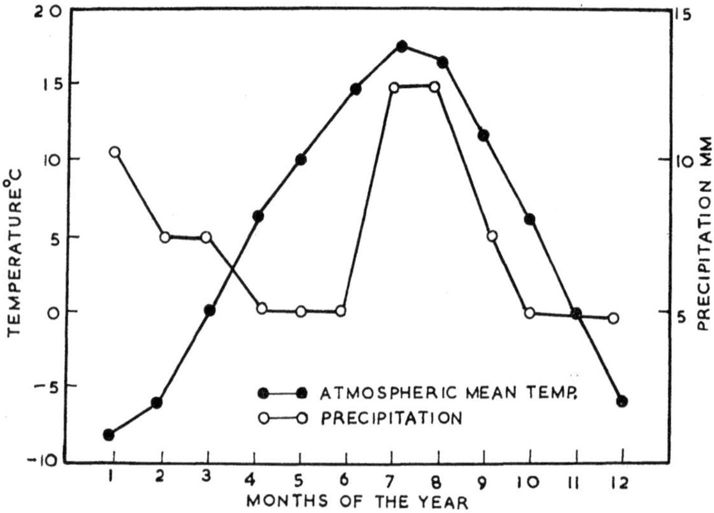

Fig. 48. The mean monthly atmospheric temperature and precipitation at an elevation of 3500 m on the Great Himalaya.

and turnip at elevations between 4270 and 4875 m is common in parts of Lahaul and Tibet. The zone between 3960 and 6000 m above mean sea-level in the East Himalaya is generally called alpine zone. The alpine zone on the Himalaya is perhaps the most extensive and also the highest in the world. The *Betula-Juniperus* belt in the west corresponds to a subalpine transitional zone. It is also generally taken as the timberline on the Himalaya. Defined in this way, the timberline on the Northwest Himalaya is much lower than in the rest of the Himalaya. The timberline in the East Himalaya is at about 4115 m on the south slope in the inner valleys and 3600-3700 m on the exposed ridges, but in the Northwest Himalaya, the timberline is somewhat above 3000 m. Even in the Northwest Himalaya, the timberline is 3600-3700 m on the south slope of the Dhauladhar Range and 3000 m on the Pir Panjal Range; the northern slopes of the Pir Panjal are barren and the ranges to the north of the Pir Panjal are far above the timberline elevations. Among the typical plants that flourish above the upper limits of the closed forest, the following may be mentioned: *Rhododendron setosum*, *Rh. anthopogon*, *Juniperus squamata*, *Gentiana*, *Primula*, *Arenaria*, *Leontopodium*, *Androsace*, *Artimisia*, *Ephedra*, etc., in addition to grasses and sedges. SWAN (1022) recognizes an aeolian zone, extending from an elevation of 6000 to 6850 m in the Nepal Himalaya. Though overlapping altitudinally, at least in part, with the alpine zone, the region included in the aeolian zone is not truly alpine in nature. This is rather a wide zone, above the limits of the growth of Phanerogams and is characterized by sparse growth of lichen, and dominance of Collembola, Anthomyiidae and Salticid spiders and by the ecological dominance of the wind-blown organic material as food resources. MANI (730) has recently pointed out the inappropriateness of the terms alpine and arctic zones in describing the biotic zonation on the Himalaya on the analogy of the conditions prevailing on the Alps. The biotic province above the upper limits of the forest extends on the Himalaya to well over an elevation of at least 6800 m and perhaps even higher.

6. THE INSECT LIFE OF THE HIMALAYA

The earliest collections of high altitude insects from the Himalaya were probably made by VON HÜGEL (550). His collections, comprising several hundred specimens, were described by KOLLAR & REDTENBACHER (643). The next important attempt at collecting insects was by the Yarkand Political Mission. To STOLICZKA, the celebrated geologist and naturalist of the Mission, should really go the credit for discovering the wealth of insect life of these inaccessible regions (1012). His collections were described by various well known specialists from Europe (264, 781, 782, 886). Some Coleoptera were also described by FAIRMAIRE (327) from the Northwest Himalaya. GUY BABAULT, who travelled through Kangra, Kulu, Lahaul, Ladak and parts of Kashmir in 1914, brought back large collections of high altitude insects. The Orthoptera from his collections were described by UVAROV (1064), the Carabidae by ANDREWES (48), Curculionidae by HUSTACHE (553) and Histeridae by DESBORDES (250). Some insect collections were also made by the Karakoram Expeditions (113, 325, 430, 431, 761, 790, 915) and by the German Nanga Parbat Expedition* (306, 307, 346, 347). HUTCHINSON of the Yale University North-India Expedition collected a number of interesting high altitude insects from Ladak and adjacent areas (50, 181, 224, 248, 449, 556). The insects collected by the different Mt. Everest Expeditions have been described by a number of specialists like ANDREWES, CAMERON, RILEY and others. MANI's Entomological Expeditions to the Northwest Himalaya collected a large number of insects at elevations above 3000 m and also made valuable field observations on the general ecology and distributional peculiarities at high elevations (76-79, 453, 454, 591, 593, 724-736, 968-978). Recently MANI (730) has given a comprehensive account of the high altitude insect life of the Northwest Himalaya.

EPHEMERIDA

Although the Ephemerida are abundant on the Himalaya, exceedingly few species have so far been described from above the forest zone. TRAVER (1045) reported a small number of species, collected by the Yale University North-India Expedition, from Ladak and other parts of the Northwest Himalaya. Recently KAPUR & KRIPALANI (606) recorded some more species collected by MANI's Entomological Expeditions to the Northwest Himalaya. There is an interesting record of an endemic genus *Ororotsia* TRAVER, represented by *O. hutchinsoni* TRAVER, from Ladak. The other species recorded from Ladak include *Ameletus primitivus* TRAVER and *Baetiella ladakae* TRAVER. *Baëtis chandra* KAPUR & KRIP. occurs at elevations of 2750-3660 m in Lahaul; the larvae of this species abound under the submerged stones in glacial torrents and near the glacier edge. *B. himalayana* KAPUR & KRIP. and *Epeorus lahaulensis* KAPUR & KRIP. occur at an elevation of 3200 m in the same area. An unnamed species of *Iron* EATON occurs at an elevation of 3970 m on the Northwest Himalaya.

*DOESBURG (DOESBURG, P. H. van, 1955. *Beaufortia, Amsterdam* 5: 47-51) has reported on the Syrphidae collected by the Fourth Dutch Karakorum Expedition of 1935.

PLECOPTERA

The high altitude Plecoptera from the Himalaya are perhaps better known than the Ephemerida. KIMMINS (614) has described a number of species collected by the Mt. Everest Expeditions. The nymphs of Plecoptera were found under submerged stones in pools and streams and the adults were taken at elevations of about 5030 m, on the Rongbuk Glacier in the Mt. Everest area. *Rhabdiopteryx lunata* KIM. was collected by HINGSTON at an elevation of 5030 m on the Rongbuk Glacier. *Kyphopteryx dorsalis* KIM. occurs on the north slope of the Mt. Everest massif, at elevations of 3353 m, in the Chumbi Valley in Nepal-Sikkim area and in Sikkim at an elevation of 3657 m. Mention should be of *Capnia pedestris* KIM. from the Rongbuk Glacier at 5030 m, *C. hingstoni* KIM. at an elevation of 3657 m in Sikkim, *C. cordata* KIM. on the north face of the Everest massif at an elevation of 3963 m, *C. tibetana* KIM. also from the same area, *C. montivaga* KIM. at an elevation of 4575 m and *Filchmeria uncata* KIM. at elevations of 2750-4360 m. JEWETT (591, 593) has recently described the Plecoptera collected by MANI's Entomological Expeditions to the Northwest Himalaya. *Rhabdiopteryx lunata* KIM. occurs on the Northwest Himalaya at elevations of 4800-5000 m. *Capnia manii* JEWETT and *Nemoura* spp. are described from material collected at elevations of 3500-4200 m. *Capnia gibbera* JEWETT and *C. bifida* JEWETT (3600 m) from the Northwest Himalaya are remarkable in that the males are apterous. SÁMAL (915) has reported on the Plecoptera collected by the Dutch Karakoram-Hindu Kush Expedition.

ORTHOPTERA

A number of interesting Orthoptera, particularly the acridids, have been observed at very high elevations on the Himalaya. In the Northwest Himalaya, the highest elevation at which these insects have so far been found is between 4575 and 4875 m. The Tettigoniid *Hypsinomus fasciata* UVAROV occurs at these elevations. The nymphs of an undescribed grasshopper have, however, been reported by UVAROV (1065) at an elevation of 5490 m in the Mt. Everest area, representing perhaps the highest altitude record for the order in the world. It is interesting to note that some typically lowland species, like *Aularches punctatus* (DRURY) that occurs widely near the sea-level in India, Malaya and Java, ascend up the valleys, to nearly 3900 m in the Himalaya. The high altitude species include *Bryodema luctuosa* (STÅL) (4600 m), *Conophyma mitchelli* UVAROV (3300-4000 m), *Gomphomastax antennatus* BRUN., *G. bolivari* UVAROV and *G. disparillis* BOL. (3000-3500 m), *Oedipoda himalayana* UVAROV (4800) m, the endemic *Dicranophyma hingstoni* UVAROV and *D. babaulti* UVAROV and *Sphingonotus rubescens* (WALK.) and *Sph. savingyi* SAUSS. (up to 3500 m) from the Northwest Himalaya. Some of these species, like *Sph. rubescens*, occur on the Turkestan mountains also. The species collected in the Mt. Everest area are *Hyboella tibetana* UVAROV (3350m), *Omocestus hingstoni* UVAROV (3400 m), *Dysanema irvinei* UVAROV (4270-4575 m), *D. malloryi* UVAROV (4875 m), *Locusta migratoria danica* (LINN.), *Bryodema tuberculata holdereri* KR. (4270-4575 m) and *Orinhippus tibetanus* UVAROV (4270-4575 m) mostly from the northern side of the main crestline of the Everest massif.

Descriptions of the Himalayan Orthoptera may be found in UVAROV (1058, 1060, 1065, 1067, 1071, 1073, 1074). UVAROV (1071) has also discussed the presence of

Mediterranean elements like *Metrioptera* WESM. and *Oedipoda* SERV. as evidence, showing that the Northwest Himalaya has had a history different from that of the rest of the Himalaya. In another contribution (1067) he has discussed the distribution of the Angaran elements like *Podisma* LATR. on the Himalaya and the Middle Asiatic mountains. The development of the alpine-zone type of Orthoptera in the autochthonous fauna of the Himalaya-Pamir complex is explained by him on the basis of the fact that the essentially thermophile forms were lifted to the alpine-zone and nival-zone elevations during the Himalayan uplift.

DERMAPTERA

The endemic *Anechura zubovskii* SEMENOV and *A. himalayana* SANTOKH (fig. 28A) are common under stones at the snow-edge and near melt-water streams, at elevations of 3500-3700 m on the Northwest Himalaya. *A. pirpanjalae* SANTOKH occurs at an elevation of about 4500 m on the Pir Panjal Range (79).

HETEROPTERA

Some typically lowland Heteroptera have become secondarily adapted for life at high elevations, especially on the south slope of the Himalaya, within relatively recent times. These secondary species are particularly common on the East Himalaya, where also purely Indo-Malayan faunal elements predominate. The Indian lowland elements, like *Carbula indica* WESTW. and *Melamphaus faber* FABR., have been collected, for example, at elevations of 5030 m by the Mt. Everest Expedition (620). In the Northwest Himalaya, there is, however, a dominance of the Palaearctic forms and there are also some very remarkable mountain autochthonous endemic genera, characteristic of high elevations. Nearly 65% of the species so far known from above the forest-zone on the Northwest Himalaya are, for example, endemic from the Middle Asiatic Palaearctic region. There are also two remarkable endemic genera, viz. *Dolmacoris* HUTCHINSON and *Tibetocoris* HUTCHINSON in the Northwest Himalaya.

The Pentatomidae are represented by an endemic species *Phimodera rupshuensis* HUTCHINSON at an elevation of 4520 m on the Northwest Himalaya. *Eusarcocoris trigonnus* KIRITSHENKO (3050 m), *Dolycoris baccarum* LINN. (4575 m), *Pentatoma (Neso-liogaster) hingstoni* KIRITSHENKO (3050 m) and *Acanthosoma chinanum* KIRITSHENKO (3050 m) were collected near the Mt. Everest area. The Coreidae comprise an undescribed species of *Stictopleurus* STÅL at an elevation of 4752 m from the Northwest Himalaya and *Coriomeris nebulivagus* KIRITSHENKO at an elevation of 5030 m from the Rongbuk Glacier area and at an elevation of 3965 m from the Rongshar Valley area (all these localities are in the East Himalaya). The Lygaeidae are represented by seven species in the Northwest Himalaya: *Bianchiella adelungi* REUTER at an elevation of 3400 m in the R. Indus Valley and also widely distributed in Mongolia, North China and Siberia; *Lamprodema brevicollis* FIEBER at an elevation of 4267 m on the Ladak Range and also known from Dalmatia (Europe); *Microplax hissarensis* KIRITSHENKO at an elevation of 4575 m on the Ladak Range and from the Hissar Mountains in Middle Asia; *Nysius ericae* (SCHILL.) at an elevation of 5200 m, but also abundant at an elevation of about 4800 m and sparsely at 3400 m on the Ladak Range and widely distributed in Middle Asia; *N. e. alticola* HUTCHINSON at elevations of 5000–5300 m, breeding on the grass *Pagaephyton scapifolium*; *Dolmacoris deterrana* HUTCHINSON at an elevation of 5200 m and *Emblethis horvathiana* HUTCHINSON at an elevation of 5200 m. The species collected by the Mt. Everest Expeditions at elevations of about 3660–4420 m include *Spilostethus oreophilus* KIRITSHENKO (3660 m) and *Geocoris chinai* KIRITSHENKO (4420 m). No high altitude species of Miridae from the Northwest Himalaya occurs at elevations below 4000 m; some of the species have indeed been collected at an elevation of 5400 m, representing perhaps the highest altitude record for Heteroptera at present. *Tibetocoris margaretae* HUTCHINSON, an endemic form, occurs at this extreme elevation. *Chlamydatus pachycerus* KIRITSHENKO, brachypterous in the female and macropterous in the male, occurs at ele-

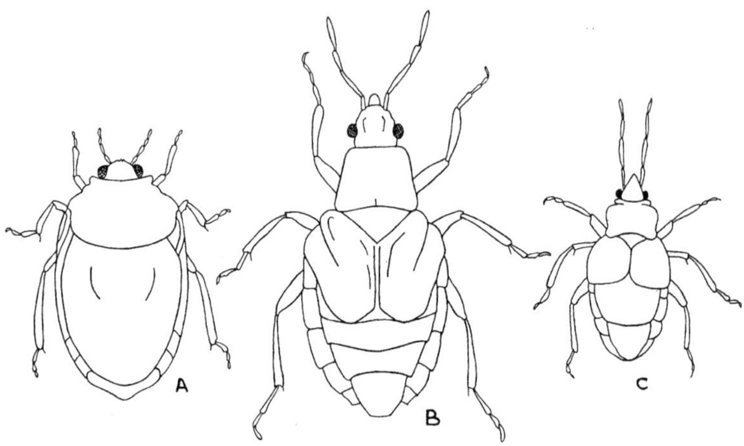

Fig. 49. The high altitude Heteroptera from the Himalaya. A. *Phimodera rupshuensis* HUTCHINSON. B. *Dolmacoris deterrana* HUTCHINSON, C. *Bianchiella adelungi* REUTER.

vations of 5100–5335 m, but sometimes it may be found at an elevation of 4530 m on the Northwest Himalaya and at elevations of 4270–5030 m in the Mt. Everest area. *Dicyphus physochlaenae* HUTCHINSON, which may perhaps prove eventually to be only a local subspecies of *D. orientalis* REUT. previously known from the Turkestan mountains, is reported on *Physochlaena praecalta* HOOKER (Solanaceae), at an elevation of 4600 m on the Northwest Himalaya. *Chiloxanthus alticola* KIRITSHENKO is the only Saldidae known so far from the Northwest Himalaya; it occurs under submerged stones in the melt-water torrents and ponds at elevations of 4200–4300 m.

HUTCHINSON (556) has pointed out the significance of the existence of a number of endemic genera and species of Heteroptera at high altitudes on mountains, which were covered by ice during the Pleistocene. Though no Heteroptera seem to have been collected at elevations above 5030 m in the East Himalaya, a number of species have been found commonly at much higher elevations in the Northwest Himalaya. HUTCHINSON refers to the four interesting endemic forms *Nysius ericae alticola* HUTCHINSON (5270 m), *Dolmacoris deterrana* HUTCHINSON (5300 m), *Chlamydatus pachycerus* KIRITSHENKO (5335 m) and *Tibetocoris margaretae* HUTCHINSON (5180 m) as representing two distinct ecological groups. *Nysius ericae alticola* and *Chlamydatus pachycerus* occur among grasses and other small herbaceous plants, particularly Cruciferae, and seem to be also widely distributed on the Northwest Himalaya and the Karakoram. They have most probably survived the Pleistocene glaciations on the mountains, in the immediate vicinity of ice. *Dolmacoris deterrana* and *Tibetocoris margaretae* are, on the other hand, confined to the more arid habitats characterized by *Artemisia minor* JAQ. as the dominant plant. Both the species belong to monotypic genera, most closely related to the forms reaching their maximum development in Middle Asia. Almost half the species of the area are peculiar to the Himalaya-Karakoram. Though richer in species, the Pamirs are poorer in peculiar forms. It seems that at the end of the Pleistocene, the Pamirs had open routes for the re-colonization from the rest of the Turkestan area (620, 878). The endemic elements of the Himalaya-Karakoram are believed by HUTCHINSON to be migrants from unglaciated areas in Tibet.

COLEOPTERA

With about 300 species, contained in some 20 families, Coleoptera represent a little more than one half of the high altitude insect life of the Himalaya. Carabidae, Staphylinidae, Tenebrionidae and Curculionidae are the dominant families above the forest-zone. Nearly 60% of the high altitude species are endemic to the Himalaya and 96% of them are Palaearctic forms. The Mediterranean elements amount to about 3% and the Indo-Malayan elements are also about the same. The Mediterranean elements are strongly developed in the Northwest Himalaya and the Indo-Malayan forms are more strongly developed in the eastern parts of the Himalaya than in the Northwest Himalaya. Nearly 76% of the high altitude species are endogeous terricoles and 78% are predators, carrion feeders and debris feeders (730, 734).

Carabidae

The high altitude Carabidae (fig. 50) of the Himalaya are remarkable for the high endemism of species; over 50% of the species are endemic to the Himalaya and a much larger proportion of the species are endemic to the vast elevated area of the Northwest Himalaya and Pamirs Region. There are 20 endemic species of *Bembidion* LATR., about five endemic species each of *Amara* BON. and *Cymindis* LATR. and two of *Nebria* LATR. The dominant genera are *Bembidion* LATR., *Amara* BON., *Carabus* LINN., *Cymindis* LATR., *Calosoma* WEBER, *Nebria* LATR., *Harpalus* LATR., *Trechus* CLAIRV., *Calathus* BON., *Bradytus* STEPH., *Clivina* LATR., *Broscus* PANZ., *Chlaenius* BON., *Dyschirius* PANZ., *Leistus* FRÖL. and *Tachys* STEPH.

The highest altitude at which Carabidae have so far been found on the Himalaya is 5300 m. *Amara brucei* ANDR. has been collected at this elevation on the Northwest Himalaya, but the same species occurs only at an elevation of 5030 m in the Mt. Everest area. A number of species like *A. ambigene* BATES and *A. lamia* BATES are common at elevations of 3400–4700 m. Over 25 species of *Bembidion* LATR. are known from above the forest-zone on the Northwest Himalaya. *B. hutchinsoni* ANDR. occurs at the edge of hot springs at an elevation of 4735 m and others like *B. ladas* ANDR. and *B. fuscicrus* MOTSCH. are typical of localities near the large glacial lakes. *B. pictum*

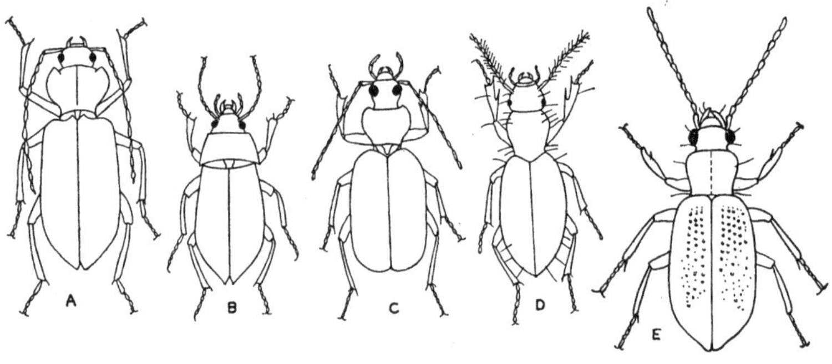

Fig. 50. High altitude Coleoptera from the Himalaya. A. *Nebria limbigera babaulti* ANDR., B. *Amara nila* ANDR., C. *Cymindis babaulti* ANDR., D. *Chaetobroscus anomalus* (CHAUD.), E. *Bembidion pamirense* ANDR.

(FELD.), *B. hutchinsoni* ANDR., *B. irregulare* NET., *B. ixion* ANDR. and *B. livens* ANDR. are confined to the elevated areas north of the main crestline of the Great Himalaya. *B. nivicola* ANDR. occurs at an elevation of 5030 m in the Northwest Himalaya, but only at an elevation of 4730 m in the Mt. Everest area. Three subspecies of the Mediterranean *Calosoma maderae* F. are found in the Northwest Himalaya: *C. m. auropunctatus* (HERB.) is common about the timberline, but may often ascend to nearly 4000 m above mean sea-level. *Carabus* (*Imaibius*) *barysomus* (BATES), *C.* (*I.*) *epipleuralis* SEM. and *C.* (*I.*) *rostianus* SEM. are endemic to the Northwest Himalaya. *Chaetobroscus* SEM. is an endemic genus, with *Ch. anomalus* (CHAUD.) at elevations of 2700-4000 m on the Northwest Himalaya. *Cymindis alticola* GEB., *C. babaulti* ANDR. and *C. glabrella* BATES occur at elevations of 3500 m and *C. rubriceps* ANDR. occurs near the snow-edge, at much higher elevations of 5200 m in Ladak. *C. mannerheimi* GEB., known also from the Pamirs and other Middle Asiatic high mountains, occurs at elevations of 3000-5000 m on the Northwest Himalaya. *Nebria limbigera* SOL., known also from the Turkestan mountains and from China, occurs at an elevation of 4600 m on the Northwest Himalaya. *N. l. babaulti* ANDR. occurs at an elevation of 4500 m on the Northwest Himalaya. *Trechus cameroni* JEANNEL is widely distributed on the Himalaya, at elevations between 2100 and 4000 m. The following species have been reported from the Kumaon Himalaya: *Carabus wagae* FAIRM., (4420 m), *Nebria pindarica* ANDR. (2500-3350 m), *Loricera aparupa* ANDR. (2500-3660 m), *Trechus pumilio* JEANNEL (2500-3050 m), *Asaphidion obscurum* ANDR. (1830-3050 m) and *Bembidion persephone* ANDR. (4800 m). *Nebria himalayica* BATES, found on the Northwest Himalaya, occurs as low as 1800 m in the Kumaon Division, but may ascend to an elevation of about 3500 m. The Carabidae from the Mt. Everest area of the Himalaya include the following: *Carabus indicus* FAIRM. (3050 m), *C. everesti* ANDR. 2895-3350 m), *C. sanchari* ANDR. (4875 m) from the north side of the Everest massif, *Paropisthius indicus* CHAUD. (2500-4875 m), *Nebria pharina* ANDR. (4875 m), *N. superna* ANDR. (5030 m) from the Rongbuk Glacier, *Loricera stevensi* ANDR. (3050 m), *Broscosoma ribbei* PUTZ. (1830-3050 m), *Trechus imaicus* JEANNEL (3050 m), *T. thibetanus* JEANNEL (5181 m), *Bembidion elatum* ANDR. (4420-4725 m) from the northern side of the Everest massif, *B. hingstoni* ANDR. and *B. livens* ANDR. (3660-4270 m).

Dytiscidae

Three endemic species *Agabus* (*Anagabus*) *jucundus* GUIG., *A.* (*Dichonectes*) *nitidus* FABR. and *A.* (*Gaurodytes*) *adustus* GUIG. occur in glacial lakes, at elevations of 3000-4870 m, on the Northwest Himalaya. *Coelambus flaviventris* MOTSCH., also endemic, occurs at an elevation of 4000 m. *Potamonectes* (*Potamonectes*) *griseostriatus* (DEG.), widely distributed in Siberia, North America and Europe, occurs at elevations of 3900-4400 m on the Northwest Himalaya and in Tibet.

Hydrophilidae

The Hydrophilidae seem to be largely restricted to the glacial lakes and ponds, at elevations above 4500 m. *Helophorus* (*Atractelopusorus*) *montanus* D'ORCH. occurs in sulphur springs at an elevation of 5400 m on the Northwest Himalaya; the spring contains sulphur dioxide dissolved in water, with temperatures ranging from 24° to 27° C. *H.* (*Helophorus*) *splendidus immensis* D'ORCH. occurs at elevations above 4200 m in Ladak and Tibet. *Hydrophilus pallidipalpus* M'LEAY, known to be widely distributed in China, Japan, Formosa, Burma and Sumatra, ascends to elevations of nearly 3900 m on the Northwest Himalaya (277).

Staphylinidae

The Staphylinidae are remarkable in that over 80% of the species, occurring above the forest-zone on the Himalaya, are endemic Palaearctic elements. The genus *Atheta* THOMS. is the dominant form, but other common genera include *Aleochara* GR., *Geodromicus* REDT., *Lesteva* LATR., *Ocyusa* KR., *Oxypoda* MANNERH., *Philonthus* CURT., *Pseudocyusa* CA., *Tachinus* GR., etc. While a number of species occur immediately above the forest-zone, some like *Atheta* (*Bessobia*) *submetallica* CAM., *A.* (*Microdota*) *ladakiana* CAM. and *Aleochara* (*Coprochara*) *bilineata* GYLL. are found at elevations of 4875 m. *Atheta* (*Dimetrota*) *hutchinsoni* CAM. is remarkable for occurring at elevations of 5600 m on the Northwest Himalaya—the maximum altitude record for Coleoptera in the world. Like other Coleoptera, the Staphylinidae exist at much higher elevations on the Northwest than on the other parts of the Himalaya. In the eastern parts of the Himalaya, the highest altitudes reached do not exceed 4800 m. *Geodromicus brevitarsis* CHAMP. and *Atheta* (*Paraconota*) *dissociata* CAM. occur at this elevation on the Kumaon Himalaya, but above 4575 m in the Mt. Everest area. *Aleochara* (*Coprochara*) *bilineata* GYLL. is widely distributed on the Kumaon Himalaya and on the Caucasus. *Atheta triangulum* (KR.), a Mediterranean form, occurs immediately above the forest-zone on the Northwest Himalaya. The species characteristic of the Kumaon Himalaya are *Amphichorum pindarensis* CHAMP. (2500-3660 m), *A. monticola* CAM. (3350 m) north of the main crestline of the Great Himalaya, *Geodromicus brevitarsis* CHAMP. (4800 m), *G. amnicola* CHAMP. and *G. crassipalpis* CHAMP (3660 m), *Hygrogeus cyanipennis* CAM. under submerged stones in melt-water torrents (3500-4575 m), *H. kumaonensis* CHAMP. (4420-4575 m), *Coprophilus* (*Zonoptilus*) *burphuensis* CHAMP. (3500 m), *Atheta* (*Paraconota*) *dissociata* CAM. (4575-4800 m), *A.* (*Metaxya*) *nimborum* CAM. (4420 m), *A.* (*Microdota*) *scabriventris* CAM. (4575 m) north of the main crestline of the Great Himalaya, *A.* (*Liogluta*) *nimborum* CAM. (4420 m), *A.* (*Dimetrota*) *sapiana* CAM. (4575 m) on the north side of the crestline of the Great Himalaya, *A.* (*D.*)

altivagans CAM. (4420 m) and *Oxypoda (Paroxypoda) morosa* CAM. (4575 m). Relatively fewer species have so far been reported from the Mt. Everest area: *Philonthus poephagus* CAM. (4270-4575 m), *Ph. azuripennis* CAM. (3660-4575 m), *Ph. nigricoxis* CAM. (3050-4420 m) and *Coprophilus (Zonoptilus) subplagiatus* CAM. (3960 m).

Tenebrionidae

Over 35 endemic high altitude species and three endemic high altitude genera of Tenebrionidae are known from the Himalaya. Descriptions of Himalayan Tenebrionidae may be found in BATES (92), GRIDELLI (431) and BLAIR (133-135). The endemic genera *Bioramix* BATES and *Chianalus* BATES are restricted to the Northwest Himalaya. *Ascelosodis assimilis* F. BATES, *A. ciliatus* BATES, *A. grandis* BATES and *A. intermedius* F. BATES are confined to elevations of about 3400 m, north of the main crestline of the Great Himalaya in the Northwest Himalaya. The genus *Ascelosodis* REDTB. is endemic to the Pamirs-Himalaya Region. *Blaps ladakensis* BATES and *B. perlonga* BATES are typically glacier species that occur at elevations of about 4300 m. *Cyphogenia* SOL., distributed in the steppes of Middle Asia and on the Afghanistan mountains, is represented by *C. plana* BATES at an elevation of 4267 m on the Northwest Himalaya. Four species of *Syachis* BATES occur at elevations of 3000-3500 m; the genus is restricted to the Northwest Himalaya and to the Middle Asiatic mountains. The Mt. Everest Expeditions collected some interesting species, often from the north side of the crestline of the main range. *Ascelosodis everestina* BLAIR was collected from under stones in the area of the Rongbuk Glacier, at an elevation of 5030 m. Two other species, *A. nitida* BLAIR (4270m) and *A. longstaffi* BLAIR (4570-5180 m) are also reported from the eastern side of the Everest massif. Other species collected by the Mt. Everest Expeditions include *Gnaptorina bruci* BLAIR (4420-4570 m) from the Rongbuk Glacier and at 4875 m on the north side of the massif, *Blaps tibetana* BLAIR (4270-4570 m), *B. apicecostata* BLAIR (3350-4570 m), *B. moerens* ALL. (4270 m), *B. himalaica* BLAIR (4575-4880 m) also from the Rongbuk Glacier (5030 m), *Platyscelis (Leptopleura) aenescens* BLAIR (3660-5030 m), *Stenillus monticola* BLAIR (3960 m) with the genus endemic to the Himalaya, *Laena parallelocollis* SCHUST. (4575 m), *L. alticola* BLAIR (5030 m) and *L. hingstoni* BLAIR (3660 m). The reader should also consult the next chapter for further accounts of the Tenebrionidae from the Pamirs-Himalaya Region.

Chrysomelidae

SHUKLA (968) has described a number of high altitude species collected by MANI's Entomological Expeditions from the Northwest Himalaya. *Chaetocnema alticola* MAULIK, an apterous species, is found under stones at elevations of 4000 m on the Northwest Himalaya and at 4575 m on the Kumaon Himalaya. *Leptosonyx* WSE., widely distributed in Middle Asia, Siberia, Mongolia and Caucasus, is represented by *L. octocostatus* WSE. at an elevation of about 4575 m on the Northwest Himalaya. *Longitarsus rohtangensis* SHUKLA, from elevations of 3600-4000 m on the Pir Panjal Range, is apterous. *Apaksha himalayensis* MAULIK (the genus is endemic) is also apterous and occurs at an elevation of about 3500 m on the Kumaon Himalaya. From the Mt. Everest area we have the records of *Swargia nila* MAULIK (4880 m) and *Shaira krishna* MAULIK (4270-4575 m). The highest altitude record for the family, as at present known, is 4880 m. *Longitarsus cyanipennis* BRYANT, widely distributed in the

Himalaya, is interesting in that the specimens from within the forest-zone are winged, but those collected from above the forest limits are apterous on the Northwest Himalaya.

Curculionidae

About 75% of the high altitude species are endemic and over 87% are Palaearctic forms. Most species seem to be typical of elevations between 3000 and 3500 m and none has so far been found above 4550 m. Nearly all the high altitude species are apterous. *Anchlaenomus babaulti* HUSTACHE and *A. squamulosus* HUSTACHE occur at elevations of 3200-4550 m on the Northwest Himalaya. The largely Indo-Malayan *Blosyrodes* JEKEL is sometimes found just above the upper limits of the closed forest. *Catopionus* SCHÖNH., widely distributed from Japan through Siberia, the Altai and the Tien Shan Mountains, the other Turkestan mountains, Karakoram, Tibet, Sikkim, Amur, etc., is represented by an endemic species *C. basilicus* SCHÖNH. at an elevation of 3200 m on the Northwest Himalaya. *Heteronyx ferus* FAUST and *H. f. loevier* HUSTACHE are confined to Lahaul-Spiti areas of the Northwest Himalaya (4000 m). *Lagenolobus lineolatus* HUSTACHE (3500-4500 m), *Otiorrhynchus russicus* STIERL., *Scepticus nubifer* (FAUST), an endemic from 3200-4550 m, and *Sitones callosus* GYLL. (known from Southwest Europe and Middle Asia also) are the other important high altitude species from the Northwest Himalaya.

HYMENOPTERA

About 47% of the high altitude species are endemic and nearly 90% are Palaearctic elements. Formicidae and Bombidae comprise the great bulk of the high altitude Hymenoptera of the Himalaya. The Formicidae from the Northwest Himalaya have been described by EIDMAN (307), MENOZZI (761) and MAYER (750). In 1919 FOREL (354) has given an account of the Himalayan ants and DONISTHORPE (276) has described the ants collected by the Mt. Everest Expedition of 1924. The endemic species *Formica* (*Serviformica*) *picea* NYL. (2800-4800 m), boreal forms like *F.* (*Formica*) *gagates* LATR. and *F.* (*F.*) *trunicola* NYL. are other important records from the Himalaya.

Subterraneobombus melanurus subdistinctus (RICH.) and *Bombus longiceps* SMITH are endemic and *B. alticus* EVERM., *B. vallestris* SMITH, *Mendacibombus margaretieri* (O. VOGT) SKORIKOV, *Lapidariobombus separandus* O. VOGT and *L. alagesianus pamirus* SKORIKOV are the common Himalayan high altitude Bombidae, some of which may be found at elevations up to 4000 m. For further accounts of the Bombidae from the Pamirs-Himalaya Region reference should be made to the next chapter, and to the papers by RICHARDS (883) and SKORIKOV (984).

TRICHOPTERA

There is a high proportion of endemic forms among the species, occurring at elevations between 3000 and 4000 m, on the Himalaya. *Pseudohalesus kashmirus* MART. occurs at an elevation of 4650 m on the Northwest Himalaya. SCHMID (925) has recently recorded *Pseudostenophylax micraulax* (MCL.) at an elevation of 4750 m. The larvae of Trichoptera have been collected at elevations of about 5000 m on the Northwest Himalaya.

LEPIDOPTERA

The Lepidoptera amount to about 25% of the high altitude insects so far known from the Himalaya. Almost half the high altitude species occur at elevations between 3000 and 4000 m. The highest altitude record for Lepidoptera in the world is 6100 m on the Himalaya. The dominant families on the Himalaya are the Papilionidae, Parnassiidae, Nymphalidae, Satyridae, Pieridae and Lycaenidae. Over 45% of the species of Rhopalocera are endemic. Although a considerable proportion of the species from the eastern Himalaya are either typically Chinese-Tibetan or also Indo-Malayan elements, the true Palaearctic forms predominate at higher elevations in the eastern Himalaya and even within the upper reaches of the forest-zone on the Northwest Himalaya. Most species from the Northwest Himalaya are common to the Pamirs and other Middle Asiatic mountains. Some remarkable Heterocera are also known at high elevations on the Himalaya. The Noctuid genus *Cteipolia* STAUD. is, for example, widely distributed on the Northwest Himalaya and the Middle Asiatic high mountains. *Ct. acrophila* HAMPS. occurs at an elevation of 4270 m on the Northwest Himalaya, *Ct. isotoma* PUNG is known from the Tien Shan and recently KAPUR (604) recorded *Ct. lithophila* KAPUR at an elevation of 4877 m from the Nepal Himalaya. *Agrotis tenuis* BUTL. is found at elevations of 4267 m on the Northwest Himalaya.

Papilionidae

The family Papilionidae is represented by two interesting subspecies of the well known Holarctic *Papilio machaon* LINN. at relatively high elevations. *P. m. asiatica* (MÉN.) occurs practically throughout the Himalaya and parts of Tibet and has been collected at elevations up to 4267 m. The second subspecies *P. m. ladakensis* (MOORE) is of darker colour than the typical from, with shorter tail in the hind wing and with the upper side of the wings of paler yellow than in *asiatica*. It is commonly found at elevations of 3500-4267 m north of the main crestline of the Great Himalaya in Ladak and in the Pamirs Region.

Parnassiidae

About a dozen species of *Parnassius* LATR. are known so far from the Himalaya. Their larvae feed mostly at elevations of 4000-5000 m, but the adults of some species may often be observed at an elevation of 6100 m. Descriptions of *Parnassius* LATR. from the Himalaya, Tibet and the Pamirs Region may be found in AVINOFF (67-70), BANG-HAAS (82, 83), FILIPJEW (345), MENZBIER (762, 763) and STAUDINGER & BANG-HAAS 1000).

P. acco (GRAY) is distributed in Ladak, the Nepal Himalaya and the adjoining parts of Tibet, at elevations of about 4890-5800 m. *P. a. acco* (GRAY) extends from the Himalaya into Tibet near Sikkim and is usually found at elevations above 5000 m, in the vicinity of ice. *P. a. tagalangi* BANG-HAAS is restricted to Ladak and *P. a. pundjabensis* BANG-HAAS to Spiti in the Northwest Himalaya. The former subspecies is intermediate between the typical form and *P. a. hampsoni* AVINOFF from the Karakoram. *P. a. baltorana* BANG-HAAS was described originally from the Baltoro Glacier area. *P. a. gemmifer* FRUHSTORFER occurs on the Sikkim Himalaya.

P. actius EVERSMANN ranges from the Trans-Alai mountains to the Northwest Himalaya. The three subspecies found within the Himalayan area (*sensu stricto*) are *P. a. sulla* BRYK & EISNER (3796 m), *P. a. catalina* EISNER & PESCHKE on the Doubounni Mountain near Gilgit and *P. a. yelyangi* BANG-HAAS on the Zaskar Range. The last mentioned subspecies marks the southernmost limit of the range of the species.

P. charltonius GRAY ranges from the Turkestan mountains to the Northwest Himalaya, Tibet and parts of southeastern China. It occurs usually near glaciers, at elevations of 4260-5800 m. *P. ch. charltonius* GRAY extends from the Pamirs, across the Northwest Himalaya, eastwards to the Kumaon Himalaya, at elevations of 3600-4000 m. *P. ch. bryki* HAUDE is peculiar to the Spiti area of the Northwest Himalaya and occurs at elevations of 4500-5000 m, but sometimes descends to even 3050 m. *P. ch. deckerti* VERITY is found at elevations of 3600-4000 m and *P. ch. ducalis* BOULLET & LE CERF occurs at an elevation of 4260 m in the Chitral area.

P. delphius EVERSMANN ranges from the Altai, south Turkestan mountains, the Pamirs, China, Northwest Himalaya and the Sikkim-Himalaya. *P. d. delphius* EVERSMANN is distributed on the Pamirs and other Turkestan mountains, the Karakoram and the Ladak Ranges and also on the mountains of north China. It has also been reported from the Sikkim-Bhutan area of the East Himalaya. *P. d. ladakensis* AVINOFF is peculiar to Ladak (5500 m). *P. d. mamaievi* BANG-HAAS is restricted to the eastern parts of Ladak and does not occur at elevations above 4000 m. *P. d. stenosomus* HOUR., restricted to Ladak, occurs as high as 5000 m above mean sea-level. *P. d. rupshuana* AVINOFF occurs at elevations of about 5300-5400 m. *P. d. workmani* AVINOFF at 5790 m on the Saltoro Glacier and *P. d. shigarensis* BANG-HAAS is restricted to the Baltoro Glacier in the Northwest Himalaya. *P. d. chitralica* VERITY, *P. d. hunza* GRUM-GRSHIMAILO from the East Hindu Kush mountains (1525 m above mean sea-level), *P. d. kafir* AVINOFF from the Safed-Koh mountains, *P. d. affinis* PESCHKE & EISNER from the Doubounin Mountains near Chitral, and *P. d. cardinalia* AVINOFF are the other forms common to the Northwest Himalaya and parts of the Middle Asiatic mountains. *P. d. lampidius* FRUHSTORFER and *P. d. lathonius* BRYK are known from the Sikkim-Bhutan area of the East Himalaya.

Parnassius epaphus OBERTHÜR ranges from the Turkestan mountains, over the Northwest Himalaya, to the Bhutan area in the East Himalaya and the southeast China mountains. It is generally not found at elevations above 4000 m on the Northwest Himalaya. *P. e. cachmiriensis* OBERTHÜR extends from the mountains of Chitral to the Kumaon Himalaya. *P. e. hinducucica* BANG-HAAS occurs at an elevation of 5660 m on the Karamar Pass of the Hindu Kush and extends also to the mountains of North Kashmir. *P. e. hillensis* BANG-HAAS from Spiti, *P. e. sikkimensis* ELWES from the Sikkim-Himalaya (4880 m) and *P. e. phariensis* AVINOFF from Bhutan are the other Himalayan forms.

P. hannyngtoni AVINOFF occurs at elevations of 4575-4880 m on the Sikkim-Himalaya. *P. hardwickei* GRAY ranges from the Northwest Himalaya over the Nepal Himalaya to Sikkim (2290-4575 m). *P. h. hardwickei* GRAY extends from Chitral to the Kumaon Himalaya (2300-4575 m). The specimens of this subspecies collected at lower elevations are conspicuously lighter-coloured than those from higher elevations. *P. h. albicans* FRUHSTORFER is recorded from the Sikkim-Himalaya.

P. imperator OBERTHÜR extends from Nanchan to Sikkim and the southeast China mountains. *P. i. augustus* FRUHSTORFER occurs at 4575-4880 m on the Sikkim Himalaya. *P. jacquemontii jacquemontii* BOISDUVAL occurs on the Northwest Himalaya,

Tibet, in China and on the Middle Asiatic mountains. *P. j. chitralensis* MOORE and *P. j. shandura* TYTL. are restricted to the Northwest Himalaya. *P. maharaja* AVINOFF is described from Ladak and from the Karakoram range at elevations of 5180-5490 m.

P. simo GRAY ranges from the Tien Shan (Issyk-Kul area) to the Hindu Kush, Northwest Himalaya, Nepal-Sikkim Himalaya and Amdo. The subspecies so far recognized include *P. s. avinoffi* VERITY from the Hindu Kush (5490 m), *P. s. lorimeri* TYTLER from Gilgit (4270 m), *P. s. peteri* BANG-HAAS (3050 m), *P. s. acconus* FRUHSTORFER from the Sikkim-Himalaya (4880 m), *P. s. colosseus* BANG-HAAS from the Saltoro and Baltoro Glaciers in the Northwest Himalaya (3960-4575 m), *P. s. simo* GRAY from Nepal, Tibet and Lahaul-Spiti, *P. s. saserensis* BANG-HAAS (5000 m), *P. s. simonides* AVINOFF, *P. s. zarensis* (BANG-HAAS) and *P. s. simoides* (BANG-HAAS) from Lahaul and Ladak (4000-5400 m).

P. stoliczkanus stoliczkanus FELD. occurs in Ladak and Lahaul, *P. st. rileyi* TYTL. on the Rupshu Glacier, *P. st. zojilaica* TYTL. from the Zojila Pass between Kashmir and Ladak (4000 m) and *P. st. spitiensis* BANG-HAAS is restricted to Spiti (4500 m). *P. st. florensciae* TYTL. is the only form that extends from the Karakoram through Ladak, Rupshu in the Northwest Himalaya to the Sikkim-Himalaya.

Nymphalidae

Argynnis aglaia vitatha MOORE is confined to elevations of 5000 m, north of the main crestline of the Great Himalaya. It is also known from the Pamirs Region. *A. clara* PLANCH. occurs up to an elevation of 4800 m on the Pir Panjal Range of the Northwest Himalaya. *A. altissima mackinnoni* DE NICEVILLE, widely distributed on the Himalaya, occurs at elevations of 4300-4500 m. *A. clara manis* FRUHSTORFER and *A. altissima* ELWES occur at elevations of 3050-4575 m in the Chumbi Valley in the Nepal-Sikkim area. *A. gemmata* BUTL. extends from the Kumaon Himalaya to the Nepal Himalaya. *A. hegemone* STAUD. occurs at elevations of 3800-4000 m in Ladak and extends also to Chitral and Turkestan mountains. *A. jainadeva* MOORE recorded from the Kumaon Himalaya at an elevation of 2800 m was collected by MANI's Entomolotical Expeditions at elevations of 4300 m on the Pir Panjal Range of the Northwest Himalaya. *A. pales generator* STAUD., distributed in Afghanistan, the Pamirs and other Middle Asiatic mountains, is common north of the main crestline of the Great Himalaya, up to elevations of 4200 m. *Melitaea sindura* MOORE is found throughout the Himalaya and occurs also in Tibet. *M. s. balbita* MOORE is restricted to the Northwest Himalaya (3500 m). *Vanessa cashmirensis* KOLLAR, widely distributed on the Himalaya, is typically a forest-zone form that is sometimes found above the forest-line on the Northwest Himalaya and ascends to elevations of 5500 m. *V. ladakensis* MOORE is also widely distributed on the Himalaya and occurs up to elevations of 4870 m in Ladak. *V. rizana* MOORE, also widely distributed on the Himalaya, does not occur at elevations below 3000 m. *V. urticae* LINN. occurs at an elevation of 4200 m on the Pir Panjal Range of the Northwest Himalaya.

Satyridae

The high altitude species of the Holarctic *Erebia* DALM. from the Himalaya are mostly brown-coloured above and tinged red on the underside. *E. mani* DE NICEVILLE extends from the Pamirs to Ladak and is confined to elevations of 4200 m on the

Himalaya. *E. kalinda kalinda* MOORE occurs at elevations of 3660-4270 m in the Garhwal-Kumaon Himalaya. *E. shallada shallada* MOORE & DE NICEVILLE ranges from the Alai-Pamirs through the Northwest Himalaya to the Kumaon Himalaya. In the Northwest Himalaya it is restricted to elevations of 3700-4000 m. *Oeneis buddha garhwalica* TYTL. occurs at an elevation of 4115 m on the Kumaon Himalaya. *Oe. pumilus bicolor* (SEITZ) occurs at elevations of 3960-5030 m on the Nepal Himalaya and *Oe. p. pumilus* (C. & R. FELD.) at 4880 m on the Northwest Himalaya. *Lethe confusa* AURIV., widely distributed on the plains in Java, Sumatra, Borneo, Burma and Assam and in the forest-zone on the outer Himalayan ranges, particularly in the east, has penetrated through the river valleys deep into the Northwest Himalaya, where specimens have been collected at elevations of 4300 m on the Pir Panjal Range. We have also records of *L. margaritae* ELWES at elevations of 1525-4880 m from the Sikkim-Himalaya, *L. serbonis teesta* TALB. (2500-4880 m), *Rhaphicera moorei mantra* TALB. (2750-4575 m) and *Oeneis palaearcticus sikkimensis* STAUD. (3050-5181 m) from the Nepal-Sikkim Himalaya.

Maniola pulchella pulchella (FELD.) from the Northwest Himalaya (4300 m), *M. pulchra pulchra* (FELD.) from the Northwest Himalaya (4300 m) and from the Kumaon Himalaya, *M. p. sylvia* HEMMING from the Pamirs to the mountains near Chitral (2500-4270 m), *M. p. astoria* TYTL. from Ladak (3350 m), *M. lupinus* (COSTA) *cheena* (MOORE) on the Nepal Himalaya (3660 m), *M. coenonympha* FELD. and *M. davendra brevistigma* MOORE from Ladak (3050-4270 m) and *M. d. davendra* MOORE from the Spiti-Lahaul sectors of the Northwest Himalaya are the other interesting records.

Hipparchia (= *Karanasa*) *hübneri* (FELD.) is widely distributed on the Northwest Himalaya, the Pamirs Region, the Tien Shan, Altai Mountains and on the mountains of Afghanistan. *H. hübneri hübneri* (FELD.) occurs in Ladak and Lahaul. *H. boloricus* (GRUM-GRSHIMAILO) from the Pamirs is represented by two subspecies on the Northwest Himalaya, at elevations of about 3050 m. *H. actaea pimpla* (FELD.) extends from the Northwest Himalaya to the mountains of Afghanistan and occurs up to an elevation of 4000 m on the Himalaya. *H. mniszechii lehana* MOORE occurs at elevations of 3660-4270 m in Ladak.

Pieridae

Aporia leucodice (EVERSMANN), from the Kumaon Himalaya, has also been reported from the mountains of Turkestan and Iran. *A. l. balucha* MARSHALL is known at elevations of 2500-3050 m and *A. l. soracta* MOORE in the upper reaches of the forest-zone on the Northwest Himalaya. *A. l. sora* EVANS is reported from the upper limits of the forest-zone on the Kumaon Himalaya. *A. nabellica hesba* EVANS and *A. n. nabellica* (BOISDUVAL) are known from the Northwest and Kumaon Himalaya (2700-3960 m). *Aporia pelorica* (HEWISTON) extends from the Kumaon Himalaya (3960 m) to the west China mountains. *Baltia butleri butleri* (MOORE) occurs on the Northwest and Kumaon Himalaya (4575-5500 m). *B. b. sikkima* FRUHSTORFER on the Sikkim Himalaya (3660-4880 m) and *B. shawii baitalensis* MOORE from the Pamirs are other interesting records. *Aporia agathon* (GRAY) is represented by several subspecies on the Kumaon and Northwest Himalaya, mostly within the upper reaches of the forest-zone. The Holarctic *Colias* FABR. is widely distributed on the Himalaya. *C. cocandica* ERSCH. ranges from the Turkestan mountains to west China, the Hindu Kush, Ladak, Tibet and the rest of the Himalaya, at elevations of about 3660 m.

C. cocandica hinducucia VERITY from the Hindu Kush Range, and *C. c. thrasibulus* FRUHSTORFER are important subspecies so far known. *C. electo* (LINN.) from China, Africa, Tibet and the Himalaya, is represented by the subspecies *C. e. fieldi* MÉN. at an elevation of 4500 m on the Northwest Himalaya, Sikkim and Bhutan and extends also to the mountains of North Burma. *C. eogene eogene* FELD. and *C. e. franscensca* WALK. from the Northwest Himalaya occur up to elevations of 5000 m. *C. berylla* FAWCETT is reported from the Sikkim-Himalaya. *C. nina nina* FAWCETT comes from the Mt. Everest area of the Nepal Himalaya (3960-4260 m). *C. erate* (ESP.), ranging from southern Russia to Japan and Formosa, is represented by *C. e. erate* (ESP.) on the Alai-Pamirs, Northwest and Kumaon Himalaya (mainly within the forest-zone in the latter area). *C. stoliczkanus* MOORE from the Turkestan mountains and the Northwest Himalaya and also from the Sikkim Himalaya is an interesting form. *C. st. stoliczkanus* MOORE occurs at elevations above 3960 m in Ladak in the Northwest Himalaya and *C. st. miranda* FRUHSTORFER at elevations of 4575-5181 m in the Mt. Everest area. *C. leechi* GRUM-GRSHIMAILO from the Karakoram (4575-5181 m), *C. dubia* ELWES from the Sikkim-Himalaya (4575-5800 m) and *C. croceus fieldi* MÉN. at elevations of 4260 m on the Northwest Himalaya and the Sikkim-Himalaya and at 650 m on the north Burma mountains are other interesting records. *C. wiskotti* STAUD. & BANG-HAAS from the Turkestan mountains is represented by *C. w. wiskotti* STAUD. & BANG-HAAS at elevations of 3350 m on the Northwest Himalaya.

Pontia chloridice (HÜBN.), distributed from southern Russia to Turkestan and Siberia, is represented by *P. chl. alpina* (VERITY) on the Northwest Himalaya and on the high mountains of Baluchistan. *P. daplidice moorei* (RÖB.) occurs at elevations of 4300 m in Lahaul in the Northwest Himalaya and is also recorded from Tibet, Yunnan and Baluchistan.

There are nine species of *Pieris* SCHR. on the Himalaya: *P. callidice* (HÜBN.) from the high mountains of Europe, Turkestan, west China and the Himalaya, with the subspecies *P. c. kalora* MOORE from the edges of glaciers at elevations of 2700-4575 m from the Safed-Koh Mountains, Northwest Himalaya and the Kumaon Himalaya; *P. dubernardi* OBERTHÜR from the Sikkim-Himalaya, Tibet, north and west China, North Yunnan, with the subspecies *P. d. chumbiensis* (DE NICEVILLE) from the Mt. Everest area (4115 m); *P. krueperi* STAUD., extending from Greece to the Turkestan mountains and the Himalaya, is represented by the subspecies *P. kr. devta* (DE NICEVILLE) at elevations of 2700 m on the Northwest Himalaya; the Holarctic *P. napi* (LINN.) with the subspecies *P. n. ajaka* MOORE from within the upper-forest zone of the Kumaon Himalaya, *P. n. montana* VERITY from the Sikkim-Himalaya and from the mountains of the southern Shan States and *P. n. melaina* RÖB. from the Mt. Everest area (4115 m).

P. extensa POUJADE, from southwest China, is represented by *P. e. bhutya* TALB. in Bhutan (2500-3050 m). *P. deota* (DE NICEVILLE) is common to the Northwest Himalaya (3350 m) and the Pamirs. *P. canidia* (SPARRMAN), extending from the Turkestan mountains to China and Malaya, is represented by the subspecies *P. c. indica* EVANS within the forest-zone on the Kumaon Himalaya. *P. brassicae* (LINN.) with *P. br. nepalensis* DOUBLEDAY, distributed from the Northwest Himalaya to Assam, occurs even in the Sub-Himalayan plains. *P. rapae* (LINN.) with *P. r. iranica* LE CERF. extends from Iraq to Iran, Baluchistan and Northwest Himalaya (1220-2500 m).

Lycaenidae

Lycaena davanica MOORE and *L. icarus* (ROTT.) from Ladak; *L. i. icadius* GRUM-GRSHIMAILO from the Pamirs; *L. orbitulus jaloka* MOORE and *L. pheretes lehana* MOORE from Ladak (3660 m); *L. omphista* MOORE from Ladak and Lahaul (all in the Northwest Himalaya); *L. pheretes asiatica* ELWES from the Sikkim-Himalaya (3660-4575 m); *L. stoliczkana* FELD., extending from the Northwest Himalaya to the Sikkim-Himalaya and *L. youngshusbandi* ELWES from the Mt. Everest area are the important records.

DIPTERA

Although Diptera are abundant at high altitudes on the Himalaya, the group as a whole has been sadly neglected by the entomologists who have so far explored the region. Some of the species occur at the highest elevations at which the higher insects have been found. The common high altitude Diptera from the Himalaya are Tipulidae, Chironomidae, Simuliidae, Bibionidae*, Blepharoceridae, Deuterophlebiidae, Culicidae, Stratiomyiidae, Syrphidae and the muscoid families, but particularly the Anthomyiidae. ALEXANDER (14) has described a number of the Himalayan Tipulidae from relatively lower elevations, but the following species occur at high elevations: *Tipula (Bellardina) arjuna* ALEXANDER at elevations of 1700-3660 m on the Kumaon and Sikkim-Himalaya; *T. (B.) hypsiostos* ALEXANDER at elevations of 4800-5181 m on the Sikkim-Himalaya; *Dicranota (Rhaphidolabis) lacteipennis* ALEXANDER and *D. (Rh.) inda* ALEXANDER at 3660 m on the Gahrwal-Himalaya. *Bibio hortulanoides* BRUN. and *B. fuscitibia* BRUN. occur at elevations of 3300-4000 m on the Northwest Himalaya. The remarkable *Deuterophlebia mirabilis* EDW. occurs at elevations of 3000-4000 m on the Northwest Himalaya (*vide* Chapter V). SANTOKH SINGH (973) has recently decribed a number of interesting high altitude Chironomidae from the Northwest Himalaya. *Brillia kultia* SANTOKH, *Metriocnemis (Heterotrissocladius) chandra* SANTOKH, *Orthocladius (Pseudorthocladius) virendri* SANTOKH and *Chironomus (Paratendipes) lahaulensis* SANTOKH occur at elevations of 3500-3900 m. *Aëdes (Ochlerotatus) pullatus* COQ. and *Theobaldia niveitaeniata* (THEOB.) are found breeding at elevations of 3600-3900 m during the summer on the Northwest Himalaya. The Syrphid *Lasiopticus selenticus* (MEIG.), widely distributed in Europe, occurs at an elevation of 3660 m on the Northwest Himalaya. *Syrphus albostriatus* (FALLEN), known also from Europe, and *Eriozona himalayensis* BRUN. occur on the Kumaon and Northwest Himalaya, at elevations of about 3300 m. The other high altitude Syrphidae so far known from the Himalaya include *Criorhina imitator* BRUN., *Arctophila simplicipes* BRUN. and *Chrysotoxum convexum* BRUN. at elevations of 2500-3350 m on the Kumaon and Northwest Himalaya. *Lucilia illustris* (MEIG.) occurs at elevations of 3000-4000 m on the Northwest Himalaya and is also known from Europe, North America, Mongolia, Manchuria and China. *Ephydra glauca* MEIG. occurs in hot springs, with the water temperature at 49° C, at elevations of 4575 m on the Northwest Himalaya. *Holmatopoda hutchinsoni* CRESS. is another Ephydrid from the Northwest Himalaya, recorded at elevations of 4575 m in the Lake Tso Kar.

* HARDY HARDY, E. 1965. *Bull. Brit. Mus. (Nat. Hist.) Ent.* 16(1): 1-23) has recently described the Bibionidae from the Nepal Himalaya.

THYSANURA

Machilinus hutchinsoni SILV. occurs at elevations of 5330 m on snow-fields and glaciers and *Ctenolepisma longicauda* ESCH. occurs at elevations of 3500-3600 m on the Northwest Himalaya. According to the observations of SWAN (1021, 1023), *Machilinus* sp. was found active on the rock surface at elevations of 5335-5800 m on the Mt. Makalu area of the Nepal Himalaya.

COLLEMBOLA

Collembola flourish in great abundance on ice and snow, on glaciers, in soil and on moss, at elevations of up to 6800 m on the Himalaya. BAIJAL (76-78) has described the species collected by MANI's Entomologcial Expeditions to the Northwest Himalaya. *Drepanoseira subornata* (DENIS) occurs at an elevation of 4700 m, *Entomobrya hutchinsoni* DENIS and other species occur at elevations of 3500-5500 m, *Friesea maxima* BAIJAL occurs at 3500-4500 m. *Fr. excelsa* DENIS at 5500 m, *Hypogastrura sonapani* BAIJAL at 4500-5000 m on the Seri Icefall and *Isotoma pallustris* (MÜLLER), found at 4000 m on the Northwest Himalaya, is widely distributed even on the plains of Europe, Canada, the U.S.A., Bear Island, Spitzbergen, Siberia and Java. The other interesting records are *Orchesella boraoi* BONET (5300-5600 m), *Proisotoma himalayanus* BAIJAL (3200-4000 m) and *Tomocerus ocreatus* DENIS (3000-4760 m). *Sminthurides aquaticus* (BOURLET), widely distributed in Europe and Australia, occurs at elevations of 4760 m in hot springs, *Sm.* (*Stenacidia*) *violaceus* (REUTER) at elevations of 4000-4700 m and *Sm. hamtaensis* BAIJAL at 4500 m on the Pir Panjal Range in the Northwest Himalaya.

7. THE HINDU KUSH RANGE

A number of interesting high altitude insects were collected by the Danish Expedition and by the Kyoto University Scientific Expedition to the Karakoram and the Hindu Kush mountains. KAWAI (610) has described the Plecoptera collected by the Kyoto University Expedition. The nymphs of *Perlodes* (*Skobeleya*) sp. were found in streams at a temperature of 4° C, at an elevation of 4000 m on the Northwest Himalaya. BRINCK (153) has described a small collection of Plecoptera from the Middle Asiatic mountains, including also parts of the Hindu Kush. *P.* (*Perlodinella*) *nuritica* BRINCK was collected by the Third Danish Expedition to Middle Asia, at elevations from above the forest-zone, (2500 m), in the Chigasarai area in Nuristan and within the forest-zone in the *Quercus balout* zone at elevations of 1400 m. He observes that the specimens collected by the Dutch Expedition to the Karakoram and previously identified as *Megarcys juenluensis* ŠÁMAL are true *Perlodes* BANKS. *P. lepneva* ŠÁMAL from the Altai mountains is really *Acrynopteryx lepneva* (ŠÁMAL). According to BRINCK, the genus *Perlodinella* KLAPÁLEK replaces *Perlodes* BANKS on the high mountains of Middle and East Asia. The nymph of *Perlodes* BANKS, collected at an elevation of 3400 m on the Karakoram, is probably identical with *Perlodinella kuenluensis* (ŠÁMAL). For further details on the Plecoptera of the region reference may be made to ŠÁMAL (915, 916).

An extremely interesting collection of Coleoptera, especially Tenebrionidae, made by HAARLOV and PALURDAN of the Danish Expedition, is described by GRIDELLI (430, 432) and KASZAB (608). KLAPPERICH also collected Coleoptera from the mountains of Afghanistan during 1952-1953. KASZAB (*op. cit.*) reports also on the Tenebrionidae of the Hindu Kush mountains. *Adesmia (Adesmia) jugalis* REITT. *gridellii* KASZAB (608) occurs at elevations of 1900 m on the Hindu Kush and at about 2000 m on the Jebel Seraj, about 90 km to the north of Kabul. SHIRÔ & SAIGUSA (964) have reported on the Lepidoptera, collected by the Kyoto University Expedition. The species listed by them include *Papilio machaon asiatica* MÉN. (3400 m), *Pieris callidice* (ESP.), *P. chrysidice* (HERRICH-SCHÄFFER) on Mt. Demavend (3500-3800 m), *Pontia daplidice persica* (BINERT) on Mt. Demavend (1400 m), *Aporia leucodice balucha* MARSHALL, *Colias eogene shandura* EVANS and *Polyommatus eros* (OCHS.) on the Nazbar Pass (3400 m), *Aulocera swaha swaha* (KOLL.) on Mt. Demavend (3500-3800 m), *Hipparchia heydenreichi shandura* (MARSHALL) at 3500 m on the Safed Koh Mountains, *Hipparchia hübneri* (TYTL.) on the Safed Koh Mountains, *Hyponephele amardaea amardaea* (LED.) on Mt. Demavend (3500-3800 m) and also known from Buchara and Ferghana, *H. lycon collina* (RÖB.) on Mt. Demavend (3500-3800 m), *Clossiana jedroni chitralensis* (MOORE) (3400 m) and *Mesoacidalia charlotta vitatha* (MOORE) from the Nuzbar Pass (3400 m) and near Dumial (3700 m). It may be mentioned that *Aulocera swaha swaha* and *Aporia leucodice* occur also on the Northwest and Kumaon Himalaya. YOSHII (1131) has reported on a small collection of Collembola from the Hindu Kush Range. *Lobella yoshibai* YOSHII occurs at an elevation of 4000 m. *Isotoma* sp. was found under stones in snow at elevations of 5000 m on the Naltar Pass and *I. innominata* B. was collected from under stones on snow-covered glacier on the Naltar Pass at an elevation of 5000 m.

8. BIOGEOGRAPHICAL CHARACTERS

The high altitude regions of the Himalaya belong to the Turkmenian Subregion of the Palaearctic Realm. The high altitude insect life of the Northwest Himalaya contains about 4% of the Mediterranean elements that extend as far east as the great defile of the R. Sutlej and ascend sometimes even to an elevation of about 4000 m. Earlier workers (1067, 1072) believed that these elements terminate in the Kashmir Valley, particularly the Mt. Nanga Parbat area. There is also a small proportion of Indo-Malayan elements but these are particularly more conspicuous in the eastern Himalaya than on the Northwest Himalaya. These elements represent ecologically transitional forms that have penetrated deep through the river valleys, in recent times. The boundary altitude between the Indo-Malayan and the Mediterranean elements in the Northwest Himalaya has been taken to be at an elevation of 3000 m by some workers, but EIDMANN (307) considers 2000 m to be the boundary altitude.

The high altitude insect life of the Himalaya as a whole is remarkable for the very high species endemism in all groups. Over 70% of the species restricted to high elevations are strictly endemic (fig. 51). The researches of SKORIKOV (984), REINIG (878) and MANI (730) have shown that the region of the Tertiary mountains must be recognized as an independent centre of faunal origin and differentiation within the Palaearctic. The lowland boreal and Middle Asiatic ancestral stock have evolved into the typically endemic high altitude elements *pari passu* with the uplift of the Himalayan System. Pleistocene glaciations and the climatic and other changes at the end of

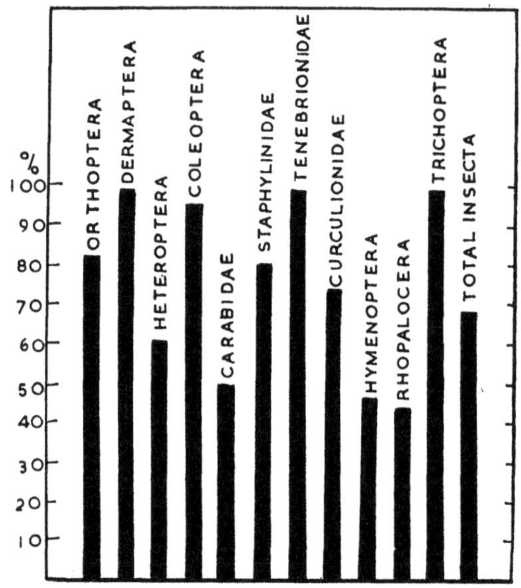

Fig. 51. Endemism in the high altitude insects from the Himalaya.

the Pleistocene glaciations have served as important factors in this evolution. The Pliocene endemic forms have survived on the nunataks during the Pleistocene glaciations and have since the last glaciation, given rise to numerous subspecies. While the evolutionary changes during the Pliocene involved largely the modification of relatively few species to true high altitude types, there is at the present time a very pronounced tendency towards an increase in the number of species by rapid process of subspeciation and isolation on single high massifs.

Biogeographically the Northwest Himalaya must be separated from the rest of the Himalaya. The Northwest Himalaya, Karakoram and the Alai-Pamirs constitute a single centre of faunal origin and differentiation. The endemic elements of this complex region have risen largely during the Pliocene, but some are also of Pleistocene and Post-Pleistocene origin.

The origin of the high endemism and the development of the ecological specializations of the high altitude insects of the Northwest Himalaya are both intimately bound up with the rise of the Himalaya itself. Essentially thermophile, lowland-forest forms were lifted up, in the course of the Himalayan orogeny, to high elevations by the rise of the ground they inhabited and thus simultaneously came to be also modified into cryophile and cryobiont mountain autochthonous types. The endemic species are the descendents of the ancestral stock of Middle Asiatic origin, which inhabited the region when its elevation was not high. While SKORIKOV (984) believes the Himalaya as a whole to be a single centre of faunal differentiation and distribution, MANI (730) has recently shown that the Northwest Himalaya has had a different history from the rest of the Himalaya and together with the Pamirs region constitutes an independent centre.

CHAPTER X

THE ALAI-PAMIRS AND THE TIEN SHAN

The elevated regions considered in this chapter belong, at least in part, to the Himalayan System, so that this chapter is, in a sense, a continuation of the foregoing. Orographically, ecologically and biogeographically, the Northwest Himalaya, the Pamirs, the Tien Shan and the other principal mountains of Middle Asia (fig. 52) constitute together an immense locality of faunal development and differentiation. The region of the Altai Mountains is likewise another independent faunal centre of Central Asia. In view of the prevailing confusion in the existing English language literature, it is necessary, at the very outset, to clearly distinguish between Middle Asia and Central Asia. Middle Asia comprises the Soviet Socialist Republics of Turkmenistan, Uzbekistan, Tadjikstan and Kirghiz (the Turkestan of old), the South Kazakstan and the North Afghanistan (north of the Hindu Kush Range). Central Asia includes the northern parts of the Kazak SSR, the Mongol Peoples' Republic, the western provinces of China (Singkiang, Tsinghai), Tibet, parts of Kansu and the Inner Mongolia. The principal mountain ranges and high elevated regions of these areas are the Alai-Pamirs, the Tien Shan and the Altai. The Altai Mountains belong to the older mountain system and the others are Tertiary formations. The reader will find excellent accounts of the tectonics of the region in BURRARD & HAYDEN (175), MUSCHKETOW (786), A. VON SCHULTZ (939) and C. C. SCHULTZ (940).

I. THE ALAI-PAMIRS REGION

There are two main areas, viz. the Alai Mountains and the Pamirs. The Alai Mountains belong to the Tien Shan System and the Pamirs are related to the mountain system of Buchara, Ladak and the Northwest Himalaya. Between the two regions is the Alai Valley.

The Pamirs (from the Persian *pai* = foot, *mir* = mountain peaks; *Bam-i-dhunya* = the roof of the world; *Hara berezaite* = the mountain of the old Iranian myth) are an elevated rectangular mass of uplands, bestrewn with boulders and other mountain debris and surrounded by high snow-capped mountain chains of the Hindu Kush, the Kashgar and the Trans-Alai, rising to elevations of 6000-7000 m. Its elevated level plateau, like those of the Tien Shan and Tibet, contains accumulations of gravel deposited in rocky troughs. V. HUMBOLDT conceived of the Pamirs as a great meridional range, connecting the Tibetan-Himalayan and the Tien Shan Systems, but SEVERSTOFF and FEDSCHENKO have shown that the fundamental mass of the Pamirs is in reality a series of parallel ranges that stretch east-west. According to STOLICZKA (1012), the Pamirs are really not a plateau at all, but a congregation of chains of mountains. The Pamirs are essentially high steppes, without outlet and with brackish lakes or they are flat broad valleys, through which flow rivers that may end in a lake. The mean elevation of the Pamirs alluvial high-level plateaux is 3660 m and that of the mountains dividing them is 5180 m.

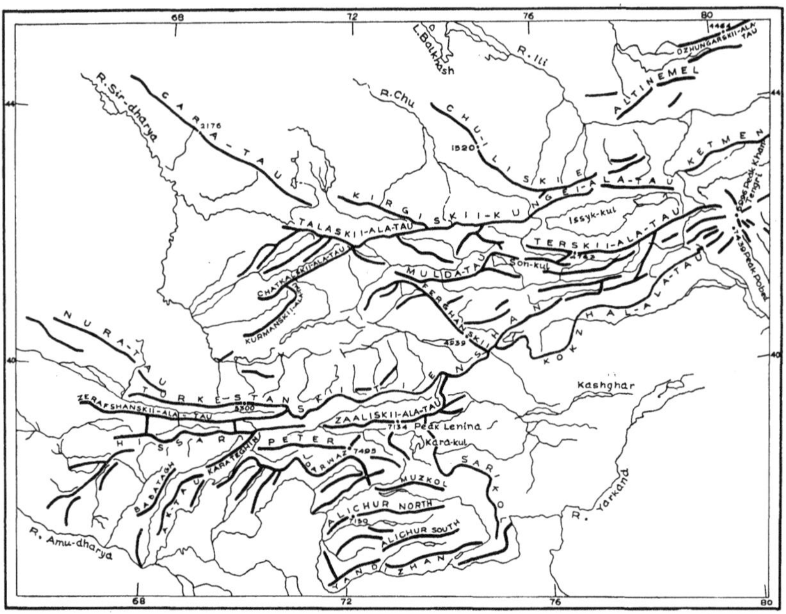

Fig. 52. Sketch-map showing the principal mountain ranges of the Alai-Pamirs Region and the Tien Shan System.

The Pamirs area forms the principal water-parting between the inland drainage systems of Asia; one ending in the Aral Sea and the other ending in Lob Nor. The snow-fields and glaciers of the mountains of the Pamirs are the principal feeders of the R. Oxus (*Ardvisura* or *Amu darya* or the *Ambu darya* of Sanskrit). The Pamirs are also the source of the R. Punj (Pundsch) or the Five Rivers, viz. the R. Burgat, Bakdjir, Bai-kara, Ab-i-Shorshil and the Pamir Darya, all having their sources at elevations of about 4600 m above mean sea-level. Garm-chasma or geysers are situated to the west of Kislak of Shund, on the northern bank of the R. Garm-chasma Darya*. The geysers are located in a line of large numbers of small craters, in some of which the water is flung to a height of 12 cm. They all gush out yellowish-green sulphurous water, which also contains dissolved salts of sodium, potassium, lithium, calcium and zinc, thus imparting a distinct sulphurous stench in the area. Most of these springs are situated about 25 m above the R. Garm-Chasma, at an elevation of about 2590 m above mean sea-level. Red, gray and green algae grow in the water at a temperature of 39-42° C. The hot spring at Zunk is situated at an elevation of 2970 m and runs into an artificial basin, in which the hot water becomes mixed with cold water, which has a resulting temperature of 44.5° C. Other springs like the Sirgyn Spring, with a temperature of 32.5° C, are situated at elevations of 2650 m. Some of the commonest plants of the Pamirs are the thorny bush *Hippophaë*, *Tamariscus*, *Salix*, etc. (*vide infra*).

* *Garam chasma* = hot spring in Persian.

2. SUBDIVISIONS OF THE PAMIRS

The Pamirs rectangle is subdivided into an eastern region of valleys, with a mean elevation of 4000 m—the Pamirs proper—and a western area of valleys, only 2400 m above mean sea-level, viz. the Western Valleys. In the east, at elevations of 4350-4500 m, the Pamir is characterized by extreme low temperatures, extreme xerophytic vegetation and poor fauna; in the west at an elevation of 3030 m, the summer is

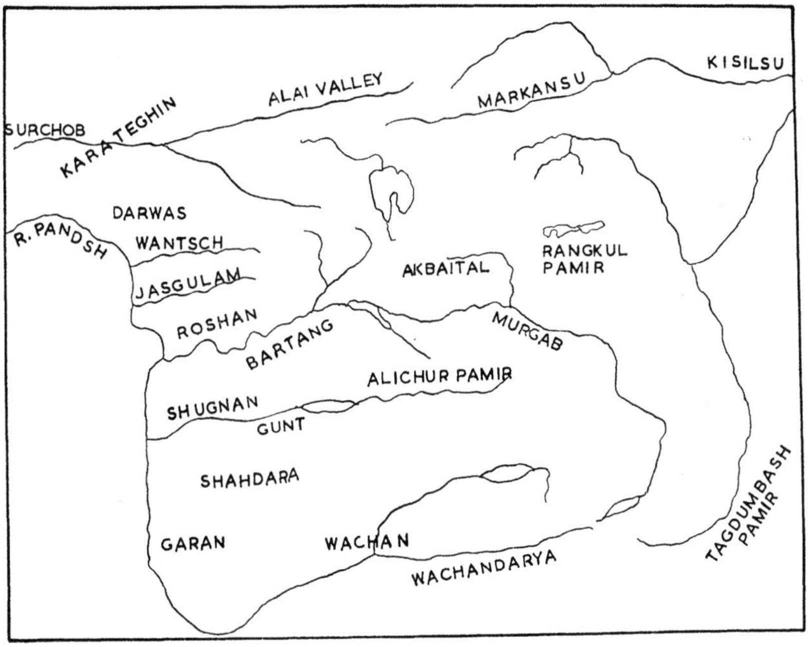

Fig. 53. Sketch-map of the Pamirs Region, showing the principal geographical subdivisions.

warm and the vegetation is luxuriant and varied and the fauna rich. The Pamirs are divided by the great meridional mountain chain of Sarikol Range (with the Muz-tagh-Ata Peak, rising to an elevation of 7422 m), mentioned in the beginning of the last chapter. The Kashgar and the Sarikol Ranges are separated by a high-level trough of a trend, not exactly meridional, but from the northwest to the southeast; these ranges form the eastern portion of the Pamirs mass. The Kashgar Range is in reality a continuation of the Kuen Lun Range. The following subdivisions of the Pamirs are generally recognized (fig. 53, 54): i. the Great Pamir, ii. the Little Pamir separated from the Great Pamir on the north by the Nichols Range, iii. the Pamir-i-Wakhan, iv. the Alichur Pamir or the Pamir of Yeshil-kul and Gund to the north of the Great Pamir and v. the Sarez Pamir or the Valley of the R. Murgab*. North of

* *Murg-ab* = Fowl-water; *ab* corruption in Persian for Sanskrit *āpah*; often also *ob*.

the Alichur Range are the Rang-kul and the Kara-kul Pamirs. They are all hemmed in and separated by snow-covered high mountains and ridges. To the east of the Sarikol Range is the Taghdumbash Pamir, with many characteristics of the western Pamirs. The Taghdumbash Pamir (3000-4000 m) and the Little Pamir (4000 m) fall respectively within China and Afghanistan. The Ak-su Valley, with extensive semi-desert terraces reveals also the typical Pamirs character even east of Pamiriskii Post and higher in the end of Ak-bura. The central sector of the Ak-su Valley is known as the Ak-su Pamir (about 20 km long, 15-20 km wide and about 500 sq. km in area). The Kara-kul* Pamir (about 30 km long, 20 km wide, 600 sq. km in area) is half lake and the Rang-kul Pamir (40 km long, 5-10 km wide and 300 sq. km in area) has a quarter of its area covered by the lakes Rang-kul and Shor-kul. The Kara-kul and the Rang-kul Pamirs are both bounded on the west by the Sulumar Mountain chain, Ak-baital Mountain chain, etc. The Great Pamir is 70 km long, 10-25 km wide and about 800 sq. km in area, of which about one-sixth is occupied by the Sor-kul lake. The Ak-su Pamir is 20 km long, 5-20 km wide and 500 sq. km in area. The Alichur Pamir and the Great Pamir pass, without sharp boundary, into the Western Valleys. The low valley portion of both these southern Pamirs, the Middle and the Upper Pundsch, the Bartang Valley, the valleys of Jas-gulam (in which end the Middle Pundsch), Wantsh, Ching-ob, the lower course of the R. Kisil-su (the upper end of which is in the Trans-Alai) are known as the Western Valleys, in contradistinction to the Pamirs. The region between the northern and the southern Pamirs is at elevations of 3500-3700 m, with comparatively mild climate and richer flora and is a sort of oasis in the Pamirs desert. This is the Murg-ab Basin and includes the Jaman Valley, the lower Ak-su, the lower Ak-baital and the area near Pamiriskii Post. Though strictly speaking outside the main Pamirs Region, Turum-tai Valley (4350 m) may also be considered as a part of the Pamirs in a certain sense. Some workers include the Alai Valley and the south slopes of the Alai Mountains in the area of Katta-kara-muk within the Pamirs Region. The Alai Valley resembles, in its wide expanse, a Pamir and is filled with river-sand and moraines. Through it flows the R. Kisil-su, from the east to the west. The valley depression, about 15 km wide and 3400 m above mean sea-level (Taun-murun Pass is 3415 m), sinks westward to an elevation of 2200 m, near Katta-kara-muk. The East Alai slopes somewhat southward to the Alai Valley, but on the north it passes gently into the Ferghana Plain. The area of the Pamirs within the USSR comprises the following six provinces: i. Wachan and Garan (with the lower affluents of the R. Pundsch), ii. Shugnan (with a part of the Middle Pundsch and the lower reaches of the R. Gund and Shah-dara), iii Roshan (with the lower reaches of Bartang, Jas-gulam and part of the northern Pundsch), iv. Bartang (including the middle and upper Bartang Valley), v. Darwas (the valleys of Wantsh, Ching-ob and the lower Pundsch) and vi. Karateghin (with the lower Kisil-su Valley, the lower Muk-su Valley and the Surch-ob Valley). All these areas are surrounded by high, snow-covered mountain ranges, with a mean elevation of 5000 m. Of the large east-west valleys, only Ching-ob, Jas-gulam and Wantsh are not connected with the high steppes of the Pamirs. The high steppes of Tuptschek are of a mean elevation of 3150 m and are bounded in the south by the steep eastern chain of Peter-the-Great Range and in the north by a continuation of the western chain of the Peter-the-Great Range (fig. 54). The

* *Kul* = tank, lake, allied to the Tamil *Kulam*.

Taghdumbash Pamir is traversed meridionally by the Sarikol Mountain chain at its western boundary. Between the northern outletless Pamirs and the southern Pamirs there is a relatively low area (3500-3700 m above mean sea-level) viz. the Murg-ab Basin, distinct climatically, floristically and faunistically as a sort of an oasis in the desert of the Pamirs Region. This comprises the Jamal Valley, the lower Ak-su, the lower Ak-baital and the environs of the Pamiriskii Post.

The Western Valley passes gradually in the north into the South Alai and the Hindu Kush Range forms its boundary in the south. The R. Pundsch that flows in the south-north direction marks its western boundary. The shrubline indicates the eastern limits of the Western Valleys. The Western Valleys comprise the Wachan (with the valleys of the Upper Pundsch, Pamir-darya and the Wachan-darya),

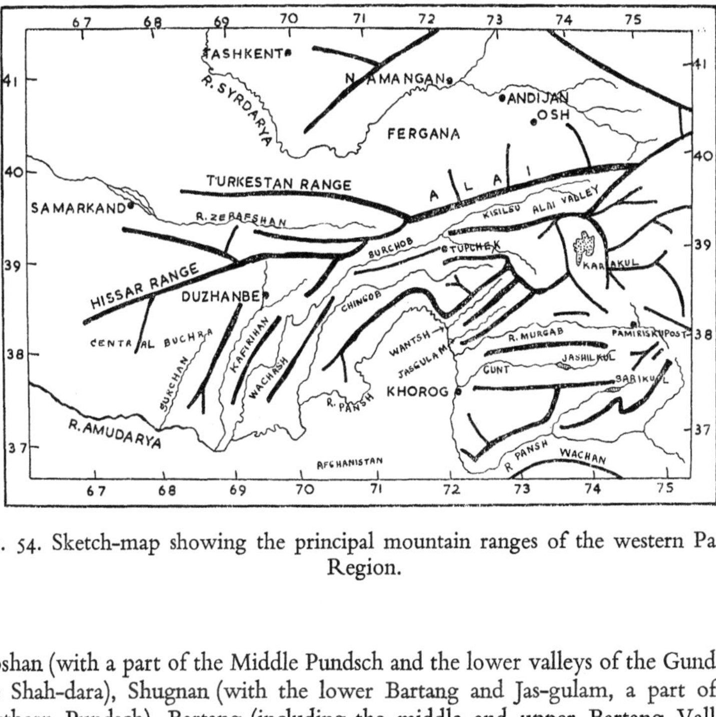

Fig. 54. Sketch-map showing the principal mountain ranges of the western Pamirs Region.

Roshan (with a part of the Middle Pundsch and the lower valleys of the Gund and the Shah-dara), Shugnan (with the lower Bartang and Jas-gulam, a part of the northern Pundsch), Bartang (including the middle and upper Bartang Valleys), Darwas (with the valleys of Wantsh, Ching-ob and the lower Pundsch) and Karateghin (with the lower Kisil-su Valley and the Surch-ob Valley). The Surch-ob Valley has a distinct northern character, more than Wantsh and the Ching-ob Valley of Darwas. The upper course of Surch-ob (the Alai Valley) is pronouncedly a high steppes-valley and it exerts its influence on the Surch-ob by its extreme winter-cold climate. A second connection of the Surch-ob with the high steppes is formed through Muk-su, so that the area has a winter colder than in Khorog, which is likewise influenced by high steppes. Ching-ob and the Wantsh are not, however, connected in this manner with the high steppes, so that at the same elevations, different ecological conditions prevail on the Alai-Pamirs, depending on whether a valley is a steppe-valley or not. Within the western Pamirs there are thus three distinct ecological floristic and faunistic zones, viz. the Kara-teghin, Darwas* and the Western Valleys.

3. ECOLOGICAL CHARACTERS OF THE PAMIRS REGION

The Pamirs ecology is essentially that of an elevated area in the middle of the largest continental mass and surrounded on all sides by high mountains. The Pamirs are high altitude steppes, characterized by very pronounced differences between the mean summer and winter atmospheric temperatures. The diurnal atmospheric temperature variations are also wide. Precise meteorological observations for a number of years are available for several stations like Osh (1023 m above mean sea-level, on the northern margin of the Alai), Khorog at the confluence of the R. Gund and R. Pundsch (2105 m) and Pamiriskii Post (3640 m in the Upper Murg-ab) (344).

The mean monthly atmospheric temperatures for Alai, Pamir and Western Valleys are shown in fig. 55. Taking the monthly mean of 7.5° C as ecologically optimal for insect life in the area, REINIG (878) has shown that the Murg-ab Basin has four months from June to September, the Western Valleys have six months from April to October, the north margin of the Alai eight months from March to October and Kara-kul only two months July-August of optimal conditions of mean temperatures of 8.2° C and 8.1° C (fig. 55). The atmospheric temperatures in different stations during the summer are summarized in table 28.

TABLE 28

Mean atmospheric temperatures in ° C during the summer months on the Pamirs Region

Station	Elevation in m	June	July	Aug.	Sept.
Osh	1025	22.0	24.6	23.9	16.4
Khorog	2160	19.2	22.5	22.2	18.1
Irkestam	2850	10.8	13.6	13.3	9.1
Pamiriskii Post	3640	10.5	13.9	13.8	7.7
Alai Val.	3145	—	9.7	10.4	7.7
Kara-kul	3960	—	8.2	8.1	2.2

In Pamiriskii Post, June-August are generally frost-free. The mean atmospheric minimum temperature during July is 4.7° C and during August 3.6° C. In Kara-kul, the minimum temperature for July is 5.5° C and for August —14.5° C. The diurnal temperature fluctuation is large and averages to 16° C. Intense night frost is, however, common in all the Pamirs, except perhaps in the Rang-kul Pamir (3780 m above mean sea-level). The surface of streams and ponds is often frozen to a thickness of 1-1.5 cm in the morning hours even during mid summer. REINIG (874) found, for example, the atmospheric temperature to be —15° C and —17° C soon after sunrise during August (10th and 11th) in Sor-kul (4350 m above mean sea-level) in the

* *Darwas* = door in Persian.

Great Pamir. The lagoons of the lakes are generally covered by ice, 1–1.5 cm thick and this melts only by midday. Even within the forest-zone at an elevation of 3400 m in the Tanimas Valley, the atmospheric temperature during July (in the shade) is only 3.0° C at sunrise; it rises to 16.0° at 8 a.m., to 29.0° C at 1 p.m. and 31° C at 2 p.m., but rapid cooling occurs at sunset and at 8 p.m. the temperature drops to 10.0° C.

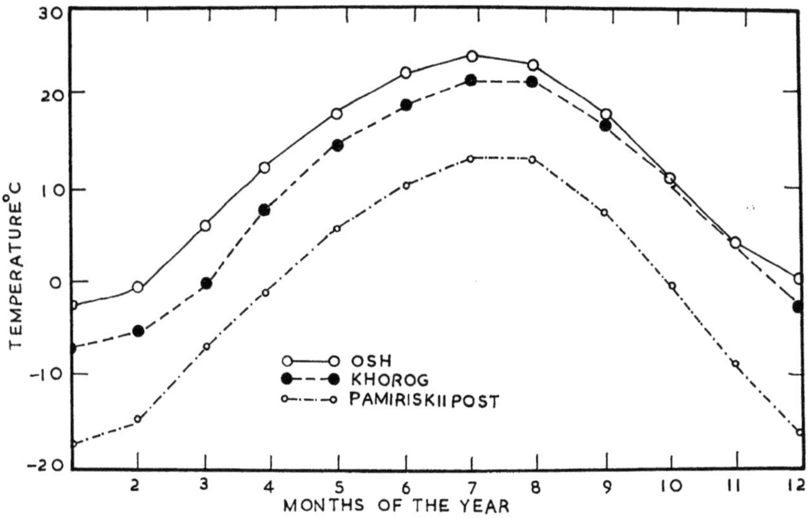

Fig. 55. The mean monthly atmospheric temperatures in the Pamirs Region.

The monthly atmospheric temperature means for July–August in the Alai Valley are 9.7° lower for July and 10.4° for August than those for Pamiriskii Post and only 1.5°–2.0° C above those of Kara-kul. The temperature maxima are not also very different; the Alai Valley has respectively 22.1° C and 24.8° C and Kara-kul has 21.2° C and 19.2° C. In Katta-kara-muk (the Western Alai Valley) (2240 m above mean sea-level) the mean atmospheric temperature for July is 17.3° C and the summer mean is 16.3° C. For Daraut-kurgan (2470 m), the means for July and for summer are respectively 16.0° C and 15.0° C. Owing to the high elevation and the pronounced atmospheric aridity of the Pamirs Region, the effect of insolation is intense. The intensity of insolation in the Alai Valley, at an elevation 3145 m, is 1.54 g-cal per sq. cm per minute. The intensity of insolation for Kara-kul in July, August and September are respectively 1.54, 1.57 and 1.67 g-cal per sq. cm per minute. Radiation is also strong during the night.

In many localities in the Central Pamirs, the ground is permanently frozen at depths of 80–100 cm and in certain places, as for example, on the southeast shore of Kara-kul and the south shore of Rang-kul, the permafrost reaches even to the surface. At greater soil depths the temperature fluctuation was between 7° and 8° C, but at lesser depths between 12° and 17° C. The range of the soil-surface temperatures was at the same time 7–33.5° C (fig. 57). At Kara-kul the maximum surface-soil tempersture for July, August and September are 55.2° C, 55.5° C and 54.0° C respectively and the corresponding minima are —7.7° C, —8.6° C and —16.5° C respectively. The values for the Alai Valley are maxima 48.9° C, 49.6° and 46.0° C and —6.3° C, —8.4° and —12.0° C respectively for the minima. The monthly fluctuations of the surface-soil temperatures at Kara-kul amount, therefore, to 70.5° C and for the Alai Valley 58.0° C. In some localities, with the ground temperature reading 28.0° C, the atmospheric temperature, at a height of 1.5 cm above the level of the ground, may amount to 14-15° C. The ground temperature rises to a maximum of 45.0° C, and sometimes even 50.0° C, at 11 in the morning. The hourly temperatures of the atmosphere (in the shade) and of the soil and the intensity of insolation during July in the Alai Valley and in Jashil-kul are summarized from REINIG (875) and from OLUFSEN (802, 803) in table 30.

The mean monthly precipitation in Osh, Khorog and Pamiriskii Post is shown in fig. 56. The annual precipitation in Pamiriskii Post is only 58.7 mm, the bulk of which occurs in summer and is distributed as follows: June 10.7 mm, July 12.6 mm, August 5.7 mm and September 2.1 mm (totalling 31.1 mm, or in other words about 53% of the total annual precipitation). In Kara-kul the precipitation in July is 0.5 mm, August 2.5 mm and in September 1.8 mm. On account of the extremely low winter precipitation, the snow-cover is scanty. In sharp contrast, however, to the scanty precipitation, atmospheric humidity is relatively high. The monthly mean relative humidity at Pamiriskii Post for July is 44% and for August 42%; for Kara-kul 45% (minimum 12%) in July and 39% (minimum 1%) in August. The mean relative humidity of the atmosphere at different stations during July-August are summarized in table 29.

TABLE 29

Relative humidity in the Pamirs

Station	July mean %	August mean %
Osh	47	48
Alai Valley	63	54
Khorog	35	32
Pamiriskii Post	44	42
Kara-kul	45	39
Jashil-kul	38	21

The protective effects of snow-cover on the soil are thus absent in most of the Pamirs. Local heavy snow-cover on the southwest slope of Naisa-tash Pass, south slope of the Djol-boeruljuk Pass and in the moraines below the Maz Pass is associated with relatively rich vegetation. The Alai Valley has also a heavy snow-cover. The Alai has, however, no glaciers in the east; only a few areas above elevations of 4500 m on the north slope have small patches of snow during summer. The snowline on the mountains of the Pamirs Region thus lies at elevations between 5000 and 5500 m, depending upon the exposition—in other words 1000-1500 m above the high-steppes of the Pamirs. The snow-cover diminishes from the east to the west, but reaches again a maximum on the Tagarma Mountain Chain. Due to the scanty precipitation in the Central Pamirs, glaciers do not descend below an elevation of 4800 m. Glaciers are, however, abundant in the Western Valleys, especially on the Sel-tau, which rises steeply from the Pamir. There are also some of the longest

TABLE 30

Diurnal fluctuations of the atmospheric and soil temperatures in the Alai Valley (3145 m) and in the Jashil-kul (3982 m) in July

Area	6	7	8	9	10	11	12	13	14	15	16	17	18	19	20	21 hours
Alai Valley																
Atm. temp.	7	9	11	14	16	18	17	19	18	17	15	14	13	19	9	—
Soil temp.	10	15	22	28	36	45	44	42	38	34	28	23	19	13	11	—
Insolation	—	—	1.45	—	—	1.54	—	—	—	—	—	—	1.16	—	—	—
Jashil-kul																
Atm. temp.	—	—	10.1	12.8	15.6	19	20	21	20.8	21.2	20	19.8	17.8	14.5	8.5	8
Soil temp.	—	—	10.4	11.2	21.8	30.8	25	23.5	23.8	22	21.1	19	17	15	13.1	12.8

extra-arctic glaciers in this area. The Fedtshenko Glacier is, for example, 77 km long and its tongue reaches down to an elevation of 2900 m. This glacier normally moves at the rate of about 50 m annually, but during May 1963 it moved this distance almost within a single week, with roaring noise, down the Wantsh Valley. After this unexpectedly extra movement, the glacier abruptly slowed down to its normal

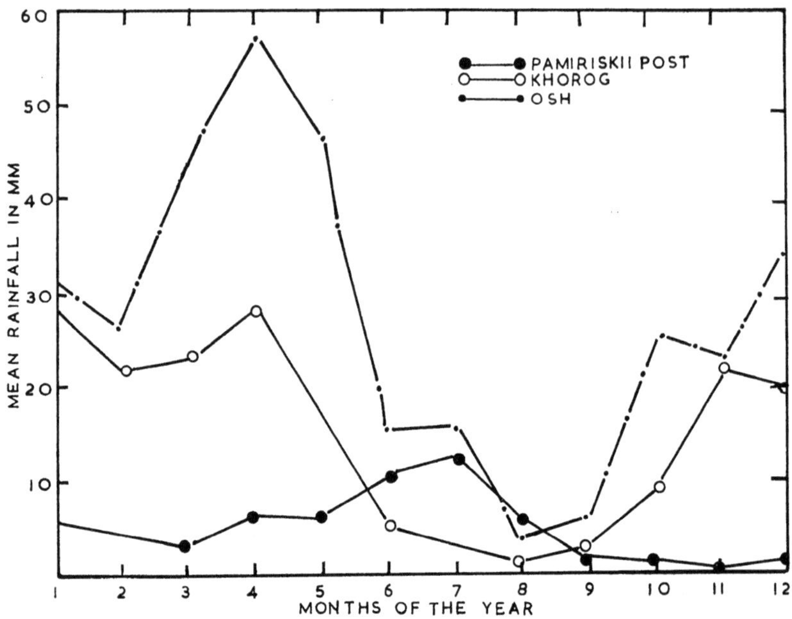

Fig. 56. Mean monthly precipitation in the Pamirs Region.

speed and then again resumed its thunderous movement, thus giving rise to gigantic crevasses, often 30-32 m deep. The snowline in this area is at elevations between 4000 en 4500 m.

To summarize, we may conclude that the general climate of the Pamirs is characterized by high atmospheric aridity, low atmospheric temperatures, intense insolation, great seasonal and large diurnal temperature fluctuations, scanty or no snow-cover and cold storms. In the region of the Pamiriskii Post only four months have temperatures above 7.5° C, viz. June 10.5° C, July 13.9° C, August 13.8° C and September 7.7° C. In the Kara-kul area, only July and August have temperatures above 7.5 C. The mean minimum temperature during July in Pamiriskii Post is 4.7° C and for August 3.6° C and the absolute minimum is — 1.3°C for July and —4.6° C for August. In the Kara-kul Pamir area, these figures are —5.5° C for July and —14.5° C for August. According to the observations of FICKER (344), the annual atmospheric mean temperature of areas at elevations of 4000-4300 m lies between —3.6° and —5.0° C. The mean of the warmest month, viz. August, lies between 6.0° and 8.0° C.

The soil temperatures show greater diurnal fluctuations than the atmospheric temperatures. The superficial-soil maximum temperature in Kara-kul is 55.2° for July, 55.5° C for August and 54° C for September and the respective minima are —7.7° C, —8.6° C and —16.5° C. In many localities the ground is permanently frozen at depths of 80-100 cm for the whole year. The ice-cold springs of the Rang-kul and the fossil-ice on the shores of the Kara-kul and Shor-kul are evidence of permafrost. In other localities like Mus-kol Valley, Ak-baital Valley and in the Upper Shah-dara, west of the Maz Pass, snow lies thick the whole year round. The mean intensity of insolation in July, August and September for Kara-kul is 1.54, 1.57 and 1.65 g-cal per sq. cm per minute and the maxima are 1.68, 1.65 and 1.67 respectively. Daily between 14 and 15 hours, violent winds break out and soon turn into storms between 17 and 18 hours and are accompanied by a rapid fall in atmospheric temperature. The wind often stops at night. Gusts of wind often reach velocities of 15-16 m/sec; this *burane* (Pamir wind) is still more violent in the southern Pamirs than elsewhere. The mountain-wind sweeps down from the crests of the mountains, as soon as the sun passes behind the peaks, often with hurricane violence. The evening wind arrives a few minutes after the sun sinks behind the mountains and the morning wind follows close on the warming up of the slopes and the bottoms of the valleys by the sunshine. The calm in Garan causes oppressive heat more than in Wakhan and the cold air of the winter is also more severe. The cold air sinks down stagnant into the narrow Garan Valley, which thus lies deep in snow, but Wakhan, though on a higher altitude, is mostly snow-free.

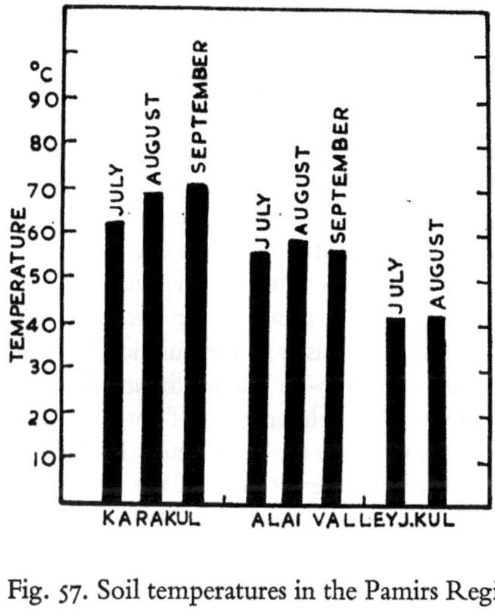

Fig. 57. Soil temperatures in the Pamirs Region.

In the Western Valleys, both precipitation and atmospheric temperatures are high. At the Khorog Meteorological Station (2105 m), the mean annual precipitation is 183.2 mm, or about thrice that of the area of Pamiriskii Post. The bulk of the precipitation does not also occur during the summer, but only between November and April, with a maximum in January and in April. About 78% of the annual precipitation occurs in these six months. June, July, August and September, during which 53% of the precipitation occurs in Pamiriskii Post, are, however, dry months in

Khorog, with the precipitation amounting to only 11.5 mm (= 6.3%) or only one-third of that of the Pamiriskii Post. This results in a very pronounced winter snow-cover and lower fluctuations of soil temperatures. Atmospheric humidity is, however, less in the Western Valleys: July 35%, August 32% in Khorog. The mean annual atmospheric temperature for Khorog is 8.8° C and six months from April to October have thus temperatures above 7.5° C. These six months have a mean monthly atmospheric temperature of 9.0° C. At Khorog the atmospheric temperature means are April 9.3° C, May 15.1° C, June 9.2° C, July 22.5° C, August 22.2° C, September 18.1° C and October 10.6° C. Jashil-kul, in the transitional area between the Pamir proper and the Western Valleys, has a monthly atmospheric mean maximum of 19.3° C for July, 21.2° C for August and 24.9° C for September and the corresponding minima are —3.5° C, —10.2° C and —3.4° C. The soil temperature maxima for July-August are 33.5° C for July and 30.8° C for August and the respective minima are 7° C and 10.4° C. Night frosts are not, however, rare events. There is sometimes frost on eight days in July and for 23 days during August. The updraft wind blows each afternoon, but never reaches the intensity of the burane of the Pamirs. Dust storms are frequent in the afternoon in the Muk-su (Alten-masar) Valley and Kudara Valley.

The Alai Valley is bounded on the east by the Surch-ob and Kisil-su regions and is about 100 km long and 15 km wide (20-30 km wide in the east) and has a mean elevation of 2400-3700 m. The monthly mean atmospheric temperatures for July and August in Sari-tash (3145 m) are less than those of Pamiriskii Post and only 1.5 to 2.0° C more than those of Kara-kul. The July mean temperature is 9.7° C and the August mean 10.4° C. The maximum and minimum are only slightly above those of the Kara-kul Pamir, which is only 800 m higher in elevation. The July maximum is 22.1° C and the minimum —3.5° C; August maximum 24.8° C and minimum —6.5° C. The soil temperature maxima and minima for July are respectively 48.9° C and —6.3° C, for August 49.6° C and —8.4° C. Rainfall is higher than in the Kara-kul Pamir; the relative humidity is 15% and 18% higher than in the Kara-kul Pamir. Considerable precipitation occurs even in winter in the Alai Valley and the snow-cover locally attains a thickness of over three metres. Insolation is estimated at 1.5 g-cal per sq. cm per minute. The afternoon winds blow from the west and are comparable in their intensity to the burane of the Pamirs.

The ancient land surface in Surch-ob has a climate characteristic of Tuptshek—Kisil-su—Muk-su triangle. The high-steppe of Tuptshek has a mean elevation of 3150 m and is bounded in the south by the steep Peter-the-Great Range, in the north by the low continuation of the Western Chain of the Peter-the-Great Range. The mean temperature for July-August is about 11.9° C and in Kulika (2822 m) 13.5° C. The mean summer temperature is 10° C and the annual mean lies between 1° C and 1.5° C.

4. VEGETATION AND BIOTIC ZONES OF THE PAMIRS REGION

From a distance, the vegetation of the Pamirs presents a characteristically brown colour. There are extensive terraces of hard, flat, stunted cushions of *Eurotia ceratoides* (LINN.), *Acantholimon diapensoides* BOISS. and in sandy localities *Astragalus* spp., *Hedysarum cephalotes* FRANCH and *Solenanthus nigricans* and often also abundant growths of *Dracocephalum heterophyllum*. FEDSCHENKO (338,338a) classifies the Pamirs vegetation as below:

I. Meadow vegetation
  1. Alpine mats, at elevations of 4000-5000 m
  2. Moist meadows
  3. Salt moor-meadow of river and lake shores at elevations of 3000-4200 m

II. Stony desert vegetation
  1. *Eurotia*-desert
  2. Slope vegetation

III. Trees
  1. *Myricaria*-bush
  2. *Salix*-bush

As may be understood, the low annual atmospheric temperature, the permanent ground frost at 80 cm depth in many places, the extreme winter aridity and the regular diurnal storms are among the major extreme unfavourable factors for the growth of trees. According to FICKER (344), an elevation of 3000 m marks the tree-line in Kara-teghin. At this elevation, the mean atmospheric temperature is 10.8° C for several months and the annual mean is 1.4° C. This treeline and the shrubline altitude is applicable perhaps only to the northern slopes; on the southern slopes and in the Alai Valley, the limits rise, however, up to 3900 m for shrubs, coinciding with the summer isotherm of 10.8° C. *Hippophaë rhamnoides* LINN. and *Salix oxycarpa* ANDERS. ascend high in the east, but *Populus* and *Betula* do so to a much less extent. *Juniperus pseudosabiana* FISCH. is common in the northern valleys and on the eastern slopes, up to elevations of 3500 m (for example, in the Trans-Alai). It may be remarked that on Sel-tau and Peter-the-Great Range, the shrubline ascends to 3500 m. The Fedtshenko Glacier descends, as already mentioned, to 2900 m. On Tanimas Valley, the glacier limit is 3600 m. The shrub vegetation thus approaches close to the glacier tongues. The maximal interval between the shrubline and the permanent snowline in the west is about 1000 m, but the glaciers do intrude into the shrub-zone or at least come into close contact with the shrubline. The conditions are, however, very different on the Pamir; the highest shrub of *Salix*-bush, at an elevation of 3700 m (in the Jaman Valley), is at least 1500 m below the snowline and more than 1000 m below the glacier tongues. This wide altitudinal zone is occupied by high montane steppes, the slopes of the Central Region and the high montane meadows.

Of the species listed by PAULSEN (815) from the alpine-mats, *Myosotis sylvatica*, *Sedum gelidum*, *Swertia marginata* SCHRENK, *Primula nivalis* PALL., *Potentilla sericea* LINN., *Saxifraga cernua* LINN., *S. flagellaris* WILLD., *S. hirculus* LINN., *Pedicularis dubia* B. FEDTSH., *P. pulchra* and *Lagotis borealis* are characteristic.

According to KOROVIN (649-651), the number of species of mountain plants on the different Middle and Central Asiatic mountains, including the Pamirs Region, are as follows: Zailiskii Ala-tau 1264 species, Chu-Iliskii Mountain 765 species, Syr Darya drainage area 2000 species, Dzhungarskii Ala-tau 787 species and the Pamirs 656 species. KOROVIN has given an excellent account of the fundamental characters, the general appearance and adaptations of the typical mountain species of plants on the Middle Asiatic Mountains, particularly the Central Tien Shan and the Pamirs Region. There is also an interesting discussion of the general ecology and characteristic vegetation of the high mountain steppes, with special reference to the

Alai-Pamirs Region. He has divided the mountain steppes into two fundamental types, viz. the steppes subaridae (*steppe subaridnie*) and the steppes frigidae (*kholodnikh steppe* or the *steppe kholodnie*), readily recognized by their characteristic vegetation. The arid steppes are characterized by *Festuca artimisiae*, *Stipeta festucae*, *Stipeta artimisiae*, *Andropogoneta*, etc. On the frigid steppes we find *Stipeta mixtoherbosa*, *Stipeta avenastri*, *Stipeta depauperata*, *Stipeta festuca alpina*, *Avenastreta*, *Festuceta*, *Ptilagrostideta*, *Atropideta* and *Leucopoeta*. The frigid steppes are subdivided into different subtypes on the basis of the characteristic vegetation. KOROVIN has also given an excellent description of the general ecology and the vegetation of montane meadows of Middle Asia, with particular reference to the Tien Shan System.

i. The Pamir Proper

Treeless steppe and semi-desert are the general formations. Extensive areas are covered with only *Eurotia ceratoides* (LINN.) and flat cushions of *Acantholimon diapensoides* BOISS. Small patches of grass may be found isolated on the arid terraces and extensive areas, especially east of Kara-kul, are wholly without a trace of vegetation, but simply covered with dust. In the furrows in which melt-water flows, there is some sparse vegetation that also presents a parched brown appearance. In the vicinity of rivers and glaciers and melt-water streams, we may find the violet-red *Astragalus* spp., *Hedysarum cephalotes* FRANCH and *Solenanthus nigricans*. *Cousinia rava* and *Saussurea crassifolia* grow in saline soil in the Murg-ab Basin. *Polygonum pamiricum* and *Carex pseudofoetida* KÜK. also occur in saline localities. In the dried-up melt-water channels and fissures may be found *Saxifraga hiculus* LINN., *Taraxacum leucanthum*, *Hedysarum cephalotes* FRANCH, *Astragalus* and *Oxytropis*. Moist localities in the neighbourhood of springs and streams have a relatively luxuriant vegetation, consisting of *Primula sibirica* JACQ., *Pedicularis cheilanthifolia* SCHRENK. *P. uliginosa*, *P. lobellata*, *Saxifraga hirculus* LINN., *Leontopodium leontopodinum*, *Lagotis decumbens* VENT., *Arnebia guttata*, *Taraxacum leucanthum*, *Potentilla bifurca* LINN., *P. multifida*, *P. sericea* LINN., *Erigeron* sp., *Ranunculus* sp., *Swertia marginata* SCHRENK, *Gentiana prostrata* HAENKE, *Inula* sp., *Smelovskia annua*, *Sedum gelidum*, *Cerastium trigynum* VILL., etc. Dry beds of streams are grown with *Dracocephalum heterophyllum*. The luxuriant moist meadows of the Murg-ab Basin are not met with anywhere else on the Pamirs proper. In the meadows of the lower Alichur Pamir and Bulun-kul there, however, occur *Myricaria squamosa* DESV., *Clematis tangustica*, *Astragalus ophiocarpus*, *Nepeta spathulifolia*, *Senecio coronopifolius parvulus*, *Chenopodium vulvaria*, etc. The vegetation of the Jaman Valley in the inner Murg-ab Basin seems unusual for the inner Pamirs. *Scrophularia incisa pamirica* grows here nearly as high as a man. *Gentiana barbata* FROEL. and *Orchis turcestanica* occur here at elevations of 3700 m. There are significant differences between the vegetation of the northern and the southern Pamirs. The Alichur Pamir has a relatively rich flora. Being open in the west, there are many species like *Cousinia rava*, *Acantholimon alatavicum*, *Papaver radicatum*, *Ligularia sibirica* and *Sedum gelidum*.

ii. The Western Valleys

In its distinctive treeline and shrubline, the area of the Western Valleys differs fundamentally from the Pamir proper. The upper limits of tree growth are at elevations of about 3000 m, coinciding with the mean monthly 10.8° isotherm from June to September. In the southern valleys, the limits are somewhat higher, by perhaps about 500 m. In the transition zone of the upper levels of the southern valleys, there are dense bushes of *Salix oxycarpa* ANDERS., *S. zygotemon, S. coerulea* E. WOLFF and *S. repens rosmarinifolia* LINN., *Hippophaë rhamnoides* LINN., *Rosa webbiana* WALL., *R. xanthia, Ribes* and *Lonicera semenovii. Juniperus pseudo-sabiana* FISCH. ascends locally to elevations of 3500 m. The herbs at elevations of 3000-3800 m include *Lagochilus paulsenii, Mulgedium tataricum* (LIPSKY), *Reganus harmala, Nepeta kakandica, N. longibracteata* BENTH., *Delphinium dasyanthum* KAR & KIR and Umbelliferae like *Archangelica songorica* and *Heracleum olgae* RGL. & SCHMALH. *Acantholimon diapensoides* BOISS., thorny spherical cushions of *A. alatavicum* BGE. *Eurotia certaoides* (LINN.) and *Ephedra.* Localized patches of *Betula, Populus, Verbascum* and *Epilobium* also occur. The apricot (*Armeniaca vulgaris*) grows at elevations of 3000 m in the Pamir Darya Valley.

iii. The Alai Vegetation

Typical tree- and shrub-steppes may be found in the east Alai Valley. The moraines at the base of the Trans-Alai mountain chain have a relatively luxuriant vegetation; the plants are, however, characteristically dwarfed, especially at higher elevations of the Alai, north of the Taldik Pass. In marked contrast to the pronounced xerophytic character of the Pamirs flora, the Alai Valley, with its relatively heavy precipitation, is noted for its mesophyll vegetation. Character species of the moraines are *Anemone tschernaewi, Astragalus danieli kochi, A. tibetanus* BENTH., *Pulsatilla albana, Ranunculus flexicaulis, Taraxacum paludosum, Primula algida* AD., *Ligularia sibirica, Androsace villosa* LINN., *Trollius, Allium, Myosotis, Alchemilla* and *Draba media. Artemisia*-steppes occur in the western parts, with sparse growth of *Salix* and *Hippophaë rhamnoides* LINN. The shrubline is at an elevation of 2700 m near Dschptik; broad-leaved trees often grow up to elevations of 2450 m.

iv. Altitudinal Biotic Zonation

The Western Valleys are below the treeline. In the transition zone from the treeline to the shrubline on the Pamirs, there are a number of species of plants and insects that cannot, strictly speaking, be described as typically high altitude mountain elements and some of them are indeed widely distributed in the Turkestan lowland steppes also. Parts of the Murg-ab Basin, such as, for example, Jaman Valley and Pamiriskii Post, cannot also be strictly described as true high altitude zones. The truly elevated areas of the Pamirs Region are subdivided into i. meadow zone, ii. subnival zone and iii. nival zone. A typical meadow zone characteristic of the Alps is not, however, found anywhere on the Pamirs Region, except perhaps on the northern chains of the Alai Mountains, at elevations of 2000-3000 m. Islands of vegetation in favourable patches, above the upper limits of the shrubline in the Pamirs Region represent perhaps the alpine-meadow zone. The altitudinal zonation on the Western Valley, proceeding eastwards, is as follows (875):

1. Cultivated zone up to an elevation of 3200 m
2. Middle montane zone or bush(Knieholz) zone, at elevations between 3200 and 3800 m

3. High montane zone from 3800 to 7500 m
   i. Lower meadow zone 3800-4000 m
   ii. High meadow zone 4000-4300 m
   iii. Mountain slope "Gehänge" steppe zone 4300-4700 m
   iv. Upper meadow zone 4700-4800 m
   v. Subnival zone 4800-5000 m
   vi. Nival zone above 5000 m

As may be expected, the boundary altitudes of the various zones are variable. The treeline is, for example, at an elevation of 2400 m in Dshailgan, the shrubline ascends to 2800 m and the continuous-meadow zone reaches up to 3400 m. The width of the subnival zone, often separated from the meadow zone by a vegetationless stretch, varies considerably, but it is usually a narrow belt, at elevations between 3400 and 3800 m. The meadow zone and the subnival zone are often also absent on the south slopes. The East Alai Valley (2800-3200 m) is above the limits of the tree- and shrublines. The steppes-vegetation represents the lowest stage of the meadow zone in some localities. This passes into a typical meadow zone in moraines on the north base of the Trans-Alai (3200-3400 m). The subnival zone lies here at elevations of 3600-3800 m and the nival zone at 3800-4000 m. The conditions are, however, very different on the south slope of the Alai, where a typical meadow zone is absent and in its place we find an arid steppe-vegetation, comparable to the *Eurotia ceratoides* formation of the Pamirs. The subnival zone is found only on the north slope and the nival zone is absent in the east. It is only in localities, where optimal amounts of moisture are available on the south slope during the summer, that we find a definitive subnival zone. The alpine zone on the Alai Mountains does not descend below an elevation of 2750 m and often lies at an elevation of 3050 m. On the Trans-Alai Mountains, the limit is not reached below an elevation of 3500 m and may often be as high as 4400 m. On the Kounjout it is even higher and the lower limits are at an elevation of about 3960 m.

The following list of the typical alpine-zone plants of the Pamirs was made available to me by my friend Mr. IGOR V. BELOLIPOV of the Botanic Garden, Academy of Sciences Uzbek SSR, Tashkent:

Gramineae: *Poa glabriflora* ROSHEV, *P. zaprjaganevi* OVCZ., *P. bucharica* ROSHEV, *P. alpina* LINN., *P. pamirica* (LITW.) ROSHEV, *P. calliopsis* LITW., *P. dschungarica* ROSHEV; several other species at elevations of 3000-4000 m on alpine meadows: *P. disiecta* OVCZ., *P. gorbunovii* OVCZ., *P. hissarica* ROSHEV, *P. litwinoviana* OVCZ., *P. marginata* OVCZ., *P. densissima* ROSHEV, etc.; *Festuca ganeschini* DROB., *F. coelestis* (ST. YVES.) KRECZ., *F. kirilovii* STEUD., etc.

Liliaceae: *Gagea pamirica* GROSSH., *G. capillifolia* VVED., *G. vedenskyi* GROSSH., *G. hissarica* LIPSKY, *G. emarginata* K. & K., etc.

Polygonaceae: *Rumex pamirica* RECH.

Caryophyllaceae: *Silene samarkandensis* PREOR., *S. conformifolia* PREOR. *S. graminifolia* OTTH., *S. pamirica* (H. WNCL.) PREOR., *S. karaczkuri* FEDTSCH., *S. trajectorum* KOM.

Ranunculaceae: *Ranunculus trautvetterianus* RGL., *R. rubrocalyx* ENGL., *R. rufosepalus* FRANCH., *R. paucidentatus* SCHRENK, *Anemone obtusiloba* DON., *A. oligotoma* J.L., *A. kostyczwii* KORSH.

Saxifragaceae: *Saxifraga hirculus* LINN.

Rosaceae: *Potentilla pamiroalaica* JUZ., *P. evestita* TH. WOLF, *Alchemilla retropilosa* JUZ.

Leguminosae: *Lotus frondosus* FREYN., *Trifolium seravschanicum* OVCZ., *Astragalus olgae* BGE., *A. chionanthus* M. POP., *A. colateocarpus* BOISS., (3100-3900 m), *A. kuschakevitschii* B. FEDTSCH. (3000-4900 m), and *A. melanostachys* BENTH. (3000-4000 m).

Geraniaceae: *Geranium sophiae* AN. FED., *G. meeboldtii* BRIGUET, *G. saxatile* K. & K., *G. ferghanense* BOBR.

Primulariaceae: *Primula eugeniae* FED., *P. lactiflora* FURKEV., *P. dorsocalyx* P. POL. & LINCZ., *P. iljinskii* FED., *P. pamirica* FED., *P. warshenevskiana* B. FEDTSCH., *P. olgae* RGL., *P. turkestanica* (RGL.) E. A. WHITE, *P. pulverea* FED.

Gentianaceae: *Gentiana kaufmanniana* ENGL. & SCHMALH., *G. olivieri* GRISEB., *G. karelinii* GRISEB., *G. vvedenskyii* GROSSH., *G. pamirica* GROSSH.

Labiatae: *Thymus seravschanicus* KLOK., *T. ashurbajevii* KLOK., *T. diminutus* KLOK.

Scrophulariaceae: *Veronica ciliata* FISCH., *V. rubrifolia* BOISS. (3850 m) and *V. michauxii* LAM. (2000-3600 m).

Campanulaceae: *Campanula evolvulacea* ROYLE.

Compositae: *Erigeron servaschanicus* M. POP., *E. sogdianus* M. POP., *E. pallidus* M. POP. and *E. vicarius* BOTSCH. (subalpine and alpine, 2200-4000 m).

5. INSECT LIFE OF THE PAMIRS REGION

The early explorations undertaken by ALCOCK (11), ALPHÉRAKY (37, 39), FEDCHENKO (337-339), SEVERTZOV in 1878, ALBERT REGEL (864, 865) and V. SCHULTZ (939) in the Alai-Pamirs Region were largely geographical. GRUM-GRSHIMAILO (446, 447) was perhaps the first to collect insects from the Pamirs, especially from Kara-teghin, Darwas and the Kara-kul Pamir, during 1886-1887 and from the eastern Pamirs and the Taghdumbash Pamir in 1889. The Government of India Boundary Commission of 1896 also undertook some zoological and botanical investigations and the material collected by its members was described by ALCOCK (11) and DUTHIE. PAULSEN was a botanist of the Olufsen Expedition of 1898-1899 to the Pamirs (815). KORB (647) collected on the Alai and in the Upper Alai Valley in 1905. CONRADT made extensive collections in Kara-teghin, Darwas, Sor-kul and in the upper course of Ak-su during 1889-1890. He travelled from Ak-su over the Beik Pass to the Kuen Lun Range and returned by Kasghar in 1891 to the East Alai and Osh. The most important work was, however, by FEDCHENKO, who explored nearly the whole of the Pamirs in 1901-1904. The German-Austrian Alpine Verein Expedition of 1913 collected considerable zoological material from the east Kara-teghin. In more recent years, the region, expecially the Buchara area, has been explored by a number of Russian naturalists like AVINOFF (67-70), KOTSCHUBEI, LAZDIN and the Pamir Expedition of 1928 (811), etc. The geographer KORSCHNEWSKY (653) explored the region for nearly half a century and also collected many specimens. The Soviet Socialist Republic Government encouraged and collaborated with many foreign scientific expeditions, to the Pamirs Region. Our knowledge of the insect life at high elevations in the Alai-Pamirs Region is largely based on the labours of the Russian-German (811), the Russian and, the Italian Pamir Expeditions and particularly to the contributions of REINIG (872-878).

The insect life of the high elevations in the Pamirs Region and on the high mountains that form the characteristic knot in the region is extra-ordinarily peculiar in many respects. These peculiarities are largely correlated with the pronounced ecological monotony of the whole of the inner Pamirs, the extreme climatic conditions, the marked paucity of vegetation, the generally arid steppe-like high mountain biotopes of the region and the orogeny of the high Himalaya and other Tertiary mountains in the area. Tube-dwelling spiders, scorpions, Diplopoda and diverse other stenotherm elements, so characteristic of the Turkestan lowlands, are completely absent on the Pamirs. Insects, the larvae of which inhabit the upper layers of the soil among the roots of plants in Turkestan, expecially Elateridae, Curculionidae, Lamellicornia, etc., are also absent on the Pamirs. Only in the Alai Valley and in the Lower Kisil-su Valley (Kitschik-kara-muk) do we find some small-sized Elateridae such as *Lacon crenicollis* MÉN., *Cardiophorus* ESCH. in Kisil-su and Atshik-alina. *Agriotes caspicus* HEYD. and *A. reticulosus* CAND. occur further west. *Sphenoptera laticeps* JAKOWL. is the only Buprestid known from the Trans-Alai moraines of the Upper Alai Valley. Owing to the absence of trees, wood-boring Cerambycidae, Ipidae, etc., are naturally not met with in the Pamirs Region. Diptera, especially Culicidae, *Scatophaga* spp., Hemiptera, Coccinellidae, Chrysomelidae and coprophagous forms are, however abundant. In Sari-tash (3200 m) in the east Alai Valley, REINIG (877) reports, for example, *Saprinus aeneus turkestanicus* SCHM., *Geotrupes impressus* GEBL., *Aphodius nigrivitis* SOL., *A. przevalskyi grombczevskii* B. KOZH., *A. vittatus mundus* REITT., *A. distinctus* MÜLL., *Onthophagus sibiricus* HAR., etc., in camel and yak dung. In the Chargush Pamir, he also found *Aphodius nigrivittis* SOL., *A. distinctus* MÜLL. and *A. pamirensis* MEDV. *Necrophorus sepulator* CHARP., *Saprinus aeneus turkestanicus* SCHM. and *Geotrupes impressus* GEBL. occur at elevations of 3200 m in the Alai Valley, but are absent in the Pamirs. Isolated examples of Silphidae, like *Aclypea pamirensis* JAKOWL., occur at an elevation of 4100 m in Djol-boeruljuk. Odonata and Myrmeleontidae are wholly absent on the Pamirs; the latter group reaches its upper altitudinal limits of 2800 m in Muk-su. *Aeschna* sp., reported from an elevation of 3700 m in the Maz Stream, represents perhaps the highest altitude record for Odonata at present.

ORTHOPTERA

A number of remarkable endemic species and some very unique endemic genera of grasshoppers occur at high elevations in the Pamirs Region. *Pamiracris reinigi* RAMME, for example, is an apterous endemic grasshopper that occurs in sandy localities at elevations of 4200 m. Other species include *Metrioptera pamirica* RAMME, with the elytra shorter in the female than in the male, at elevations of 3500 m in Kara-teghin; *M. intermedia* SERV. from the Alai; *Tettigonia viridissima* LINN., *Decticus albifrons* (F.) and *Gomphocerus sibiricus* LINN. from the Alai; *Chorthippus scalaris* F.W. from the Alai Valley; *Ch. dorsatus loratus* F. W. from the north slope of the Alai Range; *Dociostaurus crucigerus brevicollis* EV. from the Alai Range; *Conophyma birulai* MIRAM from the Karak-kul Pamir; *Sphingonotus coerulans coerulans* LINN. from Murg-ab; *Sph. rubescens* WALK., *Sph. mecheriae* KR., *Sph. birulai* MIRAM and *Sph. pamiricus* RAMME (4100-4200 m) from the Pamirs and *Conophyma miramae* UVAROV from the Alexander Mountains. Some of these genera and species are, as pointed out in the foregoing chapter, common to the Northwest Himalaya also. For descriptions of Orthoptera from the Pamirs Region reference may be made to RAMME (855), TARBINSKY (1028), UMNOV (1057) and UVAROV (1063, 1066, 1068, 1069).

DERMAPTERA

Anechura fedtshenkoi SAUSS., reported from an elevation of 1600 m on the Alai and at 1800 m from Kara-teghin, is also known from the Kara-kul Pamir. *A. bipunctata* (FABR.) is recorded at elevations of 1800-2140 m on the Pamirs Region; it is also known from the Northwest Himalaya.

HETEROPTERA

The Heteroptera from the Pamirs Region were described largely by KIRITSHENKO (617, 619).

Pentatomidae

Odontoscelis fuliginosa LINN. from the Kara-kul Pamir (4400 m); *Carpocoris fuscispinus* BOH. from Sari-kol (3200 m); *Graphosoma consimile* HORV., widely distributed from the eastern Caucasus to the Kara-tau Range (Dzhuleka), Talaskii Ala-tau, Nikolaipol, Bish-tash, the Chirchik and Chatkalskii Mountains: Kisil-kum, Kara-kum, Golodnie Steppe, Kopet-dagh Range, Zerafshan, occurs also in Kara-teghin and Darwas. The genus *Mimula* JAK. is localized at elevations of 2500-4300 m on the Alai-Pamirs, Tien Shan and the Tarbagatai Ranges. Within the Pamirs Region, we have several interesting species like M. *nigrita* JAK. (fig. 58B) from the Sari-kol Pamir (4300 m) and Naisha-tash (3700 m); *M. dungana* KIRIT. (2460-2700 m); *M. scutellaris* KIRIT. from the Koitus Mountains; *M. alatavicus* KIRIT. from the Talaskii Ala-tau; *M. hohlbecki* KIRIT. from the Zerafshan Range (3400 m); *M. vicina* KIRIT. from the Turkestan Mountains (3320 m); *M. anthracina* KIRIT. from Darwas (3770 m) and *M. maureri* KIRIT. (fig. 58A) from the Pamirs (3000 m) and from the Dzhungarskii Ala-tau mountains.

Coreidae

Of the Coreidae so far known from the Pamirs Region, *Corizus limbatus* REUT. occurs in the Kara-kul, Kara-kurgan and Rang-kul Pamirs and is also known from the Karakoram Range of the Northwest Himalaya. The other interesting species from the region include *Coriomeris scabricornis* PANZ. (3200 m), *Stenocephalus ferghanensis* HORV. (3200 m), *Liorhyssus hyalinus* FABR., *Corizus hyosciani flavicornis* PT. and *C. limbatus subsimilis* HORV.

Lygaeidae

The typical Pamirs Lygaeids include *Spilostethus ubriceps* HORV. from Sari-tash (3270 m), *Sp. melanostolus* KIRIT. (3900 m) and *Sp. alaicus* KIRIT. in moraines of the Trans-Alai; *Pionosomus opacellatus* HORV. (3100 m); *Lamprodema maurum* FABR. (3200 m); *Gonionotus marginepunctatus* WOLFF from the Kara-kul Pamir (4400 m); *Geocoris arenarius albidus* JAK. from Jashil-kul; *Jakovleffia setulosa* JAK., *Microplax interrupta* FIEB. and *Emblethis verbasci* FABR. from the Alichur Pamir; *Gonionotus reichardti* KIRIT. (fig. 59A); *Rhinocoris monticola trochantericus* REUT., widely distributed in Middle Asia, especially the Alai-Pamir and Turkestan Mountains, Alexander Mountains (2700 m). *Rh. monticola* OSH. occurs on the Tien Shan, Dzhungarskii Ala-tau, Zailiiskii Ala-tau and Terskei Ala-tau and the Pamirs Region, Zerafshan Range, Tchatkalskoi Range, Hissar Mountains (2857 m), Kara-teghin, Peter-the-Great Mountains, etc. up to an elevation of 3800 m. *Nysius ericae obscuratus* HORV. has also been recorded from the Pamirs.

Miridae

The Miridae from the Pamirs Region include the interesting *Stenodema turanicum* REUT., *St. crassipes* KIRIT., *Dicyphus orientalis* REUT., *Myrmecophyes orbicularis* KIRIT. (fig. 60C) and *M. aeneus* REUT. in addition to the Alai-Pamirs also known from the Zerafshan and Hissar Mountains and the Tien Shan System; *Scirtetellus brevipennis* REUT. from the Alai and from the Zailiiskii Ala-tau; *Sc. seminitens* HORV. widely distributed from the Tien Shan System and the Alexander Mountains and Ak-su and Kara-kul Pamirs; *Microsynamma basalis* REUT. endemic on Turkestan Mountains, Shah-i-Mardan, the Alai, Zerafshan and the Tchatkalskoi Mountains.

Saldidae

Chiloxanthus poloi KIRIT. (fig. 59B) is known from Sari-kol (3400 m) and other high mountains of Middle Asia. *Saldula jakovlevi* REUT. is known from Sari-tash and is endemic on the Turkestan Mountains, the Alai, Hissar Range, the Tien Shan, Alexander Mountains and the Bolshoi Chingan Mountains. *S. ortochila* FIEB. has been recorded from the Sari-kol Pamir and *S. melanoscela* FIEB. occurs up to elevations of 4000 m on the Pamirs.

The Heteroptera from the Pamirs Region comprise the following faunal component elements: North African, Iranian, Turanian, Siberian, Palaearctic, boreal European, south European, Caucasus and Armenian elements (619). The principal endemic genera from the Pamirs and from the other Turkestan mountains are *Melanosoma* JAK. (also known from Iran), *Promecocoris* PUT. (also from Armenia and Iran), *Periphyma* JAK. (also Iran), *Ophistochilus* JAK., *Tshingisella* KIRIT., *Hybocoris* KIRIT., *Pheocoris* JAK. (also from the south Altai, Mongolia and Tibet), *Oressibia* KIRIT., *Mormidella* HORV. (also from the Northwest Himalaya), *Ochyrotylus* JAK. (also from Iran), *Mimula* JAK., *Cercinthinus* KIRIT., *Microtelocerus* REUT., *Corizomorpha* JAK. (also from Armenia and Iran), *Limacocarcinus* KIRIT. (also from Iran), *Mecistocoris* REUT., *Aspidocanthus* REUT., *Scirtetellus* REUT., *Boopidocoris* REUT., *Pleuroxynotus* REUT., *Voruchia* POPP., *Stenaropsis* POPP. and *Heterobates* BIANCHI. The Mediterranean elements include *Byrsinus* FIEB., *Odontotarsus* LAP., *Ellipsocoris* MAYR, *Leprosoma* BAER, *Polyphema* JAK., *Microplax* FIEB. (also from Bohemia), *Auchenodus* HORV. (also from Armenia and Iran) and *Lamprodema* FIEB. (also from Armenia, Bohemia and Baikalia).

COLEOPTERA

Carabidae

The important contributions on the Carabidae from the Pamirs Region are the papers by LUTSCHNIK (714), TCHICATCHEFF (1033) and ZNOJKO (1135). In addition to

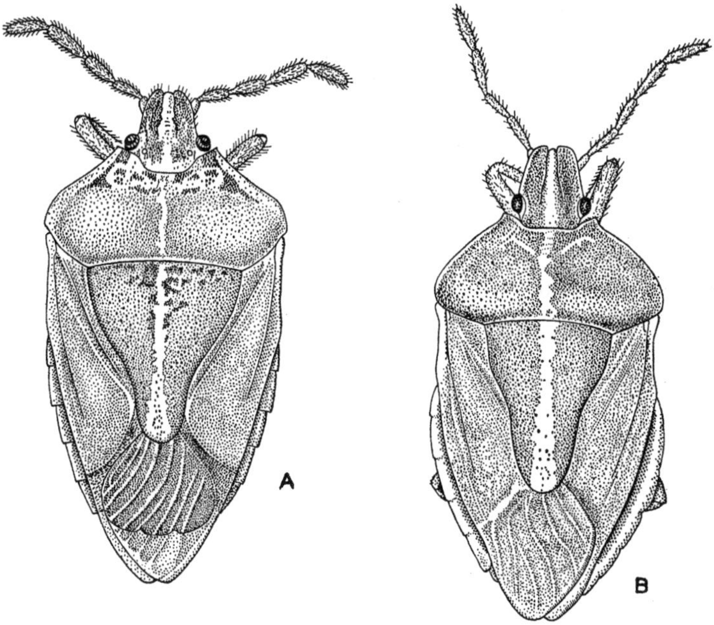

Fig. 58. The high altitude Heteroptera from the Pamirs Region. A. *Mimula maureri* KIRITS., B. *Mimula nigrita* JAK.

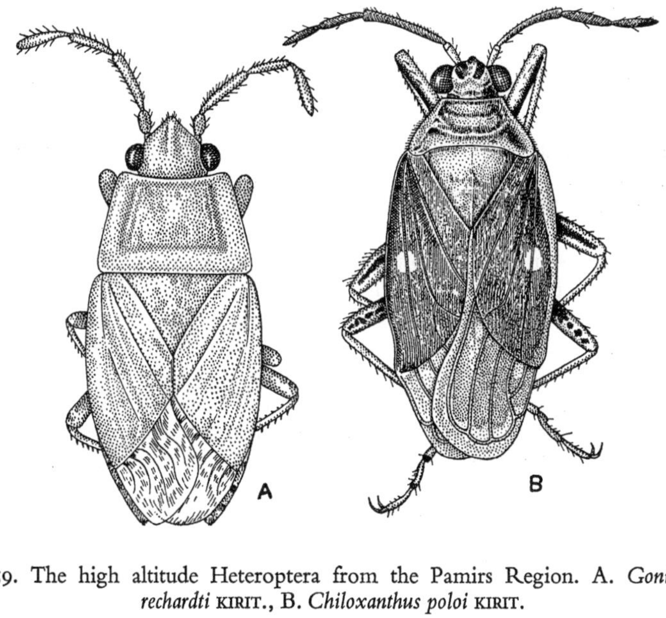

Fig. 59. The high altitude Heteroptera from the Pamirs Region. A. *Gonionotus rechardti* KIRIT., B. *Chiloxanthus poloi* KIRIT.

the high species endemism within the Pamirs Region, the Carabidae discussed in these pages are also remarkable for their high endemism to the vast area of the Pamirs and Northwest Himalaya. A great many of the high altitude species discussed here are common to the Pamirs and the Northwest Himalaya.

Agonum (*Agonum*) *ladakensis* BATES occurs at elevations of 3000 m in the region. *Amara* (*Amara*) *bamiduniya* BATES is widely distributed at elevations of 1240-4200 m. *A.* (*Celia*) *ambulans* ZIMM. (1850-3100 m), *A.* (*Leiocnemis*) *frivola* BATES (1850 m), *A.* (*L.*) *himalaica* BATES (1850 m), *A.* (*Cryptonotus*) *misera* TCHIH. (3400-4400 m) and *A.* (*C.*) *pamirensis* BATES, widely distributed from 2100 to 4200 m on the Pamirs are interesting records. A number of *Bembidion* LATR. may also be mentioned: *B.* (*Chlorodium*) *almum* SAHLB. (2010-2800 m); *B.* (*Peryphus*) *pamirense* BATES (fig. 76E) at an elevation of 3145 m in the Alai Valley, at elevations of 3960-4400 m in the Kara-kul Pamir and at 3000 m or even down to 2200 m in the Western Valleys; *B.* (*P.*) *terminale* HEER at elevations 2200-3850 m in the Western Valleys; *B.* (*P.*) *pamiricola* LUTSCH. at elevations of 4400 m in the Kara-kul Pamir and at 3200 m in the Western Valleys; *B.* (*Pamirium*) *punctulipenne* BATES (3580-3800 m); *B.* (*Daniela*) *bracculatum* BATES (2200 m); *B.* (*Emphanes*) *minimum* FABR. (2800 m); *B.* (*Notaphus*) *varium heptapotamicum* TSCHIH. from moraines of the Trans-Alai (3400 m); *B.* (*Testedium*) *bipunctatum capito* MÜLL. (3100-3300 m); *B.* (*Testediolum*) *kokandicum* SOL., *B.* (*T.*) *kokandicium fortius* NETOL. at an elevation of 3400 m in the moraines of the Trans-Alai and at an elevation of 3145 m in Kisil-su and other areas; *B.* (*T.*) *kokandicum validum* NETOL. (3100-3350 m). *Calathus melanocephalus* LINN. has been reported at elevations of 2800-3300 m. *Calosoma* (*Callistriga*) *maderae masaricum* REINIG is reported at an elevation of 2800 m. *Bradytus apricaria* LAYK. occurs at an elevation of 3400 m in moraines before Kurundi, but at lower elevations (1800 m) elsewhere. Four species of *Carabus* LINN. may be mentioned: *C.* (*Ophiocarabus*) *successor* REITT. from Ak-su (1200-2500 m); *C.* (*Ulocarabus*) *stschurovskyi* SOLS. at 1200-3000 m on the Alai-Pamirs, Zerafshan and Hissar Ranges; *C.* (*Plesius*) *dokhtouroffi* GGLB. (2500-3300 m) and *C.* (*Goniocarabus*) *grombczewskii* SEM. (3360 m). Two species *Cymindis picta pamirensis* REINIG (3600-3800 m) and *C. mannerheimi* GEBL. (3600-4400 m) may also be mentioned. *Dichirostris ustulatus* DEJ. has been recorded at an elevation of 3730 m from the Rang-kul Pamir and at an elevation of 4050 m from Djol-boeruljuk. Most species of *Harpalus* LATR. occur at lower elevations between 1700 and 3000 m, but *H. fuscipalpis* STURM. occurs at elevations of 3400-4350 m in Kisil-su, Kara-kul, Sor-kul and other localities and *H. masoreoides* BATES at 2140-4200 m. Of the two species of *Nebria* LATR., so far known, *N. limbigera* SOLS. is common to the Pamirs and the Northwest Himalaya. *N. limbigera limbigera* SOLS. occurs on the Tal-dik Pass (3300 m), Djol-masar (3850 m), Maz (3580 m), Pamir Darya at Maz (3580 m), Kisil-zu (3600 m), Utsh-kol (3800 m) and Sor-kul (4350 m). *N. l. kandshutica* GLAS. occurs at elevations of 3850-4100 m in the Western Valleys. *N. l. alaiensis* GLAS. occurs at 2470-2800 m. *N. psammophila* SOLS. ranges from an elevation of 1500 to 2800 m. *Platysma* (*Poecilus*) *liosomus* CHAUD. and *Pl.* (*Poecilus*) *cupreus matthiesseni* LUTSCH. are recorded at an elevation of 3100 m from the Mukur Valley.

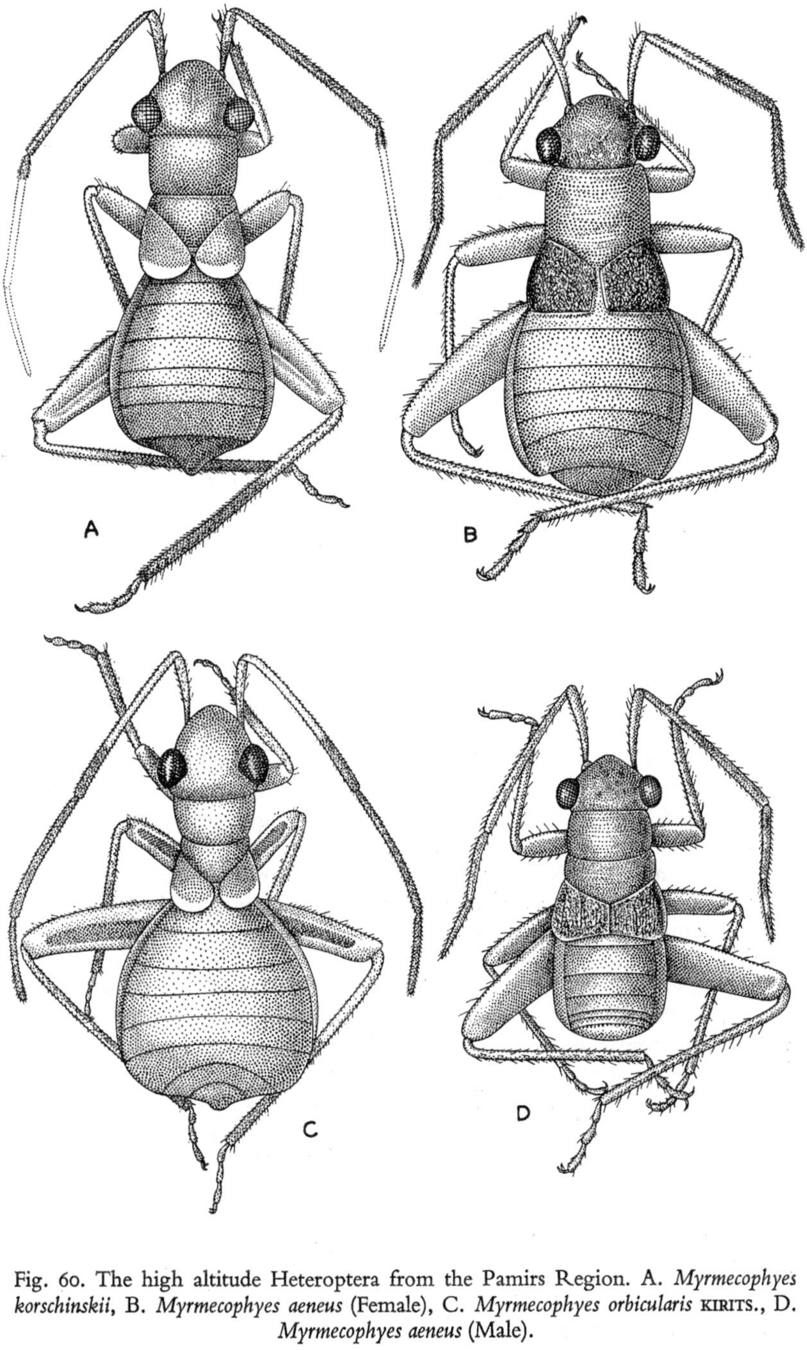

Fig. 60. The high altitude Heteroptera from the Pamirs Region. A. *Myrmecophyes korschinskii*, B. *Myrmecophyes aeneus* (Female), C. *Myrmecophyes orbicularis* KIRITS., D. *Myrmecophyes aeneus* (Male).

Staphylinidae

The Staphylinidae collected by the Pamir Expedition were described by BERNHAUER (112, 811). From the Western Valleys we have the records of *Aleochara* (*Coprochara*) *reinigi* BERNHAUER at an elevation of 3850 m. From Kara-teghin (3300 m) are reported *Quedius fulvicolleis nigricollis* KOLBE and *Tachinus rufitarsis* HOCHH. The other species known from the region include *Paederus fuscipes* CURT. (2470-3500 m), *Philonthus rotundicollis* MEN. (2750-3000 m), *Ph. r. nigropolitus* BERNH. at 2000-3900 m, *Ph. sordidus* GRAV. at 4100 m in north Pamirs, *Ph. sordidus coloripennis* BERNH. at 2900-4100 m and *Tachyporus nigrinus* LUZE at an elevation of 3100 m from Kara-teghin.

Tenebrionidae

The most important contribution on the Tenebrionidae from the Pamirs Region is by REINIG (874, 877, 878). The Platyscelinae and the Tentyriinae are represented by *Platyscelis margellanica* KR. and *Ascelosodis concinnus* BATES respectively on the Pamirs. The latter species is alone widely distributed in the region and *Platyscelis* LATR. seems to be characteristic of only the Western Valleys, including Darwas and Kara-teghin. From this area, REINIG records five species of *Platyscelis* and one species each respectively of *Trichomyatis* REINIG and *Trichoplatyscelis* REINIG. *Blaps* FABR. and *Prosodes* ESCHTZ. are only extremely sparsely represented and do not occur at elevations above 3000 m. *Dila* FISCHER DE WALDHEIM seems to be confined to the Alai, where it occurs up to an elevation of 2000 m, but is absent in the valleys of Kara-teghin. *Ascelosodis concinnus* BATES cannot strictly be described as a typical high mountain Tentyriinae. REINIG considers *Anatolica paphia* REITT., *Prosodes* (*Uroprosodes*) *costifera* KR. and perhaps also the somewhat rare *Platyscelis margellanica* KR. as character species of elevations of 3000-3200 m on the Alai Valley. These are the more important of the high-steppes forms. Under relatively more optimal conditions in the moraines of the Alai Valley in the north (than the conditions prevailing on the high-steppes), he reports the occurrence of *Blaps caraboides* ALL. and *Platyscelis modesta* KR. Comparing the Tenebrionidae from the Alai Valley with those from the Kara-teghin area, REINIG finds 13 species in the latter area, of which 11 occur in the region of the R. Surch-ob (1700-2300 m), as well as on the old landscape of the triangle formed by Tuptshek, Surch-ob and Muk-su (3000-3200 m). The altitudinal limits are, however, different further eastwards in the Alai Valley. *Itagonia* REITT. is, for example, absent in the Alai Valley. *Prosodes fallax* KR. and *Pr. alaiensis* KR. do not occur at elevations above 2250 m in the east. The eastern limit of *Platyscelis constricta* SEIDL. in the Daraut-kurghan area is at an elevation of 2250 m. *Scythis alticola* REINIG also occurs at about the same elevation in the Alai Valley, but ascends to 3000 m above mean sea-level on the northern slope of the Trans-Alai Mountains. *Gnathosia crypticola* REITT. is not so far known from the Alai Valley. We have therefore five species in the Alai Valley, only three of which occur on the high-steppes and the others occur sparsely in the moraines in the Trans-Alai, but more commonly at elevations of 3000-3200 m westwards. Leaving aside the species from the narrow biotope of the Alai Valley, only three out of eleven species (occurring at elevations of 1700-3200 m in the Kara-teghin area) are found in the Alai Valley, above an elevation of 2500 m and only two of them reach up to an elevation of 3200 m.

The following are the more important records of Tenebrionidae from the Alai-Pamirs Region: Blaptinae: *Itagonia gnapterinoides* REITT. at elevations of 2300–3500 m from Kara-teghin; *I. deplanata* REINIG at an elevation of 2500 m from Darwas; *Prosodes* (*Uroprosodes*) *costifera* KR. at 3000–3200 m in the Upper Alai Valley; *Pr.* (*U.*) *costifera neethi* REINIG (1800–3400 m); *Blaps caraboides alaiensis* REINIG, *Bl. c. emarginata* REINIG and *Bl. c. aberrans* REINIG at elevations of 2140–4100 m; *Platyscelis* (*Platyscelis*) *modesta* KR. at 2100–3200 m in the Alai area and at 3000 m in Kara-teghin; *Pl.*(*Pl.*) *constricta* SEIDL. at 1800–3100 m in Kara-teghin, at 2400–3000 m in Darwas and extending also to the lowlands of Buchara; *Pl.* (*Pl.*) *margellanica* KR., a variable species, widely distributed in the Pamirs Region at elevations of 2000–4200 m; the endemic *Trichomyatis* REINIG, with two species ranging from elevations of 2000 to 4100 m; *Myatis variabilis* BOT. (4200 m). Tentyriinae: *Ascelosodis concinnus* BATES, widely distributed on the Pamirs and a character species of the Pamirs fauna; the genus is also restricted to the Pamirs and the Northwest Himalaya; the altitudinal range of the species lies between 3000 and 4500 m; *Gnathosia crypticola* REITT. (1800–3200 m); *Anatolica paphia* REITT. widely distributed at elevations between 2000 and 3300 m; *Scythis alticola* REINIG (2800–3100 m).

TABLE 31

Distribution of Tenebrionidae in the Pamirs Region

Western Valleys	Alai Valley	Pamirs proper
Itagonia gnapterinoides		
I. deplanata		
Prosodes alaiensis		
Pr. costifera	+	
Pr. fallax		
Blaps caraboides	+	
Bl. bucharensis		
Platyscelis margellanica	+	+
Pl. constricta		
Pl. conradtii		
Pl. modesta	+	
Pl. lucidicollis		
Trichomyatis conradtii		
Myatis variabilis		
Trichoplatyscelis pamirensis		
Trigonoscelis affinis		
Stalagmopter sp.		
Ascelosodis concinnus		+
A. marginatus		
Gnathosia crypticola		
Anatolica paphia	+	
Scythis alticola		
Cabirus obtusicollis		
Stenomacidius reinigi		
Lobothorax samgarensis		
L. dilectus		

REINIG (877) has discussed the interesting reduction in the number of species of Tenebrionidae in the Pamirs Region. Of the 27 species found in the Western Valleys, Kara-teghin and Darwas, Alai Valley and the Pamir, 26 occur in the first three areas; only five occur in the Alai Valley and only two species in the Pamirs proper. Table 31 (see page 253) summarizes the data on the distributional peculiarities of the Tenebrionidae from the Pamirs Region.

Chrysomelidae

About half a dozen species of Chrysomelidae are known at present from the Pamirs Region and the highest elevation at which the family has been found so far is 4050 m. *Gynandrophthalma 11-notata* WSE. occurs in moraines in the Trans-Alai, at an elevation of 3400 m. *Leptosonyx* sp. prox. *octocostatus* WSE. occurs at elevations of 3800-3900 m. *Galeruca vittatipennis* BALY occurs at an elevation of 4050 m in Djol-boeruljuk and at 3900 m in Sor-mardansha in the Western Valley.

Curculionidae

Otiorrhynchus (Arammichnus) zumpti LONA and *O. (A.) russicus* STIERL. occur at an elevation of 4200 m, marking the extreme upper limits for the family in the Pamirs Region. The other species that deserve special mention are *Thylacites noxius alaiensis* FAUST at elevations of 2200-3100 m and *Th. n. noxius* FAUST from 3600-4100 m. The maximum altitude record 4400 m for the family in the Pamirs Region is reached by *Phacephorus vilis* FABR. In the moraines of the Trans-Alai Mountains we find *Catapionus gebleri* FAUST and *Stephanocleonus simulans* FAUST (3400 m). *Bathynoderes foveicollis* GEBL. is recorded at an elevation of 4100 m from Djol-boeruljuk.

For descriptions and records of other Coleoptera from the Pamirs Region reference may be made to BAROVSKII (84), KRYZHANOVSKII (659-661) and MEDVEDEV (755).

HYMENOPTERA

Some interesting endemic wasps are known at high elevations from the Pamirs (491). The genus *Tachygetes* HAUPT is, for example, endemic in Kara-teghin. *Tricho-syron* HAUPT is similarly endemic in the Western Valley. *Ammophila (Psammophila) hirsuta* SCOP. occurs at elevations of 4400 m in the glacier valley in the north Pamir and in other localities in the Western Valley and in the South Pamir. *A. (Ps.) caucasica* MOCS. is also widely distributed on the Pamirs, at elevations of 3600-4100 m. Another interesting record is of *A. (Ps.) alpina* KOHL at an elevation of 4100 m on the Alichur Pamir. *Crabro mocsaryi* KOHL occurs at 4100-4200 m in Djol-boeruljuk.

Formicidae

The occurrence of *Formica (Serviformica) picea* NYL. at an elevation of 3850 m in the Western Valleys and at an elevation of 1800 m in Kara-teghin is of considerable interest. In the foregoing chapter, we have already mentioned this species as occurring on the Northwest Himalaya. The highest altitude at which ants occur in the Pamirs Region is 4100 m; we find *F. fusca rufipes* SITZ. at this elevation, in the Kara-kul Pamir. *F. rufibarbis caucasicus* RUZSKY occurs at an elevation of 2800 m in Kara-teghin.

Apidae

A number of remarkable endemic subspecies, races and varieties are described by ALFKEN (26) from the Pamirs Region. At elevations of 4100-4200 m we find *Prosopis plurifasciata* ALFK., *Pr. hungarica* ALFK., *Halictoides pamirensis* MOR., *Nomada flavoguttata* W.K., *Osmia rickmersi* ALFK. in glacier valley east of Kara-kul (2400-4500 m), *Chalicodoma pyrenaicum pamirense* ALFK. at an elevation of 4300 m in the southern Pamirs and *Coelioxys rufescens tricarinatus* MOR. HEDICKE (491) has described three species of *Anthophora* LATR. as peculiar to the region; seven of the species from the Pamirs are also known from Europe and five of these have vicarious forms on the Pamirs. Four species are typical of Middle Asia. The more interesting records include *A. lacteifrons* HEDICKE from the Pamir Darya (South Pamirs, 4300 m), *A. vulpina alticola* HEDICKE at elevations of 3700 m in the Murg-ab Basin, 4050 m in the southern Pamirs and 3600-3900 m in the Western Valleys. *A. parietina pamiricola* HEDICKE is widely distributed in the region at elevations of 3400-4200 m.

Bombidae

The most important contributions on the Bombidae from the Pamirs Region are by REINIG (878) and SKORIKOV (984). With increase in elevation on the Pamirs, there is often a significant fall in the abundance of individuals, which is closely correlated with the diminution in the duration of the warm period and of the hours of bright sunshine, especially at elevations of 3800 and 4000 m. There is also a pronounced scarcity of workers, analogous to the condition found among the Arctic species, in some of which the workers are totally absent. The flight habits of *Bombus* spp. on the Pamirs present certain interesting peculiarities (878). The mature females fly from the end of June to the end of August, the workers appear from the end of July to September, the drones from early August to September and the young females from the middle of August. In each case the flight periods are shorter than in the case of the Alps, where on an average a bumblebee is on the wing from May to the end of October. The flight period on the Alai is also less by about a month. The adults of *Lapidariobombus separandus* (VOGT) are described as going into concealment on a slight snowfall at an elevation of 4000 m on the Pamirs. The most favourable hours for the bumblebees for pollen gathering on the Pamirs appear to be between 10 and 12 in the morning and the intense insolation after midday seems to be unfavourable. The cold wind that starts regularly at about 2-30 in the afternoon on the Pamirs also puts an end to the flight of the bees. The principal pollen plants for the bumblebees on the Pamirs are *Astragalus*, *Pedicularis labellata*, *P. uliginosa*, *P. cheilanthifolia*, *Solenanthus nigricans*, *Dracocephalum heterophyllum*, *Hedysarum cephalotes*, *Saussurea crassifolia*, *Primala sibirica* and *Leucanthemum leucanthemum*.

The following are some of the more important species of bumblebees known so far from the Pamirs Region: *Mendacibombus defector* (SKOR.), also widely distributed on the Trans-Alai; *M. makarjini* (SKOR.), *M. turkestanicus* (SKOR.) also from the Alexander Mountains, the Peter-the-Great Mountains and the Zerafshan Range; *Agrobombus laesus* (MOR.) known also from the Alexander Mountains and the Kungei Ala-tau; *A.* (*Adventitiobombus*) *derhamellus sartus* SKOR. from the Alexander Mountains, Trans-Alai and Peter-the-Great Mountains; *Cullumanobombus serrisquama* (MOR.) also from the Terskii Ala-tau and other mountains of the Tien Shan System; *Sibiricobombus miniatocaudatus* (VOGT) and *S. morawitzi* (RAD.); the typical Caucasus *Lapidariobombus alagesianus* with the subspecies *pamirus* SKOR. from Murg-ab; *L. separandus* (VOGT) also known from the Tien Shan and widely distributed on the Pamirs; *Subterraneobombus melanurus* (LEP.) from the Alai Valley, Trans-Alai and also known from the Alexander Mountains, Dzhungarskii Ala-tau and Peter-the-Great Mountains; *S. difficillimus* (SKOR.) on the Trans-Alai and the Pamir; *S. pamirus* (SKOR.) widely distributed in the Pamirs, especially the Alichur Pamir and the Kara-kul Pamir and in the Alai Valley; *S. fedtshenkoi* (MOR.) also from the Alexander Mountains, Peter-the-Great Mountains, Alai Valley and Kara-teghin; *Pratobombus biroi* (VOGT) widely distributed in the Pamirs and on the Peter-the-Great Range; *Pr. lapponicus relictus* SKOR. at elevations of 4000 m on the Beik Pass and *Pr. leucopygus* (MOR.) from the Alai and the Pamirs.

The following hypsometric analysis of the component faunal elements of the Bombidae from the Pamirs Region is summarized from SKORIKOV (*op. cit.*):

I. Trans-Palaearctic elements
   Agrobombus muscorum up to 2100 m
   Soroensibombus soroensis laetus (300-3600 m)
   Subterraneobombus subterraneus latreillus (560-3500 m)
   Bombus lucorum (90-3000 m)
   Hortobombus hortorum (500-2100 m)

II. Steppes elements
   Agrobombus laesus (650-3000 m)
   Cullumanobombus serrisquama up to 3000 m
   Fervidobombus demoucheli (1200-2100 m)
   Subterraneobombus fragrans (760-1700 m)

III. Mountain autochthonous elements
   Mendacibombus defector (1800-3000 m)
   M. makarjini (2600-3600 m)
   M. margaretieri (2100-2900 m)
   M. turkestanicus (2100-3800 m)
   M. marussinus up to 3350 m
   Sibiricobombus miniatocaudatus (3600-4600 m)
   S. asiaticus (1200-4600 m)
   S. morawitzi (3350-4600 m)
   Lapidariobombus alagesianus pamirus up to 3650 m
   L. anargumentosus (1820-3350 m)
   L. keriensis (1430-3960 m)
   L. separandus (1800-4900 m)
   Subterraneobombus difficillimus up to 3750 m
   S. fedtshenkoi (1370-3870 m)
   S. melanurus (1340-3960 m)
   Pratobombus biroi (790-3700 m)
   Pr. leucopygus (900-3870 m)
   Pr. lapponicus up to 900 m
   Agrobombus derhamellus sartus (2100-3660 m)

It may be observed from the analysis presented here that the mountain autochthonous elements predominate in the Pamirs Region.

LEPIDOPTERA

The Lepidoptera from the Alai-Pamirs were described by ALPHÉRAKY (36, 37, 39), AVINOFF (67), FILIPJEW (345), FORSTER & ROSEN (361), GRUM-GRSHIMAILO (445-447), PAGENSTECHER (809) and STAUDINGER (998, 1000, 1001). The Rhopalocera from the Pamirs generally extend over a relatively wide range of altitudinal zone. *Pontia (Pieris) chloridice* (HÜBN.) occurs, for example, at elevations from 300 to 3350 m. *Lycaena fergana* STGR., *L. zephyrus* FRIV., *Argynnis lathonia* (LINN.), *Melitaea sibina* ALPH. and other Satyridae occur usually at elevations of 600-2200 m. GRUM-GRSHIMAILO found, for example, specimens of *Parnassius delphius* EV., *Colias eogene* FELD., *Lycaena amor* STGR. and *Argynnis pales* SCHIFF. at elevations of about 900 m on the Alai and at about 4880 metres on Kounjout.

FILIPJEW recognizes among the Lepidoptera from the Alai-Pamirs a number of typical boreal elements, Middle Asiatic mountain elements, species widely distributed, species extending to the Northwest Himalaya, species known also from Tibet or closely related to those occurring there, Eurasiatic mountain elements, Mediterranean forms and some Kirgiz-Turanian steppes species. Exceedingly few typically Tibetan elements are met with among the Lepidoptera of the Pamirs Region.

The following are some of the more interesting records of Lepidoptera from the Pamirs Region:

Nymphalidae

Argynnis pales generator STGR. is known from an elevation of 3955 m from the Zor-tash-kol and Kara-kul Pamirs. *A. aglaia vitatha* MOORE occurs at an elevation of 3000 m in Sari-tash.

Papilionidae

Papilio machaon LINN. occurs in Sari-kol and Sari-tash (4300 m); the females from the Alai resemble very closely *P. m. ladakensis* MOORE from the Northwest Himalaya.

Parnassiidae

The Parnassiidae from the Northwest Himalaya and from the Tien Shan Mountain Systems seem to meet in the Pamirs. The typical Middle Asiatic *Parnassius tianschanica* OBERTHÜR and *P. charltonius* GRAY are widely distributed in the Pamirs. *P. tianschanica superba* GR.-GR. occurs in Roshan and *P. t. romanovi* GR.-GR. on the Trans-Alai to Kitshik-kara-muk. On the southern Alai, Issyk-kul and Kuruk-tag we find *P. t. insignis* STGR., *P. t. minor* STGR. (on the Ala-tau, Tarbagatai and Ferghana mountains), *P. t. olympius* STGR. on the Tien Shan and the eastern subspecies *P. t. erebus* VERITY from Lob-nor and the Boro-choro Mountains. *P. apollonius* EV. is distributed on the Zerafshan, East Kara-teghin, Trans-Alai, north Ferghana, Alexander Mountains and the Tien Shan. *P. actius* EV. occurs from the southern Altai Mountains to the eastern Tien Shan, the north Alai, the Trans-alai, along the eastern margin of the Pamir to Kisil-jur, Muztagh-ata (represented in the last two areas by the subspecies *brutus* BRYK and *pseudobrutus* BRYK). *P. charltonius* GRAY is absent on the Tien Shan-Semirjetsh Region, but its range superimposes on that of the two other species in the Kara-teghin-Trans-Alai area. Its principal area is the Pamirs, the Northwest Himalaya and the East Himalaya. *P. ch. vaporossus* AVINOFF occurs in Darwas. *P. ch. romonowi* GR.-GR. on the Trans-Alai and the Pamirs. *P. ch. ducalis* BOUL. & LE CERF is known from the Chitral Mountains and *P. ch. deckerti* VERITY from the Northwest Himalaya. *P. ch. bryki* HAUDE occurs in the Central Himalaya. The southeast form *P. jacquemontii* BOISD. occurs in southwest China (subspecies *tibetanus* LEECH), Himalaya (subspecies *himalayensis* ELWES), Northwest Himalaya (subspecies *hunzaica* TYTL.) to the Pamirs, Trans-Alai and the Alai (subspecies *pamira* BANG-HAAS, *rubicunda* STICHEL). *P. delphius* EV. is distributed from the Northwest Himalaya, the Hindu Kush, south Pamirs, Trans-Alai, Peter-the-Great Range, Darwas, Hissar Mountains, Alai, Ala-tau, Issyk-kul area and the Tien Shan up to Juldus. The distribution of *P. simo* GRAY is also similar; it occurs on the Karakoram and Ladak Ranges (subspecies *avinoffi* VERITY). It then goes round the Pamirs to the east, with the subspecies *reinigi* BRYK from the Taghdum-bash Pamir and the subspecies *simplicatus* STICHEL from Lob-nor, Trans-Alai, Alai, Alexander Mountains and the Tien Shan.

Satyridae

Erebia mani DE NICEVIL. is perhaps the only species that is definitely known from the Pamirs (the Alichur Pamir) and as already mentioned in the foregoing chapter, it is distributed in Ladak, northwestward on the Trans-Alai and Kokan up to the Tien Shan. *E. radiansusgentensis* RÜHL. occurs at an elevation of 3200 m at the base of the Ulu-kan Mountain. Other species, *E. meta* STGR., *E. macracandica* ERSCH., *E. hades* STGR., *E. ocnus* EV., *E. radicans* STGR. and *E. nero* STGR. are restricted to the region of the Alai, Trans-Alai, Hissar Mountains and Kara-teghin and extend eastward to the Tien Shan and to the Altai Mountains. The genus *Satyrus* FABR. also does not penetrate deep in the Pamirs, but a number of species are known from the Alai, Kara-teghin and Darwas. *S. staudingeri* BANG-HAAS is reported from the base of the Ulu-kan Mountains. *S. mniszechii sagina* RÜHL. (3000 m); *S. regeli tanceri* GR.-GR. from Kok-dzhar (3300 m); *S. cordula cordulina* STGR.; *S. hübneri leechi* GR.-GR. from Kara-kul, Mus-kul, Alai, Trans-Alai and Alichur Pamir; *S. boloricus* GR.-GR. from Kandshut and *S. alpherakyi* AVINOFF from the Pamiriskii Post are the other records. *S. hübneri leechi* GR.-GR. is perhaps the most widely distributed species on the Pamirs; it occurs on the Northwest Himalaya, from where it extends to the mountains of Afghanistan (*S. h. vogti* BANG-HAAS), Trans-Alai (*S. h. intermedia* GR.-GR.), Alai (*S. h. pamira* STGR.), Buchara (*S. h. decorata* STGR.-REBEL.), Tien Shan (*S. h. talastuana* BANG-HAAS), and Alma Ata (*S. h. dublitzkyi* BANG-HAAS). *Pararge eversmanni* STGR. and *Coenonympha alexandra* RÜHL. are known from the base of the Ulu-kan Mountains.

Pieridae

Baltia shawii baitalensis MOORE occurs at an elevation of 4500 m on the Fedtshenko Glacier. The genus *Colias* FABR. is represented by five species viz. *C. cocandica* ERSCH., *C. eogene* FELD., *C. erate* ESP., *C. wiskotti* STGR., and *C. macropolo* GR.-GR., from the upper and lower meadow zones, at elevations above 4600 m, but is absent or rare on the true high-steppes. The genus is further distributed on the Tien Shan, where it is represented by *C. wiskotti* STGR. that occurs on the Alexander Mountains also and up to the Issyk-kul area. *C. eogene* FELD. occurs, as already pointed out, on the Northwest Himalaya, south Amdo, Lob-nor, Kuku-nor, Sikkim, but is absent on the Tien Shan. *C. cocandica* ERSCH. extends over Ladak to the Pamirs, Trans-Alai, Tien Shan, Altai Mountains and northwest Mongolia. *C. macropolo* GR.-GR. is restricted to the whole Pamirs and the southeast Alichur area. *C. erate* ESP. is distributed in Ak-baital, Trans-Alai, Ak-su, Murg-ab, Alichur Pamir, Kandshut, etc. and also widely over the whole of the Palaearctic Asia. Other species like *C. christophi* GR.-GR., *C. thisar* MÉN., *C. staudingeri* ALPH., etc. from the Trans-Alai, Alai Valley, Kara-teghin, Alai and the Hissar Mountains, also extend eastward to the Tien Shan System. *C. hyale* LINN., *C. sieversi* GR.-GR., *C. romanovi* GR.-GR., *C. regia* GR.-GR. and *C. alpherakyi* STGR., all widely distributed, are, however, restricted in the Pamirs to the north and northeast mountain chains. At relatively lower elevations of 1500-3000 m may be found *Pieris brassicae ottonis* RÖB., *P. canidia palaearctica* STGR., *P. rapae debilis* ALPH., *Pontia callidice orientalis* ALPH. and *P. daplidice* LINN.

Lycaenidae

Lycaena eros erygone GR.-GR. occurs in the Murg-ab area, Tshat-shakty (4145 m) and Bash-Murg-ab (3730 m) and is also known from the mountains of southern Europe up to the Alps, the mountains of Middle and Near Asia, Boro-choro Mountains and on the Northwest Himalaya. *L. pheretes* HÜBN. is a typical boreo-alpine species that extends from the Pyrenees to the Amur region. *L. stoliczkana* FELD. is represented by the subspecies *L. st. hunza* GR.-GR. on the southeast Pamir and in the Great Pamir, at elevations of 3870-4725 m. *L. venus* STGR. from the base of the Ulu-kan Mountain and *L. icarus icadius* GR.-GR. are the important records from the Pamirs.

Heterocera

A number of Heterocera have also been recorded at high elevations from the Pamirs Region. *Macroglossa stellatarum* LINN., *Agrotis squalida* GN., *Euxoa conspicua* HÜBN., *Heliothis peltigera* SCH. and *Syngrapha circumflexa* LINN. occur at elevations of about 4500 m on the Fedtshenko Glacier. *Dipsosphecia lactea* FILIPJEW and *Staudingeria pamira* FILIPJEW are reported at an elevation of 3955 m on the northern shore of the Kara-kul, *Evergestis obsoletalis* FILIPJEW and *Alucita decipiens* LD. from Kok-dzhar (3300 m) and the interesting endemic *Reichardtiella grisea* FILIPJEW from the Alai Valley and Sari-kol at an elevation of 4300 m.

DIPTERA

The Diptera from the Pamirs Region are remarkable for the endemic genera *Reinigellum* END. (Thereviidae), *Pamirella* END. (Sarcophagidae) and *Reinigia* END. (Tachinidae), all of which occur at elevations of 4000-4400 m. The greatest bulk of the Diptera so far reported belong to the Brachycera and include the following:

Tabanidae: *Sziladynus pamirensis* END. (3400 m), *Tylostypia reinigiana* END. Shor-kul (3700 m) and Uruss-tugai (3400 m).

Asilidae: *Stenopogon flavibarbis* END. (3800 m).

Bombyliidae: *Anthrax paniscus* ROSSI (3400-3700 m), *A. maurus* LINN. (4100 m), *Argyromoeba tripunctata* WIEDEM. (4100-4200 m), *Bombylius cinerascens* MIKAN (4100-4200 m), *Systoechus nitidulus* FABR. (4100 m) and *S. niveicollis* END. at 4400 m from the Kara-kul Pamir.

Thereviidae: *Reinigellum speculiferum* END. (4100-4400 m).

Syrphidae: *Volucella bombylans* LINN. (3400 m), *Helophilus continuus* LÖW (3900 m), *Catambya pyrastri* LINN. (3400-3700 m) and *Platycheirus nigripes* END. (4100-4200 m).

Scatophagidae: *Scatophaga ampulipennis* PORTSCH. (4200 m).

Sepsidae: *Sepsis pamirensis* END. (4100-4200 m).

Ortalidae: *Melieria sareptae* PECK. and *Timia desparsata* END. at 4400 m from the Kara-kul Pamir.

Muscidae: *Chortophila turkestanica* END. and *Linnaemya nigrifacies* END. (4050-4150 m).

Sarcophagidae: *Pamirella karakulensis* END. (4400 m).

Tachinidae: *Servillia persica* PORTSCH. (3580 m), *Reinigia pamirica* END. (4050-4150 m), *Germaria violaceiventris* END. (3700 m) from the Shor-kul, *Hystriomyia fatisowi* PORTSCH. (4500 m) and *Ocyptera reinigi* END. (3900-4150 m).

6. THE TIEN SHAN MOUNTAINS

Tien Shan (= celestial mountains) is an extensive mountain system of Asia, stretching from the Aral-Caspian Depression (67° EL) to the bend of the R. Hwang-ho (103° EL). It is often usual to speak of only the part between the Khan Tengri Peak (80° 11′ EL and 42° 13′ NL) and the Barkul Depression (92-93° EL), where the northern ranges abut on the Ek-tagh Altai, as the Tien Shan. The principal divisions of the Tien Shan Ranges are: Khan Tengri and the Central Tien Shan. The Central Tien Shan is situated between 40° 19′ and 42° 25′ NL and 73° 20′ and 80° 16′ EL. The high Khan Tengri Peak (6940 m, but according to the Soviet authorities 6695 m above mean sea-level) is situated on a spur from the main crestline of the Central Tien Shan toward the southwest (fig. 61). The loftiest peak of the main crest is Mt. Nichols Mikalovich (6180 m). East of the Khan Tengri Peak is Khalyk-tau. To the west are three diverging lines of elevations, viz. the Terskii Ala-tau, Kokzhal-tau (continued in the Terez Mountains) and the Ranges of Sary-jas, Kulu-tau and Ak-shiry-col (with the peak Pabiedi, 7439 m). The snowy chain of the Khalyk-tau Mountains is the highest in the west.

From the Issyk-kul area there is a general rise in the mean elevation of some 1280 m to 2740 m on the snow-covered ridge of the Terskii Ala-tau, the peaks of which rise to elevations of 4575-4880 m or even 5180 m. Here the mountain system is about 240 km wide. It is continued westward as the Son-kul (2895 m), the Kara-kul and the Suza-mir-tau to where it abuts on the Talaskii Ala-tau. The ridges rise from 3950 m to 4880 m and are flat-topped and snow-covered. Being uplifted from a peneplain and with little or no erosion since the Tertiary, though highly mountainous, the general appearance is, however, that of a high plateau.

The eastern and northern Tien Shan or the Bar-kul Mountains (known also as the Karlyk-tagh) are snow-capped in the middle part (the snow reaching down to an elevation of 3110 m on the north) and rise to elevations of 3960-4480 m. A stony desert slopes south to the Chol-tagh, which marks the northern escarpment of the Gobi Desert. The western and the southern Tien Shan on the north side of the Issyk-kul are parallel twin ranges of the Trans-Ili Ala-tau and the Kunghei Ala-tau,

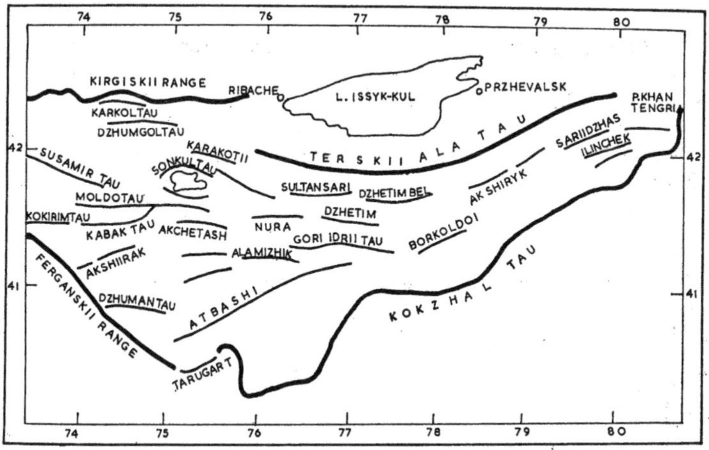

Fig. 61. Sketch-map of the area of the Central Tien Shan, showing the principal mountain ranges.

2440 m above the level of the Lake Issyk-kul and about 3960 m above mean sea-level. The Alexander Mountains terminate over against the town of Auli-Ata (71° 20′ EL) at an elevation of 750 m. Further east, they rise to elevations of 3960-4270 m and even 4680 m in Mt. Semenov. The north slopes are steep and rugged. Near the western end of the Alexander Mountains, (about 71° EL) the Kara-tau stretches northwest between Syr Darya and the Chu rivers. Stretching in the same north-westerly to southeasterly direction are the Ferghana Mountains, with a mean elevation of 3050 m, but rising to 3885 m in Suyuj. South of the Ferghana is the lofty Alai, backed by the parallel range of the Trans-Alai; both the Alai and the Trans-Alai abut on the Pamirs. The Alai is a well defined mountain range, with steep slopes, and with a mean elevation of 4880 to 5790 m. The Trans-Alai is a true border range, with the Kaufmann Peak rising to an elevation of 7128 m. The ascent of the Trans-Alai is gentle from the Pamir side on the southeast. The Alai mountains are continued westward in the radiating ranges of Kara-teghin, Zerafshan and the Hissar Ranges and the Turkestan Range of a mean elevation of 5486 m. The Trans-Alai Ranges are continued into the Peter-the-Great Range, ending in the Sandal group, at an elevation of 7500 m. For a fuller account of the general tectonics of the Tien Shan System reference may be made to SCHULTZ (940).

7. GENERAL ECOLOGICAL AND VEGETATIONAL CHARACTERS OF THE TIEN SHAN

Detailed accounts of the general physiography, ecology and the peculiarities of the vegetation of the Tien Shan mountains may be found in GOLOVKOVA (417-419). There are striking differences in the general environmental conditions and the ecological characters of the biota of the Pamirs and of the Tien Shan. The general climatic conditions at different elevations on the Central Tien Shan may be summarized as below.

TABLE 32

Atmospheric pressure and temperature on the Central Tien Shan

Station	Elevation in m	Mean atm. press. mm	Atmospheric temp. ° C Max.	Min.	Range
Kara-kudzhur	2800	267	32.7	−37.9	70.6
Central Tien Shan Physico-geographical Station	3672	303	19.5	−42.9	62.4
Chatir-kul	3619	193	20.8	−39.8	60.6
Suza-mir	2091	365	30.5	−45.6	76.1
Kuil-yu	2800	251	26.1	−38.6	64.7
Ak-shiiryuk	2780	174	28.8	−34.4	63.2
Kara-kul	3100	182	24.8	−38.7	63.5
Ak-sai	3135	246	22.0	−46.4	68.4
Arpa	3000	263	23.4	−47.4	70.8

The difference between the mean summer and winter temperatures is thus 76.1° C. The soil temperatures on the Central Tien Shan are summarized in table 33.

TABLE 33

Soil temperatures on the Central Tien Shan

Depth of soil in cm	Mean Max.	Mean Min.	Range of fluctuation
0	43.0	−2.5	45.2
25	21.5	1.5	20.0
50	19.8	4.6	15.2
100	18.8	5.8	13.0
150	17.0	7.4	9.6

In GOLOVKOVA (419) we find a beautiful review of the earlier biological explorations, especially the geobotanical investigations in the Tien Shan region by pioneer naturalists like E. P. KOROVIN, M. G. POPOV, B. A. BIKOV, N. I. RUBSHOV, K. B. STANYUKOV, A. N. KRASHNOVA, B. B. SAPOZHNIKOVA, M. M. SOVIETKINOI, E. B. NIKITINOI, B. I. VANDISHEVOI, D. N. KASHKAROV, M. M. KATOC, L. I. POPOV and others. Most of these explorations were undertaken in the region between 40° 19′ and 42° 25′ NL and 73° 20′ and 80° 16′ EL. Thanks to the labours of these Soviet biologists, our knowledge of the general ecology, vegetation and phytogeography of the Central Tien Shan may be described as far more advanced than our knowledge of the insect life of the Tien Shan. KOROVIN and KOROTOV, who have recognized 16 principal vegetation types from Middle Asia, include the following in the Central Tien Shan: i. xerophytes, ii. halophytes, ii. steppes, iv. steppe-savanna and subtropical steppes, v. montane xerophytes, vi. sub-alpine meadows, vii. taiga, vii. alpine meadow, ix. mountain tundra and x. cryophytes. GOLOVKOVA describes the subalpine communities of the Central Tien Shan under i. *Phlomis oreophila* formation, ii. *Geranium saxatile* formation, iii. *Dracocephalum diversifolium* formation, iv. *Iris brevituba* formation, v. *Festuca tianschanica* formation, and vi. *Polygonum nitens* formation. The alpine meadow communities of the Central Tien Shan comprise the following formations: *Alchemilla retropilosa* formation, *Ligularia alpina* formation, *Allium astrosanguineum* formation, *Poa alpina* formation, *Polygonum viviparum* formation, *Leontopodium ochroleucum* formation, *Taraxacum pseudoalpinum* formation, *Carex stenocarpa* formation, *Carex orbicularis* formation, *Carex griffithi* formation and *Allium semenovii* formation.

According to KOROVIN (649-652), the cryophyte vegetation of the Alai-Pamirs and of the Tien Shan, with special reference to the drainage area of the Syr Darya river system (at elevations of 3500-3600 m), the eastern Pamirs (3900-4000 m) and the Zailiiskii Ala-tau (3200-4000 m) forms a fundamental part of the general ecology of these elevated regions. The minimum temperature on the Tien Shan is — 17.8° C, but on the Pamirs —22.8° C. The maximum ground temperature is 30.7° C on the Pamirs, but only 19.5° C on the Tien Shan during August. The characteristic external appearance of the cryophyte vegetation is typified by that of *Marotomia euchroma*, *Oxytropis tianschanica*, etc., that grow at elevations of 4000 m. On the eastern Pamirs, at an elevation of about 4900 m, there are typical associations of *Oxytropis poncinsii* FRANCH., *Potentilla pamiroalaica* JUZ., etc. The following are the principal cryophyte-zone plants of the Pamirs and Tien Shan: *Acantholimon hedini* OSTENF., Pamir-Alai and Central Tien Shan (4400 m); *Potentilla pamiroalaica* JUZ. on the Pamir-Alai (4400 m) and Central Tien Shan; *Gypsophila capituliflora* RUP. at 4400 m on the Pamir-Alai and Central Tien Shan; *Potentilla pamirica* TH. WOLFF on the Pamirs; *Oxytropis immersa* (BAKES) BGE. on the Pamir-Alai and Central Tien Shan; *Oxytropis chionobia* BGE. on the Tien Shan (3600 m); *Smelovskia calycina* on the Pamirs, Tien Shan and the Altai Mountains; *Saussurea sorocephala* SCHRENK on the Alai-Pamirs and Central Tien Shan (5000 m); *Astragalus myriophyllus* BGE. on the Alai-Pamirs; *Draba korshinskyi* (O. FEDTS.) on the Alai-Pamirs; *Braya pamirica* on the Pamirs (5000 m), *Erysimum humillimum* (LDB.) N. BUSCH. on the Alai-Pamirs and Tien Shan; *Astragalus kuschakevitschii* B. FEDTS. on the Alai-Pamirs and Tien Shan and *Hedysarum cephalotes* FRANCH. on the Alai-Pamirs and Tien Shan.

In the glacier-zone (4700-5500 m) vegetation of the Pamirs Region, KOROVIN reports the following species, at an elevation of 4800 m, on Pan-chaz-Dzhilga: *Smelovskia calycina*, *Draba korshinskyi* (O. FEDTS.), *Saussurea sorocephala* SCHRENK, *S. pamirica* C. WINKL., *Oxytropis poncinsii* FRANCH., *Poa litwinowiana* OVCZ., *Tanacetum xylorrhizum* KARSCH and *Roegneria schugnanica* NEVSK. The species found at an elevation of 4600-4800 m on the Sarez Lake are *Saussurea gnaphaloides* SCH., *Nepeta longibracteata* BENTH., *Waldheimia tridactylites* KAR. & KIR. and *Rhodiola gelida* SCHRENK. The glacier-zone species at an elevation of 4000 m on the Baidulla-tau in the Central Tien Shan include *Ranunculus alberti* REG. & SCHMALH., *Callianthemum alatavicum* FREYN, *Oxygraphis glacialis* (FISCH.) BGE., *Hymenolaena pimpenellifolia* RUP., *Sedum quadrifidum*, *Smelovskia calycina*, *Oxyria digyna* (LINN.), *Saussurea sorocephala* SCHRENK, *Draba alberti* REGL. & SCHMALH. and *Waldheimia korolkovii* REGL. & SCHMALH. On the Zailiiskii Ala-tau, we find at an elevation of 3600-3900 m *Waldheimia tridactylites* KAR. & KIR., *Psychrogeton olgae* NOVOP., *Richteria leontopodium* C. WINKL., *Melandrium apetalum* LINN., *Thylacopsermum caespitosum* (CAMB.) SCHISCHK., *Pastinacopsis glacialis*

GOLOSK., *Saxifraga setigera* PURSCH., *S. cernua* LINN., *Dryadanthé tetrandra* (BGE.), *Saussurea sorocephala* SCHRENK, *Draba oreodes* SCHRENK and *Cerastium cerastoides* BRITT. On the Dzhungarskii Ala-tau, the glacier-zone plants at an elevation 3600-3700 m are *Carex nivalis* LITW., *Oxygraphis glacialis* (FISCH.) BGE., *Saxifraga oppositifolia* LINN., *S. flagellaris* WILLD., *Chorispora excapa* BGE., *Thylacospermum caespitosum* (CAMB.) SCHISCHK., *Drydanthe tetrandra* (BGE.) and *Melandrium apetalum* LINN.

There is a typical subalpine zone on the Zerafshan and the Ferghana Ranges, at elevations of 2600-2700 m, on the Zailiiskii Ala-tau at 2500-2600 m, and on the Dzhungarskii Ala-tau at 2400-2500 m. In the eastern parts of the Alai Valley, the subalpine zone is met with at elevations of 3100-3200 m. The absolute maximum temperature of the soil, at an elevation of 2200 m, on the Dzhungarskii Ala-tau, during June is 48° C and the corresponding minimum is -5.0° C. The mean diurnal fluctuation in the soil temperature is 42.5° C. The atmospheric temperature maximum is 30° C and the minimum —3° C at a height of 45 cm above the ground. The characteristic vegetation of the subalpine zone on these mountains includes *Phlomis oreophila* KAR. & KIR. and two other species, *Scabiosa alpestris* KAR. & KIR., *Pachypleurum mucronatum* (SCHRENK) SCHISCHK., *Aconitum rotundifolium* KAR. & KIR., *Delphinium confusum* M. POP., *Ranunculus grandifolius* C.A.M., *Semenovia transiliensis* REGL. & HERD., *Gentiana tianschanica* RUP., *Allium polyphyllum* KAR. & KIR., *Corydalis gortschakovii* SCHRENK, *Artemisia ashurhaevii* C. WINKL., *A. frigida* WILLD., *A. lehmanniana* REGL., *Poa relaxa* OVCZ., *Festuca rubra* LINN., *Carex* spp., *Agropyrum tianschanicum* DRAB., *Hedysarum flavescens* REGL. & SCHMALH., etc.

The typical plants of the alpine meadows, at elevations of 3500-3800 m and upwards, on the Tien Shan are as follows: *Poa alpina* LINN., *Phleum alpinum*, *Festuca* spp., but particularly *F. tianschanica* ROSHEV, *Poa litwinowiana* OVCZ., *Carex* spp., but especially *C. stenocarpa* TURCZ., *Inula rhizocephala* SCHRENK, *Schultzia crinita* (PALL.), *Astragalus alaicus* FREYN., *A. tibetanus* BENTH., *Saussurea soroscephala* SCHRENK, *Cerastium trigynum* VILL., *Polygonum biaristatum* AITCH. & HEMS., *Crepis multicaulis* LDB., *Oxytropis savellanica* BGE., *Potentilla flabellata* REGL. & SCHMALH., *P. gelida* C.A.M., *Erigeron azureus* REGL., *Allium monadelphum* LESS. and *Ranunculus songoricus* SCHRENK.

8. INSECT LIFE OF THE TIEN SHAN

The mountain autochthonous Coleoptera from the Tien Shan System include a number of common hypsobiont genera like *Carabus* LINN., *Nebria* LATR., *Trechus* CLAIRV., *Bembidion* LATR., *Pterostichus* BON., *Laena* LATR., *Otiorrhynchus* GERM., etc. SEMENOV-TIANSHANSKII (956) has recorded a number of interesting species like *Carabus* (*Ophiocarabus*) *politus* MOR. at elevations of 2440-2740 m, *C.* (*O.*) *aeneolus* MOR. at an elevation of 3050 m on the Alma-tinka Pass, *C.* (*Cratocechenus*) *elisabethae* SEM. at elevations of 1700-2200 m from the zone of the relict plant *Picea schrenkiana* on the Ala-tau. In a recent contribution, KRYZHANOVSKII (659) has reported a number of other interesting species, including *Carabus* (*Semnocarabus*) *transiliensis* SEM. at elevations of 1800-3800 m on the Zailiiskii Ala-tau; *C.* (*Anthracocarabus*) *erosus* MOTSCH. from the Kirghiz area of the eastern Tien Shan (3000 m) and from the Dzhungarskii Ala-tau; *C.* (*A*) *erosus erosus* MOTSCH. from the Terskii Ala-tau, Issyk-kul area (2000-3000 m); *C.* (*A.*) *e. korolkovi* SEM. from the Zailiiskii Ala-tau (2200- 2500 m); *C.* (*A.*) *cicatricosulus* A. MOR. from Kuld-kuruk-tag, Bar-kul, Terskii Ala-tau, etc.; *C.* (*Ophio-*

carabus) aeneolus MOR. from the north slope of the Zailiiskii Ala-tau (1800-2700 m); *C.* (*O.*) *striatus* GEB. from the Tien Shan, Kungei Ala-tau, Lake Son-kul, Zailiiskii Ala-tau, etc.; *C.* (*Cryptocarabus*) *kirghisorum* KRYZH. from the Central Tien Shan; *C.* (*C.*) *subparallelus* BALL. from the Zailiiskii Ala-tau (1700 m); *C.* (*Carabus*) *granulatus* LINN., widely distributed at elevations of 2500-2800 m on the Tien Shan System and in the north Kazakstan, in North China, north Asia, Korea, north Japan and Europe; *C.* (*Cyclocarabus*) *pullus* SEM. & ZN. at elevations of 2500-2700 m in Kimbel; *C.* (*Leptoplesius*) *gracilipennis* BREUN. at an elevation of 4000 m on Sintszin; *C.* (*Cryptocarabus*) *kaufmanni* SOLS. on the Tien Shan, the north Alai-Pamir, Talaskii Ala-tau; *C.* (*Cr.*) *grigoryi* KRYZH. at elevations of 2500-2700 m on the Talaskii Ala-tau; *C.* (*Pantophyrtus*) *turcomanorum* THIEME on the northeast Tien Shan, Alai-Pamir, Kazhan-tau, Talaskii Ala-tau and Padzha-tau (Arkit) (1800-2000 m); *C.* (*Cratocechenus*) *elisabethae* SEM. from the Zailiiskii Ala-tau (1600-2500 m); *C.* (*Cr.*) *chan* BREUN. from the Terskii Ala-tau (2000 m); *C.* (*Cr.*) *balassogloi* DOHRN from the Issyk-kul area, Terskii Ala-tau, Katmen Mountains, Khan Tengri (1500-2200 m); *Carabus* (*Cratocarabus*) *jakobsoni* SEM. at elevations of 1800-2350 m from the north slope of the Zailiiskii Ala-tau; *C.* (*Cr.*) *redikotzevi* SEM. & BREUN. at 2170 m on the Kirghiz Ala-tau; *C.* (*Alipaster*) *infantulus* MOR. (2000 m); and *C.* (*Goniocarabus*) *banghaasi* REITT. at an elevation of 3140 m on the Gun-dari Mountains.

The Tien Shan Lepidoptera, with discontinuous distribution on the Altai Mountains and often also on the Ural Mountains, the Carpathians, Alps, Apennines and the Pyrenees, include *Pieris bryoniae* O., *Anthocharis euphenoides* STGR., *Colias phicomene* ESP., *Erebia eriphyle* FRR.*, *E. manto* ESP., *E. claudina* BKH., *E. pharte* HÜBN., *E. melampus* FUESSL., *E. sudetica* STGR., *E. triarius* PRUN., *E. alberganus* PRUN., *E. pluto* PRUN., *E. gorge* HÜBN., *E. mnestra* (HÜBN.), *E. ottomana* H.-SCHÄF., *E. cassioides* R. & HOCHW., *E. pronoe* ESP., *E. styx* FRR., *E. stirius* GOD., *E. montanus* PRUN., *E. oeme* HÜBN., *E. meolans* PRUN., *Oeneis aello* HÜBN., *Ceononympha satyrion* ESP., *Ephydrysichnea wolfensbergeri* FREY, *E. cynthia* HÜBN., *E. glaciegenita* VERITY, *Melitaea varia* H.-SCHÄF., *M. asteria* FRR., *Boloria napae* HFFMGG., *Maculina rebeli* HIRSCHKE, *Polyommatus eros* O., *Pyrgus calcaliae* RBR., *P. carlina* RBR. and *P. alpinus alpinus* (ERSCH.).

A number of the Lepidoptera occurring on the Tien Shan mountains are also found to extend to the Altai, the Caucasus and parts of the Northwest Himalaya. This is, for example, the case with *Parnassius apollo* LINN., *P. mnemosyne* LINN., *Colias chrysotheme* ESP., *Satyrus ferula* FABR., *Melitaea trivia* SCHIFF., *Clossiana titania* HÜBN., *Cupido serbus* HÜBN. and *Pyrgus carthami* HÜBN.

Colias cocandica ERSCH. is common in the area of the Lake Issyk-kul in the Central Tien Shan. *Boloria pales generator* STGR. occurs on the Ala-tau, the Pamirs and the Hindu Kush. *B. p. korla* FRUHST. on the Tien Shan and *B. p. juldissica* WGN. on the Tien Shan and the Hindu Kush are other interesting forms. *Pieris callidice* ESP. occurs on the Ala-tau, the Tien Shan and the Zerafshan Range and on the Pamirs. *Plusia hochenwarthi* HOCHW. occurs on the Ala-tau, the Alai and the Tien Shan.

The ecology and biogeography of the torrential biota of the Tien Shan System present a number of interesting problems. In his valuable contribution on the torrential fauna of the Middle Asiatic mountains, BRODSKII (160) has given an account of

* There is at present certain amount of confusion about the synonymy in the genus *Erebia* DALM. We have, however, followed the general practice in the use of the specific names and have left the problem of synonyms to the specialists.

the Ephemerida, Plecoptera, Coleoptera, Trichoptera and Diptera from the Issyk river system of the Central Tien Shan mountains.

The Ephemerids *Rhitrogena tianschanica* BRODSKII, *Iron montanus* BRODSKII, *I. rheophilus* BRODSKII, *Ecdyurus* sp., *Ephemerella submontana* BRODSKII, *Baëtis issyksuvensis* BRODSKII and *Ameletus alexandrae* BRODSKII occur in the torrential feeder streams of the R. Issyk. The Plecoptera found by him include *Chloroperla capnoptera* MCLACH., *Chl. grammatica* SCOP., *Chl. avicularis* MORT. and *Amphinemura standfussi* RIS. He collected the adults of the Trichoptera *Dinarthrum reductum* MART., *D. pugnax* MART., *Agapetus tridens* MCLACH., *A. kirgisorum* MART., *Oligoplectrodes potamini excisa* MART., *Apatelia arctica* BOH. and *Dolophilodes ornata* MART. The larvae and pupae of *Rhyacophila extensa* MART., *Rh. gigantea* MART., *Apatelia copiosa* MCLACH., *Dolophilodes tinode* ULMER and *Brachycentrus montanus* KLAP were also found in the same area. The occurrence of the Coleoptera *Agabus femoralis* PAYK., *Ochthebius exsculptus* GERM. and a new variety of *Gyrinus colymbus* ER. is of considerable ecological and biogeographical importance. The torrential Diptera from the Tien Shan (Trans-Ili Ala-tau) include the remarkable *Deuterophlebia mirabilis* EDW., *Blepharocera fasciata* WESTW., *Bl. asiatica* BRODSKII and the endemic *Tianschanella monstrosa* BRODSKII. The occurrence of the Simuliid *Nevermannia* ENDER. is also of considerable interest.

BRODSKII has described the micro-environmental conditions under which some of these species occur in the Issyk river system. *Rithrogena* sp. occurs usually, for example, at elevations of about 1780-2230 m, in water at temperatures of 7.5-6.0° C to 11.0-7.0° C on the Tien Shan. *Iron* spp. occur at elevations between 1320 and 2230 m, in water with the temperature ranging from 11° C to 7° C. *Ecdyonurus* spp. are recorded by BRODSKII as common at elevations of 1760-2130 m on the Tien Shan, in water with temperatures of 9-7° C. *Ephemerella* spp. were collected by him at elevations of 1330-1560 m in water having a temperature range of 11-10° C. The great majority of the immature forms of *Baëtis* spp. was found on submerged stones, at elevations between 1330 and 2230 m, in water having a temperature range of 11.5° C to 5.0° C. *Ameletus* EATON occurs at elevations of 1720-2230 m. The altitudes and the water temperatures for Plecoptera on the Tien Shan show that *Chloroperla* NEWM. occurs at 1982 m in water with a temperature of 5.7° C. The Nemuridae were found at an elevation of 1800 m in water having a temperature of 7.5° C. The larvae of *Dinarthrum* MCLACH. feed in water with temperatures of 11-10° C at an elevation of 1550 m, but one specimen was also taken at an elevation of 1737 m, from water at a temperature of 8° C. *Agapetus* CURT. occurs at an elevation of 1550 m, in water at 11° C. *Dolophilodes* ULMER larvae occur in association with those of *Agapetus* CURT. and *Dinarthrum* MCLACH. at elevations of about 1750 m, in water having temperatures of about 8-9° C. *Rhyacophila extensa* MART. and *Rh. gigantea* MART. were collected at an elevation of 1805 m in water at 7.5° C. The former species descends to elevations of 1320 m, where the water temperature is 11-12° C. *Rh. gigantea* is typical of elevations between 1780 and 2210 m. BRODSKII found the maximum abundance of the larvae of *Deuterophlebia mirabilis* EDW. in water at a temperature of 5.7° and the pupae in water at 7.5° C and the range in altitude is 1000 to 4000 m, but in the R. Issyk it occurs usually at elevations of 1338-2230 m. Blepharoceridae like *Liponeura* LÖW, *Blepharocera* MACQ. and *Hapalothrix* LÖW. usually occur in association with *Deuterophlebia* EDW. up to an elevation of 4000 m on the Kungei Ala-tau. *Liponeura cinerascens* LW. occurs in water at 5-9° C and *Blepharocera* MACQ. occurs even at a temperature of 11° C.

9. BIOGEOGRAPHICAL CHARACTERS OF THE PAMIRS TIEN SHAN COMPLEX

Biogeographically the Pamirs, the Tien Shan and the Northwest Himalaya are closely related. The biogeographical peculiarities of this vast elevated region are correlated with the ecological and historical characters of the area.

The insect life of the Pamirs Region is composed essentially of the following elements: i. steppecole species of the Mediterranean and Turkmenian Regions, ii. montane species, iii. boreal and boreo-alpine species and iv. endemic species (878).

i. Steppecole elements

The steppecole elements occupy a prominent place in the Pamirs fauna. As examples of typical steppecole species we may mention the Coleoptera *Platyscelis* LATR., *Blaps* FABR., *Tentyria* LATR., *Anatolica* ESCHS. and *Scythis* SCHAUM (up to the shrub-zone on the Pamirs, but higher in the Alai Valley). *Platyscelis margellanica* KR. occurs, for example, at an elevation of 4200 m on the Char-gush Pamir and is also distributed in the Turkestan plain. *Chalicodoma pyrenaicum pamirensis* ALFK. occurs at an elevation of 4300 m on the Naisa-tash Pass, *Buthus kaznakovi* BIR. (Scorpionida) at elevations of 3800 m in the Pamir Darya Valley and *Locusta migratoria danica* LINN. at an elevation of 3000 m at Kara-kurgan.

ii. Montane elements

The montane elements, an important constituent of the fauna of the high-steppes of the Pamirs, are characteristic of the large areas of the Middle Asiatic mountains and must be considered as strictly alpine-meadow forms. Typical montane elements occur even on the high-steppes that are not in any sense montane biotopes. The montane elements are characteristically restricted to the inner Asia and the hygrophile montane elements of the Pamirs are confined to the lower meadows and shrub-zones. Some species, like, for example, *Nebria limbigera* SOLS. and *Cymindis mannerheimi* GEBL. have spread along the moist ground of the steppe rivers and streams up to the upper meadow zone. We have then the typical montane forms like *Parnassius* LATR., *Colias* FABR., *Erebia* DALM., etc. The apterous grasshopper *Pamiracris* RAMME, *Conophyma kusnezovi* UVAROV, *C. zubovski* UVAROV and *C. birulai* MIRAM are also montane forms on the Pamirs. The boreo-alpine *Podisma frigida* BOH. and *Aeropus sibiricus* LINN. are, however, absent on the Pamirs.

Most earlier workers like SEVERTZOV, OSHANIN and KUSNEZOV confined their analysis almost exclusively to altitudinal analysis, especially to the altitude zone between 4000 and 4500 m, an area in which the montane elements have penetrated and intermingled with the steppes elements. They entirely failed, therefore, to recognize the true montane character of the insect life of the Pamirs Region. Furthermore, KUSNEZOV ignored also the true high-steppe, the upper meadow zone, and has also confused with the montane steppe zone.

iii. Boreal and Boreo-alpine elements

The presence and the distributional pattern of the boreal and boreo-alpine elements on the inner Asiatic high mountains are of considerable biogeographic and ecological importance. The boreo-alpine elements are completely absent on some of these high mountains (*vide* Chapter XII) and the greatest majority of the boreo-alpine species seem to curiously avoid the Alai-Pamir System towards the north. While some boreo-alpine species are found on the Altai Mountains, they become progressively sparse on the Dzhungarskii Ala-tau, the Tien Shan and are almost absent on the Alai-Pamirs, particularly on the Pamirs. **Parnassius apollo* LINN. is perhaps the commonest boreo-alpine butterfly in Asia, which is widely distributed in north Europe and on the mountains of Central and south Europe, but is completely absent on the Alai-Pamirs. It is however represented on the Tien Shan by *P. a. merzbacheri* FRUHST. and *P. a. mongolica* STGR. and *P. a. alpherakyi* KRUB. on the Altai Mountains. *P. mnemosyne* LINN. is likewise found only on the northern marginal mountains of the Alai and the Hissar Range, but is absent on the Pamirs. Similarly *Erebia lappona* ESP., distributed on the Fennoscandian mountains, the Altai Mountains, the Carpathian, Rilo-Dagh, Prenj, Durmitor, the Alps and the Pyrenees, is absent on the Pamirs. Only three boreo-alpine species, viz. *Lycaena pheretes* HÜBN. from the Pamirs Northwest Himalaya area, with the subspecies *L. ph. lehana* MOORE from the Northwest Himalaya, *L. orbitulus* PRUN. with the subspecies *L. o. pheretiades* EV. from the Tien Shan, Tarbagatai, the Pamirs and the Ferghana mountains and *Caloplusia* (*Plusia*) *hochenwarthi* HOCHW. are known from the Pamirs Region at present. *Megachile analis cana* ALF., from above the shrubline at an elevation of 3850 m in the Ak-baital Valley, is considered by REINIG (*op. cit.*) to be another example of boreo-alpine species on the Pamirs. *Osmia* (*Melanosmia*) *pamirensis* GUSS. belongs to the boreal subgenus *Melanosmia*. *Halictodes pamirensis* MOR., from Djol-boeruljuk and Tshatir-tash (4100-4200 m) may also be considered a boreal form. The inner Asiatic form *Apatelia arctica* BOH. occurs in the Alai Valley, at an elevation of 3200 m in Sari-tash; this species is also widely distributed on the Altai Mountains, Changai and Ussuri Regions.

iv. Endemic elements

In addition to the large number of endemic species in almost all groups, the vast elevated area of the Northwest Himalaya, the Pamirs and the Tien Shan has a considerable number of endemic genera. Some of the more important examples include the following:

Ephemerida: *Ororotsia* TRAVER (Northwest Himalaya). Orthoptera: *Conophyma* ZUB. (Northwest Himalaya, Pamirs), *Dircanophyma* UVAROV (Northwest Himalaya) and *Pamiracris* RAMME (Pamir). Heteroptera: *Dolmacoris* HUTCHINSON (Northwest Himalaya), *Tibetocoris* HUTCHINSON (Northwest Himalaya) and *Mimula* JAK. (Alai-Pamirs and Tien Shan). Coleoptera: *Chaetobroscus* SEM. (Northwest Himalaya), *Trichoplatyscelis* REINIG (Pamirs), *Bioramix* BATES and *Chianalus* BATES (Northwest Himalaya), *Trichomyatis* SCHUSTER (Pamirs), *Ascelosodis* REDTB. (Northwest Himalaya, Pamirs), *Cyphogenia* SOLS. (Northwest Himalaya, Afghanistan, Pamirs), *Syachis* BATES (Northwest Himalaya, Pamirs), *Apaksha* MAULIK (Himalaya) and *Swargia* MAULIK (Himalaya). Hymenoptera: *Trichosyron* HAUPT (Pamirs) and *Tachygetes* HAUPT (Pamirs). Diptera: *Pamirella* END., *Reinigellum* END., *Reinigia* END. (Pamirs) and *Tianschanella* BRODSKII (Tien Shan).

* Some workers (1100) do not, however, consider this species as boreo-alpine.

v. Faunal subdivisions

The outstanding character of the insect life of the Pamirs Region is its connection with the steppe biotope. The essentially steppe character of the Pamirs explains also the relative poverty of the mountain elements in certain groups and the absence of the boreo-alpine elements. On orographical, climatic, floristic and faunistic grounds, the Pamirs Region may be subdivided into i. the Pamirs proper, ii. the Alai Valley and iii. the Western Valleys. The Pamirs proper are extensive high valleys, about 3700-4500 m above mean sea-level, with an annual mean atmospheric temperature of —1° C or even less, summer mean below 10° C, total rainfall below 100 mm and mostly not exceeding 50 mm and that also in summer. The winter snow-cover is extremely scanty or absent. The vegetation is typically an arid steppe flora of *Eurotia ceratoides* formation on the slopes and river terraces and *Trigonella* formation on the valley floors. The insect life is remarkable for the character species *Ascelosodis concinnus* BATES, *Leptosonyx* sp., *Pamiracris reinigi* RAMME, *Conophyma birulai* MIRAM, *Subterraneobombus melanurus griseofasciatus* REINIG and *S. pamirus* SKOR. The Pamirs proper comprise three parts. The outletless Northern Pamirs of Kara-kul and Rang-kul, at elevations of 3700-4200 m, are surrounded on all sides by high mountain ranges. The Kara-kul area has saline soil, with *Eurotia ceratoides* formation and poor fauna and not connected with the Western Valleys. The Southern Pamirs are high-valleys at elevations of 3800-4500 m, that pass westward gradually into the Western Valleys, with relatively mild climate, more humid than in the Northern Pamirs, but characterized by intense west-east storms. The vegetation is relatively rich in localized patches. The fauna is also relatively rich and is related to that of the Western Valleys. The Murg-ab Basin is the geologically young alluvial ground of Ak-su, Murg-ab and sheltered side valleys, at lower elevations of 3600-3800 m. The vegetation is richer, locally even with trees like *Salix oxycarpa* and *Myricaria davurica* and bushes and mesophytes. The non-steppe type of fauna is related to that of the Alai Range or of the Western Valleys.

The Alai Valley is a Pamir-like extensive high area, at about 2400-3400 m above mean sea-level, draining westward and passing gradually in the west into the Western Valleys. The area is not, however, sharply marked off from the southern ranges of the Alai and it also connects in the west with the Alai and the Western Valleys. The steppe flora is developed but locally, especially on the northern slopes of the Trans-Alai, strongly into mesophytic vegetation. Woody plants and *Artemisia*-steppes are present in the west. The fauna is on the whole generally richer than on the Pamirs proper and is related closely to that of the Alai and of the Western Valleys, but rather poor only in the east. Here we find also genera not represented on the Pamirs proper, such as, for example, *Cicindela* LINN., *Necrophorus* FABR., *Geotrupes* LATR., *Onthophagus* LATR., etc.

The Western Valleys, often only 2000-3800 m above mean sea-level, are characterized by summer temperatures above 10° C and annual rainfall of more than 100 mm, with considerable winter snow. The vegetation is richer than that of the Pamirs proper, with woody plants and *Clematis semenovii*, *Rosa webbiana*, *Rosa xanthia*, *Betula*, *Populus*, *Salix*, *Hippophaë* and *Juniperus*. The floristic boundary with the Pamirs proper and the Alai Valley coincides almost with the isotherm of 10.7° C. The fauna of the Western Valleys is rich in species and individuals and many of the species not found on the Pamirs proper occur here. *Bembex* LATR., *Sphex* LINN.,

Xylocopa LATR., *Chrysis* LINN., *Carabus* LINN., *Calosoma* WEBER, *Blaps* FABR., not known from the Pamirs proper, ascend above the shrubline in the Western Valleys. The area of the Western Valleys is subdivided into smaller subregions, nearly all of which are connected together by the Pundsch Valley. Areas with the summer temperature mean reaching to 20° C show pronounced faunistic affinities to the Turkestan lowlands. Connected with the Western Valleys are the southern valleys of Wachan, Garan, Shugnan, Roshan and Bartang, which are in their turn connected directly with the Pamirs high-steppes by the Wachan Darya, Pamir Darya, Shah Darya, Gund and Bartang Murg-ab valleys and are, therefore, climatically influenced.

vi. Faunistic affinities

The close faunistic affinity of the Pamirs Region to the areas to the east has been recognized by a number of workers like MENZBIER (762, 763). The Pamirs have indeed been included by some authors in the Mongol-Tibetan Province of the Middle Asiatic Subregion. Western Tibet and Ladak are particularly related to the Pamirs Region. KOBELT (631) has summarized the earlier views on the faunistic affinities of the Pamirs. Recently SKORIKOV (984) has discussed the faunistic similarities of the Alai-Pamirs to the Tien Shan and the Tarbagatai area.

Most earlier workers like FEDCHENKO, OSHANIN (804), SEVERTZOV, etc. believed that the fundamental elements of the insect fauna above the limits of forest on the Middle Asiatic mountains and elevated areas like the Pamirs are "Biotopfremde" elements—not boreal forms, but thermophile southern derivatives. OSHANIN indeed considered the Hemiptera to be of Mediterranean origin. KIRITSHENKO has, however, shown that the endemic elements among the Turkestan Heteroptera amount to 9.7%, while the Mediterranean elements are only 0.8% and the boreal forms are 56%. REINIG (878, 879) has also stressed the importance of Mediterranean elements, although he feels that these are also typically steppe forms, secondarily adapted for the extreme temperature fluctuations, better than other ecological forms. He has even concluded that the inner Asiatic mountain elements are much less abundant than the Mediterranean steppes forms. KUSNEZOV (669-671) believed the boundary altitude for the boreal forms to be at 450-1830 m. Exclusively on the basis of hypsometric analysis and without reference to the biotope characters of Formicidae, he concluded in 1925 that the fundamental faunal elements above the limits of forest on the Middle Asiatic mountains are not boreal, but comprise essentially southern forms, secondarily ecologically well adapted for life on the high areas. The extremely isolated examples of boreal and boreo-alpine forms found on the high-steppes apparently justified this view and indeed REINIG (878) seems to have laid considerable emphasis on the fact that boreo-alpine elements "go around" the Pamirs Region in the north and their abundance falls off steadily from the Altai Mountains towards the Pamirs.

It is no doubt true that as we proceed from the Altai Mountains towards the Pamirs Region, we observe a most significant disappearance of the boreo-alpine elements. It must not, however, be overlooked that at the same time there is a corresponding increase in the abundance of endemic elements. We have already shown that there are many endemic genera, species and subspecies and local races that are common to the Pamirs and the Northwest Himalaya. There is, therefore, considerable faunistic and biogeographic affinity of the Pamirs to the mountains of the Northwest Himalaya. Another part of the Pamirs fauna is also related to the Tien Shan and the Altai Mountains. We find more frequently that some species extend from the Altai over the Tien Shan, Alai, Trans-Alai, the Pamirs and the Northwest Himalaya and sometimes into Tibet and the Kuen Lun Ranges. The insect species complex of the Western Valleys is also related to some extent with that of the Turkestan lowlands, but such a connection is not observed between the Alai Valley and the Western Valleys.

REINIG and other workers believed that the Pamirs have received considerably from the east and southeast, but for the Alai Valley and the Western Valleys the main sources are the Tien Shan and the Turkestan lowlands. They have assumed that most species found in the Pamirs Region have apparently had their origin on the mountains to the south and the high-steppes of the Himalaya, the Karakoram and the Kuen Lun. Small numbers of species extend over to the Tien Shan System and the majority of them have not reached the Pamirs, but only up to the Trans-Alai or at the most to the upper valley ends of the Western Valleys. On the evidence of such forms like *Conophyma* ZUB. and *Gomphomastax* BRUN. that belong to an otherwise basically tropical group and on the affinity of Pamir-endemic *Pamiracris* RAMME to the Himalayan *Hypernephia everesti* UVAROV, REINIG has concluded that the Pamirs Region is inhabited by elements from the southeast, especially west Tibet and the Himalaya and to a much lesser extent only from the Tien Shan and the Altai Mountains. It is interesting to recall in this connection that FILIPJEW (345) has shown, however, that at least in the case of the Lepidoptera, MENZBIER's views of the Tibetan origin of the Pamirs fauna is untenable and the inclusion of the Pamirs Region with the Mongolian-Tibetan Province is not justified. He has demonstrated that the Tibetan element is really insignificant on the Pamirs.

The source and origin of the present-day insect fauna of the Pamirs Region are evidently closely interlinked with the orogeny of the Middle Asiatic mountains and of the Himalaya and the vast climatic and other changes during the Pleistocene. UVAROV (1067-1069) has in fact recognized the connection between the orogenic history and the faunal history in the region. While KUSNEZOV agrees with this fundamental idea in so far as the west is concerned, he emphasizes, however, the rôle of the Pleistocene climatic changes and Miocene uplift as of primary importance. The discussion of KORSCHNEWSKY (653) is also of interest in this connection. MENZBIER (762) assumed, however, that the Pleistocene glaciers descended to elevations of 2000 m and therefore the Pamirs faunal elements spread to the Turkestan lowlands and thus came to be intermingled with the Siberian-Arctic forms. KUSNEZOV did not accept this position.

It seems, however, possible that the area that is now the Pamirs was inhabited by a lowland subtropical type of fauna before the intensive mountain uplift of the region. It is not, however, clear whether this fauna retained the original character after the Pliocene phase of uplift of mountains. The Tertiary fauna, at least in the region of the Pamirs, is presumed to have been destroyed as a result of the Pleistocene glaciations and REINIG and others have therefore supposed that the southern elements are Post-glacial in origin. Recent Russian workers have, however, shown that these Tertiary elements have actually survived the glacial times in certain areas on the Pamirs.

REINIG also believes that the extent of the Pleistocene glaciations in the Pamirs Region has generally been over-estimated by some authors. It is understandable that the area was not fully covered by an inland ice mass, yet we must not at the same time overlook the possibility of high areas remaining ice-free during the whole of the Pleistocene. While parts of the Pamirs proper remained, no doubt, above the ice or were perhaps only partially covered by ice, the Western Valleys were under ice mass. The Post-Pleistocene ice recession was thus more long-drawn out in the Western Valleys than on the Pamirs proper. The Pamirs could, therefore, be conceivably be repopulated from the eastern refugium, viz. the mountains more than from the Western Valleys. After repopulation from the east, the Western Valleys elements are also assumed to have eventually pushed into the Pamirs proper. The Tien Shan elements are believed to have invaded only in the advanced stage of the Post-Pleistocene ice-melting and that also only by way of the Western Valleys to the Pamirs proper. The last glaciation seems to have been a greater determining factor for the recent fauna of the Pamirs than the earlier glaciations. The mean temperature depression, at the time of the last glaciation, seems to have been about 4° C and large-scale ice did not descend below an elevation of 2000 m. There was very little or no intermingling of the arctic elements with those of the Turkestan lowlands and of the west inner Asiatic mountains, but such an intermingling seems to have been more possible on the Tien Shan and the Altai Mountains. Later on some boreo-alpine forms seem to have spread from here to the Alai-Pamirs. The origin of the fauna of the Murg-ab Basin is somewhat obscure at present. The present-day forms of this basin may not have been derived, after having survived the Pleistocene glaciations, from the Pamirs.

Although REINIG thus seems to be largely in favour of the Post-Pleistocene origin from the west, the possibility that the general climate and ecological conditions in the whole Pamirs having been favourable for a relatively rich fauna during the whole of the last glaciation also must not be overlooked.

The available ecological and biogeographical evidence would appear to indicate that the different regions of Tertiary mountains of the Palaearctic Realm must be considered as independent centres of faunal development and differentiation, distinct from the numerous well known lowland centres, and these mountain regions must be considered to represent a sort of second alpine zone between the temperate and the subtropical zones of the earth. It seems more appropriate to employ the term faunal developmental amphitheatre (Entwicklungsschauplatz of SKORIKOV) rather than developmental centre. In place of the term alpine, we should also prefer SKORIKOV's expression Tertiary mountain zone. From his biogeographical analysis of the Bombidae, SKORIKOV (984) recognizes within the Palaearctic Realm i. a region of Tertiary mountains and ii. the adjacent boreal region. The latter comprises three distinct zones, viz. the conifer forest, the forest-steppe and the steppe-zone. This subdivision is more in consonance with the available evidence and the known distributional peculiarities of the high altitude insects of the Middle Asiatic mountains than the usual divisions into the Mediterranean and High Asiatic regions. The region of the Tertiary mountains consists of two subregions: i. the Euro-Anatolian (as a parallel to the Mediterranean subregion in the narrow sense of the term) and ii. the High Asiatic subregion, consisting of (a) the Pamirs district and (b) the Turkestan district under the Turanian Province. SKORIKOV considers the Himalaya as separate from the Turanian province. Soviet phytogeographers like KOROVIN (649) have combined the Pamirs-Tibet areas as a separate province of Middle Asia, but SKORIKOV combines the Himalaya with the Tibet into the Tibeto-Himalayan province. We have, however, already shown that the Himalaya consists of two distinctive areas, viz. the Northwest Himalaya and the rest of the Himalaya to the east of the great defile of the R. Sutlej. The Pamirs Region and the Northwest Himalaya constitute together a single amphitheatre of faunal differentiation.

In his recent discussions on the sources and the development of the Turkmenian fauna, KRYZHANOVSKII (660) has shown that the generally current views of the Post-Pleistocene migrations from the Mediterranean and Central Asiatic areas are wholly untenable. The insect fauna of the Turkmenian and Middle Asiatic mountain regions is remarkable for the large numbers of endemic forms. In addition to this strong endemic element, there are Central Asiatic, southwest Asiatic and Saharo-Sindhian elements also. The abundance of endemic genera is strong proof of the greater antiquity of the Middle Asiatic (Turanian) centre of faunal evolution and distribution and of the intensity of autochthonous speciation in the centre. The insect fauna of the mountainous desert zone of Kopet-dagh and of Badghys is closely related and the species endemism is more pronounced than in the case of the lowland fauna. There is close affinity to the fauna of other low mountains of Middle Asia, Iran and north Afghanistan. The Tien Shan, Alai-Pamirs, the Karakoram, the Hindu Kush and the Northwest Himalaya together constitute a single centre of differentiation of Tertiary mountain autochthonous hypsobiont insect fauna.

As indicated even in the foregoing chapter, we are justified in concluding that the

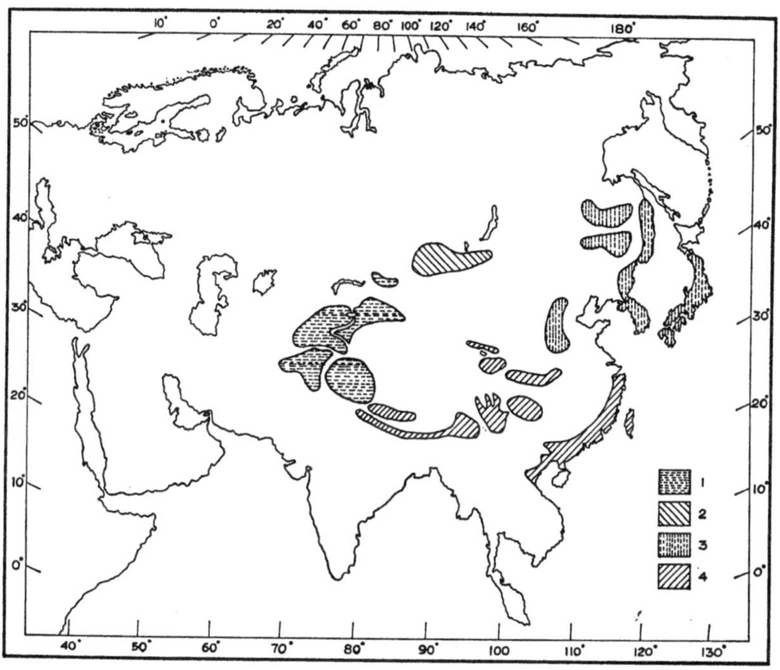

Fig. 62. Centres of differentiation of Palaearctic faunas. 1. Turkestan centre, 2. Mongolian centre, 3. Manchurio-Japanese centre, 4. Himalaya-Chinese centre (excluding the Northwest Himalaya). (Modified from GROSS).

vast elevated area of the Northwest Himalaya, the Pamirs-Alai and the Tien Shan constitute together an important Tertiary Mountain amphitheatre of faunal differentiation, dispersal and evolution within the Turkmenian Region of the Palaearctic. The rest of the Himalaya, together with Tibet, form another independent centre of faunal origin and dispersal. The distributional evidence of Lepidoptera of Eurasia, recently dicussed by GROSS (438-440), lends support to this conclusion. On the basis of the systematics and distribution of the Palaearctic Satyridae, GROSS recognizes a number of major and secondary centres of differentiation and dispersal. He also finds a close parallel between the vast geological evolution of the area and the hierarchial orders. Larger centres of faunal origin arose in an earlier period and the smaller and secondary centres arose comparatively recently. Two major Tertiary centres are recognized, viz. one in Fennoscandia and the other in the Angaraland. These centres were already advanced even during the early Tertiary, since the Tethys had already largely receded. GROSS suggests that excluding Satyridae, the rest of the allied groups had their developmental and dispersal centre in the Angaran area, but Fennoscandia was the centre of development of the original stock for Satyridae. The late Tertiary and Pleistocene faunal exchanges were far more intensive in the north of the Holarctic than in the south. The primary Pleistocene faunal dispersal centre in Asia had several sub-centres. The Turkestan centre has over 50% of the species autochthonous and 20% of the species are of Asiatic Tertiary origin; 70% are of the west Palaearctic Tertiary origin and about 10% are of uncertain evidence at present. The Asiatic Tertiary elements are found particularly in the southern parts of the Turkestan centre, especially in the region of the Northwest Himalaya. The Turkestan centre is subdivided into i. the Tarbagatai secondary centre, with predominantly Eremian biotope and rich Mongolo-Pleistocene elements; ii. the Ala-tau Tien Shan secondary centre, especially the area of the Lake Issyk-kul and not sharply demarcated from iii. the South Turkestan secondary centre. We have then the Kashmir-Chitral secondary centre (Northwest Himalaya) and the Afghanistan secondary centre. This is largely in agreement of the views expressed earlier about the affinity of the Northwest Himalaya, Pamirs and the Tien Shan (fig. 62).

CHAPTER XI

THE ALPS AND THE CARPATHIANS

1. THE ELEVATED AREAS OF EUROPE

The principal elevated areas in Europe are i. the Central Uplands "Mittelgebirge" and Plateaux, ii. the northwestern highlands and iii. the southern mountain ranges. The Central Uplands and the Plateaux comprise a continuous highland of hilly or plateau-like area, extending from south Central France to northern Czechoslovakia, with elevation ranging from 150 to 610 m above mean sea-level. Topographically, they stand intermediate between mountains and hill-lands. The Northern Highlands comprise the mountainous terrain of Scandinavia, the northern parts of the British Isles and Iceland. These areas are higher than the Central Uplands, but the general configuration of the surface is more plateau-like than mountainous. The southern mountain ranges are truly mountainous in character and comprise the Alpine System, rising to mean elevations of 3000 m and often much higher. The mountain ranges are continuous over great stretches, with deeply incised valleys and extensive basins.

The principal mountains of Europe belong to two major groups. There are several mountains, the uplift and folding of which seem to have been initiated even during the palaeozoic Era (80). In contrast to these ancient mountains, there are younger mountains of Tertiary origin. The Altai Mountains of Asia (*vide* chapter, XIII) belong, for example, to the older system of mountains, which are, therefore known as the altoid mountains. The younger Tertiary mountains like the Alpine System are post-altoid. Old Tertiary sediments form, on the younger mountain chains, a broad zone of intensive folding. In many parts of the Alps, the Miocene and even the Pliocene formations are disturbed and pushed up. In the area of the socalled Rumpfgebirge or the older mountains of Europe, not only the Old Tertiary, but even the Mesozoic sediments remain undisturbed, in horizontal position, over vast areas. To the older Rumpfgebirge belong the Bohemian Massif, the Sudeten, the Erzgebirge, the Mittelgebirge of western Germany (Fichtelgebirge, the Thüringerwald, the Harz Mountains, Rheinisches Schiefergebirge, the Schwarzwald), the Vosges, etc., the French Central Plateaux, besides the Montagne Noire, the mountain ranges of northwest France and Belgium, of England, Ireland and Scandinavia. In south Europe, we have several extensive fold mountain chains and the whole of the western and central Spain (the Iberian Meseta), the Balkan Peninsula, the Corsadinic Mass (Tyrrhenis) including the Rhodope Massif and the archaic mountains of Serbia. The Catalonian Mountains are also older mountains. To the younger mountain system belong the Balkan Mountain Ranges, the Carpathian Range, the Alps, the Apennines, the Limestone Mountains of North Sicily, the northwest African mountain chain from Tunis to Gibraltar and the Baetic Cordillera in South Spain (the tectonic continuation of which with the Balearian is submerged by the sea) and the Pyrenees.

The following orogenies are generally recognized in Europe:

1. The Pre-Cambrian Orogeny dates back to perhaps over a billion years and is

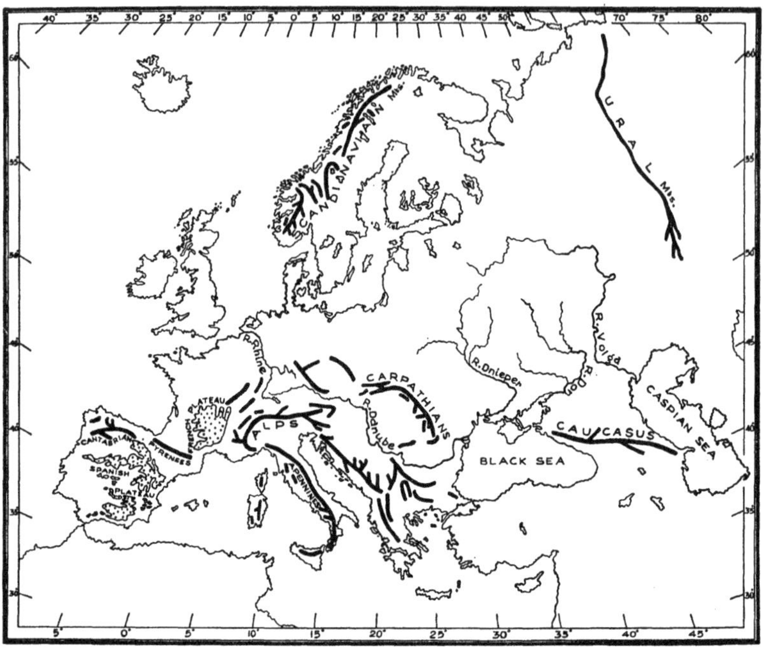

Fig. 63. Sketch-map of Europe, showing the principal elevated areas and crestlines of the mountain ranges.

at present largely obscured in most places. Extensive granite and metamorphic rocks of the Fennoscandian or the socalled Baltic Shield, the Central Siberian or the Angaran Shield, the Canadian Shield and large parts of Greenland, represent evident remnants of an old northern continent, the Laurasia, and is the core of the European mass. It forms the rigid block around which the younger mountains have been thrust up.

2. The Caledonian Orogeny was initiated perhaps during Silurian times, but presumably not later than the Devonian. The Palaeozoic marine sediments were folded and overthrust and also highly metamorphosed.

3. The Hercynian Orogeny was preceded by marine transgressions and widespread submergence during the Carboniferous Epoch, the uplift extending partly into the Permian times. The main axis of the Hercynian uplift developed south, along the northern flank or within the Hercynian Mountains. The Hercynian structure in the Alps may be observed today in the granitic massifs. Three major Post-Hercynian developments are generally believed to have given rise to the discontinuity of the Hercynian Mountains.

4. The Alpine Orogeny is of Tertiary age. The principal arch of the Alps projects northwards between the Hercynian foreland, the French Central Massif and the Bohemian Massif. The Carpathians bulge against the Bohemian Massif and a buried spur of the Russian Shield. The bend of the Jura mountains is anchored on to the two ends of the French Central Massif and the Vosges. The Mediterranean mountain

ranges are squeezed in between the Iberian, the Tyrrhenian and the Rhodope Massifs. The Tertiary orogeny involved all the major continents of the world simultaneously, so that we find today continuous Tertiary mountain chains along the southern margin of Asia and the west coast of America. We have already shown in an earlier chapter that the Himalayan System belongs to the Tertiary orogeny.

The evolution of the Alpine System is an interesting chapter in the origin of great elevated regions and of high altitude insect life. During the greater part of the Mesozoic Era, the Tethys Sea covered most of what is at present Europe. This sea stretched, as already indicated, between Laurasia and Gondwanaland. Vast amounts of sediments, mostly calcareous, were deposited in the Tethys. The seafloor of the Tethys underwent continuous subsidence, simultaneously with the deposition of sediments. This Alpine geosyncline was in a weak zone, between the stable masses of Laurasia and Gondwanaland and such subsidence could not naturally continue indefinitely. The African landmass began to push northward, eventually encroaching upon the sedimentary trough. The Hercynian mass acted, however, as a rigid block against this northward movement from the south, with the result that the sedimentary weaker layers yielded to the enormous tangential pressure and ridges began to appear perhaps even in Cretaceous times. The deformation reached, however, a peak during the Oligocene and Miocene. The sedimentary layers and often also the igneous rocks were lifted up several kilometres from the seafloor and thrown into truly gigantic folds of anticlines and synclines. The upper parts of these folds often overrode the lower or they were also dragged above the earlier mountains. These mighty crustal movements were arrested along the so-called Alpine Front in the French, Swiss and Austrian Alps, forming the sheer cliffs of today. Intrusions, volcanic eruptions, upheavals and subsidences accompanied and followed the alpine orogeny. In contrast to this view of large-scale latitudinal displacement of the Gondwana Mass, another view postulates an *in situ* development of the Alps, with the geosyncline being an underthrust from both north and south.

The principal branches of the alpine orogeny system of mountains are the Alps, the Pyrenees, the Sierra Nevada of Spain, the Atlas, the Apennines, the Dinarids (collectively the mountains of Greece and Jugoslavia), the Toros Mountains and the Anatolian Ranges (fig. 65). The Alps form, together with the Apennines, the Atlas, the Baetic Cordillera and the Carpathians, a single orogenic unit called the Alpid. The Alpids, though later formations, are European tectonic systems. Another great system is formed by the Dinarids, the Hellenids, the Taurus Mountains and the Himalayan System, which represent the marginal folds of Asia. The European Alpids and the Asiatic Dinarids unite morphologically into a single system in the Alps. KOBER (632, 633) has suggested that fold mountains are fundamentally bilateral and are, as a rule, composed of two bordering chains of mountains, separated by a median mass, the so-called "Zwischengebirge" or the between-mountains. It may be observed that the Zwischengebirge are absent in the Alps proper, where the two chains are crushed against each other and the Alpids are overridden by the Dinarids. The folded mountain chains stretch from the east to the west in great curves and extend also into the Atlas Range of North Africa and into the peninsula of Asia Minor. The following areas of curves to the west with the inner concavity marked by a depression, are generally recognized by geologists: i. the Gibraltar area, where the Atlas Mountains are separated from the Baetic Cordillera of Spain by the Strait of Gibraltar; ii. the West Alps, passing into the Apennines, with the plains of North

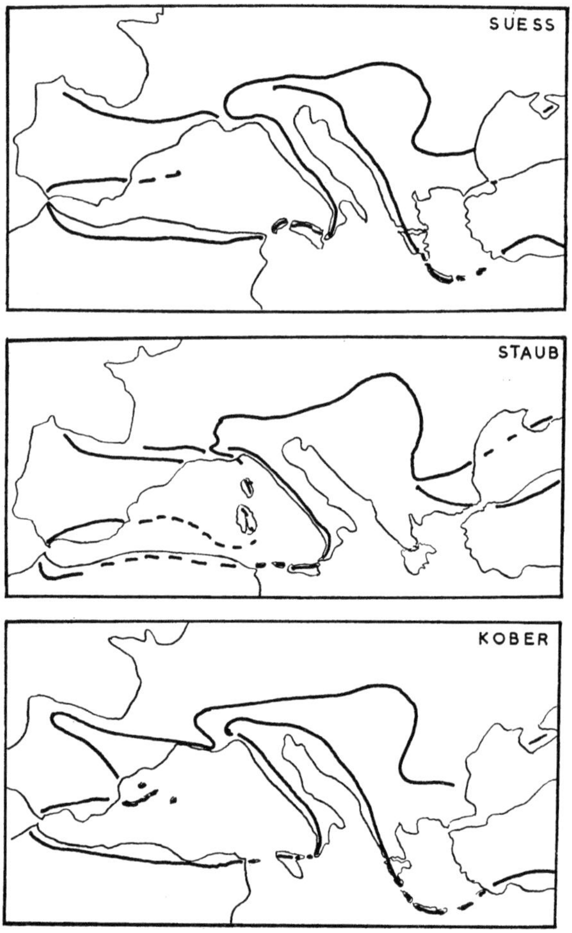

Fig. 64. Sketch-map of Europe, showing the relation of the Alpine orogeny to the other mountain systems, according to different authors.

Italy forming a depression and iii. the Transylvanian Alps curving round into the Balkan Mountains, with the depression area of the Wallachian Plain of Rumania. KOBER (633) believes, however, that the Apennines are not a continuation of the Alps, but must represent part of the Dinarid System, folded towards the east. The West Alps, represented by the outer zone and the Pennine Nappes zone, reach the sea, but are separated from the Apennines proper by a narrow belt of East Alps that extends into Corsica. Overthrusting has occurred in the Pennines, as in the Alps, with the formation of characteristic nappes. The earlier view that the Baetic Cordillera (the Sierra Nevada of Spain) are continued to the Atlas Ranges, across the Straits of Gibraltar, is not now widely accepted. The Pyrenees and the Cantabric

Mountains are not also believed to represent part of the true alpine folding, but only the chain of the Baetic Cordillera is the true Alpine element in Spain (995, 996) (fig. 64).

2. THE ALPS

The Alps are the greatest fold mountain system of Europe. Along its inner edge, it is nearly 755 km long and about 1300 km along the outer edge; the chord of the arc is nearly 555 km long. The width of system ranges from about 150 km near Mt. Blanc to about 250 km north of Verona.

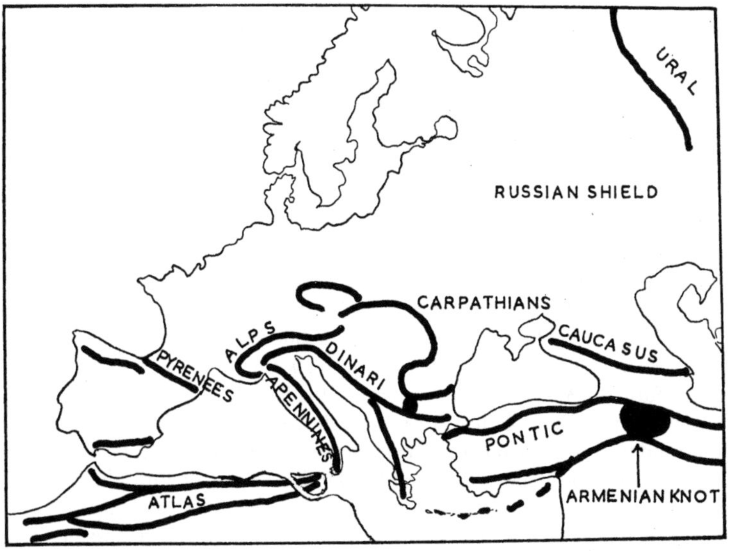

Fig. 65. Sketch-map showing the crestlines of the Tertiary mountain systems of Europe, near Asia and Northwest Africa.

The Alps comprise numerous mountain ranges, with longitudinal valleys in between them and the whole forming a broad mass, convex on the north and the valleys stretching west-east or southwest-northeast. The Coll di Altare or di Cadibona near Savona marks the limit in the west, or rather strictly speaking, on the south; on the east, the limit is over the Semmering Pass. On the north, it is marked by the R. Rhine, from Basle to Lake Constance. The Alps form an arc from the Altare Pass on the Gulf of Genova to an imaginary line near the railroad to Vienna-Graz-Maribor-Ljubljana. The Alps are connected with the Apennines, southeast of the Altare Pass. Near the Lac du Bourget in France, the Jura Mountains are detached from the Alps, so as to form a separate spur, with a plateau in between. On the interior of the arc is the Po-Adriatic Depression, towards which the Alps slope abruptly. A number of old massifs are situated along the outer edge of the great arc, against which the Alps seem to have been pushed during the gigantic folding process. The Alpine arc has numerous high peaks, rising to elevations above 3000 m; some of them situated in Switzerland attain elevations of 4575 m. Though situated at a relatively low latitude (43-48° NL), the most extensive system of glaciers on the mainland of Europe is found on the Alps. The Alps extend into Italy, France, Switzerland, Liechtenstein, Germany, Austria and Jugoslavia.

No other mountain range in the world has perhaps been so extensively explored by geologists and biologists and yet few other mountain ranges present so many unsolved problems as the Alps. The main trunk of the Alps naturally forms a geological whole, but both in the east and in the west, there are bifurcations, so that it looks as if there are two separate mountain ranges. The Dinarids and the Carpathians (separated by the Hungarian "Zwischengebirge"), unite with the Alps in the east. This union is best observed in the profile from Trieste to Vienna, where there are two distinct ranges, viz. the alpids and the dinarids. A similar condition may also be observed at the Ligurian Sea-Break in the Alps. It is thus only in the west and in the east, that the Alps are clearly bifurcated into the alpids and the dinarids. For fuller discussions on the Alpine tectonics reference may be made to HERITSCH (509), TORNQUIST (1040), HEIM (497), KOBER (632, 633), LEITMEIER (684), BUDAY *et al.* (171) and others listed in the Bibliography.

The major topographic divisions of the Alps are based on an imaginary line, extending south from Lake Constance on the R. Rhine Valley, across the Splügen Pass (2110 m above mean sea-level) to Lake Como. The West Alps that lie west of this demarcation are more narrow, compact, higher, with more and longer glaciers and more incised valleys than the East Alps. Mt. Blanc (4807 m) is the highest peak in the West Alps and Piz Bernina (4028 m) in Graubünden in Switzerland is the highest point in the East Alps. Most of the Swiss Alps are within the area of the West Alps and all of the Austrian Alps lie within the area of the East Alps. The West Alps are subdivided into i. the Franco-Italian Alps and ii. the Swiss Alps. The Franco-Italian Alps lie south and west of an imaginary line, from the eastern end of Lake Geneva, through the upper R. Rhone Valley, the Great St. Bernhard Pass (2472 m), the Dora Baltea Valley to the Piemont Lowland and the upper Po river. The Swiss Alps lie between the Great St. Bernhard Pass and an imaginary line through Lake Constance-Rhine-Lake Como. The East Alps comprise the Ortler Alps, the Ötztaler Alps, Stubaier Alps, the Hohe and the Niedere Tauern, Lavanttaler Alps, Karawanken, Steiner Alps, the Gurktaler Alps, Gailtaler Alps, Julian Alps, Carnic Alps, the Dolomites, the Brescianer Alps, etc. The Rhone-Rhine longitudinal trough divides the Swiss Alps into the Northern Alps and the Southern Alps. The Northern Alps include the Bernese Alps, the Vierwaldstätter Alps, the Glarner Alps and the Appenzell Alps. The Southern Alps include the Valais Alps, Ticino Alps and the Lepontine Alps pierced by the St. Gotthard Tunnel, the Adula Alps and the Grison or the Graubünden Alps. The subdivisions of the West Alps include the Maritime Alps, the Cottian Alps, Dauphiné Alps, etc.

The geological structure of the Alps is largely the result of the folding of sediments between the Eurasian foreland and the African landmass. The older mountains had been denuded considerably by river erosion and enormous amounts of sediments had been carried and deposited in adjacent areas, long before these folding movements commenced. During the Pleistocene the whole mountain system was repeatedly covered by ice and only small areas in the southeast and southwest apparently remained ice-free. The ice-sheet cut through and dug away large masses of the mountains and the Post-Pleistocene warming gave rise to characteristic moraines, U-shaped valleys, cirques and pyramidal peaks. In the East Alps, geologists generally recognize the following longitudinal zones: i. the Central Alps of crystalline rocks (granites, schists and gneiss) that form the principal chain of the Alps, from the edge of the Carpathian Basin to the Pennine Alps, northwest of the R. Po Basin. Bands of more readily eroded shale and slate intersperse the crystalline rocks in the East Alps. North of the Central Alps are the Northern Limestone Alps or the Calcareous Alps (die nördlichen Kalkalpen), a belt of generally lower ranges of sedimentaries, especially limestones, with deeply incised valleys. A Southern Limestone zone, not met with in the West Alps, contains volcanic extrusions of reddish-black rocks. North of the Northern Limestone Alps is the rolling foreland of Alps—hills of narrow, discontinuous Flysch zone of slate, clay and sandstone. In front of the Flysch zone in the Alpine Foreland there are stream-deposited beds of the so-called Molàsses of sandstones and conglomerates.

The following six major structural units are generally recognized: i. the igneous core or the massif along the main axis of the western and central Alps, buried under the sedimentaries-fold in the East Alps; ii. the northern limestone Alps of the Helvetides, forming the frontal ranges throughout the entire system, with several folds one above the other and the roots of the Helvetides within or south of the massifs, separated by nappes by erosion; iii. The Eastern Alps or the Austrides that pushed over all other tectonic units and covered perhaps much larger areas than visible at the present time and characterized by a series of simple, east-west trend limestones, with the central zone of complex igneous and metamorphic rocks, iv. the West Alps or the Pennides advanced against the massifs, but did not override them and were thus subjected to complex plications; v. the Southern Alps or the Dinarids, the separation of which from the Austrides and the Pennides is due to differences in the material and the mode of folding, with well developed calcareous rocks in the Karawanken; and vi. the molasses trough, in which sediments accumulated (conglomerates and sandstones deposited by primeval rivers of the Alps) overriden and tilted southward by the last Alpine front advance.

The East Alps comprise i. the southern calcareous Alps or the Dinarids and ii. the northern belt of crystalline rocks, as for example, in Silvretta, Ötztaler and Mur Alps. The Dinarids, narrow in the eastern parts of the West Alps, form in the East Alps a broad belt of the Bergamo Alps, the Dolomites and the Carso behind Trieste. The North Limestone Alps of Mesozoic sedimentation include thick limestone beds. The Tirolid or the crystalline core is absent in the West Alps. The structures seen in the West Alps are concealed by the Tirolid rocks in the East Alps, but in the Lower Engadine and the Hohe Tauern the so-called "windows" show through to the west-structure below. These windows are associated with deep-seated uplift of the Hercynian massifs, which are exposed to the surface to undergo sufficient denudation so as to reveal the Pennine Nappes. The window in the Tirolid enables us thus to see the upper Grisonid and the lower Pennides. On the basis of this hypothesis, the Swiss Alps as a whole must be described as a sort of gigantic window. It is assumed that the whole is situated on a submerged Hercynian hill. The region of the Hohe Tauern is structurally and geologically well differentiated from the rest of the East Alps and deep valleys separate high mountain ranges that consitute an effective faunal barrier. The geology of the area is discussed by a number of workers like HOTTINGER, COR-

NELIUS and CLAR, BRAUMÜLLER and others, but the reader will find a comprehensive summary of these discussions in KOBER (*op cit.*), CLOOS (213) and COLLET (219).

In the West Alps six belts are recognized: i. the Pre-Alps belt of moderate thickness in front; ii. the high Limestone Alps, with permanent snow and glaciers; iii. the discontinuous crystalline Hercynian massifs, with Mt. Blanc; iv. the axial belt of crystalline schist and gneiss of the Cottian (Mte Viso), Gran Paradiso, Pennine (Mte Rosa) and the Lepontine Alps; v. the root belt; and vi. the southern Limestone Alps (dinarids). In the Jura Mountains the sedimentary rocks from the Permian to the Tertiary rest upon a crystalline foundation. The folding seems to have affected only the surface beds up to the Middle Trias (Muschelkalk) and the older beds are unaffected.

The longest glacier on the Alps at present is the well known Aletsch Glacier. There are other glaciers at the sources of the rivers, which have often lakes in their course. The largest lakes include Lake Geneva, Neuchatel, Thum-Brienz, Lucerne and Zürich (Swiss Alps), Lake Constance, Maggiore, Lugano, Como, Garda, the Chiemsee, Wörthersee, etc. The principal rivers are the Rhine, rising on the southern Alps and flowing northward into the North Sea; the tributaries of the Danube like the Inn, Salzach, Enns, Mur and Drau; Po from the southern slopes and the Rhone in the Bernese Alps. Ignoring a small advance of about 42 m during 1917-1922, most glaciers on the Alps have been steadily receding by an average of about 30 m annually, especially in the Hintereisferner Vorfeld. The Niederjochferner Vorfeld, north of the Kreutzspitz Ridge, with a north-east-southwest trend, forms the south slope of the Ötztaler Alps, about 3000 m above mean sea-level. The glacier seems to have receded at the rate of about 12 m annually during 1932-1937. The glacier now lies at an elevation of about 2530-2550 m. During each of the Pleistocene glaciations, the general climate of the Alps was characterized by a mean temperature depression of 8.3° C and the ice accumulated and flowed as piedmont glaciers. The Alpine valleys were filled with a net-work of glaciers, which reached southern Germany and the Jura Mountains to form a continuous piedmont glacier in the northern Alpine Foreland. The valleys were sculptured into typical U-shaped cross-sections, the mountains into sharp crests and the surface of the rocks was exposed, grooved and polished. Considerable local differences existed in the extent of glaciation in different parts of the Alps. In the East Alps, for example, the Northern Limestone Alps (Kalkalpen), east of Salzach, were more intensely glaciated in the west than in the east. The Salzach Glacier, Traun Glacier and the Steyr Glacier overflowed the northern margin of the Alps. West of Salzach the Pleistocene snowline lay at an elevation of 1100-1200 m. The Enns Glacier stretched the whole length of the northern margin of the Niedere Tauern, but the Mur Glacier ended at Judenburg on the south margin. Karawanken, Steiner Alps and the Julian Alps were partly unglaciated. At the time of the Würm glaciation (18 000 years B.C.) in the Alps, the snowline was about 1200 m lower than at the present time. The Bühl stage marked the first retreat of the ice and the snowline was lower by 1100-1200 m than at the present and heavy melting of the ice had already begun. The Gschnitz and the Schlern stages were the final glaciation (8500 years B.C.), when the snowline sank by 600-800 m below the present limits. In the Daun-Egessen stage, the maximum of Daun Glacier dates back to 8000 B.C. and the snowline was 400-500 m below the present limit. In the Egessen moraines formation period, the snowline was 300 m lower than now. Then followed the

TABLE 34
Mean monthly atmospheric temperatures at different elevations on the Grossglockner Alps

Elevation in m	Jan.	Feb.	Mar.	Apr.	May	June	July	Aug.	Sept.	Oct.	Nov.	Dec.	Annual mean
2000	− 6.8	− 7.0	− 4.6	− 1.7	3.1	6.5	8.6	8.2	5.4	1.7	−2.6	−5.5	0.4
2500	− 9.5	− 9.8	− 7.7	− 5.0	− 0.1	3.0	5.2	5.0	2.3	−1.1	−5.8	−8.3	−2.7
3000	−12.2	−12.6	−10.9	− 8.2	− 3.3	−0.4	1.8	1.7	1.8	−0.8	−4.0	−9.6	−11.0

Post-Pleistocene warming up. The Fernau stage was characterized by strong retreat of glaciers (about 1520-1650 B.C.) and the snowline was 100-200 m lower than at present. About 1820-1856 A.D., the snowline was 50-100 m lower than during the fifties of the present century. For fuller information on the Pleistocene glaciations on the Alps and their influence on the biota of the region reference may be made to GAMS (396), HOLDHAUS (535), KOLBE (642), LICHTENECKER (691), ZSCHOKKE (1139) and to chapter XII. The snowline at present on the Alps ranges from 2780 m in the northern Swiss Alps to 3200 m in the central parts of the Alps. Over 13 440 sq. km of the Alps region are covered by ice and snow at present.

3. THE ECOLOGICAL CHARACTERS AND THE BIOTIC ZONATION OF THE ALPS

The general ecological conditions differ widely in the West and East Alps and also in the southern and northern Alps. The total amount of precipitation does not, for example, increase as we proceed deep into the inner Alps, where the general climate is pronouncedly more continental than in the marginal parts of the Alps (393). In the Grossglockner area the mean monthly atmospheric temperatures at different altitudes are summarized in table 34. The wide valleys are warmer during the summer and reveal a curious temperature inversion during the winter and are, therefore, colder than the nearby slopes of the mountains. The extreme summer and winter temperatures at different elevations are summarized in fig. 67.

The mean atmospheric precipitation increases with elevation up to 2440-2740 m above mean sea-level and sometimes even up to 3050 m, but thereafter decreases at higher elevations. The interior valleys are much drier and also receive much less rainfall than 750 mm annually. In winter the precipitation is mostly as snow, which often lies nine metres thick. A series of local winds are common on the Alps. In addition to the westerlies, most valleys, especially the transverse valleys, have also a regular alternation of upslope and downslope winds. Warm dry "föhn" is responsible for the sudden melting of snow and for avalanches in many parts of the Alps. The mean annual precipitation on the East Alps is summarized in table 35.

TABLE 35
Mean annual precipitation at different elevations on the East Alps

Elevation in m	Mean annual precipitation in mm
2127	1510
2560	1552
2810	1837
3080	2278
2242	2150

The duration of the winter snow-cover at different elevations on the Alps is summarized in table 36. Despite heavy precipitation, the general climate of the elevated area is dry. The relative humidity of the areas at higher elevations is less

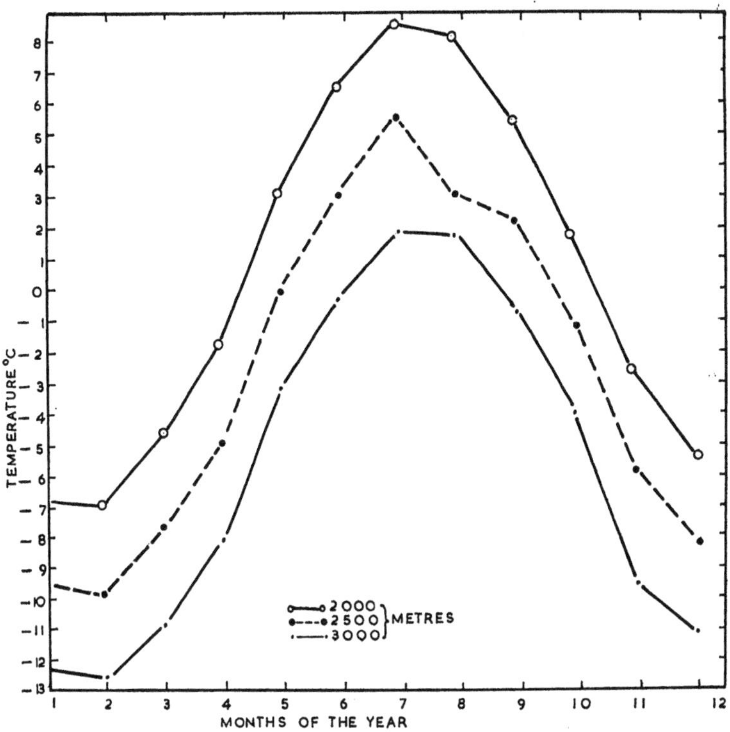

Fig. 66. Mean monthly atmospheric temperature at different elevations on the Grossglockner Alps.

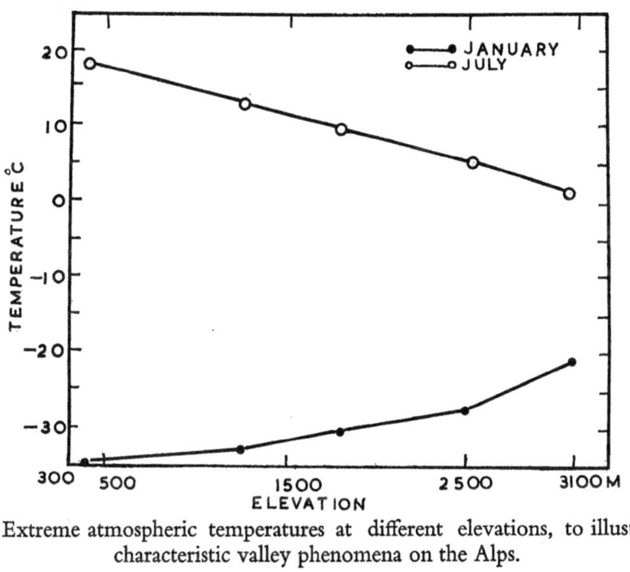

Fig. 67. Extreme atmospheric temperatures at different elevations, to illustrate the characteristic valley phenomena on the Alps.

throughout the year than at lower elevations. The East Alps are more continental in their general climate than the West Alps.

TABLE 36
The duration of the winter snow-cover at different elevations on the Alps

Elevation in m	Number of days of snow-cover
836	111
1120	156
1625	208
1950	208

The vegetation of the Alps is characterized by a number of peculiar archaic types, such as for example, *Saxifraga arachnoides*, the great age of which is correlated to i. the isolated position of the Alpine System, ii. the morpholocigal characters, iii. the ecological specializations, unilateral development and adaptations, iv. absence of dispersal and variation capacity and v. discontinuous and scattered distribution (146, 147, 149). As distinct from these archaic or the so-called Palaeogen-conserved endemics, there are also relatively younger endemic species of plants, particularly in polymorphic genera like *Hieracium*, *Alchemilla*, *Euphrasia*, etc., which are still largely in course of progressive development. The neo-endemics are accumulated characteristically in areas, which were heavily glaciated during the Pleistocene. The vegetation of the Alps contains the following floral elements: i. Tertiary basic stock of Mediterranean and Eurasiatic source, ii. boreal elements, sarmatic elements, weakly developed atlantic elements. The reader will find interesting accounts of the vegetation of the Alps in BECK V. MANNAGETTA (96-97), BRIQUET (154), BROCKMANN-JEROSCH (156), GAMS (393-397) and LÜDI (709).

The altitudinal biotic zonation on the Alps has engaged the attention of a number of workers like GAMS, HOLDHAUS, FRANZ and others. In 1929, HOLDHAUS (531) suggested the following biotic zones on the Alps: the lower forest zone or the colline zone, including the montane forest of mixed forest; the upper tree or the subalpine zone from the treeline to elevations of about 1100-1500 m and corresponding to GAMS' (393, 395) subalpine and upper montane forest zone: a transitional or intercalary zone extending from the line of snow-flakes during July to the timberline and corresponding roughly to the lower alpine zone of GAMS; and above these limits the high-alpine zone, extending from the lowest limit of snow patches in July to the upper limits of animal life. The lower forest zone is a belt of forests of the lowest mountains, nearly to the margin of the plains. Intensive cultivation has introduced many secondary features in the original faunal complex of this zone. It must be remembered that this zone is significantly deficient in true mountain autochthonous elements. The upper forest belt or the subalpine zone comprises the higher reaches of the forest, from elevations of 1100-1500 m to nearly the upper limits of the forest. Even in this zone, there is evidence of pronounced disturbance of the original mountain fauna due to cultivation, but the influence is much less than at lower elevations. There are several typical subalpine species, some of which do indeed ascend higher into the high-alpine zone. The transitional zone stretches from the forest limit to the lowest limits of snow during July. The absence of forest in this zone is correlated with the absence of many typically forest species. Owing, however, to the absence of summer snow-cover and also due to the moisture requirements, the exclusively high-alpine species do not as a rule penetrate into the intercalary zone. The fauna of the intercalary zone is, therefore, very much poorer than the zones at higher elevations. HOLDHAUS points out, for example, that the transitional zone on the Alps lacks a characteristic species-complex among Coleoptera. The intercalary zone is rather marked by negative characters. Among the few species that have their maximum abundance in the intercalary zone on the Alps, mention may be made of *Silpha tyrolensis* LAICH. The transitional zone has normally an altitudinal width of about 200 m. The high-alpine zone extends from the altitude of the lowest snow in July to the highest limit of existence of animal life. The zone is generally referred to in literature as the "alpine" zone. As pointed out by HOLDHAUS, the term alpine is applied in two wholly different contexts, viz. firstly as belonging to the Alps and secondly as pertaining to the alpine zone. In order to avoid the resulting confusion, HOLDHAUS proposes the term high-alpine (Hochalpin) for this biotic zone.

The following vegetation zones on the Grossglockner area of the East Alps are proposed by GAMS (395): i. the lower montane forest zone of mixed forest of Conifers and broad-leaved trees, reaching down to the valley floors; ii. the upper montane forest zone of *Picea excelsa* and *Larix* up to elevations of 1150-1600 m; iii. the sub-alpine zone up to elevations of 1500-1600 m on the north and 1800 m on the south slope; iv. the lower alpine (shrub) zone, varying in altitude limits and often ascending as high as 2400 m and descending to 1900 m above mean sea-level; v. the upper alpine or the grassy-meadow zone, between the closed meadows and the shrub zone, about 2040-2100 m in Pinzgauer Täler, 2100-2150 m in the Mölltal, 2150-2230 m in Kals, most alpine communities belong here and this is also the zone of the climax community of the *Caricion curvulae* (149); vi. the subnival zone or the vegetation-mat zone up to the closed meadow zone, 2400-2600 m on the north and 2600-2700 m on the south slope, with the snowline hanging down within this zone and with festoons of the timberline ascending also into it; and vii. the nival zone from the highest peaks to the level of the abundant growth of vegetation mats, 2900-3100 m on the north and 3180-3330 m on the south slope.

FRANZ (369) combines, however, the subnival and nival zones into a single nival-subnival zone, with a plant cushion fauna, the species composition of which decreases with increase in altitude and is characterized by the absence of Orthoptera, Hemiptera, Formicidae and Isopoda. Only Acarina, Tardigrada and Collembola are found at the highest elevations. Below this lies the high-alpine meadow-zone, to which belong the greatest majority of the high altitude insects. The transitional zone of shrubs lies between the meadows and the dwarf-trees. The subalpine zone is of dwarf-tree belt and below this lies the lower forest zone, from the upper limits of the mixed forest to the lowest valley floor.

The boundary altitudes between the various biotic zones mentioned above are naturally not sharply defined, but show considerable variations. As we find in the case of the Himalaya, in addition to the north-south slope differences, there are differences between the East and the West Alps also. Though the subnival zone lies at elevations of 2500-2800 m, these limits do not, however, hold good from the extreme West Alps to the last peaks in the extreme East Alps. A similar situation is also found in case of the nival zone, recognized by BÄBLER (74) at elevations of 2800-

4800 m. The nival zone is indeed a belt that is found invariably in the isohypse of 2800 m, but the limit oscillates considerably both upwards and downwards. BÄBLER considers the actual snowline as the most satisfactory lower altitudinal limit of the nival zone. As, however, differences exist between the orographic snowline and the climatic snowline, the latter is to be preferred for determining the limits of the nival zone. The altitudinal biotic zones, described here on the Central European mountains, are more or less readily applicable for other mountain ranges of the Balkan Peninsula, on the Apennines, on the mountains of Sicily and Corsica, etc. On the Pyrenees a subalpine zone may be distinctly recognized, with characteristic subalpine insect species complex, separate from the forest belt. In the Mediterranean area there are also high peaks with summer snow and these high-alpine zones have in south Europe a rich alpine-zone insect life.

The biotic zones at lower elevations are on the whole poorer in typical mountain autochthonous insects than the zones at higher elevations. In the lower biotic zones on the East Alps we find, for example, *Leistus piceus* FRÖL., *Trechus palpalis* DEJ., *Pterostichus metallicus* FABR., *Licinus hoffmannseggi* PANZ., *Platynus scrobiculatum* FABR., *Bythinus longulus* KIESW., *Cephennium* spp., *Euconnus* spp., etc. At higher elevations of 1100-1500 m, within the forest zone, there is a remarkable group of species, some of which descend to lower elevations locally in certain sheltered situations. Above these limits are the greatest bulk of the species that are completely absent at lower elevations. Among the species that are typical of the forest-zone we may mention the following Coleoptera from the East Alps: *Carabus fabricii* PANZ., *C. silvestris* PANZ., *Leistus nitidus* DUFT., *Nebria dejeani* DEJ., *N. fasciatopunctata* MILL., *Trechus limacodes* GGLB., *Tr. alpicola* STURM., *Pterostichus* spp., *Orina intricata* GERM., *O. gloriosa* FABR., *O. alpestris* SCHUM., etc. In the zone immediately above the upper limit of the forest we come across a wholly new type of species complex. There is a significant decline of the subalpine elements, though there may be a few forest types. At elevations of 200 m above the forest limits we have the approximate boundary of a wholly new type of biota of the high-alpine zone. There are several subgenera like *Oreonebria* DAN., *Leirides* PUTZ., *Leiromorpha* GGLB., etc., which are exclusively high-alpine zone forms. According to well known authorities, there are over 75 species of exclusively high-alpine Coleoptera on the Alps.

4. THE INSECT LIFE OF THE WEST ALPS

ORTHOPTERA

The most important contributions to our knowledge of the Orthoptera from the Alps are by GRABER (421), FRUHSTORFER (390), EBNER (291) and RAMME (856). While we find a number of endemic genera on the Pamirs and the Northwest Himalaya, the Alps are remarkable for the presence of a number of characteristically widely distributed hypsobiont genera like *Gomphocerus* THUNB., *Podisma* LATR. and *Bryodema* FIEB., which are often represented only by endemic subspecies and by boreo-alpine species. *Br. tuberculata* (FIEB.), found on the West Alps, is, for example, a Siberian form that extends from North China to the Caspian Sea area, Finland and Denmark. *Chorthippus viridulus* (LINN.) occurs up to an elevation of 2730 m. There are several flightless species of *Podisma* LATR., like *P. alpina* (KOLL.) with the subspecies *P. a. alpina* (KOLL.) and *P. a. subalpina* FISCHER occurring at elevations of about 2000 m and sometimes even on the Aletsch Glacier, at an elevation of 2800 m. Other species include *P. frigida* BOH. at elevations of 2100-2600 m and usually associated with the boreo-alpine *P. pedestris* (LINN.) at elevations of 2200-2500 m and sometimes even at 3000 m. *Aeropus* (= *Gomphocerus*) *sibiricus* (LINN.) is a typical Siberian-alpine element, abundant at elevations of about 1800 m and also widely distributed on the Apennines, Pyrenees, the Cordillera Carpentata, Schieferalpen (2800 m), Harz, Bavarian Alps (2000 m) and the Caucasus Mountains.

HETEROPTERA

The maximum altitudinal limit for most Heteroptera on the West Alps is the treeline (1800-2200 m). As the mean of the atmospheric and soil temperatures at higher elevations is much lower than the mean thermal requirement minima of 10-15° C for these insects, there is an abrupt fall in their abundance at the treeline. Very few species are typically subalpine or alpine and exceedingly rarely do we find any truly resident Heteroptera in the subnival and nival zones of the Alps. There are hardly a dozen species recorded so far from above the treeline. *Eurydema oleraceum* LINN. often occurs at an elevation of 2600 m and *Chlorochroa juniperina* LINN. has also been reported from about the same elevation. The record of *Therapha hyosciami* LINN. on a nunatak at an elevation of 3100 m, in association with *Dichroscytus rufipennis* FALL., is extremely interesting. The strictly alpine forms are *Eurydema fieberi rotundicolle* DOHRN., *Geocoris megalocephalus* COSTA and *Systelnotus alpinus* F.-G. Most species have a single generation and lay eggs in spring, complete development in 6-8 weeks during the summer and the adults hibernate during the winter. HOFFMÄNNER (524) has given an interesting account of the ecology of Heteroptera from the Swiss Alps and the reader should refer to his paper for further details.

COLEOPTERA

Nearly fifty percent of the Coleoptera at high elevations, particularly in the subnival and nival zones of the West Alps, are Carabidae. The species mostly occur in vegetation terraces, strewn with boulders and stones. *Bembidion glaciale* HEER and *Trechus glacialis* HEER occur, however, even on barren stony moraines and sometimes also on glaciers. The former species was collected by BÄBLER (74), for example, at an elevation of 3237 m on glaciers. *Nebria germari* HEER occurs on the Silvretta Glacier area and *N. bremii* GERM. on the Bündnerberg Glacier. From the Aletsch Glacier we have the records of *N. frontalis* DAN. and *N. castanea* RON. According to BÄBLER, there are about ten endemic species in the Finsteraarhorn area and five in the Silvretta Glacier area. The other nival-zone Carabidae include *Carabus concolor amplicollis* KRATZ at elevations of 2700-2750 m, *Nebria germari* HEER at 2500-2900 m, *Bembidion bipunctatum* R. at 2760-3000 m and all found under stones; *B. glaciale* HEER on sand at an elevation of 3020 m and under stones at 2500-3237 m; *B. pyrenaeum* DEJ. under stones at elevations of 2900-3000 m; *Trechus glacialis* HEER among moss in rock crevices at elevations of 2600-3000 m; *Tr. pertyi langobardus* PUTZ under stones at an elevation of 2900 m; *Amara familiaris* DEFT., under stones at an elevation of 2500 m and *Cymindis vaporariorum* LINN. under stones at elevations of 2800-3000 m.

The Staphylinidae include *Leptusa globulicollis* MULS. & REY at an elevation of about 2890 m, *Atheta leonhardi* BRB. on humus under stones at an elevation of 2700 m, *A. raettgeni* BRB. on vegetation debris at an elevation of 3030 m and *Aleochara bipustulata* LINN. at an elevation of about 2890 m.

Other Coleoptera from the nival zone include the Byrrhidae *Byrrhus pilula* LINN. (2500-2800 m), Hydrophilidae *Helophorus schmidti* VILLA and *H. glacialis* VILLA (2700-2800 m), Chrysomelidae *Orina speciosissima viridescens* SUFFR. on vegetation at elevations of 2800-2900 m and *O. sp. troglodytes* KIESW. under stones at 2700-2800 m, *Phytodecta flavicornis nigra* SUFFR. under stones and in moraines (2500 m) and the Curculionid *Otiorrhynchus pupillatus subdentatus* BACH. under stones at elevations of about 2800 m.

The important subalpine and alpine Coleoptera from the Swiss Alps include *Cychrus cordicollis* CHAUD., *Carabus fabricii* PANZ., *C. latreillei* DEJ., *Nebria crenatostriata* BASSI, *N. cordicollis* CHAUD., *Trechus strigipennis* KIESW., *Tr. montis-rosae* JEANNEL, *Tr. artemisiae* PUTZ., *Tr. caprai* JEANNEL, *Tr. ceresai* BIN., *Tr. pertyi* HEER, *Tr. glacialis* HEER, *Pterostichus parnassius* SCHAUM., *Platynus depressum* DEJ. and *Amara doderoi* BAL.

LEPIDOPTERA

Widely distributed lowland forms like *Vanessa urticae* LINN. are often common among the high-alpine zone flowers at high elevations on the West Alps. The most characteristic species of the nival zone in the West Alps is *Erebia glacialis* ESP. that occurs in great abundance among snow patches, among grass, under stones, etc. at elevations of 2800-3250 m. Unlike the adults from the subnival zone, those from the nival zone have brighter spots on the forewings. *E. gl. pluto* ESP. occurs at elevations of about 2800 m. The other important species are *E. gorge* ESP., *E. alecto* ESP. (2800-2960 m) and *Argynnis pales* SCHIFF. (2800 m). *Pieris callidice* ESP. occurs on grass under stones at elevations of 2700-2800 m. A number of moths like *Dasydia tenebraria* ESP. on the lichen-incrusted rock and under stones at elevations of 2500-2800 m, *Psodos alticolarius* MÉN. on lichen-incrusted stones at 2700-3237 m, *Gnophos caelibaria spuricaria* LATR. on lichen-incrusted rock at elevations of 2700-3000 m and *Gn. zelleraria* FABR. under stones at elevations of 2700-2800 m (Geometridae); *Setina andereggia* HS. (Arctiidae) under stones at 3237 m and *Anarta melanopa* THUNB. (Noctuidae) on vegetation-cushions at elevations of 2700-2800 m have also been found in the subnival and nival zones of the West Alps.

DIPTERA

Diptera are extremely abundant and include a number of endemic species. Most species have the capacity to tolerate rapid and wide changes of temperature and swarms of flies often settle down on warm rock surfaces during the hours of bright sunshine, but get under shelter in cloudy weather. According to BÄBLER (74) and other workers, there does not seem to be a well defined upper altitudinal limit for Diptera on the West Alps. Adult flies have indeed been found on some of the highest peaks, at elevations of 3300-4275 m. The larvae of *Simulium* LATR. were observed by BÄBLER in melt-water streams below glacier snouts. Some remarkable Chironomidae from the Alps have been described by THIENEMANN (1035, 1036). LINDNER (697) has given an interesting account of the distributional and ecological peculiarities of the high altitude Diptera, particularly the diurnal periodic activities of the adult Anthomyiidae from the Alps. While most Tipulidae from the Alps are apterous only in the female, LINDNER records the apterous condition in the males of *Tipula sexspinosa* STROBL from the Koralps. The following are some of the more typical high altitude Diptera from the West Alps: Chironomidae: *Orthocladius stercorarius* DG. on vegetation mats at elevations of 2700-2900 m, Mycetophilidae: *Sciara quinquilineata* MACQ. *nigripes* STROBL in rock crevices at 2850 metres, *Sc. silvatica* MEIG. on vegetation mats at 2800-2900 m, *Sc. affinis* ZETT. under stones at 2765-3020 m, and *Sc. pallipes* T. on vegetation mats at 2800-2850 m, in rock crevices at 2970 m and under stones at 3316 m. Limnobiidae: *Dactilobis denticulata* BERG. on moss at 3100-3200 m and *Chionea alpina* BEZZI in rock crevices at elevations of 2700-2800 m. Syrphidae: *Platycheirus podagratus* ZETT. in rock crevices at 2765 m and *Hencostomis vivax* LÖW on vegetation mats at elevations of 2800-2900 m. Empidae: *Clinocera trinotata* MCK. on moist stones (2690 m), *Cl. beckeri* MCK. on moss under moist stones (2700-2800 m) and *Cl. undipes* LÖW on moss under stones at 2700 m. Syrphidae: *Syrphus toparius* MEIG. on vegetation, on wings, at 3000 m and among moss at 2950-2970 m. Sarcophagidae: *Sarcophaga carnaria* LINN. under stones at 3020 m. Muscidae: *Cystoneura pascuorum* MEIG. under stones (2780 m), *Pollenia atramentaria* MEIG., *Daryphora piatorum* MEIG. and *D. versicolor* MEIG. under stones at 2970 m, *Lucilia cornicia* MEIG., *Calliphora erythrocephala* MEIG. and *Myospila meditabunda* FABR. (2970 m). Anthomyiidae: *Hoplogaster obscuricula* ROND. (2850 m), *Chortophila coerulescens* STROBL on vegetation at 2800-2975 m, *Ch. c. grisella alpina* STROBL on moss at 2870 m and on moist stones at 3000 m, *Anthomyia aestiva* MEIG., *Mydaea depuncta* FALL. and *Pogonomyia alpicola* ROND. (2750-2800 m). The Anthomyiidae are dominant Diptera at the highest elevations.

COLLEMBOLA

BÄBLER (74) recorded the Collembola as occurring at the highest elevations, in nearly all the biocenoses, in moraines, glaciers, snowfields, on stones and under stones and in glacial streams. He found the Entomobryiidae to be dominant. An excellent account of the general ecology and habits of the Collembola from the Swiss Alps is given by GISIN (409). The following important species deserve mention: *Hypogastrura bengtssoni* AGR., known also from Scandinavia, occurs on moss at an elevation of 2500 m; *H. armata* NIC., a cosmopolitan species found throughout Europe, North and South America, Sumatra and New Zealand, occurs at an elevation of 2500 m, on the Desor Glacier and at an elevation of 2800 m on the Ewigschneehorn in the Swiss Alps; *H. manubrialis* TULLB., widely distributed in north and Central Europe, Siberia and South America, occurs at elevations of 2600-3400 m near glaciers on the Swiss Alps; *H. vernalis* CARL. occurs under stones near snowfields at an elevation of 2500 m and is also known from Fennoscandia; *H. sahlbergi* REUT., a widely distributed species in Russia, Finland and Germany, occurs at elevations of 2600-3300 m on the Swiss Alps and

H. frigida AXELS. occurs on the Unteraar Glacier at an elevation of 2400 m. *Onychiurus armatus* TULLB., widely distributed in Europe, Siberia, Iceland, Greenland and other Arctic Islands, North America and Chile, occurs at elevations of 2600-3400 m in the immediate vicinity of glaciers on the West Alps; *O. a. inermis* AXELS. occurs at elevations of 3200-3400 m; *O. zschokkei* HANDSCHIN occurs under stones at an elevation of 2650 m; *O. ambulans* NIC. collected at an elevation of 2890 m on the West Alps is previously known from North America and is also widely distributed in Europe, except in the extreme north; and *O. tuberculatus* MON. is known from elevations of 2600-3000 m on the West Alps and occurs also in caves in North France. *Tetracanthella alpina* CARL and *T. afurcata* HANDSCHIN occur at elevations of 2900-3400 m. *Proisotoma schoetti* D.T., found at elevations of 2220-2500 m on the West Alps, is known also to occur in north and Central Europe, Spitzbergen and North America and *Pr. crassicauda* TULLB., known from Sweden, Russia, the Shetland Islands and Hungary, occurs at elevations of 3000-3100 m. *Isotoma (Vertagopus) westerlundi* REUT., found at elevations of about 2500-4257 m, is a typical winter form in Finland; *I. (V.) sensibilis* TULLB., from north and Central Europe, Spitzbergen and North America, occurs at elevations of 2100-3400 m; *I. (Isotoma) saltans* AG., a character species of snow and glaciers on the Alps, is dominant at elevations of about 2500-2800 m, but may often ascend to 3800 m and occurs under stones in crevasses; *I. nivalis* CARL occurs under stones and inside rock crevices at elevations of 2700-2850 m and *I. hiemalis* SCHÖTT., a true boreo-alpine species occurs at elevations of 2600-3100 m. *Agrenia bidenticulata* TULLB. is another boreo-alpine species that occurs at elevations of 2200-2500 m. *Isotomurus palustris* MÜLL., a widely distributed species, occurs at elevations of 2500-3000 m. *Entomobrya nivalis* LINN., widely distributed in Europe and North America, occurs up to an elevation of 3200 m. *Lepidocyrtus lanuginosus* GMEL., widely distributed throughout Europe and North America, occurs up to an elevation of 3200 m and *L. cyaneus* TULLB. occurs at elevations of 2700-3100 m. *Orchesella bifasciata* NIC., found in north and Central Europe, ascends to elevations of 2500-3000 m and *O. alticola* UZEL is perhaps one of the commonest species at elevations of 2400-3100 m on the West Alps. *Bourletiella pruinosa* TULLB. and *B. lutea* LUBB., known from Fennoscandia and parts of Central Europe, occur at elevations of 2400-3300 m.

5. INSECT LIFE OF THE EAST ALPS

The literature on the high altitude insects of the East Alps is very extensive. The Coleoptera are described by a number of workers like AMANN & KNABL (44), BREUNING (150), FRANZ (367, 369-374), HÖLZEL (520), HORION (542), GREDLER (422), HEBERDEY & MEIXNER (490), HEER (492), HOLDHAUS (526), KRAUSS & GANGLBAUER (656), MANDL (723), MILLER (774), MOOSBRUGGER (783), SCHWEIGER (944, 945, 947) and WÖRNDLE (1121). Descriptions of Lepidoptera may be found in CARNELUTTI (187), HOFFMANN (523), HORMUZAKI (543), KITSCHELT (625) and PAX (817). The Collembola are described by HANDSCHIN (469), LATZEL (677) and STEINBÖCK (1003-1007). AUSSERER (60) and SCHWOERBEL (949) have described the Arachnida and BIGLER (124-126) and ATTEMS (58) have dealt with the Diplopoda. FRANZ (375) and JANETSCHEK (564) have discussed various aspects of the ecology and HOLDHAUS (534) and JANETSCHEK (567) have made valuable contributions to the biogeography of the region.

On the East Alps, the Middle Hohe Tauern have been a centre of intensive entomological and ecological investigations since the botanist FRANZ XAVIER VON WOLFEN visited the Glockner area in 1775. The most important biological explorations in the area were made by SIGMUND VON HOCHENWARTH, who also collected a large number of alpine-zone Lepidoptera. SCHULTES and the botanist HOPPE made large collections of diverse high altitude insects. In 1844 the entomologist NICKOL made important collections of Lepidoptera. Most valuable observations were also made by the SCHLAGINTWEIT brothers during 1848-1851 (924). In recent years HOLDHAUS and FRANZ have made valuable contributions to our knowledge of the general ecology and biogeography of the high altitude insects form the East Alps.

ORTHOPTERA

The commonest high altitude grasshoppers on the East Alps are *Acrydium bipunctatum* (LINN.) at elevations of 2200-2400 m, *Podisma alpina alpina* (KOLLAR) at an elevation of about 2200 m, *P. frigida strandi* FRUHST. (2300-2400 m), the boreo-alpine *Aeropus sibiricus* (LINN.) (1000-2400 m) and *Metrioptera brachyptera* (LINN.) at elevations of 2000 m and above.

DERMAPTERA

Anechura bipunctata (FABR.) is confined in the East Alps to elevations of about 1400-2500 m, on the southern slopes of the Hohe Tauern.

COLEOPTERA

Nearly 70 high altitude species of Coleoptera have so far been described from the East Alps and many exclusively alpine-zone species are sporadic on the East Alps and occur in greatly isolated and localized concentrations. There are also many boreo-alpine species. The Carabidae account for over 85% of the high altitude Coleoptera from the East Alps.

Carabidae

The following are some of the important records: *Cychrus schmidti* CHAUD. and *C. cylindricollis* PINI; *Carabus depressus* BON., with the subspecies *C. d. bonellii* DEJ. at elevations of 1600-2000 m on the Hohe Tauern; *C. fabricii* PANZ. as low as the subalpine zone; *C. alpestris* STURM. common locally in the alpine zone, but descending to an elevation of 2000 m; *C. bertolinii* KR. exclusively in the alpine zone; *C. carinthiacus* STURM. both in the subalpine and alpine zones and often descending to an elevation of 1200 m. *Bembidion bipunctatum* LINN. is common in the alpine zone and often at the snow-edge at an elevation of 2500 m; *B.* (*Peryphus*) *nitidulum alpinum* DEJ. near melt-water streams; and *B.* (*Testediolum*) *pyrenaeum glaciale* HEER occurs at elevations of 2000-2700 m, often also at 3000 m. There are several species of *Nebria* LATR. The widely distributed and exclusively alpine-zone *N. hellwigi* PANZ. never occurs below an elevation of 1800 m. The subspecies *N. h. stigmata* DEJ. is endemic to the Central Hohe Tauern. *N. bremii* GERM. ascends often to the snowline and *N. germari* HEER also occurs at the snow-edge. The subspecies *N. g. norica* SCHAUB. is endemic on the Central Hohe Tauern and occurs at the glacier margin, at elevations of 2000–2500 m. *N. castanea* BON. is common at the edge of melting snow and the subspecies *N. c. brunnea* DUFT. occurs at an elevation of 2000 m on the Hohe Tauern. *N. schusteri* GGLB. is restricted to the Koralps and *N. lombarda* DAN. is widely distributed in the alpine zone, but sometimes occurs as low as the forest-edge and is common at the snow-edge at an elevation of 2700 m. There are two other exclusively alpine-zone species, viz. *N. diaphana* DAN. and *N. angustata* DEJ.

Three species of *Broscosoma* PUTZ. occur under stones in the alpine zone; of these *Br. relictum* WEISSM. is localized on the summit of Mte Alben in the Bergamasker Alps. A number of species of *Trechus* CLAIRV. are confined to the alpine zone of isolated summits on the East Alps. Some of these species are, however, widely distributed throughout the Alps and occur both in the subalpine and alpine zones. The exclusively alpine-zone species include *Tr. pallidulus* GGLB., *Tr. meschniggi* JEANNEL (restricted to only the Mte Raut, Mte Verzegnis and Cima Monfalcone in the Venetian Alps), *Tr. wagneri* GGLB., *Tr. ovatus* PUTZ. (predominantly found at the edge of the melting snow), *Tr. glacialis* HEER, *Tr. ochreatus* DEJ., *Tr. rudolphi* GGLB., *Tr. hampei* GGLB. (often coming down to the subalpine zone) and *Tr. regularis* PUTZ. (restricted to the alpine zone of the Koralps only). While *Pterostichus ziegleri* DUFT., *Pt. kokeili* MILL. and *Pt. lineatopunctatus* DEJ. are exclusively alpine-zone species, others like *Pt. cognatus* DEJ., *Pt. illigeri* PANZ., *Pt. schachli* GGLB. and *Pt. multipunctatus* DEJ. are found both in the subalpine and alpine zones. *Pt. aurus* DUFT. occurs in the alpine zone on the East Alps and in the upper reaches of the forest on the Carpathians. *Amara spectabilis* SCHAUM. is exclusively alpine, but *A. alpestris* VILLA occurs sometimes also in the upper reaches of the forest.

Staphylinidae

Atheta alpestris HEER and *Aleochara* (*Coprochara*) *bilineata* GYLL. occur under stones on the high-alpine grassy-meadows and are widely distributed on the East Alps.

Chrysomelidae

Orina melanocephala DUFT. and *Crepidodera simplicipes* KUTSCH. are exclusively alpine-zone forms and the first mentioned species ascends often to the snowline. *Chrysomela lichenis* RICHT. is an alpine-zone form that occurs sometimes in the subalpine zone on the East Alps.

Curculionidae

The exclusively alpine-zone species include *Otiorrhynchus chalceus* STIERL., *O. hadrocerus* DAN., *O. cadoricus* DAN., *O. costipennis* ROSCH., *O. schmidti* STIERL. and *O. planiceps* DAN. These species occur mostly under stones. A number of other typically subalpine species of the genus often occur in the lower levels of the alpine zone also.

HYMENOPTERA

The ecology and biogeography of the Bombidae and Formicidae from the East Alps present a number of interesting features. *Bombus (Hortobombus) hortorum* LINN., known from Middle Asia, also occurs within the subalpine zone on the East Alps. *B.* (*Pomobombus*) *elegans* SEIDL., widely distributed on the Alps, ascends to an elevation of 2600 m on the East Alps. Though *B.* (*Soroensibombus*) *soroensis proteus* GERST. may often be found at an elevation of 2800 m, it generally nests at much lower elevations and is also widely distributed. *B.* (*Alpinobombus*) *alpinus* LINN. is a boreo-alpine species that occurs up to an elevation of 2000 m on the East Alps. *B.* (*Lapidario-bombus*) *lapidarius* LINN., widely distributed in Europe, occurs up to an elevation of 2000 m. *B.* (*L.*) *alticola* KRIECHB. is widely distributed at elevations above 1600 m on the Alps, the northern Apennines, Pyrenees and the mountains of the Balkan Peninsula. The boreo-alpine *B.* (*Pratobombus*) *lapponicus hypsophilus* SKORIKOV occurs both within the subalpine and in the alpine zones and often ascends to elevations of 3000 m on the East Alps. The high altitude ants from the East Alps include *Myrmica* (*Myrmica*) *lobicornis* NYL., *Leptothorax* (*Mycothorax*) *acervorum* (FABR.) up to an elevation of 2400 m on the East Alps and 2600 m on the West Alps. *Formica* (*Serviformica*) *fusca fuscogagtes* FOREL occurs at elevations of 1300-2600 m on the East Alps.

LEPIDOPTERA

The following are the common alpine and high-alpine zone species: *Parnassius delius* ESP., with the larvae feeding on *Saxifraga aizoides*; *Colias phicomene* ESP., sometimes found even within the subalpine zone; *Melitaea asteria* FRR., *Argynnis pales* SCHIFF., *Erebia epiphron cassiope* FABR., a boreo-alpine form that extends from the subalpine to the high-alpine zone, *E. gorge* ESP., *E. lappona* ESP., *Coenonympha arcania satyrion* ESP., *Lycaena eros* OS., *L. corydon* PODA and *L. semiargus* ROT. The common Heterocera include *Gnophos caelibarius intermedius* KANTZ at elevations of 2300-3100 m, *Gn. zellerarius* FRR., an endemic species characteristic of the zone between 2300 and 2500 m, *Dasydia tenebraria innuptaria* HERR.-SCH. at elevations of 2200-3200 m, *Psodos alticolarius* MÉN. not found below an elevation of 2400 m, *Ps. alpinatus* SCOP., *Ps. noricanus* WAGN. and *Arctia quenseli* PAYK., the caterpillars of which occur under stones at elevations of about 2700 m.

DIPTERA

Nearly 600 species of Diptera have so far been described from the East Alps and some of them are endemic to the area, but the others are boreo-alpine and still others extend to the West Alps also, or to the Carpathians.

The more important records include the following: Tipulidae: *Pachyrhina excisa* SCHUMM., a widely distributed high-alpine zone species, that occurs particularly in the northern parts of the Alps, up to elevations of 2800 m and *P. irregularis* POK. endemic to the Alps and also restricted to the altitude zone of 2500-2900 m. Sciaridae: *Neosciara auripes* WINN., *Orinosciara brachyptera* LENGERSDORF (the genus is endemic to the East Alps), *Caenosciara ignava* LENGERSDORF and *Neosciara diversiabdominalis* LENGERSDORF found at an elevation of 2650 m, in soil at a depth of about 5 cm. Other records are *Bibio johannis* (LINN.), *Hydrina nubeculosa* STREBL. (Ephydridae) (3000 m), *Eriozona syrphoides* FALL. and *Lasiopticus pyrastri* LINN. (Syrphidae), and the Anthomyiid *Phaonia morio* ZETT., a boreo-alpine species, found at elevations of 2000-2100 m. *Rhynchocoenops obscuricola* ROND. is found up to an elevation of 2900 m; this species occurs also on the Apennines and the Pyrenees. *Rh. subrostratus* ZETT., a boreo-alpine species, occurring up to an elevation of 3000 m on the East Alps and *Rh. villosus* HENDEL, an endemic species on the Alps, occurring at elevations of 2500-3000 m are interesting records. *Mydaea fuscula* FALL. (3000 m) and the Sarcophagid *Brachicoma devia* FALL. (2600-3000 m) may also be mentioned.

COLLEMBOLA

While the Collembola from the West Alps have been intensively studied by several workers, our knowledge of the Collembola from the East Alps is derived largely from the studies of LATZEL (676, 677) and STEINBÖCK (1003-1007). About 70 species are subalpine forms, but extremely few seem to be truly alpine-zone elements in the East Alps. In 1948, FRANZ (370) made a valuable contribution to our knowledge of the soil Collembola from the East Alps, with special reference to their ecology and distribution. The cosmopolitan ubiquit *Hypogastrura armata* (NIC.) occurs in soil and on moss at elevations of 2800 m on the East Alps. The Holarctic *Lepidocyrtus lanuginosus albicans* REUT. occurs at elevations of 2350-3200 m. *Orchesella alticola* UZEL is widely distributed and occurs under stones, at elevations of about 3100 m. *Entomobrya nivalis* LINN. is another well known Holarctic species that occurs at elevations of 1630-3200 m. The other records include *Friesea emucronata* STACH, *Onychiurus alpinus* STACH on and near snow, at elevations of 2200-3100 m; *Isotoma (Pseudisotoma) sensibilis* TULLB., *Lepidocyrtus cyaneus* TULLB., *Orchesella bifasciata* (NIC.), *O. montana* STACH under stones, at elevations of 2300-3100 m and *O. viridilutes* STACH at an elevation of 2500 m on the East Alps and known also from the Carpathians.

6. OTHER ARTHROPODA

ARANEIDA

Hilaira montigena L. KOCH is perhaps the most widely distributed high altitude spider on the Alps. A number of other species occur above the snowline, on some of the highest summits. BÄBLER (74) found, for example, spiders at elevations of 4500 m on the Punta Gnifetti. *Drassodes* spp. are common at elevations of about 2950 m and *Pardosa* spp. occur in a variety of biocenoses, but most usually under stones, at elevations up to 3500 m.

The following species are known from the West Alps: *Drassodes heeri* PAVESI, under stones, at elevations of 2700-3000 m; *Dr. troglodytes* L. KOCH under stones at elevations of 2820-3000 m, *Prosthessima clivicola* L. KOCH under stones at elevations of 2820-3000 m, *Gnaphosa petrobia* L. KOCH under stones at elevations of 2735-3000 m, *Diplocephalus eborodunensis* CAMB. in rock crevices at elevations of 2700-3000 m,

Stylotector brocchus L. KOCH under stones at elevations of 2740-3030 m, *Cornicularia karpinskii* CAMB. under stones at an elevation of 2760 m, *Hilaira montigena* L. KOCH under stones at elevations of 2500-3320 m, *Macrargus adipatus* L. KOCH under stones at an elevation of 2770 m, *Microneta glacialis* L. KOCH under stones at 2725 m, *Micryphantes gulosus* L. KOCH in rock crevices and under stones at elevations of 2700-3350 m and *M. nigripes* SIMON inside rock crevices at an elevation of 2893 m and under stones at elevations of 2780-2825 m. *Lephthyphantes baebleri* DE LESSERT occurs under stones at elevations of 2500-2825 m and inside rock crevices at 3030 m. *Xysticus glacialis* L. KOCH occurs under stones at elevations of 2700-2900 m. Under stones may also be found *Pardosa giebeli* PAVESI (2500-2800 m), *P. nigra* L. KOCH (2800-3000 m) and *P. pedestris* SIMON (2700-2800 m). From the East Alps is recorded *Titanocera obscura* (WALCK.) that is widely distributed in the plains also and occurs up to the high-alpine zone (2300 m). *Erigone remota* L. KOCH is a boreo-alpine species that occurs at the snow-edge and under stones at elevations of 2350-3100 m. Other records from the East Alps include *Heliophanus dubius* L. KOCH (2200-2300 m), *Xysticus desidiosus* SIMON (2200 m), *Lycosa hyperborea pusilla* THORELL (2200 m) and *Gnaphosa badia* (L. KOCH), an endemic form from the Alps found at elevations of about 2300 m.

ACARINA

A number of species are widely distributed on the Alps and occur in diverse habitats and communities, but others seem to be characteristically restricted to isolated high peaks like the Finsteraarhorn (4275 m) and the Silvrettahorn (3248 m) on the Swiss Alps. In association with Collembola, the Acarina may be said to be character species on some of the highest isolated summits. The highest elevation at which BÄBLER (74) collected mites is between 4500 and 4600 m on the Mte Rosa. The following are some of the interesting records:

Trombididae: *Caeculus echinus* DUFOUR under stones, at elevations of 2500-2800 m and only rarely at an elevation of 1600 m on the East Alps; *Trombidium bicolor* L. KOCH and *Tr. pusillum* HERM. under stones, at elevations of 2735-2820 m; *Erythræus regalis* C. L. KOCH under stones at elevations of 2820-2850 m; *Rhyncholophus unidentatus* TADB. under stones at 2820 m; *Rhagidia miniatus* HERM, *Rh. nemorum* C.-L. KOCH and *Rh. gigas* CAN. under stones at 2500-2970 m; *Pentathodes ovatus* C.-L. KOCH under stones and on moss at elevations of 2800 m; *Bryobia speciosa* C.-L. KOCH inside rock crevices at elevations of 2760-3000 m and *Tarsotomus hercules* (BERL.) under stones at elevations of 2770-3000 m. Oribatidae: *Damarus clavipes* HERM. under stones at 2820 m; *Ceratophila bipilis* HERM. among roots of vegetation and among moss at elevations of about 2800 m; *Cepheus* sp., *Oribatella* sp. and *Neoliodes concentricus* SAY under stones at elevations of about 2750-2800 m; *Oribata orbicularis* C.-L. KOCH under stones at elevations of 2770-4090 m; *O. fuscipes* C.-L. KOCH and *O. setosa* C.-L. KOCH in soil and under stones at elevations of 2770-3000 m; *Oribatula tibialis* NIC. and *O. exilis* BERL. under stones at elevations of 2900-3000 m. Gamasidae: *Pergamasus quisquiliarum* G.R.S. on moss and under stones at an elevation of 2800 m and *Eugamasus* sp. in rock crevices at an elevation of about 3316 m. Bdellidae: *Bdella capillata* KRAM. under stones at elevations of 2500-2800 m and in rock crevices at elevations of 2890-

3300 m; *Bd. vulgaris* HERM. under stones at elevations of 2500-2800 m, in rock crevices at elevations of 2900-3000 m and on moss at an elevation of 3000 m and *Cyta (Ammonia) latirostris* HERM. under stones at 2800 m and on moss at 3000 m.

The records of Acarina from the East Alps include the interesting *Pergamasus parvulus* BERL. that occurs from the surrounding lowlands to elevations of nearly 2500 m and *Rhagidia terricola* C.-L. KOCH occurring both in the lowlands and on the mountains up to an elevation of nearly 2400 m and sometimes even up to 3100 m in parts of the Swiss Alps. Other records are *Bdella subulirostris* BERL. (3100 m), *Neomolgus capillatus* KR. (2200-3000 m), *N. monticola* (2600-2900 m), *Bryobia praetiosa* (C.-L. KOCH) from the Ötztaler Alps (2250-3775 m) and Swiss Alps (3000 m) and *Microtrombidium sucidum* BERL. exclusively from the snow-edge and from the snow surface.

The true subterranean water-mites from the Alps include *Feltria phreaticola* SCHWOERBEL, *Lethaxona pygmaea* SZALAY, *Stygomomonia latipes* SZALAY and *Atractides magnirostris* MOTAS. The hyporheic water-mites from the Alps belong to the following ecological groups (949): i. Hyporheobiont species, found exclusively in subterranean water, like for example, *Stygothrombium chappuisi* SCHW., *St. bispinosum* SCHW., *Wandesia thori* SCHW., *Lebertia hausmanni* SCHW., *Torrenticola jeanneli* MOTAS & TAN., *Atractides latipalpis* SCHW., *A. pygmaea* MOTAS & TAN., *A. szalayi* SCHW., *Feltria subterranea* VIETS, *Ljania macilenta longissima* SCHW., *Stygomomonia latipes transversalis* SCHW., *Stygohalacarus subterraneus* SCHW. and *Arrenurus haplurus* SCHW.; ii. Hyporheophile species that occur regularly in underground waters, but often also above ground on the Alps, such as, for example, *Bandakia concerta* SCHW., *Feltria cornuta longispina* SCHW., etc.; iii. Autohyporheoxene species occur less regularly as members of underground water communities, may be found in torrents and do not exist above ground: Examples include *Protzia eximia* (PROTZ) and *Torrenticola elliptica* MAGLIO; and iv. Tychohyporheoxene species occur only sparsely and occasionally in underground waters: Example *Atractides gibberipalpis* PIERS.

DIPLOPODA AND CHILOPODA

The most important contribution on the high altitude Diplopoda from the Alps is by BIGLER (124-126). We know of at least four typically mountain autochthonous high-alpine endemic species, two predominantly high-montane endemics and four valley endemics. The high-alpine endemic species are *Polydesmus monticola* LATZ. (1700-2400 m), *Orotrechosoma alticum dormeyeri* VERH. (2100-2750 m), *Trimerophorella nivicomes* VERH. (2100-2700 m) and *Leptoiulus sarasini* BIG. (2350-2600 m). The second and the third species, mentioned above, belong also to typically mountain genera. The highest elevation at which BÄBLER (74) found Diplopoda on the Swiss Alps is 2900 m, but on the south slopes of Mte Rosa he found them at least 300 m higher. *Leptoiulus (Leptoiulus) alemannicus alemannicus* VERH. and *L. simplex simplex* VERH. occur under stones in the alpine and high-alpine zones. Among the high altitude Diplopoda from the Alps we find eight western elements, three southern elements, four boreal elements, five eastern elements and three widely distributed species.

The Chilopoda seem to occur at much higher elevations on the Alps than the Diplopoda. *Lithobius lucifugus nivalis* ROTHENB. was recorded by BÄBLER at elevations of 2500-3237 m. Some species, like *Hyposoiulus alpivagus* VERH. (Juloidea), occur at the snow-edge, at elevations of 2500-2800 m.

7. COMMUNITIES AND ECOLOGICAL SUCCESSION ON THE EAST ALPS

i. Some typical Communities

The principal high altitude communities are grouped as follows (369):

1. The communities of the subnival-nival zones, mainly *Nebria atrata-Gnophos caelibarius* association in plant-cushions and on rock surfaces covered by vegetation-mats above the snowline. 2. The subnival "Kalkphyllite" debris mounds association of *Caeculus echinus-Chrysomela crassicornis norica.* 3. The young moraines and the high-alpine zone slopes, strewn with stones and boulders, with association of *Lycosa nigra-Machilis alpestris.* 4. The sandy glacier fore-field with association of *Bembidion nitidulum alpinum-Amara quenseli-Bledius erraticus bosnicus.* 5. The high-alpine grassy-meadows with association of *Carabus concolor-Zygaena exulans.* 6. The snow valleys with association of *Nebria hellwigi* in damp localities and at the snow-edge.

1. *Nebria atrata-Gnophos caelibarius* association contains the following character species: *Nebria atrata* DEJ. (2400-2500 m), *Gnophos caelibarius intermedius* KUTZ., the larvae of which occur on the high-alpine zone grassy-meadows, *Bombus alpinus* LINN., *Erigone remota* LINN. a boreo-alpine form found in the subnival zone, *Pergamasus franzi* WILLM., *Bdella iconica* BERL., *Rhagidia intermedia alpina* WILLM., *Microtrombidium sucidum* BERL. (boreo-alpine), *Nebria germari* HEER, *Scythrio glacialis* FREY (2400-2900 m), *Sphaleroptera alpicolana* HÜBN. (1800-3500 m), *Olethreutes spuriana* HERR.-SCH. (2100-3000 m), *Tmeticus graminicolus* SUND., *Isotomurus palliceps* UZEL, *Orchesella bifasciata* NIC., *O. mantana* STACH and *Lepidocyrtus lanuginosus* GML. (3200 m). There are, in addition to the above mentioned species, a number of other associated species like, for example, *Nebria castanea* RON., *N. hellwigi* PANZ. and *N. austriaca* GGLB.; *Notiophilus aquaticus* LINN., *Bembidion bipunctatum* LINN., *B. pyrenaeum glaciale,* HEER, *Amara quenseli* SCHÖNH., *Pieris callidice* ESP., *Erebia gorge* ESP., *Gnophos zellerarius* FRR., *Dasydia tenebraria, Psodos alticolarius* MÉN., *Ps. alpinatus, Tipula excisa* SCHUM., *Thereva brevicornis* HW., *Thynchopsilops villosus, Podothrombidium bicolor, P. montanum* and *Lycosa nigra.* The Anthomyiidae amount to about 50-80% of the Diptera and the Syrphidae stand next in abundance. This community becomes gradually impoverished at elevations above 3000 m.

2. *Caeculus echinus-Chrysomela crassicornis* association has the character species *Caeculus echinus* DUF. widely distributed on the South European mountains and confined particularly to pioneer vegetation mats, *Chrysomela crassicornis* HELLIS, a boreo-alpine species that occurs exclusively in sandy localities. The subspecies *Chr. cr. norica* occurs on the southern slopes of the Hohe Tauern, from the Brenner Pass eastward to the Sonnblick Group. *Amara quenseli* SCHÖNH. is confined to the high-alpine zone on the East Alps. *Cymindis vaporariorum* LINN. is widely distributed in sunny localities on the Alps. *Otiorrhynchus chalceus* STIERL. is restricted to the Alps, particularly in localities above the shrub-zone. *Gnophos caelibarius intermedius* KAUTZ and *Dasydia tenebraria innuptaria* HERR.-SCH. The other character species of this association are *Psodos alticolarius* MÉN., *Bombus alpinus* LINN., *Tachista interrupta* LÖW, *Drassodes lapidosus* (WALCK.), *Xysticus bifasciatus* C.-L. KOCH, *Neomoglus monticola* WILLM., *Erythraeus regalis* C.-L. KOCH, *Bdella iconica* BERL., *Orchesella montana* STACH and *Pyramidula rupestris* DRAP. In addition to these character species, the other associated species include *Hypnoidus dermestoides* HBST., *Erebia gorge* ESP., *E. lappona* ESP., *Anarta melanopa, Gnophos zellerarius* FRR., *Endrosa roscida melanomos, Lycosa giebeli,*

Xysticus desidiosus, Bdella longicornis, Orchesella bifasciata (NIC.), *Isotomurus palliceps,* etc. This community is remarkable for the conspicuous absence of myriapods, Orthoptera and Heteroptera.

3. *Lycosa nigra-Machilis alpestris* association. The large dark coloured spider *Lycosa nigra* occurs in association with *Machilis alpestris* on stones among the glacier ice and snow in moraines. The other associated species include *Nebria germari, Tmeticus graminicolus, Linopenthaleus irki, Bdella iconica* and *Isotoma saltans.*

4. The communities of the sandy foreground of glaciers. In the area uncovered from the glaciers that started receding about 1856 and on the sandy beds of the melt-water streams on the high-alpine meadows stage and in the shrub-zone stage the association is composed of *Bembidion alpinum* DEJ., *Amara quenseli* SCHÖNH., *Bledius erraticus bosnicus* BERNH., *Malthodes trifurcatus atramentarius* KSW., *Hypnoidus maritimus* CURT., *H. dermestoides* HBST., *Otiorrhynchus rugifrons* GYLL., *Tachista interrupta* LÖW, *Otiorrhynchus chalceus* STIERL., *Listrocheiritium cervinum* VERH. and *Arianta arbustorum* LINN.

5. *Carabus concolor- Zygaena exulans* association is characteristic of the high-alpine grassy-meadow zone, extending from the upper limits of the closed forest to the upper limits of the continuous shrub zone. The character species are *Carabus concolor* FABR., *Staphylinus ophthalmicus hypsibatus* BERNH., *Quedius alpestris* HEER, *Qu. a. spurius* LOK., *Dasytes alpigradus* KSW., *Rhagonychus maculicollis* MÄRCK., *Malthodes trifurcatus* KSW., *M. tr. atramentarius* KSW., *Byrrhus alpinus* GORY, *Corymbites cupreus* FABR., *C. rugosus* GERM., *Chrysochloa frigida* WSE., *Chr. viridis* DUFT., *Otiorrhynchus alpicola* BOH., *Melitaea asteria* FRR., *M. cynthia* HÜBN., *M. arinia meropa* PRUN., *Argynnis pales* SCHIFF., *Erebia gorge* ESP., *E. lappona* ESP., *E. tyndarus* ESP., *Plusia hochenwarthi* HOCHW., *Gnophos zellerarius* FRR., *Psodos coracinus* ESP., *Thanatus alpinus* KULCZ., *Xysticus desidiosus* SIM., *Lycosa ferruginea* L. KOCH, *Gnaphosa dadia* L. KOCH, etc. A very large number of associated species including *Pterostichus jurinei* PANZ., *Amara quenseli* SCHÖNH., *Cymindis vaporariorum* LINN. and *Gnophos caelibarius* occur in this association.

6. *Nebria hellwigi* PANZ. association contains the character species *Nebria castanea brunnea* DUFT., *Bembidion pyrenaeum glaciale* HEER, *B. bipunctatum* LINN., *Helophorus schmidti* VILLA, *Atheta tibialis* HEER, *Quedius punctatellus* HEER, *Aleochara rufitarsis* HEER, *Aphodius mixtus* VILLA, *Phytodecta affinis* GYLL., *Otiorrhynchus dubius* STRÖM., *Microthrombidium sucidum* BERL., etc. Other species which are generally associated with the character species include *Nebria germari, N. austriaca* GGLB., *Otiorrhynchus chrysocomus, Rhagidia terricola, Isotomurus palliceps* and *Orchesella montana.*

The associations in the vegetable-mats and shrub-zone, a transition between the high-alpine grassy-meadow zone and the subalpine forest zone, is remarkable for the presence of large numbers of ants like *Formica fuscogagates, Myrmica lobicornis, M. sulcinodis, Leptothorax acervorum,* etc. A number of other character species include *Carabus depressus bonellii* DEJ., *C. violaceus neesi* HOP., *Calathus erratus* SHLB., *C. melanocephalus* LINN., *Harpalus laevicollis* DUFT., *Pterostichus lepidus* LESKE, *Amara erratica* DUFT., *Anthophagus alpinus* FABR., *Quedius paradisianus* HEER, *Malthodes trifurcatus* KSW., *Corymbites cupreus* FABR., *C. melancholicus* FABR., *Evodinus interrogationis* LINN., *Cryptocephalus hypochoeridis* (LINN.), *Anechura bipunctata* FABR., *Aeropus sibiricus* (LINN.), *Podisma pedestris* LINN., *Colias phicomene* ESP., *Argynnis pales* SCHIFF., *Erebia epiphron cassiope* FABR., *E. tyndarus* ESP., *E. pharte* HÜBN., *Coenonympha arcania, Lycaena optilete cyparissus* HÜBN., *L. pheretes* HÜBN., *Bombus derhamelus* KIRBY, *B. pyrenaeus* PER., etc.

ii. Some Peculiarities of Ecological Succession

The ecological succession of species and the peculiarities of colonization of the ground, newly exposed from under the receding glacier-edge, at elevations of 2250-2500 m, on the sunny slope of the Ötztaler Alps were recently studied by JANETSCHEK (564). Depending upon their relative age, three major zones of associations are recognized by him. The zones in front of the glacier snout represent the areas expoxed successively in the years 1850, 1860, 1890 and 1920, by the receding glacier. The zone I of JANETSCHEK lies between the present limit of the glacier and a distance of about 80 m away, where the glacier stood in 1920. The soil maturity in this zone is about 1.5 decades. The II zone comprises IIa and IIb sub-zones. The sub-zone IIa extends from a distance of 80 m to 250 m from the present glacier snout, in other words, to the point where the glacier snout remained in 1890. The soil age in this area is about 4.5 decades. The sub-zone IIb extends to 600 m from the glacier snoüt, to the point where it stood in 1860. The soil age in this sub-zone is 6-7 decades. The zone III is beyond this point and extends to 900 m, up to the point where the glacier extended in 1850. The soil age is 8 decades.

Though no locality remains really unoccupied, there are, however, significant differences in the total numbers of individuals, in the absolute and in the percentage abundance of species of different groups and in other ecological characters. The total abundance of individuals increases with the increasing maturity of the biotope. The succession follows the series: spiders, Collembola, Diptera. The speed of the succession series is largely governed by the conditions of elevation, the microclimatic characters, the speed of the soil changes, etc. With the increase of distance from the glacier-edge and the soil age, we observe a significant increase in the mean density and height of the vegetation as a whole, with the eventual development of the high-alpine meadow communities as climax formations. In the young zones and for the rest above the timberline, spiders are among the larger types and are dominant both in absolute numbers and relative proportions. The zones I and IIa have a relative abundance of about 98%. Above the timberline, the peak of spider abundance is to be found in the zone II. The Post-Pleistocene recolonization of the glacier front appears to have progressed relatively rapidly along the margin of the Alps. The climax communities or the communities approaching climax conditions are typically characterized by a dominance of Acarina over Collembola. The succession series follow first the geoxene, then the geophiles (hemi-edaphon) and finally the geobiont (euedaphon), depending upon the soil changes brought about by the zoocenoses and atmobios.

As the succession of the vegetation and insects are closely interlinked, we may present the following synopsis of the vegetational succession (Table 37, p. 302).

The general ecology of the soil arthropods, with special reference to the Collembola, on the East Alps and in the Pannonic area, has been described by FRANZ (374). Certain species stand out as character species for a given habitat type. The influence of diverse climatic factors and other environmental conditions on the species composition of different soils on the East Alps are also discussed by him.

TABLE 37
Synopsis of vegetational succession on the Ötztaler Alps

Zone	Limit of glacier edge	Distance from glacier snout in m	Soil maturity (decades)	Dominant associations
I	1920	80	1.5	*Poa laxa-Cerastium unifolium* association.
IIa	1890	250	4.5	*Agrostis rupestris-Polytrichum juniperinum.*
IIb	1860	600	6-7	*Trifolium palescens-Polytrichum juniperinum*
III	1850	700	8	*Festuca halleri.*

The ecological succession of animals may be summarized as follows:

Vegetation zone 0 = animal zone Ia. *Pardosa waglerei nigra—Nebria jockischi—Proisotoma crassicauda.*

Vegetation zone I = animal zone Ib. *Plaesiocraerus helleri—Pardosa wagleri nigra—Hypnoidus maritimus—Proisotoma schoetti.*

Vegetation zone IIa = animal zone IIa. *Lithobius lucifugus—Robertus truncorum—R. arundineti—Simplocaria semistriata—Notiophilus biguttatus—Monochus muscorum.*

Vegetation zone IIa = animal zone IIb. *Lithobius—Meioneta gulosus—Oxypoda tirolensis.*

Vegetation zone IIb = animal zone IIIa. *Lithobius—Lephthyphantes fragilis—Quedius alpestris—Leptoiulus simplex—Euconulus trochiformis.*

Vegetation zone IIb = animal zone IIIb. *Lithobius—Haplodrassus signifer—Pterostichus subsinuatus—Epichnopteryx pulla—Scolioplanes acuminatus—Oxypoda tirolensis* (direct from IIb or also through IIIa).

High-alpine grassy-meadow. *Deltocephalus striatus—Dasytes alpigradus—Lithobius—Chorthippus parallelus—Oxypoda* (succession from IIIb).

8. THE CARPATHIANS

Together with their forelands, the Carpathian Region comprises i. the Moldavian Hill-lands extending from the east and south, ii. the Wallachian Plain, iii. the Carpathian (Pannonian) Basin and iv. the Carpathian Ranges. The R. Danube flows through the Carpathian Basin. The arc of the Carpathian Ranges extends nearly 1600 km, from the Vienna Basin to the Eisernes Tor (the Iron Gate). Though nearly as long as the Alps, the Carpathians do not exceed an elevation of 2652 m even in the Vysoké Tatry (Hohe Tatra) in the West Carpathians. There is a continuous outer zone of Tertiary sandstones and discontinuous central zone of crystalline rocks and limestones. There is also a volcanic inner belt towards the basin of the central zone. The Carpathians are divided into i. the Northwest Carpathians, ii. the Central Carpathians, iii. the East Carpathians and iv. the South Carpathians. The Northwest Carpathians extend from the R. Danube and the R. Morava to the Dukla Pass (450 m above mean sea-level) in Czechoslovakia; the Carpathians are widest in this area. The R. Vah separates the Vysoké Tatry and the Nízke Tatry (Niedere Tatra);

the former is south of the main chain of the West Carpathians and is a zone of crystalline rocks, which was covered by ice during the Pleistocene. The mean elevation varies from 1830 to 2440 m and the Stalin Peak (2663 m) is the highest on the Carpathian Range.

The Nízke Tatry, composed of gneiss and granite, is situated south of the Vysoké Tatry. The Central Carpathians, about 90-100 km wide, extend from the Dukla Pass to nearly the R. Tisza. The East Carpathians lie within the Rumanian territory and extend southward to Brasov (Orasul Stalin). The East Carpathians are subdivided into three zones of parallel ridges of an eastern sandstone zone, a central limestone and crystalline zone and western zone of young volcanic material. Rounded mountain tops, weathered from crystalline schists, are familiar features in this area. The peaks here have a mean elevation of 1525-2130 m. The Pleistocene glaciations appear to have been scanty in this area. The Transylvanian Alps or the South Carpathians extend from Brasov to the Kazan Gorge of the R. Danube, a point where the mountains curve south and southeastward. The Transylvanian Alps are continued as the Banat Mountains, west of the R. Timis. Many peaks in the Transylvanian Alps exceed 2440 m in elevation. Flat-topped plateaux and terraces are characteristic features of this area. The Bihor Mountains are in the eastern portion of the Carpathian Basin.

Like the Balkan Mountains, the Alps, the Pyrenees and the Apennines, the Carpathians belong also to the Tertiary mountain system. The principal folding movements appear to have occurred about the middle of the Tertiary times, but there are indications of perhaps an earlier folding even during the Cretaceous Period in the south and the east. The Upper Cretaceous folding movements gave rise to the uplift of the Transylvanian Alps, which like parts of the East Carpathians, probably projected as islands from a Palaeocene sea. The tectonic activity seems to have continued into the late Tertiary and even Pleistocene times, after the main folding movements were over. Folds and overthrusts of the late Tertiary sediments at the foot of the Carpathians, extensive and partly projecting the uplift areas, river terraces of Pleistocene and Pliocene age, etc. are clear evidence of such continued recent upheavals. The recent movements seem to have attained great intensity especially in the South Carpathians, where it continued right down into Pleistocene times, but in the North Carpathians the movements apparently slowed down relatively early. The conditions are not very clear in the East Carpathians. The movements of the Pleistocene upheavals in the Transylvanian Alps attained 300-400 m. The Carpathians are separated from the Alps by the Pannonic and Vienna Basins, from the Sudeten by the Oder-Betschwalfurche, from the mountains of the Balkan Peninsula by the R. Danube at the Iron Gate.

While there are no glaciers at the present time on the Carpathians, numerous glaciers existed during the Pleistocene, but the extent of glaciation seems to have been much less than on the Alps. Evidence of the Pleistocene glaciers has so far been found in the Beskids; on the highest peak Babiagora (1723 m) a small hanging glacier has left its marks on the northwest side of the mountain. The Vysoké Tatry carried most intensive glaciation in the whole of the Carpathians during the Pleistocene. An ice-ring, about 4.5 km wide, surrounded the base of the Vysoké Tatry, from Krivan to Weisswassertal. The Carpathians are at present remarkable for their moraine amphitheatres, covered with needle-forests. On the north of the Tatra, the Pleistocene ice-flows extended in great beds, filling the transverse valleys of the

Javorinka, Bialka, Panszyca and Suchwod. The snowline during the Pleistocene period on the Vysoké Tatry was 1500 m above mean sea-level. The Nízke Tatry also carried glaciers, especially the Dumbier, where the moraines reach down to 900 m and indicate a Pleistocene snowline of 1500 m above mean sea-level. Even in the eastern Nízke Tatry there is evidence of Pleistocene glaciers. In the Czornahora area, indications of past glaciers have been found on Goverla (2058 m). The highest parts of the Swidowiec Mts. also seem to have been ice-covered. In the Rodna Mts., on the north slope of Kuhorn, there is evidence of Pleistocene glaciers, which filled the Lala Valley and left moraines at an elevation of 1620 m. The summit of the Rodna Mts. was also under Pleistocene ice. On the Transylvanian Alps, from the massive Godeanu in the Banat area eastwards to Bucsecs, all peaks above an elevation of 2000 m were covered by ice, but Ciucas (1925 m) was not covered. The Pleistocene snowline was at an elevation of 1900 m and the lowest moraines are found at an elevation of 1100-1200 m. The autochthonous glaciers of the Carpathians seem to have been independent of the northern inland ice-mass. During the maximum glaciation, the inland ice-mass seems to have reached up to Galizia and the Austria-Silesia area, along the Oder Valley up to the Weisskirchner Wasserscheide. The ice margin was directly at the foot of the Carpathians and the terraces left by it may be found locally at an elevation of about 400 m. Tongues of the inland ice seem to have extended to over 30 km in the Beskid valleys, open to the north. The part of the Carpathians in contact with the inland ice-mass extends from the area of Neutitschein and Stramberg eastwards to Santal. The outer margin of the Carpathians, east of Santal, was not, however, reached by the inland ice-mass. In comparison to the Alps, the Carpathians seem to have offered better environmental conditions for plants and animals during the height of the Pleistocene glaciations. The distance from the snowline to the timberline was about 800 m and the upper limit of the *Fagus*-forest on the Carpathians lies only 300 m below what it was in the past. It seems probable that not only needle-forest, but also extensive broad-leaved forests flourished on the Carpathians during most of the Pleistocene. The reader may also consult the collected papers of BUDAY *et al.* (171) for fuller accounts of the tectonics of the Carpathians.

9. THE ECOLOGICAL CHARACTERS AND BIOTIC ZONATION OF THE CARPATHIANS

As may be expected from the geographical situation, the general climate of the Carpathians is far more continental than that of the Alps. The difference between the mean atmospheric temperatures in the summer and winter is more intensified and the mean precipitation is less on the Carpathians than on the Alps. Furthermore, the climate becomes more and more pronouncedly continental eastward on the Carpathians. All parts of the Carpathians have, however, sufficient precipitation to ensure requisite moisture for a rich montane fauna. The mean annual precipitation rises with the increase in elevation up to the forest-line.

As on the Alps, the maximum precipitation occurs about the upper limit of the forest, but above these limits there is a distinct fall in the mean annual precipitation with further increase in elevation. The summer is the rainiest season on the Carpathians. On the Vysoké Tatry, July is the rainy month of the year, but June and August have also considerable rainfall. On the East and South Carpathians, the maximum monthly precipitation occurs during June, but July also receives much rain and the autumn months are relatively dry. The snow-flakes persisting in the high-alpine zones, during summer, provide adequate moisture for the high altitude organisms, Kis-krivan (1711 m), the summit of which projects a little above the stunted-tree zone, has relatively extensive snow-fields during July. The Vysoké Tatry bears, in the high-alpine zone, large and perennial snow-fields. Though the highest peak projects over the climatic snowline, the gradient of the peak prevents the formation of glaciers. Snow-fields occur on Dumbier (2045 m) on the Nízke Tatry. On Czornahora (2058 m) snow was reported on July 8th in 1875. On the Rodna Mts. snow has been reported on August 11th in 1882 in the area of the Kuhorn Peak (2280 m). With their extensive and massive elevation, the Transylvanian Alps have an extensive high-alpine region, in which the winter snow remains unmelted far in mid summer. As at an elevation of 200-300 m above the timberline on the Alps, the Carpathians also have an extensive high-alpine zone, in which the winter snow remains unmelted till late in summer and all peaks, at least 200-300 m above the timberline, thus offer optimal conditions for nival elements. Perennial snow is found, however, only on the Vysoké Tatry.

Extensive *Quercus*-forests flourish at elevations of about 800 m in many parts of the Carpathians, but the *Fagus*-forest extends over the whole Carpathians. The upper limit of the continuous *Fagus*-forest is 1250 m on the Tatra, 1300 m on the Central Carpathians and between 1300 and 1500 m on the South Carpathians, depending upon the slope exposure. The upper forest limit is lowest on the North Carpathians, but rises towards the East Carpathians and South Carpathians. On Babiagora the forest rises to only 1300 m above mean sea-level, on the Tatra Mts. (Central Carpathians) to 1500 m, in the Czornahora, Rodna Mts. and in Rareu it rises to 1600 m, in the Caliman, Hargita and Nagy-Hagymas ranges to 1700 m. In the South Carpathians the upper limit of the forest is at an elevation of about 1700-1800 m. The upper limit of the forest on the Transylvanian Alps is about 100 m more on the south and east slopes than on the north slope.

Botanists like PAX (816) have recognized the following biotic zones on the Carpathians:

I. The zone up to the upper limit of the forest
   i. The cultivated zone, 1000 m in the Central Carpathians and 1100 m on the Transylvanian Alps.
   ii. The lower forest-zone, from the upper limit of the cultivated zone to the upper limit of the *Fagus*-forest.
   iii. The upper forest-zone, from the upper limit of the *Fagus*-forest to the forest-line (spruce forest).

II. The zone above the upper forest limit
   i. The subalpine zone or the elfin-wood zone, from the upper limit of the forest to the upper limit of elfin-wood, about 200-300 m wide and characteristic for the subalpine Strauch-formation, (green alder or *Alnus glutinosa*), and by the formation of alpine meadow and snow-flora, while the rock flora is still in the background.
   ii. The alpine zone, lacking the shrub vegetation formation; in place of the alpine meadow we have here the alpine-mats, boulder stone flora; the part of the mountain above the elfin-wood zone.

As shown by HOLDHAUS & DEUBEL (536), this terminology is different from that generally used by others and is also likely to lead to some confusion, especially the use of the term lower-forest zone for the narrow belt above the isohypse of 1000 m. The term subalpine zone is generally widely applied in botanical and zoological literature for the upper forest-zone. MARTONNE suggested the vertical zonation on the basis of distribution of plants, summarized below:

I. "La zone subalpine (zone de la forêt)" from an elevation of 600 m to the timberline
   i. "La zone subalpine inférieure" from the lower limit of the forest to the upper limit of the continuous *Fagus*-forest.
   ii. "La zone subalpine supérieure" or the "zone du sapin", from the upper limit of the *Fagus*-forest to the timberline.

II. "La zone alpine", above the timberline
   i. "La zone alpine inférieure", from the forest limit to the upper limit of the elfin-wood, corresponding to the subalpine region of PAX.
   ii. "La zone alpine proprement dite", above the elfin-wood.

With reference to insect life, we may recognize the following biotic zones on the Carpathians:

1. The forest zone from the lowest elevations of occurrence of true montane terricole insects up to the upper limits of the forest, characterized by a dominance of forest-insects, with an abundance of planticole and terricole species that are exclusively typical of the forest. The insect fauna of the montane forest zone of the Carpathians shows significant qualitative differences as the altitude increases, so that we may recognize a lower forest zone and an upper forest zone—the latter is in fact the subalpine zone. The subalpine types descend often to 600-700 m above mean sea-level, in deep sheltered gorges, but ascend higher on the sunny and more arid slopes.

2. The transitional zone, lacking the forest elements and also without summer snow-cover, is unfavourable for the occurrence of typical alpine-zone forms. There are therefore few characteristic types in this zone. The zone extends from about 200-300 m immediately above the upper limits of the forest and its upper limit is recognized by the presence of dwarf plants. When compared to the forest zone and the high-alpine zone, the transitional zone is strikingly poor in insect life.

3. The high-alpine zone differs from the lower zones by the complete absence of trees and by the presence of small patches of unmelted winter snow during summer and by a combination of diverse other climatic and edaphic conditions. The zone is rich in character species. In addition to the typically high-alpine zone authochthonous forms, there are also some species, particularly terricole elements, typical of the forest-zone, occurring in certain localities within the high-alpine zone of the Carpathians. A number of interesting Coleoptera of the forest seem to have thus become more or less re-adapted for life at higher elevations on the Carpathians. The typically high-alpine forms occur only on the peaks that have unmelted winter snow during the summer (middle of July).

10. CHARACTERISTICS OF THE HIGH ALTITUDE INSECTS FROM THE CARPATHIANS

The literature on the high altitude insects from the Carpathians is extensive. The comprehensive account of HOLDHAUS & DEUBEL (536) on the Coleoptera may be considered as generally illustrative of the peculiarities of the ecology and biogeography of the region.

The outstanding character of the high-alpine zone insects from the Carpathians is the fact that in sharp contrast to the Alps, there is a pronounced dominance of terricole species typical of the forest-zone and a corresponding insignificance of the exclusively high-alpine character species. Most of the terricole Coleoptera, known so far from the Carpathians, occur within the forest zone. The terricole species complex of the broad-leaved forest zone differs very much from that of the needle-forest zone. A number of these typically forest-terricole species ascend higher to the zone above the upper limit of the forest. Depending upon diverse local conditions, some of the exclusively montane-terricole species from the lower-forest zone also descend right down to the foot of the mountains. Of the total montane Coleoptera from the Carpathians, 25% are Carabidae, 16% Staphylinidae, 15% Chrysomelidae, 22% Curculionidae, 4% each of Pselaphidae, Scydmaenidae and Silphidae, 16% Hydrophilidae, 1.3% Tenebrionidae and the rest belong to the Dryopidae, Byrrhidae, etc. About 12% of the total montane-terricole Coleoptera occur in the high-alpi-

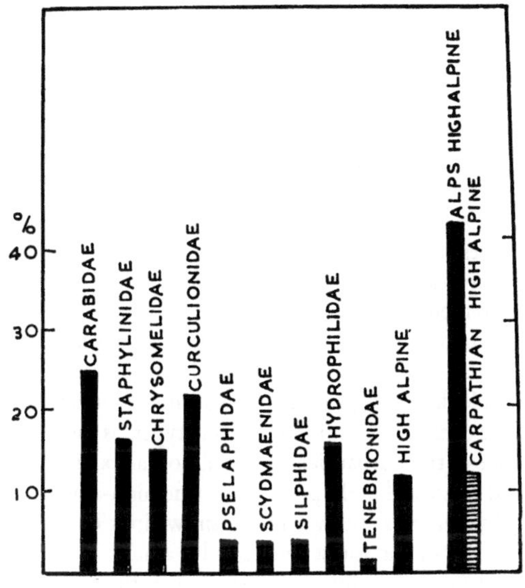

Fig. 68. The taxonomic composition of the montane Coleoptera from the Carpathians.

ne zone and are exclusively characteristic of the zone. The number of montane species of Coleoptera from the Carpathians as a whole is very large, but the greatest bulk of the species are generally confined to lower elevations, within the forest zone. The montane-terricole Coleoptera from the region show a gradual, but at the same time, pronounced impoverishment from the south to the north, a peculiarity that cannot be readily explained purely on the present climatic and edaphic grounds. The Banat Mountains seem to be the most densely populated in the entire region.

The Transylvanian Alps stand intermediate between the Reyezat and Bucses. Compared to the South Carpathians, the East Carpathians are relatively poor, particularly in the blind terricole Coleoptera. The Coleoptera from the Tatra, West Beskids and Trenčin Komitat are poor and monotonous in comparison to the Coleoptera from the East Carpathians. The northward impoverishment is perhaps most pronounced in the case of the species of the lower-forest zone and least in the case of upper-forest zone.

While the high-alpine zone of the Alps is rich in Coleoptera, the Carpathians are very poor in the exclusively high-alpine types. We know, for example, at least 80 species of the high-alpine zone Coleoptera from the East Alps, but the Carpathians have at the most only 22 exclusively and typically high-alpine zone species. Most of these species are also sparsely found and in a few localities. We do not also find on the Carpathians such well known exclusively high-alpine forms like *Cychrus* FABR., *Carabus* LINN., *Trechus* CLAIRV., *Amara* BON., *Orina* MOTSCH., *Crepidodera* CHEVR., etc., and we do not also come across the subgenus *Oreonebria* of *Nebria* LATR. (so very characteristic of the Alps). The following are some of the exclusively and typically high-alpine montane-terricole Coleoptera from the Carpathians: *Nebria carpathica* BIELZ., *N. tatrica* MILL., *Leistus gracilis* FUSS., *Deltomerus tatricus* MILL. (the closely related *D. carpathicus* MILL. occurs in the subalpine zone of the East Alps), *Oxypoda nimbicola* FAUV., *Atheta carpathica* MILL. (the closely related *A. alpicola* MILLER is subalpine), *Mycetoporus oreophilus* BERNH., *Coryoides deubeli* BERNH., *Niphetodes redtenbacheri* MILL., *Bolitophaga alpicola* KÜST., *Choleva oreositropha* GGLB. (all the other species of the genus occur within the forest-zone on the Carpathians and *Ch. nivalis* KR. occurs both in the forest and the high-alpine zones), *Rybinskiella magnifica* RYB., *Chrysomela schneideri* WSE., *Otiorrhynchus alpicola* BOH., *O. fusciventris* FUSS., *O. alpigradus* MILL., *O. granicollis* BOH. (the closely related *O. proximus* STIERL. occurs both within the forest and the high-alpine zones), *Brachynodontus reitteri* WSE. (the closely related *Br. deubeli* GGLB. is subalpine) and *Aphodius montanus* ER. About 50 other species are also mentioned in the literature as having been collected in the high-alpine zone of the Carpathians, but we do not know at present whether these records refer to specimens accidentally transported from lower elevations.

It seems remarkable that though at the present time favourable conditions exist for an abundance of the exclusively high-alpine zone types, as much as on the Alps, and the high-alpine zone of the Carpathians is also almost as extensive as on the Alps, the Carpathians should be so strikingly poor in the true high-alpine forms. The Vysoké Tatry is covered by perennial snow and the snow limit elsewhere lies 200-300 m above the upper limits of the forest during July, almost as we find on the Alps. HOLDHAUS & DEUBEL (536) have attempted to explain this curious condition on the basis of the conditions that prevailed during the Pleistocene in the area. There was, during the late Tertiary, an essentially warmer climate in Central Europe than at present. This was correlated with a much higher upper limit of the forest on mountains. The relatively low Carpathians were far below this high upper limit of the forest and had thus very little or perhaps no high-alpine area at all. The relatively high Alps had, even during the late Tertiary, an extensive high-alpine zone, so that a diversified high-alpine fauna was able to evolve there. As a result of the Post-Tertiary climatic changes, the upper limit of the forest sunk even on the Carpathians and the areas that were formerly densely forested, but are now above the forestline, could not, therefore, be populated by the subalpine elements from below, although such a tendency may be observed today to some extent. Strictly speaking, there does not seem to have been enough time since the retreat of the last Pleistocene ice for the development of an abundant and well diversified high-alpine zone species complex on the Carpathians. Many species that now occur within the forest zone on the Carpathians seem to have quite recently become more or less partially adapted, secondarily, for life in the high-alpine zone, so that they are actually subalpine types on the high-alpine zone. This is, for example, the case with *Carabus fabricii* PANZ., *Nebria transsylvanica* GERM., *Trechus procerus* PUTZ., *Pterostichus blandulus* MILL. *Niphetodes eppelsheimi* GGLB., *Carpathobyrrhulus transsilvanicus* SFFR., *Orina viridis* DUFT., *Chrysomela lichenis* RICHT., *Chr. carpathica* FUSS., *Otiorrhynchus graniventris* MILL., *O. riessei* FUSS., etc. A number of other subalpine and even some lowland species that ascend to the forest-zone on the Carpathians have developed local races on the high-alpine zone. We have, for example, the high-alpine races of *Carabus scheidleri* PANZ.

The studies of PRÜFFER (846) on the Lepidoptera from the Polish Tatra contain an interesting review of the earlier investigations, the high altitude localities and the ecology and geography of the region. He has listed the following species as characteristic of the nival zone (2250-2663 m): *Pieris brassicae* LINN., *P. rapae* LINN., *P. napi* LINN., *Vanessa urticae* LINN. and *Plutella maculipennis* CURT. On the high-alpine zone, at elevations of 1920-2250 m, he records over 30 species, with notes on the plants on which the larvae feed at high elevations. The more interesting records include *Erebia mnestra* HÜBN., *E. gorge* ESP., *E. pandrose* BKH., *E. p. nowickii* PRÜFFER, *Gnophos operaria* HÜBN., *Psodos alpinatus* SCOP. and *Ps. coracina* ESP.

In an interesting contribution, KRZYWICKI (662) has summarized the outstanding features of the Rhopalocera from the Vysoké Tatry. He has discussed 64 species from the Polish Tatra, with details of their phenology, the ecological correlations of the species and the environmental conditions and the character species of different biotic zones. The Rhopalocera of the Vysoké Tatry studied by him belong to two broad groups, viz. i. the species that are permanent residents and ii. the species that occur only sporadically. The species endemic to the Polish sector of the Tatra Mts. are *Parnassius apollo niesiolowskii* KRZ., *Erebia manto praeclara* NIES., *E. aethiops magdalena* KRZ., *E. gorge rudkowskii* O.B.H., *E. pandrose robertsi* PESCHKE, and *Boloria pales insignis* KRZ. These species occur at elevations of 1500-2000 m. In addition to these endemic elements, other species like *Pieris bryoniae* (HÜBN.), *P. napi* LINN., *Erebia pronoë* (ESP.), *Coenonympha tullia* (MÜLLER), *C. iphis* (SCHIFF.), *Pararge petropolitana* (FABR.), *P. maera* (LINN.), *Palaeochrysophanus hippothoë* (LINN.), etc. were also collected by him. The typical alpine zone species are *Parnassius apollo* (LINN.), *Erebia manto* (SCHIFF.), *E. epiphron* (KNOCH.), *E. pharte* (HÜBN.), *E. gorge* (HÜBN.), *E. pronoë* (ESP.) and *E. mnestra* (HÜBN.). The boreo-alpine elements include *Pieris bryoniae* (HÜBN.), *Erebia euryale* (ESP.), *E. pandrose* (BKH.) and *Boloria pales* (SCHIFF.). The Euro-Caucasus elements include *Maniola jurtina* (LINN.), *Pararge aegeria egerides* STGR., *P. megera* (LINN.), *P. petropolitana* (FABR.), *P. maera* (LINN.) and *Melitaea athalia* (ROTT.). The Mediterranean elements are represented by *Colias australis* VERITY, *Vanessa atalanta* (LINN.), etc. KRZYWICKI considers *Parnassius apollo* (LINN.), *Pieris bryoniae* (HÜBN.), *Erebia ligea* (LINN.), *E. euryale* (ESP.), *E. manto* (SCHIFF.), *E. epiphron* (KNOCH.), *E. pharte* (HÜBN.), *E. æthiops* (ESP.), *E. gorge* (HÜBN.), *E. mnestra* (HÜBN.), *E. pronoë* (ESP.), *E. pandrose* (BKH.), *Pararge petropolitana* (FABR.) and *Boloria pales* (SCHIFF.) as eualpine autochthonous types on the Polish Tatra. *Coenonympha pamphilus* (LINN.),

Aglais urticae (LINN.), *Pieris brassicae* LINN., *P. rapae* LINN., *Colias hyale* LINN., *Coenonympha tullia* (MÜLLER) and *Argynnis paphia* LINN. as tycho-alpine types. *Vanessa cardui* LINN. and other species that occur only sporadically in the alpine zone are classified as xeno-alpine types; to this group belong *Colias croceus* (FOURC.) and *Pararge aegeria egerides* STGR. The percentage abundance of the different ecological elements on the Polish Tatra are tabulated as follows by KRZYWICKI:

TABLE 38

Ecological types of Rhopalocera on the Polish Hohe Tatra (percentage abundance), according to KRZYWICKI

Family	Eualpine	Tycho-alpine	Xeno-alpine	Total
Papilionidae	—	—	50	50
Parnassiidae	50	—	—	50
Pieridae	7	50	29	86
Satyridae	32	12	12	56
Nymphalidae	2	29	12	43
Lycaenidae	—	7	18	25
Total	9	17	15	41

II. BIOGEOGRAPHICAL CHARACTERS OF THE ALPS AND THE CARPATHIANS

The biogeographical peculiarities of the Alps and the Carpathians may be traced, in part at least, to the alpine tectonics and orogeny and in part to the Pleistocene glaciations. The general distributional pattern and the faunal affinities of these elevated regions have been greatly influenced by the Pleistocene glaciations in particular. While some species have doubtless disappeared, numerous others have survived the glacial times (*vide* next chapter). The Postglacial recolonization of the former ice-covered regions does not seem to have ended, but continues even today, both on the Alps and on the Carpathians.

The north-south discontinuous distribution or the well known boreo-alpine phenomena of the Central and south European mountains is dealt with in the next chapter. HORION (542) has recently drawn attention to a no less interesting east-west discontinuity in the distribution of the high altitude Coleoptera of the Alps. The distributional pattern of *Staphylinus chloropterus* PANZ., *Tachinus discoideus* ER., *Quedius vexans* EPPELSH., *Omalium validum* KRAATZ and *Aphodius conjugatus* PANZ. shows two distinctly separated ranges, one of which lies in southeast Central Europe and the other in southwest Central Europe. This discontinuity, like the north-south one, is also attributed by HORION to the effects of Pleistocene glaciations. It is assumed that before the Pleistocene, these species had a continuous distribution. They are supposed to have survived in both these widely separated refugial areas and to have spread again more or less through Central Europe at the end of the glacial epoch. Above all, the southern slopes of the Alps are in the zone in which these species do not seem to occur.

The distribution of the blind montane Coleoptera on the Alps and the Carpathians is of considerable interest. The blind montane Coleoptera, occurring on the

Austrian Alps, include, in addition to the strictly cavernicolous species, blind forms that occur outside the cave environment. The latter type of species are exclusively found south of the Drautal. For a long stretch of the Alps to the west, the blind montane species are restricted to the south strip of the marginal zone of the Alps. The area of the blind species in the south Alps is believed by HOLDHAUS (535) to be approximately coincident with the zone of his "massifs de refuge" (*vide* chapter XII). In the area of the Alps, the blind montane species penetrate into the mountains of Kärnten and Südsteiermark, least towards to north. The northern limit of the blind montane Coleoptera (the Blindekäferlinie) may be defined by the distribution of the following species: Dobratsch: *Anophthalmus mariae* SCHATZM (cavernicole) and *Troglorrhynchus anophthalmus* SCHMIDT (also free-living in Dobratsch, but otherwise cavernicole). The Kärnten slope of the Karawanken: *Anophthalmus bernhaueri* GGLB. (free-living in Obir), *A. sp.* (cavernicole in Eisenkappel), *Scotoplectus capellae* REITT. (free-living), *Bathyscia celata* HAMPE (free-living also in Satnitz), *Troglorrhynchus anophthalmus* SCHMIDT, Bachergebirge and Tatra Hügel on the north slope of Drau in Marburg: *Bathyscia celata* HAMPE, *Anophthalmus bielzi* SEIDL. (free-living and cavernicole), (from the Transylvanian Alps over the East Carpathians-Rodna Mts., Nagy-Hagymas, Caliman, Rareu, Marmaros, etc.—to the Vysoké Tatry and in the West Beskids). The largest blind species is *Molops* (*Typhlochoromus*) *stolzi* MOCZARSKI from the subalpine and alpine zones (1330-1700 m) of the Venetian Alps. The absence of the blind montane Coleoptera on the mountains of Germany and on the Bohemian Massif and Sudeten is explained by HOLDHAUS on the assumption that these became extinct, because of the Pleistocene climatic changes and their postglacial rewandering and recolonization have not been possible. On the Alps, a large number of species have an extremely restricted range and are also pronouncedly discontinuous in distribution. On the Carpathians also we find the species with very restricted range or with pronounced discontinuity, but the total area of distribution of these species is on the whole much less on the Carpathians than on the Alps. Most of the species occurring on the Carpathians extend to nearly the whole of it or at least to the major part of it. In other words, they are more widely distributed on the Carpathians than on the Alps.

The high altitude insects of the Carpathian Region, with its much less intense glaciation during the Pleistocene, have become extremely slightly modified and in any case, much less than those from the Alps. The distributional and other peculiarities that we see on the Alps are also repeated, but only in a very much weakened condition on the Carpathians. Of the few true high-alpine species from the Carpathians, *Oxypoda nimbicola* FAUV., *Otiorrhynchus alpicola* BOH. and *Aphodius montanus* ER. are also found on the Alps. The Collembola *Hypogastrura monticola* STACH and *Onychiurus alpinus* STACH (992, 993) occur for example, both on the Vysoké Tatry and on the Hohe Tauern (Alps), at elevations of 2100-2300 m and 2200-3100 m respectively. The terricole montane Coleoptera from the Alps and Carpathians include *Carabus fabricii* PANZ., *Bembidion viridimicans* DAN., *Pterostichus kokeili* MILL., *Pt. maurus* DUFT. (known also from the Apennines), *Atomaria grandicollis* BRIS., *Crepidodera cyanipennis* KUTSCH and *Cr. cyanescens* DUFT. A number of species are common to the Carpathians, the Variscischengebirge and often also to the Sudeten mountains, like for example, *Trechus montanellus* GEMM., *Tr. latus* PUTZ., *Tr. amplicollis* FAIRM., *Pterostichus negligens* STURM., *Pt. rufitarsis* DEJ., *Choleva nivalis* KR. and *Chrysomela carpathica* FUSS. Some species like *Carabus concolor* FABR., *Trechus splendens* GEMM. and *Chrysomela lichenis* RICHT., are common to the Alps, Carpathians and the Variscischengebirge. The species that are common to the Carpathians and the mountains of the Balkan peninsula are *Cychrus semigranosus* PALL., *Nebria reichi* DEJ., *Bembidion balcanicum* APFB., *Geodromicus puncticollis* WSE., etc. A number of other species like *Bembidion glaciale* HEER, *Platynus scrobiculatum* FABR., *Pterostichus transversus* DUFT., *Orina vittigera* SUFFR., *Crepidodera melanostoma* REDTB., *Otiorrhynchus pulverulentus* GERM., etc. are common to the Alps, Carpathians and the mountains of the Balkan peninsula.

The ecological and distributional peculiarities of the montane-terricole Coleoptera on the Alps and Carpathians are best explained on the basis of their Tertiary origin. The close affinity of the mountains, separated at the present time by effective dispersal-barriers, indicate that the fauna of these mountains must be traced back to a time, when such barriers did not exist. We have, for example, the great affinity of the montane-terricole species of the mountains of Corsica and the Alps, of the Tyrrhenian Islands among themselves and with those of the Apennines and the northwest African mountain chain, the affinity of the montane species of the Balaeares to those of south Spain, etc.

It seems thus evident that the mountain autochthonous insects, especially the Coleoptera, of the Alps and Carpathians are predominantly Tertiary derivatives. The mighty tectonic movements of the Tertiary times gave rise to these young mountain chains and these uplifts produced the mountain-fauna also. The conclusion also seems to be justified that the roots of this Tertiary mountain fauna must be sought in an older fauna, which had evolved for a longer period on older mountain systems. These roots should not, however, be sought only in the older mountain systems of Europe. The fauna of the European high mountains shows a most remarkable similarity to the high altitude types of the high mountains of Asia, as far as Siberia. In East Siberia lies indeed the ancient mountainous area of the Angaraland, from where the westward migration to Europe was possible along the central and west Asiatic young mountain ranges (*vide* chapter XIII). Even in the present fauna, we see undoubted evidence of the contributions of the older mountain fauna to the fauna of the younger mountains. The East Alps and Carpathians are, for example, related faunistically, since the faunae of these two regions have been derived from the same older Bohemian Massif and the Sudeten on one side, and the Central Mass of the Balkan Peninsula on the other side. The insect fauna of the West Alps is, on the other hand, radically different from that of the East Alps, since the West Alps were doubtless populated by roots from the Central French Plateau and to a much less extent from the Tyrrhenian Massif, by way of the north Apennines.

In conclusion, we may observe that the high altitude insects of the Alps comprise the following faunal elements: i. the endemic Alpine and widely distributed species like *Trechus alpicola* STURM., *Carabus concolor hoppei* GERM., *Nebria hellwigi* PANZ., etc., ii. the Euro-Siberian valley elements that have penetrated into the Alps since the Pleistocene times, iii. the steppes relicts and iv. the boreo-alpine species.

CHAPTER XII

THE BOREO-ALPINE INSECTS

I. WHAT ARE BOREO-ALPINE INSECTS?

The boreo-alpine insects (including the arctic-alpine or boreo-montane insects) represent a most interesting component element of the mountain autochthonous insects of the Central and South European mountains. They are of considerable ecological and biogeographical importance and present complex problems of origin, evolution and distribution.

The boreo-alpine species are characterized by their discontinuous distribution in the northern parts of Europe and at high elevations on the Central European and partly also southern European mountains, but absence in the intervening areas (fig. 69). In their northern range, many boreo-alpine species extend eastwards to Siberia and to a limited extent occur on the Altai Mountains, but only very exceptionally on the Pamirs-Himalayan system (*vide* chapter X). The north range of a number of boreo-alpine species is characteristically circum-polar, so that these species occur in the extreme north of Europe, Asia and North America. The grasshopper *Podisma frigida* BOH. occurs, for example, in Fennoscandia (vide chapter XV), northern Russia, Siberia, the Altai Mountains, northern Mongolia, Manchuria, Kamchatka, Alaska and the Alps. In addition to the typical boreo-alpine species, others are more or less sparsely but widely distributed in the intervening vacant areas, but attain their maximal abundance only in the north and on the elevated areas of Central Europe. The distributional pattern of these species represents a transition between the continuous distribution and the extreme boreo-alpine mode of discontinuous distribution. *Trechus rubens* FABR., *Cymindis vaporariorum* LINN. and *Hypnoidus riparius* FABR. have, for example, this type of transitional distribution.

The boreo-alpine insects are generally known as glacial relicts or Pleistocene relicts. It was generally assumed until recently that the Pleistocene glaciation brought about whole-sale destruction of flora and fauna and the relict species were assumed to have survived only in the marginal zone of the Alps. At the end of the Pleistocene, the relict species were assumed to have rewandered to the interior of the Alps. As we shall see, it has, however, been recently shown that some of the boreo-alpine species have actually survived the Pleistocene glaciations in the so-called "ice-devastated" area in the heart of the Alps, on high peaks that have remained as rock islands above the ice mass.

Considerable attention has been devoted to the study of the boreo-alpine forms by biologists even during the last century. It is not, however, OSWALD HEER (492), as is widely asserted, who was the first to suggest an explanation of the boreo-alpine distribution on the basis of the Pleistocene climatic conditions, but the botanist EDWARD FORBES (353). As early as 1846, FORBES recognized the phenomenon, with particular reference to the occurrence of typical boreal plants on the Central European mountains and attributed their presence to the Pleistocene glaciations. He conceived of the plants having migrated on moving icebergs and ice-masses.

The boreo-alpine distribution of plants and animals is now widely admitted to be the result of Pleistocene glaciations. An enormously continuous mass of ice and glaciers, representing the northern inland ice-mass, extended during the Pleistocene times over the areas which are now northern Europe, great parts of England, Ireland, Holland, northern Germany, Poland and many parts of Central Russia. The ice extended also from Kiev eastwards to the Urals. In the area of the Central European mountains, the Alps were the most extensively glaciated part. There were, however, relatively small ice-free islands. In western Europe, ice covered the Jura, Vosges, the highest summits of the French Central Plateau, the lower Plateau de Millevaches, the Pyrenees and other higher plateaux of the Iberian Peninsula. On the Carpathians the Pleistocene ice covered the Babiagora, the Tatra Mts., the Czornahora and the Rodna Mts. Evidence of Pleistocene glaciers is found even on the Spanish Sierra Nevada, on the Monte Sirino in south Italy, Monte Pollino (Sierra di Dolcedorme), Greece, etc.

2. SOME TYPICAL BOREO-ALPINE INSECTS

ORTHOPTERA

The boreo-alpine Orthoptera from Europe are listed by ANDER (45). *Aeropedellus variegatus* FISCH.-WALDH., strongly brachypterous in the female and wholly brachypterous in the male, is restricted to grassy-meadows, at elevations of 1500-2600 m, on the Basses Alpes, Hautes Alpes, Graubünden, Monte Cimone (Apennines) and the Caucasus. The north area: Finland, the Altai Mountains, northern Mongolia, Lena, Kamchatka and Alaska. *Podisma frigida* BOH., brachypterous, but sometimes macropterous in Siberia, occurs sporadically on the West Alps. The north area: Fennoscandia, northern Russia, northern Siberia, Kokshetov Mountains in Akmolinsk, the Altai Mountains, Irkutsk, Trans-Baikal, northern Mongolia, Manchuria, Kamchatka, Alaska and northern Canada.

HETEROPTERA

Arctocorisa carinata SAHLB. occurs in ponds in the subalpine and alpine zones of the Alps and Pyrenees. The north area: Iceland, Faeroes, Shetland, Scotland, Fennoscandia, Kola Peninsula and Kamchatka. *Glaenocorisa cavifrons* THOMS. occurs in the subalpine and alpine zones of the Alps and Carpathians. The north area: Hebrides, Scotland, England, Fennoscandia and Kola Peninsula.

COLEOPTERA

The most important papers on the boreo-alpine Coleoptera are those of ROUBAL (902), HOLDHAUS (529, 533, 534), HOLDHAUS & LINDROTH (537) and HORION (542). On the whole, 43 species of boreo-alpine Coleoptera are so far known from the world. The bulk of the species belong to the Carabidae and Staphylinidae.

Nebria rufescens STRÖM. (= *N. gyllenhali* SCHÖNH.) occurs on snowfields in the alpine zone and near ponds and streams in the subalpine and alpine zones of the Pyrenees, Auvergne, Alps, the Swiss Jura, Sudeten, Carpathians, Apennines, Montenegro, northern Albanian Mountains and the Rila-planina in Bulgaria. In the northern area it occurs on stony ground and on snowfields, up to an elevation of 1000 m. The north area: Fennoscandia, Iceland, Faeroes, Shetlands, St. Kilda Is., Scotland, England, Ireland, northern Russia, North America and Greenland.

Bembidion fellmanni MANNERH. occurs on the eastern parts of the Transylvanian Alps and in the north area in Fennoscandia and West Siberia. *B. difficile* MOTSCH. occurs on the Tatra Mts. The north area: Fennoscandia, North Russia and northwest Siberia.

Patrobus assimilis CHAUD. occurs under stones on the Tyrolean and Salzburg Alps, the Erzgebirge, the Harz Mountains and the Sudeten, partly above the timberline and partly also within the forest, especially in moors. In the north area, it is predominantly found in the taiga forest, particularly in moors and moraines and in the alpine zone, up to an elevation of 800 m. The north area: the Faeroes, Shetlands, St. Kilda Is., Ireland, the Isle-of-Man, Scotland, England, Fennoscandia, Denmark, Mecklenburg, north Brandenburg, northern Russia and northwest Siberia.

Pterostichus blandulus MILL. occurs under stones in moist localities and mostly in the alpine zone on the Tatra Mts. The north area: northern Siberia, from the R. Yenissi to the R. Lena and perhaps also arctic Russia. *Pt. kokeili* MILL. occurs under stones in the alpine zone, rarely on grassy-meadows in the upper reaches of the forest on the Transylvanian Alps, the central chain of the Alps, the Rodna Mts. (East Carpathians).

Amara erratica DUFT. is widely distributed in the south area on the Pyrenees, Cantal Mts., the French and Swiss Jura, Vosges, the Alps, Schwarzwald, Thüringerwald, Harz Mountains, Sudeten, Carpathians, the Bosnian high mountains, Montenegro, Kopaonik, Stara-planina, Rila-dagh, the Central Rhodope Range and the Caucasus. It is found mainly above the timberline on the Central European mountains, particularly on grassy-meadows, in the vicinity of snowfields and under stones or even in soil. It comes to the open surface near snowfields on warm and sunny days. In the north area, it is mainly an alpine-zone form, but may occur sometimes in the tundra zone and subalpine zone in Fennoscandia; further eastward it may descend to the taiga forest. The north area: Fennoscandia, north Russia, Siberia eastwards to Kamchatka and northern North America. *A. quenseli* SCHÖNH. always occurs above the timberline on the East Alps; it is also known from the West Alps, Pyrenees, Vysoké Tatry, South Carpathians, the Bosnian high mountains, Herzegovina and Montenegro, north Albanian mountains, the mountains of Serbia, Bulgaria and on the Caucasus. It is conspicuously xerophile and occurs in sandy moraines in the *Betula*-zone up to an elevation of 1000 m in Lappland and 1200 m in Iceland.

Trichocellus mannerheimi SAHLB. is found just below the timberline on the Central East Alps and under stones in the tundra zone in the north. The north area: Kola Peninsula, Siberia eastwards to Ochotsk and Baikal.

Ilybius crassus THOMS. is sparsely distributed in the alpine zone, up to an elevation of 900 m, on the Hohe Venn, Schwarzwald, the Alps, Böhmerwald, Brdy in Central Bohemia, Erzgebirge, Sudeten, Vysoké Tatry, Czornahora (East Carpathians).

Mannerheimia arctica ER. occurs in the lower levels of the alpine zone, especially in sheltered and moist localities, on the Alps. The north area: Fennoscandia, north Russia and west Siberia.

Arpedium brachypterum GRAV. occurs almost exclusively above the timberline on the Alps, especially in moist localities under stones, near the snow-edge. It is also known from the Carpathians and the Caucasus. In the north area, it is generally found within the forest zone, but sometimes ascends to the subalpine and alpine regions, to elevations of 1000 m, and occurs among moss in moist localities in Sweden. The north area: Scandinavia, Faeroes, Shetlands, St. Kilda Is., Scotland, Ireland, the Isle-of-Man, England, Bear Is., Denmark, North Germany, Estonia, north Russia and Siberia.

Geodromicus globulicollis MANNERH. occurs exclusively above the timberline on the Alps, near the margins of ponds and streams in the vicinity of summer snow patches; it is also known from the Pyrenees, Haute Auvergne, the French Jura and the Caucasus. In the north area, covering Iceland, Scotland, northern England, Fennoscandia and the Kola Peninsula, the species occurs mainly in stony or sandy margins of streams, in subalpine and lower levels of the alpine zone, at elevations of 900-980 m.

Anthophagus alpinus FA. is found under stones, on moss and in flowers of *Primula glutinosa* WULF. in the subalpine and alpine zones of the Central Austrian Alps, the French Jura, Vosges, Sudeten, Carpathians and the Apennines. It is a planticole predator in the north area and occurs particularly in the lower levels of the alpine zone and often also within the tundra, but sparsely in the taiga. The north area: Fennoscandia, north Russia, Scotland, Ireland and England. *Anthophagus omalinus* ZETT. occurs under stones and on diverse plants in the subalpine zone and sometimes also above the timberline on the Alps, the Krainer Schneeberg, Böhmerwald Sudeten and the Carpathians. In the north area of Fennoscandia, north Russia and the R. Yenissi region, it occurs up to an elevation of 800 m in the subalpine zone and is also widely distributed within the taiga.

Boreaphilus henningianus SAHLB., from the Rhöngebirge in Central Germany, occurs in the north area in the lower reaches of the alpine zone and is abundant in the *Betula*-zone, up to an elevation of 1300 m, in semi-moist localities, under moss, grass and fallen vegetable debris. The north area: Fennoscandia, north Russia and Siberia.

Autalia puncticollis SHARP occurs in the lower reaches of the alpine and in the subalpine zone on the Alps, the French Jura and Carpathians. It occurs above the timberline in the north only in Norway. The north area: Fennoscandia, Iceland, Faeroes, the Orkney Is., Scotland, Estonia, Denmark and north Russia.

Atheta laevicauda SAHLB. is found on the upper reaches of the forest zone and above the timberline on the Central European mountains and often up to an elevation of 2600 m (Ötztaler Alps); it is also known from the Carpathians, Sudeten and Böhmerwald. In the north area, covering Fennoscandia and north Russia, the species occurs in moss in the subalpine alpine regions.

Pteroloma forsstroemi GYLLH. occurs near streams on the Harz Mountains, Thüringerwald, Erzgebirge, Brdywald, Sudeten, North Carpathians and the East Alps. In the north area of Fennoscandia, Kola Peninsula and Estonia, it is found mostly in the subalpine regions, and above the timberline only in Norway, where it is specially common in moss among stones in sheltered and moist localities.

Silpha tyrolensis LAICH. occurs on the open ground or also under stones in the subalpine and alpine zones, in cadavers, etc. on the Alps. Other distribution in the south: the Sierra Nevada, Sierra de Guadarrama, Cuenca Mountains, Serra do Gerêz. the Cantabric Range, Pyrenees, Salvaget près Castres, Monts d'Aubrac, Haute Auvergne, Montagne de l'Aigoual, the French Jura, Vosges and Sudeten. The north area: Ireland, England, Scotland, Hebrides.

Agathidium arcticum THOMS. occurs in moss in the subalpine areas on the Alps and Carpathians, but is high boreal and occurs in alpine regions in the north area of Fennoscandia, north Russia, Siberia and Scotland. It is recorded at an elevation of 2100 m on the Stubai Alps.

Neuraphes coronatus SAHLB. is a subalpine form that ascends to about 100 m above the timberline on the Alps and Carpathians; it is also known from Auvergne, Schwarzwald, Böhmerwald, the Harz Mountains and Sudeten. The north area: Fennoscandia, Russian Karelia and Kola Peninsula; under moss in the spruce-birch zone, often subalpine.

Coccinella trifasciata LINN. is found on the Graubünden and north Tyrolean Alps. The north area: Fennoscandia, north Russia (southwards to Leningrad), Siberia eastwards to Kamchatka and east Siberia to the Altai Mountains, northern Mongolia, Manchuria, North China, North Tibet, Canada, northern U.S.A. on the Cordillera to Central California and south Colorado.

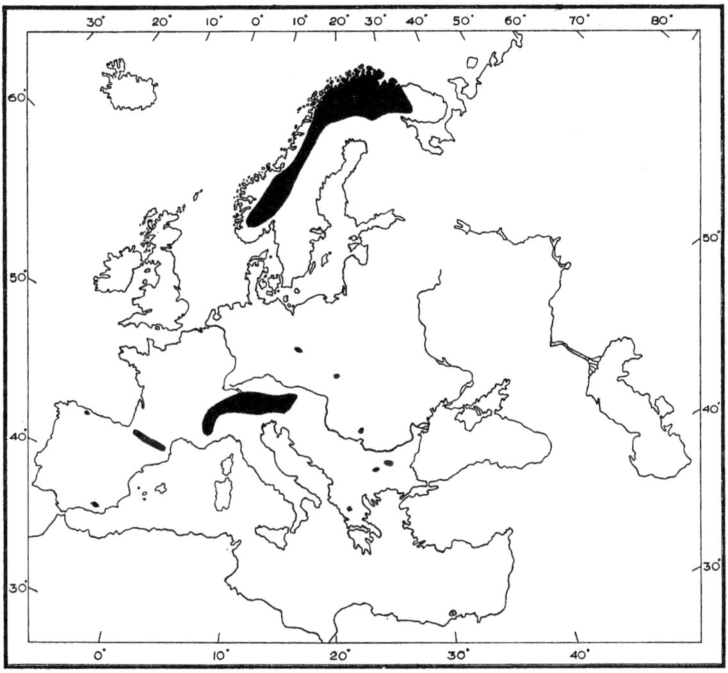

Fig. 69. The boreo-alpine distribution of *Helophorus glacialis* VILLA.

Helophorus glacialis VILLA (Fig. 69) is found exclusively in stagnant waters in the upper reaches of the forest and just above the timberline, sometimes at the snow-edge in the alpine zone and only rarely above the snowline on the Alps. It is also known from the Cantabric Range, Pyrenees, the Spanish Sierra Nevada, Corsican mountains, the higher Apennines, Krkonose (Riesengebirge), the Vysoké Tatry, Transylvanian Alps, Schipka-Balkan, Rila-planina, Muss-Alla on the Rhodope Range, Peristeri east of Janina and Lycian Toros. In the north area of Fennoscandia and Kola Peninsula, the species is much less bound to water, but occurs even in moist vegetable debris. In Lappmark, it is reported in stagnant waters in the alpine regions and sparsely in the subalpine region, at elevations of 1060 m.

Simplocaria metallica STURM. is sparsely found on moss in the subalpine zone and immediately above the timberline on the Gailtaler Alps, Krkonose (Riesengebirge) and Carpathians. In the north area of Fennoscandia and Greenland, it occurs in the alpine region, in sandy localities, near streams.

Podabrus obscuripes SAHLB. occurs on the north Tyrolean Alps and on the Karawanken. It is found under stones, on *Empetrum*-meadows, above the forest, at elevations of 900-1100 m, on Mt. Noulja in the north area. The north area: Fennoscandia, Finnish Lappmark, Kola and Kanin Peninsulas, northwest Siberia and Kamchatka.

Corymbites cupreus FABR. occurs on vegetation in the subalpine and alpine zones and under stones on the grassy-meadows on the Alps. It is also known from the Sierra de Guadarrama, Pyrenees, Montagnes Lyonnaises, Auvergne, Jura, Vosges, Schwarzwald, Böhmerwald, Brdywald, Erzgebirge, Sudeten, Carpathians, Bihor Mountains, the mountains of Mal i Gjalices in North Albania, Rila-dagh, Schipka-Balkan and the Apennines. The north area: Ireland, Scotland, England, parts of Norway, Swedish Lappmark, Finland, north and central Russia, Siberia eastwards to the Altai Mountains and the Yenissi Region. *C. rugosus* GERM. is found under stones, exclusively above the timberline, at elevations of 2900-3000 m, on the Alps; it is also known from the Vosges. The north area: Petschora Region, North Urals, Siberia southwards to the Altai Mountains and Trans-Baikal and eastwards to Sachalin, Mt. Daisetsu in the Hokkaido Is. of Japan and perhaps also Alaska. *C. affinis* PAYK. is essentially a forest-form that occurs in the subalpine zone on the Carpathians; it is also known from the Auvergne, l'Aigoual, Lyonnais, Alps, Vosges, Schwarzwald, Hohes Venn, Harz Mountains, Thüringerwald, Erzgebirge, Sudeten and the Bihor Mountains. In the north area, it often occurs in the regio alpina in Fennoscandia, Russia, Siberia and Sachalin.

Hypnoidus rivularius GYLLH. occurs in the subalpine and alpine zones at the edge of stagnant waters and in the margins of streams, but often also under stones near the snow-edge on the Alps, Krkonose, Tatra Mountains, the high summits of the East and South Carpathians, Caucasus, Montagnes de la Lozère and Mont Mézenc. In the north area, it is predominantly found in moist localities, at the edges of ponds and streams, in the lower regio alpina and descends also to the taiga zone, but occurs up to an elevation of 900 m on Abisko. *H. hyperboreus* GYLLH. occurs on the French Alps, Walliser Alps, Gran Paradiso and the southern Dolomites. The north area: Fennoscandia, Kola Peninsula, Petschora Region and Siberia; in the regio alpina and tundra, but even in the regio subalpina and within the *Betula*-forest zone.

Aphodius piceus GYLLH. is a subalpine form on the Central European mountains and occurs in dung above the forest in Sudeten; it is also known from the Alps, Harz Mountains, Tatra Mountains. Unlike the other species, it does not occur in dung in the north area, where it is common in the regio alpina, at elevations of 1000 m in North Sweden and descends up to the southern limits of the high boreal taiga forest. In addition to Fennoscandia, the north area includes northern Russia and Siberia southwards to the Altai Mountains.

Bius thoracicus FABR. is a subalpine form on the Central European mountains, especially on the Hautes Pyrenees, the Swiss and French Alps, North Tyrol and southern East Carpathians. In the north area, it occurs under bark. The north area: Fennoscandia, Russia southwards to nearly the 57th north parallel and west Siberia.

Evodinus interrogationis LINN. is a typically subalpine form on the Alps. It occurs also on the Auvergne, Jura, Vosges, Siebengebirge, Erzgebirge, North Carpathians and the Caucasus. The north area: Fennoscandia, Denmark, Estonia, north and central Russia, Siberia, Tarbagatai, northern Mongolia, Manchuria, Korea and Sachalin.

Acmaeops septentrionis THOMS. from the French Alps is a meadow-form in the north area and occurs in the flowers of *Chamaenerium angustifolium*, *Angelica*, *Spiraea ulmaria*, *Sorbus aucuparia*, *Prunus padus*, *Alchemilla millefolium*, etc. *Acmaeops smaragdula* FABR. occurs in the subalpine zone on the French Alps and in the *Betula*-forest zone in Fennoscandia, north and central Russia, Siberia southward to the Altai Mountains and eastward to Amur and Korea.

Chrysomela crassicornis HELLIES is found on the Cantabric Mountains and sparsely on the Alps. The north area: Ireland, Scotland, Orkney and Shetland Is., southwest Norway, Auli-Ata in the Syr-Darya drainage basin and the east Siberian coast. It is found under stones in the *Armeria maritima*-zone on the mountains of Norway. On the East Alps, it is restricted to the Kalkphyllite and Mesozoic limestone areas and occurs under stones in the alpine zone, at elevations of 2040–2800 m.

Phytodecta affinis GYLLH. occurs under stones up to an elevation of 1100 m on the Alps, but may sometimes ascend above the timberline also. The north area: Fennoscandia, Kola Peninsula, Kanin, Petschora, Siberia, Alaska, the Hudson Bay Territory and Colorado Mountains.

Otiorrhynchus morio FABR. from the Cantabric Mountains, Pyrenees, Haute Auvergne, Cévennes, Savoye, Dauphiné, Swiss Alps, northern East Alps, Jura, Vosges, Mittelgebirge of Germany, Schwarzwald, Bohemian Forest northwards to the Carpathians, is found under stones, in the low-subalpine zone, and sometimes also in the alpine zone. *O. dubius* STRÖM. occurs in moss and under stones in the subalpine zone of the Alps, Carpathians, Vosges, Schwarzwald, Bohemian Forest, Erzgebirge, Thüringerwald, Harz, Sudeten, etc. In the north area, it inhabits the not-too-moist, open ground in the southern boreal-taiga forest. The north area: Greenland, Iceland, Faeroes, Shetlands, Ireland, Isle-of-Man, Scotland, northern England, north Jutland, Fennoscandia, north Russia, Estonia and north Prussia. *O. arcticus* FABR. occurs under stones, generally above the timberline, on the Pyrenees, Haute Auvergne, Sudeten, Vysoké Tatry and Czornahora (northeast Carpathians). The north area: Fennoscandia, Greenland, Iceland, Faeroes, Shetland, Ireland, Isle-of-Man, St. Kilda Is., Scotland, Kola and Kanin Peninsula, etc. *O. salicis* STRÖM. occurs exclusively in the subalpine zone on the Central European mountains, from the Swiss Alps to the Carpathians, the mountains of Bosnia and Bulgaria. The north area: southern Norway, central Sweden and East Prussia.

Barynotus squamosus GERM. occurs on the Pyrenees, Cévennes and Mont-Dore; it is an alpine zone form on the Pyrenees. The north area: northeast North America, Iceland, Faeroes, Shetlands, Orkney Ls., St. Kilda Is., Ireland, Scotland, England,

Jutland, Norway and Sweden; it was probably introduced into North America by human agency.

HYMENOPTERA

PITTIONI (837) has listed the boreo-alpine Hymenoptera, particularly the Bombidae. *Bombus alpinus* LINN. from the Glockner area in the Alps (up to an elevation of 3100 m) and from the north area of Norway-Sweden and Kola Peninsula and *B. lapponicus* FABR. from the Pyrenees, Alps (up to an elevation of 3000 m), the high mountains of the Balkan Peninsula and the Macedonian Mountains are the two important boreo-alpine species. The north area: Scotland, Fennoscandia, north Russia, Novaya Zemlya, north Siberia eastward to Kamchatka and Tschukot Peninsula, the Altai Mountains, Tien Shan and Pamirs, the Aleutian Is., Alaska, Sitka, northern Canada eastwards to Labrador, the Cordillera of Canada southwards to Oregon, Utah, Colorado southwards to the mountains of Arizona.

LEPIDOPTERA

Parnassius phoebus FABR. from the Alps breeds on *Saxifraga aizoides* and *Sempervivum montanum* at elevations of 1500-2500 m. The north area: Petschora, the northern Urals, the Altai Mountains, Tarbagatai, the Sayanskii Range, Irkutsk, eastern Tien Shan, northwest Mongolia, Kamchatka, the Rocky Mountains of Alaska, British Columbia, Montana southwards to New Mexico and the Californian Sierra Nevada.

The north area of *Melitaea iduna* DALM. (Fig. 70), from the northwest Caucasus, embraces the Swedish Lappland, Norway, Finland, Russia, the Altai Mountains, Tarbagatai, the Sayanskii Range and northern and southern East Siberia. The larvae feed in *Bartsia alpina*, *Vaccinium uliginosum* and *Vaccinium myrtillus*. *Argynnis pales* SCHIFF. from the subalpine and alpine zones of the Alps (3000 m on the Swiss Alps), Pyrenees, Abruzze Mountains, the Vysoké Tatry and the mountains of northern Balkan Peninsula, and the Himalaya, extends in the north in Fennoscandia, the Russian Lappland, polar Siberia and perhaps also Alaska and north Canada. *A. thore* HÜBN. occurs at elevations of 800-2000 m on the Alps. The north area: north and central Scandinavia, north Finland, Kola Peninsula, Ladoga-Karelia, Wiatka, Kasan West Siberia, the Altai Mountains, east Siberia, Amur, Ussuri and Kamchatka. *Erebia pandrose* BKH. (= *E. lappona* ESP.) (Fig. 70) occurs on the Pyrenees, Alps, Carpathians, the Rila and Rhodope Mountains, Macedonian mountains, north Albanian mountains, etc. It always occurs above an elevation of 1500 m on the Pyrenees, including the Spanish side on the Massiv of the Mont Perdu. On the Alps it occurs at elevations of 1600-3000 m and on the Rila Mountains at 2200-2900 m. The larvae feed on *Festuca ovina* and other grasses. The north area: Fennoscandia, Kola and Kanin Peninsulas, the Altai and the Sayanskii Ranges. *E. epiphron* KNOCH is a subalpine and alpine zone species on the Alps and occurs at elevations of 1200-2600 m; on the French side of the Pyrenees at elevations of 1500-2000 m; on the Cantabric Mountains, Auvergne, Cevennes, Vosges, the Harz Mountains, Sudeten, Carpathians, Apennines and the high mountains of the Balkan Peninsula. The north area: Ireland, Scotland and northern England.

Pieris callidice ESP. breeds on the Alps on *Cardamine resediflora*, *Erysimum pumilum* and *Sempervivum arachnoideum* at elevations of 1500-3333 m. On the Stubaier Alps it occurs up to an elevation of 3500 m. On the Middle Asiatic Mountains, the adults

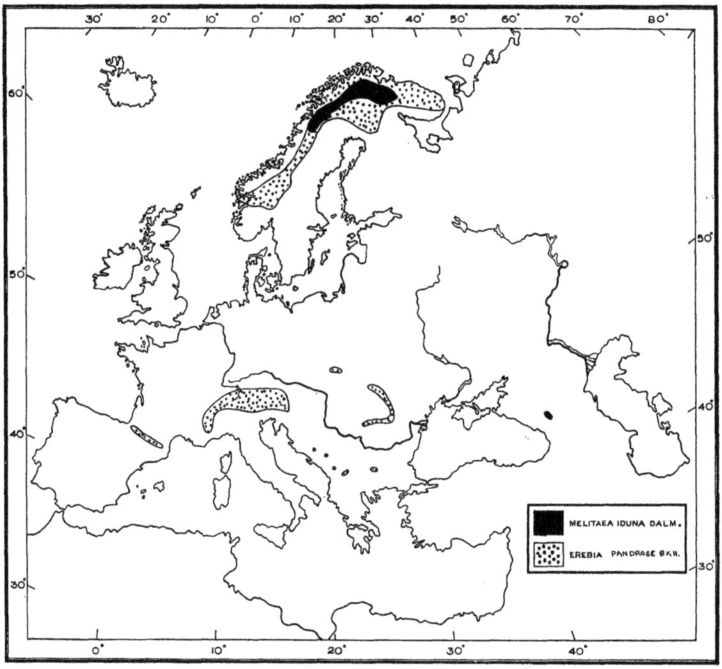

Fig. 70. The boreo-alpine distribution of *Melitaea iduna* DALM. and *Erebia pandrose* BKH.

occur at elevations of 3000 m and on the Northwest Himalaya between 3000 and 5000 m. The south area: Pyrenees, Caucasus, Mt. Ararat, Perli-dagh, Mt. Elbrus, etc. The north area: north Urals, northwest Siberia, the Altai Mountains, Ala-Tau, Tien Shan, Ferghana and Zerafshan Ranges, the Pamirs, Tibet, Hindukush and Tarbagatai.

Two Lycaenidae are known to have boreo-alpine distribution. *Polyommatus glandon* PRUN. (= *Lycaena orbitulus* of authors) occurs on the Cantabric Mountains, Pyrenees, Alps, Herzegovina, Alibutusch Range in Macedonia and perhaps also the Spanish Sierra Nevada, Mt. Ararat and the Caucasus. The north area: Fennoscandia, northern parts of Asia, northern North America and Greenland. *P. orbitulus* PRUN. (= *Lycaena pheretes* HÜBN.) breeds on *Astragalus* and *Trifolium* at elevations of 1000-3000 m on the Swiss Alps. It occurs also on the central Scandinavian mountains, the Altai Mountains, East Siberia, Issyk-kul area of the Tien Shan, West China, Tibet, the Sikkim-Himalaya, Northwest Himalaya and the Pamirs.

Hesperia andromedae WALLGR. occurs at elevations of 1250-2500 m on the Swiss Alps and is also known from the Pyrenees, Transylvanian Alps and the high mountains of Bosnia and Herzegovina. It occurs in the upper *Betula*-zone and above the timberline in Fennoscandia, Irkutsk and Amur.

A number of Heterocera are known to have boreo-alpine distribution. *Rhyacia alpicola* ZETT. (= *Agrotis hyperborea* ZETT.) from the Alps, Krkonose and the Nízke Tatry feeds on *Empetrum nigrum*, *Arctostaphylos uvaursi*, *Vaccinium myrtillus*, *Betula nana*, etc. in the subalpine and alpine zones of the Alps. It occurs at an elevation of 2500 m on the Swiss Alps. The north area: Scotland, Orkney and Shetland Is., Fennoscandia, Estonia and north Russia. *Agrotis speciosa* HÜBN. occurs on the Alps, Auvergne, the Swiss Jura, Vosges, Schwarzwald, Thüringerwald, Harz Mountains, Bohemian Forest, Erzgebirge, Sudeten, Carpathians, north Albanian mountains, the Rila Range and the Macedonian mountains. The north area: northern North America, the Adirondack Mountains, the White Mountains, Fennoscandia, Estonia, north Russia, the Urals, northern Siberia, the Altai and the Sayanskii Ranges, the Dzhungarskii Ala-Tau, northern Mongolia, the Changai Mountains, etc. *A. faticida* HÜBN. occurs on the Pyrenees, Alps, the Albanian mountains and the Caucasus. The larvae feed on grass roots at elevations of 2700 m on the Alps. The north area: Norway, southern Urals and Siberia. *Crymodes* (=*Hadena*) *maillardi* (H.-S.) occurs on the Alps, Pyrenees, Carpathians, Rila Range and Macedonian mountains. The north area: Fennoscandia. On the Alps, it is a subalpine and alpine zone form that ascends to an elevation of 3000 m in South Tyrol. The larvae feed on grasses like *Nardula stricta* and *Poa alpina*. *Anarta melanopa* THUNB. ascends up to the snowline on the Alps and also occurs on the Gran Sasso, the Rila Mountains and the mountains of Macedonia and northern Albania. The north area: Fennoscandia, Shetlands, Scotland, the Kola Peninsula, Labrador, White Mountains, Colorado and the Taos Peak in New Mexico. It is a polyphagous species in these areas and feeds on *Vaccinium uliginosum*, *V. myrtillus*, *Polygonum*, *Salix*, *Betula*, etc. It also occurs at high elevations in Lappland. *Sympistis funesta* PAYK. (=*Anarta funebris* HÜBN.) occurs at elevations of 2000-2500 m on the Alps. The north area: Fennoscandia up to the arctic region, north Siberia and Labrador. *Plusia hochenwarthi* HOCHW. occurs in Fennoscandia up to the arctic region, the Kola Peninsula, West Siberia, East Siberia, the Altai Mountains, Tarbagatai, the Ala-Tau, the Alai, Tien Shan, the Ferghana Ranges, Tibet, Northwest Himalaya, northwest Caucasus, the Armenian mountains, the Alps (1700-2700 m), the arctic North America and the White Mountains. *Cidaria* (=*Larentia*) *nubiliaria* (H.-S.) from the Alps, the southern East Carpathians, Herzegovina, Montenegro and the Albanian mountains, occurs in north and central Scandinavia, north Finland and Siberia. The larvae feed in *Saxifraga oppositifolia*, *S. moschata* and *S. bryoides*. While in the north area it occurs mostly on grassy-meadows, it ranges from an elevation of 470 m to over 2500 m on the Swiss Alps. *Cidaria* (=*Larentia*) *munitata* (HÜBN.) occurs in the subalpine and alpine zones of the Pyrenees and the Alps and is also known from the Caucasus. On the Swiss Alps the larvae feed on *Alsine media* and *Bellis perennis*. The north area: Fennoscandia up to the arctic region, Estonia, Kurland, north Russia, Urals, Siberia eastward to Kamchatka, Amur, Ala-Tau and Issyk-kul area of the Tien Shan, Shetlands, Scotland, north England, Wales, St. Kilda Is., Ireland, Faeroes, Iceland and northern parts of the Nearctic regions. *Cidaria* (=*Larentia*) *turbata* (H.-S.) occurs in the subalpine and alpine zones (1500-2500 m) on the Alps, Pyrenees, Auvergne, Carpathians and the Bosnian and Macedonian mountains. The north area: Fennoscandia, northwest Russia, the Altai Mountains, and the Sayanskii Range and Kamchatka. *C.* (=*L.*) *flavicinctata*

(HÜBN.) occurs on the Cantabric Mountains, the Pyrenees, the Spanish Sierra Nevada, the French Central Plateau, Swiss Jura, Alps, the Apennines (Abruzze), the Vysoké Tatry, the mountains of Bosnia and Herzegovina, Montenegro, northern Albania, Schar Planina in Serbian Macedonia, the Rila Mountains, etc. It breeds in *Saxifraga*, *Sedum*, *Alchemilla montana* and *Salix* in the subalpine and alpine zones, up to an elevation of 2400 m on the Swiss Alps. The north area: Ireland, north England, Scotland, Hebrides, Orkney Is., Norway and northwest Finland.

Six species of Geometridae are boreo-alpine. *Eupithecia scriparia* H.-S. (=*Tephroclysia undata* FRR.) occurs in the subalpine and alpine zones and ascends to an elevation of 2300 m on the Swiss Alps and is also known from the Pyrenees, Auvergne, Apennines, Vysoké Tatry, Transylvanian Alps, Bosnian mountains, Herzegovina, Macedonian mountains, Armenian mountains and northeast Asia Minor. The north area: Fennoscandia, east Siberia and Labrador. *Gnophos sordarius* THUNB. is a predominantly subalpine zone species from the Central European mountains, but it also occurs in the alpine zone, up to elevations of 600-2300 m, on the Alps. It is distributed on the Jura (Swiss), Vosges, Schwarzwald, Harz Mountains, Sudeten, Bohemian Forest eastwards to the western Waldviertel in Lower Austria, the Carpathians and the mountains of Herzegovina and Albania. The north area: Fennoscandia, Kanin Peninsula, the Altai Mountains and the southern East Siberia. *Gn. myrtillatus* THUNB. is distributed on the Cantabric Mountains, Pyrenees, Central Spanish mountains like the Sierra de Albarracin, Sierra de Gredos and the Sierra Nevada, the Apennines, Alps, Transylvanian Alps, the mountains of Herzegovina, Durmitor, northern Albania, Rila and Rhodope mountains, Pirin Mountain, Schar Planina, Caucasus and the Armenian mountains. The north area: Scotland, Fennoscandia, Estonia, north Russia, the Altai Mountains, Tien Shan (Issyk-kul area). *Ps. coracina* ESP. from the Pyrenees and Alps (1800-2600 m), occurs in the north in Fennoscandia, the Kola Peninsula and the Altai Mountains. *Pygmaea fusca* THUNB. occurs at elevations of 1800-2500 m on the Swiss Alps and at 3000 m in south Tyrol, and it is also known from the Pyrenees and Schar Planina. In the north area it breeds on *Viola calcarata*, *Vaccinium uliginosum* and *Arctostaphylos uvaursi* in Fennoscandia and extends up to the arctic area. *Isturgia* (=*Fidonia*) *carbonaria* (CL.) ranges at elevations from 400 to nearly 2500 m on the Alps and Pyrenees. The north area: Fennoscandia, Scotland, Denmark, Schleswig, northern East Prussia, north Poland, north Russia and east Siberia.

Arctia flavia FUES. occurs at elevations of 1400-3320 m on the Swiss Alps and has also been reported from the Rila Mountains. The north area: southern and central Urals, the plains of Russia to the west of the Urals, the Altai and the Sayanskii Mountains, Tarbagatai, Irkutsk, northern Mongolia, Amur and the Chingan mountains. The larvae feed on *Cotoneaster vulgaris* and *Aconitum napellus*. *Arctia quenseli* PAYK. occurs at 2000-2700 m on the Alps as far as the Hohe Tauern; it is also known from the Transylvanian Alps. The north area: arctic Fennoscandia, northern East Siberia, Amur, Tarbagatai and the arctic North America. *Lithosia cereola* HÜBN. occurs at elevations of 1200-1900 m on the Alps and Fennoscandia, Estonia and eastern central Russia in the north. *Zygaena exulans* HOCHW. occurs on the Alps (1600-3000 m), Pyrenees, Apennines, Transylvanian Alps, etc. The north area: Fennoscandia, Scotland, Murmansk, the Kola Peninsula, the Altai and the Sayanskii mountains and Tarbagatai. *Sterrhopteryx standfussi* H.-S. is a subalpine and alpine zone species on the Alps, South Carpathians, Sudeten, Bohemian Massif, etc. The north area: Fennoscandia, Harz Mountains.

3. SOME TYPICAL BOREO-ALPINE ARACHNIDA

Very few boreo-alpine spiders appear to have been recorded so far. *Tiso aestiva* KOCH from the Vysoké Tatry and Iceland, *Cornicularia karpinskii* CAMBR., *Erigone remota* KOCH, *Tarentula* (*Arctosa*) *alpigena* DOL. (known from Greenland also), and *Pellenes lapponicus* SUDEV (also from the Pyrenees) are perhaps the only true boreo-alpine spiders known at present.

THIENEMANN (1036) considers *Hydrovolzia placophora* MONTI, *Lebertia glabra* THOR., *Hygrobates foreli* LEB. and *Feltria minuta* KOEN. to be boreo-alpine Hydracarina. *Eutrombidium frigidum* BERL., *Microtrombidium sucidum* BERL., *Tarsolacrus arctulosus* S.T., *Podothrombium curtipalpe* BERL. and *Parazerkon sarekensis* WILLM. are probably boreo-alpine terricole mites.

4. THE BOREO-ALPINE INSECTS IN THE NORTH AREA

Owing to the greater extent and the relative uniformity of vegetation within the limits of the north area, the range of the boreo-alpine species in the north is uninterrupted and also generally more extensive than in the south area. Only some species like *Otiorrhynchus morio* FABR., *O. salicis* STRÖM and *Erebia epiphron* KNOCH. have a smaller range in the north area than in the south. The European part of the north area contains 32 species of boreo-alpine Coleoptera, of which only five are absent in Fennoscandia. The position of the southern limits of the north area differs in different species and is governed by a complex set of ecological conditions and past distribution. The boreo-alpine species of Europe are generally characterized by the high endemism on the mainland of north Europe. As we proceed from the central north Europe, we see a gradual impoverishment both to the west and to the east.

Some boreo-alpine species appear to have a wider altitudinal distribution in the north area rather than in the south area. A number of species occur, for example, both within the subalpine and alpine regions in the north area, but predominantly or even exclusively within the alpine zone in the south area. *Mannerheimia arctica* KR., *Hypnoidus hyperboreus* GYLL., *Phytodecta affinis* GYLL. and *Otiorrhynchus arcticus* FABR. are, for example, exclusively alpine-zone species in the south area and *Geodromicus globulicollis* MANNERH. is also found above the forestline on the Alps. *Corymbites rugosus* GERM. occurs within the forest-zone in Siberia, but on the Alps it is confined to the alpine zone, though sometimes it may be found sparsely in the subalpine zone on the Vosges. *Amara erratica* DUFT., *A. quenseli* SCHÖNH. and *Arpedium brachypterum* GRAV. are predominantly alpine-zone forms on the mountains in the south area, but true forest-zone forms in the north. *Hypnoidus hyperboreus* GYLL. has perhaps minimal range within the forest zone in the north area and is confined in Fennoscandia to the *Betula*-zone. In the north area, *Geodromicus globulicollis* MANNERH. occurs largely above the forestline and also within the *Betula*-zone, but it does not descend far in the taiga forest, except perhaps in valleys. A complex set of ecological factors seems to underlie these differences in the north and south. In the north area the uppermost limit of the forest is characterized in most localities by more or less open *Betula*- forest, but such a *Betula*-zone is absent on the mountains in the south. The needle-forest, with abundant rhododendron, ascends to the forestline on the mountains in the south area.

There are, in addition, a number of mountain autochthonous species in the south area, near the forestline, but such species are generally absent in the north area and the boreo-alpine species have, therefore, apparently a much wider area for distribution in the north area. It is now generally accepted that the boreo-alpine species are mostly forest forms and practically none that are exclusively confined to the alpine zone and thus represent true tundra elements. It seems, therefore, that the boreo-alpine mode of distribution could only have risen during the Pleistocene times, when the ice-free area of Central Europe did not have the tundra character, but exhibited subarctic parkland characters, in other words, forest with isolated treeless meadow ground, somewhat like what we find today in the *Betula*-zone in the north area. It is interesting to recall in this connection that most boreo-alpine species occur in open grassy localities and never within the closed forest area.

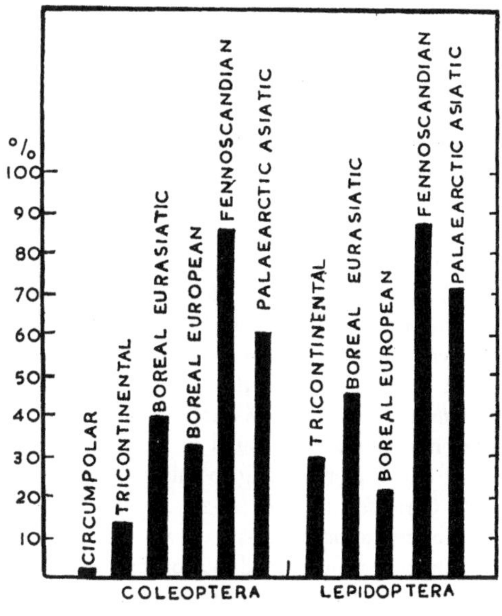

Fig. 71. The component elements of the boreo-alpine Coleoptera (43 species) and Lepidoptera (33 species) in the north area.

Depending upon the differences in the extent of the north area, the boreo-alpine insects belong to the following groups:

1. Boreal tricontinental elements that have a circumpolar distribution and thus occur on all the three northern continents, viz. Europe, Asia and North America. The circumpolar species, in the strict sense of the term, is *Nebria rufescens* STRÖM. that occurs not only on the mainland of Europe, Asia and North America, but is also widely distributed in Greenland, Iceland, Faeroes and Great Britain. *Amara erratica* DUFT., *Coccinella trifasciata* LINN., *Hypnoidus hyperboreus* GYLL. and *Phytodecta affinis* GYLL. occur in the three northern continents, but are absent in the Atlantic islands and in the islands of the northern ice-seas. *Corymbites cupreus* FABR. also belongs to

this group, but it is known from Alaska. The tricontinental boreo-alpine Lepidoptera are *Parnassius phoebus* FABR., *Agrotis speciosa* HÜBN., *Cidaria munitata* (FABR.), *Sympistis funesta* PAYK., *Plusia hochenwarthi* HOCHW., *Eupithecia scriparia* H.-S. and *Arctia quenseli* PAYK. None of these species occur in the north Atlantic islands. *Argynnis pales* SCHIFF., *Polyommatus glandon* PRUN. and *Anarta melanopa* THUNB. also belong to this group.

2. Boreal Eurasiatic species occur in the north of Europe and Asia, but are absent on the mainland of North America. Examples: *Podabrus obscuripes* SAHLB., *Trichocellus mannerheimi* SAHLB., *Bembidion difficile* MOTSCH., *Boreaphilus henningianus* SAHLB., *Mannerheimia arctica* ER., *Agathidium arcticum* THOMS., *Arpedium brachypterum* GRAV., *Anthophagus omalinus* ZETT., *Hypnoidus rivularius* GYLL., *Corymbites affinis* PAYK., *C. cupreus* FABR., *Aphodius piceus* GYLL., *Bius thoracicus* FAB., *Evodinus interrogationis* LINN., *Acmaeops septentrionis* THOMS., *A. smaragdula* FABR. and *Chrysomela crassicornis* HELLIES. The following 15 boreo-alpine Lepidoptera have Eurasiatic distribution in the north area: *Pieris callidice* ESP., *Melitaea iduna* DALM., *Argynnis thore* HÜBN., *Erebia pandrose* BKH., *Polyommatus orbitulus* PRUN., *Hesperia andromedae* WALLGR., *Agrotis faticida* HÜBN., *Cidaria nubiliaria* (H.-S.), *C. turbata* (H.-S.), *Psodos coracina* ESP., *Gnophos myrtillatus* THUNB., *Isturgia carbonaria* (CL.), *Arctia flavia* FUES., *Zygaena exulans* HOCHW. and *Hepialus ganna* HB.

3. Exclusively European species, extending sometimes only to Greenland, but never occurring on the mainland of North America. Examples: *Ilybius crassus* THOMS., *Geodromicus globulicollis* MANNERH., *Anthophagus alpinus* FABR., *Autalia puncticollis* SHARP, *Atheta laevicauda* SAHLB., *Neuraphes coronatus* SAHLB., *Pteroloma forsstroemi* GYLL., *Silpha tyrolensis* LAICH., *Helophorus glacialis* VILLA, *Simplocaria metallica* STURM., *Otiorrhynchus dubius* STRÖM., *O. morio* FABR., *O. arcticus* FABR. and *O. salicis* STRÖM. As already indicated, *Barynotus squamosus* GERM. seems to have been introduced by human agency into North America. Only *Simplocaria metallica* STURM., *Otiorrhynchus dubius* STRÖM. and *O. arcticus* FABR. are endemic in Greenland. Some of these species may perhaps be found in future in north Asia also. Seven boreo-alpine species of Lepidoptera belong to this group: *Erebia epiphron* KNOCH., *Phycita alpicola* ZETT., *Crymodes maillardi* (H.-S.), *Phygmaena fusca* THUNB., *Lithosia cereola* HÜBN. and *Sterrhopteryx standfussi* H.-S.

Six of the 43 boreo-alpine Coleoptera and four of the 33 boreo-alpine Lepidoptera from Europe have not so far been found in Fennoscandia, but appear either in Great Britain and other north Atlantic islands or also in north Russia. The extreme eastern limit of many boreo-alpine Coleoptera, like for example, *Geodromicus globulicollis* MANNERH., *Helophorus glacialis* VILLA, *Otiorrhynchus salicis* STRÖM., *Barynotus squamosus* GERM. and *Simplocaria metallica* STURM., is within Fennoscandia itself. *Otiorrhynchus salicis* STRÖM. and *Barynotus squamosus* GERM., though Fennoscandian elements, are absent in Finland proper. The boreo-alpine Lepidoptera *Crymodes maillardi* (H.-S.) and *Sterrhopteryx standfussi* H.-S. occur exclusively in Fennoscandia in the north area. The southernmost limit of many boreo-alpine Coleoptera, like *Bembidion fellmanni* MANNERH., *Anthophagus alpinus* FABR., *Geodromicus globulicollis* MANNERH., *Helophorus glacialis* VILLA and *Phytodecta affinis* GYLL., in Fennoscandia is on the southern Norwegian mountains. The other boreo-alpine species extend in the north area further south than Fennoscandia.

The Palaearctic Asia has 26 boreo-alpine Coleoptera and 24 boreo-alpine Lepidoptera, but their distribution within Asia is at present only very imperfectly known.

Some species like *Evodinus interrogationis* LINN. and *Acmaeops smaragdula* FABR. extend eastward to Korea and Lepidoptera extend further southwards also. *Pieris callidice* ESP. occurs, for example, as far south as the Hindu Kush and the Northwest Himalaya. *Polyommatus glandon* PRUN. extends south to the Pamirs Region and the Northwest Himalaya and the Sikkim-Himalaya. *Arctia flavia* FUES. extends to the Chingan Mountains. Some of the boreo-alpine species, like *Corymbites rugosus* GERM. and *Arctia quenseli* PAYK., in the Palaearctic of Asia, have also been recorded from the Mt. Daisetsu in Hokkaido Island (Japan). The following species reach their extreme eastern limit in the north area in West Siberia: *Patrobus assimilis* CHAUD., *Amara quenseli* SCHÖNH., *Mannerheimia arctica* ER., *Anthophagus omalinus* ZETT., *Agathidium arcticum* THOMS., *Corymbites cupreus* FABR. and *Bius thoracicus* FABR. *Nebria rufescens* STRÖM., *Amara erratica* DUFT., *Corymbites rugosus* GERM., *C. affinis* PAYK., *Evodinus interrogationis* LINN., *Acmaeops septentrionis* THOMS. and *A. smaragdula* FABR. are, on the other hand, widely distributed in West Siberia.

The distribution of the boreo-alpine insects in the North Atlantic islands is peculiar in many respects. Only the species that are found in Britain occur in Ireland, the Isle-of-Man, Hebrides, Orkney and Shetlands, the Faeroes and Iceland. This fact is all the more remarkable, when it is recollected that only a small number (15 out of 37 from Scandinavia) of species are endemic in Britain. *Simplocaria metallica* STURM., a species that is absent on all the other North Atlantic islands (but occurs in Fennoscandia on the European mainland) is known from Greenland. Almost all the North Atlantic islands have also only those species that are capable of flourishing in the alpine-zone, but all forest species are absent. *Coccinella trifasciata* LINN., *Corymbites affinis* PAYK., *Bius thoracicus* FABR., *Otiorrhynchus salicis* STRÖM., *Ilybius crassus* THOMS., *Neuraphes coronatus* SAHLB. and *Pteroloma forsstroemi* GYLL. are absent on all the North Atlantic islands. The distributional pattern of the boreo-alpine Lepidoptera in these islands is also similar to that of the boreo-alpine Coleoptera. Six boreo-alpine Coleoptera *Nebria rufescens* STURM., *Amara erratica* DUFT., *Coccinella trifasciata* LINN., *Phytodecta affinis* GYLL. and *Hypnoidus hypereroreus* GYLL. have so far been found on the mainland of the Nearctic Realm. Except the last mentioned species that is found only in Alaska, the others are widely distributed in North America. Seven boreo-alpine Lepidoptera, three of which are found on the three northern continents, occur in North America. *Anarta melanopa* THUNB., occurring on the White Mountains, is characterized by its discontinuous distribution in Labrador. *Agrotis speciosa* HÜBN. and *Plusia hochenwarthi* HOCHW. are also known from the White Mountains.

5. THE BOREO-ALPINE INSECTS IN THE SOUTH AREA

The abundance of boreo-alpine species on the mountains of the south area depends primarily on the massiveness of the mountain range and its distance from the southern margin of the Pleistocene northern inland ice-mass. With their relatively larger size, the Alps surpass all the other mountains of Europe in their wealth of boreo-alpine species. Though considerably less than that of the Alps, the number of boreo-alpine species on the Carpathians is not small. The Harz Mountains, the Thüringerwald, Erzgebirge, Bohemian Forest, Schwarzwald, Vosges, the French Central Plateau and the Jura mountains are, however, relatively poor in boreo-alpine insects. The

Pyrenees and the mountains of the Balkan Peninsula are far too distant from the southern edge of the northern inland-ice of the Pleistocene and support, therefore, a considerably lesser number of boreo-alpine species than the Alps. The distance of the mountain range from the southern margin of the Pleistocene northern ice-mass has a far greater influence on the distribution pattern of forms like Coleoptera, with poor capacity for dispersal than in the case of insects with well developed powers of extensive locomotion and migration. The southern mountain ranges have, therefore, a much larger number of boreo-alpine Lepidoptera than Coleoptera (fig. 72). The mountains of the Balkan Peninsula are, for example, particularly rich in boreo-alpine Lepidoptera.

Seven species of boreo-alpine Coleoptera from the Alps, viz. *Trichocellus mannerheimi* SAHLB., *Mannerheimia arctica* ER., *Coccinella trifasciata* LINN., *Podabrus obscuripes* SAHLB., *Hypnoidus hyperboreus* GYLL., *Acmaeops smaragdula* FABR. and *Phytodecta affinis*

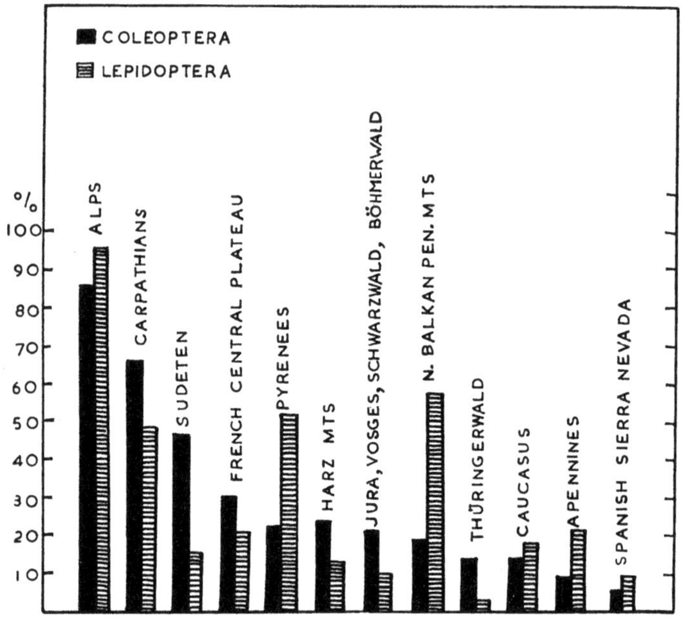

Fig. 72. Analysis of the boreo-alpine Coleoptera and Lepidoptera on different mountains in the south area.

GYLL. are absent on the other Central European and south European mountains. In their south area, the three species, *Bembidion difficile* MOTSCH., *B. fellmanni* MANNERH. and *Barynotus squamosus* GERM. are restricted to the Carpathians. The southern limit of the last mentioned species extends only to the Pyrenees and the French Central Plateau and of *Corymbites rugosus* GERM. to the Alps and the Vosges. *Pterostichus kokeili* MILL. occurs on the central chain of the East Alps and parts of the East and South Carpathians. The south area of *Boreaphilus henningianus* SAHLB. is restricted to the Hohe Rhön only. All the other boreo-alpine Coleoptera have in the south area a more or less wide distribution on the mountains of Central and southern Europe and occur often on very widely separated mountains. *Parnassius phoebus* FABR., *Argynnis thore* HÜBN., *Polyommatus glandon* PRUN., *Sympistis funesta* PAYK., *Lithosia cereola* HÜBN. and *Hepialus ganna* HB. in the south area occur exclusively on the Alps and *Melitaea iduna* DALM. on the Caucasus. All the other boreo-alpine Lepidoptera in the south area occur on more than one mountain. The distribution of the boreo-alpine species in the south area may be summarized as follows:

TABLE 39

Distribution of boreo-alpine insects in the south area

Mountain range	Number of species of Coleoptera present	%	Number of species of Lepidoptera present	%
Alps	37	86	32	97
Carpathians	29	67	16	49
Sudeten	20	47	5	15
French Central Plateau	13	30	7	21
Pyrenees	10	23	17	52
Harz Mountains	10	23	4	12
Vosges	9	21	3	9
Schwarzwald	9	21	3	9
Jura	9	21	3	9
Erzgebirge	9	21	1	3
Bohemian Forest	8–9	21	3	9
North Balkan Pen. mountains	8	19	19	58
Thüringerwald	6–7	14	1	3
Caucasus	6	14	6	18
Apennines	4	9	4	21
Sierra Nevada	2	5	2–3	9
Total number of boreo-alpine species	43		33	

Of the 43 boreo-alpine Coleoptera, *Bembidion fellmanni* MANNERH., *B. difficile* MOTSCH., *Pterostichus blandulus* MILL., *Boreaphilus henningianus* SAHLB., *Otiorrhynchus arcticus* FABR. and *Barynotus squamosus* GERM. are absent on the Alps. With 37 species, the Alps have, therefore, a much larger number of boreo-alpine elements than any other mountain in Central and southern Europe. Only a small number of the 37 species of boreo-alpine Coleoptera are, however, widely distributed on the Alps. Others, though widely distributed, are often completely absent from large areas of the Alps and are sometimes restricted to relatively small isolated localities. Some species, like *Amara quenseli* SCHÖNH., *Geodromicus globulicollis* MANNERH., *Arpedium brachypterum* GRAV. and *Phytodecta affinis* GYLL., are widely distributed at higher elevations on the West Alps and in the western parts of the East Alps, but are almost wholly absent in the eastern parts of the East Alps. *Evodinus interrogationis* LINN., *Hypnoidus rivularius* GYLL. and *Otiorrhynchus morio* FABR. are likewise absent in the greater part of the East Alps. *O. dubius* STRÖM. and *O. salicis* STRÖM. are common on the East Alps and Swiss Alps, but are totally absent on the French Alps and the Italian Alps, near the Franco-Italian border. There is, however, no significant difference in the abundance of the boreo-alpine Coleoptera in the north and south Alps. The southernmost Dolomites have, for example, no fewer species than the northern North Tyrolean Alps (Limestone Alps). It is only on the lower hills near the southern margin of the Alps that we may observe a fall in the abundance of the boreo-alpine Coleoptera. Of the 33 boreo-alpine Lepidoptera, only *Melitaea iduna* DALM. does not occur on the Alps. As in the case of the boreo-alpine Coleoptera, many boreo-alpine Lepidoptera also do not occur at higher elevations throughout the Alps. *Parnassius phoebus* FABR., *Pieris callidice* ESP., *Polyommatus glandon* PRUN., *Polyommatus orbitulus* PRUN., *Arctia flavia* FUES., *A. quenseli* PAYK. and *Zygaena exulans* HOCHW. are, for example, widely distributed in the West Alps and in the western parts of the East Alps, but are absent from the rest of the East Alps. *Isturgia carbonaria* (CL.) and *Hepialus ganna* HB. are also absent in the eastern parts of the East Alps. *Argynnis thore* HÜBN. is widely distributed throughout the Alps, but is not found in the French Alps.

In the general composition of the boreo-alpine species, the Carpathians present many interesting similarities with the Sudeten, but are richer in species. Of the 20 boreo-alpine Coleoptera, known from the Sudeten, only *Patrobus assimilis* CHAUD. and *Silpha tyrolensis* LAICH. do not occur on the Carpathians. In addition to the other 18 species that are common to the Sudeten, there are on the Carpathians the following species also: *Bembidion fellmanni* MANNERH., *B. difficile* MOTSCH., *Amara quenseli* SCHÖNH., *Pterostichus blandulus* MILL., *Pt. kokeili* MILL., *Autalia puncticollis* SHARP, *Agathidium arcticum* THOMS., *Bius thoracicus* FABR., *Evodinus interrogationis* LINN., *Acmaeops septentrionis* THOMS. and *Otiorrhynchus morio* FABR. There are, therefore, on the whole no less than 29 boreo-alpine species of Coleoptera on the Carpathians; 24 of these species occur on the Tatra Mts. and 13 on the Transylvanian Alps. Most of the species from the Carpathians are characterized by discontinuous distribution and are also restricted to small localized patches. The following boreo-alpine Lepidoptera are known from the Carpathians: *Argynnis pales* SCHIFF., *Erebia pandrose* BKH., *E. epiphron* KNOCH., *Hesperia andromedae* WALLGR., *Rhyacia hyperborea* ZETT., *Agrotis speciosa* HÜBN., *Crymodes maillardi* (H.-S.), *Cidaria nubiliaria* (H.-S.), *C. turbata* (H.-S.), *C. flavicinctata* (HÜBN.), *Eupithecia scriparia* H.-S., *Gnophos sordarius* THUNB., *Gn. myrtillatus* THUNB., *Arctia quenseli* PAYK., *Zygaena exulans* REIN. & HOHENW. and *Sterrhopteryx standfussi* H.-S.

6. THE ORIGIN OF THE BOREO-ALPINE INSECTS

SCHARFF (920) and HULTÉN (551) suggested that the area of north Asia is the place of origin of the biota of Arctic mountains, including also the boreo-alpine species. From here the parental stock dispersed partly towards the arctic areas and partly as the mountain-branch in Europe and North America. SCHARFF and SAINTE-CLAIRE DEVILLE (909) believe that the high altitude insects of the Central European mountains show very close affinity to those of Central Asia and the Siberian mountains. The boreo-alpine distribution was explained by these workers on the assumption that the boreo-alpine species migrated to Europe from the high mountains of Siberia and Central Asia by two principal routes. One route brought the boreo-alpine elements even during the Miocene times from the heart of Asia over the chain of the Toros-Dinar Mountains of the northeast Mediterranean region to the Alps. The second route brought the boreo-alpine species to the extreme north on the Palaearctic mainland, from Siberia to Fennoscandia and partly also further to Scotland and Ireland. According to these views, the boreo-alpine distribution was not due to the Pleistocene glaciations and the concomitant climatic and other changes. There are, however, many difficulties in accepting these views. All European species cannot be considered as relicts of such an ancient distribution, originating from northeast Asia. Though the arctic and boreo-alpine biota of Europe may have originally come largely from this area, this must have evidently been so much earlier, that all the species, recognized at the present time, have been built up on the European soil. The explanation of the boreo-alpine mode of distribution must indeed be sought within the limits of Europe. If an exchange between the north and south has been possible within Europe, circumpolar distribution does not alter the position. The existence of a continuous range of high mountains between Caucasus-Jaila-dagh and the mountains of the Balkans Peninsula has not, however, been established either by geological or biological evidence. The recent distribution of boreo-alpine species cannot, therefore, be satisfactorily accounted for on the basis of a hypothesis that takes for granted such a mountain connection. If the boreo-alpine species had migrated from Asia, in the early Tertiary or even during the Pleistocene, by the northeast Mediterranean mountain chain into Europe, we must then expect to find some evidence of this old migration on the mountains of Asia Minor and the Balkan Peninsula. On the other hand, the Sudeten, which is obviously very difficult to reach by the migrants from the Taurus mountains and the Caucasus, has a very much larger number of boreo-alpine Coleoptera than the Caucasus. Even the Balkan Peninsula, despite the relatively large alpine features, higher than on the Sudeten, has very few boreo-alpine Coleoptera. The abundance of boreo-alpine Coleoptera on the Central and southern European mountains and their poverty on the Caucasus is in no way in agreement with the hypothesis of migration over the Toros-Dinar route, but corresponds throughout to the view that the boreo-alpine species are Pleistocene relicts. Furthermore, it may be observed that the boreo-alpine species on the Central and southern European mountains are completely identical with those in the high north of Europe and Asia. Except in the case of *Bembidion fellmanni* MANNERH., *Geodromicus globulicollis* MANNERH., *Phytodecta affinis* GYLL., *Otiorrhynchus dubius* STRÖM. and *O. morio* FABR., none of the boreo-alpine species show any evidence of development of local races within the limited area of the south. Even among these five species, most individuals from the south area are completely identical with those from the north area. This complete identity of specimens from the north and south areas is perhaps the strongest proof that the origin of the boreo-alpine elements does not really go beyond the Pleistocene times. If the boreo-alpine species had already migrated to Europe during Miocene times, the prolonged isolation in the widely separated north and south areas must have led to race formation, advanced enough at the present time to be readily observed. Because, however, the boreo-alpine forms have risen in relatively recent times, viz. the Pleistocene, race formation has had no time to take place as yet. Further, the Asiatic species are also often different from the European forms. It may also be recalled in this connection that many of the boreo-alpine species do not occur in Asia at all.

Even as early as 1846, FORBES (353) attempted to explain the boreo-alpine phenomena on the basis of the Pleistocene climatic changes. He emphasized that, as suggested by LYELL in the case of the Scandinavian rocks present in Great Britain, even organisms could have drifted on icebergs from the north to the south, though apparently LYELL was himself not completely satisfied with this drift theory, in at least so far as the boreo-alpine plants are concerned. The drift theory was eventually overthrown in 1875 by TORELL. A number of other well known authors like CHARLES DARWIN, WALLACE and HEER attempted to explain the boreo-alpine distribution on the basis of the influence of the Pleistocene glaciations.

Considerable literature exists on the Pleistocene glaciations, their extent, the thickness of the ice, the cause and the effects of glaciations, the influence of glaciation on the topography and geology of the glaciated areas and the associated climatic changes. The reader will find useful summaries of the available information in COLEMAN (217, 218), ANTEVS (51), DALY (231), DE TERRA & PETERSON (256), FLINT (350), MIREINK (778), PENK & BRÜCKNER (824), WRIGHT (1126), ZEUNER (1133) and others listed in the bibliography. The influence of the Pleistocene glaciations and the associated changes on the vegetation and animal life have engaged the attention of several workers (286a, 343, 383, 394, 396, 397, 422, 424, 489, 490, 642, 674, 691-700, 701, 858, 944). In 1912, ZSCHOKKE (1140) published a critical review of the results of researches on the biogeographical effects of Pleistocene glaciations. There is also an excellent recent discussion on the biogeographical importance of Pleistocene times by DEEVEY (240). The effects of glaciations on the vegetation were discussed by CHANDLER (198). A number of workers have investigated the effects of Pleistocene glaciations on the high altitude insect life of the Central and southern European mountains and summaries of these discussions may be found in HOLDHAUS (529, 533, 534, 535), HARNISCH (478), FRANZ (368, 373), GAMS (396) and RAND (858).

In addition to diverse other effects on the general ecology, habits and organization and evolution of fauna, the Pleistocene glaciations have most profoundly influenced the geographical distribution of the mountain autochthonous insects in Europe. The effect of glaciations on the insect life of the Alps is seen, for example, not only in the occurrence of numerous high-alpine glacial relicts on the high mountains, but also in a series of peculiarities of distribution of the mountain autochthonous forms in general. Though some of the mountain autochthonous species inhabit the forest-zone, most others are restricted to the life zones above the timberline. Their most striking character is the profound localization and the feeble and greatly limited powers of movement; a great many of them occur in soil and are apterous and often also blind. This limited capacity for migration of the great majority of the mountain autochthonous species explains many of the traces left by the Pleistocene glaciations on their geographical distribution.

The absence of exclusively torrenticole and petrophile Coleoptera on the mountains of Fennoscandia is also explained by HOLDHAUS as an effect of the Pleistocene glaciations. The glaciations are supposed to have annihilated the older forms and the Post-Pleistocene repopulation of the Fennoscandian mountains by the petrophile fauna from the south or from the east has not been possible, because the north German and Russian lowlands have served as effective barriers. The few petrophile elements that occur in Fennoscandia have no doubt survived in the region during the Pleistocene.

It has generally been taken for granted that as a result of the widespread glaciations of the Pleistocene, the mountain autochthonous insects on the Alps and other mountains in the northern hemisphere were either totally destroyed or decimated to mere relicts. The southern mountains must therefore have a much richer petrophile fauna than the northern mountains. When the Pleistocene ice receded, the species capable of migration and already adapted to the Pleistocene conditions repopulated the so-called "ice-devastated" areas. The species with limited migration capacity, such as, for example, the small apterous and blind Coleoptera, numerous small apterous Coleoptera of the high-alpine snow-edge, etc. must, however, have been left behind and are even today restricted in their distribution to larger or smaller parts of the marginal areas of the Alps, which were not covered by ice during the Pleistocene. It was believed that in the marginal area of the Alps we have a remarkable mountain relict fauna, richer in species than in the inner ice-devastated areas. The condition of the marginal area of the Alps has been termed by botanists as the "massifs de refuge". HOLDHAUS (533) adopted this term in explaining the distribution of the boreo-alpine insects, particularly Coleoptera. He has assumed that the southern marginal area of the Alps has remained ice-free during the whole of the Pleistocene and has served as the survival refugium for the autochthonous species. A number of species appear to be exclusively confined to the massifs de refuge. Some of these species also occur on isolated single summits or on a few adjacent ones. The re-immigrant species include those that occur not only on the massifs de refuge, but also on other neighbouring areas, which were ice-covered during the Pleistocene. HOLDHAUS believed that these species have repopulated the ice-devastated areas in post-glacial times from the massifs de refuge. According to him, *Cychrus schmidti* CHAUD. and *Trechus pallidulus* GGLB. have migrated in this way over short distances, but others like *Cychrus attenuatus* FABR. have migrated over long distances.

HOLDHAUS believes that the undisturbed mountain types are rich in the massifs de refuge on the Alps, in the southeast and south fringe of the Alps. In contrast to the intensely glaciated central and northern parts of the Alps, the area of the massifs de refuge contained, during the Pleistocene, numerous unglaciated areas and the high altitude of the snowline during the glacial times in the south fringe and the southeast margin of the Alps permitted the occurrence of forests at much lower elevations in the region of the marginal mountains. As at the present time, even during the Pleistocene, the southern margin of the Alps had xerothermic localities, with protected and south-exposed conditions, thus enabling the thermophile species to survive. It is because of this that the massifs de refuge were supposed by HOLDHAUS to have at present a rich petrophile fauna, while the other localities, outside the massifs de refuge, are poor in them. HOLDHAUS compares the flightless terricole Coleoptera of the Koralps with those of the Radstädter Tauern or of the Hohe Tauern and the species complex of the Venetian Alps (Mte Cavallo) on the one hand with those of the Dolomites or the Gailtaler Alps on the other hand. He believes that on the East Alps the character of the massifs de refuge is best observed in the Bergamasker Alps Mte Baldo, Monti Lessini, Venetian Alps, Julian Alps, Steiner Alps, the east Karawanken, etc.

SCHWEIGER (944) has also assumed the survival of species of *Trechus* CLAIRV. on the massifs de refuge during the Pleistocene glaciations. The Trechinae are indeed well suited to demonstrate the traces of the glacial epoch in the fauna of the Alps. All the high altitude and even the subalpine species of Trechinae are apterous and have thus very little capacity for migration, both because of their small size and also because of the peculiarities of their life-cycles, so that their distribution pattern has changed very little since the end of the glacial epoch. Even the pattern of their present-day distribution is therefore of considerable importance in our discussions. SCHWEIGER thinks that not a single high-alpine zone species is present now in the central area of the Alps, which was covered by the Pleistocene ice, while in the marginal zone or the periphery of the Alps (the massifs de refuge) a considerable number of these species are still sheltered. The conditions of the subalpine species are, however, different and being much less ecologically specialized than the high-alpine zone species, they are better fitted for migration. They were, therefore, able to repopulate the areas of the inner Alps. *Trechus alpicola* STURM. and *Tr. limacodes* DEJ., which occur both in the subalpine and alpine zones, are widely distributed throughout the East Alps. The repopulation of the ice-devastated inner Alps by these two species was possible, because of their survival on the massifs de refuge.

A number of workers like FRANZ (370), HEER (494), BÄBLER (74), HANDSCHIN (468) and STEINBÖCK (1003, 1004) have shown that *nunatak* rocks projecting over the snowline support insects even in the roughest climate at the present time. A number of the cold-adapted and glacier species of Collembola, Acarina, Araneida, etc. thrive for example, on the nunatak rocks, far above the glaciers and above the permanent snowline. FRANZ doubts whether the repopulation, as suggested by HOLDHAUS, from the massifs de refuge, could adequately explain the alpine endemic species. The extensive investigations of FRANZ (373) on the soil biota of the Alps seem to show that the so-called devastation by the Pleistocene glaciations in the inner Alps has not been really as complete or effective as has been generally assumed by most earlier workers. He suggests, therefore, that as at the present time, here and there high elevated parts had remained ice-free during Pleistocene times also; these ice-free areas supported, as now, diverse organisms during the Pleistocene. The relict species that survived on the nunatak are termed as the nunatak relicts. The extremely restricted distribution of these nunatak relict species must be interpreted as due to their survival on unglaciated localities in the midst of heavily glaciated areas. The distributional pattern of the nunatak relict species can at once be distinguished from that of the species which have rewandered in post-glacial times.

JANETSCHEK (567) has recently brought together considerable evidence on the survival of insects and other related terrestrial arthropods on the inner Alps nunatak system. The inner Alps nunatak system embraces all the rocks in the interior heart of the Alps, in contrast to the marginal zone, projecting above the general surface of the Pleistocene ice-mass on the heavily glaciated areas of the Alps. The highest summits and ridges of the Central Alps and the Limestone Alps were largely glaciated and covered by snow, but locally there were snow-free and ice-free areas, depending upon the gradient, exposition and the prevailing wind. The ice-free localities served as more or less small refugial niches for the survival of diverse species during the Pleistocene. It is generally admitted that in the innermost and higher Alps, the Pleistocene glacier area did not exceed the present-day heights much and the ice-mass did not also overflow the highest peaks. On the other hand, most peaks and ridges often rose to some 1000-2000 m above the general level of the ice-mass. Cryptogams and even diverse Phanerogams flourished on these nunatak systems and survived the Pleistocene glaciations (769).

The possibility of the Pleistocene survival of species on the inner Alps nunatak system has been recognized by nearly all students of nival fauna. Even HOLDHAUS (535) is aware of the fact that the subnival *Nebria atrata* DEJ. and *N. angustata* DEJ.

survive on nunataks in the heavily glaciated areas. The species with their present distributional range restricted wholly to the inner Alps or at least showing close affinity to the Pleistocene inner Alps nunatak system are strong proof of the idea of their Pleistocene survival on the nunatak system.

The possibility of a secondary restriction of the range on the nunatak system due to post-glacial warmer times is excluded in the light of not only the distribution data, but also on ethological considerations, especially the constancy of the biotope, the homing behaviour and the poor dispersal capacity and even the ecological peculiarities of the species. The strongest proof is provided by the strictly terricole insects that lack the capacity for rapid dispersal over the vast net-work of glaciers and ice-mass, from the glacier foreground on the fringe of the Alps to the inner Alps nunatak system. These are the species that were widely distributed in the pre-glacial and inter-glacial times, but came to have their range broken up into the irregular and restricted nunatak system, as a result of glaciation and have survived on these high massifs. The restriction of the species exclusively to the refugial nunatak system is not, however, necessarily the sole evidence of their Pleistocene survival on the nunatak system. At the same time, the view of HOLDHAUS that the species now found on the deep inner Alps represent post-glacial rewanderings from the marginal massifs de refuge is, however, precluded by consideration of the vast distances involved. Additional evidence is also provided by the peculiarities of the distribution of the high altitude Arachnida, Diplopoda and Apterygota.

JANETSCHEK has stressed the following facts as strongly proving the idea of Pleistocene survival by boreo-alpine insects on the inner Alps nunatak system:

1. Refugiocavale species that have free high-alpine-nival zone populations on the inner Alps.
2. Species that occur free (not cavernicole) in the inner Alps and their distribution here is closely bound up with the inner Alps nunatak system. In contrast, other species of the same genera occur exclusively or nearly so in caves of the Alps marginal area or in caves of unglaciated or slightly glaciated areas.
3. High-alpine and nival zone species, the inner-Alps area of which is at present closely connected to the Pleistocene nunatak system.
4. Species with discontinuous distribution in the Alps and other mountains or in the north, the alpine area of which is greatly restricted.
5. Restricted distributional range of superspecific ranks to the inner Alps nunatak system.
6. Adaptations of the nival species to their present-day habitats.
7. Relation between the areal types of the high-alpine to the nival zoocenoses.
8. Comparative historical zoogeographical studies of the North European and Arctic regions.

It is concluded that the inner Alps nunatak system offered optimal conditions during the Pleistocene, no less than the marginal refugial areas, for the survival of species. It seems reasonably certain that not only the strictly boreo-alpine but likewise the other mountain autochthonous insects have survived the Pleistocene glaciations on the Alps and perhaps also on other Tertiary mountains.

CHAPTER XIII

OTHER MOUNTAINS OF THE NORTH TEMPERATE REGIONS OF THE OLD WORLD

This chapter deals with the Altai Mountains, the Pyrenees, the Spanish Sierra Nevada, the Apennines, the Caucasus and other lesser mountains of the north temperate regions of the Old World. Compared to the great Alps-Himalayan Systems, the ecology and biogeography of the high altitude insects of these mountains are only very imperfectly understood at present. Strictly speaking from the biogeographical point of view, we should perhaps begin the chapter with the Altai Mountains, but with a view to maintain the continuity with the foregoing chapters, we shall take first the Pyrenees and proceed eastwards to the Altai region.

1. THE PYRENEES

The Pyrenees Mountains, separating the Iberian Peninsula from France, stretch about 385 km, from the Bay of Biscay to the Cape Creus. Three sections, viz. the Central Pyrenees, the Atlantic or the Western Pyrenees and the Mediterranean or the Eastern Pyrenees, are generally recognized. The highest summits like the Aneto or the Pic de Methou (3405 m) and the Mont Perdu or the Monte Perdido (3368 m) are situated in the Central Pyrenees. The mean elevation is maintained in the eastern section, but in the Atlantic Pyrenees the elevation diminishes westwards. Geologists recognize the following zones: i. the Central Massif zone comprises the Primary rocks, Archaean, Cambrian, Ordovician, Silurian, Devonian and Lower Carboniferous, with great masses of granites, and forms the highest summits. On the French side of the Pyrenees are ii. the Ariège zone of Lower Cretaceous and Jurassic beds and granites and iii. the Petites Pyrenees zone of Upper Cretaceous and Eocene and thin outcrop of the Jurassics; and iv. the Aragon zone of Eocene and Primary rocks. On the Spanish side are v. the Mont Perdu zone of Upper Cretaceous and Eocene, with thin outcrop of the Jurassic and vi. the Aragon zone of Eocene and the Sierras zone of Trias, Cretaceous and Eocene. Although thus the same number of zones lies on the two flanks of the Pyrenees, the zones do not really correspond exactly everywhere. Large masses of rocks seem to have been brought forward upon the thrust planes, over the edges of others. Denudation has also occurred on the Pyrenees to a much greater extent than on the Alps. The axis of the Pyrenees was outlined by the Hercynian movement and the folding took place along it at the close of the Dinarian Epoch and again before the Permian times. The chain was submerged later, up to the early Cretaceous, and the final uplift began and continued till the Oligocene and thus completed well before that of the Alps. Unlike in the Alps, high altitude lakes are absent on the Pyrenees and the mountain torrents on the Pyrenees locally form also the so-called characteristic "gaves" of lofty waterfalls.

The general climatic conditions on the Pyrenees differ considerably from those prevailing on the Alps. Precipitation is, for example, much greater on the West Pyrenees and on the East Pyrenees. There are no glaciers on the East Pyrenees and

West Pyrenees. The glaciers on the Central Pyrenees are also confined to the north (French) slope. The snowline varies from 2680 to 2800 m above mean sea-level. There is an interesting account of the general tectonics of the Spanish mountains by STAUB (996).

The literature on the general natural history and fauna of the Pyrenees is much scattered and is also mostly inaccessible. The reader will find, however, interesting accounts of the insect life of the Pyrenees in DESLANDERS (251), DUFOUR (285), HARCOURT-BATH (475), KITSCHELT (626) and SEITZ (955). RONDOU (894, 895) published a general catalogue of the Lepidoptera from the Pyrenees. JEANNEL (574, 576, 586) has given interesting accounts of a number of remarkable cavernicolous Coleoptera from the Pyrenees. There is a striking abundance of blind cavernicolous Coleoptera like *Anophthalmus* STURM., *Adelops* TELLK., etc. in the caves of Ariège. JEANNEL's monograph on the *Trechus* CLAIRV. from the Pyrenees is a most valuable contribution to our knowledge of the insect life of the region (557). In recent years, BERTRAND (114-119) and BERTRAND & VERRIER (121) have, in a series of papers, described the Ephemerida, Chironomidae, aquatic Coleoptera and other aquatic insects and BERTRAND & AUBERT (120) have similarly dealt with the Plecoptera from the Pyrenees.

Some of the outstanding peculiarities of the altitudinal zonation of life on the Pyrenees and other Spanish mountains have been discussed by FRANZ (378). As on other mountains of the world, the altitudinal faunal zonation on the Pyrenees is closely related to the altitudinal vegetational zonation. It is not only the present-day vegetational zonation, but the effects of the vegetational zonation during the late Pleistocene and the post-glacial climatic and other changes have a very profound influence on the faunal zonation on the Pyrenees and other Spanish mountains. The sharp contrast that we see between the biocenoses of the forest and the forest-free areas of the Central European mountains is not, however, directly applicable to the mountains in the Mediterranean region. Here the two areas interpenetrate each other in a most complex manner, so that the animal communities of the forest and the forest-free areas are spatially not really sharply demarcated. The light character of the arid forest of these regions seems to be the principal reason for this peculiarity. The altitudinal biotic zonation, characteristic of the Alps, may be seen more or less on the Apennine of Toscana and on the East Pyrenees. On Mont Canigou, the subalpine belt, with *Pinus uncinata*, is at an elevation of 2200-2500 m and meets the dwarf-shrub zone here. Only above an elevation of about 2500 m lies the high-alpine grassy-meadow zone, rich in endemic forms. The snow-valleys are poor in species and have only *Nebria lafresnayi* SERV., *Haptoderus* (*Pyreneorytes*) *glacialis* BRIS. and the somewhat less characteristic *Bembidion* (*Testediolum*) *pyrenaeum* DEJ. The dwarf-shrub zone on Monte Perdido (between 1800 and 2200 m) in the Central Pyrenees is characterized by numerous heliophile species like *Timarcha altimontana* BECHYNE, *Galeruca monticola* KIESW., *Dorcadion molitor* FABR., etc. These peculiarities are even more pronounced on the Cantabric Mountains than on the Pyrenees. The dwarf-shrub zone is thus at a much lower level, viz. 1600-2000 m. Here the character species are *Pseudaptinus pecoudi* PUEL, *Harpalus* sp., *Cymindis melanocephala kricheldorffi* FUENTE, *Heliopathes parceioveolatus* REITT., *Dorcadion heydeni* KRAATZ, *Timarcha geniculata* GERM., etc.

Between the subalpine-forest belt and the alpine grassy-meadow zone, there must have existed for a long time a broad zone, which was partially forested or had at least dwarf-shrubs flats, with open grasslands. This condition must have been conspicuously developed on the Central Spanish high mountains, on the Sierra de Guadarrama, Sierra de Gredos, Sierra del Moncayo, etc. Most of the inhabitants of the grassy-meadows descend here to an elevation of about 2300 m, a belt that is dominated by *Festuca indigesta*, so that these mountains seem to have had no typically high-alpine forms. This is not, however, strictly correct, because only the boundary between the high-alpine grassy-meadow fauna and the subalpine-forest belt is not so sharply defined either climatically or distributionally as on the Central European mountains. That the Spanish mountains possess, however, a true mountain autochthonous fauna is evident from the typically valley elements, despite their poverty of species. On the Sierra de Guadarrama, to this group belong *Nebria vuillefroyi* CHAUD. and *Bembidion* (*Testediolum*) *pyrenaeum carpetanum* SHARP and on the Sierra de Gredos *Nebria pazi* SEIDL. Similar conditions prevail even on the mountains to the northwest of the Sierra de Gredos, viz. on the Monte Teleno, on the Peña Trevinca and on the Sierra de Queija. *Nebria belloti* FRANZ is the most dominant species at the snow-edge on these mountains. The arid localities of the high elevations are covered predominantly by associations of Ericaceae, *Genista* and *Juniperus nana* and by grassy-meadow only to a limited extent. The heliophile Coleoptera here belong to the genera *Zabrus* CLAIRV., *Harpalus* LATR., *Cymindis* LATR., *Dinodes* DEJ., *Heliopathes* MULS., *Dorcadion* DALM., *Cyrtonus* LATR., *Otiorrhynchus* GERM., which sometimes descend very low. They are found not only above the timberline (i.e. the zones which were formerly grown with *Quercus pyrenaica-Betula verrucosa* associations), but also within the *Quercus pyrenaica* zone, down to elevations of 1500 m. This is particularly observed on the Serra da Estrela. All these mountains seem to have, within the forest zone, extensive open forest-free patches, inhabited by heliophile species.

The French side of the Pyrenees has been somewhat better explored by entomologists than the Spanish side. The Lepidoptera and Carabidae of the French Pyrenees have also been more extensively studied than the other groups. RONDOU (894, 895) has compared the Lepidoptera of the Pyrenees with those from the Alps. According to him, the species from the Pyrenees are on the whole larger and more vividly coloured than those from the Alps. These striking differences seem to be correlated, at least in part, with the differences in the latitudes of these mountain systems—the Pyrenees Orientales is one degree latitude south of the southern end of the French Alps. This difference, ecologically important even on the lowlands, becomes very pronounced at high altitudes on mountains. Melanism is reported to be more pronounced on the Pyrenees than on the Alps in many Lepidoptera, particularly Boarmiinae, *Dasydia* GU., *Psodos* TR., *Cleogenea* DUP., etc. There are only two species of *Erebia* DALM. viz. *E. lefebvrei* DUP. and *E. gorgone* BDV., which are, strictly speaking, restricted to only the Pyrenees (313). Nine species are common to the Pyrenees and the Alps: *E. melampus* (FUESS.), *E. mnestra* (HÜBN.), *E. christi* RAETZ., *E. gorgophone* BELL., *E. pharte* (HÜBN.), *E. ceto* HÜBN., *E. scipio* (BOSID.), *E. glacialis* ESP., and *E. montanus* (PRUN.) (=*E. goante* ESP.). *Lycaena pyrenaica* BOSID. is the only endemic Lycaenid from the Pyrenees. *L. optilete* KNOCH., *Polyommatus orbitulus* PRUN. and *Lycaena sebrus* BK. from the Alps occur also on the Pyrenees. Three species of *Cidaria*, viz. *C. multistrigaria* HAW., *C. dissimulata* RBR. and *C. flavofasciata* THUNB. are true Pyrenees forms, but the Alps-species like *C. cupressata* HÜBN.-GEYER, *C. kollariaria* H.-SCH., *C. cambrica*, *C. quadrifasciata* Q., *C. pomaerafia* EV., etc., are often found on the Pyrenees. Some of these differences are closely correlated with the differences in the flora of these regions. The Lepidopterophile flora of the Pyrenees is only 4.6%

of the total vegetation of the region, but on the Alps it amounts to 9.3%.

The following are some of the more interesting Lepidoptera from the Pyrenees: *Papilio podalirius* LINN., East Pyrenees (2600 m), *Parnassius apollo* LINN. with a number of local subspecies (*P. a. pyrenaica* HARC.-BUTL., *P. a. chrysophorus* FRUHST., etc.), *P. mnemosyne* LINN. in humid localities at elevations between 1500 and 2000 m, *Aporia crataegi* LINN. on meadows at elevation of 2000 m, *Pieris brassicae* LINN., *P. rapae* LINN., *P. manni* MAYER., *P. napi* LINN., *P. callidice* ESP. (2000-2800 m, breeding on *Sempervivum montanum* and *S. arachnoidum*), *Euchloë crameri* BUTL. at elevations of 1000-1620 m on the East Pyrenees, *E. cimplonia oberthüri* VERITY breeding on *Sisymbrium erucastrum* at elevations of over 2000 m, *Anthocharis cardamines* LINN. at an elevation of 1800 m, *A. euphenoides* STGR. at elevations of 1600-1750 m on the East Pyrenees, *Colias phicomone* ESP. widely distributed and particularly in the *Rhododendron*-zone (1500 m), *C. hyale* LINN. (with the subspecies *C. h. obsoleta* TUTT. and *C. h. vernalis* VERITY), *C. croceus vernalis* VERITY, *C. cr. helicina* OBERTHÜR, *C. cr. aubuissoni* CARADJA, *C. cr. helice* HÜBN., *C. cr. albissima* RAG., *Erebia epiphron* KNOCH. (1500-2500 m), *E. e. cassiope* FABR., *E. e. pyrenaica* H.-SC., *E. e. melanus* BDV., *E. e. albinescens* OBERTHÜR, *E. manto* ESP. in the *Rhododendron*-zone, *E. oeme* HÜBN. (2000 m), *E. gorge* ESP. widely distributed up to an elevation of about 3000 m and rarely below 2000 m (with several varieties), *E. euryale* ESP. and several varieties and *E. tyndarus* ESP. with numerous varieties at elevations of 1500-2500 m and sometimes even at 3000 m. There are also some other interesting species like *Pararge egeria* LINN., *Ephinephele janira* LINN. *E. pasiphae* ESP., *Coenonympha pamphilus* LINN. (2000 m), *Melitaea* sp. (2000 m). *Argynnis selene* SCHIFF. (2170 m), *A. euphrosyne* LINN. (2500 m), *A. pales* SCHIFF. (2500 m) and *A. niobe* LINN. (2600 m) on the Pyrenees.

BOURGOGNE (142, 143) has recently given an interesting account of some insects collected by him on the French Pyrenees and parts of the French Alps. He reports finding *Hesperia* (=*Pyrgus*) *andromedae* (WALLGR.), the small Noctuid *Omia cymbalariae* HÜBN. and the Pyralid *Pyrausta nigralis* F. at elevations of 2190 m on the Rocher de Villeneuve. A number of other species like *Pieris bryoniae* (HÜBN.), *Colias phicomone* ESP., *Clossiana titania* HÜBN., *Erebia alberganus* PRUN., *E. melampus* FUES., *E. aethiops* (ESP.), *Lycaena irregularis* TUTT., *Eumedonia chiron* ROT., *Erebia pharte* (HÜBN.), etc. are also recorded. Although scarce during 1961, the Arctiid *Endrosa aurita ramosa* FAB. was very common in 1962. There was a pronounced abundance of *Erebia mnestra* (HÜBN.), *E. pandrose* (BKH.), *E. epiphron* (KNOCH.), *E. pluto* PRUN., *Euphydryas cynthia* HÜBN., *E. aurinia glaciegenita* VERITY, *Oeneis glacialis* SCHRANCK, *Polyommatus glandon* (PRUN.) and some examples of *Papilio machaon* LINN. at elevations of about 2000 m. He collected a number of other butterflies at elevations of about 2000 m on Mont Chevrier. On the Petit Mont Blanc (2680 m), situated to the south of the Massif du Plassa, he found *Pyrgus cacaliae* RBR., *Pyrausta nigralis* F., etc. The other areas from which he reports collections of butterflies include the Cirque du Grand Marchet (1800 m), Mont Bochor and Col de la Vanoise. According to him, the butterflies are common on their wings on these mountains, between June and July, at elevations of 2500-3000 m. The same author (143) has also reported the interesting find of *Melanargia russae* ESP. from the French East Pyrenees; this species had earlier been known only from the West Pyrenees.

Ten boreo-alpine Coleoptera are known from the Pyrenees: *Nebria rufescens* STRÖM., *Amara quenseli* SCHÖNH., *A. erratica* DUFT., *Geodromicus globulicollis* MANNERH., *Silpha tyrolensis* LAICH., *Helophorus glacialis* VILLA, *Corymbites cupreus* FABR., *Otior-*

rhynchus morio FABR., *O. arcticus* FABR. and *Barynotus squamosus* GERM. *Geodromicus globulicollis* MANNERH. is known only from the Hautes-Pyrenees, but the rest of the species are more or less widely distributed on the higher regions of the Pyrenees. The boreo-alpine Lepidoptera from the Pyrenees are *Hesperia andromedae* (WALLGR.), *Pieris callidice* ESP., *Argynnis pales* SCHIFF., *Erebia pandrose* (BKH.), *E. epiphron* (KNOCH.), *Polyommatus glandon* PRUN., *Agrotis faticida* HÜBN., *Crymodes maillardi* (H.G.), *Cidaria munitata* (HÜBN.), *C. turbata* (H.-S.), *C. flavicinctata* (HÜBN.), *Eupithecia scriparia* H.-S., *Psodos coracina* ESP., *Gnophos myrtillatus* THUNB., *Pygmaena fusca* THUNB., *Isturgia carbonaria* (CL.) and *Zygaena exulans* HOCHW. Intensive explorations in future may perhaps bring to light some more boreo-alpine insects on the Pyrenees.

2. THE SPANISH SIERRA NEVADA AND OTHER SPANISH MOUNTAINS

Some account of the general geology, climate and vegetation of the Spanish Sierra Nevada may be found in MATEU (745, 746), who has also described a number of interesting Coleoptera from the region. ROUDIER (903, 904) has also described a number of Coleoptera from the Spanish Sierra Nevada. There is an interesting account of the Pleistocene glaciations in the region by PASCHINGER (813). JANETSCHEK & STEINER (568, 569) recently undertook a zoological exploration of the region and collected a large number of insects and other related terrestrial arthropods. The Diptera from their collections are described by LENGERSDORF (688) and SCHMITZ (926), the Curculionidae by ROUDIER (904), Heteroptera by WAGNER (1090) and the spiders by DENIS (247). The Staphylinidae from the Sierra Nevada were described by JARRIGE (573), the Tenebrionidae by ESPAÑOL (323), the Elateridae, Buprestidae, Chrysomelidae, etc. by COBOS (214) and the Coccinellidae by CAPRA (183). In his recent discussions on the altitudinal biotic zonation on the Spanish mountains, FRANZ (378) has drawn particular attention to the significant differences between the Alps and the Sierra Nevada. On the southernmost mountain range of Europe, rising to an elevation of nearly 3500 m in the Mulhacen and the Veleta, the timberline was formerly at about 2000 m. The forest has at present given place to the shrub-patches in the low valleys. About 2000-2600 m above mean sea-level, we find thorny plants, with the general landscape highly reminiscent of African character. The Tenebrionidae are conspicuously much richer on the Spanish Sierra Nevada than on any other Spanish mountain. There are a number of character species like *Tentyria incerta* SOL., *T. platyceps* STEV., *Dichillus subcordatus* SOL., *Alphasida parallela morensi* KOCH, *Asida oblonga frigida* ESC., *A. pygmaea* ROSCH., *Pimelia monticola strobli* REITT., *Opatrum baeticum gregarium* BOH., etc. There are also a number of typically heliophile species like *Timarcha insparsa* ROSH., *Dorcadion lorquini* FAIRM. and *Rhytirrhinus nevadensis* DESBR. It may be observed that these species constitute essentially a steppe-fauna, with the unmistakable characters of African arid plains.

While, inspite of the large-scale deforestation, the typically forest elements persist at lower elevations on the northern mountains of the Iberian Peninsula, we find that the typical heliophile elements dominate on the Sierra Nevada. These elements have apparently occupied areas in all altitudinal zones, perhaps even before the disappearance of the forest and also appear to have spread from these areas to larger ones, with the progressive deforestation. Again, while in Central

Europe the rewandering of the heliophile mountain elements during the post-glacial times seems to have been confined to the refugial niches, above the timberline, in the southern Mediterranean region the forest seems to have dominated even during the late Pleistocene times.

In association with STEINER, JANETSCHEK (568, 569) studied the animal communities at high elevations on the Spanish Sierra Nevada, during the summer of 1954. The importance of the region lies in part in the nearness to North Africa and in part in the relatively short distance from the Alps. Although they were able to collect insects only for a short time of about 20 days during July-August, the results of their studies are nevertheless very interesting. The collections were largely on the northern slopes of the Sierra Nevada, at elevations of about 2400 m in the Veleta area and also in the region of Corral de Veleta and closeby to the southeast slope, opposite the Laguna del Rio Veleta. The collections were made mostly at elevations of 2400–3500 m, in the crystalline zone of the Sierra.

The interesting endemic species of Tettigoniid *Baetica ustulata* EBNER occurs at elevations of about 2440–3350 m on the Sierra Nevada. *Omocestus bolivari* EBNER is also endemic on the Sierra. A distinct local race of the Central Spanish *Myrmeleotettix maculatus* THUNB. was found at elevations of about 2800 m. *Eumigus monticola* RAMB., also known from the Sierra de Alcaraz, was found at elevations of 2500–2600 m on the Sierra Nevada. *Sphingonotus coerulans corsicus* CHAP. was collected on the north slope, at an elevation of about 2720 m on Veleta. *Dociostaurus maroccanus* THUNB. was taken at an elevation of 2800 m. The Dermaptera are abundant in damp and humid localities and the endemic *Lithinus analis* RAMB. is dominant at elevations of 2400–2900 m and the cosmopolitan *Forficula auricularia* L. occurs up to an elevation of 3170 m.

The Heteroptera are interesting for the significant dominance of Miridae. Some of the interesting records include *Macroplax fasciata* H.S. (2550–3350 m), *Apterola iberica* (2440–3460 m), the widely distributed Mediterranean *Microplax interrupta* FIEB. (2600–3350 m), *Ischnocoris punctulatus flavipes* SIGN. (2700–3170 m), *Sehirus melanopterus* HERR.-SCHÄFF. (2700–3350 m), *Parahypsitylus nevadensis* (2500–2600 m), *Trapezonotus montanus* WAG. (2690–2760 m), *Dicyphus albonasutus* WAG. (3060 m), *Systellonotus putoni* REUT. (2600 m), *Tinicephalus hortulans* MEY. (3170 m), *Pachytomella alutacea* WAG. (2500 m), *Dichroscytus algiricus* WAG. (2400–2600 m) and the brachypterous females of *Globiceps picteti* FIEB. at an elevation of about 2600 m. Most of the Homoptera collected by them belong to widely distributed lowland forms, but *Tettigometra hispanica* WAG. was found as high as 3130 m above mean sea-level. The ants *Tetramorium* MAYR and *Formica* LINN. were found at moderate elevations, but the highest elevation at which ants were collected is 3170 m.

The Coleoptera from the Spanish Sierra Nevada are characterized by the endemism of species in a number of mountain groups. Nearly 30% of the Carabidae and 40% of the Tenebrionidae are, for example, endemic. The endemic Carabidae include *Trechus planipennis* RASH. (3060–3170 m), *Ocys (Oreocys) andreae* JEANNEL in humid localities at elevations of 2600–3160 m, *Harpalus nevadensis* K. & J. DANIEL (2600 m), *Zabrus angustatus* RAMB. dominant at elevations of about 2700–3300 m. *Trymosternus cordatus* RAMB. dominant at an elevation of about 2240 m, *Bembidion pyrenaeum montanum* RAMB. (2700–3160 m), *B. bipunctatum glaciale* RAMB. (2600–3160 m), *Asaphidion cyanicorne* PAND. (2700 m), *Brachynus (Pseudaptinus) baeticus* RAMB. endemic in the Iberian Peninsula and sub-dominant at elevations of 2700-3130 m, *Asaphidion pallipes* DUFT (2700 m), *Harpalus decipiens* DEJ. (2600-2900 m) and *Microlestes luctuosus chobauti* JEANNEL (2400-3380 m). The Staphylinidae comprise *Tachyporus nitidulus* F., *T. hypnorum* F., *Aleochara bipustulata* L., *Leptusa* (*Micropisapoda*) sp., and a microphthalmous *Oxypoda* (*Bessopora*) sp. in soil at an elevation of 2700 m. The Tenebrionidae include *Phylan indiscretus* MULS., *Tentyria incerta incerta* SOL. and *Alphasida parallela* SOL. The endemic Chrysomelidae are *Cyrtonus gratiosus* ROSENH. (3020 m), *Timarcha lugens* ROSENH. (2400-3350 m), *T. rugosa* L., (2700-3170 m), *Orina* (*Cobosorina*) *colasi* COBOS (2800-3060 m), *Labidostomis nevadensis* DAN. (2950-3270 m), *Aphthona pyrenaea* HEIKERTINGER (2550 m), *Coptocephala apicalis* LACORD. (2700 m) and known also from the Caucasus and *Aphthona euphorbiae* SCHRANK (2400-3380 m). Curculionidae are *Otiorrhynchus sulcogemmatus* BOH. (2440-2760 m), *O. mesnili* HOFFMANN from an elevation of 2700 m to the highest summits, *O. nevadensis* STIERL. (2620-3380 m), *Strophosomus alticola* SEIDL. (2550-2800 m), *Homapterus nevadensis* ROUDIER (2400-2800 m), *Rhytirrhinus nevadensis* DESBR. (2400-2600 m), *Rh. mateui* SOL. (2700-3170 m), *Cathormiocerus janetscheki* ROUDIER (2600 m), *Dichotrachelus janetscheki* ROUDIER (2720-2900 m), and *Ceutorrhynchus steineri* ROUDIER (2700-2900 m). The boreo-alpine Hydrophilid *Helophorus glacialis* VILLA is represented by the subspecies *H. gl. insularis* REICHE at an elevation of about 2800 m.

The Diptera include several interesting species of Sciaridae, collected at elevations of 2600-3350 m and Phoridae mostly like *Megaselia* (*Megaselia*) *evecta* (2900-3050 m), *M.* (*M.*) *teneripes* SCHM. (3060 m), *Metopina nevadae* SCHM. (3060-3160 m), etc. It is remarkable that *Megaselia* (*Aphiochaeta*) *euryprocta* SCHM., known to have occurred on the northeast Alps during the Pleistocene, was found at an elevation of 3170 m on the Sierra Nevada. *M.* (*Megaselia*) *teneripes* is also known from an elevation of 2500 m on the Dolomites. Many terrestrial Chironomidae and brachypterous Simuliidae are found at elevations of 2400-2900 m.

Of the other terrestrial arthropods, mention may be made of the endemic Pseudoscorpion *Neobisium* (*Neobisium*) *nivale* BEIER, occurring only at elevations of 3000-3460 m. *Chthonius hispanis* BEIER, widely distributed from Austria to Portugal, was found at elevations of about 3000 m. The Phalangid *Odiellus duriusculus* (SIMON), found at an elevation of 3300 m on the Sierra Nevada, is interesting, because it is also known from North Africa and the south Spanish Coast. *O. troguloides* (LUCAS), occurring at elevations of 2400-2500 m on the Spanish Sierra Nevada, is also known from Algeria and Central Spain. The spiders taken at high elevations include *Aelurillus tristis* (3380-3400 m), *Euophrys patellaris* (3050 m), *Harpactocrates cantabricus* SIMON (2700-3270 m), *Thanatus fuscipes concolor* DENIS (2400-3460 m) (the typical form occurs in Algeria), *Theridion pyrenaeum* DENIS (2500-3380 m), *Pardosa blanda* (L.-KOCH) (2400-3170 m) (known also from the south slopes of the Pyrenees), *Gnaphosa tigrina* SIMON (2440-3460 m) (known also from the Alps, Pyrenees and the Moroccan High Atlas Mountains), *Gn. inconspecta* SIMON known also from the Pyrenees and *Tegenaria pallidula* SIMON, previously known from the Alps and the East Pyrenees. Some of the typically Mediterranean and austral transgressive species like *Liocranum majus* SIMON (3460 m), *Drassodes albicans* (SIMON) (3050 m), *Haplodrassus dalmatensis* (L. KOCH) (2400 m) and *Berlandina plumalis* DALMAS (2400 m) (circum-mediterranean, tropical Africa, Turkestan, Mongolia, India and Burma) occur at high elevations on the Sierra Nevada. Typically mountain autochthonous spiders are, however, only poorly represented on the Sierra Nevada. The widely distributed *Scutigerella imma-*

culata (NEWPORT) is reported at an elevation of 2760 m and *Scolopendra cingulata* NEWPORT at 2700–2800 m on the Sierra Nevada.

Although a number of species characteristic of the Pyrenees, Alps and other Central and north European areas are found on the Spanish Sierra Nevada, the high altitude insects of the Sierra Nevada must be considered as essentially mediterranean-montane in composition. Though the development of the present fauna has been greatly influenced by historical factors, the present-day ecological conditions also play a most important rôle in the composition of the high altitude insect fauna of the Spanish Sierra Nevada. The relatively scanty precipitation and the resultant higher atmospheric aridity and the poor snow-cover should perhaps explain the absence of hygrophile elements, so that the snow-edge forms are only very scantily found in the region. In correlation with the general physiographic conditions, there is a peculiar zonal inversion of the Mediterranean forms of the plains with the high montane forms. While the former inhabit the open arid sunlit localities in summit areas, the latter occur only in the sheltered, shady and deeper levels. The high montane character, except in some groups, is thus much less pronounced than on the principal mountain ranges of Europe. In addition to the presence of the typical high altitude forms, we also find characteristically melanistic forms, especially among the Heteroptera. The lower latitude of the Sierra Nevada has the result of pushing back the lower altitudinal limit of the thermophile species.

The Cordillera Cantabrica are to the west and a little south of the Pyrenees (43° NL), south of the Bay of Biscay and have likewise an east-west trend. The higher reaches of these mountains have not so far been adequately explored by entomologists, but the following species are known at present: *Silpha tyrolensis* LAICH., *Helophorus glacialis* VILLA., *Chrysomela crassicornis* HELLIES, *Otiorrhynchus morio* FABR., *Erebia epiphron* (KNOCH.), *Polyommatus glandon* PRUN., *Cidaria flavicinctata* (HÜBN.) and *Gnophos myrtillatus* THUNB.

The mountains of Central and South Spain and of Portugal are very imperfectly explored and of the species so far reported, *Silpha tyrolensis* LAICH. is perhaps the most widely distributed. This species is known from the Sierra de Guadarrama, the Cuenca Mountain, Sierra Nevada and the Serra do Gerêz in Portugal. *Helophorus glacialis* VILLA. occurs widely on the Sierra Nevada. *Corymbites cupreus* FABR. is also found on the Sierra de Guadarrama. *Otiorrhynchus estrellaiensis* ZUMPT from the Serra da Estrela is considered by some authors to be only a subspecies of *O. morio* FABR. *Cidaria flavicinctata* (HÜBN.) and *Gnophos myrtillatus* THUNB. occur on the Sierra Nevada; the latter species is also known from the Sierra de Albarracin and Sierra de Gredos. *Lycaena nevadensis* ZÜLLICH from the Sierra Nevada is only a race of the boreo-alpine *Polyommatus glandon* PRUN.

CHAMPION & CHAPMAN (195) made two interesting entomological excursions to Central Spain, especially to the Sierra de Bejar in 1902. The hill-sides are covered by *Genista florida*, but at higher elevations *Cytisus furgans* is abundant. Large patches of unmelted snow may persist in July on the summit of the Sierra de Bejar. At higher elevations, near the summit, they found the Coleoptera *Asida castellana* GRAELLO, *Dorcadion dejeani*, *Corymbites* spp., *Cymindis melanocephala* DEJ., *Otiorrhynchus dentipes* GRAELLO, *Nebria* spp., *Amara ooptera* PUTZ., *Trechus pandellei* PUTZ., *Bembidion glaciale carpetanum* SHARP, etc. They believe that the Coleoptera of the Sierra de Bejar are very similar to those of the Sierra de Guadarrama.

3. THE ATLAS MOUNTAINS

The Atlas Mountains are more or less parallel to the coast of northwest Africa and stretch nearly 2400 km, from the Cape Nun, west to the Gulf of Gabes. The Atlas Region is bounded on the north by the Mediterranean Sea and on the south by the vast Sahara desert. The range consists of two principal chains of mountains, viz. the Maritime Atlas and the inner and more elevated ranges, extending south of the Maritime Ranges and separated from them by a plateau. The western Atlas ranges are known as the Moroccan Range of the Idraren-Draren (=mountain of mountains) and are composed of five distinct chains of mountains. The main range or the Great Atlas is the central and the longest and also the loftiest of all the ranges and has a mean elevation of 3305 m. The slopes are steep on the Atlantic side. One or two peaks rise above the permanent snowline, but a number of summits are generally covered by patches of snow during most of the summer months. Moraines indicate that glaciers existed in many places during the Pleistocene in the Atlas Region. The Atlas Mountains were uplifted during the Alpine period, but the uplift commenced perhaps during the Jurassic Epoch, was renewed during the Upper Cretaceous and continued during the Miocene.

Within the forest-zone of the Atlas Mountains, many European genera of insects are commonly represented, especially the typical terricole blind Coleoptera. The Bathysciinae are, however, absent and the genus *Carabus* LINN. is also poorly represented and there are indeed no mountain-endemic species of this genus. *Cychrus* FABR. is also absent on the Atlas. Two cavernicole blind Coleoptera from the Atlas mountains are of exceptional interest, viz. *Trechus jurjurae* PEYER. and *Aphaenops iblis* PEYER. from the caves of the Djur-djura Mountain. *Nebria* BON. from the caves on the Atlas has curiously elongate antennae and legs and forms the subgenus *Spelaeonebria* PEYER. *Sp. nudicollis* PEYER. has minute compound eyes. Two blind cavernicole Staphylinidae, *Paraleptusa cavatica* PEYER. and *Apteraphaenops longiceps* JEANNEL and the blind Curculionid *Troglorrhynchus mairei* PEYER. occur in the Djur-djura Cave. The blind troglophile Carabid *Laemostenus fezzensis* BON. comes from the Beni-Add Cave of the Atlas Mountains, near Ain-Fezza. Among the high-alpine zone Coleoptera from the Atlas Mountains, we find a number of genera typical of the high-alpine zones of the European mountains, but there are exceedingly few endemic genera on the northwest African mountains. The Carabids *Trechopsis* PEYER. and *Oreocys* PEYER. are perhaps the only endemic genera so far known from the Atlas Mountains. For further accounts of the high altitude insects from the Atlas Mountains reference may be made to PEYERIMHOFF (833).

4. THE FRENCH CENTRAL PLATEAU, THE JURA MOUNTAINS, VOSGES, THE SUDETEN, ETC.

The French Central Plateau is much lower in elevation than the Pyrenees, but a remarkably larger number of boreo-alpine Coleoptera are known from the region than from the Pyrenees. The following 13 species have so far been recorded: *Nebria rufescens* STRÖM., *Amara erratica* DUFT., *Geodromicus globulicollis* MANNERH., *Silpha tyrolensis* LAICH., *Agathidium arcticum* THOMS., *Neuraphes coronatus* SAHLB., *Corymbites affinis* PAYK., *C. cupreus* FABR., *Hypnoidus rivularius* GYLL., *Evodinus interrogationis* LINN.,

Otiorrhynchus morio FABR., *O. arcticus* FABR. and *Barynotus squamosus* GERM. Some of these species, like for example, *Amara erratica* DUFT., *Geodromicus globulicollis* MANNERH., *Neuraphes coronatus* SAHLB. and *Evodinus interrogationis* LINN. occur only at the highest elevations of the Central Plateau, in the volcanic area of the Haute Auvergne. *Hypnoidus rivularius* GYLL. occurs on the Montagnes de la Lozère and Mont Mézenc. *Agathidium arcticum* THOMS. is known only from the Pionsat in Puy de Dome. The other species are widely distributed at higher elevations of the Central Plateau. Of the boreo-alpine Coleoptera from the Pyrenees, *Amara quenseli* SCHÖNH. and *Helophorus glacialis* VILLA. are absent on the French Central Plateau. In marked contrast to the boreo-alpine Coleoptera, fewer Lepidoptera occur on the French Central Plateau. The following species alone seem to have been recorded so far: *Erebia epiphron* (KNOCH.), *Agrotis speciosa* HÜBN. *Crymodes maillardi* (H.-S.), *Cidaria turbata* (H.-S.), *C. flavicinctata* (HÜBN.), *Eupithecia scriparia* H.-S. and *Gnophos myrtillatus* THUNB. (known only from Cévennes).

The Jura Mountains have only 9 species of boreo-alpine Coleoptera: *Nebria rufescens* STRÖM., *Amara erratica* DUFT., *Anthophagus alpinus* FA., *Geodromicus globulicollis* MANNERH., *Autalia puncticollis* SHARP, *Silpha tyrolensis* LAICH., *Corymbites cupreus* FABR., *Evodinus interrogationis* LINN. and *Otiorrhynchus morio* FABR. The number of boreo-alpine Lepidoptera is even less; only three species *Agrotis speciosa* HÜBN., *Cidaria flavicinctata* (HÜBN.) and *Gnophos sordarius* THUNB. are known at present.

The Vosges have relatively been better explored than the Jura Mountains. Nine boreo-alpine Coleoptera, viz. *Amara erratica* DUFT., *Anthophagus alpinus* FA., *Silpha tyrolensis* LAICH., *Corymbites cupreus* FABR., *C. affinis* PAYK., *C. rugosus* GERM., *Evodinus interrogationis* LINN., *Otiorrhynchus morio* FABR. and *O. dubius* STRÖM., are recorded so far from the region. *Erebia epiphron* (KNOCH.), *Agrotis speciosa* HÜBN. and *Gnophos sordarius* THUNB. are the only boreo-alpine Lepidoptera known at present from the Vosges.

From the Sudeten Region the following 20 boreo-alpine Coleoptera are known: *Nebria rufescens* STRÖM., *Amara erratica* DUFT., *Patrobus assimilis* CHAUD., *Ilybius crassus* THOMS., *Anthophagus alpinus* FA., *A. omalinus* ZETT., *Arpedium brachypterum* GRAV., *Atheta laevicauda* SAHLB., *Pteroloma forsstroemi* GYLL., *Silpha tyrolensis* LAICH., *Neuraphes coronatus* SAHLB., *Helophorus glacialis* VILLA., *Simplocaria metallica* STURM., *Corymbites affinis* PAYK., *C. cupreus* FABR., *Hypnoidus rivularius* GYLL., *Aphodius piceus* GYLL., *Otiorrhynchus dubius* STRÖM., *O. arcticus* FABR. and *O. salicis* STRÖM. The boreo-alpine Lepidoptera from the Sudeten are *Erebia epiphron* (KNOCH.) from the Praded (Altvatergebirge) westwards to the Góry Sniezne (Glatzer Schneeberg), *Rhyacia hyperborea* ZETT. from the Krkonose, *Agrotis speciosa* HÜBN., *Gnophos sordarius* THUNB. and *Sterrhopteryx standfussi* H.-S.

On the Central Mountains of West Germany and the Bohemian Massif, in close correlation with their low elevations, there are very few boreo-alpine Coleoptera, the bulk of which is also restricted to small isolated elevated localities. Only four species, *Corymbites affinis* PAYK., *C. cupreus* FABR., *Otiorrhynchus morio* FABR. and *O. salicis* STRÖM. seem to be relatively widely distributed in the area.

5. THE APENNINES

The Apennine mountain ranges traverse the Italian Peninsula. Originally the name (from the Celtic *pen* = mountain top) was applied to only the northern part of the range, from the Maritime Alps to Ancona, but at present, it embraces the entire range. The Apennines have a length of nearly 1280 km and a width of 112-130 km. Three divisions are generally recognized: the North Apennines, the Central Apennines and the South Apennines. The North Apennines extend from the Maritime Alps to the Bocchetta di Altare, about 8 km to the west of Savona, on the main highroad to Torino. It includes the Ligurian Apennine, the Toscan and the Umbrian Apennines. The south sides of the Apennines are generally steep and rise abruptly from the sea and are well known as the Riviera. Monte Bue (1862 m) is situated on the Ligurian Apennine, Monte Cimone (2165 m) is on the Toscan Apennine and Monte Nerone(1525 m) is on the Umbrian Apennine. The Alpi Apuane are a detached chain of the Valley of Serchio and rise to an elevation of 1859 m and contain the well known marbles of Carrara. The Central Apennines stretch as far as the Valley of the Sangro. To the north are the Monti Sibillini, with the highest point Monte Vettore, rising to an elevation of 2478 m. The Central chain has Monte Terminillo (2211 m) and Monte Velino (2487 m). The East Chain has Monti Maiella (M. Amaro) (2779 m) and the highest summit of all, viz. the Gran Sasso d'Italia is 2913 m above mean sea-level. The South Apennine, south of the Sangro Valley, has three parallel chains that break into smaller groups like Monte Miletto (2080 m), the isolated Mte Gargano and Mte Pollino (2263 m). The slopes of the Apennines are generally densely wooded, but winter snow lies unmelted long during the summer on the highest parts of the mountains.

As mentioned earlier, the Apennines form a part of the Alpine-Himalayan Tertiary mountain systems. The Apennines and the Atlas Mountains are believed by some workers to form a system, distinct from the Alpine system and to lie in between the Alpine and the Dinarid systems. Others consider, however, the Apennines as part of the Dinarid branch, separated from the Alpine by the Corso-Sardinian Massif (Tyrrhenis) and the exposed part of the Hercynian Range. The Apennines consist mostly of Triassic, Jurassic, Cretaceous, Eocene and Miocene beds, as in the outer zones of the Alps. The folding that gave rise to the existing chain of mountains occurred during the Eocene and continued perhaps up to the Miocene. Then followed subsidence and the Pliocene sea overflowed the lower parts. Subsequent elevation, without folding, raised the Pliocene beds to as much as 900 m and they now lie in undisturbed horizontal position on the older beds. The last uplift led to the formation of numerous lakes, now filled with Pleistocene deposits. The North Apennines appear to have become elevated earlier than the rest. There are at present no glaciers on the Apennines, but the numerous Post-Pleistocene moraines of Basilicata are evidence of former glaciers.

As the Apennines are situated much further south than the Alps and are thus climatically different, the subalpine and the alpine vegetation and insect life are also different. The high boreal elements that we find in such great abundance on the Alps are, however, extremely sparsely represented on the Apennines. On the Central Apennines, the principal species of plants include *Salix herbacea*, *Saxifraga aizoides*, *Cerastium ceratoides*, *Veronica fruticosa*, *Silene acaulis*, *Sibbaldia procumbens*, etc. Unlike on the Alps, the Ericaceae (*Loiseleuria procumbens*, *Empetrum nigrum* and *Arctostaphylos alpina, Geum triviale*), etc. are extremely sparsely found in certain localities only on the Apennines. The typically cryophile alpine-zone plants like *Ranunculus glacialis* are naturally absent. In marked contrast to the subalpine zone, commencing at an elevation of 1800 m on the Alps, this zone lies above an elevation of 2300 m on the Apennines, so that there is practically very little of the true alpine zone. Except at elevations above 2000 m, the summer is relatively dry on the Apennines.

FRIDÉN (386) recently collected subalpine and alpine zone insects from the Apennines, especially from the Gran Sasso, Campotosto, Rovere and Scanno. BINAGHI (128) records a number of interesting Coleoptera from the Corno Grande (2914 m) and Corno Picolo (2687) on the Apennines. Snow-species and petricole species were collected mostly at elevations of 2700 m on the north slope of the Corno Grande. *Cychrus attenuatus latialis* LINGIONI from under stones at an elevation of 2500 m is somewhat darker than *C. a. attenuatus* FABR. *Leistus (Orebius) glacialis relictus* BINAGHI (2500-2600 m), *Nebria (Helobia) orsinii* VILLA from the snow-edge on an alpine meadow at an elevation of 2300 m, *Trechus italicus* DANIEL and *Bembidion (Testediolum) magellense* SCHAUB. are among the other interesting Carabidae collected on the Corno Grande. *Trechus italicus* DANIEL from elevations of 2200-2300 m are 3.8-4.0 mm long, but those collected at elevations of 2400-2600 m are 4.0-4.2 mm long. *Mannerheimia aprutiana* GRIDELLI from under stones at an elevation of 2500 m, *Chrysolina lurida obscurefacta* BECHYNE from under stones near *Salix retusa* and *Otiorrhynchus consentaneus alticola* SOLARI from an elevation of 2500 m are the other interesting Coleoptera collected by FRIDÉN. *Stenus biguttatus* LINN. and *Asaphidion flavipes* LINN. were collected near streams, at the edge of the Lake Compotosto (1425 m). From the same locality are recorded *Clivina fossor* LINN. and *Pterostichus nigrita* LINN. *Bembidion bipunctatum* LINN. was found on snow patches. Other records include *Aclypea opaca sammitica* FIORI, *Cymindis humeralis* GEOFFR., *Amara praetermissa* SAHLB., *Helophorus glacialis* VILLA., etc.

Only four boreo-alpine Coleoptera are known at the present time from the Apennines: *Nebria rufescens* STRÖM., *Anthophagus alpinus* FA., *Helophorus glacialis* VILLA. and *Corymbites cupreus* FABR. The first mentioned species occurs on the Flysch Apennine of Emilia and also in the neighbourhood of Toscana. *Anthophagus alpinus* FA. occurs on the Abruzze and *Corymbites cupreus* FABR. on the high Apennine of Toscana and Aspromonte. The boreo-alpine Lepidoptera from the Abruzze are *Argynnis pales* SCHIFF., *Erebia epiphron* (KNOCH.), *Anarta melanopa* THUNB., *Cidaria flavicinctata* (HÜBN.), *Eupithecia scriparia* H.-S., *Gnophos myrtillatus* THUNB. and *Zygeana exulans* HOCHW.

We may also mention at this stage the insects from the Corsican Mountains. The Mte Cinto rises to an elevation of 2710 m. The Carabid *Percus* BON. is widely distributed on the mountains of Corsica, Sardinia, Maritime Alps, Ligurian Alps, Apennines, Mte Gargano, etc. *Percus villae* KR. is endemic on the southwest Alps. *Scotonomus* FAUV. from the Central Apennines and *Trogaster* SHARP are other interesting genera. Corsica and Sardinia are remains of an ancient area, viz. the Tyrrhenis, the central mountainous area of which existed even from the Palaeozoic Era. As the Tertiary mountains of the Apennines, north Sicilian chains and the Atlas Mountains arose, many Corsican-Sardinian faunal elements populated these mountains. The high-alpine endemic Coleoptera of this area are typified by *Trechus varendorffi* DEV., *Amara bickhardti* DEV., *Atheta varendorffi* DEV. and *Otiorrhynchus corsicus* FAIRM. *Agrotis haverkampfi* STANDF. is found exclusively in the alpine zone.

In the region of the Adriatic Bridge (the eastern margin of the Italian Peninsula) there are two mountain folds, predominantly of Mesozoic calcareous sediments. The Mte Gargano is different geologically from the Apennines, but belongs to the Dinarid System. Some interesting high altitude insects are known from the Mte Gargano. *Carabus cavernosus* FIRV. from the Mte Gargano is also known from the mountains of Bosnia, Gran Sasso, etc. and is a subalpine to alpine zone form that ascends only up to an elevation of 1000 m. *Nebria kratteri* DEJ. is a typically forest form that is common on the mountains of Greece, Albania and a terricole form on the South Apennines, descending to an elevation of 800 m. *Leptomastax emeryi* SIMON occurs on the Mte Gargano, down to 700 m. *Bathyscia (Phaneropella) lesinae* REITT. is a terricole species found in caves at elevations down to 800 m on the same mountain. Many typically mountain autochthonous forms like *Cychrus* FABR., *Abax* BON., *Byrrhus* LINN., *Orina* MOTSCH., *Melitaea* FABR., *Erebia* DALM., *Parnassius* LATR., etc. so characteristic of the Alps-Carpathians are totally absent on the mountains of Corsica-Sardinia. It may be remembered, however, that many of these forms occur on the Spanish mountains on the west, on the Apennines and sometimes even on the Atlas Mountains.

6. THE BALKAN MOUNTAINS

The high mountains of the Balkan Peninsula have been more or less extensively explored by a number of entomologists like APFELBECK (53), REBEL (862), REBEL & ZERNY (863) and others. The Albanian Alps, largely of massive limestones, have a general west-east trend and have a mean elevation of 1820–2590 m. In the northwest there is predominance of limestones and they are considered as continuations of the high karst of Montenegro. In the northeast we find Palaeozoic schists. Considerable areas of plateaux have summer pastures. There is no permanent snow, but in sheltered localities there are often isolated patches of the unmelted winter snow. During the Pleistocene the area was heavely glaciated. The Albanian Alps end aburptly in the west, about 45 km from the Adriatic Coast.

The following are interesting records from the Balkan Peninsular mountains: *Nebria rufescens* STRÖM., *Amara erratica* DUFT., *A. quenseli* SCHÖNH., *Arpedium brachypterum* GRAV., *Neuraphes coronatus* SAHLB., *Helophorus glacialis* VILLA., *Corymbites cupreus* FABR. and *Otiorrhynchus salicis* STRÖM. Except *Arpedium brachypterum* GRAV (restricted to only Vitosa and Rila-planina in Bulgaria) and *Neuraphes coronatus* SAHLB. (from Bosnia only), the other species occur in the northern parts of the Peninsula and are widely distributed on the mountains. There are almost twice as many boreo-alpine Lepidoptera as Coleoptera on the mountains of the Balkan Peninsula. Most of these species occur also on more than one mountain. The exact southern limits of these boreo-alpine Lepidoptera on the mountains of the Balkan Peninsula do not seem to have been determined satisfactorily so far, but the southernmost mountain mass on which boreo-alpine species occur is Perister in Monastir (Bitola) and the Ali-butusch Range in southeast Macedonia.

On the West Rhodope Mountains in Bulgaria, the following boreo-alpine Lepidoptera have been recorded: *Argynnis pales* SCHIFF., *Erebia pandrose* (BKH.), *E. epiphron* (KNOCH.), *Polyommatus glandon* PRUN., *Hesperia andromedae* WALLGR., *Agrotis faticida* HÜBN., *A. speciosa* HÜBN., *Crymodes maillardi* H.G., *Anarta malanopa*

THUNB., *Cidaria nubiliaria* (H.-S.), *C. turbata* (H.-S.), *C. flavicinctata* (HÜBN.), *Eupithecia scriparia* H.-S., *Gnophos sordarius* THUNB., *Gn. myrtillatus* THUNB., *Psodos coracina* ESP., *Pygmaena fusca* THUNB., *Arctia flavia* FUESS. and *Zygaena exulans* HOCHW. Most of these species are widely distributed on the different mountains.

7. THE CAUCASUS

The Caucasus are nearly parallel ranges that stretch 1400 km, northwest to southeast, from the Strait of Kerch (between the Black Sea and the Sea of Azov) to the Caspian Sea. The Caucasus has on the whole greater similarities to the Pyrenees than to the Alps. Like the Pyrenees, the mean elevation of the Caucasus is maintained over long distances without diminution. The higher summits like Mt. Elbrus, Mt. Kazbek, Mt. Dykh-tau, Mt. Koshtan-tau, Janga-tau and Shkara, are situated on spurs and not on the ranges themselves. The Caucasus is divided into i. the West Caucasus, ii. the Middle Caucasus and iii. the East Caucasus. The West Caucasus extends nearly 670 km from the Strait of Kerch to Mt. Elbrus (42° 40′ EL), parallel and close to the Black Sea Coast. Between it and the sea coast, there are other minor ranges like the Bzybi Khrebet, Kodorskii Khrebet, Egrissi Khrebet, etc. (*khrebet* = range in Russian). The Bolshoi Kabkasa or the main range has a mean elevation of 3050-3700 m and is composed of crystalline rocks. The permanent snowline on this range is 2740 m, particularly on the summits of the high peaks. East of Oshten the crest of the main range is snow-capped and there are also hanging glaciers. The southern slopes are more luxuriantly forested than the northern slopes. The Middle Caucasus has an elevation exceeding 3050 m, but it is surpassed by the secondary Bokovoi Khrebet. The permanent snowline here is between 2890 and 3050 m on the north and about 300 m higher on the south slope. There are nearly 900 glaciers, but none of them descends low. The best known of these glaciers are the Bezingi or the Ulu Glacier between Dykh-tau and Janga-tau (about 16.8 km long and descending to an elevation of 1937 m) and the Karagom Glacier from Adai-khok (about 15 km long and descending to an elevation 1770 m). The eastern part of the Middle Caucasus, from Kazbek and Darial Gorge to Baba-dagh (48° 25′ EL), is about 390 km long. The maximum width of the Caucasus Region is attained in this area. The mean elevation is 3050 m, but many peaks rise to elevations of 3660-4575 m. The general elevation increases eastward. The principal peaks are Tsmiakom-khok (4136 m), Shan-tau (4429 m), Kidenaismagali (4217 m), Zilga-khok (3852 m), Julti-dagh (3786 m), Alakhun-dagh (3867 m), Maghi-dagh (3795 m), etc. on the spurs and on the subsidiary ranges. On the main range we have the peaks Borbalo (3103 m), Great Shavi-kildeh (3756 m), Murov (3385 m), Ansal (3604 m), Ginor-roso (3390 m), etc. Flanking the main range on the north is the remarkable highland area of Daghestan. The important peaks in this area are Tebulos-mta (4506 m), Tugo-mta (4130 m), Komitotavi or Kachu (4270 m), Donos-mta (4130 m), Diklos-mta (4187 m), Kavvlos-mta or Kolos-mta (3986 m), Motschekh-tsfari (4004 m), Galavanas-tsfari (4040 m), etc. Nevé and glaciers occur on some of them; the glacier on Diklos-mta descends to an elevation of 2345 m on the north and 2545 m on the south. The East Caucasus diminishes in elevation east of the Baba-dagh (3626 m) and no peaks are higher than 2743 m above mean sea-level.

Treeless valleys are common features of the Caucasus. There are numerous ancient and extinct volcanoes on the intersections of the main and transverse ranges. In the centre of the Caucasus Region, we find crystalline rocks, beneath on both sides are folded Palaeozoic rocks and Jurassic rocks. The final folding of the mountains of the Main Caucasus Range occurred probably during the Miocene.

There is a great deal of literature, especially in the Russian language, on the general natural history of the Caucasus Region. One of the earliest accounts of the general natural history of the region was by SCHNEIDER (929). SCHNEIDER & LEDER (930) have also described a number of interesting Coleoptera and RAMME (857a) has given an account of the Orthoptera collected by the Polish Alpine Expedition to the Caucasus. The Lepidoptera collected by the same expedition were described by WOJTUSIAK & NIESIOLOWSKI (1122); these authors have also discussed the ecology and biogeography of the Caucasian Lepidoptera. A short review of the literature on Lepidoptera from the region may be found in WARNECKE (1101). For a general account of the ecology of the Caucasus Region, reference may be made to the recent studies by SAKHOKIA *et al.* (910). GADZHIEV (392) has given a beautiful and detailed account of the subalpine vegetation of the Caucasus. Phytogeographical investigations were carried out by GROSSHEIM & SOSNOVSKII (441) on the Caucasus. An account of the general landscape-hydrological zones of the region may be had in KAVRISHVILI (609).

Among the better explored areas of the Caucasus Region the following are included: i. The Alazani Valley, near Signakhi, about 30-35 km wide, with steep slopes of the Great Caucasus Range. The mean elevation of the main range in the area is between 3000 and 3500 m. There are no glaciers on the south slope of the main range. Small glaciers existed, however, towards the end of the Pleistocene, as low as 2700 m. The upper limits of the alpine zone, below the subnival zone, are clothed with the characteristic *kar* vegetation. The winter snow-cover persists nearly to the middle of June. The forest-zone ascends here up to an elevation of 2000-2300 m. The mean annual rainfall in the area is about 800-950 mm. There is a subalpine zone above the well known Lagodekhi Reservation on the south slope of the main range, above the forestline, at elevations between 2000 and 2300 m. The Meteorological Station of the Lagodekhi Reservation is located here at an elevation of 1950 m. ii. The Bordzhomi-Bakuriani-Tzkhra-Tzkharo. The Bordzhomi Gorge is of considerable interest because of the overlying lava of Pliocene times. Subalpine and alpine zones are found on the Tzkhra-Tzkharo-Kodiani (2800 m). The subalpine meadows commence at an elevation of about 2000-2300 m, with communities of *Anemone fasciculata* LINN., *Scabiosa caucasica* W., *Betonica grandiflora* W., *Geranium ibericum* CAV., *Veratrum lobelianum* BERNH., etc. The alpine zone is relatively a narrow strip, above 2600-2700 m on the Tzkhratzkhrao-Kodiani Range. *Festuca ovina* LINN., *F. varia* HAENKE, *Carex tristis* M.B., *Nardus glabriculmus* SAKALO, etc. are some of the common plants in this belt. The High Altitude Botanic Garden of the Institute of Botany of the Academy of Sciences of the Georgian SSR, with its fine alpinarium, is situated on the Bakuriani.

GADZHIEV's account (392) of the subalpine vegetation of the Caucasus Region deals with the area in the Azerbaidjan SSR in particular, viz. the main Caucasus, Northeast Bokovoi Range, and the Bodorazdel Range, in which are situated the high peaks Mrov-dagh (3545 m), Maki (3488 m), Salavat (2852 m), Shailkhan (3545 m), Bazardynzi (4485 m), Tfan-dagh (4206 m), Baba-dagh (3632 m), Shah-dagh (4250 m), etc. The mean elevation of Bodorazdel Range is 3000 m and some of the peaks rise to higher elevations: Tipovrassa (3385 m), Guton (3695 m), Akhbai (3485m),

Tsaplakhan (3541 m), Malgamul (3875 m), Kharut (4033 m), Tfan (4206 m), etc. A brief account of the hydrological peculiarities of the region is also given by GADZHIEV.

At elevations between 500 and 1200 m, temperate climate prevails, between 1200 and 1800 m cool climate prevails and the subalpine zone commences at elevations between 1800 and 2000 m. At the Ali-bek Meteorological Station, situated at an elevation of 1750 m, the mean annual minimum temperature is —5° C and the annual maximum is 16° C. Subtropical climate prevails on the coastal ranges, up to an elevation of nearly 2000 m on the south slope. The mean temperature ranges from +3° C to —5° C. The cold temperate climate prevails on the Main Caucasus Range. The altitudinal zonation of the subtropical vegetation is up to 300 m; above this lies the *Quercus*-forest up to an elevation of 700 m; the *Fagus*-Conifer forest extends from about 600 m to 2000 m and the high-montane meadow zone lies between 1800 and 3000 m (fig. 73).

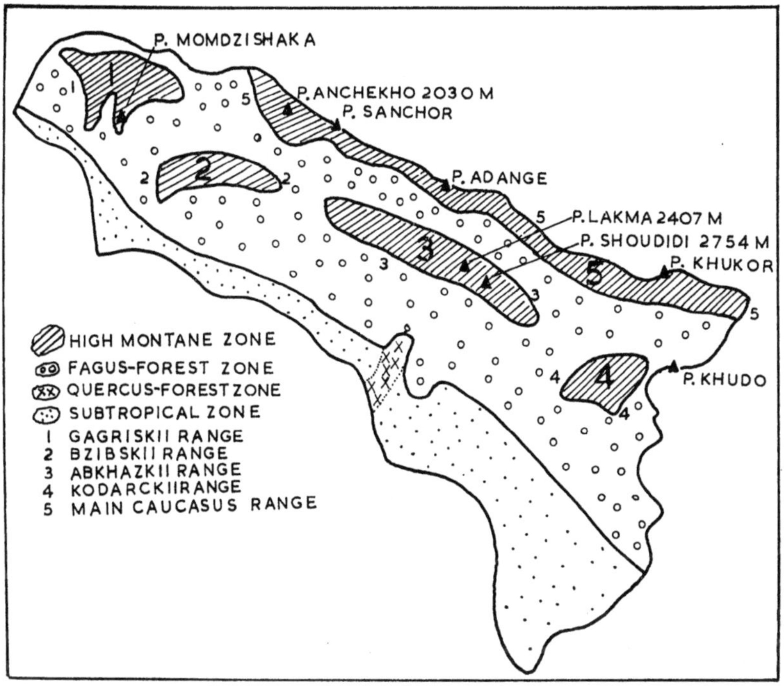

Fig. 73. Sketch-map of the West-Caucasus Region, showing the altitudinal zonation of vegetation.

Biogeographically, the Caucasus Region is subdivided into the following horizontal divisions by UVAROV (1059): i. The Kuban-Terek District, belonging to the steppes subregion, embraces the northern slopes of the mountains. The southern limits are formed by the forestline on the main Caucasus Range. ii. The Daghestan District, in direct contact with the Kuban-Terek District, Somkheto-Kakhetian, Caspian, Trans-Caspian and Eastern Caucasus, but with affinities to the steppes subregion or to the mediterranean subregion. The northwestern and southwestern limits coincide with the lower limits of the alpine zone on the East Caucasus. The boundaries between the Daghestan and the Kuban-Terek District are not sharp. iii. The West Caucasus District, in which UVAROV lists a number of subalpine and alpine-zone grasshoppers, of which nearly half are boreal forms. The northern and southern boundaries coincide with the forestline on the corresponding slopes of the Caucasus. iv. The East Caucasus has a much poorer Orthoptera fauna than the West Caucasus and has also fewer boreal elements. v. The Minor Caucasus District, with pronounced Armenian affinity and considerable numbers of boreal elements, which are, however, absent on the Armenian mountains, and derived from the main Caucasus Range. The area embraces the high tablelands of Alexandropol, Kars and Akhalkalaki. The northern boundary coincides with the upper Somkheto-Kakhetan forestline. vi. The Trans-Caspian-Caucasus, on either side of the R. Kura, from west of Tbilisi to the Caspian Sea.

The high altitude insects of the Caucasus Region are remarkable for large numbers of endemic species, but unfortunately relatively little attention seems to have been paid to the insect life of the higher elevations in the region. Valuable information about the subalpine and alpine zone Orthoptera may be found in ADELUNG (7). Some aspects of the geographical distribution of the Orthoptera, especially the Acrididae, of the Caucasus Region were discussed by UVAROV (1059). SCHNEIDER (929) has also described the general natural history of the region. STOLJAROV (1013) has recently dealt with some of the peculiarities of the geographical distribution, ecology and biology of the grasshoppers from the Caucasus.

The alpine-meadow Orthoptera of the Caucasus include *Acryptera fusca* PALL. and *Nocarodes cyanipes* FISCHER V. WALDH. at an elevation of 2745 m from the Jalanus-Tscham, *Poecilimon tschorochenosis* ADELUNG and the boreo-alpine *Aeropedellus variegatus* FISCHER V. WALDH. The remarkable Tettigoniid *Schizonotinus* RAMME was described by RAMME (854) from the Caucasus. STOLJAROV (1013) has listed nine species of Tettigoniids from the high altitude meadows of the Abkhaskii Range, five of which do not occur at an elevation below 1800 m. *Poecilimon similis* RET. occurs from the sea-level to an elevation of 3000 m and thus ranges from the subtropical forest zone to the high montane meadow zone on the Abkhaskii Range. It is particularly abundant in the southwest in the *Quercus-Fagus*-zone and in the northeast it is common in the *Fagus*-zone. *Isophya kalishevskii* ADELUNG occurs on the subalpine meadows, at elevations of 800-3000 m on the Bziebskii Range, on the Abkhaskii Range in the northeast area and on the southern slopes of the Main Caucasus Range. Two species, *Poecilimon djakonovi* MIR. and *Schizonotinus fortificalis* BEJ-BIENKO, are found at elevations of 1600-3000 m. The former species occurs in the subalpine zone of the Gagrskii Range, Bziebskii Range, the southern slopes of the Main Caucasus Range and southwest slope of the Abkhaskii Range. The latter species occurs on the Kodorskii Range, Abkhaskii Range, Bziebskii Range, the south slope of the Main Caucasus Range in the *Fagus*-zone and in the subalpine zone. *Schizonotinus crassicerus* TARB., *Semenovites specularis* FISCHER V. WALDH., *S. inflatus* UVAROV, *Polysarcus zacharovi* STSCHELK. and *Decticus verrucivorus* RAMME are found at elevations of 1800 and 3000 m. *Schizonotinus crassicerus* TARB. occurs on alpine meadows, during August, on the Gagrskii Range and on the southern slopes of the Main Caucasus Range, but also descends into the subalpine zone during July. *Semenovites specularis* FISCHER V. WALDH. is common on the southern slopes of the Main Caucasus Range. *S. inflatus* UVAROV is known from the subalpine meadows on the Kodorskii Range and Abkhaskii Range, but does not seem to extend to the west. *Polysarcus zacharovi*

STSCHELK. is also confined to the Kodorskii Range (alpine-meadow zone). *Decticus verrucivorus verrucivorus* RAMME is known from the alpine-meadow zone of the Kodorskii Range. The boreal European *Decticus annaelisae* RAMME, considered by STOLJAROV to be a race of *Decticus verrucivorus* RAMME, is confined to the high-montane meadows in the Caucasus Region. Of about 40 species of subalpine and alpine grasshoppers listed by UVAROV from the West Caucasus, the boreal forms include *Stenobothrus* spp., *Gomphocerus sibiricus caucasicus* MOTSCH., *Mecostethus* sp., *Podisma* sp., *Locusta* sp., etc. In the Minor Caucasus District, UVAROV recognizes two endemic species *Stenobothrus werneri sviridenkoi* RAMME and *Platycleis iljinskii* RAMME. *Podisma lezgina* RAMME and *Phlocerus menetriersi* FISCH.-WALDH. are endemic in the East Caucasus District.

In a recent contribution, KOPANEVA* has described the seasonal and altitudinal changes in the habitats of grasshoppers in the upper reaches of the R. Teberda in the northern Caucasus. The region lies mostly at elevations of 2000-3000 m, some parts at elevations of 3000-4000 m. Of the five distinct groups of habitats-biotopes, the fourth group embraces the subalpine belts, at elevations above 2400 m and the fifth group comprises the alpine-zone areas, up to elevations of 3500 m. The character species of the subalpine belt are *Chorthippus longicornis* LATR. and *Decticus verrucivorus* LINN. The species characteristic of the alpine zone meadows are *Gomphocerus sibiricus caucasicus* MOTSCH., *Podisma teberina* RAMME, *Nocaracris cyanipes* F.-W., *Chorthippus apricarius* LINN. and *Decticus verrucivorus* LINN.

The following subalpine Coleoptera are known from the West Caucasus: *Carabus stjernvalli* MANNERH., *Bembidion ellipticircutum* NETOL., *B. zaitzevi* LUTSCHN., *Hydroporus koslovskii* ZAITZ., *Hydnobius puncticollis* REITT., *Oxytelus strigifrons* HOCHH., *O. gibbulus* EPP., *Helophorus kerimi* GGLB., *H. armeniacus* GGLB., *Athous vulpeculus* REITT., *Helops detarus* ALLARD., *Semiadalia schelkovnikovi* DBZH. and *Otiorrhynchus kirschi* STIERL. The typical subalpine Coleoptera from the East Caucasus are *Cymindis intermedia* CHAUD. and *Otiorrhynchus cinereus* STIERL. A number of species are widely distributed over the whole of the Caucasus Region and the more important of these are *Carabus calleyi* FISCH., *Harpalus raddei* TSCHITCH., *Zabrus aurichalceus* ALLARD, *Metabletus humeralis* MOTSCH., *Blaps luctuosa* MÉN., *Dorcadion* spp. and *Otiorrhynchus carbonarius* HOCHH. In the alpine zone of the West Caucasus we find *Trechus armenus* KHNZ., *Pterostichus chydaeus* TSCHITCH., *Pt. armenus* FALD. and *Agonum rugicolle* CHAUD. The typical alpine-zone Coleoptera, widely distributed in the Caucasus Region, include *Nebria schlegelmilchi* ALLARD, *Bembidion armeniacum* CHAUD., *B. caucasicum* MOTSCH., *Pristonychus mannerheimi* KOL., *Gaurodytes glacialis* HOCHH., *Olophrum caucasicum* FAUV., *Aphodius asphaltinus* KOL. and *Otiorrhynchus foveicollis* HOCH.

Despite its massiveness, the Caucasus has only a relatively small number of boreo-alpine species. As mentioned in the preceding chapter, *Amara erratica* DUFT., *A. quenseli* SCHÖNH., *Geodromicus globulicollis* MANNERH., *Arpedium brachypterum* GRAV., *Hypnoidus rivularius* GYLL. and *Evodinus interrogationis* LINN. are the only boreo-alpine Coleoptera on the Caucasus. The boreo-alpine Lepidoptera are *Pieris callidice* ESP., *Melitaea iduna* DALM., *Agrotis faticida* HÜBN., *Plusia hochenwarthi* HOCHW., *Cidaria munitata* (HÜBN.), and *Gnophos myrtillatus* THUNB. As it may be recalled, *Melitaea iduna* DALM. is the only boreo-alpine species that is restricted to the Caucasus in the south area.

* KOPANEVA, L. M. 1963. *Entomologicheskoe Obozrenie*, 42 (3): 564.

8. THE ARMENIAN HIGHLANDS

Armenia is largely a plateau, about 1830-2440 m above mean sea-level, with short ridges and isolated volcanic mountains. Mt. Alagoz (4095 m) was active during the Tertiary. Lake Goktcha (Sevan) (1382 m above mean sea-level) is surrounded by volcanic mountains and the neighbourhood of Leninakan is covered by volcanic ash. The general climate of the region varies from place to place, with differences in altitude and aspect; the winters are, however, generally severe, dry and thus the climate is on the whole extreme. Leninakan (1529 m) has a January temperature mean of 11.1° C and July mean of 18.3° C. The annual mean rainfall is 405 mm. The general trend of the mountain ranges is almost north-south in southern Armenia and the well known high peaks are Alagoz, Quzei Dara (3447 m), Arazhin (3220 m), Armaz (3385 m) and Kopuzhah (3918 m). The area of the Armenian Highlands falls within the Armenian District of the Caucasus Region (1059), with the northern boundary on the mountain chain of Pontus, to nearly Batumi in the east and also embracing a narrow part of the Azerbaidjan Mountain chain, extending south on the Arsian Chain to Arex, further east to Mt. Ararat. The western boundary is the margin of the Anatolia Plateau, the southern boundary along the Cylician Toros and Anti-taurus Mountains (Güneydogu Toroslar).

LUPPA *et al.* (713) have given an interesting account of their recent explorations in Armenia, where they collected a number of Carabidae, Coccinellidae and other insects. RAMME (854) reports the Orthoptera *Stenobothrus werneri sviridenkoi* RAMME at an elevation of 2200 m from Tschbuchy on Goktcha Lake, *Omocestus demokidovi* RAMME at an elevation of 2800 m on Alagoz and *Metrioptera armeniaca* RAMME at an elevation of 2000 m on the Goktcha Lake and from the Mt. Alagoz at elevations of 2000-2800 m. The most important work on the Coleoptera from the Armenian highlands is by YABLAKOV-KHNZORYAN (1128). He has recorded *Aphodius quadrinaevulus* REITT., *Onthophagus basipustulatus* HEYD., *Mannerheimia kirschenblatti* KHNZ., *Selostomus pecircanus* REITT., *Helops constrictus* SEIDL., *H. humeridens* REITT., *H. eligius* REITT. and *Otiorrhynchus sunicensis* ARNOLDI from the subalpine zone, at elevations of about 3500 m, on the Armenian mountains. The West Caucasus elements on the subalpine zone include *Carabus stjernvalli* MANNERH., *Bembidion ellipticircutum* NETOL., *B. zaitzevi* LUTSCH., *Hydroporus koslovskii* ZAITZ. *Hydnobius puncticollis* REITT., *Oxytelus strigifrons* HOCHH., *O. gibbulus* EPP., *Helophorus kerimi* GGLB., *H. armeniacus* GGLB., *Athous vulpeculus* REITT., *Helops detarus* ALLARD., *Adalia schelkovnikovi* DBZH. and *Otiorrhynchus kirschi* STIERL. *Cymindis intermedia* CHAUD. and *Otiorrhynchus cinereus* STIERL. are East Caucasus elements. A number of Pan-Caucasus forms like *Carabus calleyi* FISCH., *Harpalus raddei* TSCHITCH., *Zabrus aurichalceus* ALLARD, *Metabletus humeralis* MOTSCH., *Blaps luctuosa* MEN., *Dorcadion* spp., *Otiorrhynchus carbonarius* HOCHH., etc. are also found in the subalpine zone of the Armenian mountains. The species from Mt. Ararat are *Dailognatha pumila* BAUDI, *Haltica globicollis* WSE., *Otiorrhynchus punctirostris* STIERL., *O. angustifrons* STIERL. and *O. cribripennis* STIERL. The Central European-Caucasus elements are *Bembidion starcki* DAWSON, *B. redtenbacheri* DAN., etc. The Euro-Siberian elements on the Armenian mountains are *Harpalus rufitarsis* DUFT., *Amara municipalis* DUFT., *Coelambus marklini* GYLL., *Platystethus nodifrons* SAHLB., *Quedius nitidipennis* STEPH., etc. The typical Palaearctic species on the subalpine zone include *Amara famelica* ZIMM., *Necrophorus investigator* ZETT., *Philonthus micans* GRAV. and *Quedius boops* GERM.

There are a number of endemic species of Coleoptera on the alpine zone: *Nebria gotschi* CHAUD., *Trechus oresitrophus* KHNZ., *Tr. dzermukensis* KHNZ., *Amara monticusta* KHNZ., *Agonum brachyderum* CHAUD., *Olophrum aragatzense* KHNZ., *Philonthus rufimargo* REITT., *Bryoporus abdominalis* LUZE, *Tachinus gracilicornis* LUZE, *Aphodius inclusus* REITT., *Olibrus ovalis* KHNZ., *Dorcadion cineriferum* SUV., *Longitarsus alpigradus* KHNZ., *Otiorrhynchus richteri* KHNZ., *O. nairicus* KHNZ., *O. arnoldi* KHNZ. and *O. sulcibasis* REITT. The West Caucasus elements on the alpine zone are *Trechus armenus* KHNZ., *Pterostichus chydaeus* TSCHITCH., *Pt. armenus* FALD. and *Agonum rugicolle* CHAUD. The Pan-Caucasus elements of the alpine zone are *Nebria schlegelmilchi* ALLARD, *Bembidion armeniacum* CHAUD., *B. caucasicum* MOTSCH., *Pristonychus mannerheimi* KOL., *Gaurodytes glacialis* HOCHH., *Olophrum caucasicum* FAUV., *Aphodius asphaltinus* KOL. and *Otiorrhynchus foveicollis* HOCH. Eight boreo-alpine Coleoptera *Bembidion bipunctatum* LINN., *Amara cursitans* ZIMM., *A. quenseli* SCHÖNH., *Potamonectes griseostriatus* DEG., *Gaurodytes solieri* AUBÉ, *G. congener* THUNB., *Geodromicus globulicollis* MANNERH. and *Atheta tibialis* HEER are known from the Armenian mountains. The boreal Palaearctic species common in the alpine zone of the Armenian mountains include *Hydroporus memnonius* NIC., *Eriglenus labiatus* BRAHM, *Copelatus haemorrhoidalis* FABR., *Ilybius subaeneus* ER., *Liodes dubia* KUG., *Aclypea undata* LINN. and *Chaetocnema arida* FOUDR.

In a recent contribution, SCHWEIGER (948) has discussed some of the striking characters of the Coleoptera from the mountains of West Anatolia. He finds them to be fundamentally different from those of the Alps and this difference is all the more striking, when we compare them with the species from the Pontic Mountains and from the Caucasus Region. The apterous hygrophile snow-field species of *Carabus* LINN., *Nebria* LATR., *Trechus* CLAIRV., *Pterostichus* BON., *Stenus* LATR., etc. so well represented on the Alps, Caucasus and the Pontic Mountains, are absent at comparable elevations of the mountains of West Anatolia. There are also no montane species of *Nebria* LATR. on the West Anatolian Mountains; the genus is confined to the lower altitudes, within the forest-zone. *Trechus quadristriatus* SCHR. is perhaps the commonest species that extends from the low elevations ot 2200 m on snow on Ulu-dagh. He recognizes three faunal zones, viz. the Pontic mountains, the Taurus mountains and the isolated West Anatolian high summits.

9. THE URAL MOUNTAINS

The Ural Mountains extend north-south from the Arctic Ocean to the Caspian Sea and separate Asia from Europe. They represent separate series of upheavals, some with north to west strike and others with north to east strike. Paé-khoy or the Coast Ridge (Stony Ridge) is independent of the Urals proper and has a north-northwest and northwest trend. Obdorsk or the North Urals begin near the Kara Bay and extend southwest to 64° NL as a stony and craggy mass, in which the highest peaks are Khard-yues (1131 m) and Paé-yer (1440 m). The highest peaks of the Urals are, however, of sedimentary rocks, viz. Sablya (1646 m) and Töll-poz-ir or the Muraichakhl (1689 m). Dense forests of fir, pine and larch clothe the sides of the hills on the south. The forest disappears at the base of the mountains near the Arctic Circle and near 65° NL. The sections between 64 and 61° NL have distinctive broad marshy valleys, with isolated peaks, mostly below an elevation of 941 m. Only Yang-tump rises to an elevation of 1271 m (62° 43′ NL). The Middle Urals are densely forested and the South Urals (55° 30′ to 51° NL) consist of three parallel chains of mountains, northeast to the southwest and constitute an independent part of the Ural System. The Urals proper are low sinuous chains of hills, hardly 850 m above mean sea-level. Towards the west, there is a parallel chain of mountains that rise to elevations of 1595 m. The Urals are the western edge of a broad belt of folding, the greater portion of which is buried beneath the Tertiary deposits of West Siberia. The chain of granites, diorites, peridotites, gneisses, etc. are covered by Silurian, Devonian, Carboniferous, Permian and Trassic strata, which are thrown into numerous folds.

The Ural mountains seem to have been sadly neglected by entomologists and we really know very little of the subalpine and alpine elements of this region (85). From the low southern mountains, the boreo-alpine Lepidoptera *Agrotis faticida* HÜBN., *Cidaria munitata* (HÜBN.) and *Arctica flavia* FUES. are known. *Corymbites rugosus* GERM. has also been reported from the northern Urals and *Pieris callidice* ESP. from the arctic parts of the Urals.

10. THE ALTAI AND OTHER MOUNTAINS OF THE ANGAR-BERINGEA

The principal mountains of the Angar-Beringean area are the Altai, the Sayanskii Mountains to the northeast of the Altai, the Baikal Mountains, the Yablonovy Mountains, Stanovoi Mountains, Sikhote-Alin Ranges, Cherskogo Mountains,

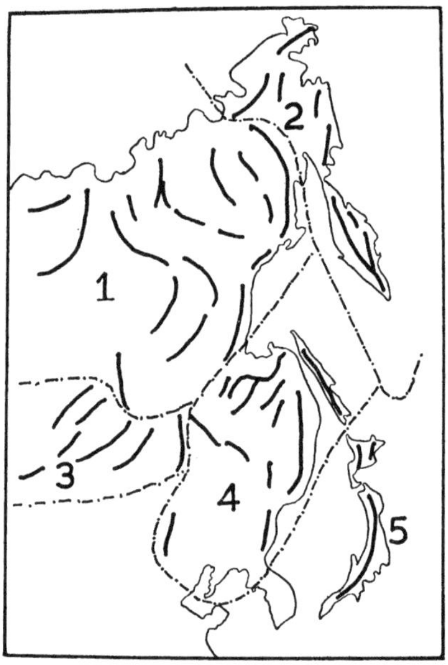

Fig. 74. Sketch-map of Angar-Beringea, showing the crestlines of the principal mountain ranges. 1. Angaran area, 2. Beringea, 3. Baikal-Mongol area, 4. Pri-Amur area, 5. Peripheral mountains area. (Modified from KURIENTZOV).

Anadyr Mountains, the peripheral ranges of Kamchatka Peninsula and Japan. Part of this area lies within the subarctic and arctic regions, but these mountains are best dealt with here.

The Altai (from the Mongolian *altain-ula* = mountain of gold), situated in West Siberia and Mongolia, covers a much wider area of the earth's surface than the Alps, but it belongs to the older system of mountains. It extends from the R. Irtysh and the Dzhungarskii Depression (between 46 and 47° NL) northwards to the Sayanskii Mountain Range. The Sailughem or the Kolyvan Altai Mountain forms a sort of principal backbone of the region and stretches northeastward from 49° NL and 86° EL to the west end of the Sayanskii Mountains (51° NL and 89°EL). The mean elevation of the Altai Mountains is about 2050 m in the north and about 2377 m in the south. Many rugged peaks rise, however, to more than 900 m above the permanent snowline. The Altai Range is flanked by the Mongolian Plateau in the east and southeast, with several other lesser plateaux in between, such as, for example, the Ukök Plateau (2378 m), Chenya (1829 m), Juvlu-kul (2410 m), etc. There are a number of lakes, like Ubsa-nor (715 m above mean sea-level), Kirghiz-nor, Durga-nor, Khodo-nor (1158 m above mean sea-level), etc. The Altai Ranges are crossed by smaller mountain ranges, like Tannu-ola, parallel to the Sayanskii Range, as far east as Kosso-gol (100-101° EL) and the Khan-khu Mountain, with east-west trend.

The Altai proper, also called the Ek-tagh or the Altai Mountain, the Yuzhnovo Altai (southern Altai) or the Great Altai, comprises two parallel chains of mountains that separate the Kobdo Basin (north) from the Irtysh Basin (south) and rises from the Dzhungarskii Depression (460-914 m above mean sea-level), but descends on the north by a short slope to the plateau of North Mongolia, at an elevation of 1220-1677 m. East of the 94th meridian the range continues by a double series of mountain chains: the southern chain is Kara-adzirga and Burkhan-ola ends near the 99th meridian. The northern chain is Artsi-bogdo and Saikhat that extends nearly to the bend of the R. Hwang-ho. The western Ek-tagh Altai rises above the snowline, but the eastern double range rises barely above this elevation and is forested right up to an elevation of 1900 m.

II. GENERAL ECOLOGICAL CHARACTERS OF THE ALTAI

Despite the size, location and altitude, the Altai Mountains lack glaciers almost completely; the only glaciers that may perhaps be mentioned are at the source of the R. Katuna on the higher and steeper mountains and some smaller ones near the Kurai Pass. This peculiarity is no doubt related to the extreme atmospheric aridity and the deep continentality of the Altai Region and the relatively light winter snowfall. Though it is extremely cold on the Upper Tchuja, there is relatively very little snowfall. The higher regions of the Altai are, however, liable to heavy rains and frequent snow-storms. The lower areas are densely forested and the valleys are marshy, intersected by numerous streams and rivers.

The flora and the fauna of the Altai Mountains are greatly influenced by the extreme climatic conditions, especially the extreme heat and cold, the frequent thunder storms accompanied by hail and snow at higher elevations almost daily during the summer. The storms arrive mostly from the east. Large patches of unmelted snow may be found during July, at elevations of 2150 m. Up to an elevation of nearly 300 m on the north side and 600 m on the south, the plants are generally reminiscent of the European flora, with abundance of *Larix* at the foot of the mountains. The steppe-flora ascends up to elevations of 250-360 m and in the valleys often to 1675 m. The alpine vegetation is met with from above this elevation. Trees like *Betula, Populus, Pinus cembra*, etc., grow often at elevations between 1900 and 2475 m (the latter limit on the north). Firs do not grow above an elevation of 760 m. The Tchuja-Steppes are a treeless low mountainous area, about 1980 m above mean sea-level, with isolated patches of *Salix*, up to an elevation of 2290 m. The mountains are steep, but also with grassy meadows. The alpine-zone flora of the Altai Mountains include *Primula nivalis, Dracocephalum grandiflorum, Polemonium pulchellum, Gentiana altaica, Pedicularis verticillata, P. foliosa, P. comosa, Allium sibiricum, Linum caeruleum, Iris tigridia, Pyrethrum pulchellum, Corydalis, Aquilegia glandulosa, Astragalus, Lloydia serotina, Papaver alpinum, Draba ochroleuca, Saxifraga oppositifolia* (up to an elevation of 2595 m), *Dryas octopetala, Rhubarb rhaponticum*, etc. In the area of Lake Teletskoi, the vegetation is characterized by the presence of dense growths of *Aconitum, Delphinium*, etc. The only tree *Larix sibirica* often grows up to an elevation of 2130 m.

12. THE INSECT LIFE OF THE ALTAI

Our knowledge of the insect life of the Altai Region must be described as extremely fragmentary. The outlying spurs of the Altai and the Sayanskii Ranges were explored in part by PALLAS during the eighteenth century and by a number of Russian geologists and botanists during the last century. The names of HELMERSEN, LEDEBOUR (681) and TCHICATCHEFF (1033) stand foremost among the early explorers on the Altai. Considerable literature exists in the Russian language on the general natural history of the Altai Region. KINDERMANN collected a number of interesting Lepidoptera-Rhopalocera from the Western Altai during 1852-1853 and his material was described by LEDERER (682, 683). The next important attempts at collecting insects on the Altai were by RUDOLF TANCRÉ in the vicinity of Lake Saisan, Katun-karagai and Tchingistai and by ELWES, FLETCHER, BEREZOWSKY and ALEXIS JACOBSON. ELWES (314) has given an interesting account of the general zoological and botanical observations made by him during his travels. Nearly all the entomological explorations in the region have so far been confined to the part within the USSR, but the Mongolian area has not been explored at all.

HILDÉN (515) has listed the Coleoptera collected by K. HILDÉN & J. J. GRANÖS during the summer of 1914. They found 49 species, varieties, aberrations, etc., mostly from the steppes area of the R. Tscholyschman, Tchuja Katu, Kygy near the Lake Teletskoe. The interesting records include *Entelocarabus arvensis rufipes* GEB. from Tchuja, *Pancarabus aeruginosus aereus* FISCH. from Katun, *Corymbites cupreus* FABR. from Kygy and Tscholyschman, *C. pectinicornis* LINN., *Evodinus interrogationis* LINN., *E. interrogationis schamoysi* KR., *E. variabilis* GEB., *E. v. nova* and *E. v. semipratensis obscuripennis* PIC. from Kygy.

The Lepidoptera from the Altai are better known than any other group of insects. Among the Parnassiidae, we have the interesting *Parnassius apollo sibiricus* NORDM. commonly at elevations of 300-1400 m and its larvae feed on *Sedum ewersi* in Bashkaus, Tscholyschman and Bija Valleys. *P. delius intermedius* MÉN. occurs at elevations of 1220-2500 m and the larvae feed on *Saxifraga aixoides* in the Tchuja Valley. Other species include *P. nomion* F.D.WALD. at an elevation 1700 m, but occurring also at lower elevations (1200 m) near Lake Teletskoi; *P. actius* EV. and *P. tenedius* EV. from Tchingistai; *P. stubbendorfi* MÉN. on marshy alpine meadows on the Kurai Pass (1830-2135 m) and *P. eversmanni* MÉN. breeding on *Corydalis gigantea* on marshy alpine meadows near Darkoti (2135 m). The Nymphalidae include *Neptis lucida* FABR. (1830 m) from the Tchuja Valley; *Melitaea iduna* DALM. from the Tchuja Steppes (1830-2500 m); *M. cinxia* LINN. from the Tchuja Valley (2135 m) *M. arcesia minor* STGR. on grassy slopes (2000-2500 m) and *M. aurelia britomartis* ASSM. in the Tchuja Valley (1230 m).

Over 30 species of Satyridae have so far been recorded from the Altai Region. While some of the Satyridae are endemic, others are Palaearctic or even Holarctic forms. A few species extend to the Himalaya. The following are some of the important records: *Argynnis aphirape* HÜBN. on grassy meadows and slope (2100-2500 m); *A. selene* SCHIFF. from Upper Tchuja Valley; *A. selenis* EV. from the Lower Tchuja Valley (900-2135 m); *A. euphrosyne* LINN. from the Tchuja Valley (900-1230 m); *A. freya pallida* ELWES on grassy meadows (2135-2500 m) and extending also to the East Sayanskii Mountains; *A. pales* SCHIFF. common on marshy meadows above the timberline, at elevations of 1830-2135 m; *A. dia alpina* ELWES from alpine meadows; *A. amathusia sibirica* STGR. from the Tscholyschman Valley (1525 m) and from the Kentei Mountains in Mongolia; *A. angarensis* ERSCH. in the Tscholyschman Valley (1830 m); *A. frigida alpestris* ELWES on wet alpine meadows (2100-2500 m) on the Tchuja Mountain; *A. ino* ESP. below an elevation of 1230 m near Lake Teletskoi and Bashkaus; and *A. aglaia* LINN. at an elevation of 1830 m from Lake Teletskoi; *Erebia theanus* TAUSCH. (900-1830 m); *E. maurisius* ESP. (1525-2500 m); *E. kefersteini* EV. (1675-2290 m); *E. tyndarus* ESP. (1830-2500 m); *E. euryale* ESP. on the north side of the Kurai Pass, in marshy localities (1525-1830 m); *E. lappona* ESP. (2500 m) on the Tchuja Mountain; *E. rossii* CURT. on the mountains south of the Tchuja Steppes (2135-2750 m); *E. edda* MÉN. on the Tchuja Mountain; *Oeneis norma altaica* ELWES from the Tchuja Valley (1225 m); *Oe. nanna* MÉN. from the Upper Tchuja Valley (1850 m); *Oe. sculda* EV. (2135-2500 m); *O. bore ammon* ELWES (2135-2750 m); *Satyrus hippolyte* ESP. from the Tchuja Steppes (1830 m); *Coenonympha iphis iphicles* STGR. from Tchuja and Bashkaus (900-1830 m); *C. hero persei* LED., *C. amaryllis* CRAM. and *C. phryne* PALL. from the Kurai Steppes and mountains, up to elevations of 2500 m.

The Pieridae comprise widely distributed forms like *Aporia crataegi* LINN. on stony localities (2000 m), *Pieris napi* LINN., *P. rapae* LINN., *P. daplidice* LINN., *P. chloridice* HÜBN., *P. callidice chrysidice* H.S. (on alpine zone), *Colias palaeno* LINN. in swampy-flats, *C. nastes mangola* ALPH. near Kuch-Agatch (2135 m), *C. melinos* EV. in subalpine valleys and on alpine meadows (2500 m) on the high Tchuja Mountain, *C. hyale* LINN. and *C. chrysotheme* ESP. from the Tchuja Valley.

The Lycaenidae comprise three species of *Thecla* FABR., viz. *Th. prunoides* STGR. from the Tscholyschman Valley and from the southwest Altai, *Th. rubi* LINN. from the Tchuja Valley and *Th. frivaldszky* LED. from the Katuna and Tchuja Valleys (1220-1530 m). The genus *Lycaena* FABR. is richly represented. The occurrence of *L. pheretes* HÜBN. at elevations of about 1830-2500 m is of considerable biogeographical importance. The other species so far recorded include *L. fischeri* EV. from the Tchuja Valley, *L. aegon* S.V. from the Tchuja and Bashkaus Valleys and from the Kentei Mountain,

L. argus LINN., *L. lucifera* STGR., *L. optilete* KNOCH., *L. cyane deserticola* ELWES from the Tchuja Steppes, *L. orion* PALL., *L. astrache* BG.-STGR., *L. eros* O., *L. damone altaica* ELWES from the Tchuja and Bashkaus Valleys (1525-1830 m) (the species is also known from the Urals and from the Armenian Mountains), *L. donzelii* B., *L. argoli* ELWES on the mountains to the south of Kuch-Agatch and *L. arion cyanecula* EV. to the south of Lake Teletskoi.

The high altitude insect fauna of the Altai Mountains is essentially boreal and not Middle Asiatic. Except a single species of Middle Asiatic origin, all the species of *Parnassius* LATR. known at present from the Altai Mountains are Siberian elements. One species of *Colias* FABR. is of Middle Asiatic origin, but the rest are either European forms or are closely related to them. Only a single species of *Thecla* FABR. is non-European and only 7 out 34 species of *Lycaena* FABR. are similarly non-European. Thirteen species are common to the Altai and the Pamirs. While only a single species of *Argynnis* FABR. is common to the Altai and the Tien Shan System, none of the species of *Erebia* DALM. are common to the Altai and the Tien Shan. Only a single species, out of about half a dozen, of *Oeneis* HÜBN. from the Altai is of Middle Asiatic origin.

The reader may also refer to the interesting observations on the diurnal moths from the Altai by SHTANDEL (967) and TSCHETVERRIKOV (1051).

13. OTHER MOUNTAINS OF THE ANGAR-BERINGEA

In a recent contribution, KURIENTZOV (667) has given an interesting outline of the principal biogeographical characters of the subalpine and alpine insect fauna of the mountains of the Angaris, Beringea and the Amur area, the Baikal-Mongolia Region and other peripheral mountains, including the Sikhote-Alin and the Kamchatka mountains. Some of these high mountains, like the Sikhote-Alin Ranges, reach elevations of 1400-1500 m. The Pri-Amurskii Range rises to an elevation of 1000-1200 m. KURIENTZOV has discussed the distribution of the typical steppe elements like *Triphysa phryne tscherskii* GR.-GR., *Colias meinos* STGR., *Synchloe callidice orientalis* ALPH.; arcto-alpine elements of the Angaris-centre endemics like *Hyperborea czekanowskii* GR.-GR., *Hyalocoa diaphana* EV., *Brenthis erda* CR., *Colias hyperborea* GR.-GR., *Erebia kozhantchikovi* SHEL., *Pterostichus* (*Cryobius*) *jakobsoni* POPP., *Pt.* (*Cr.*) *poppiusi* SEM.; northeast Siberian *Colias melinae herzi* STGR., *C. viluensis* MÉN., *Argyroploce czekanowskii* KUSN., *Pterostichus* (*Cryobius*) *parviceps* POPP., *Pt.* (*Cr.*) *frigidus* DEJ., *Bembidion breve* MOTSCH. and the high altitude Altai elements like *Erebia kefersteini* EV., *Parnassius eversmanni altaica* VERITY, *Erebia tyndarus sibirica* STGR., *E. kindermanni* STGR., *Oeneis ammon* ELWES, *Colias thiosa* MÉN. etc. The Angaran mountain elements are of considerable biogeographical importance. The Angaran centre of faunal origin and differentiation is very rich. The Pleistocene glaciations, with associated changes in the sea-level, have very profoundly influenced the alpine-zone insects of the Beringean Region. Discussing the different phases of faunal differentiation, on an ecological-biogeographical basis, in the Angaran high altitude insect fauna during the Tertiary, KURIENTZOV traces the oreophily of the insects to the Pliocene, representing the first phase. The origin of the cryo-xerophile character in these insects marks the second phase. During the Pleistocene the Angaran elements moved southward and came to be widely distributed and they even penetrated the high mountain areas of Central and Middle Asia and west North America and gave rise in these areas to centres of endemic species.

i. The Sikhote-Alin Mountains

The Sikhote-Alin Mountains are a nearly meridional range of coastal mountains, stretching almost north-south, from about 51° NL to 44° NL, east of Vladivostock and are apparently part of the Dzhugdzhur Mountains and the Kelyma Mountains further north. The biogeographical importance of the Sikhote-Alin Mountains has been recently emphasized by the researches of KURIENTZOV (664, 665) on the ecology, biology and distribution of the relict species found in the region. There is even now considerable trace of the ancient elements that connected the fauna of Europe and East Asia and came to be broken up by the Pleistocene conditions. There are also affinities of the fauna of the mountains to those of the North American. There are a number of endemic genera like *Grylloblattina* BEJ-BIENKO (99), *Hypsopedes* BEJ-BIENKO, *Eobroscus* KR., *Eonebria* ZSNOIKO, etc. and numerous endemic species on the Sikhote Mountains. The high mountain and Chinese-Himalayan elements on the Sikhote Mountains are typified by *Eolimenitis eximia* MOLT., *Rosalia coelestis* SEM., etc. Ecologically the relict species are represented by local varieties, adapted to diverse environmental conditions. The largest number of relicts are found in the fir-cedar zone, at elevations of 700-1200 m and above on the fir-taiga zone they fall in their abundance.

ii. The Mountains of Kamchatka

The mountains of Kamchatka stretch nearly southwest to northeast, between 52° and 60° NL. The ecological and biogeographical investigations on the montane insects from this region, discussed by KURIENTZOV (666), are of considerable importance. The zoogeography of the Kamchatka region was discussed to some extent earlier by AVERIN (66), but the most important contribution to our knowledge of the insect life of the region was by the Swedish Kamchatka Expedition of 1920-1922 (75, 221, 358, 463, 646, 794).

The principal mountains of Kamchatka discussed by KURIENTZOV are the Central Range, the East Range and the volcanic area. The Central Range has a mean elevation of 1300-1400 m and the highest peaks reach to elevations of about 2000 m. The Ichiskii Volcano rises to an elevation of 3621 m above mean sea-level. The East Range mountains rise to elevations of 1400-2000 m. The volcanic area has a mean elevation of 4850 m. In the east, the mean atmospheric temperature during February is —10.2° C and during August 12° C. The mean atmospheric precipitation is 400-500 mm. In the west, the precipitation is 700 mm; the August mean atmospheric temperature is 10° C and the February mean temperature is —21° C. The maximum atmospheric temperature in the region is 35° C. The mean precipitation of the region is 450 mm. In the southwest, the minimum winter temperature is —20° C and the lowest temperature is —55° C, the maximum temperature is between 10 and 12° C.

The essential features of the vegetation of Kamchatka mountain region have been described by KOMAROV (646). The taiga zone extends up to elevations of 750-1300 m and the montane-tundra zone to elevations of 2000-3000 m. The alpine meadows have *Epilobium* spp., *Aster consanguineus* LDB., *Pedicularis oederi* VAHL., *Primula cuneifolia* LDB., *Oxytropis revoluta* LDB., *Anemone narcissiflora* LINN., *Gentiana glauca* PALL., *Thalictrum alpinum* LINN., *Senecio kamschatkicus* KOM., *Pedicularis verticillata* LINN., etc. Ecologically the fauna is classified as follows: the high mountain alpine-meadow and the montane-tundra. The high altitude insect life above elevations of 2000–2800 m includes the following species:

Coleoptera: *Dyschirius jezoensis* BATES, *Nebria carbonaria* ESCH., *Amara alpina* FABR., *Notiophilus reitteri* SPAETH., *Agonum alpinum* MOTSCH., *Pterostichus* (*Cryobius*) *middendorffi* SAHL., *Pt.* (*Cr.*) *septentrionis* CHAUD. and *Pt.* (*Cr.*) *homalonotus* TCHITSCH.

Lepidoptera: *Brenthis eugeniae* EV., *Br. euphrasyne kamtschadcalis* STGR., *Argynnis pales arsilache* ESP., *Hyperborea czsekanowskii* GR.-GR., *Erebia rossi* CURT., *E. fasciata* BUTL., *Parnassius eversmanni* MÉN., *P. delius kamtschaticus* MÉN., *Euchloë belia orientalis* BREM., *Colias villuensis* MÉN., *Pieris callidice orientalis* ALPH., *Lycaena argus* LINN., *L. optilete sibirica* STGR., *L. orbitulus wosnesenskyi* MÉN., *Cidaria citrata* LINN., *C. kamtschatica* DJAK., *C. incursata derzhavini* DJAK., *C. frigidaria* GEN., *Anarta melanopa* THUNB., *Sympistis melaleuca* THUNB., *Orodemnias quenseli daisetsuzana* MATS., etc.

Diptera: *Tabanus lapponicus* LINN., *Syrphus lapponicus* LINN., etc.

The Orthoptera *Aeropedellus variegatus borealis* MISCH. and *Melanoplus frigidus kamtschatkae* SJÖST. have been also recorded.

iii. The Japanese Mountains

The earliest collections of insects from the mountains of Japan, especially the Mt. Fuji, were made by Abbé DAVID. MATSUMURA later collected a number of species peculiar to the mountains. At elevations above 1220 m, the ground is covered with volcanic ash and there are no plants. In this barren zone, MATSUMURA found some interesting Geometrids like *Elphos latiferaria* WALK. At higher elevations, he reports finding *Calosoma mikado* under stones. Near the summit he found *Argynnis paphia* (LINN.), *A. nerippe* FLD., *Musca corvina* FABR., *M. domestica* LINN., *Leucorrhina fujisana*, etc.

Mt. Atusanupuri is an active volcano, situated near the east side of Lake Kutcharo (510 m at the top and about 200 m above mean sea-level at the foot), with numerous small craters and deposits of sulphur and ash near the summit. The slopes are covered by alpine plants like *Pinus pumila*, *Ledum palustre yezoense*, *Empetrum nigrum*, *Sorbus sambusifolia*, etc., inspite of the low elevation. According to HAYASHIDA (483), the typical alpine ants from Japan are *Formica fusca* LINN., *Camponotus herculeanus sachaliensis* MEYR., *Myrmica kurokii* and *Leptothorax acervorum* FABR. in Hondshu and *Formica fusca* LINN., *F. picea kurokii*, etc., in Hokkaido. The typical alpine species on Mt. Atusanupuri are *Formica fusca* LINN. and *Leptothorax acervorum* FABR., although *Camponotus herculeanus herculeanus* LINN. has hitherto been recorded from only the boreal subalpine regions in North America. The higher levels of the mountain are reported to be dominated by *Formica fusca* LINN.

14. BIOGEOGRAPHICAL AFFINITIES OF THE EUROPEAN AND ANGARAN MOUNTAINS

It is now generally recognized that the fundamental elements of the European fauna are of Angaran origin. Europe has indeed been most appropriately described as a sort of zoogeographical appendage of Asia (641). In marked contrast to Asia, Europe lacks a proper distribution centre for its fauna. The immediate relatives of the genera and species of north and Central Europe occur in Siberia and very often the genera and species are identical. Fennoscandia was connected by land to the ancient land mass and many species appear to have evolved there, so that we find a number of endemic forms in north Europe. The rest of Europe appears to have been composed of islands throughout the Mesozoic Era. The European fauna was, however, enriched by migration from outside. During Jurassic times, North America, Greenland and Iceland are believed to have been connected together and extended eastward as far as Europe and were only separated from Fennoscandia by a narrow gap in the Shetland Strait. Europe was thus closely connected with Asia. The Angaran area is an ancient landmass, with abundant evolutionary elements that developed on an extensive scale. Central Asia, Middle Asia, especially Turkestan and inner Asia contained the reserve of highly differentiated fauna. According to KOLBE (641), here was the centre of distribution of the fauna of Europe. In an earlier chapter, we have referred to the recent works of GROSS (438-440) on the source and origin of the Eurasiatic fauna. The affinities of the Angaran area and of Fennoscandia are also emphasized by him. FRIESE (388) has recently grouped the insects of the mountains of the Palaearctic Realm as follows: In addition to considerable endemic elements, we have i. Boreo-Altaic-alpine elements, ii. Altaic-alpine elements, iii. European elements, iv. Holo-mediterranean elements, v. Atlanto-mediterranean elements, vi. Ponto-mediterranean elements, vii. Euro-mediterranean elements, viii. Euro-Near-Asiatic elements, ix. Euro-west-Asiatic elements, x. Eurasiatic elements, xi. Holarctic elements and xii. circumpolar elements. The Boreo-Altaic-alpine elements are characterized by discontinuous distribution in the north Palaearctic and elevated regions of the Eurasiatic mountain ranges, with different geographical subspecies in the north and south areas. Some of these species are circumpolar and most species occur on the Alps and many of them are also restricted to Central Europe. Some species occur on the Vysoké Tatry and the Carpathians and a few on the Sudeten and Vosges. The Altaic-alpine elements are characterized by their distribution on the Alps, Pyrenees, Apennines, Carpathians, the Altai, Tien Shan and the Urals, but the species are absent in north Europe and north Asia. Some of these species are restricted in Europe to the Alps, but all species occur on the Alps. The European elements are present on the mountains of the Balkan Peninsula and the Iberian Peninsula, both of which are the areas of their maximum distribution. The Holo-mediterranean elements are characterized by the Mediterranean refugium, with only a very limited Post-Pleistocene extension. The northern limit is on the Pyrenees and extends eastward along the southern margin of the Alps to Bulgaria. The Atlanto-mediterranean elements belong to the West Mediterranean refugium as the main centre of radiation and are also present on the mountains of northwest Africa. The Ponto-mediterranean types have their centre in the east Mediterranean region, particularly the Balkan Peninsula and extend to the fringe of the Pyrenees and their northern limit is the southern margin of the Alps. The Euro-Mediterranean and the Euro-Near-Asiatic types are relatively minor constituents. The Euro-West-Asiatic types extend across Near Asia to Middle Asia and Central Asia, nearly up to the Altai Region and West Siberia, also Northwest Himalaya and the Tien Shan Mountains. The Eurasiatic types occur throughout. The Holarctic types occur in the whole of Eurasia, North Africa and North America. The circum-polar elements must not be confused with the boreo-alpine elements; their refugium is believed to be Siberian-Mongolian border mountains, but the species are both montane and moorland forms.

CHAPTER XIV

MOUNTAINS OF THE NORTH TEMPERATE REGIONS OF THE NEW WORLD

The mountains of the north temperate regions of the New World embrace the principal mountains of Canada, the United States of America and of Mexico. The elevated regions of Canada are i. the Eastern Highlands or the Laurentian Highlands (the Labrador Plateau), ii. the Appalachian Highlands and iii. the Western Highlands or the Cordilleran mountain system. The Appalachian Highlands and the Cordillera continue southward into the United States of America. In marked contrast to the great Alpine-Himalayan Mountain Systems, the North American mountains are mostly meridional in their general trend.

While the other branches of entomology have made great advances in America in recent years, our knowledge of the insect life of the North American mountains is, however, extremely fragmentary. The Appalachian Ranges may perhaps be said to be somewhat better explored than the more important and massive and much higher Cordillera. The ecology and insect life of the White Mountains on the Presidential Range in the east are, for example, much better known those of Mt. Rainier in the western highlands.

I. THE APPALACHIAN HIGHLANDS

The Appalachian Highlands in Canada include the area, about 242 000 sq. km, to the east of an imaginary line from Lake Champlain to Quebec and south of the St. Lawrence. Together with Newfoundland, they form the Appalachian Mountain System, with a width of about 240-640 km and length of about 2800 km, extending southwestward, nearly to the Gulf of Mexico. Many of the higher peaks of the Appalachian Mountains, rising to elevations of over 1800 m, are situated in the U.S.A. and the Canadian part of the Appalachian is much lower. The greatest height is attained in a rather discontinuous range, extending in a wide curve from the Canada-United States boundary, northeastward towards Quebec and thence more easterly to Gaspé. The southern part consists of three parallel ranges. The Notre Dame Mountains do not rise to elevations over 600 m above mean sea-level. The Shickshock Mountains in Gaspé rise to an elevation of 1060 m and the Tabletop Mountain is 1220 m above mean sea-level. The Notre Dame and the Shickshock Mountains slope abruptly northwestward, but to the southeast the descent is irregular up to an elevation of about 50 m along the Atlantic coast of Nova Scotia.

Within the U.S.A., the Appalachian Highlands include the Blue Ridge, the Great Smoky Mountain National Park, the Appalachian Ridges from Alabama to the Hudson Valley, the Catskill Mountains, the White Mountains and the Green Mountains, associated with extensive dissected plateaux. The Green Mountains of New England Acadian Region are also known as the Berkshire Hills and rise to an elevation of 1330 m. The Taconic Range west of the Berkshire Hills and the Green

Mountains, north-south ridges, rise to an elevation of 550-610 m. The White Mountains of New Hampshire, especially Mt. Monadnock, are about 965 m above mean sea-level. The Presidential Range is the most picturesque feature of the White Mountains in the central part of New Hampshire. It includes the well known landmarks like Mount Jefferson, Mount Madison, Mount Adams and Mount Monroe. The best known group is, however, Mt. Washington (1918 m above mean sea-level). The southwestern division of the Appalachian rises to an elevation of 1830 m, with the gorges at elevations of 610-900 m. Here is included the Piedmont Upland. From here follow the Older Appalachian Mountains northwards, from the Hudson Valley to Georgia, composed mostly of high metamorphosed Pre-Cambrian rocks. The Smoky Mountain National Park is situated within this area and contains Mt. Mitchell (2045 m), the highest peak in the eastern United States. The Newer Appalachian Mountains, west and northwest of the Older Appalachians, are nearly parallel hills, not higher than 900-1200 m above mean sea-level.

The Appalachian Plateau borders the Newer Appalachian Mountains on the west and northwest; the eastern boundary is formed by the Allegheny Front in the north and the Cumberland Escarpment in the south. The mean elevation of the Appalachian Plateau is 1067 m. Mount Katahdin rises somewhat abruptly 450-900 m from the sloping table-land, 1300-1430 m above mean sea-level and is about 5 km long and 0.75 km wide.

The Adirondack Mountains belong to the Laurentian Upland and consist of crystalline and metamorphic rocks, with the highest peaks rising above the peneplain. Mt. Marcy (1630 m) and Mt. McIntyre (1560 m) are two well known peaks of the Adirondack Mountains. The Appalachian Mountains in Canada were first elevated about the Devonian Epoch and have since then undergone continuous erosion, combined with oscillations in elevation. The Pre-Cambrian is represented in the Appalachians by crystalline and greatly deformed rocks and the Cambrian is represented by shales, slates and sandstones of the Upper Cambrian times. Sedimentation seems to have continued through Ordovician, Silurian and Devonian times. The reader will find interesting accounts on the White Mountains in ANDERSON & MOORE (46), BAILEY (80), BILLINGS (127), LAHEE (675), MATTHES (748), WILLIS (1112) and others.

2. ECOLOGICAL CHARACTERS OF THE WHITE MOUNTAINS

The Mount Washington Range (44° 16′ 15″ NL and 71° 18′ 13″ WL) or the Presidental Range, as it is also called, is primarily a narrow ridge, about 22 km long, extending north-south. With its minor northern and southern peaks, the range should measure nearly 33 km long. The range is deeply dissected by a number of basins and deep round-bottomed valleys, with semi-circular and precipitous headwalls. The important peaks on the Presidential Range, from the north to the south, are the Pine Peak (744 m), the Madison Peak (1640 m), the Adams group of peaks with the Peak John Quincy Adams (1722 m), Adams (1770 m) and Sam Adams (1702 m), the Peak Jefferson (1745 m), the three peaks of Clay, of which the highest is 1687 m above mean sea-level and the Washington Peak (1917 m). Southwestward from Mt. Washington are the peak Monroe (1643 m), peak Franklin (1532 m),

Pleasant (1555 m), Cinton (1303 m), Jackson (1223 m) and Webster (1181 m). The principal basins and cirques (413), which dissect the range, include the Madison Gulf, Jefferson Ravine, Great Gulf, Huntington Ravine, the Tuckerman Ravine and the Gulf of Slides on the east; Oakes Gulf on the south; King Ravine, Cascade Ravine, Castel Ravine, Burt Ravine and the Ammonoosuc Ravine on the north and west.

The Presidential Range is composed largely of metamorphic rocks of sedimentary origin. Alternate layers of Devonian quarzites and schists underlie the higher peaks. The Silurian granulites or narrow calcareous layers occur south of Mt. Washington. The Ordovician beds on the Mt. Franklin, Mt. Monroe and Mt. Clay represent the so-called injection gneiss. The oldest beds, perhaps the Ordovician of volcanic origin, occur on the lower slopes of northern peaks. Granites occupy, however, a very minor place in the main body of the Presidential Range.

The summit of Mt. Washington has been most appropriately described as an "arctic island in the temperate zone". Although situated about the 44th north parallel, the general climatic conditions on the Presidential Range are virtually those that prevail at the 60th north parallel in Labrador, more than 1600 km to the north-east. The winter on the mountain generally corresponds to the climate of west Greenland, on the 70th north parallel. The mountain lies within the isotherm of —3.3° C. The mountain is also characterized by frequent coastal storms, storms from the Great Lakes area and from the Gulf of Mexico and the semi-stationary Newfoundland storms. The duration and severity of the diverse storms are generally greatly accentuated owing to the elevation on the mountain. The isolated location in the middle of a range of north-south trendline has also a barrier-effect on the prevailing westerly winds. Atmospheric condensation is generally heavy and frequent. The mean atmospheric temperature on the summit of Mt. Washington is below the freezing-point from October to April and above this limit from May to September, so that less than half the total precipitation is as snow and dense wet fogs are therefore frequent. Low fogs surround the summit, for example, for nearly 300 days in the year. The mean annual rainfall is nearly twice that of the surrounding lowland country and is maximum during July and minimum during February. The maximum total rainfall for the year 1937 was 2634 mm and the minimum rainfall of 1718 mm was in 1933. The lowest atmospheric temperature so far recorded is —32.7° C in February of 1871 and the highest temperature of 22.2° C was recorded in July of 1881. The annual mean zero isotherm elevation is 1322 m, but the zero isotherm in July should lie at an elevation of 3446 m, in other words fully 1530 m above the summit of the mountain. The result is that the mountain is ice-free during summer. It is particularly interesting to observe that on the Pikes Peak, about 105 km south of Denver on the Western Highlands, the zero isotherms are at elevations of 3200 and 4970 m respectively and the winter is also much less severe than on Mt. Washington. The prevailing winds are from the west and northwest, reaching velocities of over 160 km/h and sometimes even 275 km/h.

The mean atmospheric temperature on the summit corresponds to that at sea-level in Labrador; during July-August it is 8.4° C. The July maximum is 17.7° C and the minimum is —1.1° C. The maximum diurnal range is 11.1° C and the annual fluctuations of the atmospheric temperature correspond closely to those observed in Labrador. The mean monthly atmospheric temperatures on the summit and base of the Presidential Range and in Labrador are summarized in table 40.

TABLE 40

The monthly atmospheric mean temperatures on the Presidential Range and in Labrador

Months	Presidential Range Summit	Presidential Range Base	Labrador
January	−13.5	−8.4	−14.4
February	−14.7	−7.7	−14.7
March	−11.1	2.0	−14.9
April	− 5.3	8.3	−12.9
May	1.8	12.6	− 0.2
June	7.5	16.0	2.6
July	9.6	22.4	7.0
August	9.1	17.2	8.2
September	5.2	15.0	5.1
October	− 1.7	8.0	0.0
November	− 6.6	−1.4	− 4.9
December	−13.1	−8.4	−14.4

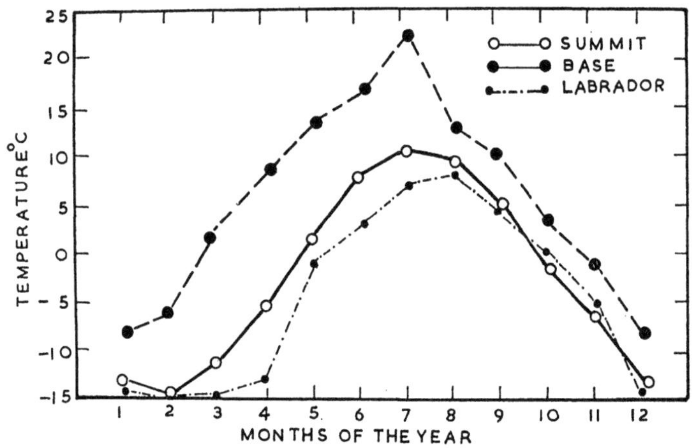

Fig. 75. The mean annual atmospheric temperatures on the Presidental Range and near sea-level in Labrador.

The mean lapse rate of temperature with increase in altitude on the Presidential Range is 5.24° C for every increase of altitude of 1000 m. With the atmospheric temperatures during the winter going down to – 13.7° C, the snow-cover does not exceed 27 cm and must, therefore, be described as relatively scanty. This has the result of ground frost to nearly 0.75–1.3 m or sometimes even 2.0 m depth.

3. ALTITUDINAL BIOTIC ZONES ON THE NORTH AMERICAN MOUNTAINS

The altitudinal biotic zonation on the North American mountains corresponds roughly to that recognized by American biologists, as one proceeds from the north to the south on the lowlands, viz. i. the arctic zone, ii. the Hudsonian zone, iii. the Canadian zone, iv. the Alleghenian zone and the Vancouverian zone, v. the Austroriparian zone and vi. the Sonoran zone. The Hudsonian zone is of the barren ground, south of the bleak arctic regions, with some trees and characterized by severe winters. The summer days are hot and long, the winter days are correspondingly short and cold. Many arctic species of insects occur within the Hudsonian zone also. South of the Hudsonian zone lies the Canadian zone or a broad belt across Canada and the northern parts of the United States, from the Atlantic Coast to the Rocky Mountains. It extends as a narrow tongue southwards along the higher levels of these mountains and on the more eastern slopes of the Cascade Mountains and the Sierra Nevada. This is generally a heavily forested region, chiefly with true firs, *Abies*, *Picea*, *Pinus*, etc. The Canadian zone is characterized by abundant summer rains, lakes, rivers, swamps and rich aquatic fauna. South of the Canadian zone is the Alleghenian zone in the east United States, typically in the Appalachian Region, but extending more or less throughout the eastern parts. The Alleghenian zone is sometimes restricted to the more mountainous area and the term Carolinan zone is applied to the lowlands in the east. The Vancouverian zone corresponds to the Alleghenian zone in the west; it extends from the Aleutian Island and the south-eastern parts of Alaska, southwards along the Pacific Coast and west of the Shelkirk and the Cascade Mountains, through western Columbia, Washington and Oregon and in a narrow belt south along the coast of California to Monterey.

Diverse attempts have been made to define the biotic zones on Mt. Washington from the early days of exploration. The highest open area is generally recognized by botanists as the alpine zone and a subalpine is often recognized beneath it. In 1874 SCUDDER published detailed coloured maps, showing the altitudinal limits of the following zones: i. mountain-forest zone or the Canadian zone, ii. subalpine-scrub or the Hudsonian zone and iii. the alpine (Labrador or the subarctic) zone. The subalpine zone is characterized by dwarf-spruce and includes also the heads of the deepest ravines. The alpine zone is characterized by naked, broken masses of rock, except on the level spots, where sedges conceal them. The alpine zone is subdivided by SCUDDER into an upper alpine and a lower alpine zone. In 1885 HAYWARD extended the Merriam system of biotic zonation, generally followed in America, to the Presidential Range: i. the alpine zone, as restricted to the summits of the higher mountains, corresponds to the arctic-alpine zone of MERRIAM; ii. the subalpine zone confined to the neighbourhood of the timberline corresponds to the Hudsonian zone; iii. the greater part of the region below this, with fauna typical of lowlands of northern New England, represents the Canadian zone and iv. the immediate vicinity of the large river bottoms with fauna identical with that of the eastern Massachusetts, is the transitional zone of MERRIAM. ANTEVS (52) has given detailed descriptions of the habitats, the general environmental conditions and the biotic zones on the Mt. Washington Range. In 1938 DICE combined the transition zone with the Canadian zone, approximately reaching to an elevation of 1220 m, thus limiting the Hudsonian zone to a relatively narrow belt between 1220 and 1525 m. There is an excellent general review of the biological environment of the Presidential Range by ALEXANDER (12, 13). ALEXANDER (12) believes in a somewhat lower border, approximately 916 m above mean sea-level on the east side, between the Hudsonian and Canadian zones on the Presidential Range.

According to the observations of ANTEVS (52), the belt above the continuous scrub-forest on Mt. Washington really represents the alpine zone. An alpine zone can be recognized on the White Mountains of New Hampshire, on Mt. Katahdin of Maine, the Green Mountains of Vermont and on the Adirondack Mountains of north New York. The scrub-forest line and the scrub limit on the White Mountains are interesting in a number of respects. On the White Mountains *Picea rubra* grows up to elevations of nearly 1220 m and just above this limit are tall scrubs of *Abies balsamea* as the dominant species, *Betula papyrifera cordifolia, B. p. minor, B. glandulosa sibirica, Pyrus americana, P. mariana*, etc. (349). The scrub is finally characterized by crooked stems, truncated tops and impenetrable hedge-like growths, up to an altitude of about 1463-1586 m. The forest limit on the White Mountains is thus really a scrubline. On the mountains of the western parts of North America, it is, however, a true treeline. The scrub-forest represents, therefore, the subalpine or the Hudsonian zone.

The upper limit of the continuous scrub is really an irregular zone, ascending higher on the cones and ridges than in the gaps, and is largely influenced by the gradient of the slope, exposure to the prevaling winds, the nature of the debris cover and other complex sets of factors. On the east side of the ridge, between Mt. Franklin and Mt. Monroe, the scrub-forest zone ascends to elevations of 1490-1525 m, but on the west side on Mt. Monroe only to an elevation of 1490 m. On the west side of Mt. Washington Range the scrub-forest extends to elevations of 1464-1525 m. On the Adirondack Mountains of northern New York, the forest-line on Mt. Marcy is just below an elevation of 1465 m, on Mt. McIntyre about 1426 m, on Mt. Wright at about 1295 m. There is therefore no absolute treeline on Mt. Marcy, since trees grow in isolated patches even on the summit. ADAMS *et al.* (6) have summarized the general ecology and altitudinal zonation of life on Mt. Marcy. The transition from the closed forest to the open vegetation marks the commencement of the alpine-zone vegetation of *Empetrum, Rhododendron lapponicum, Diapensia lapponica, Solidago cutleri* and *Vaccinium caespitosum*, at elevations of 1460-1490 m. Above these limits, the Mt. Marcy is subjected to exposure on all sides, rapid drainage, frequent precipitation, etc., so that the alpine-zone vegetation is subjected to destruction by climatic changes, sliding of boulders, local erosions, etc. Local and well established associations may thus disappear and pioneers invade the area, to be followed by climax vegetation.

The higher limits of the forest and scrub on the east side of the White Mountains are correlated with the prevailing west and northwest winds. The alpine zone extends about 15 km along the crestline, from Mt. Franklin to Mt. Madison and varies in width from a few metres to about one kilometre, at elevations between 1460 and 1585 m. Along the entire crest, the zone is recognized by *Loiseleuria procumbens* (LINN.) DESV., *Phyllodoce coerulea* (LINN.) BAB., *Cassiope hypnoides* (LINN.) D.DON., *Diapensia lapponica* LINN., etc. The barren and exposed summits of the ranges have often been described as tundra, but as has been pointed out by ANTEVS, the term tundra should, strictly speaking, be applied to the treeless, wet moorland, with moss and lichen in an arctic climate. ANTEVS has classified the alpine-zone plant formations of the Presidential Range into i. the alpine heath, ii. the heath-meadow, iii. the bog, iv. the bog-meadow and v. the scrub thickets, where one or more of the alpine species dominate, with admixture of other types. The alpine-heaths are further subdivided into i. the *Juncus trifidus-Carex concolor* heath; ii. *Juncus trifidus-Carex concolor-Cladonia*-moss heath; iii. *Carex concolor-Cladonia* heath; iv. *Carex concolor-*

Cladonia-moss heath; v. *Diapensia* heath; vi. *Salix-Carex-Arenaria-Cladonia*-moss heath; vii. *Vaccinium-Carex-Cladonia*-moss heath; and viii. *Cladonia*-moss heath. The heath communities are thus predominantly dwarf shrubs and xerophile rushes, sedges and grasses, with undergrowth of lichen and moss. The heath-meadow, with dwarf shrubs and meadow plants side by side, is intermediate between heath and meadow. The bogs, of heath composition, are distinguished by the predominance of sphagnum mosses. The bog meadows are recognized by the sphagnum and dwarf shrubs being more or less equally developed.

The flora of the alpine zone on the White Mountains includes both arctic and subarctic species of plants (343). About 57 species of arctic plants reach their maximum southern limits on the White Mountains and occasionally some of these species descend here even below the limits of the alpine zone. 48 of the 63 species of alpine plants of the Presidential Range are also known from Europe, 45 from Asia and 44 are common to Eurasia.

4. CHARACTERS OF THE HIGH ALTITUDE INSECT LIFE OF THE APPALACHIAN REGION

Though the Presidential Range was climbed by DARBY FIELD in 1642, the first scientific exploration of the range was undertaken by the Rev. Dr MANASSEH CUTLER-JEREMY BELKNAP in 1784. The early entomological studies on the range initiated by WILLIAM DANDRIGE PECK in 1804 were followed by many explorations by NUTTALL, PICKERING, HARRIS and others. Extensive collections of insects were made by SCUDDER (952) during 1859 and the Cambridge Entomological Club organized a camp on the mountains during the summer of 1874. A little later, MORRISON collected about 20 000 specimens of insects. Most of these collections by the early students consist of Coleoptera and Lepidoptera. During the late eighties of the nineteenth century, Mrs. ANNIE TRUMBULL SLOSSON made extensive collections of insects from the alpine region of Mt. Washington (985). Although the majority of her specimens were collected at elevations above 1525 m, it does not necessarily follow that all the species listed by Mrs. SLOSSON are true alpine-zone forms. Her lists include a great many species that are accidental visitors from lower elevations and also many other species which have been passively blown by wind from the lowlands. Some collections were also made by JOHNSON during 1913-1916. MARION E. SMITH and staff of the Mount Washington Observatory have also made frequent collections of insects from the region.

The alpine-zone insects from Mt. Washington are characterized by the dominance of many typically boreal elements. Some genera are also common to the high mountains of the Palaearctic Realm. The majority of the species are no doubt Nearctic, but some species from the summit area, like for example, *Trechus rubens* FABR. and *Simplocaria metallica* STURM., are also found on the mountains of Central Europe and are often widely distributed in the high northern latitudes. Some of the species represent arctic or Pleistocene relicts and boreal species from the surrounding conifer forests. There are also some boreo-alpine elements. *Brenthis montinus* SCUDDER, *Oeneis semidea* (SAY), *Anarta melanopa* THUNB., *A. schoenherri* ZETT. and *Zubovskya* (=*Podisma*) *glacialis glacialis* (SCUDDER) are boreo-alpine species from the Presidential Range. *Oeneis semidea* (SAY) is restricted to the alpine zone, but *Brenthis montinus*

SCUDDER is subalpine and *Anarta melanopa* THUNB. is circumpolar. *A. schoenherri* ZETT. is also known from Labrador, Greenland, Arctic Asia and Fennoscandia. *Zubovskya glacialis glacialis* (SCUDDER) is the well known apterous White-Mountain locust that occurs among the dwarf *Betula*, but often also near snowbanks at the head of the Tuckerman Ravine at an elevation of 1220 m and on the summit of Mt. Madison, in the alpine zone on Mt. Marcy, the area of Lake Chateaugay on the Adirondacks and in the upper limits of the scrub-forest on Katahdin and also in Labrador.

The alpine-zone Coleoptera have been described by AUSTIN & LE CONTE (65), GARDINER (398), BOWDITCH (144), SHERMAN (962), etc. Almost half the Coleoptera so far recorded from the region are Carabidae, with the Staphylinidae standing next in importance. The following are some of the important species: Carabidae: *Carabus chamissonis* FISCH. from under stones on the summit; *Amara hyperborea* DEJ., the most common carabid on the mountains, from under stones on the summit; *Calosoma frigidum* KB. near the summit; *Notiophilus aquaticus* (LINN.), *N. borealis* HARRIS, *N. sylvaticus* ESCH. and *N. sibiricus* MOTSCH. from under stones; *Nebria sahlbergi* FISCH., *N. suturalis* LEC., *Loricera caerulescens* (LINN.), *Bembidion (Diplocampa) cautum* LEC., *B. (Lopha) mutatum* G. & H., *B. (Notaphus) versicolor* (LEC.), *B. (Peryphus) nitens* LEC., *B. (P.) scopulinum* (KB.), *Amerizus oblongulus* (MANNERH.), *Patrobus septentrionis* DEJ., *Trechus borealis* SCHFFR., *Bothriopterus adstrictus* (ESCH.), *B. luczotti* (DEJ.), *Cryobius hudsonicus* LEC., *Cr. mandibularis* KB., *Celia erratica* STURM., *Curtonotus cylindricus* LEC., *C. brunnipennis* DEJ., *Calathus ingratus* DEJ., *Platynus stygicum* LEC., *Cymindis unicolor* KB., *Harpalus laticeps* LEC., *H. varicornis* LEC. and *Trichocellus cognatus* GYLL.

The Staphylinidae include *Tachinus* spp., *Lesteva pallipes* LEC. on moss, *Acidota quadrata* ZETT., *Quedius fulvicollis* STEPH., etc. The Byrrhidae include such interesting forms like *Simplocaria metallica* STURM., *Byrrhus cyclophorus* KBY., *B. geminatus* LEC., etc. A number of Dytiscidae like *Deronectes griseostriatus* (DEG.), *Agabus congener* PAYK. and Chrysomelidae like *Pytodecta pallida* (LINN.) and *Crepidodera robusta* LEC. are also known from the alpine zone on Mt. Washington.

The Lepidoptera from the alpine zone on Mt. Washington include the Satyridae like *Oeneis semidea* (SAY), the Geometrid *Eupithecia cretaceata* PACKARD and a number of Noctuidae like *Anarta quieta* (HÜBN.), *A. melanopa* THUNB., etc.

Although Diptera are abundant, we do not know what proportion of the species so far recorded really belong to the alpine-zone. The following species may, however, be mentioned: *Tipula (Nitidotipula) pachyrhinoides* ALEX. mostly from the alpine, but often also from the Hudsonian zone. *T. (Oreomyza) insignifica* ALEX. at elevations of 1525-1675 m from Mt. Washington and from Mt. Katahdin. *T. (O.) nebulipennis* ALEX. common at elevations of 1630 m on the Presidential Range and also known from Gaspé and Labrador. *Nephrotoma penumbra* ALEX. from the Presidential Range and Katahdin. *Limonia (Dicranomyia) halterella* (EDW.) at an elevation of 1525 m on the Presidential Range and known earlier from Scotland also. *Dicranota (Plectromyia) petiolata* (ALEX.) at an elevation of 1525 m on the Presidential Range and recorded also from Gaspé and the higher Rockies of Colorado. *Phyllolabis lagganensis* ALEX. at an elevation of 1280 m on Mt. Madison and from the Canadian Rockies, a low alpine-zone form. In addition to Tipulidae, the other dominant Diptera are Anthomyiidae, Empididae, Dolichopodidae, etc.

Mention should be made of the following Orthoptera from the White Mountains: *Mecostethus gracilis* (SCUDDER), *Melanoplus borealis* (FIEBER), *Acrydium granulatum incurvatum* HANCOCK, *A. hancocki* MORSE, the apterous White Mountain locust

Zubovskya glacialis glacialis (SCUDDER) (typically restricted to the high Hudsonian zone and also widely distributed on the White Mountains), *Nemotettix cristatus cristatus* (SCUDDER) restricted to high elevations in the Appalachian Region and said to become intermediate between *N. cr. cristatus* and *N. cr. compressus* MORSE southwards of the region and *N. femoratus* (SCUDDER) occurring up to an elevation of 1675 m on the southern Appalachians and known also from Mt. Roan in North Carolina.

A small number of Plecoptera from the White Mountains include *Dictyopterygella washingtoniana* HANSON, *Perlodes margarita* ALEXANDER from the foot of the glacier in the Tuckerman Ravine and the Hudsonian-zone species *P. slossonae* BANKS.

Among the other terrestrial arthropods from the Appalachian Region mention may be made of the spiders *Lycosa albohastata* EMERTON, *L. pictilis* EMERTON, *Pardosa glacialis* THORELL, *P. groenlandica* THORELL, *P. luteola* EMERTON, *P. muscicola* EMERTON, *P. uncata* THORELL, *P. tachypoda* THORELL, *Araneus carbonarius* (L. KOCH) and *Cochlembolus alpinus* BANKS, found at the highest elevations. The species characteristic of the Hudsonian zone are *Gnaphosa brumalis* THORELL, *Theridion montanum* EMERTON, *Tmeticus bidentatus* EMERTON, *Tm. montanus* EMERTON, *Zilla montana* C. KOCH and *Pardosa* sp.

5. THE WESTERN HIGHLANDS

The western side of North America is buttressed by a system of mountains that constitute the Cordillera, far more extensive and imposing than the Appalachians on the Atlantic Coast (341, 342). The cordilleran system extends, with a width of 640-1600 km, from the extreme northwest corner of Alaska to the Isthmus of Panama, a distance of over 6900 km. The Appalachian Ranges had long been elevated and were already in an advanced stage of demolition, when the uplift of the Cordilleran Mountains commenced. The orogenic movements that led to the uplift of the Cordilleran System have continued right to the present time. The Cordilleran Mountains are, therefore, loftier and far more rugged than the Appalachian hills.

The mountains of western North America are part perhaps of the longest continuous highland region on the Earth. Starting from the far northwest Alaska, practically within sight of Siberia, the mountain ranges extend south and east along the Pacific Coast. They cover all of the southern and eastern parts of Alaska and vast areas of western Canada, broadening southward. The mountains then continue through the U.S.A., Mexico and Central America to Panama and thence to merge with the northernmost spurs of the Andes to continue along the whole length of South America. The whole highland region is almost 16 000 km long; Mt. McKinley in Alaska and Sarmiento beyond the Magellan Strait are literally half a world apart, but belong nevertheless to the same highland system. The mountains of Alaska are the highest of the Cordilleran System and they are true arctic mountains; their rocky masses lie buried under enormous masses of ice and snow. The Canadian Cordillera comprise two, nearly parallel, principal mountain systems, separated by a linear belt of lower mountains and rugged plateaux. There are from the east to the west: i. the Canadian Rockies, ii. the Interior Plateau and iii. the Canadian Coast Mountains. The Western Mountains comprise the Coast Mountains to Mt. St. Elias (5490 m) in Alaska, the Middle Range consisting of the Gold Mountains and the Eastern Mountains form the Rockey Mountains. These constitute together the western highlands of Canada, about 600 km broad and with numerous ridges that rise to elevations of 3050 m and in between the mountains there is the plateau of British Columbia.

The Rocky Mountains begin south of the Canada-United States border and extend to the Arctic Ocean, in a series of overlapping ranges that become progressively wider and lower, generally 80-160 km wide and rising to elevations of 3050-3660 m. North of the R. Liard, the Mackenzie Mountains overlap the Rockies on the east, thence northward and westward between the Yukon and Mackenzie rivers. The Franklin Mountains are an overlapping range, not over 1525 m in elevation. These mountains descend in the east through a series of foothills that form a belt 16-20 km wide. The Coast System of mountains likewise comprises several ranges. The Coast Range is 160 km wide and borders the Pacific Coast from the United States frontier, to nearly 1500 km to Skagway and thence turns inland. Though older and thus much less rugged than the Rockies, these mountains rise 1500-1900 m directly out of the sea and 2400-2750 m further inland. Part of these mountains are at present submerged in the Pacific Ocean. The Purcell and Selkirk Mountains, east of the Coast Range and adjacent to the Rockies, are believed to have been formed at the same time as the Coast Range. Between the Coast Range and the Rockies there is an irregular belt of 240-320 km and at elevations of 900-1500 m above mean sea-level, representing the Interior Plateau, broken by valleys, 300-900 m deep.

The term Rocky Mountains is generally applied to the eastern chain of the Cordillera within Canada and the United States, but it seems appropriate to include the entire chain from the northernmost parts of Alaska to the southern Mexico. The Brooks Range of Alaska in the northern margin of the Yukon Territory, extending eastward into Canada, is also part of the Rocky Mountains. In the Pacific Coast area, the Sierra Nevada is the eastern belt and the Coast Ranges are the western belt. The Cascade Mountains are east in Oregon and Washington and the Olympic Mountains and the Coast Range are west. The western Sierra Madre forms the eastern boundary in Mexico. The Sierra Nevada has many well known landmarks like Mt. Whitney (4422 m), the highest peak. The Sierra Nevada is 650 km long, 80-130 km wide and gently tilted westward. The famous Yosemite Valley is situated in the Sierra Nevada. The Cascade Range is west of the Columbia Basin and north of the Sierra Nevada. The range is remarkable for the large number of volcanoes, of which there are at least 120. The Lassen Peak is indeed a volcano that rises to an elevation of 3180 m. Mt. Shasta (4316 m) in the southern Cascade Range has perennial snow and glaciers; the Whitney Glacier descends to an elevation of 2896 m. Mt. Mazma is situated in the middle of the south Cascade Range and is famous for its Crater Lake of Oregon. The northern Cascade contains Mt. Rainier (4390 m), covered with ice that descends to an elevation of 1830 m (927). The Glacier Peak is about 160 km north of Mt. Rainier; it is indeed a volcano (3180 m) that has 50 glaciers. Mt. Baker (3300 m) is an isolated volcanic cone in the northwest corner of the Cascade Range. The northern Cascade Range has hundreds of small glaciers. Mt. Whitney is in the neighbourhood of the well known Death Valley (44.1 m below mean sea-level). Unlike in the case of the Canadian Rockies, we do not find a single unbroken chain of mountains in the United States, but a complex of many chains and ranges, often widely separated by great rolling plateaux. The northern Rockies form the mountains of Montana and North Idaho, where we find the famous Glacier National Park. The Beartooth Mountains are northeast of the Yellowstone Park in the west corner of Wyoming. South of this geyser basin are the Teton Mts. The Colorado Mountains are the highest and the most extensive of the system, with the peaks rising to an elevation of 4200 m; Mt. Elbert rises to an elevation of 4395 m.

The Cordilleran Plateaux include the Mexican Plateau from the Sierra de Sur northward into New Mexico and Arizona, to the margin of the Colorado Plateau, bounded on the east by the Sierra Madre Oriental and on the west by the Sierra Madre Occidental. The Colorado Basin contains the Grand Canyon, the Craters of the Moon are in the Snake River Plains. The Utah Plateau also forms a part of the system.

The sedimentation has coincided since the earliest times with the present mountain systems. The Selkirk, Purcell and the Rocky Mountains contain an old series of conglomerates, sandstones, argillites and impure limestones, often 9750 m thick, known as the Belt Series. The Belt Series appear to have been continuously laid in shallow waters. The thickening and coarsening of the deposits in the west are assumed to have been derived from land. The Belt Series seem to have been formed about the close of the Pre-Cambrian times and to represent the time of commencement of sedimentation that continued without interruption to the close of the Palaeozoic Era. Progressive submergence is also indicated by the gradual and conformable overlap of the Cambrian to Devonian beds over the Belt Series areas in the Purcell Mountains. The region was submerged throughout the Carboniferous Epoch. Permian and Triassic times saw perhaps the uplift and igneous activity. With the uplift of the Coast Range, most of the Cordilleran Mountains in Canada were also raised. In the early Eocene there was again renewed activity of folding, faulting and uplift of the Palaeozoic and Mesozoic strata in the eastern part of the Cordillera, to give rise to the Canadian Rocky Mountains. Compressive movements were confined largely to the Rocky, Mackenzie and Franklin Mountains. The Rocky Mountains are believed by geologists to have been raised by at least four separate and marked periods of upheavals, of which the first was from the close of the Carboniferous, the second was at the close of the Triassic, the third about the close of the Cretaceous and the fourth during the Tertiary. During and since the mountain-building movements, the Cordilleran Mountains have been subjected to vigorous and continuous erosion.

Three main centres of Pleistocene glaciations are of importance, viz. the Cordilleran ice-sheet covering the area of the western mountains, the Keewatin sheet on the plains of the west Hudson Bay and the Labrador sheet. The Labrador-Keewatin ice-sheet is estimated to have covered an area of some 9 000 000 sq. km to a thickness of about 2400 m; the ice thus had a volume of nearly 21 600 000 cu. km. The Cordilleran ice cap was over 800 m thick and extended to nearly 2 500 000 sq. km, amounting to a mass of 2 000 000 cu. km (51, 350).

6. ECOLOGICAL CHARACTERS OF THE CORDILLERAN MOUNTAINS

The general ecological characters of the Cordilleran Mountains depend to a large extent on their meridional trend, on the massiveness of the mountains and their proximity to the Pacific Coast and the Pleistocene glaciations. The conditions on the coastal ranges differ from those on the interior mountains. The general ecological conditions in the northern Cordillera approach more and more those of subarctic and arctic mountains. While the ecological peculiarities of the Appalachian Highlands have been more or less extensively studied by a number of workers, our knowledge of the high altitude ecology of the Cordilleran mountains is limited to the observations of FLETT (349), JONES (599), TAYLOR (1031), TAYLOR & SHAW (1032) on Mt. Rainier. JONES has also made valuable observations on the Olympic Mountains. Mt. Rainier is a massive eminence of basaltic material, characterized by humid climate, extremely heavy snowfall and with numerous glaciers. The climatic altitudinal biotic zones on Mt. Rainier are i. a transitional zone (600-900 m), ii. the Canadian zone (900-1370 m), iii. the Hudsonian zone (1370-1830 m) and iv. the alpine zone up to the summit (4437 m). The transitional zone ascends up to 1066 m on the western side of the mountain and sometimes even to 1220 m. The altitudinal limit of the Canadian zone often also extends up to 1525 m and the Hudsonian zone reaches to 1980 m. The Hudsonian zone on Mt. Rainier is characterized by the climax community of *Abies-Tsuga-Chamaecyparis* and the Canadian zone by *Tsuga-Abies-Pinus* community. The principal plant and animals habitats on Mt. Rainier have been described in considerable detail by a number of workers and are closely correlated with the conditions of atmospheric moisture. The dominant plants of the arctic-alpine zone, snow and ice-covered upper slopes of the mountains are *Arenaria, Castillejo, Erigeron, Empetrum, Erigonum, Gilia, Oreostemma, Pedicularis, Pentastemon, Saxifraga*, etc. The timberline on Mt. Rainier is at an elevation of 2316 m, but trees seldom actually grow to this altitude. It has been assumed by a number of workers that above an elevation of 3050 m, practically life ceases on Mt. Rainier. The timberline on the Rocky Mountains, at a latitude of 56° NL, is 1220 m. In Montana the timberline is 2900 m, but 3700 m at 39° NL. On the Olympic Mountains, the transitional zone extends from sea-level to an elevation of about 450 m, the Canadian zone from 450 m to 900 m, the Hudsonian zone from 1000 to 1525 m, often with tongues descending to even 760 m and the arctic-alpine zone extends from 1525 to 2130 m.

7. INSECT LIFE OF THE CORDILLERAN MOUNTAINS

Some of the characters of the insect life on the Cordilleran Mountains are correlated to the peculiarities of the forest-line. The relatively high timberline of the Cordillera results in the generally higher altitudinal limits of the subalpine insects. The zone between elevations of 2750 and 3350 m on the Colorado Mountains approximates the subalpine conditions on the Central European mountains. According to CARPENTER (189), the insect life on the Colorado Mountains is very rich at an elevation of about 3350 m, a little above the timberline, but the abundance of species falls off a few hundred metres higher and becomes sparse at an elevation of about 4260 m.

A number of interesting grasshoppers occur at high elevations on the Cordilleran mountains. GORDON ALEXANDER (21) has published an interesting account of the high altitude Orthoptera from the Colorado Mountains. He collected the following typically alpine-zone forms, mostly at elevations of 1525-3960 m: Acridinae: *Aeropedellus clavatus* (THOMAS). Oedipodinae: *Xanthippus corallipes altivolus* (SCUDDER), Cyrtacanthacrinae: *Melanoplus alpinus* SCUDDER, *M. borealis monticola* SCUDDER. *M. bruneri* SCUDDER, *M. dodgei dodgei* (THOMAS), *M. d. bohemani* (STÅL), *M. fasciatus* (F. WALK) (?), *M. kennicotti nubicola* (SCUDDER), *M. spretus* (WALSH) (now extinct) and

M. oregonensis marshallii (THOMAS). A number of species that really belong to much lower elevations and often found in the alpine zone on the Colorado Mountains, represent the accidental elements carried by strong updraft air currents from below the forest-line. The majority of the true alpine-zone forms are typically brachypterous. *Aeropedellus clavatus* (THOMAS) has the widest latitudinal and altitudinal range among the alpine-zone grasshoppers of the Cordillera and occurs from elevations of about 1525 to 4115 m in the Boulder Region in Colorado. *Melanoplus dodgei dodgei* (THOMAS) is a small-sized subspecies in northern Colorado and the larger subspecies *M. d. bohemani* (STÅL) is characteristic of the southern Colorado Mountains. *M. oregonensis marshallii* (THOMAS) is a brachypterous subspecies from the Mosquito Range and the Loveland Pass and from below the timberline north of this pass. *M. alpinus* SCUDDER from the northern Colorado Mountains, *M. borealis monticola* SCUDDER from the southern Colorado Mountains and *Xanthippus corallipes altivolus* (SCUDDER) from the north and south Colorado Mountains have somewhat reduced wings. In the female, the wings are sometimes too short to be of much use in flight. *Melanoplus bruneri* SCUDDER is macropterous. *Xanthippus* overwinters as juveniles and reappears as adults before midsummer, at elevations of 3350 m. *Aeropedellus clavatus* (THOMAS) and *Melanoplus dodgei dodgei* (THOMAS) appear as adults during the middle of July. The macropterous *M. alpinus* SCUDDER and *M. borealis monticola* SCUDDER appear as adults about the middle of August, by which time the adults of *Xanthippus corallipes altivolus* (SCUDDER) have already disappeared.

ALEXANDER (22) records finding *Pedioscirtetes nevadensis* THOMAS and *Capnobotes fuliginosus* (THOMAS) at elevations of 1525-2618 m, near the Yampa Canyon, on Colorado Mountains. The following table summarizes the abundance of species of high altitude Orthoptera, at different altitude zones, on the Cordillera:

TABLE 41

Hypsometric analysis of Orthoptera on the Colorado Mts.

Altitude in m	Absolute No. of species	Percentage of species At upper limit	At lower limit
1525-1830	65	30.7	—
1830-2130	50	36.7	10.0
2130-2440	32	12.5	0.0
2440-2750	27	29.5	3.7
2750-3050	19	26.3	0.0
3050-3350	15	46.7	6.7
3350-3660	8	25.0	0.0
over 3660	6	—	0.0

According to the observations of REHN & GRANT (869), the alpine-zone forms of Orthoptera occur up to a maximum elevation of 4267 m on the Cordillera.

Tetrix bruneri BOLIVAR extends from Alaska and Canada southwards along the Cascade Mountains and on the Mountains of Montana, Wyoming, Colorado, Utah and Idaho. *Paratettix femoratus* (SCUDDER) is reported to occur up to an elevation of 2280 m near the Californian Death Valley. *Psychomastax robusta* HEBARD occurs on the southwestern Sierra Nevada, around Mt. Whitney. The other interesting species recorded by REHN & GRANT include *Psychomastax psylla psylla* REHN & HEBARD, at elevations from 1160 to 2530 m, in the Upper Sonoran and transitional zones; *Eumorsea balli* HEBARD at an elevation of about 2745 m on Mt. Santa Rita; the predominantly petrophile *Phrynotettix tschivavensis pusillus* REHN & GRANT from the east-central elevated areas of Arizona and south-central Mexico, often as low as 1830-1950 m and as high as 2580 m and *Phr. robustus* (BRUNER) at elevations of 2430-3050 m on Mt. Popocatepetl in Mexico.

Mention should be made here of the remarkable alpine rock-crawler *Grylloblatta campodeiformis campodeiformis* WALKER, found in moist localities, under deeply sunk stones, at elevations of 1700-3050 m, on the summit of the McDonald Peak, the Holland Peak on the Swan Range and other localities (201, 776, 838).

Many high altitude Coleoptera from the Cordillera belong to genera characteristic of high mountains of Asia and Europe. We have, for example, *Tachypachys* MOTSCH., *Notiophilus* DUM., *Nebria* LATR., *Bembidion* LATR., *Platynus* BON., *Harpalus* LATR., *Hydroporus* CLARIV., *Helophorus* FABR., *Stenus* LATR., *Geodromicus* REDTB., *Lesteva* LATR., *Olophrum* ERICHS., *Deliphrum* ERICHS., *Orobanus* LEC., *Anthobium* LEACH, etc. Among the Coleoptera occurring at elevations above 3960 m on the Colorado Mountains are *Carabus taedatus* FABR., *Tachypachys inermis* MOTSCH., *Notiophilus hardyi* PUTZ., *Cymindis unicolor* KIRBY, *Pterostichus surgens* LEC. (2750-3965 m), *Amara gibba* LEC., *A.* (*Curtonotus*) *cylindrica* LEC., *Aphodius terminalis* SAY, *Podabrus lateralis* LEC. (2900-3965 m), *Collops hirtellus* LEC., *Acmaeops atra* LEC., *Chrysomela montivagans* LEC. (3350-3960 m), *Adimonia externa* SAY, *Epicauta pruinosa* LEC. and *Lepyrus gemellus* KIRBY. *Pterostichus surgens* LEC. is closely related to *Pt. fatuus* LEC. from Alaska. *Geodromicus ovipennis* LEC. occurring at elevations of 2750-3050 m on the Colorado Mountains is believed by competent authorities to correspond to the European *G. plagiatus* F. Many of these species occur within the forest-zone, often immediately below the timberline and it is not always possible at the present time to give anything like a complete list of the exclusively alpine-zone Coleoptera from the Rockies. It may, however, be remarked that *Carabus* LINN., which is widely distributed both in the subalpine and alpine zones on European mountains, is rather poorly represented on the Cordillera. We do not indeed know of any typical mountain autochthonous Nearctic species of the genus at present. The tribe Cychrini is, however, represented by several species, both in the subalpine and alpine zones of the North American mountains, but there are perhaps no true alpine-zone species of *Cychrus* FABR. in the region. Relatively few species of *Trechus* CLAIRV. are known at present and none of them also seems to be an exclusively high-alpine zone form. The total absence of *Otiorrhynchus* GERM. on the Cordilleran mountains and on the Appalachian Region is of considerable biogeographical and ecological significance, particularly in view of the fact that this genus is commonly found in the North American lowlands, where a number of species have been recently introduced by human agency. The Chrysomelid genus *Orina* MOTSCH. also is not represented on these mountains.

The high altitude Lepidoptera from the Cordilleran System include *Parnassius* LATR., *Colias* FABR., *Oeneis* HÜBN., *Erebia* DALM., *Argynnis* FABR., etc., in other words, genera which are widely distributed on the mountains of Eurasia. While *Parnassius* LATR. occurs in Alaska up to the Arctic Circle, in the rest of North America, the genus is found only on the Cordilleran mountains, southwards up to the 35th north parallel and ascends here usually to an elevation of about 4000 m. Even *Erebia* DALM. and *Oeneis* HÜBN. are found at high elevations only. *Erebia epipsodea* BUTL. occurs, for example, at an elevation of 3660 m on the Colorado Mountains, *E. magdalena* STECK. at elevations of 3050-4270 m also on the Colorado Mountains, *Oeneis beani* ELWES at 2440-2750 m in Alberta (Canada) and *Oe. semidea* (SAY) and *Oe. brucei* EDW. at elevations of 3660-4270 m on the Colorado Mountains. The distribution of *Oe. bore* ESP. (*Oe. taygete* HÜBN.), recently discussed by DOS PASSOS (281), is of considerable interest. *Oe. bore gaspeensis* DOS PASSOS occurs at an elevation of 915 m on the Canadian Rockies, but the Colorado Mountains appear to have perhaps the largest populations of *Oeneis* HÜBN. in North America. *Oe. bore edwardsi* DOS PASSOS occurs on the mountains of Montana and Colorado. Further lists of the Lepidoptera from the Yellowstone Park and from other regions of the Rocky Mountains may be found in MCDUNNOUGH (753), REAKIRT (861) and GEDDES (400). CARPENTER (189) found *Parnassius smitheus* DOUBL. and *Oeneis semidea* (SAY) to be the most abundant species on the Colorado Mountains. *Arctia quenseli* PAYK., also known from Mt. Washington on the Appalachians and from the Alps, is widely distributed on the Colorado Mountains. CARPENTER found several other species like *Pieris occidentalis* REAKIRT, *Argynnis freya* ESP., *Melitaea nubigena* EDW., *Agrotis islandica* STAUD., etc. at an elevation of about 3660 m. PACKARD (808) has discussed the zoogeographical significance of the Lepidoptera, collected by CARPENTER from the alpine zone of the Colorado Mountains. He points out that *Oeneis semidea* (SAY), *Arctia quenseli* PAYK., *Anarta quadrilunata* GROTE and *A. subfuscata* GROTE, *Agrotis islandica* STAUD. and *Pieris hochenwarthi* HOCHW., occurring in the alpine zone of the Colorado Mountains, are also common to Labrador and *Agrotis islandica* STAUD. occurs further in Greenland and Iceland also. The second, third and the last mentioned species occur also on the Alps. The subalpine and circumpolar forms that occur at elevations of 2133-2750 m on the Colorado Mountains include *Larentia caesiata* S.V., *Cidaria populata* LINN., *C. lugubrata* MOESCH., *C. testata* LINN., *Melanippe hastata* (LINN.), *M. tristata* (LINN.), etc. All these species are known from the high mountains of Asia and most of them occur also in Labrador. *Larentia caesiata* S.V., *Cidaria populata* LINN. and *Coremia ferrugata* (STGR.) are also known from Mt. Washington. *Cidaria populata* (LINN.) was collected by CARPENTER at elevations of 3660-3960 m on the Berthoud's Pass in Colorado, *C. lugubrata* MOESCH. and *C. testata* LINN. occur at elevations of 2440-2600 m and the latter species is also known from the Altai and Ural Mountains. *Melanippe tristata* (LINN.), found at an elevation of 2750 m on the Colorado Mountains, is also known from the Urals and from south European mountains. *Cidaria lugubrata* MOESCH. from the Colorado Mountains is larger than those from Alaska and nearer the European forms than those from Labrador and New England. *Arctia quenseli* PAYK., occurring in the alpine zone of the Colorado Mountains, is also known from an elevation of 2440 m on the Alps and is widely distributed in Labrador and on the summits of Mt. Washington and Mt. Madison.

There are interesting records of Diptera at elevations of 3660 m on the Colorado Mountains. We have here such well known types like *Bibio* GEOFFR., *Simulium* LATR., *Plecia longiceps* LÖW, *Hesperinus brevifrons* WALK. (also known from Mt. Washington), *Bibiocephala grandis* OST.-SACK., *Exoprosopa decora* LÖW, *Anthrax* SCOP., *Systœchus* spp., *Odontomyia nigrirostris* LÖW, *Sericomyia militaris* WALK., *Helophilus bilineatus* CURTIS (also from Labrador), and *Chrysotoxum derivatum* WALK. ALEXANDER (13) has recorded a number of interesting subalpine-zone Tipulidae from the Rocky Mountains; most of these species were collected at elevations of 1670-1700 m, but represent Canadian-zone forms.

8. BIOGEOGRAPHICAL CHARACTERS OF THE NORTH AMERICAN MOUNTAINS

VAN DYKE (1079) has discussed some aspects of the biogeographical characters of the North American mountains. The main component elements of the fauna of North America have been derived from Asia. The fundamental elements of the present day fauna have not recently evolved, but represent rather an old element which developed to a great extent on the highlands of Middle Asia and spread to the present territory, probably during the Pleistocene (4, 442). The species have naturally become greatly modified in diverse ways in these frigid areas and are characterized by reduction of wings and concomitant modifications of body, size, etc. ROSS (897, 899) has also shown that the Nearctic fauna in general has had a complex origin and includes many endemic types and colonizations from Asia at different times and perhaps also from Europe and South America. The complexity of the distributional pattern of insects suggests that the system of a few time-faunas, as used by zoogeographers, is an over-simplification. He believes that the Pleistocene conditions did not really cause wholesale migrations and exterminations of the insect life of the temperate regions, as is generally assumed by many workers. The chief effects of the Pleistocene glaciations on the North American insects seem to have been i. multiplication of some species due to diverse processes of speciation, ii. reduction in the area of eastern communities, but elimination of none and this reduction has progressed in an orderly fashion and iii. the spread of some groups southwestward into the Ozarkian uplift.

It may also be mentioned at this stage that the Cordilleran mountains have served as routes for the distribution of typically boreal elements of plants and insects to the tropical American and South American Andes.

CHAPTER XV

SUBARCTIC AND ARCTIC MOUNTAINS

The principal high mountains of the subarctic and arctic regions, so far explored by entomologists, include the mountain systems of Fennoscandia, the Kolyma Mountains and the mountains of the Tai Myr Peninsula (75° NL) in North Siberia, the Anadyr Mountains a little to the south of 70° NL (between 170° and 180° EL) west of the Bering Strait (part of the Beringean Region was already discussed in chapter XIII), the mountains ranges of the Alaska-Yukon Territory, parts of British Columbia, Labrador; Mt. Peterman Peak (3353 m above mean sea-level, 73° NL and 30° WL) and Mt. Rigny (2399 m) south of 70° NL on the east coast of Greenland. Mention may also be made of the arctic mountains of the Queen Elizabeth Islands of the Canadian Arctic Archipelago (north of 74° NL to nearly 83° 7′ NL), within 750 kilometres of the North Pole (164). Some of these mountains in the Archipelago attain elevations of 3000 m in the east, in the northern Ellesmere Is. and Central Axel Heiberg Is. Some mountains in the Melville Is. are only 900 m above mean sea-level. The extreme northern mountains of northern Scotland, north of the 58th north parallel, are almost on the fringe of the subarctic mountain system. The northern parts of the Ural Mountains are in the subarctic and arctic regions. FREEMAN (381) has recently given a brief historical resumé of the entomological explorations in the subarctic and arctic regions of Canada, including Alaska. The same author has also recently discussed the general distributional peculiarities of the subarctic and arctic butterflies.

1. FENNOSCANDIAN MOUNTAINS

As a biogeographical term, Fennoscandia refers to Scandinavia, Finland, the Russian part of Karelia and Lappland (east of the R. Oenga and R. Wyg). The Scandinavian mountains stretch nearly north-south, between 60° and 70° NL and about half the mountain system lies within the Arctic Circle (fig. 76). The mountains of Lappland have, however, a nearly east-west trend. The higher mountain ranges are near the Atlantic coast of Norway.

The bulk of the Scandinavian mountains are situated in Norway. The southern part of the mountainous ridge constitutes the Langfjellene or the Long Mountains of Norway and divides the country into the Östlandet and the Vestlandet. In the Jotunheimen Massif, with the highest peak Galdhøpiggen rising to 2468 m above mean sea-level, the ridge turns more east-westward and is known as the Dovrefjell (=Dovre Mountain). These mountains separate the Östlandet from Trøndelag in the north. The ridge is broken by the Trøndelag-Jämtland Gap, but is resumed as the Kjölen Range in the north. In its northern reaches, the main range of Kjölen extends through the Swedish Territory and reaches its maximum elevation of 2120 m. The highest mountain in Finland is Haltiatunturi (1324 m above mean sea-level).

The Scandinavian mountains exhibit a characteristic weather-divide effect. The maritime west coast is separated by them from the more continental eastern interior.

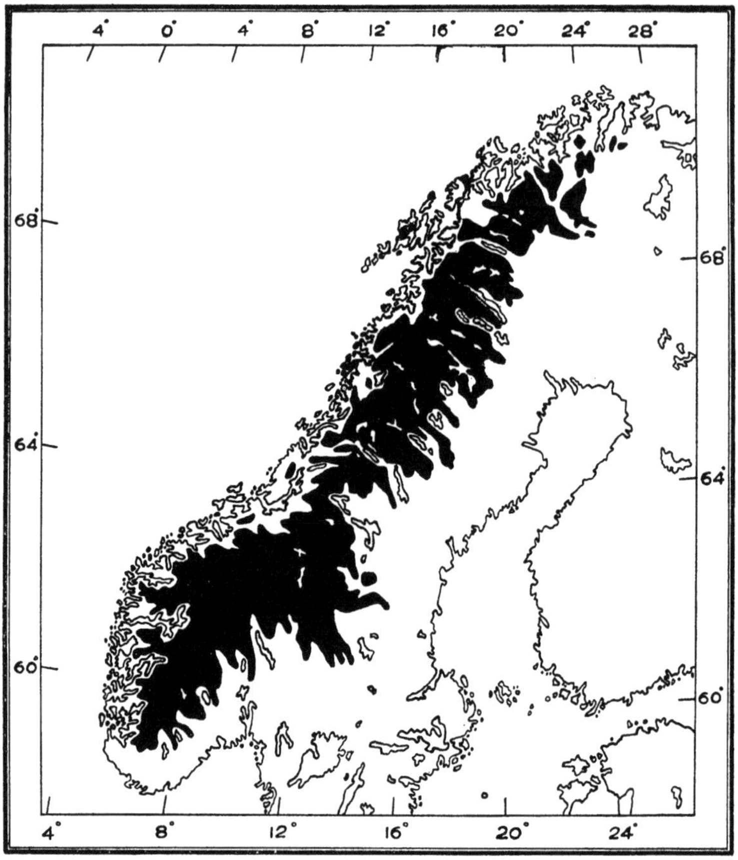

Fig. 76. Sketch-map of Fennoscandia, showing the principal elevated areas.

The atmospheric winter temperatures on the mountains are considerably low, so that deep and stable snowfall occurs for more than six to seven months annually. The mean annual precipitation on the western slopes is about 2000 mm, but more than three times this amount in the higher parts of the Vestlandet. East of the mountains, the annual mean rainfall is much below 2000 mm. The summers are warmer and sunnier in the west than in the east. In the south, the upper limit of coniferous forest is at elevations of 600 m, but in the far north, the limit is reached even at an elevation of 360 m. The *Betula*-scrub zone rises to an elevation 240 m higher up. The greater part of the mountains lie, therefore, within the alpine zone. Though the elevation of the Scandinavian mountains is not impressive in comparison to those of the Alps or the Himalaya or even of the American Cordillera, they nevertheless constitute important barriers.

The main highland mass of Scandinavia is a much eroded plateau, known as Kjölen (mean elevation about 450 m), extending from southwest Norway to the North Cape and perhaps also to Spitzbergen. The ridge-like character is, however, most marked about the 63rd north parallel. The high plateau consists of Palaeozoic rocks, representing the oldest structural elements in Europe. The Archaian zone extends from Bergen to Hammerfest, interrupted in the north by the overlying Palaeozoic beds. Gneiss predominates in most places, but other types of crystalline rocks also occur. East of an imaginary line from Nase to the North Cape, the Cambrian deposits are not also overloaded, overthrust or metamorphosed, but west of this line, there is considerable folding and metamorphosis. Near the Finnish border, the peaks rise to elevations of 1230 m; some of the better known higher peaks include Jaeggevarre (1845 m) between Lyngen and Ulls Fjords, Kistefjeld (1723) m and Sulitjelmo (1914 m) on the Norway-Sweden frontier. Nearer the coast and almost on the Arctic Circle, is the great Svartisen Icefield, through which the Snetind Nunatak (1600 m above mean sea-level) projects. South of this area are Okstindane (1912 m) and Børgefjell (1703 m). The lofty summits are situated in the Jotunheimen; Galdhøpiggen (2468 m) is, for example, the highest Norwegian mountain. Not counting about 30 m of snow and ice at the top, the Glittertind attains an elevation of 2529 m. About 30 peaks rise to elevations above 1980 m in the Jotunheimen Mass. Dovrefjell and Rondanefjell have the peaks Snøhletta (2300 m) and Rondeslottet (2100 m) respectively.

The ancient mountain range forms a sort of backbone for Sweden and the main crestline of the range constitutes also the Swedish-Norwegian boundary. The principal divisions of the range are i. the northern mountains and the lake region, covering all Norrland and western parts of Svealand, ii. Småland Highlands in the south and southeast. The Northern Mountains attain elevations of 1830 m; Kebnekajse is 2123 m and is an isolated summit; a little south, beyond Lule Älf, is the lofty Sarektjåkko (2125 m) and further west is the Sulitjelmo (1914 m), beyond which the elevation diminishes progressively southwards to Åreskutan (1419 m). The snowline near Helagsfjället (1797 m) is 914 m. There are a number of U-shaped valleys. The largest European icefield, viz. Jostedalsbre, is about 1300 sq. km in area and the glaciers from it descend to 45-60 m above mean sea-level. The somewhat smaller Svartisen Icefield sends its glaciers down almost the sea-level. The small Engabrae Glacier enters the sea in Jokulford (70° NL). The permanent snowline in Norway varies with the aspect of the mountains, but is relatively low and ranges from 900 m in Seiland to nearly 1525 m in the south. The conifer-forests extend from the sea-level to elevations of 760-900 m, but near Trondheim sink to 600 m. Above the conifer-forest belt, we find the *Betula*-zone, above which lies the lichen-belt up to the permanent snowline. As considerable parts of the mountains are within the Arctic Circle, the boundary mountains are sufficiently high to be much colder than the surrounding lowlands, but are not really high enough to act as effective barriers to the warming action of the southwest winds.

2. ECOLOGICAL CHARACTERS AND ALTITUDINAL BIOTIC ZONATION ON FENNOSCANDIAN MOUNTAINS

The ecological, floristic and faunistic peculiarities are largely correlated with the extensive range of latitude and altitude of the Fennoscandian mountains. An outstanding feature of these mountains is the fact that the largest positive temperature anomaly in the world is found on the Norwegian mountains. The mean atmospheric temperature on the mountains within the Arctic Circle is higher than in the areas further east and 20° latitudes further south (917). This interesting temperature anomaly is the result of the warm-water currents and the air-drift across the Atlantic to the Norwegian shores. At an elevation of about 600 m, the mean atmospheric temperature during February, the coldest month, is —19.9° C and the absolute minimum temperature is —26.6° C. The forestline is relatively low and alpine conditions prevail even at an elevation of about 600 m above mean sea-level. The biotic zones are (fig. 77): i. the *Fagus*-forest zone, ii. the conifer-forest zone, iii. the

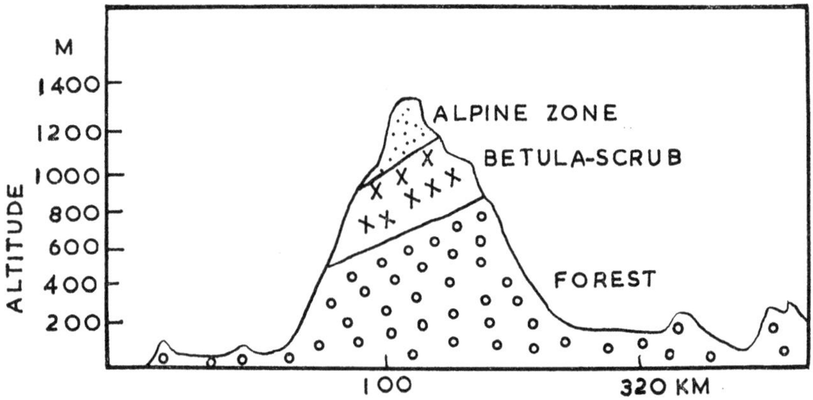

Fig. 77. Profile section of Scandinavian mountains, showing the altitudinal biotic zonation.

Betula-zone and iv. the alpine zone. The alpine zone is, however, confined to the upper levels of the spinal mountains, above an elevation of 440 m in the north and 900 m in the south. Stunted shrublike *Salix* and *Betula*, with extensive area of heather, are characteristic of the alpine areas. The reindeermoss, arctic poppy and *Saxifraga* are sometimes found on the bleak, wind-blown mountain heights. The Finnish region is subdivided into the following ecological-biogeographical zones: i. a narrow coastal strip of *Quercus*-zone; ii. broad-leaved forest zone of south Finnland and rest of the country south of the 62nd north parallel; iii. the Central Finnish zone, reaching to nearly the 64th north parallel; iv. the lower zone of north Finnish needle-forest, between the 64th and the 65° 30′ north latitudes; v. the higher zone of needle-forest of north Finland, extending somewhat into the polar circle; vi. the needle-forest zone of Lappland, partly extending to 69° 30′ NL; and vii. the alpine and subalpine zones of Lappland, including some of the meadows in the south to peaks in the northern area and along the coast. FRANZ (378) has recently shown that the altitudinal biotic zones on the Fennoscandian mountains have, in addition to the present-day climate, been influenced by the Pleistocene climatic and distributional conditions. There are fundamental differences between the conditions prevailing on the Alps and those of the Scandinavian mountains, with special reference to the altitudinal zonation. The rather obscurely defined boundaries of the altitudinal zones are closely correlated with the absence of strictly localized endemic species of insects. The fauna is on the whole poorer in character species than the Alps and is also particularly remarkable for the absence of strictly localized endemic forms. The old endemic elements are generally assumed to have perished during the Pleistocene glaciations, so that only some types have survived on the nunataks. The grassy-meadow forms of the Swedish Lappland, at elevations of 600-800 m, are poorer but otherwise equivalent to the alpine elements. The snow-valley fauna is extremely poor in species, but there is a characteristic plant-polster fauna. The subalpine elements from the forest penetrate into pockets in the arctic-alpine grassy-meadow areas and the grassy-meadow types in their turn populate open localities within the subalpine-forest zone.

3. CHARACTERS OF THE INSECT LIFE OF THE FENNOSCANDIAN MOUNTAINS

A striking feature of the high altitude insect life of the Fennoscandian mountains is the conspicuous poverty of endemic elements, when compared with the Central European mountains. There are, for example, no endemic species of the typically terricole hypsobiont genera *Cychrus* FABR., *Carabus* LINN., *Nebria* LATR., *Trechus* CLAIRV., *Pterostichus* BON., etc. on the mountains of Finland and Scandinavia. The endemic terricole hypsobiont species are also lacking in the planticole genera like *Chrysomela* LINN., *Orina* MOTSCH., *Crepidodera* CHEVR., *Alophus* SCHÖNH., *Plinthus* GERM. and *Liparus* OLIV. According to the evidence brought together by LINDROTH (698), there are indeed no endemic Carabidae on the Fennoscandian mountains. *Bembidion scandicum* LINDR., originally supposed to be endemic in north Norway and Sweden, is now believed by LINDROTH to be identical with *B. mackinleyi* FALL. from Alaska. The high altitude insects from this region are essentially either Palaearctic or Holarctic elements with circumpolar distribution. Many of the species are, therefore, widely distributed in the northern and eastern parts of Siberia, on the Altai Mountains, in Alaska, the Yukon Territory, British Columbia, Greenland and often also on the Central and south European mountains. The most important constituents of the Fennoscandian mountain insect fauna are the boreo-alpine species, which have already been discussed in chapter XII. The boreo-alpine species represent in fact the commonest subalpine and alpine forms on the Fennoscandian mountains. It may be recalled that only 6 of the 43 species of boreo-alpine Coleoptera, and only 4 of the 33 boreo-alpine Lepidoptera have not so far been found in the region. The southernmost limit of the greatest majority of the boreo-alpine species in Fennoscandia is on the Norwegian mountains. This is, for example, the case with *Bembidion fellmanni* MANNERH., *Anthophagus alpinus* FABR., *Geodromicus globulicollis* MANNERH., *Helophorus glacialis* VILLA. and *Phytodecta affinis* GYLL. The distribution of *Simplocaria metallica* STURM. on the Fennoscandian mountains presents certain interesting peculiarities; this species has conspicuous lacunae of about 3° 50′ latitude in the middle sections on these mountains. It is also interesting to observe that even the other, non-boreo-alpine Fennoscandian Carabidae are mostly either varieties or subspecies of widely distributed species or they are identical with the latter (395).

Apart from the work of LINDROTH already referred to, the literature on the mountain autochthonous insects from Fennoscandia is extremely scanty. KAISILA (600) made interesting observations on insects found on snowfields and on other alpine-zone localities of the Pallas Ounastunturi National Park in west Finnish Lappland, especially the Keimiötunturi (626 m above mean sea-level), Sammaltunturi and

Lommoltunturi Mountains (about 67° 55′ and 68° 20′ NL), the peak Taivaskero (821 m) on the Central Pallastunturit. He collected a number of interesting species of Collembola, Plecoptera, Trichoptera, Ephemerida, Odonata, Hemiptera, Megaloptera, Neuroptera, Raphidiodea, Coleoptera, Lepidoptera, Hymenoptera and Diptera. Most of the insects collected by him represent forms that have been blown by wind from lower elevations and from the forest zone. He observed *Nebria gyllenhali* SCHÖNH., *Limnophora triangulifera* ZETT. and the spider *Xysticus cristatus* (CLERK) using the snowfields as hunting-ground; they suck the juices from the wind-blown derelicts cast on the snow.

VALLE's (1076) observations on the Odonata of the subarctic and arctic mountains of Finnland are of considerable interest. The dragonfly *Ennallagma cyathigerum* (CHARP.) occurs throughout east Finland and Scandinavia and has also been reported from the north coast in the Kola Peninsula (Lapponia murmanica, about 68° 48′ NL). It is also known from the tundra-zone (regio alpina). The northernmost locality within Finland is Kryö in Ivalojoki. *Aeschna coerulea* (STRÖM) (=*borealis* ZETT.) is distributed throughout Finland up to Utsjoki in North Lappland (about 69° 50′ NL) and occurs also in the subalpine and alpine areas, though in the northernmost localities in the Kola Peninsula it is found within the regio silvatica (forest zone). The northern species *Somatochlora alpestris* SELYS occurs in Central Finland up to Jyväskylä (62° 20′ NL), the northernmost distribution extends up to Utsjoki. In Central Finland, it is restricted to open moorland localities. *S. arctica* ZETT. is widely distributed up to Enontekiö, Inari and Gavrilovo (about 69° 10′ NL) and occurs in subalpine and alpine zones and in the southern parts it is sparse and confined to moors.

LINDBERG (693, 694) studied the subalpine and alpine insects, especially Coleoptera, in the extreme northwest Finnland, near the Fenno-Swedish border, in the area of the Lake Kilpisjärvi. Lake Kilpisjärvi in northwest Enontokis is about 3 km long and 4-9 km wide and lies in the general direction of southeast to northwest. It is divided into two subdivisions by a small peninsula. Kilpisjärvi is the source lake of the R. Muonio and lies at 69° NL and somewhat west of 21° EL, at about an elevation of 475.5 m and is thus situated entirely within the subalpine area. The upper limit of the subalpine zone in the region is at an elevation of 570 m, the lower limit is, however, less clearly marked. The conifer-forests appear, however, only 100 km to the southeast. As the southern parts of Kilpisjärvi belong to Sweden, LINDBERG confined his observations entirely to the north. On the northernmost shore of the lake, the lofty projection of Saana rises to an elevation of 1055 m. The subalpine zone is chiefly characterized by the presence of *Vaccinium-Betula* associations and in isolated localities *Geranium* and *Trollius*. There are also *Salix glauca* and *Betula nana* in some places. The highest parts of the alpine zone, the peak of Fjelde, are relatively poor in vegetation, especially the higher plants. KOTILAINEN's subdivisions, "the regio alpina sterilis" and "regio alpina fertilis" are not recognized by LINDBERG. In the lower reaches of the zone, alternating with the alpine meadows, there are wide shrub patches, with *Dryas*, *Empetrum*, *Betula nana*, *Arctostaphylos alpina*, *Azalea procumbens*, etc.

The following Coleoptera are characteristic of the regio subalpina: *Nebria rufescens* STRÖM., *Bembidion bipunctatum* LINN., *B. hasti* SAHLB., *Patrobus septentrionis* DEJ., *Amara alpina* FABR., *Geodromicus globulicollis* MANNERH., etc. These species occur particularly on sandy and dry shores of Lake Kilpisjärvi. On the damp shore of the lake are found *Notiophilus aquaticus* LINN., *Stenus hyperboreus* SAHLB., *Atheta graminicola* GRAV., *A. laevicauda* SAHLB., *A. aequata* ER., *A. microptera* THOMS., etc. On the open dry and humid shore of the R. Muonio he found a number of typical forms like *Pelophila borealis* PAYK., *Notiophilus germinyi* FAUV. (=*hypocrita* PUTZ.), *Dyschirius globosus* HERBST, *Bembidion bipunctatum* LINN., *B. aeruginosum* GEBL., *B. virens* SCHILSKY, *B. prasinum* DUFT., *Stenus hyperboreus* SAHLB., *St. proditor* ER., *St. canaliculatus* GYLL., *St. fuscipes* GRAV., *Atheta frigida* SAHLB., *A. graminicola* GRAV., *Byrrhus fasciatus* LINN., etc. The species from the regio alpina include *Nebria rufescens* STRÖM., *N. nivalis* PAYK., *Pelophila borealis* PAYK., *Stenus lapponicus* SAHLB., *St. niveus* FAUV., *Atheta arctica* THOMS., *A. frigida* SAHLB., *A. graminicola* GRAV., etc. The species from under stones on the alpine-meadows are *Notiophilus aquaticus* LINN., *N. germinyi* FAUV., *Patrobus septentrionis* DEJ., *P. assimilis* CHAUD., *Amara praetermissa* SAHLB., *A. alpina* FABR., *Mannerheimia arctica* ER., *Geodromicus globulicollis* MANNERH., *Quedius fulvicollis* STEPH., *Atheta islandica* KR., *A. graminicola* GRAV., *A. microptera* THOMS., *Lepyrus arcticus* PAYK., etc. *Geodromicus globulicollis* MANNERH., *Atheta frigida* SAHLB. and *A. graminicola* GRAV. are also found under stones in the ravines of the higher parts of the regio alpina.

According to LINDBERG a biocenosis of essentially the same composition appears on similar biotopes at different levels. The significant differences observed in the composition of species are attributed to the fact that some species which occur at lower elevations are absent at higher levels. The typical alpine-species occur only in biotopes that are confined to the regio alpina. New species appear on the other hand at higher elevations. The same species exhibits different habits and modes of life under different ecological conditions. LINDBERG has also drawn particular attention to the pronounced rufinism of the alpine-zone species in Finland. There are normally both the black and reddish-brown forms in the northern region and the rufinistic forms become more abundant at higher elevations than at lower. There are also transitional forms. The distinctly rufinistic species are *Patrobus septentrionis* DEJ., *Amara alpina* FABR., *Hydroporus lapponum* GYLL. and *Agabus thomsoni* SAHLB. *Nebria rufescens* STRÖM. occurs under stones in association with *Bembidion hasti* SAHLB. and nearly 60% of the specimens collected in the alpine zone are rufinistic, but only 14% are rufinistic in the subalpine area. *Notiophilus aquaticus* LINN. occurs both in the alpine and subalpine areas, though more frequently in the former. At higher elevations this species is generally associated with *Amara alpina* FABR., *Miscodera arctica* PAYK. and *Hypnoidus rivularius* GYLL. (= *Cryptohypnus rivularius* GYLL.) under stones. *Notiophilus germinyi* FAUV. is more sparsely found than the foregoing species and occurs partly also within the subalpine zone. *Dyschirius septentrionum* MUNST. is restricted to the subalpine zone. *Miscodera arctica* PAYK. occurs under stones in association with *Notiophilus aquaticus* LINN., *Amara alpina* FABR. and *Cymindis vaporariorum* LINN. *Patrobus septentrionis* DEJ. occurs under stones and fallen leaves in humid and moist localities. About 16% of the specimens from the alpine zone are non-rufinistic, but 77.5% of the subalpine-zone specimens are typically black coloured. *Patrobus assimilis assimilis* CHAUD., not so common as the foregoing species, also occurs both in the subalpine and on grass-meadow and under stones in the alpine zone. *Amara praetermissa* SAHLB. and *A. brunnea* GYLL. occur under stones in the alpine zone. *A. torrida* ILLIG. is found under stones in the alpine zone, but also descends to a transitional belt between the alpine and subalpine zones. *A. alpina* FABR. occurs, often in association with *Nebria rufescens* STRÖM., *Bemdidion hasti* SAHLB. and *Patrobus septentrionis* DEJ. in the alpine zone and subalpine zone; in the alpine zone it is found particularly in association with *Notiophilus aquaticus* LINN., *Cymindis vaporariorum* LINN. and *Miscodera arctica* PAYK. It is curious that specimens of this species from the subalpine zone are rufinistic, but those from the alpine zone are black coloured.

Dytiscidae like *Coelambus marklini* GYLL. occur in association with other species in small ponds. *Hydroporus lapponum* GYLL. also occurs in small ponds, in association with a number of other species of the genus. *Deronectes griseostriatus multistriatus* FALKENSTRÖM belongs to the alpine zone and rarely occurs in the subalpine area. It generally occurs in association with *Helophorus glacialis* VILLA. at higher elevations. A number of other species like *Agabus elongatus* GYLL., *A. guttatus* PAYK., etc. are common in the subalpine zone and *A. thomsoni* SAHLB. is common in the alpine zone and less frequent in the subalpine zone. While *Helophorus glacialis* VILLA. and *H. fennicus* PAYK. occur in the alpine zone, *H. lapponicus* THOMS. is typical of the transitional belt between the alpine and subalpine zones.

A large number of Staphylinidae are recorded by LINDBERG and include the following: *Megarthrus depressus* PAYK. under dung, *Anthobium lapponicum* MANNERH. on moors and usually in flowers of *Rubus chamaemorus*, *Acrulia inflata* GYLL. in fungi, *Omalium brevicolle* THOMS., *Mannerheimia arctica* ER., *Olophrum boreale* PAYK. in the subalpine zone, *Arpedium quandrum* GRAV. in moist localities and under stones both in the subalpine and alpine zones, *Arpedium tenue* LEC., *A. brunnescens* SAHLB., *Geodromicus globulicollis* MANNERH. common both in the subalpine and alpine areas and particularly in localities covered by snow during the winter, *Anthophagus alpinus* FABR. in the alpine zone, *Boreaphilus henningianus* SAHLB. in association with *Olophrum boreale* PAYK. under stones near streams, *Oxytelus laqueatus* MARSH in decaying organic matter, *Stenus canaliculatus* GYLL., *St. proditor* ER., *St. ruralis* ER., *St. fuscipes* GRAV., *St. hyperboreus* SAHLB., *Philonthus* spp., *Quedius fulvicollis* STEPH. usually under stones on grassy-meadows and in the alpine areas, *Atheta islandica* KR., *A. arctica* THOMS., *A. frigida* SAHLB., *A. aequata* ER., *A. depressicollis* FAUV., *A. laevicauda* SAHLB. *A. microptera* THOMS., *A. granigera* KIESW., *A. atramentaria* GYLL., *A. muensteri* BERNH., *A. altaica* BERNH. and several species of *Mycetoporus* MANNERH., *Bryoporus* KR., *Bolitobius* MANNERH., *Tachyporus* GRAV., *Tachinus* GRAV., *Gnypta* THOMS. and *Gymnusa* GRAV.

The Curculionidae are represented by *Otiorrhynchus dubius* STRÖM. and by a number of species of *Polydorus* GERM., *Lepyrus* GERM., *Pissodes* GERM. etc.

The Fennoscandian mountain autochthonous Lepidoptera appear to belong to a relatively young fauna (830, 831). *Colias nastes* BOISD. occurs over a wide range of elevations, from 400 to 900 m in Scandinavia and above the timberline on North American mountains (544). It extends up to 69° 4′ NL in Scandinavia, Novaya Zemlya (73° NL) and Spitzbergen, and occurs on the Kentei and Sayanskii Mountains and in Alaska and Labrador. On the Scandinavian mountains, the larvae feed in *Astragalus alpinus*. *Colias hecla* LEF. extends on the Scandinavian mountains from Jämtland (64° 50′ NL) to Lakselv and Porsanger (70° NL). It is also known from Siberia, Alaska, the northern parts of Canada, Greenland and on the Rocky Mountains, up to nearly the 54th north parallel. The altitudinal range of this species is relatively narrow on the Scandinavian mountains, but wide on the North American mountains. On the Scandinavian mountains, its larva feeds on *Astragalus alpinus* but on the North American mountains on *Hedysarum americanum*. *Argynnis pales* SCHIFF. is represented on the Scandinavian mountains by *A. p. lapponica* STGR. and *A. p. subalpina* PET. above the forestline in Central Norway but in the subalpine zone in other parts. The caterpillars feed on *Viola biflora* in Scandinavia. *Boloria arsilache* ESP. occurs at lower elevations than *Argynnis pales* SCHIFF. on the Scandinavian mountains. *Boloria improba* BTL. extends from Haditjokko in the north to the montane lake Torne Träsk in the south, along a relatively narrow belt of about 160 km width. The species is a circumpolar element, but it ascends to much higher elevations on the Scandinavian mountains than the other circumpolar butterflies. In Finland, it occurs at elevations of 750-950 m, but at 1000-1100 m on Vassitjåkko and Kårsanjonnje, especially in heavily snowed localities. *B. thore* HÜBN. is sometimes found above the timberline on Scandinavian mountains and *B. polaris* BOISD. is an alpine-zone species from Lule Lappmark in the south. It occurs, however, at much lower elevations in Novaya Zemlya, northwest Siberia, the Anadyr Mountains, in Alaska, Labrador and Greenland. In the lower levels of the alpine areas on the Scandinavian mountains, *Oeneis bore* ESP. is common; as already mentioned, this species is also known from Alaska, Montana and Colorado (*vide* chapter XIV). *Hesperia andromedae* WALLGR. is common at elevations of 400-1000 m all over the Scandinavian mountains. *Melitaea iduna* DALM., known also from Central and northeast Siberia and from the Caucasus (*vide* chapters XII & XIII), occurs on north Scandinavian mountains. *Polyommatus glandon* PRUN. is a subalpine and alpine species on the north Scandinavian mountains. Another boreo-alpine species, viz. *Erebia pandrose* BKH. occurs on the mountains of Västmanland and in Finland.

The component elements of the butterflies of the Scandinavian mountains are either Palaearctic or Holarctic. The latter include three sub-types: i. the species that also occur on the taiga zone of the Middle Asiatic mountains, ii. the species that occur on the taiga zone of North American mountains and in the arctic areas in Eurasia and iii. the species that are strictly arctic forms. PETERSEN (830, 831) has suggested that the insect fauna of the Scandinavian mountains invaded the area perhaps during the late Pleistocene, after the older fauna had already disappeared. Three distinct groups, viz. i. arctic invaders, ii. the European invaders and iii. the non-arctic invaders from Asia are recognized by him. The arctic elements like *Colias hecla* LEF., *Boloria polaris* BOSID., *B. improba* BTL., *Oeneis bore* ESP., etc. came from North America and are at present confined to alpine areas of the northern parts of the Scandinavian mountains. The non-arctic elements of Asia, like *Melitaea iduna* DALM. and *Colias nastes* BOISD. (?), though likewise restricted to the northern parts of these mountains, are found both in the subalpine and alpine zones. The European elements on the Scandinavian mountains include *Hesperia andromedae* WALLGR., *Polyommatus glandon* PRUN., *Psodos coracina* ESP., etc. which occur mostly over the whole mountain chain or only on the southern parts and some of them are subalpine and others are alpine forms. PETERSEN has also recognized several distinct types of boreo-alpine species, especially the northern elements. *Boloria charicola* SCHN.-type occurs in the arctic and is perhaps a re-immigrant to Eurasia from America or the northeastern parts of Siberia. *Polyommatus glandon* PRUN.-type is an arctic-alpine invader from non-arctic Asia. The third group is from Central Europe and the fourth group is from parts of the northeastern Middle Asia.

4. TIMANSKII MOUNTAINS

The Timanskii Mountains lie southeast of the Kalin Peninsula in northern Russia. The mountains were uplifted during the Permo-Triassic, replacing an older Timan

Range that has been submerged. The region was covered by ice during the Riss and Würm glaciations, but nunatak areas existed for the survival of an extremely interesting group of relict species. The northern mountains have practically lost their orographic identity at present and are flat erosion plateaux, 150–230 m above mean sea-level. The relict species from this region have been studied recently by SEDYKH (954). The relicts reveal considerable Central Asiatic affinities and indeed occur in the highest parts of the South Timan mountains. Most species are mountain-taiga elements, with very close affinities to those occurring on the Altai Mountains, the East Sayanskii Range, Primor'ye, Sachalin or North America. The Timanskii Range separates the Komi ASSR into two halves and lies northwest to the southwest. Among the interesting relicts found here are the Mirid *Megaloceraea reticornis* GEOFFR., the Carabid *Dyschirius jakuticus* ZM. known also from East Siberia, the Lepidoptera *Brenthis angarensis* ERSCH. known also from the Altai Mountains, Primor'ye and *Sympistis funesta* PAYK. (=*funebris* HÜBN.) also known from the Alps, north and east Siberia, Labrador and the Scandinavian mountains.

5. ALASKAN MOUNTAINS

Alaska is very similar to the western North America physiographically and the general outline of the mountain ranges is a continuation of that of Canada in the south, but curved northwest to southwest toward Asia. From the south to the north, we find a double coastal system of igneous mountains, an interior plateau of ancient metamorphic rocks, a sedimentary mountain system, a sedimentary plateau and finally a coastal plain. The mountains of Alaska are the largest and the highest in North America and are true arctic mountains. The rocky cores of these mountains remain permanently buried under truly enormous masses of ice and snow. The region as a whole is essentially a vast treeless plateau, with grass and moss. The valleys lie at about 900 m above mean sea-level and are remarkable for the great abundance of flowers.

The arctic mountains in northern Alaska extend from the Cape Lisburne to the Mackenzie Delta in the Yukon Territory of Canada and the foothills pass gradually into the coastal plain that flanks the Arctic Ocean. The mountains are in some cases hardly 18–19 km from the sea. There are numerous tributaries of the rivers from the foothills and the ground-ice reaches up to varying depths. The coastal mountains extend to some of the islands of the southeast. The inner coastal mountains include the Coast Range of the southeast, the Alaska Range and the Aleutian Range. Valleys separate the inner and outer coastal mountains. Many of the valleys are indeed submerged at present and others like the Copper Valley, Gulkana Valley, Matanuska Valley and the Susitna Valley, show evidence of former heavy glaciations. The interior plateau extends over most of Alaska and the Yukon Territory and is a general peneplain of about 1830 m elevation in the east, but less in the west. The northern mountain system is relatively narrow and has undergone extensive glaciation.

Orographically three subdivisions are recognized: i. the Pacific Mountain System, ii. the Central Plateau and iii. the Rocky Mountains or the Cordilleran System.

The Pacific Mountain System consists of four principal ranges: i. the Coast Range of the Panhandle, stretching to over 160 km in Alaska and British Columbia, without, however, a well-defined crestline. The summits of the Coast Range attain a mean elevation of about 1525–1830 m. ii. Mt. St. Elias Range, separated from the Coast Range by the Shilkat River, includes the mountains between the Cross Summit, Mt. St. Elias, Chugach, Wrangell and Nutzotin Mountains. It lies in the southeast part of the continental Alaska. The well known high peaks in this range are Mt. Sanford, Mt. Fairweather (4663 m) completely under snow-cover the whole year, and Mt. Vancouver, all rising to elevations over 4575 m. Mt. Wrangell is an active volcano on the Nutzotin Mountains. Mt. St. Elias (5489 m) belongs to the Cordilleran System. Northward from the Glacier Bay is the Fairweather Range, with Mt. Fairweather. iii. The Aleutian Range constitutes a sort of backbone for the Aleutian Peninsula and rises to elevations of 900–1830 m. It is remarkable for perhaps one of the greatest volcanic eruptions in recent times. The top of Mt. Katmai, which showed no indication of its volcanic nature, blew off *en masse* and a large number of volcanic vents appeared in what is now the well-known "Valley-of-ten-thousand-smokes" (the Katmai National Monument). iv. The Alaska Range and the Nutzotin Mountains form together a great arc. The Alaska Range is 960 km long and 80–120 km wide and is south of the Yukon drainage system. It is connected with the Nutzotin and Skolai branches of the St. Elias Range. The Alaska Range has many well known peaks like Mt. Foraker (5181 m); the South Pinnacle or the Mt. McKinley (6187 m). the highest mountain in North America; the North Pinnacle is only 90 m less high). Most of the Alaska Range is under permanent snow-cover and great alpine-glaciers fill the valleys. North of the Yukon river lie the Baird Mountains and the Endicott Mts. Northernmost is the Brooks Range, extending nearly east-west, between the Arctic Circle and the 70th north parallel. The Endicott Mts. are a part of the Brooks Range. Inland to the east in the Northwest Territory (in Canada) are the Mackenzie Mountains, with northwest to southeast trend.

The Alaskan mountains are remarkable for the number and size of their glaciers. There are literally several hundreds of alpine-glaciers, ranging from 1.5 to over 22 km in length in the upper valleys. There are also numerous piedmont glaciers. The Malaspina Glacier, the largest in Alaska, is about 80 km long, along the seaward base of the Mt. St. Elias. The Muir Glacier was shattered by the disastrous earth-quake of 1899. It is known that most of the Alaskan glaciers are gradually receding at the present time and have terminal moraines many kilometres back in a number of cases.

The Alaskan mountains consist of folded and contorted sedimentary beds, with underlying igneous rocks. The intrusion of the igneous masses have led to the uplift of the sedimentary rocks. The Alaska Range was elevated during the late Tertiary. There is abundant evidence of a long series of uplifts and depressions, of erosions and deposition of sediments, of folding, faulting, metamorphism of beds and of intermittent volcanism of the Alaskan mountains. The older sedimentary beds are of Pre-Cambrian age in the Upper Yukon Valley (the socalled Birch-creek schist). During the Ordovician and perhaps also the early Devonian, there was marine submergence, with marine deposition during the Carboniferous and Permian times. The Cretaceous Epoch saw renewed submergence. Most Tertiary rocks are hard conglomerates, shales or they have basaltic lava. The Pleistocene beds are abundant in a number of places. The Pleistocene glaciers abraded the highland rocks and carried gravel, sand, silt, till, etc. and the glacial deposits along the R. Copper attain a thickness of some 800 m.

The St. Elias Range is continued into the Yukon Territory and forms the highest mountain chain in Canada in the southwest. Mt. Logan (6050 m) and Mt. St. Elias (5489 m) are the most important massifs. There is also the north extension of the Rocky Mountains, along the south and northeast. The interior of the Yukon Territory is high towards the southeast, but the elevation falls towards the northwest. This area is best described as a much dissected peneplain, with low mountains to the south. As the lofty mountains on the west act as effective rain-shadow, there is general atmospheric aridity in the Yukon Territory and rainfall and snowfall are scanty, so that the general climate is typically continental, with very cold winters extending over seven months and short but sunny summers. The permanent snow-line on the Yukon Territory mountains is at an elevation of 1220 m in the area, but the moss-covered ground is frozen, often to depths of 30-60 m.

6. ECOLOGICAL CHARACTERS OF THE ALASKAN MOUNTAINS

As already pointed out, the Alaskan mountains are typically arctic mountains, the rocky cores of which are often buried under great masses of ice and snow. On the inland side of the mountains, the climate is on the whole continental, with short, warm and sunny summers and long and cold winters. On the Pacific side, however, the summer is longer and cooler. Over most of Alaska, the mean atmospheric temperature during the summer lies between 12° and 15° C, but in the Aleutian and the Bering Sea Coast areas 10° C and much less in the area of the Arctic Coast. The winter temperatures fall below —15° C in most areas, except perhaps in the extreme south and southwest. While the climate of the interior of Alaska is continental, it is oceanic in the southwest. The moisture-laden winds from the warm Pacific Ocean cause heavy precipitation on the windward slopes of the coastal mountains.

The timberline in Alaska is different from that in Canada in a number of respects (433-436). While the timberline seems to have remained practically stationary on the Canadian Rockies, it is rapidly advancing upwards on the Alaskan mountains. In North Canada the timberline is fairly static or is perhaps slightly receding at present. It is marked by characteristically stunted, crippled and prostrate trees, so that there is an abrupt change in the flora and fauna beyond. The peculiar anomalous conditions in northwest Alaska are not due to the oceanic climate or even to the heavy precipitation, but probably to the Post-Pleistocene history of the region (744). In Central Alaska, the timberline is about 915 m, but the arctic conditions commence only at about 1060-1220 m above mean sea-level. On the Kodiak Islands and in the Alaska Peninsula, arctic conditions begin at an elevation of about 300 m. The low sheltered valleys on the north slope of the Brooks Range appear to have arctic elements, with strong boreal intrusions. The altitudinal biotic zones in Alaska are the Canadian, the Hudsonian and the arctic zones of MERRIAM. The Canadian zone shows two subdivisions, viz. the western Hemlock-Sitka-spruce in the southeastern parts of Alaska and the interior white-spruce-aspen-birch zone. Most of Alaska is typically Hudsonian zone of open, dwarf-spruce forest, with undergrowth of Ericaceae, lichen and shrubs. In the Canadian zone the undergrowth is on the other hand typically of moss and herbs. The upper limit of the Hudsonian zone or the subarctic zone is usually defined by the timberline. Deciduous *Betula* and *Populus* often grow beyond the timberline in parts of Alaska, as for example, north of the Brooks Range. The July isotherm of 10° C may also be utilized to delimit the Hudsonian zone.

7. CHARACTERS OF THE INSECT LIFE OF THE ALASKAN MOUNTAINS

In view of the enormous ecological and biogeographical importance of the region, it seems rather unfortunate that we know practically nothing of the high altitude insects of the Alaskan mountains. A large number of insects, spiders, centipedes and mites have no doubt been collected by the famous Harriman Alaskan Expedition of 1913-1918. It is, however, only exceptionally that we find records of the exact altitude of the collection in the Report of the Harriman Expedition (479). Most of these collections were also made, in fact, not on mountains at all and the exceedingly few specimens that seem to have been taken on mountains come from within the forest-zone.

A fact of great biogeographical and ecological importance is that a great many of the species reported from Alaska-Yukon Territory are also known from the Canadian Rockies, the mountains in Colorado and Montana, from Labrador and from the Appalachian Region, especially the Presidential Range. Many of them extend moreover into the region of Beringea and Siberia and occur also on the Altai Mountains and still others turn up even on the Fennoscandian mountains, in Novaya Zemlya and Greenland.

The Carabidae from the Alaskan-Yukon Territory mountains are remarkable for the presence of the characteristic small Pterostichine *Cryobius* CHAUD. *Melanoplus frigidus* (BOH.) has been reported at an elevation of 305 m on the south side of the hills of the Melville Mountains, about 60 km inland on the R. Horton. The Chironomid *Diamesa arctica* (BOHEM.) was collected on the Colville Mountains in the Wollaston Peninsula and from the Angmaloktok Lake on the Colville Mountains. Mention may also be made of the spider *Xysticus bimaculatus* EMERTON from the Rocky Mountains at Leggen and from Mt. Lincoln in Colorado, reported to have been collected at an elevation of 3355 m in the Alaska-Yukon region. The Chilopod *Cryophilus alaskanus* CHAMBERLAIN was found under stones in the high tundra areas in Alaska.

The great bulk of insects collected from the Alaskan region is, however, Lepidoptera. GIBSON (406) has listed a number of species and recently DOS PASSOS (281-283) has also given an interesting account of the Alaskan and Yukon mountain Lepidoptera. *Papilio machaon aliaska* SCUDDER occurs below the summit of the Chilkoot Pass on the Coast Range in British Columbia (60° NL). The records include *Oeneis uhleri* (REAK.) at an elevation of 760 m on the Mackenzie, Nahanni Mountains; *Oe. mackinleyensis* DOS PASSOS from the Mt. McKinley; *Oe. brucei yukonensis* GIBSON at an elevation of 2500 m on the Klutlan Glacier in the Yukon region during June; *Erebia rossii gabrieli* DOS PASSOS at an elevation of 1066 m from the Mt. McKinley National Park; *E. r. rossii* (CURTIS); *E. r. ornata* LEUSSLER; *E. r. kuskokwiima* HOLLAND from the R. Kuskokwiim Valley; *E. magdalena mackinleyensis* GUNDER from the Sabel Pass on Mt. McKinley National Park; *Boloria improba* (BTL.) (=*Erebia youngi rileyi* DOS PASSOS) from the Mt. McKinley National Park. *Boloria frigga* THUNB. occurs throughout the tundra zone to much lower elevations on the Alaskan mountains. *B. polaris* BOISD. is also found on the Alaskan mountains and the Anadyr

Mountains of eastern Siberia and occurs in Labrador and Greenland. *Oeneis bore* ESP. extends from Alaska to Labrador and as already stated, occurs also on the mountains of Colorado, Wyoming and Montana. *Oe. semidea* SAY has been reported from the R. Steward area in Yukon at 1220 m above mean sea-level; on the mountain summit above the R. Nadaleen on Yukon Mountain, near the upper R. Pelley; Mt. Eduni (1370-1830 m) and also on the mountains below the R. Natla in the Northwest Territory. *Oe. arctica* GIBSON also occurs in the same localities. The occurrence of *Erebia magdalena* STKR. on the rocks above timberline at an elevation of 1830 m in Yukon is of considerable interest. The specimen from this locality is described as being smaller than those from the Colorado mountains. *Argynnis pales alaskensis* (HOLL.) was collected from mountains below the Forty-mile and Mission Creeks. *Boloria natazhati* (GIBSON) was taken at an elevation of 2620 m on the north of Mt. Natazhat on the 141th meridion. *B. polaris* (BOISD.) is reported from Lake Angmaloktok on the Colville Mountains. *B. improba* (BTL.) comes from the Klutlan Glacier (2743 m). *B. distincta* (GIBSON) is recorded at an elevation of 1830 m on the R. Gravel on the Eduni Mountains in the Northwest Territory. Other records include *Anarta richardsoni* CURTIS at an elevation of 2743 m on the Klutlan Glacier, *Bombyx quenseli* PAYK. at an elevation of 1920 m on Mt. Natazhat, *Pieris occidentalis* REAK. at 2283 m on the Klutlan Glacier (141th meridian), *Colias nastes* BOISD. on rocky areas above the timberline on Alaskan mountains and on barren ground or alpine tundra, as low as 900 m or as high as 2000 m; *C. hecla* LEFEBRE on the Mt. McKinley Range and *C. palaeno* LINN. at 2750 m on the Alaskan mountains. It may be remembered that these species occur frequently on the Scandinavian mountains. In an interesting discussion on the general ecology and distribution of *Erebia* DALM. on the mountains of Alaska and the Yukon Territory, EHRLICH (304) has shown that in comparison to their north-south distribution in non-montane areas, the vertical distribution of these boreal forms on the subarctic mountains is more interesting in a number of ways Of about a dozen Nearctic species, *E. fasciata* BTL. is the most typically arctic and occur

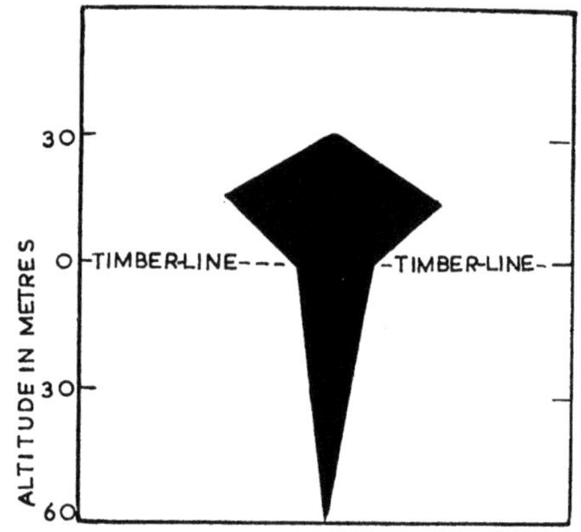

Fig. 78. The relative abundance of *Erebia* spp. immediately below and above the timberline in Alaska.

above the timberline on moist tundra on Alaskan and Yukon mountains (fig. 78). It does not, however, extend horizontally to the treeline in the south. *E. rossii* CURTIS extends south to the northern limits of the boreal forest and is also widely distributed on the mountains of British Columbia. It is also found on Alaskan mountains, in open areas, well below the treeline. *E. disa* (VILL. & GUÉR.) occurs occasionally in the open near the treeline on Mt. McKinley National Park. *E. youngi* HOLLAND occurs north of the treeline in Herschel Island, at the treeline in the Mackenzie Delta and in the Mt. McKinley National Park mainly above the timberline. It is common at an elevation of about 1070 m on the Eagle Summit in Alaska, above the timberline and at 365 m on the Richardson Mountains. *E. epipsodea* BTL. occurs in the Hudsonian, Canadian and the upper transitional zones and on the mountains near the Alaska-Yukon border within the treeline.

Some of these distributional peculiarities may be summarized as follows:

i. Many species, strictly coastal in Alaska, become strongly continental on the Canadian Rockies.

ii. Many interior Alaskan elements reach the Pacific Coast slope in southwest Alaska, but not elsewhere on the Pacific Coast.

MASON (744, 744a) recognizes two major faunal groups on the basis of the distributional pattern, viz. the species that have re-invaded Alaska from Pleistocene refugium in North America and species that have survived the Pleistocene glaciations in the Bering refugium or in eastern Siberia.

The Nearctic ranges embrace i. the coastal Coleoptera from northern California to southeast Alaska, ii. the coastal form in Alaska but occurring throughout in British Columbia, extending to Colorado and Californian mountains; many of these species extend as far as the Alaska Range. iii. The widespread western elements with Cordilleran distribution outside Alaska have general distribution in the warmer parts of Alaska. iv. Transcontinental boreal elements found on most of northern Canada and the Rocky Mountains, but only sparsely on the wet coastal British Columbia and southeast Alaska; southeastern Alaska has a more western fauna than southwestern Alaska. v. Widespread arctic, circumpolar elements.

The Bering ranges include the Aleutian elements extending to Kamchatka and southern Alaska, the interior Alaskan elements centre around the Yukon Valley and extend east to the Mackenzie Valley and Hudson Bay, southward to the Rocky Mountains up to Alberta and westward to the Aleutian and across the Bering Sea to Siberia. The Arctic-Alaskan or the Old World arctic elements on the north slope of Alaska are widely distributed at high altitudes throughout Alaska. Reference may also be made in this connection to the work of FREEMAN (381, 382), who groups the subarctic and arctic Rhopalocera into i. transcontinental elements, ii. western arctic elements, iii. eastern arctic elements, iv. northern arctic elements, v. mid-arctic elements, vi. south-arctic elements, vii. pan-arctic elements and viii. southern arctic relicts.

CHAPTER XVI

THE MOUNTAINS OF THE SOUTH-TEMPERATE, AUSTRAL AND ANTARCTIC REGIONS

The principal elevated areas, of entomological interest, in the south-temperate and antarctic regions are the mountains of Australia, New Zealand, Tasmania, the Chilean Andes, the Patagonian Steppes, the Andes of Tierra del Fuego and the mountains of Antarctica. Although parts of the high regions of northern Australia are really within the torrid zone, a number of considerations justify discussion of all the Australian mountains as part of the south temperate region. The Andes south of Bolivia, in other words the Chilean Andes, the Argentinian Andes, the Patagonian Andes and the Patagonian Steppes and the southernmost parts of the Andes in Tierra del Fuego constitute perhaps the most remarkable high altitude regions of the south-temperatre and sub-antarctic areas. As mentioned elsewhere, the main trendline of the Andes in Tierra del Fuego is almost east-west. It has recently been shown that Tierra del Fuego and Patagonia are geographically closely related. Besides the Andes, we must also consider the area of the Mesetas to the east of the Andes. As the predominant winds are westerly, the rainfall and dense forests are concentrated on the Andes; the Mesetas and the plains are typically steppes and semi-deserts. The area closest to the Andes was covered by ice during the late Pleistocene. Old glacial moraines, east of the limits of the youngest glaciation, are correlated with the penultimate glaciation that corresponds probably with the Würm Glaciation in the northern hemisphere. Among the more important high mountain peaks, situated within the south temperate region, are Llullaillaco (25° SL and 6720 metres above mean sea-level), Ojos del Salado (27° SL and 6870 metres), Mercedario (32° SL and 6765 metres) in Argentina-Chile; and Aconcagua in Argentina (32° 30′ SL and 7034 metres).

I. ECOLOGICAL CHARACTERS OF THE MOUNTAINS OF THE SOUTH-TEMPERATE REGION

The outstanding features of the high mountains of the south-temperate and sub-antarctic regions centre largely around their fundamental ecological and biogeographical differences from the north-temperate and boreal mountains. Some of these differences were apparently recognized even by v. HUMBOLDT and since his times several workers, but notably TROLL (1047-1049), have laid considerable emphasis on this aspect. The recent discussions on the biology of the southern cold-temperate zone, under the leadership of PANTIN, in the Royal Society of London, have served to focus attention on the great fundamental importance of the problem (287, 668, 1048).

While great continental masses dominate the cold-temperate and subarctic regions in the north, the region of the south-temperate is only a narrow extension of the South American continent and various scattered islands, surrounded by vast expanse of oceans. The general climate of the south-temperate area is thus oceanic, in sharp contrast to the largely continental climate of the boreal cold and temperate regions. Nowhere in the north do we find such small differences between the summer and winter atmospheric temperatures as in the sub-antarctic area. On the islands, south of New Zealand, the annual atmospheric temperature amplitude is only 3.5° C. In corresponding boreal latitudes the winters are nowhere mildenough for the dense growth of broad-leaved evergreen forest as we find, for example, south in Chile, in Tasmania, the South Island of New Zealand, the Steward Island and the Auckland Island. Tree ferns flourish in these areas, at latitudes of nearly 51° south, in other words corresponding to the north latitudes of Newfoundland and southernmost Aleutian Islands and southernmost point of the Kamchatka Peninsula. Yet the lower atmospheric temperatures of these southern islands should justify our considering them as part of the southern-cold and south-temperate regions. The presence of the evergreen broad-leaved rain-forests in this region must be attributed wholly to the mild winters, with no or very little frost at sea-level. There is on the other hand a conspicuous similarity in the general climatic, floristic, faunistic and other ecological peculiarities between tropical high mountains and the south-temperate and sub-antarctic mountains. TROLL (1046) has discussed some of the climatic and vegetational similarities of these two regions. The peculiarities of the south-temperate mountains are to be attributed largely to the peculiarities of the distribution of land and oceans in the extra-tropical latitudes. We have the greatest part of the mainland in the northern latitudes at about 60° and this landmass diminishes from the north southward to the fringe of the tropics and gives place to the polar seas at about 70°NL. We have cold winters in these high north latitudes. The boreal climate is characterized by temperature variations of about 65° C during the year, as for example, in Siberia. At corresponding latitudes in the south, there is, however, a large mass of ocean, in which lies Tierra del Fuego, South New Zealand and other scattered oceanic islands, with mean annual temperature variation of about 3.5° C only. The needle-forests, summer-green broad-leaved forests and mixed forest of the northern cold-temperate regions of Eurasia and North America have in the tropical high altitudes little relation to the tundra, with its cold winters and permafreeze of the ground. The northern vegetational zones of the needle and deciduous forests extend southwards at high elevations of 3000-4000 m on the high meridional mountains and on the Mexican mountains (fig. 79).

The climate of the north polar regions may be described as essentially seasonal, the equatorial climate as diurnal and the climate of the sub-antarctic region as isothermal. The high Andes resemble the sub-antarctic Macquarie Islands in having relatively small diurnal and annual atmospheric temperature ranges. Close below the permanent snowline, at elevations of about 4650-4800 m, on Mt. Chimborazo, the extreme temperature variation lies between 3.5° and —2.0° C. On Macquarie Island the temperature varies the whole year only between 7.6° and 2.8° C. The ecological effects of such climatic conditions are very clearly observed in the general nature of the frost action at the surface of the soil. In the polar climate, with severe winters and perennially frozen ground, the soil particles are structurally predominantly large to very large. On the tropical high mountains and in the sub-antarctic areas, the frost and the thaw cycle are characteristically diurnal and the frost does not also extend deeper than a few centimetres. At an elevation of about 4800 m on the Andes of southern Peru, alternate night frosts and daytime thaws occur, for example, on 349 days in the year and in the Kerguelen Islands at sea-level there are 236 days

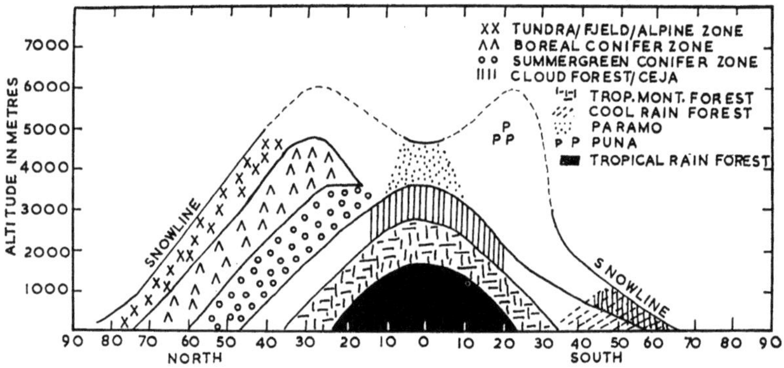

Fig. 79. Diagrammatic altitudinal vegetational zonation profile, from the Arctic to the Antarctic, showing the asymmetry between the northern and sourthern hemispheres. The boreal elements extend southwards at higher and higher elevations and reach their maximum southern limits on the margin of the tropics. The southern elements are related to the high altitudes of the tropics (Modified from TROLL).

of frost annually. The summer temperatures, the summer vegetation period, the winter snow-cover and the depth of the summer thawing of the ground, especially in regions of permafrost on the lowlands of large continental areas—decisive factors that play such an important rôle in the general ecology of the treeline zone in the northern hemisphere—are completely absent in sub-antarctic and on tropical mountains. The fact that only miniature forms of the typical frost-soil are found even in the periglacial zone of the sub-antarctic regions and not in the Antarctic, clearly shows the part played by the short periodic frost of slight depth. According to the measurements made by the German Antarctic Expedition, the soil temperature in the Kerguelen Islands, far away from the Antarctic forest-line and a few hunderd metres below the permanent snowline, is related to the fact that 235 days of the year there is frost, but not deeper than 5 cm of the soil. This climate is not absolutely free from seasonalism, but the mean amplitude of the annual temperature variation is only 6.5° C. Only a single month is frost-free, but even during the warmest month of the year, the temperature on the surface of the ground is 0° C. These are precisely the conditions that are never met with at comparable latitudes in the northern hemisphere. Under these high oceanic conditions, with lower summer temperatures and the possibility of frost during all the months of the year, the antarctic forest-line is situated at relatively low latitudes like 40-45° south in the sub-antarctic, but only at 55° NL in the North Atlantic Islands.

The typical boreal vegetation types, like the tundra, coniferous forests, are characteristic of only the continental winter-cold climates of Eurasia and North America. In the high altitude areas of the oceanic cold-temperate regions of the southern hemisphere we find, however, tussock grasslands, communities of cushion plants and luxuriant evergreen scrubs and forests. The cool-temperate forests of Patagonia and New Zealand have a striking similarity to the upper montane forests of tropical mountains, the Ceja de la Montaña of the Andes, the moss forests and the elfin forests of tropical East Africa and of Malaysia (*vide* Chapter

VIII). The tussock grasslands of the sub-antarctic region have similarly their counter-parts in the tussock and cushion formations, the paramos, of equatorial high mountains. The same zonal succession may be observed vertically between the lowlands and the mountains in New Guinea, as is encountered when we proceed from the north to the south on the lowland in New Zealand. In each case, the succession passes from the tropical rain forests, through lower montane forests with *Weinmannia*, *Podocarpus* and *Nothofagus*, to the upper montane forests with *Libocerus* and *Dacrydium* and then to the tussock grasslands with *Danthonia*. These characteristic similarities and differences are closely correlated with the climatic peculiarities described above. The tree-forms of the antarctic treeline either in the Auckland Island or on the high Patagonia are quite different from those of the arctic or alpine timberlines of the Alps and the Himalaya, in the winter-cold regions of the northern hemisphere and are curiously similar to the upper limits of the high altitude forests on the mountains of tropical East Africa. Many southern genera and sometimes even species of the cool-temperate forests extend northward to the tropical upper montane forests. The high forests of the tierra fria show the greatest affinity to the sub-antarctic rain-forests of Patagonia and Tierra del Fuego and New Zealand. This is also equally true of the paramos and the punas of the tropical high mountains and the humid or steppe-like grasslands of the sub-antarctic. In the temperate, permanently cool climates of these grasslands, with their meadows, moors, stony pastures and steppes, we find wholly similar organisms as are found on the high mountains of the tropical regions, viz. the tussock grass, hard polsters, woolly-leaved herbs, etc. A number of the sub-antarctic plants, like *Azorella*, *Acaena*, *Colobanthus*, *Lycopodium saururus*, *Oreobolus*, *Donatia*, *Astelia*, *Gaimardia*, *Abrotanella*, *Arctiastrum*, *Lomaria*, etc., and partly also species belonging here, are found on the Andes of the tropics. In addition to the sub-antarctic-tropico-montane vegetation, there are also others that are evidently cases of convergence. This is, for example, the case with the tussock grass of the sub-antarctic area and the ichu grass of the Andes, the woolly herbs *Pleurophyllum* and *Culcitium*, the woolly-polster plants *Raoulia*, *Haastia* and *Pycnophyllum*, the leafy *Stilbocarpa* and *Gunnera*, etc. The tropical high mountains and the sub-antarctic areas have also special turf-building hard polster plants of the moors-type, as for example, *Distichia*-moor of West Patagonia, New Zealand and Hawaiian Islands. It has been estimated by competent authorities that at least 65% of the cushion-plants so far known from the world grow either on the Andes of the tropics or in the sub-antarctic areas. The *Azorella*-formation of the Kerguelen Islands, Tierra del Fuego and Patagonia is also found at elevations of 4600 m on the high Andes of the tropics. The woolly *Raoulia-Haastia* have similarly their counterparts in the Andean *Pycnophyllum*. *Pleurophyllum hookeri*, with dense white woolly coat from the Macquarie Island, resembles *Culcitium* occurring almost up to the snowline on the Andes of the equatorial region (fig. 80).

The trees of the timberline of the antarctic regions are different from those of the arctic and the alpine timberlines of the northern hemisphere. They are evergreen trees, mostly hard, broad-leaved, though often also needle-leaved or also with the leaves rolled; densely foliaged, spherical, dome-shaped or umbrella-shaped types, characteristic of the upper limits of the tropical cloud-forest (fog-forest). Some of the best known types are *Metrosideros lucida* (Myrtaceae) from New Zealand, *Phylica nitida* (Rhamnaceae) of the islands of the Indian Ocean and *Nothofagus* from Tierra del Fuego. The continental timberline in Patagonia has deciduous *Nothofagus*. There is, therefore, considerable similarity between the antarctic timberline and the altitudinal limits of the forests on the tropical mountains. For a review of the plant life of the Subantarctic and Antarctic see WACE 1085a.

2. THE SOUTHERN ANDES

The southern parts of the Andes are perhaps entomologically the best explored area in the south-temperate region. The treeline and snowline are, as may be expected, much lower in this area than on the northern Andes. The permanent snowline is at an elevation of 1500 m at the latitude of Chiloe, but south of the Magellan Strait, about 900 m above mean sea-level and glaciers often descend to the sea-level in many places.

The high altitude Coleoptera from this region are extremely poorly known, but the Lepidoptera are relatively better known. Most of the high altitude butterflies from the southern Andes are typically small forms, characteristically reminiscent of the high-alpine types from the Alps in their general colour and habits. The genera peculiar to the southern parts of the Andes are *Phulia* HERRICH-SCHÄFFER and *Trifurcula* ZELL. that have their close relatives among the Pierids of Asia. *Colias* FABR. and *Pieris* LINN. are also found here. The Nymphalidae and Satyridae are represented by *Vanessa* FABR. and *Argynnis* FABR., *Pseudomaniola* WEYM., *Lymanopoda* WESTW., *Pedaliodes* BUTLER, etc. in place of *Erebia* DALM. The Heterocera include *Agrotis* O., *Prodenia* GUEN., *Mamestra* OCHSENHEIMER, *Psodos* TR., *Azelina* GUEN., etc. (810).

The high altitude Lepidoptera of the southern Andes are thus characterized by extremely few typically characteristic genera. Although a number of the species from the forest zone seem to ascend to great elevations above the timberline and often even up to the permanent snowline, the true autochthonous forms above the forest-line are not many. These include *Faunula leucoglene* FELD., *Argynnis modesta* BLANCH., *Scolitantides andina* CALV., *Phulia nymphula* BLANCH and perhaps also *Colias exuxanthe* FELD. The Satyrid *Faunula lecoglene* FELD. is remarkably similar in its habits to *Erebiola butleri* FER., from New Zealand mountains.

The importance of terrestrial zoology of the alpine zones of the southern Andes has recently attracted considerable attention (668). Both above and below the lower limits of the alpine zone, the deciduous trees *Nothofagus antarctica* and *N. pumilio* flourish. The alpine zone contains, among others, the cushion plants *Azorella* and *Bolax* and also *Empetrum rubrum*. The zone is not restricted to the southern Andes, but extends the whole length of the Andes to Ecuador and Colombia, though becoming more and more restricted to the higher altitudes and at the same time retaining the same fundamental features of the general composition of the fauna.

In the region between the 36th and 48th south parallels, the dominant Curculionidae are (Cylindrorrhinini) *Adioristus* WAT., *Scotoeborus* SCHÖNH., *Cylindrorrhinus* GUÉR. and the Tenebrionidae *Epipedonota* SOL. (Nycteliini). The dominance shifts southwards to the Listroderini-Curculionidae and the Helopinae of Tenebrionidae. Above the upper limits of *Nothofagus pumilio* and *N. antarctica*, up to 100-200 m the ground is mostly snow-free during the summer but occasional snow storms may leave patches of snow on the ground. There are no ponds and the region is subject to strong winds, but the soil is relatively rich in apterous and brachypterous insects, especially *Listroderes* SCHÖNH. of the subgenus *Antarctobius* FAIRM. and *Falklandius*, ENDERL. (Curculionidae) and the Carabid *Merizodius* SOL. and the Tipulids and Empidids with atrophied wings. The Patagonian steppe-zone embraces the large unforested areas, east of the Andes in Patagonia. As it recedes from the Magellanic forests, its semi-desert character becomes accentuated. The vegetation is dominated by low, xerophilous, spiny and small-leaved shrubs and hard grasses. A small part of the Patagonian steppe zone lies within Chile and is divided into a drier sub-zone and a wetter sub-zone. The drier sub-zone, reaching to 52° SL, is characterized by tussock grasses and *Mulinum spinosum*. The wetter sub-zone has grassland, extending south from the 52nd south parallel to the islands southeast of the Beagle Canal, including the Isla de Los Estados and the Falklands. The first sub-zone is remarkable for the dominance of Tenebrionidae-Nycteliini *Nyctelia* LATR., Scotobiini *Scotobius* GERM. and *Emmalodera* SOL. and Praocini *Praocis* ESCHS. The dominant Curculionidae are *Cylindrorrhinus* GUÉR., *Adioristus* WAT., *Scotoeborus* SCHÖNH., etc. Here we also find the grasshoppers *Trimerotropis* STÅL, *Dichroplus* STÅL and *Bufonacris* WALK. The dominant types in the second sub-zone are *Listroderes* SCHÖNH., with the subgenus *Antarctobius* FAIRM., *Puranius* GERMAIN, *Falklandius* ENDERLEIN and *Falklandella* SCHELLENBERG (Curculionidae); *Migadops* WATERH., *Merizodus* SOL., *Metius* CURT. and *Abropus* WATERH. (Carabidae); *Parahelops* WATERH., *Hydromedion* WATERH. (Tenebrionidae); *Taurocerastes patagonicus* PHIL. (Scarabaeidae) and *Eriopis magellanica* PHIL. (Coccinellidae).

The terrestrial Plecoptera from the southern Andes are of considerable ecological and biogeographical importance (560, 561). The Andiperline group is remarkable for being wholly apterous and for lacking ocelli (59). *Andiperla willinki* AUBERT occurs under stones on the Argentinian Andes and other parts of the southern Andes, up to elevations of about 1000 m, at latitudes of 49° south. *Megandiperla* ILLIES, also belonging to the same group, has smaller eyes, is of darker colour and completely lacks ocelli and wings. The male of *M. kuscheli* ILLIES measures about 19-20 mm long and the female 21-23 mm long; both occur under stones on the high meadows on the Monte Jervis at 49° SL on the Chilean Andes. The group is apparently endemic to a small region of the south Patagonian mountains, characterized by relatively heavy precipitation. The well developed eyes of *Andiperla willinki* AUBERT would seem to indicate life in a rather dimly lighted biotope and the larvae are apparently normally aquatic. The first definitive record of an entirely terrestrial larva refers to *Megandiperla kuscheli* ILLIES. The larvae of this species have extremely well developed abdominal gills between the cerci, a condition correlated no doubt with the high atmospheric humidity of the region. Southwest Patagonia is indeed a region of heaviest precipitation and it rains practically every day in the year. The mountains are only exceptionally free from clouds. The species occurs at elevations of 900-1000 m, in other words in the zone with lichen, moss and some alpine vegetation of *Azorella selago*. Although there are no streams and ponds, there is abundant snow and the meadows are strewn with stones, under which the larvae are found. Like other Gripopteryidae, the species seems to feed on lichen and moss. *Apteryoperla* WIS., also belonging to the same family, occurs at an elevation of 2000 m on the alpine zones on Mt. Irene in New Zealand under similar conditions. Numerous larvae of *A. monticola* WIS. occur under stones in the vicinity of melting snow. This species is also completely apterous, but has well developed ocelli (59, 1117). It has recently been shown that the Andiperlinae are an unnatural subfamily (563). The genera

Andiperla AUBERT and *Andiperlodes* ILLIES are placed under Paragripopteryinae, *Apteryoperla* WIS. in Leptoperlinae and *Megandiperla* ILLIES in Antarctoperlinae.

3. AUSTRALIAN AND NEW ZEALAND MOUNTAINS

The mountain ranges of Australia do not rise to any great altitudes. Most of them are essentially portions of the Miocene and Pre-Miocene surfaces, largely composed of granites and folded strata, elevated by warping and isolated by dissection. This is, for example, the case with the Macdonnell, Musgrave and Everard Ranges, the Grampians of Western Victoria, the Drummond Range of Queensland. etc. The Stirling Range of Western Australia may have been uplifted with faulting, The Mt. Lofty and the Flinders Ranges of South Australia are horsts, eroded into ridges. Other mountains are dissected volcanoes or masses of Cainozoic basalt that have suffered erosion. Some of the so-called mountains of Australia, such as for example, the Darling Range of Western Australia and the Blue Mountains of New South Wales, are in reality more or less extensive plateaux, intersected by valleys. In New South Wales there are the volcanic peaks Bajimba (1525 m), Capoompeta (1555 m) and the Snowy Mountains (1525 m). The highest summit on the Stirling Range is only 1100 m above mean sea-level in the southwest. East of 126th meridian and between the 24th and 27th south parallels, there arise from the plain a series of elevations, viz. the Rawlinson, Warburton, Blackstone and Townsend Ranges. Mt. Aloysius is one of the highest and rises to an elevation of 1103 m. The Macdonnell Ranges dominate the Northern Territory and include the Macdonnell, Strangways, Ferguson, Harts, etc. The Macdonnell Ranges stretch nearly 400 km, in the form of an open sigmoid curve. Mt. Ziel, in the extreme west, rises to an elevation of 1450 m. Mt. Gillen (1940 m) and Mt. Sander (1370 m) are on the Heavitree Range (721). The South Australian mountains do not usually rise to elevations above 900 m and include the Mt. Lofty Range in the east; the highest point is about 1200 m above mean sea-level. Mt. Woodroffe (1660 m) is on the

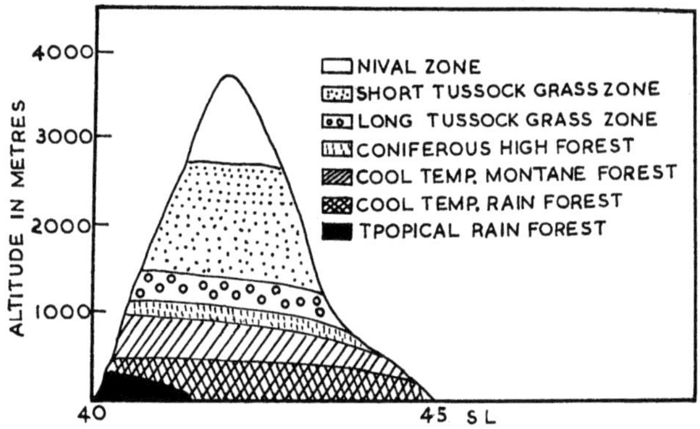

Fig. 80. Profile section of New Zealand (South Island) mountains, showing the altitudinal zonation of vegetation. (Modified from TROLL).

Musgrave Range. The eastern discontinuous belt of highlands, known collectively as the Coastal Range, are usually not higher than 1200 m above mean sea-level. Mt. Barle Frère (1610 m) is on the Bellenden Ker Range. In New South Wales, Mt. Spiriby, Bajimba and Capoompeta rise to an elevation of 1555 m and the Chandler Peak is 1565 m above mean sea-level. In the Southern Highlands are the well known Mt. Kosciusko (2227 m, 36-37° SL and 148° EL) (190), Mt. Gingera, Bimberi and Murray (all above 1800 m above mean sea-level). In Victoria we find Mt. Cobberas (1825 m), Mt. Gibbo (1755 m) and Mt. Bogong (1982 m).

The Central Plateau of Tasmania is for the most part stony and barren, sloping southwards and ranges in elevation from 720 to nearly 1525 m. The highest summit in Tasmania is the Legge Peak (1558 m) that forms a part of the triangle of highlands Mt. Arthur, Mt. Victoria and Ben Lomond.

In New Zealand, the high peaks and ridges in the area are the Cradle Mountain (1525 m), Mt. Olympus, Mt. Hugel, Mt. Roland, Mt. Black Bluff, etc. South of 42° 30′ SL are the Southern Highlands, with the Snowy Mountains, Wellington Range, etc. Mt. Tasman (3499 m) is on the Southern Alps that have a general trend of northeast to southwest, a little north of 44° SL and east of 170° EL in the South Island, New Zealand. Mt. Cook (3764 m) is situated a little southwest of this. The well known Tasman Glacier is to the southeast. The Pleistocene glaciations in Australia were confined to the higher latitudes and altitudes. About 870 sq. km area in the higher parts of the Kosciusko Plateau and more than one-third of Tasmania were covered by ice. The general geographical features of southwestern New Zealand are described by BENSON (108) and the physiography of the Western Macdonnell Ranges is described by MADIGAN (721). See also KEAST (611a).

The records of high altitude insects from the mountains of Australia and New Zealand are extremely scanty. The Satyrid *Heteronympha* WALLGR. is considered as typical of the high mountains of Tasmania and New Zealand. *Xenica* BUTL. is likewise represented by a number of species, like *Xenica orichora* MAYR. *Xenica correae* OLLIFF., etc. wich are exclusively high altitude forms that occur at elevations of 1500-1800 m on Mt. Kosciusko. Most of the species have typically localized distribution on the Tasmanian mountains. Some species of *Erebia* DALM. have also been recorded from the mountains of New Zealand and it is extremely interesting to observe that the genus is absent on the Australian and Tasmanian mountains. *Erebiola butleri* FER. and *Percnodaimon pluto* FER., closely related to the high mountain Satyridae of the Palaearctic, are exclusively mountains species that occur at elevations of 1000-1900 m on the South Island (New Zealand). *Argyrophenga antipodum* DBLD. is another Satyrid found up to an elevation of 1200 m in the South Island. Other mountain species belong to *Metacrias* MEYR., *Notoreas* MEYR., *Dasyuris* GN., etc. The adult female of the Arctiid *Metacrias* MEYR. is interesting for its greatly atrophied wings. The grasshopper *Pezotettix* BURM. is also reported to be characteristic of the mountains of New Zealand. There are interesting discussions of the origin of the flora and fauna of New Zealand and Australia by HUTTON (558, 559) and KEAST et al. (611a).

4. ANTARCTIC MOUNTAINS

Over 500 expeditions have perhaps explored the Antarctic Continent since the end of the Second World War, but only about 150 of these are of geographical and scientific importance. Important results have been archieved by the British-American Watkins-Hearst Antarctic Expedition of 1928-1929, the British-Australian-New Zealand Antarctic Expedition of 1929-1931 and the Norwegian (Maudheim) Expedition of 1949-1952. Biological observations were not part of the Geophysical-Year Antarctic Programme of 1957-1958, but the opportunity was utilized by the Special Committee for Antarctic Research and the International Council of Scientific Unions to make some preliminary biological observations also. Some attention was thus paid to the terrestrial Invertebrates and their ecology, but many details are not yet available (GRESSITT (427a), DALENIUS (223a). As a part of the recent United States Antarctic Research Programme, JANETSCHEK (570, 571, 572) studied some important aspects of the general terrestrial animals and their ecology up to the Beardmore Glacier, at an elevation of 3330 m (the Anna Heights), north of Mt. Nimrod (85° 15′ SL and 167° EL). A number of observations were made at elevations of 2200 m and up to the summit on Mt. Erebus (4040 ± 20 m), the active volcano in Antarctica.

The maximum thickness of the Antarctic ice*, at a distance of about 160 km east of the U.S. Station, is estimated to be about 4270 m. The mean thickness of the inland-ice is between 2000 and 2500 m, totalling to a quantity of ice of about 29 000 000 cu km. This would represent nearly 91% of all the ice on the surface of the Earth at present. The mean annual increase of ice is estimated at 1.32 billion tons and the mean annual loss at 0.41 billion tons. The mean sea-level thus falls by 3.6 mm and rises by 1.1 mm annually. The Antarctic ice-cover may therefore be described as fairly stable at the present time. This enormous mass of ice naturally depresses the earth's crust to an extent of about one kilometre, so that earthquakes are comparatively rare occurrences in the Antarctic Region. Recent explorations (HARRINGTON (479a)) seem to indicate that we deal with two distinct areas in the Antarctic: the East Antarctic with its structure analogous to that of large continental shell, as in the case of the adjacent southern hemisphere; and the West Antarctic separated from the East Antarctic by the Weddel Sea. In its subglacial history, the latter area is an archipelago, which is in fact a continuation of the Andes—with mountains of alpine type, reaching to elevations of over 5000 m—and connecting with South America by the backbone of Nova Scotia (the South Antiles). The thickness of the crust in the West Antarctic is about 30 km, while that of the East Antarctic is 40-48 km. Fossiliferous and coal-bearing sediments, associated with immense tillite deposits, contain *Glossopteris*-flora of the Gondwana-Series, characteristic of the other southern continents. Milder climates appear to have prevailed on the coastal strips during the past, perhaps about 10 000 yaers ago.

GOLLERBAKH & SYROECHKOVSKY (414) recognize the following subdivisions of Antarctica: i. the Glacier Cap ice-cap area, with isolated rock nunataks in the inner continent, with lichens; ii. the young Antarctic cold desert, comprising the coastal strip of the mainland, though isolated from the oceans, characterized by sparse vegetation of algae, moss and lichens; iii. the mature Antarctic oasis or the mature cold desert, low areas near the margin of the continent, connected by bays to the sea, but also separated by extensive ice-belts from it; iv. the coastal islands and the coastal ice; and v. the subantarctic islands. On the coastal islands the tropical affinities are evident and an increased dependence of animals on vegetation may also be observed

* BAUER, A. & C. LORIUS, 1964. *UNESCO Impact. Sci. Soc.* 14(4): 223-238 have given a summary of some recent observations.

The present-day climate of the region is remarkable for the extremely low atmospheric temperature and precipitation—the total annual precipitation is between 145 and 192 mm and that also falls as snow. Winds are constant and favour rapid evaporation and strong cooling. The so-called "cold pole" is situated at an elevation of about 3000 m and has a mean annual temperature of —56° C; the lowest temperature so far recorded in the region is —90° C. The "wind pole" is situated on the coast of Terre Adélie, where the maximum wind velocities reach 250 km/h and the duration of winds is also the longest in the day. In areas with rich animal life, in the East Antarctic, near the coast in the Australian-American Wilkes Station, the mean annual atmospheric temperature is —8.4° C and the mean summer temperature is —1.7° C. Insolation is remarkably intense and even at the sea-level is as strong as at an elevation of 3000 m on the Alps. The layer of air immediately above the ground thus becomes rapidly warmed up, but the low atmospheric humidity is the most decisive ecological factor in the region.

Under the extreme atmospheric conditions, exceedingly few frost-hardy forms, also capable of resisting the rapid desiccation, seem to have become adapted to survive the deep Antarctic winter. The micro-climatic conditions are, however, often comparable to those met with in temperate latitudes. Photosynthesis of the Thallophytes enclosed within the ice is facilitated by the intense insolation during the brief summer. The insolation is indeed so intense that these plants are actually enclosed in an immediate layer of melt-water, sufficient also to activate the associated microfauna. As on high mountains in other parts of the world, the intensity of light may, however, serve as an inhibitory factor at the peak of the vegetative period. The primary producers in the food chain are found among the thermophiles, but there is also a fairly wide range of terrestrial Arthropoda that are characterized by different temperature optima, cold resistance and other micro-climatic conditions. We do not, however, know the precise mode of survival of the various organisms during the winter in many localities. Some of the Collembola, especially *Isotoma* spp., occurring at elevations of 1000-1800 m, are devoid of pigment and are apparently strictly cold stenotherm elements. Another Collembolan *Gomphiocephalus hodgsoni* CARPENTER has, on the other hand, a temperature preference of 11° C and becomes inactive even at 0° C and is conspicuously heavily pigmented (572).

The conditions of atmospheric humidity are, however, much less optimal for most organisms, so that there is general predominance of encystment as an adaptation against the prolonged periods of severe desiccation. The soil moisture and the soil-air conditions are also other important limiting factors for these organisms. The nutritional behaviour of bryocole Arthropoda is more or less similar to that prevailing on high mountains in the temperate regions of the world. The Trombidid mites and Collembola of the high Antarctic feed mostly on the fungus *Aspergillus* and on moss and to a small extent on lichens.

As may perhaps be readily expected, the distribution of the Arthropoda is irregularly patchy in the region. JANETSCHEK reports, for example, that on 78th south parallel in certain localities, often as many as 300 individuals of Collembola occur per square metre on mossy ground, but in other localities, apparently equally favourable, not a single specimen may be found.

An interesting comparison is made by JANETSCHEK of the terrestrial biology of the high Antarctic with that of the extra-polar high mountains regions. The ecological relations are fundamentally very different in the two cases. In a sense, the conditions reverse to those prevailing on Antarctica are met with on the high mountains in other parts of the world. Even in apparently optimal localities there is an extreme poverty of organisms in the lowlands of Antarctica, but on the high Antarctic mountains there is relatively great abundance and variety of organisms. In Antarctica the coastal glaciers alone reveal locally isolated snow vegetation and the inner glacier mass is white. On the Antarctic mountains, however, the glaciers are dusty and covered by diverse organic particles and detritus, in addition to the presence of certain organisms in the melt-waters. These attract an enormous variety of detritus-feeding species. The ecosystems of the extra-polar high mountains are described as anabolic-allotroph systems, but those of the Antarctic mountains are katabolic autotoph systems (571). In comparison to those of the extra-polar high mountains, the ecosystems of the Antarctica are extremely simple and largely independent (autonomous) systems. Considerable floristic and faunistic affinities are also evident with Australia, Tasmania, New Zealand, New Guinea and New Caledonia, besides South America. Several considerations indicate that the Plecoptera arose in the Antarctic or at least in the southern hemisphere.

The whole of the Antarctica is far south of the southern treeline and lacks truly endemic terrestrial Vertebrates. The northernmost part of the Antarctic Peninsula has three Phanerogams, several liverworts, some higher fungi and about 75 species of mosses. In comparison to this, the arctic flora is very rich, with over 100 species of Phanerogams. In the Canadian arctic there are, for example, nearly 300 species of Phanerogams, in Novaya Zemlya 182 and in Spitzbergen 128. Further north of these there are over 150 species of mosses. Antarctica is the only landmass where lichens seem to out-dominate most other types of plants. Over 250 species of lichens have already been described from the Antarctic Continent. The lichens seem to thrive exceedingly well under the conditions of high atmospheric aridity and grow on the Antarctic mountains at or near 88° SL. Blue algae also abound near the coast and occur at elevations of 3000 m on Mt. Erebus, along with some other microphytes, soil Protozoa and Rotifera. This is the maximum altitude, at which organisms have so far been found to flourish in Antarctica. In the soil are also found some Nematoda, Tardigrada, about 15 species each of Collembola and Acarina and a couple of species of brackish-water Chironomidae. The Sub-Antarctic area has a relatively rich flora and fauna, with over 700 species of Arthropoda. In all cases the ecosystems have microphytes and Collembola at the base of the pyramid of numbers. Lichens abound on the nunatak rocks on the inland ice and are characterized by high endemism of species, especially among the Lecideaceae, Lecanoraceae and Blatsteniaceae. The southernmost record for a winged insect is *Parochlus* (*Podonomus*) *steineni* GERCKE. Acarina occur up to an elevation of 1800 m and Collembola ascend to 2000 m at 77° SL. This should perhaps be considered at present as the upper altitudinal limit for most Arthropoda on the Antarctic landmass, but Tardigrada are reported at an elevation of 3000 m on Mt. Erebus. Larvae of Lepidoptera, found at 1800 m on the 77th south parallel, represent probably wind-blown specimens from the Sub-Antarctic areas.

JANETSCHEK (570, 571) recognizes two principal types of terrestrial ecosystems: i. the chaliko system and ii. the bryo-system. The former refers to the naked rubble, derived from the weathering of the bedrock and moraine of different ages, without any apparent macrophytes. It is typically characterized by semi-edaphic mesofaunat comprising some half a dozen species of Collembola and some Acarina. The very short, simple food-chain of the mesofauna begins by the utilization of soil fungi by the Collembola. Wherever Arthropods exist, the soil microphytes also seem to be present, even at elevations of 2000 m. The mountain crest of Mackay Glacier may perhaps be the absolute limit for terrestrial animal life. At these elevations near the upper limits of distribution, the chaliko-system exhibits an impoverished facies, consisting of only one or two species of Collembola. The most important limiting factors for these forms are not perhaps the soil temperatures as such but the moisture content of the soil. The soil-air humidity and the surface are correlated with the clay content of the soil and therefore the general petrography of the bedrock is of considerable importance. The bryophyte system is composed of more or less open, rarely closed macrophytic vegetation of moss, lichen, thin algal cover, etc. The microfauna forms the dominant types: Protozoa, Nematoda, Rotifers especially bdelloid, and Tardigrada. The component elements of the bryo-system are, therefore, the same as in other parts of the world. The vertical distribution of this bryo-system is closely correlated with the upper limit of growth of moss and soil lichen, to about 1300 m (crest of Mt. Suess and Mt. England). Both the chaliko-system and the bryo-sytem often occur side by side, so as to form a mosaic.

The absolute highest altitude, at which direct evidence of plants and animals was observed by JANETSCHEK, is 3600 m on the western slopes of Mt. Erebus, in the soil near fumaroles, between the active crater and the next older one below. The soil has a pH of 4.2-4.3 and contains mixed populations of fungi and bacteria, Cyanophyta, Rhizopoda, Rotifers and perhaps also Tardigrada. The top of Mt. Erebus, the ridge of the active crater, a little more than 4000 m above mean sea-level, with the soil pH of 2.7-2.8, is wholly sterile.

The Collembolan *Gomphiocephalus hodgsoni* CARPENTER does not seem to be completely in harmony with the ecological conditions prevailing in Antarctica and may have perhaps immigrated relatively recently from some of the Sub-Antarctic islands. Both the chaliko-system and the bryo-system reach almost the 86th south parallel on the Queen Maud Range and the vertical distribution, as reported by the USSR Antarctic Expedition, is 3000 m on the coastal range and 2700 m in the region of the USSR Station Lazarev in Queen Maud Land (71-72° SL).

BIBLIOGRAPHY

1. ABSOLON, K. 1900. Vorläufige Mitteilung über einige neue Collembolen aus den Höhlen des mährischen Karstes. *Zool. Anz.*, 23: 265-269.
2. ABSOLON, K. 1911. Gletscherflöhe in den nieder-österreichischen Voralpen. *Mitt. Sekt. Naturf. Österr. Touristenkl.*, 24: 1.
3 ABSOLON, K. 1915/1916. Bericht über höhlenbewohnende Staphyliniden der dinarischen und angrenzenden Karstgebiete. *Koleopt. Rundschau*, 4: 132-151 (1915); 5: 1-18 (1916).
4. ADAMS, C. C. 1902. Postglacial origin and migrations of the life of the northeastern United States. *J. Geogr.*, 1: 303-310, 352-357.
5. ADAMS, C. C., 1915. An ecological study of the prairie and forest Invertebrates. *Bull. Illinois Lab.*, Urbana, 11: 31-280, pl. i-lxiii.
6. ADAMS, C. C., G. P. BURNS, T. L. HANKINSON, B. MOORE & N. TAYLOR, 1920. Plants and animals of Mount Marcy, New York. *Ecology*, 1: 71-94, 204-233, 274-288.
7. ADELUNG, N. 1908. Beitrag zur Kenntnis der Orthopterenfauna Transkaukasiens. *Horae Soc. ent. Rossicae*, 38 (2): 32-82, pl. i.
8. AELLEN, V. & P. STRINATI, 1956. Matériaux pour une faune cavernicole de la Suisse. *Rev. Suisse Zool.*, 63 (1): 183-202.
9. AGRELL, I. 1934. Studien über die Verteilung der Collembolen. Untersuchungen im schwedischen Lappland. *Opusc. ent.* (Suppl.) 3 (6): 1-236.
11. ALCOCK, A. 1897. Report upon the natural history of the Pamir Boundary Commission. *Rep. Proc. Pamir Boundary Commission*, Calcutta, pp. 69-70.
12. ALEXANDER, C. P. 1940. The Presidential Range of New Hampshire as a biological environment, with particular reference to insects. *Amer. Midland Nat.*, 24 (1): 104-132.
13. ALEXANDER, C. P. 1949. Records and descriptions of North American craneflies (Diptera) Part VIII. The Tipulidae of Washington. I. *Amer. Midland Nat.*, 42 (2): 257-333.
14. ALEXANDER, C. P. 1956/1957. Undescribed species of craneflies from the Himalaya Mountains (Tipulidae: Diptera) I. *J. N. Y. ent. Soc.*, 64: 137-147 (1956); II. *ibid.*, 65: 147-157 (1957).
15. ALEXANDER, C. P. 1961. New or little known species of exotic Tipulidae (Diptera). IX. *Proc. R. ent. Soc. London*, (B) 30: 167-172, fig. 12.
16. ALEXANDER, C. P. 1962. New or little known Tipulidae from Eastern Asia (Diptera) XIX. *Philip. J. Sci.*, 90 (2): 155-214, pl. i-v.
17. ALEXANDER C. P. 1962. Beiträge zur Kenntnis der Insektenfauna Boliviens XVII. Diptera II. The craneflies (Tipulidae: Diptera). *Veröff. zool. Staatssammlung München*, 7: 9-159.
17a. ALEXANDER, C. P. 1963. Blepharoceridae and Deuterophlebiidae. Guide to Insects of Connecticut VI. The Diptera or true flies 8. *Bull. State geol. nat. Hist. Surv. Connecticut*, 93: 39-80, pl. xv.
18. ALEXANDER, C. P., C. D. HARRISON & J. R. MALLOCH, 1919. Diptera. *Rep. Canadian Arctic Expedition 1913-1918*, Ottawa, 3 (C): 1-90 (1922).
19. ALEXANDER, G. 1933. Some relations between altitude and the distribution of Acrididae in Colorado. *Ann. ent. Soc. Amer.*, 26: 585-590.
20. ALEXANDER, G. 1937. Morphological variations in Acrididae correlated with altitude. (Abstract). *J. Colorado-Wyoming Acad. Sci.*, 2: 46.
21. ALEXANDER, G. 1941. Keys for the identification of Colorado Orthoptera. *Univ. Colorado Studies*, (D) 1 (3): 129-164.
22. ALEXANDER, G. 1951. The occurrence of Orthoptera at high altitudes, with special reference to Colorado. Acrididae. *Ecology*, 32 (1): 104-112.

23. ALEXANDER, G. 1951. The natural history of high altitudes. *The Biologist*, 33 (3): 91-97.
24. ALEXANDER, G. 1961. The type locality of *Gomphocerus clavatus* Thomas (Orthoptera: Acrididae). *Ent. News*, 72 (1): 107-111.
25. ALEXANDROV-MARTYNOV, O. N. & L. BIANCHI, 1931. Neuroptera. Abhandlungen der Pamir Expedition 1928, 8 (Zoology): 119-125.
26. ALFKEN, J. D. 1931. Entomologische Ergebnisse der Deutsch-Russischen Alai-Pamir-Expedition (II): Hymenoptera IV. Apidae. (excl. *Bombus* and *Anthophora*). *Mitt. zool. Mus. Berlin*, 16: 823-844.
27. ALLEE, W. C. 1931. Animal aggregations: A study in the general sociology. University Chicago Press.
28. ALLEE, W. C. 1934. Recent studies in mass physiology. *Biol. Rev.*, 9 (1): 1-48.
29. ALLEE, W. C., A. E. EMERSON, O. PARK, T. PARK & K. P. SCHMIDT, 1949. Principles of Animal Ecology. W. B. Saunders Company, Philadelphia and London, pp. 1-837.
30. ALLUAUD, CH. 1908. Les Coléoptères de la faune alpine du Kilimandjaro avec notes sur la faune du Mont Méru. *Ann. Soc. ent. France*, 77: 21-32.
31. ALLUAUD, CH. 1917. Les Carabiques de la fauna alpine des hautes montagne de l'Afrique orientale. *Ann. Soc. ent. France*, 86: 73-116, fig. 9.
32. ALLUAUD, CH. 1927. Études des Coléoptères de la famille des Carabidae. *Ark. Zool.*, 18 (A): 1-22.
33. ALLUAUD, CH. 1937. Entomological Expedition to Abyssinia 1926-27. Coleoptera, Carabidae. *Ann. Mag. nat. Hist.*, (10) 9: 272-287.
34. ALLUAUD, CH. & R. JEANNEL, 1911-12. Voyage en Afrique orientale. Result. scientifiques 1913-1925. Paris.
35. ALLUAUD, CH. & R. JEANNEL, 1912-1929. Voyage de Ch. Alluaud et R. Jeannel en Afrique orientale 1911-1912: Liste des stations avec une carte 1912; Résult. scientifique. Paris.
36. ALPHÉRAKY, S. 1887. Diagnosen einiger neuer Centralasiatischer Lepidopteren *Stett. ent. Z.*, 48: 167-171.
37. ALPHÉRAKY, S. 1889. Le Pamir et sa faune Lépidoptérologique. Noctuelites. *Mem. Rom.*, 5: 124-191, pl. vi-viii.
38. ALPHÉRAKY, S. 1897. Lepidopteren aus Kamtschatka, gesammelt von O. Herz. *Mém. Lepidop.*, 9: 301-307.
39. ALPHÉRAKY, S. 1897. Mémoire sur différents Lépidoptères, tant nouveaux que peu connus de la faune palaearctique. *Mém. Lepidop.*, 9: 185-227.
40. ALTLAND, P. D. 1946. Effects of discontinuous exposures to 25 000 ft simulated altitude on body weight and reproductive system of immature rats. *Anat. Rec.*, 96: 27.
41. ALTLAND, P. D. 1949. Effect of discontinuous exposure to 25 000 ft simulated altitude on growth and reproduction of the albino rat. *J. exp. Zool.*, 110 (1): 1-15.
42. ALTLAND, P. D. & M. PARKER, 1955. Effects of hypoxia upon the box turtle. *Amer. J. Physiol.*, 180 (2): 421.
43. ALTLAND, P. D. & B. HIGHMAN, 1957. Effects of high altitude exposures on dogs and on their susceptibility to endocarditis. *J. Aviat. Med.*, 28: 253.
44. AMANN, J. & H. KNABL, 1912-1913. Die Coteopterenfauna des Ötztals. *Coleop. Rundschau*, 1: 36-40, 57-61, 73-77, 92-96, 112-115, 143-147, 161-163, 181-185 (1912); 2: 40-42, 51-59, 71-75, 82-90 (1913).
45. ANDER, A. 1949. Die boreoalpinen Orthopteren Europas. *Opusc. ent.*, 14: 89-104.
46. ANDERSON, J. & S. MOORE, 1930. The book of the White Mountains. pp. 300.
47. ANDREWES, H. E. 1923. Coleoptera of the Second Mount Everest Expedition 1922. Part I. Carabidae. *Ann. Mag. nat. Hist.*, (9) 11: 273-278.
48. ANDREWES, H. E. 1924. Carabidae. Guy Babault Mission dans les Provinces Centrales de l'Inde et dans la région occidentale de l'Himalaya 1914. pp. 1-125, pl. i-iv.

49. ANDREWES, H. E. 1930. The Carabidae of the Third Mount Everest Expedition 1924. *Trans. R. ent. Soc. London*, 78: 1-44, map 1.
50. ANDREWES, H. E. 1934. Yale North-India Expedition, Report on Coleoptera of the family Carabidae. *Mem. Connecticut Acad. Arts & Sci.*, 10 (4): 23-28.
51. ANTEVS, E. 1929. Maps of the Pleistocene Glaciations. *Bull. geol. Soc. Amer.*, 40: 631-720.
52. ANTEVS, E. 1932. Alpine zone of Mt. Washington Range. Auburn, Maine, pp. 1-118, fig. 38.
53. APFELBECK, H. 1904. Die Käferfauna der Balkanhalbinsel. I. Caraboidea. Berlin, pp. 1-422.
54. ARMSTRONG, H. G. & J. W. HEIM, 1938. The effect of repeated daily anoxemia. *J. Aviat. Med.*, 9: 92.
55. ARNOLD, G. 1950. New Pompilidae from Kilimanjaro. *Occ. Papers Nat. Mus. S. Rhodesia*, 16: 400-401.
56. ARROW, G. J. 1909. Zoological results of the Ruwenzori Expedition 1905-1906. No. 14. Coleoptera. *Trans. zool. Soc. London*, 19 (2): 194-195.
57. ASPER, G. & J. HEUSCHER, 1855-1856 & 1887-1888. Zur Naturgeschichte der Alpenseen. *Jber. St. Gall. naturf. Ges.*, 1, 2.
58. ATTEMS, C. 1949. Die Myriopodenfauna der Ostalpen. *Sitzb. österr. Akad. Wiss.*, (math.-naturw. Klasse) (1) 158 (1/2): 79-153, map 1.
59. AUBERT, J. 1956. *Andiperla willinki* n. sp., Plécoptère nouveau des Andes de Patagonie. *Mitt. schweiz. ent. Ges.*, 19: 229-232.
60. AUSSERER, A. 1867. Arachniden Tirols nach ihrer horizontalen und vertikalen Verbreitung. *Verh. zool.-bot. Ges. Wien*, 17: 137-170, pl. vii-viii.
61. AUSTANT, J. L. 1889. Les Parnassiens de la faune palaearctique. Leipzig.
62. AUSTANT, J. L. 1899. Sur deux *Parnassius* asiatiques nouveaux. *Le Natural.*, 21 (2): 154.
63. AUSTANT, J. L. 1912. Notice sur quelques formes aberrantes de *Parnassius*. *Int. ent. Z.*, 5 (50): 359.
64. AUSTEN, E. E. 1909-1910. Ruwenzori Expedition Reports. Diptera. pp. 85-102.
65. AUSTIN, E. P. & J. L. LECONTE, 1874. Catalogue of the Coleoptera of Mt. Washington, N.H., with descriptions of new species. *Proc. Boston Soc. nat. Hist.*, 16: 265-276.
66. AVERIN, U. V. 1957. Zoogeographicheskii ocherk Kamchatki. *Bull. Mosck. Obsh. Isp. prir.*, Otd. Biol., 12 (5): 29-39.
67. AVINOFF, A. 1910. Zur Rhopalocerenfauna des östlichen Pamir. *Horae Soc. Ent. Ross.*, 1909: 225-245, pl. xiv.
68. AVINOFF, A. 1913. Butterflies of the genus *Parnassius* in the Indian Museum. *Rec. Indian Mus.*, 9: 327-331.
69. AVINOFF, A. 1916. Some new forms of *Parnassius*. *Trans. ent. Soc. London*, 1915: 351-360, pl. lii-liv.
70. AVINOFF, A. 1922. Considérations sur les Parnassiens d'Asie Centrale. *Étude Lép. Comp.*, 19: 47-70.
71. AXELSON, W. M. 1903. Beiträge zur Kenntnis der Collembolenfauna Sibiriens. *Öfv. vet. Soc. Förh.*, 45: 1-13.

72. BACK, E. A. & R. T. COTTON, 1925. The use of vacuum for insect control. *J. agric. Res.*, 31: 1035-1041.
73. BADER, F. 1960. Die Verbreitung borealer und sub-antarktischer Holzgewächse in den Gebirgen des Tropengürtels. *Nova Acta Leopaldina*, 23 (148).
74. BÄBLER, E. 1910. Die wirbellose, terrestrische Fauna der nivalen Region. *Rev. Suisse Zool.*, 18: 761-915.
75. BÄNNINGER, M., F. NETOLITSKY & E. SCHAUBERGER, 1929. Entomologische Ergebnisse

der schwedischen Kamchatka-Expedition 1920-1922. Carabidae. *Ark. Zool.*, 20 (B) (4): 1-5.

76. BAIJAL, H. N. 1955. Entomological Survey of the Himalaya. Part VI. Two new species of Collembola. *Agra Univ. J. Res.*, (Sci.) 4 (1): 175-178.

77. BAIJAL, H. N. 1955. Entomological Survey of the Himalaya. Part XI. Five new species of Collembola. *Agra Univ. J. Res.*, (Sci.) 4 (2): 531-538.

78. BAIJAL, H. N. 1958. Entomological Survey of the Himalaya. Part XXVIII. Nival Collembola from the Northwest Himalaya. *Proc. nat. Acad. Sci. India*, 28 (B) (5): 349-360.

79. BAIJAL, H. N. & SANTOKH SINGH. 1954. Entomological Survey of the Himalaya. Part III. On a collection of Dermaptera. *Agra Univ. J. Res.*, (Sci.) 3 (2): 455-456.

80. BAILEY, E. B. 1928. The Palaeozoic Mountain Systems of Europe and America. *Rep. Br. Assoc. Adv. Sci. Glasgow*, (C): 57-76, fig. 2.

81. BALDI, E. V., TONOLLI & L. I. PIROCCHI, 1953. La differente evoluzione di due Lughi Gia constituenti un unico bacino: il lago maggiore ed il lago mergazzo. *Mem. Ist. Ital. Idrobiol.*, 7: 49-107.

82. BANG-HAAS, O. 1915. Zur Kenntnis von *Parnassius delphius* Eversm. und verwandter Arten. *Dtsch. ent. Z., (Iris)* 29: 148-170, pl. iv-v.

83. BANG-HAAS, O. 1937. Neubeschreibungen und Berichtigungen der paläarktischen Macrolepidopterenfauna. XXXI. *Ent. Z.*, 2 (32): 302-304.

84. BAROVSKII, V. 1931. Coccinellidae. Abhandlungen der Pamir Expedition 1928. 8 (Zoology): 127-131.

85. BARTEL, M. 1902. Lepidopteren des südlichen Urals. *Dtsch. ent. Z., (Iris)* 15: 183-230.

86. BASILEWSKY, P. 1950. Description d'un *Orinodromus* nouveau des hauts sommets du Kilimadjaro. *Bull. (Ann.) Soc. ent. Belg.*, 86: 118-120.

87. BASILEWSKY, P. 1953. Expedition to the Gughé Highlands 1948-1949: Coleoptera Carabidae. *J. Linn. Soc. London* (Zool.), 42: 276-292.

88. BASILEWSKY, P. 1957. Journey to Northern Ethiopia (Simien) 1952-53: Coleoptera, Carabidae. *J. Linn. Soc. London* (Zool.), 43: 188-202.

89. BATES, F. 1891. Additions to the Carabideous fauna of Mexico, with remarks on some of the species previously recorded. *Trans. ent. Soc. London*, 223-278, pl. xiii-xiv.

90. BATES, F. 1891. Heteromera in Coleoptera of the Second Yarkand Mission. pp. 55-79.

91. BATES, H. W. 1891. Coleoptera from Kulu in Northwest India. *Entomologist*, 24 (Suppl.): 7-22.

92. BATES, H. W. 1891. Coleoptera collected by Mr Pratt on the Upper Yangtse and on the borders of Tibet. Second Notice of Journey of 1890. *Entomologist*, 24 (Suppl.): 69-80.

93. BATES, H. W. 1891. Geodiphaga and Longicornia in the Scientific Results of the Second Yarkand Mission. Coleoptera, pp. 1-36.

94. BAUMAN, F. 1910. Beiträge zur Biologie der Stockhornseen. *Rev. Suisse Zool.*, 18: 647-728.

95. BEARD, J. S. 1955. The classification of tropical American vegetation types. *Ecology*, 36: 89-100.

96. BECK V. MANNAGETTA, G. 1906. Die Umkehrung der Pflanzenregionen in den Dolinen des Karstes. *Sitzb. math. Kl. Akad. Wiss. Wien*, 115 (1): 3-20.

97. BECK V. MANNAGETTA, G. 1913. Vegetationsstudien in den Ostalpen. III. *Sitzb. Akad. Wiss. Wien*, **122**: 157-367, pl. i-iii.

98. BECKER, TH. 1907. Zur Kenntnis der Dipteren von Central-Asien I. Cyclorrhapha Schizophora holometopa und Orthorrhapha brachycera. *Ann. Mus. Zool. Akad. Imp. Sci. St. Petersburg*, 12 (3): 253-317, pl. i-ii.

99. BEI-BIENKO, G. R. 1951. Novie predstavitel ortopteroidnikh nasekomikh iz gruppi Grylloblattoidea (Orthoptera) v faune SSSR. *Ent. Obozr.*, 31 (3): 4.
100. BEIER, M. 1944. Über Pseudoscorpionen aus Ostafrika. *Eos*, 20: 173-219.
101. BEIER, M. 1950. Zur Phenologie einige *Neobisium*-Arten (Pseudoscorp.). *VIII int. Congr. Ent.*, 1002-1007, fig. 3.
102. BEIER, M. 1951. On some Pseudoscorpioidea from Kilimanjaro. *Ann. Mag. nat. Hist.*, (12) 4: 606-609.
103. BENSON, R. 1935. The high mountain sawflies of Britain (Hymenoptera Symphyta). *Trans. R. ent. Soc. London*, 83: 23-39, fig. 20.
104. BENSON, R. 1955. Sawflies of the high Swiss Alps (Hymenoptera Symphyta). *Mem. Soc. R. ent. Belg.*, 27: 74-81.
105. BENSON, R. 1955. On some high alpine Nematine sawflies (Hym. Tenthredinidae). *Ent. month. Mag.*, 91: 103-105, fig. 10.
106. BENSON, R. 1958. Arctic sawflies and the open habitat. *Proc. X int. Congr. Ent.*, 1: 693.
107. BENSON, R. 1960. Some more high-alpine sawflies (Hymenoptera Tenthredinidae). *Mitt. schweiz. ent. Ges.*, 33 (3): 173-182.
108. BENSON, W. N. 1935. Notes on the geographical features of Southwestern New Zealand. *Geogr. J.*, 86: 393-401.
109. BENT, A. H. 1911. A bibliography of the White Mountains. vii+114. Boston: Appalachian Mountain Club.
110. BENT, A. H. 1915. Animal life in the White Mountains. *Appalachia*, 13: 349-352.
111. BERGER, B. 1907. Über die Widerstandsfähigkeit der Tenebriolarven gegen Austrocknung. *Arch. ges. Physiol.*, 118: 607-612.
112. BERNHAUER, M. 1930. Entomologische Ergebnisse der Deutsch-Russischen Alai-Pamir-Expedition 1928. (I) Coleoptera 1. (Staphylinidae). *Mitt. zool. Mus. Berlin*, 16: 241-243.
113. BERNHAUER, M. 1935. Zoologici raccolti dalla spedizione italiana al Karakorum. Beschreibungen einiger neuen Staphyliniden-Arten (Coleoptera). *Atti Mus. Stor. nat. Trieste*, 12: 86-88.
114. BERTRAND, H. 1952. Récoltes de Diptères Chironomides dans les Pyrénées (2e note). *Vie et Mileu*, 3 (3): 314-321.
115. BERTRAND, H. 1953. Diptères Chironomides dans les Pyrénées (2e note). *Bull. Soc. ent. France*, 71-79.
116. BERTRAND, H. 1953. Récoltes de Coléoptères aquatiques (Hydrocanthares) dans les Pyrénées: Observations écologiques. (2e note). *Bull. Soc. Zool. France*, 79: 2-3, 91-105.
117. BERTRAND, H. 1955. Sur la faune entomologique des Pyrénées. *Trav. Ass. int. Limnol. theor. appl.*, 12: 634-642.
118. BERTRAND, H. 1956. Récoltes de Coléoptères aquatiques (Hydrocanthares) dans les Massifs Montagneux de l'Espagne: Observations écologiques (2e note). *Bull. Soc. Zool. France*, 81: 12-23.
119. BERTRAND, H. 1956. Récoltes de Coléoptères aquatiques (Hydrocanthares) dans les Massifs Montagneux de la Grèce; observations écologiques. *Bull. Soc. zool. France*, 81: 323-338.
120. BERTRAND, H. & J. AUBERT, 1952. Récoltes des Plécoptères dans les Pyrénées. *Ann. Soc. ent. France*, 121: 93-196.
121. BERTRAND, H. & M. L. VERRIER, 1954. Contribution à l'écologie des Ephémeroptères des Pyrénées espagnoles. *Bull. Biol. France et Belgique*, 88 (2): 146-153.
122. BEZZI, M. 1914. Ditteri cavernicoli. *Atti Soc. ital. Sci. nat. Milano*, 53: 207-230.
123. BEZZI, M. 1918. Studi sulla ditterofauna nivale delle Alpi italiane. *Mem. Soc. ital. Sci. nat. Milano*, 9: 1-164.

124. BIGLER, W. 1919. Beiträge zur Kenntnis alpiner Leptoiuliden. *Rev. Suisse Zool.*, 27 (8): 283-333.
125. BIGLER, W. 1919. Diplopodenfauna des schweizerischen Nationalparks. *Ergebn. wiss. Unters. schweiz. Nationalparks*, 5: 1-88.
126. BIGLER, W. 1925. Zur Verbreitung der Diplopoden des schweizerischen Nationalparks. *Rev. Suisse Zool.*, 32 (4): 73-76.
127. BILLINGS, M. P. 1938. Geology in the Presidential Range. *Mount Washington Observatory News Bull.*, 4: 8-11.
128. BINAGHI, G. 1959. Contributo alla conoscenza della Coleopterofauna nivale del Gran Sasso d'Italia. *Boll. Soc. ent. ital. Genoa*, 89: 6-11.
129. BISCHOFF, W. 1922. Zur Kenntnis der Blepharoceriden. *Zool. Jb.*, 46: 102.
130. BISCHOFF, W. 1924. Die Segmentierung der Blepharoceridenlarven und Puppen und die ökologische Begründung der Larvenphylogenie. *Zool. Anz.*, 60: 231-251.
131. BISCHOFF, W. 1928. Die Ökologie der paläarktischen Blepharoceridae. *Ergebn. Fortschr. Zool.*, 7: 209-278.
132. BISCHOFF, W. 1930. Entomologische Ergebnisse der Deutsch-Russischen Alai-Pamir-Expedition 1928. II. Hymenoptera VII. *Mitt. zool. Mus. Berlin*, 16 (5): 860-864.
133. BLAIR, K. G. 1922. Coleoptera of the Mount Everest Expedition 1921. *Ann. Mag. nat. Hist.*, (9) 9: 558-562.
134. BLAIR, K. G. 1923. Coleoptera of the Mount Everest Expedition 1922. Part II. Heteromera. *Ann. Mag. nat. Hist.*, (9) 11: 278-285.
135. BLAIR, K. G. 1927. Heteromera of the Third Mount Everest Expedition 1924. *Ann. Mag. nat. Hist.*, (9) 19: 241-255.
136. BODINE, J. H. 1921. Factors influencing the water content and the rate of metabolism of certain Orthoptera. *J. exp. Zool.*, 32: 137-164.
137. BODINE, J. H. 1925. Effect of temperature on the rate of embryonic development of certain Orthoptera. *J. exp. Zool.*, 42: 91-109.
138. BOEKEMÜHL, J. 1956. Die Apterygoten des Spitzberges bei Tübingen, eine faunistisch ökologische Untersuchung. *Zool. Jb.*, 84 (2/3): 113-194.
139. BOHEMAN, L. 1943. Zwei neue Otomys-Rassen aus Ostafrika. *Zool. Anz.*, 143: 153-155.
140. BOND, R. M. 1934. Report on Phyllopod Crustacea (Anostraca, Notostraca and Conchostraca), including a revision of the Anostraca of the Indian Museum. *Mem. Connecticut Acad. Arts & Sci.*, 10 (5): 29-62.
141. BORN, P. 1908. Zoogeographisch-Carabologische Studien. *Ent. Wochenbl. Leipzig*, 25.
142. BOURGOGNE, J. 1962. Une bonne localité: Pralognan-Vanoise. Rocher de Villeneuve et de la Roche. *Alexanor: Rev. Lép. franç.*, 11 (7): 255-262, pl. vii.
143. BOURGOGNE, J. 1962. Sur la présence de *Melanargia russae* Esp. dans les Pyrénées Français. *Alexanor: Rev. Lép. Franç.*, 11 (7): 276.
144. BOWDITCH, E. C. 1896. List of Mount Washington Coleoptera. *Psyche*, 7 (Suppl.) II: 1-11.
145. BRASS, L. J. 1941. The 1938-39 Expedition to the Snow Mountains, Netherlands New Guinea. *J. Arnold Arboretum*. 22.
146. BRAUN-BLANQUET, J. 1923. Über die Genesis der Alpenflora. *Verh. naturf. Ges. Basle*, 25: 243-261.
147. BRAUN-BLANQUET, J. 1931. Recherches phytogéographiques sur le massif du Grossglockner (Hohe Tauern). *Rev. geogr. alpine*, 19: 675-735, fig. 3.
148. BRAUN-BLANQUET, J., G. L. FULLER & H. S. CONRAD, 1932. Plant Sociology: A study of Plant Communities. McGraw-Hill Book Company, New York.
149. BRAUN-BLANQUET, J. & H. JENNY, 1926. Vegetationsentwicklung und Bodenbildung

in der alpinen Stufe der Centralalpen (Klimagebiet des *Caricion curvulae*). *Denkschr. schweiz. naturf. Ges.*, 43: 183-349.

150. BREUNING, ST. 1924. Beiträge zur Kenntnis der Caraben der Ostalpen. *Koleopt. Rundschau*, 11: 1-20; II. *ibid.*, 13: 10-28 (1927); III, *ibid.*, 13: 115-126 (1927).

151. BRIGHAM, W. T. 1909. The volcanoes of Kilauea and Mauna Loa on the Island of Hawaii. *Bernice Mus. Mem.*, 2 (4): 1-222.

152. BRINCK, P. 1949. Studies on Swedish stoneflies (Plecoptera). *Opusc. Ent.*, (Suppl.) 11: 1-250.

153. BRINCK, P. 1950. Plecoptera from Afghanistan. The 3rd Danish Expedition to Central Asia. Zool. Res. 1. *Vidensk. Medd. Dansk. naturh. Foren.*, 112: 133-138.

154. BRIQUET, J. 1905. Le développement des flores dans les Alpes occidentales avec aperçu sur les Alpes en général. *Wiss. Ergebn. int. bot. Kongr. Wien*, pp. 130-173.

155. BRITTON, E. R. 1959. Carabidae (Coleoptera) from New Zealand caves. *Proc. R. ent. Soc. London*, (B) 28: 103-106, fig. 4.

156. BROCKMANN-JEROSCH, H. 1919. Baumgrenze und Klimacharakter. *Beitr. geobot. Landesaufn.*, Zürich, 6: 255.

157. BRODSKII, A. & K. BRODSKII, 1926. *Deuterophlebia* dans les montagnes de l'Asie Centrale. *Bull. Asie Centrale*, 13: 23-26.

158. BRODSKII, K. 1930. Zur Kenntnis der mittelasiatischen Ephmeropteren I. Imagines. *Zool. Jb.*, (Syst.) 59: 681-720, fig. 42.

159. BRODSKII, K. 1930. Zur Kenntnis der Wirbellosenfauna der Bergströme Mittelasiens II. *Deuterophlebia mirabilis* Edw. *Z. Morphol. Ökol. Tiere*, 18: 289-391; III. Blepharoceridae. I. Imagines. *Zool. Anz.*, 90: 129-160, fig. 15.

160. BRODSKII, K. 1935. Contribution to the knowledge of Invertebrate fauna of the mountain streams of Central Asia (Materialii k poznanyu faunii bespozvochnikh gornikh potokov Srednii Azii). *Trudi Sredniiaziaticheskovo Gosud. Univ. Tashkent*, 15: 1-152, fig. 22, pl. iii.

161. BRODSKII, K. 1937. A contribution to the knowledge of the fauna of mountain streams of Central Asia IV. Blepharoceridae II. *Trav. Inst. Zool. Acad. Sci. USSR, Leningrad*, 4: 71-105, fig. 22, map 1 (In Russian with a summary in English).

162. BRUCE, N. 1953. Eine neue *Mnionomus*-Art aus Afrika. *Ann. Mag. nat. Hist.*, (12) 6: 75-76.

163. BRUES, C. T. 1929. Insect fauna of thermal springs. *Trans. IV int. Congr. Ent. Ithaca*, 2: 237-240.

164. BRUGGEMANN, P. F. 1958. Insects and environments of the High Arctic. *Proc. X int. Congr. Ent.*, 1: 695-782.

165. BRYANT, G. E. 1950. New species of Halticinae from Kilimanjaro and Ruwenzori collected by Dr. G. Salt. *Ann. Mag. nat. Hist.*, (12) 3: 163-166.

166. BRYK, F. 1911. Über eine unbekannte Neuerung in der Flügeltracht der Gattung *Parnassius* Latr. *Soc. Ent.*, 26 (15): 54-55.

167. BRYK, F. 1912. Vornehme *Parnassius*-formen. *Jb. Nass. Ver. Nat. Wiesb.*, 65: 1-35.

168. BRYK, F. 1913. Kritische Zusammenstellung der Formen von *Kailasius charltonius* Gray. *Int. ent. Z.*, 7 (22/24): 149-151, 153-154, 161-163, pl. iv-v.

169. BRYK, F. 1935. Lepidoptera: Baronidae, Teinopalpidae, Parnassidae. *Das Tierreich*, 65.

170. BRYK, F. & C. EISNER, 1934. Das Männchen von *Parnassius stoliczkanus* Feld. subsp. *tytlerianus* (n). *Parnass.*, 3 (3): 40.

171. BUDAY, T., O. KODYMSEN, M. MAHEL, M. MÁSKA, A. MATEJKA, J. SVOBODA & V. ZOUBEK, 1960. Tectonic development of Czeschoslovakia. *Praha Nakladatestvi Československé Akademie Ved*, pp. 224, map 1.

172. BURESCH, I. 1925. Sur la faune cavernicole de Bulgare et description d'un nouveau Pheggonisetes. *Bull. Soc. ent. France*, 115-120.

173. BURESCH, I. 1926. Über die Höhlen-Trechinen Bulgariens, nebst Beschreibung einer neuen Art *Paraduvalius regis-borisi*. *Koleopt. Rundschau*, 12: 156-162.

174. BURESCH, I. & W. ARNDT, 1926. Die Glazialrelikte stellenden Tierarten Bulgariens und Mazedoniens. *Z. Morphol. Ökol. Tiere*, 5: 381-405.

175. BURRARD, S. G. & H. H. HAYDEN, 1907-1908. A sketch of the geography and geology of the Himalaya Mountains and Tibet. Calcutta, pp. 1-230, charts 50.

176. BUSK, D. L. 1955. The mountains of Ethiopia. *The Mountain World*, pp. 213-222.

177. BUTLER, A. G. 1888. Descriptions of some new Lepidoptera from Kilimanjaro. *Proc. zool. Soc. London*, pp. 91-98.

178. BUXTON, P. A. 1930. Evaporation from the mealworm and atmospheric humidity. *Proc. R. Soc. London*, (B) 106: 560-577.

179. CALLONI, S. 1889. La fauna nivale, con particolare riguardo ai viventi delle alte Alpi. Pavia, 478, 20.

180. CAMERANO, L. 1909. Di alcuni Coleotteri dell Ruwenzori; apud S. A. R. il Principe Abruzzi. Il Ruwenzori, parte scientifica I. Zoologica, Milano, pp. 221-253.

181. CAMERON, M. 1934. Yale North-India Expedition. Report on Coleoptera of the family Staphylinidae. *Mem. Connecticut Acad. Arts & Sci.*, 10: 17-21.

182. CAPORIACCO, L. DE, 1934. Aracnidi dell'Himalaya e del Karakorum. *Mem. Soc. ent. Ital.*, 13: 113-160.

183. CAPRA, F. 1954. Coleotteri di la Sierra Nevada. Coccinellidae. *Arch. Inst. Aclimatación, Almeria*, 2: 103-104.

184. CARADAJA, A. 1933. Gedanken über die Herkunft und Evolution europäischer Lepidopteren. *Ent. Rundschau*, 50.

185. CARL, J. 1899. Über Collembola der Schweiz. *Rev. Suisse Zool.*, 6: 264-362.

186. CARL, J. 1901. Zweiter Beitrag zur Kenntnis der Collembolenfauna der Schweiz. *Rev. Suisse Zool.*, 9 (2): 243-278, pl. xv.

187. CARNELUTTI, J. 1958. Alpine petrophile Lepidopteren des Karstgebietes. *Verh. dtsch. zool. Ges., Leipzig*, 1957: 506-511, map 1.

188. CARPENTER, G. D. HALE, 1935. The Rhopalocera of Abyssinia: A faunistic study. *Trans. R. ent. Soc. London*, 83: 313-448, pl. viii-xiii.

189. CARPENTER, W. L. 1874. Report on the alpine insect-fauna of Colorado. *Ann. Rep. U.S. geol. geogr. Surv. of the Territories embracing Colorado*, being a report of the progress of the explorations for the year 1873. pp. 539-542.

190. CARTER, H. J. 1906. A beetle hunt on Mt. Kosciusko. *The Australian Naturalist, Sydney*, 1 (2): 17-23.

191. CASTEL, G. P. 1939. The Plecoptera of Montana. *Canad. Ent.*, 71: 208-211.

192. CAUDELL, A. N. 1909. Some insects from the summit of Pikes Peak found on snow. *Proc. ent. Soc. Wash.*, 5: 74-82.

193. CAVALHO, J. C. & T. R. E. SOUTHWOOD, 1955. Revision of the complex *Cyrtorhinus* Fieber-*Mecoma* Fieber. *Bol. Mus. para Goeldi*, 11: 1-72.

194. CAVE, C. J. P. 1926. The physical basis of insect drift. *Nature, Lond.*, 118 (2958): 50.

195. CHAMPION, G. C. 1903. An entomological excursion to Bejar, Central Spain. *Trans. ent. Soc. London*, 1903: 165-182.

196. CHAMPION, H. G. 1926. Entomological notes on a tour of the Kumaon-Tibet Border. *Ent. mon. Mag.*, 62: 271-275, pl. ii-iii.

197. CHAMPION, H. G. 1936. A preliminary study of the forest types of India and Burma. *Indian Forest Rec.*, (NS) Silviculture, 1: 1-256.

198. CHANDLER, M. E. J. (Miss), 1935. The effects of the southern extension of glaciers and ice-sheets on the preglacial vegetation. The nature of the flora as revealed by the plant remains associated with glacial and interglacial deposits. *Proc. R. Soc. London*, 118: 205-210.

199. CHANEL, J. 1899. Voyage au Kilimandjaro au 1894. *Le Tour de Monde* (NS) 5: 33-37.
200. CHAPMAN, J. A. 1924. Observations on snow insects in Western Montana. *Canad. Ent.*, 86: 357-363.
201. CHAPMAN, J. A. 1953. Ecological notes on *Grylloblatta* in Montana. *Proc. Montana Acad. Sci.*, 13: 39-41.
202. CHAPMAN, J. A. 1954. Studies on summit frequenting insects in Western Montana. *Ecology*, 35 (3): 41-49.
203. CHAPMAN, J. A. 1954. Swarming of ants on Western United States Mountain summits. *Pan-Pacific Ent.*, 30 (2): 93-102.
204. CHAPMAN, J. A., J. I. ROMER & J. STARK, 1955. Ladybird beetles and army cutworm adults as food for grizzly bears in Montana. *Ecology*, 36 (1): 156-158.
205. CHAPMAN, R. N. 1931. Animal Ecology, with special reference to insects. McGraw-Hill Book Company Inc., New York and London, pp. 1-464.
206. CHAPPUIS, P. A. 1927. Die Tierwelt der unterirdischen Gewässer. Stuttgart, pp. 1-175, pl. iv.
207. CHEESMAN, L. E. 1951. Old Mountains of New Guinea. *Nature, Lond.*, 168: 597.
208. CHODAT & PAMPANINI. 1902. Sur la distribution des Plantes des Alpes austro-orientales. *Le Globe* (Genève) 41.
209. CHOPARD, L. 1923. On some New Zealand cave Orthoptera. *Trans. N. Z. Inst. Wellington*, 54: 230-239.
210. CHOPARD, L. 1928. La faune des Orthoptères des montagnes des Etats-Unis et ses rapports avec la faune paléarctique. *Mem. Soc. Biogéogr.*, 2: 142-149.
211. CHOPARD, L. 1964. La vie des insectes sur les hauts sommets de l'Himalaya. *Sci. Progrès*, (La Nature), 3353: 353-360, fig. 14.
212. CLENCH, H. & N. SHOUMATOFF, 1956. Lepidoptera Rhopalocera (Insecta) from Afghanistan. *Vid. medd. Dansk. naturh. Foren.*, 118.
213. CLOOS, H. 1936. Einführung in die Geologie. Berlin: Gebrüder Bornträger. pp. 503.
214. COBOS, A. 1954. Coleopteres de Sierra Nevada (Familias Buprestidae, Elateridae y Eucnemidae) *Arch. Inst. Aclimatación*, Almeria, 2: 81-88; (Familias Cerambycidae y Chrysomelidae). pp. 139-155.
215. COCKERELL, T. D. A. 1893. The entomology of the midalpine zone of Custer Country, Colorado. *Trans. Amer. ent. Soc.*, 20: 305-370.
215A. COE, M.J. 1967. The Ecology of the Alpine Zone of Mt. Kenya. *Monogr. Biol.* XVII. Dr. W. Junk, The Hague.
216. COLE, W. H. 1906. Bionomics of the grain weevils. *J. econ. Ent.*, 1: 63-73.
217. COLEMAN, A. P. 1920. Extent and thickness of the Labrador ice sheet. *Bull. geol. Soc. Amer.*, 31: 319-328.
218. COLEMAN, A. P. 1935. Pleistocene glaciations in the Andes of Colombia. *Geogr. J.*, 86: 330-334.
219. COLLET, L. W. 1936. The structure of the Alps. London: Edward Arnold & Sons (2dn Ed.) pp. 304.
220. COLLINS, C. W. & W. L. BAKER, 1934. Exploring the upper air for wind-borne gypsy moth larvae. *J. econ. Ent.*, 27: 320-327.
221. CORTI, A. 1929. Entomologische Ergebnisse der schwedischen Kamtchatka-Expedition 1920-22. Lepidoptera IV. Notodontidae und Noctuidae. *Ark. Zool.*, 21 (B) (1): 1-4.
222. COTTON, A. D. 1930. A visit to Kilimanjaro. *Kew Bull.*, 97-121.
223. COTTON, C. A. 1926. Geomorphology of New Zealand. Wellington, N. Z. Dominion Museum, pp. 462.
224. CRESSON, JR. ERZA, 1934. Yale North-India Expedition. Report on Diptera of the family Ephydridae. *Mem. Connecticut Acad. Arts & Sci.*, 10 (1): 1-4.
225. CZEKELIUS, D. 1929. Über das Vorkommen und die Verbreitung von *Parnassius apollo* in Spitzbergen. *X int. Congr. Zool. Budapest*, 2: 1561-1463.

226. DADAY, E. 1910. Untersuchungen über die Süsswassermikrofauna Deutsch-Ostafrikas. *Zoologica*, 59: 1-314.
227. DÄNIKER, A. 1923. Biologische Studien über Baum- und Waldgrenze, insbesondere über die klimatischen Ursachen und deren Zusammenhänge. *Vierteljahrsschr. Naturf. Ges. Zürich*, 68: 1-102.
228. DAHL, R. G. 1959. Studies on Scandinavian Ephydridae (Diptera Brachycera). *Opusc. ent.*, (Lund) (Suppl.) 15: 1-224, fig. 84.
228a. DALENIUS, P. 1965. The Acarology of the Antarctic Regions. In: P. van Oye & J. van Mieghem (eds.) Biogeography and Ecology in Antarctica. *Monogr. Biol.* XV, The Hague, Dr. W. Junk.
229. DALENIUS, P. & O. WILSON, 1957. On the soil fauna of the Antarctic and of the Sub-Antarctic Islands. The Oribatidae (Acari). *Ark. Zool.*, 11 (23): 393-425.
230. DALTON, A. J., B. F. JONES, V. B. PETERS & E. R. NITCHELL, 1945. Organ changes in rats exposed repeatedly to lowered oxygen tension with reduced barometric pressure. *J. nat. Cancer Inst.*, 6: 161.
231. DALY, R. A. 1934. The changing world of the ice age. New Haven: Yale University Press. pp. 217.
232. DARLINGTON, P. J. JR. 1936. Variation and atrophy of flying wings of some Carabid beetles. *Ann. ent. Soc. Amer.*, 29: 136-179.
233. DARLINGTON, P. J. JR. 1943. Carabidae of mountains and islands: Data on the evolution of isolated faunas and on atrophy of wings. *Ecol. Monogr.*, 13: 37-61.
234. DARLINGTON, P. J. JR. 1952. The carabid beetles of New Guinea. Part 2. The Agonini. *Bull. Mus. comp. Zool. Harvard, Cambridge*, Mass., 107: 89-252, pl. iv.
235. DAUBENMIRE, R. F. 1938. Merriam's Life zones of North America. *Quart. Rev. Biol.*, 13: 327-332.
236. DAUBENMIRE, R. F. 1942. An ecological study of the vegetation of southeastern Washington and adjacent Idaho. *Ecol. Monogr.*, 12: 53-79.
237. DAUBENMIRE, R. F. 1946. The life zone problem on the northern intermontane region. *Northwest Sci.*, 20: 28-38, fig. 1.
238. DAVENPORT, C. B. 1908. Experimental Morphology. New York: The MacMillan Company. pp. 488.
239. DEES, E. DUDAY DE, 1908. Entomostraca and Hydrarachnoidea from Tibet in Report on a collection of aquatic animals in Tibet by Captain F. H. S. Stewart, during the year 1907. *Rec. Indian Mus.*, 2: 323-341.
240. DEEVEY, E. S. 1949. Biogeography of the Pleistocene. *Bull. geol. Soc. Amer.*, 60: 1315-1416.
241. DE FILIPPI, F. 1908. Ruwenzori: An account of the Expedition of H. R. H. Prince Luigi Amedeo of Savoy Duke of the Abruzzi. London: pp. 408, maps. Appendix C. Zoology.
242. DE FILIPPI, F. 1912. Karakoram and Western Himalaya. London.
243. DE FILIPPI, F. 1932. The Italian Expedition to the Himalaya, Karakoram and Eastern Turkestan 1913-1914. London: Edward Arnold & Co. pp. 1-528, figs. 300, panoramas 17, maps 4, pl. ii.
244. DELAMARE-DEBOUTTEVILLE, C. 1953. Collemboles du Kilimandjaro récoltés par le Dr George Salt. *Ann. Mag. nat. Hist.*, (12) 6: 817-831.
245. DE LATTIN, G. 1956. Die Ausbreitungszentren der holarktischen Landtierwelt. *Verh. dtsch. zool. Ges.*, 1956: 380-410, maps 3.
246. DENIS, J. 1950. Spiders from East and Central African Mountains collected by Dr. George Salt. *Proc. zool. Soc. London*, 120: 497-502.
247. DENIS, J. 1957. Zoologisch-systematische Ergebnisse der Studienreise von H. Janetschek und W. Steiner in die spanische Sierra Nevada 1954. VII. Araneae. *Sitzb. österr. Akad. Wiss.* (math.-naturwiss. Klasse) (1) 166: 265-302.

248. DENIS, J. R. 1934. Yale North-India Expedition. Report on Collembola. *Mem. Connecticut Acad. Arts & Sci.*, 10: 261-282.

250. DESBORDES, H. 1921. Histeridae. Guy Babault Mission dans les Provinces Centrales de l'Inde et dans la Région occidentale de l'Himalaya 1914. pp. 13.

251. DESLANDERS, M. 1930. Trois formes nouvelles des Rhopalocères des Pyrénées françaises (Lep.). *Bull. Soc. ent. France*, pp. 243-244.

252. DESPAX, R. 1930. Plécoptères des Pyrénées III. Étude des espèces du genre *Lectura* Steph. appartenant au sous-genre *Pachylectura* Desp. *Bull. Soc. Hist. nat. Toulouse*, 59: 139-151, fig. 10.

253. DE TERRA, H. 1930. Zum Problem der Austrocknung des westlichen Zentralasien. *Z. ges. Erdk.*, 161-177.

254. DE TERRA, H. 1932. Geologische Forschungen im westlichen Kun-lun und Karakorum-Himalaya. In: Trinkler, E. & H. De Terra, Wissenschaftliche Ergebnisse der Dr Trinklerschen Zentralasien Expedition. Berlin: Dietrich Reimer/Ernst Vohsen. pp. 1-196, fig. 70, maps 2, pl. xxii.

255. DE TERRA, H. 1934. Physiographic results of the recent survey in Little Tibet. *Geogr. Rev.*, 24 (1): 12.

256. DE TERRA, H. & T. T. PETERSON, 1939. Studies on the Ice Age in India. Carnegie Inst. Washington, 493: 1-354.

257. DEWITZ, J. 1902. Der Apterismus bei Insekten, seine künstliche Erzeugung und seine physiologische Erklärung. *Arch. Anat. Physiol.*, (Abt. Physiol.) 1902: 61-67.

258. DEWITZ, J. 1920. Über die Entstehung rudimentärer Organe bei den Tieren. 4. Die Beeinflussung der Flügelbildung bei Insekten durch Kälte und Blausäuregase. *Zool. Jb.*, (Allg. Zool.) 37: 305-312, fig. 1-3, pl. v.

259. DICE, L. R. 1943. The biotic province of North America. pp. 78.

260. DIEM, E. 1903. Untersuchungen über die Bodenfauna in den Alpen. *Jber. St. Gall. naturw. Ges.*, 1901-1902: 234.

261. DIES, L. R. 1955. Natural Communities. Ann Arbor: University of Michigan Press. pp. 547.

262. DIGBY, P. S. B. 1958. Flight activity in the blowfly *Calliphora erythrocephala* in relation to wind speed, with special reference to adaptation. *J. exp. Biol.*, 35 (4): 776-795.

263. DILL, D. B. 1938. Life, Heat and Altitude. Cambridge: Harvard. pp. 1-183.

264. DISTANT, W. L. 1807. Scientific results of the Second Yarkand Mission. Rhynchota. pp. 1-15.

265. DISTANT, W. L. 1900. Undescribed genera and species belonging to the Rhynchotal family Pentatomidae. *Trans. ent. Soc. London*, pp. 163-174.

266. DISTANT, W. L. 1909-1910. Ruwenzori Expedition Report. Rhynchota. pp. 67-84, pl. ii.

267. DITMAN, L. P., G. B. VOGT & D. R. SMITH, 1948. Undercooling and freezing of insect. *J. econ. Ent.*, 36: 304-311.

268. DOBZHANSKY, TH. 1925. Über das Massenauftreten einiger Coccinelliden im Gebirge Turkestans. *Z. wiss. Insektenbiol.*, 20: 249-256.

269. DODDS, G. S. 1919. Altitudinal distribution of Entomostraca in Colorado. *Proc. U.S. nat. Mus.*, 54: 59-87.

270. DOODS, G. S. & F. L. HISAW, 1924. Ecological studies of aquatic insects I. Adaptations of mayfly nymphs to swift streams. *Ecology*, 5 (2): 137-148.

271. DODDS, G. S. & F. L. HISAW, 1924. Ecological Studies of aquatic insects II. Size of respiratory organs in relation to environmental conditions. *Ecology*, 5 (3): 262-271.

272. DODDS, G. S. & F. L. HISAW, 1925. Ecological Studies of aquatic insects III. Adaptations of caddisfly larvae to swift streams. *Ecology*, 6 (2): 123-137.

273. DODDS, G. S. & F. L. HISAW, 1925. Ecological Studies of aquatic insects IV. Alti-

tudinal range and zonation of mayflies, stoneflies and caddisflies in the Colorado Rockies. *Ecology*, 6: 380-390.

274. DODGE, H. R. & J. M. SEAGE, 1954. Sarcophagidae and other Diptera taken by trap and net on Georgia Mountain summits in 1952. *Ecology*, 35: 52-61.

275. DONCASTER, J. P. 1956. Two aphids from the Indian Region. *Proc. R. ent. Soc. London*, 25 (7/8): 111-116.

276. DONISTHORPE, H. 1930. Formicidae (Hymenoptera) taken by Major P. W. G. Hingston on the Mount Everest Expedition 1924. *Ann. Mag. nat. Hist.*, (10) 5: 444-447.

277. D'ORCHYMONT, A. 1943. Les Palpicornia du Tibet. *Bull. Mem. Hist. nat. Belg. Brussels*, 19 (57): 1-18.

278. DORFMEISTER, G. 1864. Über die Entwicklung verschiedener, während der Entwicklungsperioden angewandter Wärmegrade auf die Färbung und Zeichnung der Schmetterlinge. *Mitt. naturw. Ver. Steiermark*, 2: 99-109.

279. DORNO, C. 1919. Physik der Sonnen- und Himmelstrahlung. *Die Wissenschaft*, 58 (7): 1-126.

280. DORNO, C. 1927. Grundzüge des Klimas von Muottas-Muraigl Oberengadin. Braunschweig. 10: 1-177 Tables 41.

281. DOS PASSOS, C. F. 1949. The distribution of *Oeneis taygetes* Geyer in North America, with description of new subspecies (Lepidoptera: Satyridae). *Amer. Mus. Nov.*, 1399: 1-21, fig. 16.

282. DOS PASSOS, C. F. 1949. New butterflies from Mount McKinley National Park, Alaska, with a review of *Erebia rossii* (Rhopalocera: Satyridae). *Amer. Mus. Nov.*, 1389: 1-17, fig. 28.

283. DOS PASSOS, C. F. 1958. The Satyrid butterflies of Northwestern North America (Lepidoptera: Satyridae). *Proc. X int. Congr. Ent.*, 1: 673-681.

284. DOWDY, W. W. 1944. The influence of temperature on vertical migration of Invertebrates inhabiting different soil types. *Ecology*, 25: 449-460.

285. DUFOUR, L. 1851. Les zones entomologiques dans nos Pyrénées occidentales et désignation des insectes, qui les habitent. *Acta Soc. Linn. Bordeaux*, 17: 304-364.

286. DUGGAR, B. M. 1936. Biological effects of radiation. New York: McGraw-Hill Book Company Inc. vols. 2.

286a. DU RIETZ, G. E. 1935. Glacial survival of plants in Scandinavia and British Isles. *Proc. R. Soc. London*, 118: 226-229.

287. DU RIETZ, G. E. 1959. Remarks on the botany of the southern cold temperate zone. The Royal Society Expedition to Southern Chile 1958-1959. *Proc. R. Soc. London* (B) 152: 431-682.

288. DUTTON, E. A. T. 1929. Kenya Mountain. London. pp. 1-219, pl. lv.

289. DUZEE, E. P. VAN, 1919. Hemiptera. Report Canadian Arctic Expedition 1913-1918. Ottawa, 3 (F): 1-5 (1922).

290. DYTE, C. E. 1957 Journey to the Gughé Highlands 1948-1949. Coleoptera: Pselaphidae. *J. Linn. Soc. London* (Zool.), 43: 111-112.

291. EBNER, R. 1937. Orthopterologische Studien in Nordwest-Tirol. *Konowia*, 16: 28-40, pl. i.

292. EDMONDS, W. T. & G. E. HUTCHINSON, 1934. Report on Rotataria. Yale North-India Expedition. *Mem. Connecticut Acad. Arts & Sci.*, 10 (11): 152-186.

293. EDWARDS, F. W. 1922. Deuterophlebiidae. *Ann. Mag. nat. Hist.*, 9 (9): 380.

294. EDWARDS, F. W. 1928. Entomologische Ergebnisse der schwedischen Kamtchatka Expedition 1920-1922. Diptera, Nematocera. *Ark. Zool.*, 19 (A) (31): 1-3.

295. EDWARDS, G. J. 1950. Amphizoidae of the world. *Wasmann J. Biol.*, 8: 303-332, fig. 30, pl. iv.

296. EDWARDS, G. J. 1951. The natural history of high altitudes. *The Biologist*, 33 (3): 91-97.
297. EDWARDS, G. J. 1954. Observations on the biology of Amphizoidae (Coleoptera). *Col. Bull.*, 8: 19-25.
298. EDWARDS, G. J. 1956. Entomology above timberline. *Mazama Club Annual* (Portland: Oregon), 38 (13): 13-17.
299. EDWARDS, G. J. 1957. Some general observations on the ecology of Glacier National Park, Montana, with special reference to certain entomological aspects. *Wasmann J. Biol.*, 15 (1): 123-151.
300. EDWARDS, G. J. 1957. Entomology above timberline II. The attraction of ladybird beetles to mountain tops. *Col. Bull.*, 11: 41-46.
301. EDWARDS, G. J. 1960. The alpine insects of the Teton Range. *Rep. activities Jackson Hole biol. Res. Sta.*, Summer 1960: 20210 (1)-2020 (4).
302. EDWARDS, W. H. 1880. Experiments upon the effect of cold applied to chrysalis of butterflies. *Psyche*, 3: 1-6, 15-19, 75-76.
303. EHRENBERG, C. G. 1854. Bericht über die mikroskopischen Organismen auf den höchsten Gipfeln der europäischen Centralalpen und über das kleinste Leben der bayerischen Kalkalpen. K. & A. Schlagintweit, Neue Untersuchungen über die physikalische Geographie und die Geologie der Alpen. Leipzig. pp. 233-268.
304. EHRLICH, P. R. 1958. Problems of arctic-alpine insect distribution as illustrated by the butterfly genus *Erebia* (Satyridae). *Proc. X int. Congr. Ent.*, 1: 683-686.
305. EIDAM, P. 1937. Entomologische Sammelergebnisse der Deutschen-Hindukusch-Expedition 1935 der deutschen Forschungsgemeinschaft. *Arb. Morphol. Taxon. Ent.*, 4: 177-191.
306. EIDMANN, H. 1941. Zur Ökologie und Zoogeographie der Ameisenfauna von Westchina und Tibet. Wissenschaftliche Ergebnisse der 2. Brook Dolan Expedition 1934-1935 (Hymenoptera). *Z. Morphol. Ökol. Tiere*, 38: 1-43.
307. EIDMAN, H. 1942. Zur Kenntnis der Ameisenfauna des Nanga Parbat. (Hymenoptera). *Zool. Jb.*, (Syst.) 75: 239-266, fig. 2.
308. ELLER, K. 1939. Fragen und Probleme zur Zoogeographie und zur Rassen- und Artbildung in der *Papilio machaon*-Gruppe. *VII int. Kongr. Ent.*, 1: 74-101.
309. ELMERSEN, VON, 1848. Reise nach dem Altai, im Jahre 1834. St. Petersburg.
310. ELTON, C. S. 1925. The dispersal of insects to Spitzbergen ice. *Trans. ent. Soc. London*, pp. 289-299.
311. ELWES, H. J. 1882. On a collection of butterflies from Sikkim. *Proc. zool. Soc. London*, pp. 398-407, pl. xxv.
312. ELWES, H. J. 1886. On butterflies of the genus *Parnassius*. *Proc. zool. Soc. London*, pp. 6-53, pl. i-iv.
313. ELWES, H. J. 1887. On the butterflies of the French Pyrenees. *Trans. ent. Soc. London*, pp. 385-403.
314. ELWES, H. J. 1898. On the zoology and botany of the Altai Mountains. *Proc. Linn. Soc. London*, 27: 23-46.
315. ELWES, H. J. 1903. On a collection of Lepidoptera from Arctic America. *Trans. ent. Soc. London*, pp. 239-243, pl. ix.
316. ELWES, H. J. 1903. The butterflies of Chile. *Trans. ent. Soc. London*, 3: 263-301, pl. xii-xv.
317. ELWES, H. J., H. G. HAMPSON & J. H. DURRANT, 1906. On the Lepidoptera collected by the officers on recent Tibet Frontier Commission. *Proc. zool. Soc. London*, 2: 479-498, pl. xxxvi.
318. EMBERGER, L. 1930. La végétation de la région Mediterranéenne. Essai d'une classification des grouppements végétaux. *Rev. gén. Bot.*, 42: 641-662, 705-721.
319. EMDEN, F. VAN, 1951. Muscidae collected on Mt. Kilimanjaro by Dr. G. Salt. *Ann. Mag. nat. Hist.*, (12) 4: 786-793.

320. EMERTON, J. H., N. BANKS & R. V. CHAMBERLIN, 1919. Spiders, mites and myriapods. Report Canadian Arctic Expedition 1913-1918. Ottawa, 3 (H): 1-22 (1922).

321. ENGLER, A. 1925. Die Pflanzenwelt Afrikas. vols. 3.

322. ERHARD, H. 1931. Die Tierwelt der Alpen. In: Alpines Handbuch. Leipzig: F. v. Brockhaus. pp. 107-204.

323. ESPAÑOL, F. 1954. Coleopteres de Sierra Nevada. Tenebrionidae. *Arch. Inst. Aclimatación*, Almeria, 2: 111-122.

324. EVANS, G. O. 1953. On a collection of Acari from Kilimanjaro. *Ann. Mag. nat. Hist.*, (12) 6: 258-281.

325. EVANS, W. R. 1927. Lepidoptera-Rhopalocera obtained by Mme J. Visser-Hooft of the Hague (Holland) during an Exploration of the unknown country in the western Karakoram, N.W. India. *Tijdschr. Ent., Amsterdam*, 70: 158-162.

326. EVERLING, E. 1914. Einige Beobachtungen von Insekten in höheren Luftschichten. *Z. Naturw. Leipzig*, 85: 241-244.

327. FAIRMAIRE, L. 1891. Coléoptères des montagnes de Kashmir. *C. R. ent. Belg.*, 35: 88-103, 121-134.

328. FAIRMAIRE, L. 1894. Coléoptères du Kilimandjaro et des environs. *Ann. Soc. ent. Belg.*, 38: 386-395.

328a. FASSATI, M. 1964. Zoologische Ergebnisse der österreichischen Karakoram-Expedition 1958. Die Arten der Gattungen *Bembidion* Latr. und *Amara* Bon. (Col. Carabidae). *Ent. Arb. Mus. Frey*, 15 (2): 721-724.

329. FASSL, A. H. 1910. Die Erforschung des Monte Tolima. *Ent. Z. Frankfurt*, 24: 250-252.

330. FASSL, A. H. 1911. Die vertikale Verbreitung der Lepidopteren in den Colombischen Central Cordilleren. *Fauna Exotica*, Frankfurt, 1 (6/8).

331. FASSL, A. H. 1914. Tropische Reisen VI. Die Hochcordillere von Bogotá. *Ent. Rundschau*, 31: 97-100, 104-105.

332. FASSL, A. H. 1915. Die vertikale Verbreitung der Lepidopteren in den Colombischen Westcordilleren. *Ent. Rundschau*, 32: 9-12.

333. FASSL, A. H. 1920. Meine Bolivia-Reise. *Ent. Rundschau*, 32: 10-15, 15-18, 22-23 (Lepidoptera).

334. FAURE, J. C. 1943. Phase variation in the armyworm *Laphygma exempta* (Walk.). *Sci. Bull. Dept. Agric. Forest South Africa*, 234: 1.

335. FAUVEL, A. 1907. Voyage de M. Ch. Alluaud dans l'Afrique Orientale: Staphylinidae. *Rev. Ent.*, 26: 10-17.

336. FAWCETT, J. M. 1904. On some new and little known butterflies mainly from high elevations in the Northeast Himalaya. *Proc. zool. Soc. London*, ii: 134-141, pl. ix.

337. FEDCHENKO, B. A. 1902. Ocherk rastitelnosti Pamira, Shugnana e Alaiya. *Trudi S. Petersburgskovo Obshestva estestvoispitalei*, 33: 1.

338. FEDCHENKO, B. A. 1927. Vertikalnoe rasprostranenie paporotenikov Turkestaneskoe gornoi systemi. *Izv. Glavn. botanich. Sada*, 26: 1.

338a. FEDCHENKO, B. A. 1947. Botanikogeographicheskoe raionirovanie Pamira. Tezisi Dokaldu II. Sbezda Vsesoyuzhnovo Geographicheskovo Obshestva.

339. FEDCHENKO, O. 1903/1909. Flore du Pamir. *Acta horti Petropolitana*, 21 (1903); Suppl. 1, 24 (1904); 2, 24 (1905); 3, 28 (1907); 4, 28 (1909).

340. FELT, E. P. 1926. The physical basis of insect drift. *Nature, Lond.*, 170 (2952): 754-755.

341. FENNEMAN, N. M. 1931. Physiography of western United States. New York: McGraw-Hill Book Company Inc. pp. 534.

342. FENNEMAN, N. M. 1938. Physiography of eastern United States. New York: McGraw-Hill Book Company Inc. pp. 714.

343. FERNALD, M. L. 1925. Persistance of plants in unglaciated areas of boreal America. *Mem. Amer. Acad. Arts Sci. Boston*, 15 (3): 237-342.
344. FICKER, H. 1921. Untersuchungen über die meteorologischen Verhältnisse der Pamirgebiete (mit 17 Textfiguren und 1 Kartenskizze). *Denkschr. Akad. Wiss. Wien*, (math. naturw. Klasse) 97: 151-255.
345. FILIPJEW, K. 1931. Lepidoptera. Pamir Expedition 1928. Abhandlungen der Expedition, 8: 143-174, pl. i-x.
346. FINSTERWALDER, R. 1935. Forschung am Nangaparbat, Deutsche-Himalaya Expedition 1934. *Geogr. Ges. Hannover* (Sonderveröffentlichung H. Spreitzer: Hannover).
347. FINSTERWALDER, R. 1936. Die Formen der Nangaparbat-Gruppe. *Z. ges. Erdk.*, Berlin.
348. FISCHER, E. 1900. Lepidopterologische Experimentalforschungen. *Illustr. Z. Ent.*, 5.
349. FLETT, J. B. 1922. Features of the flora of Mt. Rainier National Park. U.S.D.I., pp. 50, fig. 39.
350. FLINT, R. F. 1947. Glacial geology and the Pleistocene Epoch. New York.
351. FOCARILE, A. 1963. Osservazioni preliminari sul criotropismo del genere *Oreonebria* Dan. *Boll. Soc. ent. ital.*, 93 (9/10): 159-164.
352. FOLSOM, W. J. 1919. Collembola. Report Canadian Arctic Expedition 1913-1918, Ottawa, 3 (A): 1-29 (1922).
353. FORBES, E. 1846. On the connection between the distribution of the existing fauna and flora of the British Isles and the geological changes which have affected their area, especially during the epoch of the Northern Drift. *Mem. geol. Surv. Gr. Britain*, 1.
354. FOREL, A. 1919. Les fourmes de l'Himalaya (Hymenoptera). *Bull. Soc. Vaud. Sci. Nat.*, 42.
355. FOREL, F. 1878. Faunistische Studien in den Süsswasserseen der Schweiz. *Z. wiss. Zool.*, 30: 382-391.
356. FORMOZOV, A. N. 1939. The snow covering as an environmental factor and its importance in the ecology of mammals and birds. *Bull. Soc. nat. Moscow*, 18: 60-68.
357. FORMOZOV, A. N. 1946. La couverture de neige comme facteur integrant du milieu et son importance dans l'écologie des mammifères et des eiseux (In Russian, with summary in French). Material for Fauna and Flora of USSR. Moscow (NS) (Zool.) 5: 1-152.
358. FORSIUS, R. 1929. Entomologische Ergebnisse der schwedischen Kamtchatka-Expedition 1920-1922. Tenthredinidae. *Ark. Zool.*, 20 (A) (5): 1-4.
359. FORSTER, W. 1955. Beiträge zur Kenntnis der Insektenfauna Boliviens. I. Lepidoptera. *Veröff. zool. Staatssamml. München*, 3: 81-160, pl. ix.
360. FORSTER, W. 1958. Die tiergeographischen Verhältnisse Boliviens. *Proc. X int. Congr. Ent.*, 1: 843-846.
361. FORSTER, W. & K. ROSEN, 1940. Entomologische Ergebnisse der Deutsch-Russischen Alai-Pamir Expedition 1920. Lepidoptera. *Mitt. münchn. ent. Ges.*, 30: 907-919.
362. FOX, H. M. & M. L. JOHNSON, 1934. The control of respiratory movements in Crustacea by oxygen and carbon dioxide. *J. exp. Biol.* 11: 1.
363. FOX, H. M., B. G. SIMMONDS & R. WASHBOURN, 1935. Metabolic rates of Ephemerid nymphs from swiftly flowing and from still waters. *J. exp. Biol.*, 12: 179.
364. FOX, H. M. & J. SYDNEY, 1953. The influence of dissolved oxygen on the respiratory movements of caddis larvae. *J. exp. Biol.*, 30: 234.
365. FRANZ, H. 1930. Untersuchungen über den Wärmehaushalt der Poikilothermen. *Biol. Zbl.*, 50: 158-182.
366. FRANZ, H. 1935. Beobachtungen über das Vorkommen von Koleopteren und anderen Insekten auf Schnee. *Koleopt. Rundschau*, 21: 9-14.

367. FRANZ, H. 1936. Die hochalpine Koleopterenfauna der Karnischen und Venezianer Alpen. *Koleopt. Rundschau*, 22: 230-251.

368. FRANZ, H. 1936. Der Einfluss der Eiszeiten auf die Lebewelt der Alpen. *Jb. Ver. Schutz Alpenflor. Tiere*, 10.

369. FRANZ, H. 1943. Die Landtierwelt der mittleren Hohen Tauern: Ein Beitrag zur tiergeographischen und soziologischen Erforschung der Alpen. *Denkschr. Akad. Wiss. Wien* (math.-naturw. Klasse) 107: 1-552, fig. 6, pl. i-xiv, maps 11.

370. FRANZ, H. 1948. Untersuchungen über die Kleintierwelt Ostalpiner Böden II. *Zool. Jb.*, (Syst.) 77: 81-162.

371. FRANZ, H. 1949. Erster Nachtrag zur Landtierwelt der mittleren Hohen Tauern. *Sitzb. österr. Akad. Wiss. Wien*, 158: 1-77.

372. FRANZ, H. Zur Kenntnis der Rassenbildung bei Käfern der ostalpinen Fauna. *Zbl. Gesamtgebiet Ent.*, Lienz, 3: 3-23, fig. 2, map 1.

373. FRANZ, H. 1950. Prä- und interglaziale Relikte in der Bodenfauna der Nordostalpen. *Proc. VIII int. Congr. Ent. Stockholm*, p. 382.

374. FRANZ, H. 1950. Die Artzusammensetzung ostalpiner und pannonischer Bodentiergemeinschaften in ihrer Abhängigkeit vom Standort. *Proc. VIII int. Congr. Ent. Stockholm*, pp. 401-407.

375. FRANZ, H. 1950. Tiergesellschaften hochalpiner Lagen. *Biol. Gen.*, 18: 1-29.

376. FRANZ, H. 1951. Der "hochalpine" Charakter der Felsenheidenfauna in den Ostalpen. *Biol. Gen.*, 19: 299-311.

377. FRANZ, H. 1952. Bemerkenswerte in den Nordostalpen gemachte Dipterenfunde. *Z. wien. ent. Ges.*, 37: 38-43.

378. FRANZ, H. 1957. Die Höhenstufengliederung der Gebirgsfaunen Europas. Consejo Superior de investigaciones científicas. *Publicaciones del Inst. Biol. appl. Barcelona, Simposio de Biogeográphica Ibérica*, 26: 109-116.

379. FRANZ, H. 1960. Zur Kenntnis der Bodenfauna der Südostalpen. *Z. Arb. Gem. österr. Ent.*, 12: 19-23.

380. FREEMAN, D. S. 1951. A subapterous species of *Tipula* from East Africa. *Entomologist*, 83: 61-63.

381. FREEMAN, T. N. 1958. A historical account of insect collecting in Northern Canada. *Proc. X int. Congr. Ent.*, 1: 613-617.

382. FREEMAN, T. N. 1958. The distribution of arctic and subarctic butterflies. *Proc. X int. Congr. Ent.*, 1: 659-672.

383. FRENZEL, B. 1959-1960. Die Vegetations- und Landschaftszonen Nord-Eurasiens während der letzten Eiszeit und während der postglazialen Wärmezeit I & II. *Akad. Wiss. Lit. Mainz*, Abh. math.-naturw. Klasse.

384. FREUDE, H. 1953. Bemerkenswerte Neufunde von *Mesoniscus alpicola* (Hell.) im Berchtesgadener Naturschutzgebiet. *Jb. Ver. Schutz Alpenflora und Tiere.*

385. FRIDÉN, A. 1956. Coleopterafauna i Tärna. *Opusc. Ent.* (Suppl.) 13: 1-128, fig. 71.

386. FRIDÉN, A. 1962. Betrachtungen der Käferfauna der mittleren Apenninen, nebst einigen floristischen Gedanken. *Boll. Soc. ent. ital.* 92 (5/6): 77-80.

387. FRIES, R. E. & T. C. E. FRIES, 1948. Phytogeographical researches on Mt. Kenya and Mt. Aberdare, British East Africa. *K. Svensk. Vet. Handl.*, (3) 25: 1-83.

388. FRIESE, G. 1958. Zoogeographische Betrachtungen zur Rhopalocerenfauna Mitteleuropas (Lepidoptera). *Proc. X int. Congr. Ent.*, 1: 731-736.

389. FRISON, T. H. 1929. Fall and winter stoneflies or Plecoptera of Illinois. *Illinois nat. Hist. Soc. Bull.*, 18: 345-409; The stoneflies or Plecoptera of Illinois. *ibid.*, 30: 218-471.

390. FRUHSTORFER, H. 1921. Die Orthopteren der Schweiz und der Nachbarländer auf geographischer sowie ökologischer Grundlage mit Berücksichtigung der fossilen Arten. *Arch. Naturg.*, 87 (5): 1-262.

391. FUCHS, V. E. 1935. The Lake Rudolf Rift Valley Expedition 1934. *Geogr. J.*, 86: 114-142.

392. GADZHIEV, V. D. 1962. Subalpiiskaya Rastitelnost Bolshavo Kabkaza (v Priedelak Azerbaidjanskoi SSR). *Akad. Nauk Azerbaidjan SSR, Inst. Bot.* V. L. Komarova Bak. pp. 1-171, fig. 23.
393. GAMS, H. 1923. Die Waldklimate der Schweizeralpen, ihre Darstellung und ihre Geschichte. *Verh. naturf. Ges. Basel*, 25: 262-276.
394. GAMS, H. 1933. Das Alter des alpinen Endemismus. *Ber. schweiz. bot. Ges.*, 42: 467-483.
395. GAMS, H. 1935. Pflanzenleben des Glocknergebietes. Kurze Erläuterungen der Vegetationskarte. *Z. dtsch.-Österr. Alpenver.*, 66: 157-176.
396. GAMS, H. 1936. Der Einfluss der Eiszeiten auf die Lebewelt der Alpen. *Jb. Ver. Schutz Alpenflora und Tiere*, 10.
397. GAMS, H. 1938. Die nacheiszeitliche Geschichte der Alpenflora. *Jb. Ver. Schutz Alpenflora und Tiere*, 10.
398. GARDINER, F. 1879. Coleoptera of the White Mountains. *Psyche*, 2: 211-213.
399. GAUCKLER, H. 1899. Untersuchungen über beschleunigte Entwicklung überwinternder Schmetterlingspuppen (Treiben der Puppen). *Illustr. Z. Ent.*, 4 (7): 12, 15.
400. GEDDES, J. G. 1883. List of diurnal Lepidoptera collected in the Northwest Territory and Rocky Mountains. *Canad. Ent.*, 15: 221-223.
401. GEIGER, R. 1927. Das Klima der bodennahen Luftschicht. Braunschweig. pp. xii+246.
402. GEIGER, R. 1942. Das Klima der bodennahen Luftschicht. *Die Wissenschaft*, 78 (2): xvi, 465.
403. GEIGER, R. 1950. The climate near the ground. Cambridge, Mass. Harvard University Press.
404. GEROULD, S. H. 1924. Seasonal changes in the melanic pigmentation in butterflies of the genus *Colias*. *Anat. Rec.*, 29: 83-94.
405. GERSTAECKER, A. 1873. Die Gliederthierfauna des Sansibargebietes. C. von der Decken Ost-Afrika Expedition 1862. Leipzig.
405a. GHAURI, M. S. K. 1964. A new species of *Meconema* (Heteroptera: Miridae) from Ruwenzori. *Proc. R. ent. Soc. London*, 33 (1/2): 13-14, fig. 1.
406. GIBSON, A. 1920. Lepidoptera. *Report Canadian Arctic Expedition* 1913-1918. Ottawa. 3 (I): 1-58 (1922).
407. GIESE, A. C. 1945. Ultraviolet radiation and life. *Physiol. Zool.*, 18: 223-250.
408. GILBERT, G. K. 1914. The transportation of debris by running water. *Prof. Papers U.S. geol. Surv.*, 86: 1-263.
409. GISIN, H. 1943. Ökologie und Lebensgemeinschaften der Collembolen im schweizerischen Exkursionsgebiete Basels. *Rev. Suisse Zool.*, 50 (4): 131-224.
410. GLENNIE, E. A. 1941. Supposed cannibalism among spiders in high altitudes. *J. Bombay nat. Hist. Soc.*, 42 (3): 667.
411. GLICK, P. A. 1939. The distribution of insects, spiders and mites in the air. *U.S. Dep. Agric. Tech. Bull.*, 673: 1-151.
412. GODMAN, F. D. 1885. List of the Lepidoptera collected by Mr H. H. Johnston during his recent Expedition to Kilimanjaro. *Proc. zool. Soc. London*, pp. 537-541.
413. GOLDTHWAIT, J. W. 1913. Glacial cirques near Mount Washington. *Amer. J. Sci.*, 35: 1-19.
414. GOLLERBAKH, M. M. & E. H. SYROECHKOVSKY, 1958. Biogeographical observations in Antarctica 1957. *Izv. Akad. Nauk SSSR*, (geogr.) 6.
415. GOLOSKOLOV, V. P. 1949. Flora i rastitelnost visokogornikh poysov Zailiskovo Ala-Tau. Alma-Ata.

416. GOLOVKOVA, A G. 1957. Visokigornaya rastitelnost Centralnovo Tian-Shanya. *Sbornik Kirghizkovo Nauchno Issledovatelskovo Instituto Pedagogiki.* Frunze.

417. GOLOVKOVA, A. G. 1959. Rastitelnost Centralnovo Tian-Shanya. Kirghiz-Gosudarst. Universitet Kafedra Bot. Frunze. pp. 1-456, fig. 128.

418. GOLOVKOVA, A. G. 1961. O flore Centralnovo Tian-Shanya. Kirghiz. Gosudarst. Universitet. Frunze.

419. GOLOVKOVA, A. G. 1962. Geobotanicheskoe Raionirovanie Centralnovo Tian-Shanya. Kirghiz-Gosudarst. Universitet. Kafedra Bot., Frunze. pp. 1-139.

420. GOUGH, W. G. 1935. Some butterflies of Nepal. *J. Bombay nat. Hist. Soc.*, 38: 258-265.

421. GRABER, V. 1867. Die Orthopteren Tirols mit besonderer Rücksicht auf ihre Lebensweise und geographische Verbreitung. *Verh. zool.-bot. Ges. Wien*, 17: 251-280.

422. GREDLER, V. 1856. Die Alpenkäfer und die Eiszeit. *Verh. Mitt. Siebenbürg. Ver. Naturw.*, pp. 15-19.

423. GREGORY, J. K. 1921. The Rift Valleys and Geology of East Africa.

424. GREGORY, J. W. 1894. The glacial geology of Mt. Kenya. *Quart. J. geol. Soc.*, 1: 515-530.

425. GREGORY, J. W. 1900. The geology of Mt. Kenya. *Quart. J. geol. Soc.*, 56: 223-229.

426. GRESSITT, J. L. 1956. Entomological investigations in New Guinea Mountains. *Proc. Hawaii. ent. Soc.*, 16 (1): 47-69.

427. GRESSITT, J. L. New Guinea and insect distribution. *Proc. X int. Congr. Ent.*, 1: 767-778.

427a. GRESSITT, J. L. 1965. Biogeography and Ecology of Land Anthropods of Antarctica. In: P. van Oye & J. van Mieghem, Biogeography and Ecology in Antarctica. *Monogr. Biol.* XV, The Hague, Dr. W. Junk.

428. GRESSITT, J. L. & M. E. PRYOR, 1960. Supplement to "Bibliographic introduction to Antarctic-Subantarctic Entomology". *Pacific Insects*, 3 (4): 563-567.

429. GRESSITT, J. L. & N. A. WEBER, 1959. Bibliographic introduction to Antarctic-Subantarctic Entomology. *Pacific Ent.*, 1 (4): 441-480.

430. GRIDELLI, E. 1935. Materiali zoologici raccolti dalla spedizione italiana al Karakorum, Coleoptera: Staphylinidae. *Atti Mus. Stor. nat. Trieste*, 12: 69-85, fig. 4.

431. GRIDELLI, E. 1935. Materiali zoologici raccolti dalla spedizione italiana al Karakorum, Coleoptera: Tenebrionidae. *Atti Mus. Stor. nat. Trieste*, 12: 37-68, fig. 1, pl. i-ii.

432. GRIDELLI, E. 1955. Tenebrionidae (Insecta) from Afghanistan. The Third Danish Expedition to Central Asia. Zool. Res. 15. *Vidensk. Meded. Dansk. naturh. Foren.*, 177: 23-74.

433. GRIGGS, R. F. 1934. The edge of the forest in Alaska. *Ecology*, 15: 80-96.

434. GRIGGS, R. F. 1936. Vegetation of Katmai District. *Ecology*, 17: 380-417.

435. GRIGGS, R. F. 1938. Timberlines in northern Rocky Mountains. *Ecology*, 19: 548-564.

436. GRIGGS, R. F. 1946. The timberlines of North America and their interpretations. *Ecology*, 27: 275-287.

437. GRINNELL, J. 1908. The biota of San Bernadino Mountain. *Univ. California Publ. Zool.*, 5: 1-170, pl. i-xxiv.

438. GROSS, F. J. 1958. Zur Schmetterlingsfauna Ostasiens I. Gattung *Satyrus* Latr., Untergattung *Aulocera* Butl. (Lepidoptera: Satyridae). *Bonn. zool. Beitr.*, 9 (2/4): 261-293.

439. GROSS, F. J. 1960. Zur Geschichte und Verbreitung der europäischen Satyriden (Lepidoptera). *Verh. dtsch. zool. Ges. Bonn/Rhein*, pp. 513-528 fig. 7.

440. GROSS, F. J. 1961. Zur Evolution europäischer Lepidopteren. *Verh. dtsch. zool. Ges. Saarbrücken*, pp. 461-478, fig. 5.

441. GROSSHEIM, A. A. & D. I. SOSNOVSKII, 1928. Opit botanichesko-geographicheskovo raionirovaniya Kavkazkovo Kraya. *Izv. Tbilisi Polytech. Inst.*, 1: 48-51.
442. GROTE, A. R. 1875. The effect of glacial epoch upon the distribution of insects in North America. *Ann. Mag. nat. Hist.*, (4) 16: 440-442.
443. GROTE, A. R. 1875. On the insect fauna of the White Mountains. *Psyche*, 1: 76-77.
444. GROTE, A. R. 1876. Arctic Lepidoptera in the White Mountains. *Amer. Nat.*, 10: 129-132.
445. GRUM-GRSHIMAILO, G. 1885. Bericht über meine Reise in das Alai-Pamir Gebiet. *Mem. Rom.*, 2: 214-247.
446. GRUM-GRSHIMAILO, G. 1890. Le Pamir et sa faune Lepidopterologique. *Mem. Rom.*, 4: 17-577, pl. i-xxi.
447. GRUM-GRSHIMAILO, G. 1899. Novae species et varietates Rhopalocerarum e Pamir. *Horae Soc. ent. Rossae*, 22: 303-307.
448. GRUNNE, X. DE (with L. HAUMANN, L. BURGEON & P. MICHOT) 1937. Le Ruwenzori. Vol. 1, maps 17.
449. GUINOT, F. 1954. Dytiscidae recuillies par la Yale North-India Expedition au Kashmir et au Pendjab (Coleoptera). *Opusc. ent. Lund*, 19: 221-224.
450. GUNN, D. L. 1942. Body temperature of poikilotherm animals. *Biol. Rev.*, 17 (4): 293-324.
451. GUPPY, H. B. 1897. On the summit of Mauna Loa. *Nature, Lond.*, 57 (1462): 20-21.
452. GUPPY, H. B. 1925. Dispersal of butterflies and other insects. *Nature, Lond.*, 116: 543.
453. GUPTA, V. K. 1954. Entomological Survey of the Himalaya X. On a collection of Ichneumonidae. *Agra Univ. J. Res.*, (Sci.) 4 (2): 513-530.
454. GUPTA, V. K. 1960. Entomological Survey of the Himalaya Part XXII. Notes on some Hymenoptera collected by the Third Entomological Expedition to the Northwest Himalaya. *Rec. Indian Mus.*, 55 (1/4): 23-27 (1957).
455. GUPTA, V. K. 1962. Taxonomy, zoogeography and evolution of Indo-Australian *Theronia* (Hymenoptera: Ichneumonidae). *Pacific Ins. Monogr.*, 4: 1-142.
456. GURNEY, A. B. 1953. Grasshopper Glacier of Montana. *Ann. Rep. Smithsonian Inst.*, 1952: 303-325.

457. HACKMAN, W. 1964. On reduction and loss of wings in Diptera. *Not. Ent. Helsinki*, 46: 73-93.
458. HAECKER, V. 1925. Klima und tierische Pigmentierung. Verh. Klimatol. Tagung in Davos.
459. HAEMPEL, O. 1923. Zur Kenntnis einiger Alpenseen, III. Millistüttersee. *Int. Rev.*, 15: 345-400.
460. HAEMPEL, O. 1926. Zur Kenntnis einiger Alpenseen, IV. Der Altersee. *Int. Rev.*, 15:273-322.
461. HAEMPEL, O. 1932. Zur Kenntnis einiger Alpenseen, V. Der Irrsee. *Int. Rev.*, 26: 337-387.
462. HALL, F. G. 1922. The vital limits of exsiccation of certain animals. *Biol. Bull.*, 42: 31-56.
463. HAMFELT, B. 1930. Entomologische Ergebnisse der schwedischen Kamtchatka-Expedition 1920-22. Lepidoptera V. Microlepidoptera. *Ark. Zool.*, 21 (A) (13): 1-6.
464. HAMMOND, E. C. 1938-1939. Biological effects of population density in lower organisms. *Quart. Rev. Biol.*, 13: 421-438; 14: 35-59.
465. HAMPSON, G. F. 1909-1910. Ruwenzori Expedition. Lepidoptera-Rhopalocera, pp. 103-140.
466. HANCOCK, G. L. & W. W. SOUNDRY, 1931. Notes on fauna and flora of northern Bugishu and Masaba (Elgon). *J. E. Afr. nat. Hist. Soc.*, 36: 165-183, pli-iv.
467. HANDSCHIN, E. 1909. Über Relikte. *Verh. zool.-bot. Ges. Wien*, 59: 183-207.

468. HANDSCHIN, E. 1919. Beiträge zur Kenntnis der Wirbellosen terrestrischen Nivalfauna der schweizerischen Hochgebirge. Leistal. pp. 151.
469. HANDSCHIN, E. 1919. Über Collembolenfauna der Nivalstufe. *Rev. Suisse Zool.*, 27 (4): 65-98.
470. HANDSCHIN, E. 1920. Die Onychiurinen der Schweiz. *Verh. naturf. Ges. Basel*, 32: 1-37.
471. HANDSCHIN, E. 1924. Ökologische und biologische Beobachtungen an der Collembolenfauna des schweizerischen Nationalparks. *Verh. naturf. Ges. Basel*, 35 (2): 71-101.
472. HANDSCHIN, E. 1924. Die Collembolenfauna des schweizerischen Nationalparks. *Denkschr. schweiz. naturf. Ges.*, 60 (2): vi+89-174, pl. vii.
473. HANDSCHIN, E. 1925. Ziele und Probleme der zoologischen Erforschung der Hochalpen. *Rev. Suisse Zool.*, 32: 65-71.
474. HANITSCH, R. 1900. An Expedition to Mount Kina Balu, British North Borneo. *J. Straits Branch R. Asiatic Soc.*, 34: 49-88, pl. iv (pp. 76-84 deal with insects).
475. HARCOURT-BATH, W. 1896. On the vertical distribution and derivation of the Rhopalocera in the Pyrenees. *The Entomologist*, pp. 320-325.
476. HARDY, A. C. & P. S. MILNE, 1938. Studies in the distribution of insects by aerial currents. Experiments in aerial tow-netting from kites. *J. anim. Ecol.*, 7: 199-229.
477. HARKER, J. E. 1950. The effect of temperature on the final instar nymphs of three species of Australian Ephemeroptera. *Proc. R. ent. Soc. London*, 25 (10/12): 111-114.
478. HARNISCH, O. 1925. Die Beziehungen der mitteleuropäischen Tierwelt zur Eiszeit. *Arch. Hydrobiol.*, 15.
479. HARRIMAN, 1904. Alaska Expedition. 8 Insects. Part I. vol. 9, Part II. New York.
479a. HARRINGTON, H. J. 1965. Geology and Morphology of Antarctica. In: P. van Oye & J. van Mieghem (eds.), Biogeography and ecology in Antarctica. *Monogr. Biol.* XV, The Hague, Dr. W. Junk N.V.
479b. HARZ, K. 1968-1969. Die Orthopteren Europas – The Orthoptera of Europe. *Series ent.* Vols. 5 and 6. Dr. W. Junk N.V., The Hague.
480. HASEMAN, J. D. 1912. Some factors of geographical distribution in South America. *Ann. N. Y. Acad. Sci.*, 22: 9-112.
481. HAUFE, W. V. 1954. The effects of atmospheric pressure on the flight responses of *Aëdes aegypti* (Linn.) (Diptera). *Bull. ent. Res.*, 45: 507-526.
482. HAWKES, O. A. 1926. On the massing of ladybird *Hippodamia convergens* (Coleoptera) in the Yosemite Valley. *Proc. zool. Soc. London*, 2: 693-705.
483. HAYASHIDA, K. 1959. Ecological distribution of ants in Mt. Atusanupuri, an active volcano in Akan National Park, Hokkaido. *J. Fac. Sci. Hokkaido Univ.*, Sapporo, (6) Zool., 14: 252-260, fig. 3.
484. HAYDEN, H. H. 1907. Notes on certain glaciers in Northwest Kashmir. *Rec. geol. Surv. India*, 25: 127-137.
485. HAYDEN, H. H. 1916. Notes on the geology of Chitral, Gilgit and the Pamirs. *Rec. geol. Surv. India*, 14.
486. HEBARD, M. 1928. Orthoptera of Montana. *Proc. Acad. nat. Sci. Philadelphia*, 80: 211-306.
487. HEBARD, M. 1929. The Orthoptera of Colorado. *Proc. Acad. nat. Sci. Philadelphia*, 81: 303-425.
488. HEBARD, M. The Orthoptera of the Upper Rio Grande Valley and the adjacent mountains in northern New Mexico. *Proc. Acad. nat. Sci. Philadelphia*, 87: 45-82.
489. HEBERDEY, R. F. 1935. Die Bedeutung der Eiszeit für die Fauna der Alpen. *Zoogeographica*, 1: 353-412.
490. HEBERDEY, R.F. & J. MEIXNER, 1933. Die Adephagen der östlichen Hälfte der Ostalpen. Eine zoogeographische Studie. *Verh. zool.-bot. Ges. Wien*, 87: 1-164, map 1.

491. HEDICKE, H. 1930. Entomologische Ergebnisse der Deutsch-Russischen Alai-Pamir-Expedition 1928 (11). Hymenoptera V. *Anthophorus* Latr. *Mitteil. zool. Mus. Berlin*, 16 (6): 845-857.

492. HEER, O. 1834. Geographische Verbreitung der Käfer in den schweizer Alpen, besonders nach ihren Höhenverhältnissen. I. Teil: Canton Glarus; II. Teil: Rhätische Alpen. *Mitt. Gebiet. theor. Erdk. Froebel und Heer*, Zürich, 1: 36; 2: 131.

493. HEER, O. 1836. Einfluss des Alpenklimas auf die Farbe der Insekten. *Mitt. Gebiet. theor. Erdk. Froebel und Heer*, Zürich, pp. 161-170.

494. HEER, O. 1840. Die Käfer der Schweiz mit besonderer Berücksichtigung ihrer geographischen Verbreitung II. *Neue Denkschr. schweiz. Ges.*, 4: 67.

495. HEER, O. 1845. Die obersten Grenzen des tierischen und pflanzlichen Lebens in den schweizer Alpen. *Neujahrsbl. naturf. Ges. Zürich*, Meyer & Zeller, 47.

496. HEERE, W. 1953. Tierwelt und Eiszeit. *Biol. Gener.*, 19.

497. HEIM, A. 1918/1922. Geologie der Schweiz. Leipzig. vols. 3.

498. HEIM, A. 1938. The Himalayan Border compared with the Alps. *Rec. geol. Surv. India*, 72: 413-421.

499. HEIM, A. 1956. The geological structure of the Himalaya compared with the Alps. *Proc. nat. Inst. Sci. India*, 22 (A): 228-235.

*500. HEIM, A. & A. GANSSER, 1939. Central Himalaya, geological observations of the Swiss Expedition 1936. *Denkschr. schweiz. naturf. Ges.*, 73 (1).

501. HEINIS, F. 1921. Über die Mikrofauna alpiner Polster- und Rosettenpflanzen. *Festschr. Zschokke*, 9: 22.

502. HEINIS, F. 1937. Beiträge zur Mikrobiocönosis in alpinen Pflanzenpolstern. *Ber. geobot. Forschungsinst. Rübel*, Zürich, 1936.

503. HEINRICH, G. H. & V. K. GUPTA, 1957. Entomological Survey of the Himalaya. Part. XX. On a collection of Ichneumonidae (Parasitic Hymenoptera) from Northwest Himalaya. *Agra Univ. J. Res.*, (Sci.) 5 (2): 367-368 (1956).

504. HELLER, C. & K. W. VON DALLA TORRE, 1881. Über die Verbreitung der Tierwelt im Tiroler Hochgebirge. *Sitzb. Akad. Wiss. Wien*, (math.-naturw. Klasse) 83 (1): 8-53; 83 (2): 103-178; 83 (6); 86 (2): 8-83 (1883).

505. HEMMINGSEN, A. M. 1956. Deep-boring ovipositor of some cranefly species (Tipulidae) of the subgenus *Vestiplex* Bezzi and *Oreomyza* Pok. and some associated phenomena. *Vidensk. Meded. dansk. naturh. Foren. Kobenhavn*, 118: 243-315.

506. HENDRIKSEN, K. L. & W. LUNDBEK, 1917. Groenland Landarthropoder (Insecta et Arachnida Groenlandica) conspectus faunae Groenlandica 2. Kobenhaven. *Meded. Groenland*, 22 (2): 481-821.

507. HEPTNER, V. G. 1945. Desert and steppe fauna of the Palaearctic Region and centres of its development. *Bull. Soc. nat. Moscou*, Biol., 50: 1-2, 17-38.

508. HERING, M. 1932. Morphologische Untersuchungen in der Gattung *Parnassius* (Lepidoptera) als Beitrag zu einer Kritik am Begriff der Unterart. *Mitt. zool. Mus. Berlin*, 18: 275-317, fig. 150, map 1.

509. HERITSCH, FR. 1923. Die Grundfragen der alpinen Tektonik. Berlin: Gebrüder Bornträger. pp. 259, fig. 33.

510. HERON, F. 1909-1910. Ruwenzori Expedition Results. Rhopalocera. pp. 141-176, pl. v.

511. HERRERA, A. L. & L. D. VERGARA, 1899. La vie sur les hauts plateaux. Mexico. pp. 790, tables 18.

512. HIGHMAN, B. & P. D. ALTLAND, 1949. Acclimatization response and pathogenic changes in rats at an altitude of 25 000 ft. *Arch. Pathol.*, 48: 503.

513. HIGHMAN, B. & P. D. ALTLAND, 1955. Effect of altitude and cobalt polythemia hypoxia and cortisone on susceptibility of rats to endocarditis. *Circulation Res.*, 3: 351.

* Since this was sent to press, a book by GANSSER "Geology of the Himalayas", pp. 300, has been published by John Wiley-Interscience: London.

514. HIGHMAN, B., P. D. ALTLAND & H. EAGLE, 1954. Experimental bacterial endocarditis in altitude rats, development and regression in cardiac lesions, including lesions in rat treated with penicillin. *A.M.A. Arch. Pathol.*, 58: 241.

515. HILDÉN, I. 1924. Zur Kenntnis der Käferfauna im Altai. *Not. entom.*, 4: 94-95.

516. HILL, L. & A. EIDNOW, 1923. The biological action of light. *Proc. R. Soc. London*, (B) 95: 163-180.

517. HINCKS, W. D. 1950. Some Dermaptera from Kilimanjaro, Tanganyika. *Ent. mon. Mag.*, 86: 179-181.

518. HINGSTON, R. W. 1925. Animal life at high altitude. *Geogr. J.*, 65: 186-198.

519. HINGSTON, R. W. 1925. In: E. Norton's The fight for Everest. London: Edward Arnold & Sons.

520. HÖLZEL, E. 1958. Die Koleopterenfauna des östlichen Teils der Karnischen Nordkette. Faunistik und zoologische Darstellung. *Mitt. münch. ent. Ges.*, 48: 1-50, fig. 2, pl. ii, map 1.

521 HÖVERMANN, J. 1954. Über die Höhenlage der Schneegrenze in Äthiopien und ihre Schwankungen in historischer Zeit. *Nachr. Akad. Wiss. Göttingen*, iia (math.-physik. chem. Abt.) 6 : 111-137.

522. HOFENEDER, K. 1928 Über die Larven der Blepharoceriden und ihren merkwürdigen Anheftungsapparat. *Verh. zool.-bot. Ges. Wien*, 77: 82-98, fig. 1-9.

523. HOFFMANN, FR. 1908-1909. Beitrag zur Lepidopterenfauna des Glocknergebietes. *Jber. ent. Ver. Wien*, 19: 63-84.

524. HOFFMÄNNER, B. 1925. Beitrag zur Ökologie und Biologie der schweizerischen Hemipteren. *Rev. Suisse Zool.*, 32 (5): 181-206.

525. HOLDHAUS, K. 1904. Ergebnisse einer koleopterologischen Exkursion in das Gebiet des Mte Cavallo in den venetianer Alpen. *Münsch. Kol. Z.*, 2: 215-228

526. HOLDHAUS, K. 1906. Über die Verbreitung der Coleopteren in den mitteleuropäischen Hochgebirgen. *Verh. zool.-bot. Ges. Wien*, 629-641.

527. HOLDHAUS, K. 1911. Über die Abhängigkeit der Fauna vom Gestein. *I. Congr. int. Ent.*, 2: 321-344.

528. HOLDHAUS, K. 1911. Über die Coleopteren- und Molluskenfauna des Monte Gargano. *Denkschr. Akad. Wiss. Wien*, (math.-naturw. Klasse) 87: 342-465; *Wien. ent. Z.*, 34: 349-352 (1915).

529. HOLDHAUS, K. 1912. Kritisches Verzeichnis der boreoalpinen Tierformen (Glazialrelikte) der mittel- und südeuropäischen Hochgebirge. *Ann. naturh. Hofmus. Wien*, 26: 399-440.

530. HOLDHAUS, K. 1924. Spuren der Eiszeit im Faunenbild von Europa. Wien, pp. 1-23, fig. 9.

531. HOLDHAUS, K. 1929. Die geographische Verbreitung der Insekten. In: Schröder Handbuch der Entomologie, 2: 592-1058.

532. HOLDHAUS, K. 1932. Die boreoalpinen Arten der Gattung *Bembidion* Latr., nebst Bemerkungen über die Genese der boreoalpinen Verbreitung. *Soc. Ent. Fr. Paris* (Livre cent.) 353-368.

533. HOLDHAUS, K. 1933. Das Phänomen der Massifs de Refuge in der Coleopterenfauna der Alpen. *V. Congr. int. Ent.*, 2 (Trav.) 1932: 397-406.

534. HOLDHAUS, K. 1938. Verschiedenartige Verbreitungsbilder unter den boreoalpinen Insekten Europas. *VII. int. Kongr. Ent. Berlin*, 1938: 211-224.

535. HOLDHAUS, K. 1954. Die Spuren der Eiszeit in der Tierwelt Europas. *Abh. zool.-bot. Ges. Wien*, 18: 1-493, pl.i-lii, map 1.

536. HOLDHAUS, K. & F. DEUBEL, 1910. Untersuchungen über die Zoogeographie der Carpathen (unter besonderer Berücksichtigung der Coleopteren). *Abh. k.k. zool.-bot. Ges. Wien*, 6 (1): vi+202, map 1. Jena: Gustav Fischer.

537. HOLDHAUS, K. & C. H. LINDROTH, 1939. Die europäischen Koleopteren mit boreoalpiner Verbreitung. *Ann. nat. Hist. Mus. Wien*, 50: 123-293, pl. vi-xviii.

538. HOLLAND, T. H. 1906. General Report (Himalayan Glaciers). Geological Survey of India. *Rec. geol. Surv. India*, 35: 32. A preliminary survey of certain glaciers in the Northwest Himalaya by the officers of the Geological Survey of India. *ibid.*, pp. 123-126.

539. HOLM, Å. 1962. The spider fauna of the East African Mountains. *Zool. Bidr.*, Uppsala, 35: 19-204.

540. HOPKINS, A. D. 1920. The bioclimatic law. *J. Wash. Acad. Sci.*, 10: 34-40.

541. HOPKINS, A. D. 1938. Bioclimatic law. *U. S. Dept. Agric. Misc. Pub.*, 280.

542. HORION, A. 1950. Discontinuierliche Ost-West Verbreitung mitteleuropäischer Käfer. *Proc. VIII int. Congr. Ent., Stockholm*, p. 408.

543. HORMUZAKI, C. *v.* 1900. Beitrag zur Makrolepidopterenfauna der österreichischen Alpenländer. *Verh. zool.-bot. Ges. Wien*, pp. 24-33.

544. HOVANITZ, W. 1950. The biology of *Colias* butterflies I. The distribution of North American species. *Wasmann J. Biol.*, 81 (1): 49-75, fig. 8.

545. HOWARD, L. O. 1918. A note on insects found on snow at high elevations. *Ent. News*, 29 : 375-377.

546. HUBAULT, E, J. 1927. Contribution à l'étude des invertebrés torrenticoles. *Bull. Biol. Fr. Belg.* (Suppl.) 9: 1.

547. HUBBARD, H. G. 1891. Insect life in the hot springs of the Yellowstone National Park. *Canad. Ent.*, 23: 226-238.

548. HUDSON, G. V. 1890. An entomological tour of the table-land of Mount Arthur (New Zealand). *Entomologist*, 23: 8-12, 52-55.

549. HUDSON, G. V. 1905. Notes on insect swarms on mountain tops in New Zealand. *Trans. New Zeal. Inst.*, 38: 334-336.

550. HÜGEL, C. P. VON, 1848. Kaschmir und das Reich der Siek. Stuttgart. Vols. 4.

551. HULTÉN, E., 1937. Outline of the history of arctic and boreal biota during the Quaternary period. Stockholm.

552. HUNTINGTON, E. 1906. Pongong: A glacial Lake. *J. Geol.*, 14: 599-617.

553. HUSTACHE, A. 1928. Curculionidae. Guy Babault Mission dans les Provinces Centrales de l'Inde et dans la region occidentale de l'Himalaya 1914. pp. 1-22.

554. HUSTACHE, A. 1936. Entomological expedition to Abyssinia 1926-1927: Coleoptera, Curculionidae. *Ann. Mag. nat. Hist.*, (10) 18: 353-373, 419-446, 483-505, 563-587.

555. HUTCHINSON, G. E. 1933. The zoogeography of the African aquatic Hemiptera in relation to climatic changes. *Int. Rev. Hydrobiol.*, 28: 436-468.

556. HUTCHINSON, G. E. 1934. Yale North-India Expedition. Report on terrestrial families of Hemiptera-Heteroptera. *Mem. Connecticut Acad. Art. & Sci.*, 10 (8): 119-146, pl. viii-x.

557. HUTCHINSON, G. E. 1937. Limnological studies in Indian Tibet. *Int. Rev. Hydrobiol.*, 35: 133-177.

558. HUTTON, F. W. 1884. On the origin of the fauna and flora of New Zealand. *Ann. Mag. nat. Hist.*, (5) 13: 425-448; 15: 77-107 (1885).

559. HUTTON, F. W. 1896. Theoretical explanation of the distribution of southern faunas. *Proc. Linn. Soc. N. S. Wales*, 21: 36-47.

560. ILLIES, J. 1960. Phylogenie und Verbreitungsgeschichte der Ordnung Plecoptera. *Verh. dtsch. zool. Ges.* (Zool. Anz. Suppl.), 24: 384-394.

561. ILLIES, J. 1960. Die erste auch im Larvenstadium terrestrische Plecoptera. *Mitt. schweiz. ent. Ges.*, 33 (3): 161-163, fig. 4.

562. ILLIES, J. 1961. Die Lebensgemeinschaft des Bergbaches. Wittenberg-Lutherstadt: A. Ziemsen Verlag. pp. 106, fig. 55.

563. ILLIES, J. 1964. Die Plecopteren-Unterfamilie Andiperlinae (Auflösung einer unnatürlichen Negativgruppe nach Remane). *Zool. Anz.*, 172 (1): 37-48, fig. 5.

564. JANETSCHEK, H. 1949. Tierische Succession auf Hochalpinem Neuland. Nach Untersuchungen an Hintereis-, Niederjoch- und Gepatschferner in den Ötztaler Alpen. *Schlern-Schriften*, 67: 1-215, pl. vii.

565. JANETSCHEK, H. 1955. Nunataktiere? Ein aktuelles Problem der zoologischen Heimatforschung. *Jahrb. Voralberger Landes-Mus. Bregenz.*

566. JANETSCHEK, H. 1955. Tierleben auf den höchsten Alpengipfeln. *Der Schlern*, 29 (5/6).

567. JANETSCHEK, H. Das Problem der inneralpinen Eiszeitsüberdauerung durch Tiere (Ein Beitrag zur Geschichte der Nivalfauna). *Österr. zool. Z.*, 6 (3/5): 421-506, fig. 13.

568. JANETSCHEK, H. 1957. Zoologische Ergebnisse einer Studienreise in die spanische Sierra Nevada (Vorläufige Mitteilung). Simposio de Biogeographía Ibérica. Consejo Superior de investigaciones cientificas. *Publicationes del Instituto de Biologia Applicada, Barcelona*, 26: 135-153.

569. JANETSCHEK, H. 1957. Zoologisch-systematische Ergebnisse der Studienreise von H. Janetschek und W. Steiner in die spanische Sierra Nevada 1954. I. Einführung. *Sitzb. österr. Akad. Wiss.* (math.-naturw. Klasse) (1) 166 (5/6): 215-218.

570. JANETSCHEK, H. 1963. Wissenschaft in der Antarktis. *Sitzb. Ges. naturf. Freunde Berlin*, (NF) 3 (3): 123-128.

571. JANETSCHEK, H. 1963. Zur Biologie von Antarktika. *Ber. naturw.-mediz. Verein Innsbruck*, 53 (1959-1963): 235-246.

572. JANETSCHEK, H. 1963. On the terrestrial fauna of the Ross-Sea area, Antarctica. Preliminary Report. *Pacific Insects*, 5 (1): 305-311.

573. JARRIGE, J. 1954. Coléoptères de Sierra Nevada: Staphylinidae. *Arch. Inst. Acclimatación*, Almeria, 2: 73-79.

574. JEANNEL, R. 1908. Biospeologica V. Coleoptera. *Arch. Zool. Paris*, (4) 8: 267-326, pl. iii; (5) 1: 447-532, pl. viii.

575. JEANNEL, R. 1913. Trois nouveaux Trechus des hautes montagnes de l'Afrique orientale. *Bull. Soc. Ent. Fr.*, 87-90.

576. JEANNEL, R. 1919. Sur la distribution géographique de Coléoptères cavernicoles dans Pyrénées. *Assoc. Espan. para el Progresso de las Ciencias Nat.*, Madrid (Zool.) 41: 1-25, map 1.

577. JEANNEL, R. 1921. Les Trechus des Pyrénées et de la Chaine Cantabrique. *Bull. Soc. Hist. Nat. Toulouse*, 49: 165-182.

578. JEANNEL, R. 1926-1930. Monographie des Trechinae, morphologie comparée et distribution géographique d'un groupe de Coléoptères. *L'Abeille*, 32: 221-550; 33: 1-592; 35: 1-808; Suppl. Les Trechini cavernicoles. 34: 59-122.

579. JEANNEL, R. 1928. Les *Trechus* des hautes montagnes, leur origine et leur histoire. *Soc. biogeogr. Peuplm. des Hautes Montagnes*, pp. 122-134.

580. JEANNEL, R. 1935. Carabidae: Trechinae et Perigoninae. Mission scientifique de l'Omo. 2 (Zool.) 2: 23-75.

581. JEANNEL, R. 1936. Sur les Trechinae recueillies par M. H. Scott en Abyssinie. *Rev. franç. Ent.*, 2 (4): 205-212.

582. JEANNEL, R. 1937. Les Bembidiides endogés: Monographie d'une lignée gondwaninne. *Rev. franç. Ent.*, 3: 241-399.

583. JEANNEL, R. 1940. Carabidae: Trechinae. British Museum (Nat. Hist.) Ruwenzori Expedition 1934-1935. 3: 123-127.

584. JEANNEL, R. 1950. Hautes Montagnes d'Afrique. *Mus. Nat. d'Histoire Paris* (Suppl.) 1: 1-253.

585. JEANNEL, R. 1950. Sur quelques Carabiques de la famille des Trechides recueillies par M. H. Scott dans le sud de l'Abyssinie. *Rev. franç. Ent.*, 17: 176-183.

586. JEANNEL, R. 1952. Sur les Bathysciides des Basses-Pyrénées. *Notes biospéol.*, Paris, 7: 51-52.

587. JEANNEL, R. 1952. Coléoptères cavernicoles du Devoluy (Hautes Alpes). *Notes biospéol. Paris*, 7: 35-39, fig. 9.

588. JEANNEL, R. 1954. Sur quelques Psélaphides de l'Uganda et du Mont Elgon recoltés par N. Leleup en décembre 1953. *Rev. franç. Ent. Paris*, 21: 149-170.

589. JEANNEL, R. 1954. Les *Trechus* du Simiene recoltés par Dr. H. Scott en 1952 (Coleoptera: Trechidae). *Rev. franç. Ent. Paris*, 21: 171-183, pl. ii.

590. JEANNEL, R. 1959. Coleoptera: Carabidae-Trechinae. Ruwenzori Expedition 1952. London: British Museum (Nat. Hist.) 2: 17-19, fig. 4.

591. JEWETT, JR. S. G. 1958. Entomological Survey of the Himalaya. Part XXVIII. Stoneflies (Plecoptera) from the Northwest Himalaya. *Proc. nat. Acad. Sci. India*, 28 (B) (4): 320-329.

592. JEWETT, JR. S. G. 1959. The stoneflies (Plecoptera) of the Pacific Northwest. *Oregon State Coll. Monogr. Studies in Entomology*, 3: iv+95.

593. JEWETT, JR. S. G. 1960. Entomological Survey of the Himalaya. Part XXXI. Plecoptera collected by Prof. Mani's Third Entomological Expedition to the Northwest Himalaya. *Agra Univ. J. Res.*, (Sci.) 9 (2): 229-232.

594. JOHANSEN, F. 1921. Insect life on the western arctic coast of America. *Rep. Canadian Arctic Expedition* 1913-1918. Ottawa. 3 (K): 1-61.

595. JOHNSON, C. G. & L. R. TAYLOR, 1962. High altitude migration of *Ocinella frit* (Linn.) (Diptera: Chloropidae). *J. anim. Ecol.*, 31 (2): 373-383, fig. 2.

596. JOHNSON, N. K. & E. L. DAVIS, 1927. Some measurements of temperature near the surface in various kinds of soils. *Quart. J. R. Meteorol.*, 53: 45-59.

597. JOHNSTON, H. H. 1886. The Kilimanjaro Expedition. London (Appendix 4: 372, Insects).

598. JONES, C. G. 1940. Empididae: Hybotinae, Ocydromiinae, Clinocerinae and Hemerodromiinae. British Museum Ruwenzori Expedition 1934-1935. 2 (5): 257-323.

599. JONES, N. G. 1938. The flowering plants and ferns of Mount Rainier. pp. 1-192, pl. ix.

600. KAISILA, J. 1952. Insects from arctic mountain snows. *Ann. ent. fenn.*, 18 (1): 8-25.

601. KAPUR, A. P. 1955. Occurrence of the fruitfly *Acanthiophilus helianthi* (Rossi) (Diptera: Trypetidae) on snow over the Rohtang Pass, N. W. Himalayas. *Indian J. Ent.*, 18 (2): 193.

602. KAPUR, A. P. 1955. On moths of the greasy cutworm *Agrotis ypsilon* Roth. found dead in numbers on snow over the Rohtang Pass, N. W. Himalayas. *Indian J. Ent.*, 18 (3): 289.

603. KAPUR, A. P. 1958. High altitude insects of the Himalaya. *X int. Congr. Ent. Montreal*, 1956, 1: 775-784.

604. KAPUR, A. P. 1962. On the genus *Cteipolia* Staudinger (Lepidoptera: Noctuidae), with a description of a new species from Nepal. *Rec. Indian Mus.*, 58 (2): 115-120 (1960).

605. KAPUR, A. P. 1963. The Coccinellidae of the Third Mount Everest Expedition 1924 (Coleoptera). *Bull. British Mus.* (Nat. Hist.) Ent., 14 (1): 48.

606. KAPUR, A. P. & M. B. KRIPALANI, 1961. The mayflies (Ephemeroptera) from the Northwestern Himalaya. *Rec. Indian Mus.*, 59 (1/2): 183-221, pl. vii.

607. KASHYAP, S. R. 1928. The vegetation of the Western Himalaya and Western Tibet in relation to their climate. Presidential Address Annual Meeting of the Indian Botanical Society, Benares, 1925. *J. Indian bot. Soc.*, 4 (9/10): 327-334.

608. KASZAB, Z. 1958. Ergänzungen zur Tenebrioniden (Insekten) Aufarbeitung von Dr E. Gridelli der 3. dänischen-zentralasiatischen Expedition. Zool. Res., 24. *Vidensk. Meded. Dansk. naturh. Foren.*, 120: 237-255.

609. KAVRISHVILI, V. I. 1955. Landschaftnovo hydrologicheski zonie Gruciiskoi SSR. Tbilisi.

610. KAWAI, T. 1963. Stoneflies (Plecoptera) from Afghanistan, Karakoram and Punjab Himalaya. In: Masuzo Uéno's Results of the Kyoto University Sci. Expedition to Karakoram and Hindukush 1955. 4 Insect fauna of Afghanistan and Hindukush. pp. 53-86, fig. 18.

611. KAWAMURA, T. 1932. *Deuterophlebia. Incongr. Insect. japon.*, p. 2204.

611a. KEAST, A. 1958. Biogeography and Ecology in Australia. *Monogr. Biol.* VIII. The Hague., Dr. W. Junk. N.V.

612. KEISER, F. 1950. Phänologische Beobachtungen an alpinen Dipteren. *Proc. VII int. Congr. Ent.*, pp. 418-422.

613. KENNEDY, H. D. 1958. Biology and life history of a new species of mountain midge *Deuterophlebia nielsoni* from eastern California (Diptera: Deuterophlebiidae). *Trans. Amer. micr. Soc.*, 77 (2): 201-228.

614. KIMMINS, D. E. 1946. New species of Himalayan Plecoptera. *Ann. Mag. nat. Hist.*, (11) 13 (2): 721-740.

615. KIRITSHENKO, A. N. 1910. Espece nouvelle du genre *Phimodera* Germ. trouvée dans l'Altai. *Rev. russe Ent.*, 10: 21.

616. KIRITSHENKO, A. N. 1912. K poznaiyu roda *Acanthia* F. Latr. (Hemiptera: Heteroptera). Ad cognitionem generis *Acanthia* F. Latr. (Hemiptera: Heteroptera). Ezhevodnik zoologicheskovo Muzeya Imperatskoi Akademii Nauk. *Ann. Mus. Zool. Acad. Imp. Sci. St. Petersburg*, 16: 539-549 (1911).

617. KIRITSHENKO, A. N. 1913. Hemiptera-Heteroptera turanica nova. *Rev. russe Ent.*, 13: 412.

618. KIRITSHENKO, A. N. 1916. Hemiptera-Heteroptera from the Arctic Urals 1909. *Mem. Acad. Imp. Sci. Petrograd*, Cl. Phys.-math. (8) 28 (19): 1-11, fig. 2.

619. KIRITSHENKO, A. N. 1931. Hemiptera-Heteroptera. Abhandlungen der Pamir-Expedition 1928. 8 (Zoology): 77-118.

620. KIRITSHENKO, A. N. 1931. Hemiptera-Heteroptera of the Third Everest Expedition 1924. *Ann. Mag. nat. Hist.*, (10) 7: 362.

621. KISHKO, YA. G. 1959. Species composition of the microflora in high atmosphere layers. *Mikrobiologia*, 28 (5): 713-716 (Translation from Russian in: Amer. Inst. Biol. Sci. Washington).

622. KISIMOTO, K. 1956. Effect of crowding during larval period on the determination of the wing from of an adult plant hopper. *Nature, Lond.* 178 (4534): 641.

623. KITAKAMI, S. 1929. An unusual insect from Japan (Diptera: Deuterophlebiidae). *Zool., Mag.*, 41: 379-380.

624. KITAKAMI, S. 1938. The Deuterophlebiidae of Japan. *Annot. Zool. Japon.*, 17: 487-513.

625. KITSCHELT, R. 1925. Zusammenstellung der bisher in dem ehemaligen Gebiet von Südtirol beobachteten Grossschmetterlinge. Wien, p. 421.

626. KITSCHELT, R. 1933/34. Sammelergebnisse in den spanischen Zentral Pyräneen. *Int. ent. Z.*, 26: 447-450, 483-485, 526-530; 27: 46-47, 66-69, 115-116.

627. KLEBELSBERG, R. V. 1933. Grundzüge der Geologie Tirols. Tirol Land, Natur, Volk und Geschichte. Hauptausschuss D. Ö. Alpen Verein. München: Bruckmann.

628. KLUTE, F. 1920. Ergebnisse der Forschungen am Kilimandscharo 1912. Berlin. pp. viii+136.

629. KNIGHT, H. H. 1924. On the nature of the colour patterns in Heteroptera, with data on the effects produced by temperature and humidity. *Ann. ent. Soc. Amer.*, 17: 258-272.

630. KNOLL, F. 1921-1922. Insekten und Blumen: Experimentelle Arbeiten zur Vertiefung unserer Kenntnis über die Wechselbeziehungen zwischen Pflanzen und Tieren. *Abh. zool.-bot. Ges. Wien*, 12 (1/2): 1-65,4l. i-x.

631. KOBELT, W. A. Geographichekoe rasprostranenie Zhivotnikh. Leipzig and Moscow.
632. KOBER, L. 1923. Bau und Entstehung der Alpen. Berlin: Gebrüder Bornträger.
633. KOBER, L. 1931. Das alpine Europa und sein Rahmen: Ein geologisches Gestaltungsbild. Berlin: Gebrüder Bornträger. pp. 310, fig. 33, pl iii.
634. KÖPPEN, W. 1919. Baumgrenze und Lufttemperatur. *Petermanns Mitt.*, 65: 201-203.
635. KÖPPEN, W. 1923. Die Klimate der Erde. Berlin: Walter de Gruyter & Co. pp. 1-369.
636. KÖPPEN, W. 1936. Das geographische System der Klimate. In: Handbuch der Klimatologie. Berlin: Gebrüder Bornträger. 1 (C): 1-44.
637. KOLBE, H. 1891. Aufzählung der von Herrn Dr Hans Meyer im Jahre 1889 im Gebiete des Kilimandscharo und Ugueno Gebirges gesammelten Coleopteren. *Ent. Ztg.*, 52: 18-36.
638. KOLBE, H. 1895. Über die in Afrika gefundenen montanen und submontanen Gattungen der mit *Calosoma* verwandten Coleopteren. *Sitzb. Ges. naturf. Freunde Berlin*, 4: 50-69.
639. KOLBE, H. 1897. Die Käfer Deutschostafrikas. Berlin.
640. KOLBE, H. 1912. Glazialzeitliche Reliktfauna im Hohen Norden. *Dtsch. ent. Z.*, 33-63.
641. KOLBE, H. 1913. Die Differenzierung der zoogeogeographischen Elemente der Kontinente. *Trans. II int. Congr. Ent.*, 2: 433-476.
642. KOLBE, K. 1923. Über das Klima und die Insektenwelt Mitteleuropas während der Eiszeit und der Nacheiszeit. *Dtsch. ent. Z.*, 1923 (1): 1-32.
643. KOLLAR, V. & L. REDTENBACHER, 1848. Aufzählung und Beschreibung der von Freiherrn Carl von Hügel auf seiner im Himalayagebirge gesammelten Insekten. Aus: C. v. Hügel, Kaschmir und das Reich der Siek. Stuttgart. 4: 395-564.
644. KOMÁRCK, J. 1914. Über die Blepharoceriden aus dem Kaukasus und Armenien. *Sitzb. k. böhm. Ges. Wiss. Prag.*, pp. 1-19.
645. KOMÁRCK, J. 1914. Die Morphologie und Physiologie der Haftscheiben der Blepharoceridenlarven. *Sitzb. k. böhm. Ges. Wiss. Prag.*, 9: 25.
646. KOMAROV, V. L. 1951. Flora Polustrova Kamchatki. *Izbp. Soch.*, 7: 1-506,; 8: 1-526.
647. KORB, M. 1916. Reise in den hohen Alai. *Mittl. münch. ent. Ges.*, 7: 7-24.
648. KORFF, S. A. 1954. The world's high altitude research stations. New York University Press.
649. KOROVIN, E. P. 1934. Vaprosam palaeo-ecologicheskikh smien v Srednii Asii. Vaprosi Ecologii i biocaenologii.
650. KOROVIN, E. P. 1934. Rastitelnost Srednii Asii i yuzhnovo Kazakstana. Tashkent.
651. KOROVIN, E. P. 1962. Rastitelnost Srednii Asii i yuzhnovo Kazakstana II. Akad. Nauk Uzbek SSR. Tashkent. pp. 1-548, fig. 231.
652. KOROVIN, E. P. & A. ROZANOV, 1938. Pachvi i rastitelnost Srednii Asii kak estestvennaya proizvoditelnaya sila. *Trudi Central. Asii Gosudarst. Universitet.*, (11) a: 17.
653. KORSCHNEWSKY, N. L. 1930. Das Alaital (Orohydrographie und Vereisung). Abhandl. der Pamir Expedition 1928. 3: 1-62.
654. KOZHANCHIKOV, I. V. 1938. Physiological conditions of cold hardiness in insects. *Bull. ent. Res.*, 29: 253-262.
655. KRAMER, S. 1958. Pigmentation in the thoracic musculature of cockroaches and related Orthoptera and the analysis of flight and stridulation. *Proc. X. int. Congr. Ent.*, 1: 569-579.
656. KRAUSS, H. & L. GANGLBAUER, 1902. Eine koleopterologische Exkursion auf dem Mte. Canin in den Julischen Alpen. *Verh. zool.-bot. Ges. Wien*, 52: 101-109.
657. KRÜGER, P. 1929. Die Bedeutung der Ultraroten Strahlen für den Wärmehaushalt der Poikilothermen. *Biol. Zbl.*, 49: 65-82.
658. KRÜGER, P & F. DUSPIVA, 1933. Die Einfluss der Sonnenstrahlen auf die Lebensvorgänge der Poikilothermen. *Biol. Gen.*, 19: 168-198.

659. KRYZHANOVSKII, O. L. 1953. Zhuki-zhuzhelitsi Roda *Carabus* Srednei Azii. Opredeliteli po faune SSR. Izdabanenie zoologicheskii Institutom Akademii Nauk SSSR. Izdatelstvo Akademii Nauk SSSR. Moscow-Leningrad, 52: 1-132.

*660. KRYZHANOVSKII, O. L. 1961. On the zoogeographical features of the Coleoptera fauna of the deserts of Turkmen SSR. *Beitr. Ent.*, 11 (3/4): 426-445.

661. KRYZHANOVSKII, O. L. 1962. The caterpillar hunters *Calosoma* Web. and *Callisthenes* Fisch.-w. (Coleoptera: Carabidae) in the USSR. *Ent. Oboz.*, 41: 163.

662. KRZYWICKI, M. 1963. Przyczynek do znajomosci fauny Rhopalocera Tatr Polskich (Lepidoptera): Beitrag zur Rhopalocerenfauna des polnischen Teiles des Tatra-Gebirges (Lepidoptera). *Polska Akad. Nauk Inst. Zool., Ann. Zoologici*, Warszawa, 21 (12): 151-122, pl. i-vii.

663. KULCZYNSKI, 1924. Das boreale und arktisch-alpine Element in der Mitteleuropäischen Flora. *Bull. int. Acad. Pol. Sci. Let.* (math.-natur.) (B).

664. KURIENTZOV, A. I. 1959. Relikti v faune Sikhote Alinya. *Komarovskie chteniya, Akad. Nauk SSSR* (Sibiriskoe otdelenie) 8: 19-53 (relicts of fauna in Sichote-Alin Mountains: English summary).

665. KURIENTZOV, A. I. 1961. V ubezhishakh Ussuriskikh Reliktov. Vladivostok: Primorskoe Knizhnoe Izdatelstvo. pp. 1-181, figs. numerous.

666. KURIENTZOV, A. I. 1963. Fauna Kamchatkoi oblasti Akad. Nauk SSSR (Sibiriskoe otdelenie). Inst. Vulkanologii. *Trudi Kamchatkoi Kompleksnoii Expeditsii*. Moscow-Leningrad. pp. 1-60.

667. KURIENTZOV, A. I. 1964. Visokogornaya fauna dalnevo Vostoka i ee proiskhodzhnie. *Akad. Nauk SSSR, Zool. Zhurnal*, 43: 1585-1600, fig. 6.

668. KUSCHEL, G. 1960. Terrestrial zoology in southern Chile. *Proc. R. Soc. London*, (B) 152: 540-550, map 1.

669. KUSNEZOW, N. N. 1923. Die genetischen Elemente der Ameisenfauna des russischen Turkestan. *Zool. Anz.*, 57: 82-88.

670. KUSNEZOW, N. N. 1924. Die Bedeutung der Eiszeit in der Entwicklung der Fauna Turkestans. *Zool. Anz.*, 59: 52-55.

671. KUSNEZOW, N. N. 1925. Zur Frage der vertikalen Verteilung der Faunenelemente Turkestans. *Zool. Anz.*, 62: 107-117, fig. 2.

672. KUZNETSOV, N. Y. 1935. The origin of the Lepidopterous fauna of the Arctic Eurasia. *Arctica*, 3: 115-138.

673. KUZNETSOV, N. Y. 1938. Arkticheskaya fauna Evrazii i ee proiskhzhdenie. *Trudi zool. Inst. Akad. Nauk. SSSR*, 5: 1-85.

674. LA GRECA, M. 1955. Influenza delle variazioni climatiche del Quaternario sul popolamento entomologico d'alta montagna. *Boll. Zool.*, Naples, 22: 489-562, fig. 20.

675. LAHEE, F. H. 1930. Field Geology. New York: McGraw-Hill Book Co. Inc. pp. 700.

676. LATZEL, R. 1907. Massenerscheinung von schwarzen Schneeflöhen in Kärnten. *Carinthia*, 2 (97): 54-71, 145-173.

677. LATZEL, R. 1917. Neue Collembolen aus den Ostalpen und dem Karstgebiet. *Verh. zool.-bot. Ges. Wien*, 71: 49-85.

678. LATZEL, R. 1921. Die Apterygotenfauna der Ostalpen und des anschliessenden Karstes. *Verh. zod.-bot. Ges. Wien*, 71: 49-85.

679. LAWSON, A. C. 1921. The Sierra Nevada. *Univ. California Chron.*, 130-149.

680. LECONTE, J. L. 1878-1880. Coleoptera of the alpine regions of the Rocky Mountains. *Bull. U. S. geol. geogr. Surv. Territ.*, 4 (2): 447-480 (1878); 5: 499-520 (1880).

* Since this was sent to press KRYZHANOVSKII has published a comprehensive account of the Middle Asiatic terrestrial fauna in his book "Sostovi proiskhozhdenie Nazemnoi fauni Srednei Azii". Akademiya Nauk USSR. Moscow-Leningrad (1965), pp. 1-419.

681. LEDEBOUR, C. F. VON 1820. Reise durch das Altai-Gebirge.
682. LEDERER, J. 1853. Lepidopterologisches aus Sibirien. *Verh. zool.-bot. Ges. Wien*, 1-36.
683. LEDERER, J. 1855. Weiterer Beitrag zur Schmetterlingsfauna der Altai-Gebirges in Sibirien. *Verh. zool.-bot. Ges. Wien*, 97-120, pl. i-ii.
684. LEITMEIER, H. 1928. Die österreichischen Alpen. Wien: F. Deuticke.
685. LELEUP, N. 1956. La faune cavernicole du Congo Belge et considérations sur les Coléoptères reliques d'Afrique intertropicale. *Ann. Mus. Congo Belge*, 8 (Sci.) (Zool.) 46: 1-171.
686. LELEUP, N. 1958. Les faunes Coléopterologiques humicole et terricole des montagnes de l'Afrique Continentale au sud Sahara. *Rapp. ann. Inst. Rech. Sci. Afr. Centr.*, Bukavu, 11 (1958): 151-154.
687. LENGERSDORF, F. 1937. Beitrag zur Kenntnis und Systematik der bisher bekannten paläarktischen Lycoriiden (Sciariden) bei denen Flügellosigkeit oder Flügelrückbildung mit Reduktion der Palpen gleichzeitig auftritt. *Decheniana*, 95 (B): 36.
688. LENGERSDORF, F. 1957. Zoologisch-Systematische Ergebnisse der Studienreise von H. Janetschek und W. Steiner in die spanische Sierra Nevada 1954. III. Neue Lycoriiden (Sciariden) (Ins. Diptera). *Sitzb. österr. Akad. Wiss.* (math.-naturw. Klasse) (1) 166 (5/6): 227-229.
689. LESSERT, R. DE, 1915-1926. Araignées du Kilimandjaro et du Merou. *Rev. Suisse Zool.*, 23: 99; 28: 381; 31: 429; 33: 335.
690. LESTAGE, J. A. 1916. Über Larven von Ephemeriden. *Ann. Biol. Lacustr.*, 8: 313.
691. LICHTENECKER, N. 1936. Die gegenwärtige und eiszeitliche Schneegrenze in den Ostalpen. *Verh. II int. Quatär-Konf. Wien*, 141-147, map 2.
692. LIGNAU, N. G. 1912. Neue Beiträge zur Myriopodenfauna des Kaukasus. Ezhevodnik zoologicheskovo Muzeya Imperatskoi Akademii Nauk. *Ann. Mus. Zool. Acad. Imp. Sci. St. Petersburg*, 16: 37-60, pl. i (1911).
693. LINDBERG, H. 1927. Zur Ökologie und Faunistik der subalpinen und alpinen Käferwelt in Enontokis Lappmark. *Acta Soc. Fauna Flora Fenn.*, 56 (14): 1-51.
694. LINDBERG, H. 1933. Untersuchungen im N. Petsamo über die Käferfauna hochnördlicher Biotopen. *Mem. Soc. Fauna Flora Fenn.*, 9: 103-125.
695. LINDBERG, H. 1958. The insect fauna of Cape Verde Islands. *Proc. X int. Congr. Ent.*, 1: 785-787.
696. LINDNER, E. 1930. Blepharoceridae und Deuterophlebiidae. In: Die Fliegen der paläarktischen Region, II (2): 1-37.
697. LINDNER, E. 1956. Zur Verbreitung der Dipteren (Zweiflügler) in den Hochregionen der Alpen. *Jb. Verein Schutz. Alpenpflanzen und Tiere*, München, 1956: 121-128.
698. LINDROTH, C. H. 1941. Endemismus in der Fennoscandischen Fauna, erläutert an der Fam. Carabidae. 20. *Zool. Bidr.*, Uppsala, 20: 431-442.
699. LINDROTH, C. H. 1945-1949. Die fennoscandischen Carabidae. Eine tiergeographische Studie 1-3. *Göthesborgs Kungl. Vetensk. Vitterh. Samh. Handl.*, (6) (B) 4: (1-3).
700. LINDROTH, C. H. 1953. Influence of Pleistocene climate changes on the insect fauna of northern Europe. *Trans. IX int. Congr. Ent.*, 2: 145-153.
701. LINDROTH, C. H. 1957. The faunal connections between Europa and North America. New York: London.
702. LINDROTH, C. H. 1961. The ground-beetles (Carabidae excl. Cicindelidae) of Canada and Alaska. *Opusc. ent.* (Suppl.) 20: 1-200.
703. LIST, T. 1899. Ueber den Einfluss des Lichtes auf die Lagerung von Pigment. *Arch. Entwickl. Mech. Organ.*, 8: 618-632.
704. LÖNNBERG, E. 1929. The development and distribution of the African Fauna in connection with and depending upon climatic changes. *Ark. Zool.*, 21 (A) (4): 1-33, fig. 5.

705. LONG, B. B. 1953. Effects of population density on larvae of Lepidoptera. *Trans. R. ent. Soc. London*, 104 (15): 541-591, fig. 14, pl. vi.
706. LONGSTAFF, T. G. 1920. Glacier Exploration in the Eastern Karakoram. *Geogr. J.*, 35: 622.
707. LOSINA-LOSINSKY, L. 1937. Cold hardiness and anabiosis in larvae of *Pyrausta nubilalis*. *Zool. J.*, 16: 642.
708. LUDWIG, D. 1928. Development of cold hardiness in the larva of Japanese beetle (*Popillia japonica* Newm.). *Ecology*, 9: 303-306.
709. LÜDI, W. 1921. Die Pflanzengesellschaften des Lauterbrunnentales und ihre Sukzessionen. *Beitr. geobot. Landesaufn. Schweiz. Zürich*, 9: 1-364, fig. 4, map 2.
710. LUNDBLAD, O. 1934. Report on Hydracarina. Yale North-India Expedition. *Mem. Connecticut Acad. Sci. Arts.*, 10 (7): 85-118.
711. LUNDEGÅRDH, R. 1930. Klima und Boden in ihrer Wirkung auf das Pflanzenleben. Jena: (2te Auflage) pp. 1-480, 1 map.
712. LUNDQVIST, C. 1936. Hochasiatische Binnenseesedimente. *Mem. Connecticut Acad. Sci. Arts*, 10 (1): 193-240.
713. LUPPA, H., G. MÜLLER, M. NICHT & F. TIETZE, 1960. Vorläufiger Bericht über eine 1959 durchgeführte zoologische Sammelreise nach Armenien. *Wiss. Z. Martin-Luther Univ. Halle-Wittenberg*, 9 (1): 259-262, fig. 4.
714. LUTSCHNIK, W. N. 1930. Carabidae. Abhandlungen der Pamir Expedition 1928. II. Zool. pp. 31-50.
715. LUTZ, F. E. 1929. Experiments with wonder creatures. *Nat. Hist.*, New York, 29: 160-168.
716. LUTZ, F. E. 1932. Our ignorance concerning insects. *Canad. Ent.*, 64: 56.
717. LYELL, G. 1908. Lepidoptera of the Victorian Alps. *Vict. Nat.*, Melbourne, 25: 31-35.
718. LYMAN, H. H. 1892. Can insects survive freezing? *Canad. Ent.*, 24: 1-5.

719. MACHATSCHKE, F. 1955. Das Relief der Erde. Berlin: Gebrüder Bornträger. vols. 2.
720. MACLAGEN, D. S. 1932. The effect of population density upon the rate of reproduction with special reference to insects. *Proc. R. Soc. London*, (B) 111: 437-454.
721. MADIGAN, C. T. 1931. Physiography of Western MacDonnell Ranges. *Geogr. J.*, 78: 417.
722. MALAISE, R. 1932. Entomologische Ergebnisse der schwedischen Kamtchatka-Expedition 1920-22. Tenthredinidae. *Ark. Zool.*, 23 (A) (8): 1-68.
723. MANDL, K. 1940-1944. Die Blindkäferfauna der Karawanken. *Koleopt. Rundschau*, 26: 25-36 (1940); 29: 103-108 (1944), fig. 25.
723a. MANDL, K. 1964. Zoologische Ergebnisse der österreichischen Karakoram-Expedition 1958. 4. Die Arten der Familie Carabidae (Coleoptera). *Ent. Arb. Mus. Frey*, 15 (2): 716-720.
724. MANI, M. S. 1955. Entomological Survey of the Himalaya. Part IV. Expedition to the Upper Chenab Valley 1954. *Agra Univ. J. Res.*, (Sci.), 4 (1): 157-170.
725. MANI, M. S. 1955. Entomologists in the Himalaya. *Turtox News*, 33: 1.
726. MANI, M. S. 1956. The Second Entomological Expedition to the Himalaya. *Nature, Lond.*, 177: 124-125.
727. MANI, M. S. 1956. The Second Entomological Expedition to the Northwest Himalaya. *Turtox News*, 34: 1.
728. MANI, M. S. 1958. High altitude insect life of the Northwest Himalaya. Presidential address at the Biological Section of the XXVI Annual Session of the National Academy of Sciences of India at the Muslim University, Aligarh 1957. pp. 41-62.
729. MANI, M. S. 1959. On a collection of high altitude scorpions and pseudoscorpions (Arachnida) from the Northwest Himalaya. *Agra Univ. J. Res.*, (Sci.) 8 (1): 11-16.

730. MANI, M. S. 1962. Introduction to high altitude entomology. London: Methuen & Co. pp. 304, fig. 88, pl. xvii.
731. MANI, M. S. 1963. High altitude insects. *Agra Univ. J. Res.*, (Sci.) 12 (1): 171-196.
732. MANI, M. S. & SANTOKH SINGH. 1955. Entomological Survey of the Himalaya. Part XIII. Second Entomological Expedition to the Northwest Himalaya (1955). *Agra Univ. J. Res.*, (Sci.) 4 (Suppl.): 717-740.
733. MANI, M. S. & SANTOKH SINGH. 1958. Entomological Survey of the Himalaya. Part XIX. Faunistics of the high altitude Coleoptera from the Northwest Himalaya. *Agra Univ. J. Res.*, (Sci.) 6 (2): 93-125 (1957).
734. MANI, M. S. & SANTOKH SINGH. 1961. Entomological Survey of the Himalaya. Part XXV. A contribution to our knowledge of the ecology of high altitude insect life of the Northwest Himalaya. *Proc. zool. Soc. Calcutta*, 14 (2): 61-135, fig. 18.
735. MANI, M. S. & SANTOKH SINGH. 1961-1963. Entomological Survey of the Himalaya. Part XXVI. A contribution to our knowledge of the geography of the high altitude insects of the nival zones from the Northwest Himalaya. *J. Bombay nat. Hist. Soc.*, 58: 387-406, 724-748 (1961); 59: 77-99, 360-381, 834-861 (1962); 60: 160-172 (1963); fig. 55.
736. MANI, M. S., SANTOKH SINGH, V. K. GUPTA & H. N. BAIJAL, 1955. Entomological Survey of the Himalaya. Part IX. First annotated checklist of the insects from the Northwest Himalaya. *Agra Univ. J. Res.*, (Sci.) 4 (2): 471-512.
737. MANK, E. W. 1934. Annotated list of Coleoptera of Glacier National Park, Montana. *Canad. Ent.*, 66: 73-81.
738. MANNHEIMS, B. J. 1935. Beiträge zur Biologie und Morphologie der Blepharoceriden. *Zool. Forschungen*, Leipzig: R. Noske, 2: vi+115, fig. 95.
739. MARCHAND, W. 1917. Notes on the habits of snowfly *(Chionea)*. *Psyche*, 24: 142-153, pl. xiii-ix.
740. MARSHALL, G. A. K. 1950. New Curculionidae from Kilimanjaro. *Ent. mon. Mag.*, 86: 75-77.
741. MARSHALL, G. A. K. 1954. Journey to the Gughé Highlands 1948-1949. Coleoptera, Curculionidae from the high mountains. *J. Linn. Soc. London*, (Zool.) 42: 369-381.
742. MARTYNOV, A. W. 1930. Trichoptera. Abhandlungen der Pamir Expedition 1928. II. Zool. pp. 79-82.
743. MASON, K. L. 1930. The glaciers of the Karakoram and neighbourhood. *Rec. geol. Surv. India*, 63: 214-278.
744. MASON, R. W. M. 1958. Distributional problems in Alaska. *Proc. X int. Congr. Ent.*, 1: 703-710.
744a. MASON, R. W. M. 1964. Southward extension of the Arctic zone. *Canad. Ent.*, 96 (1/2).
745. MATEU, J. 1954. Apuntes sobre la geología, vegetación y climatología de la Sierra Nevada. *Arch. Inst. Aclimatación*, Almeria, 2: 7-33; Coleopteros de Sierra Nevada Familias Dytiscidae, Hydrophilidae, Dryopidae, Catopidae, Silphidae, Pselaphidae, Histeridae, Cleridae, Cantharidae, Dasytidae, Aderidae, Anthicidae. *ibid.* pp. 89-101.
746. MATEU, J. & G. COLAS, 1954. Coleopteros de Sierra Nevada: Caraboidea. *Arch. Inst. Aclimatación*, Almeria, 2: 35-72.
747. MATSUMURA, M. 1898. Insects collected on Mt. Fuji. *Annot. zool. japon.*, 2: 113-124.
748. MATTHES, F. E. 1922. Mount Rainier and its glaciers. *U.S. Dept. Infor.*, pp. 1-48, fig. 26.
749. MAULIK, C. 1936. Fauna of British India: Chrysomelidae: Galerucinae. pp. 1-648.
750. MAYER, G. 1914. Formiciden aus Tibet und der südlichen Gobi (nach den von dem Obersten P. K. Koslov gesammelten Materialen). *Ann. Mus. Zool. Acad. Sci. Petrograd*, 20.

751. MCATEE, W. L. 1917. Showers of organic matter. *U.S. month. Weather Rev.*, 45: 217-224.

752. MCCOLLOCH, J. W. & P. M. WILLIAM, 1923. Soil temperature and its influence on white-grub activities. *Ecology*, 4: 29-36.

753. MCDUNNOUGH, J. H. 1929. Notes on some diurnal Lepidoptera from Yellowstone Park and the adjacent regions of Montana. *Canad. Ent.*, 61: 105-107.

754. MEAMY, E. S. 1916. Mount Rainier, a record of exploration. MacMillan & Co. pp. 325.

755. MEDVEDEV, S. N. 1930. Scarabaeidae. Abhandlungen der Pamir Expedition 1928. II. Zool. pp. 55-58.

756. MEDVEDEV, S. N. 1959. Gattung *Aphaleria* Rtt. (Coleoptera: Tenebrionidae) und ihre Stellung in der System. *Rev. Ent. USSR*, 38: 219-222.

757. MEINERTZHAGEN, R. 1927. Ladak, with special reference to its natural history. *Geogr. J.*, 70: 129-163.

758. MELANDER, A. L. 1923. Collecting insects on Mount Rainier. Smiths. Report for 1921: 415-422, pl. ix.

759. MELL, R. 1958. Zur Geschichte der ostasiatischen Lepidopteren I. Die Hebung Zentralasiens, das westchinesische Refugium zentralasiatischer Ankömmlinge und die Verbreitungsachse Sikkim/Kashiaberge-Zentralforma. *Dtsch. ent. Z.*, (NF) 5: 185-213, fig. 11.

760. MELLANBY, K. 1940. The activity of certain insects at low temperatures. *J. anim Ecol.*, 9: 296-301.

761. MENOZZI, C. 1939. Formiche dell Himalaya e del Karakorum raccolte dall spedizione italiana commandata da S. A. R. il duca di Spolate 1929. *Atti Soc. Ital. Sci. Nat. Milan*, 78: 285-345.

762. MENZBIER, M. A. 1914. Les districts zoologiques du Turkestan et l'origine probable de sa fauna. *Ann. Soc. encour. Sci. exp. leurs appl. nom. de Chr. Ledentroff*, Moscou, 1914: 1-144.

763. MENZBIER, M. A. 1923/1924. Ueber die Entstehung der Fauna der Tundren. *Bull. Mosk. Obsh. Isp. prir.*, Otd. Biol., 32 (1/2): 76-98.

764. MENZEL, R. 1914. Über die mikroskopische Landfauna der schweizerischen Hochalpen. *Arch. Naturg.*, (A) 80 (3): 1-98.

765. MERKER, E. 1929. Licht und Atmung, neue Ergebnisse. *X int. Congr. Zool. Budapest*, 1: 603-616.

766. MERRIAM, C. H. 1892. The geographical distribution of life in North America. *Proc. biol. Soc. Wash.*, 7: 1-44.

767. MERRIAM, C. H. 1898. Life zones and crop zones of the United States. *U.S. Dept. Agric., Div. Biol. Surv. Bull.*, 10: 1-79, map 1.

768. MERRIFIELD, F. 1892. The effects of artificial temperature on the colouring of several species of Lepidoptera, with an account of experiments on the effects of light. *Trans. ent. Soc. London*, 1: 33-44.

769. MERXMÜLLER, H. & J. POELT, 1954. Beiträge zur Florengeschichte der Alpen. *Ber. Bayer. bot. Ges.*, 30: 91-106, map 6.

770. MEYER, H. 1891. Across East African Glaciers. London. pp. xx+404.

771. MEYER, H. 1900. Der Kilimandjaro, Reisen und Studien. Berlin. pp. xvi+436.

772. MEYERICK, E. 1885. An ascent of Mount Kosciusko. *Ent. mon. Mag.*, 22: 78-82.

773. MICHAL, K. 1931. Beziehung der Populationsdichte auf Lebensoptimum und Einfluss des Lebensoptimum auf das Zahlenverhältnis der Geschlechter bei Mehlwurm und Stubenfliege. *Biol. Gener.*, 7: 631-646.

774. MILLER, L. 1868. Eine entomologische Reise in die ostgalizischen Karpathen. *Verh. zool.-bot. Ges. Wien*, 18 (Abh.): 1-34.

775. MILLER, L. 1878. Eine coleopterologische Reise durch Krain, Kärnten und Steiermark 1878. *Verh. zool.-bot. Ges. Wien*, 28: 463-470.

776. MILLS, H. B. & J. H. PEPPER, 1937. Some observations on *Grylloblatta campodeiformis* Walker. *Ann. ent. Soc. Amer.*, 30: 269-274.
777. MIRAM, E. 1931. Dermaptera, Blattodea, Orthoptera. Abhandlungen der Pamir Expedition 1928. VIII. Zoology. pp. 65-75.
778. MIRČINK, 1930. On the determination of southern boundary of the glacier of the Würmian time. Bull. Kommiss. Erforsch. Quartär. Leningrad.
779. MONAHAN, R. S. 1933. Timberline. *Appalachia*, 26: 401-426.
780. MOORE, C. & D. PRICE, 1948. A study at high altitudes of reproduction, growth, sexual maturity and organ weights. *J. exp. Zool.*, 108: 171.
781. MOORE, F. 1878. Description of new species of Lepidoptera collected by the late Dr F. Stoliczka during the Indian Government Mission to Yarkand in 1873. *Ann. Mag. nat. Hist.*, (5) 1: 227-237.
782. MOORE, P. 1879. Lepidoptera. Scientific Results of the Second Yarkand Mission. pp. 1-18.
783. MOOSBRUGGER, J. 1932. Alpine und subalpine Käfer des Stierischen Ennsgebirges. *Koleopt. Rundschau*, 18: 217-226.
784. MORRISON, H. K. 1875. On the insect fauna of the White Mountains. *Psyche*, 1: 85.
785. MÜLLER, H. 1881. Die Alpenblumen, ihre Befruchtung durch Insekten und ihre Anpassung an dieselben. Leipzig.
786. MUSCHKETOW, D. J. 1924. Die Hauptfragen der Tektonik in Turkestan. *Bull. Moskauer naturf. Ges.* (Abt. Geol.) 3. (In Russian).
787. MUTTKOWSKI, R. A. 1927. A new and unusual insect record from North America (Diptera: Deuterophlebiidae). *Bull. Brooklyn ent. Soc.*, 22: 246-249.

788. NASIMOVICH, A. A. 1964. Zhizn zhivotnikh v gorak na bolshikh visotakh (Gimalai, Kilimandzharo, Andi; obzor zarubezhnikh dannikh) The life of animals in the mountain heights (The Himalayas, Kilimanjaro, Andes; Survey of foreign data). *Bull. Moskovskovo Obshstva Ispitatelei Prirodi*, (Otd. Biol.) 69 (5): 5-15.
789. NETOLITSKY, F. 1929. Zoogeographische Untersuchungen in der Carabidengruppen Bembidini. *Koleopt. Rundschau*, 15: 31-37.
790. NETOLITSKY, F. 1935. Materiali zoologici raccolti dall spedizione italiana al Karakorum. Coleoptera: Carabidae, Bembidini. *Atti Mus. Stor. Civ. nat. Trieste*, 12: 89-100, pl. i.
791. NETOLITSKY, F. 1935. Neue Bembidini aus Vorderasien. *Koleopt. Rundschau*, 21: 165-168.
792. NEWCOMB, H. H. 1901. *Chionobas katadin* and an account of its discovery. *Ent. News*, 12: 225-231.
793. NÖTH, L. 1931. Glazialogische und morphologische Untersuchungen im Nordwest-Pamir. *Mitt. geogr. Ges. München*, 24: 154-193.
794. NORDSTRÖM, F. 1929. Entomologische Ergebnisse der schwedischen Kamtchatka-Expedition 1920-1922. Lepidoptera. II. *Ark. Zool.*, 20 (A) (12): 1-6.
795. NORIN, E. 1925. Preliminary notes on the late Quaternary glaciations of the Northwest Himalaya. *Geogr. Annaler*, 7 (3).
796. NUORTEVA, P. 1963. The influence of *Oporinia autumnata* (Bkh.) (Lep. Geometridae) on the timber-line in subarctic conditions. *Ann. Ent. fenn.*, 29: 270-277, fig. 5.

797. OBERTHÜR, C. 1870. Note sur les myriades d'insectes dans le voisinage du cratère du Vésuve. *Ann. Soc. Ent. France, Bull.*, p. xlix-l.
798. OBERTHÜR, C. 1891. Lépidoptères du genre *Parnassius*. *Etude d'Ent.*, 14: 1-18, pl. i-ii.
799. OEHLER, E. 1915. Von einer Forschungsreise am Kilimandscharo im Jahre 1912. *Z. dtsch.-österr. Alpenverein*, 46: 124-156.
800. OGILVIE-GRANT, W. R. 1909-1910. Zoological Results of the British Museum Ruwenzori Expedition 1905-1906. *Trans. zool. Soc. London*, 19: 1.

801. OLLIFF, A. S. 1889. On Rhopalocera from Mt. Kosciusko, New South Wales. *Proc. Linn. Soc. N.S. Wales*, (2) 4: 619-624.
802. OLUFSEN, O. 1903. The Second Danish Pamir Expedition. Det Nordiske Forlag. Ernst Bojesen.
803. OLUFSEN, O. 1904. Through the unknown Pamir: The Second Danish Pamir Expedition 1898-1899. London: William Heinemann. pp. 238.
804. OSHANIN, R. D. 1891. The zoogeographical character of Heteropterous fauna of Turkestan. *Zarisk. Russk. geogr. Obsch.*, 23: 56.

805. PACKARD, A. 1868. The insect fauna of the summit of Mount Washington, as compared with that of Labrador. *Proc. Amer. Ass. adv. Sci.*, 17 Meet., 1867: 154-155.
806. PACKARD, A. 1885. Occurrence of *Colias nastes* in the Pamir mountains. *Amer. Nat.*, 19: 1220.
807. PACKARD, A. 1894. On the origin of the subterranean fauna of North America. *Amer. Nat.*, 28: 727-752, pl. i.
808. PACKARD, A. S. JR. 1874. On the geographical distribution of moths of Colorado. *Ann. Rep. U.S. geol. geogr. Surv.*, pp. 543-560, fig. 15.
809. PAGENSTECHER, A. 1898. Die Lepidopteren der Hochgebirge. *Jb. Nassauisch Ver. Naturk. Wiesbaden*, 60: 91-178.
810. PAGENSTECHER, A. 1910. Die geographische Verbreitung der Schmetterlinge.
811. PAMIRISKAYA Expedition 1928. Akademiya Nauk Soyusa Sovietskikh Socialisticheskikh Republik. Trudi Expeditsii Bipusk 8. Zoology. Leningrad 1931. Pamir Expedition 1928. Akademie der Wissenschaften der Union der sozialistischen Soviet-Republiken. Abhandlungen der Expedition. Lieferung 8. Zoologie, pp. 1-247.
812. PARKER, R. N. 1921. Insects living on snow at 14 000 ft. *J. Bombay nat. Hist. Soc.*, 27: 639.
813. PASCHINGER, H. 1954. Würmvereisung und Spätglazial in der Sierra Nevada (Spanien). *Z. Gletscherk. Glazialgeol.*, 3 (1): 55-67.
814. PAULMANN, R. & A. VILLIERS, 1940. Les Coléoptères des *Lobelias* des montagnes du Cameroun. *Rev. franç. Ent.*, 7: 72-83.
815. PAULSEN, O. 1920. The studies in the vegetation of the Pamirs. Copenhagen.
816. PAX, F. 1898-1908. Grundzüge der Pflanzenverbreitung in den Karpathen. I (1898); II (1908).
817. PAX, F. JR. 1906. Die Lepidopteren der Rodnaer Alpen. *Jb. Nassau. Schles. Ges. vaterländ. Kultur*, 84: 43-53.
818. PAYNE, N. M. 1926. The effect of environmental temperatures upon insect freezing points. *Ecology*, 7: 99-106.
819. PAYNE, N. M. 1927. Two factors of heat energy involved in insect cold hardiness. *Ecology*, 8: 194-196.
820. PAYNE, N. M. 1927. Freezing and survival of insects at low temperatures. *Quart. Rev. Biol.*, 1: 270-282.
821. PAYNE, N. M. 1927. Freezing and survival of insects at low temperatures. *J. Morphol.*, 43: 521-546.
822. PAYNE, N. M. 1927. Measures of insect cold hardiness. *Biol. Bull.*, 52: 449-457.
823. PAYNE, N. M. 1929. Absolute humidity as a factor in insect cold hardiness with a note on the effect of nutrition on cold hardiness. *Ann. ent. Soc. Amer.*, 22: 601-620.
824. PENK, A. & E. BRÜCKNER, 1909. Die Alpen im Eiszeitalter. Leipzig: Tauchnitz, vols. 2.
825. PENNAK, R. W. 1941. An introduction to the limnology of northern Colorado. *Univ. Colorado Studies*, (D) 1: 203-220.
826. PENNAK, R. W. 1945. Notes on mountain midges (Deuterophlebiidae) with a

description of the immature stages of a new species from Colorado. *Amer. Mus. Nov.*, 1276: 1-10.

827. PENNAK, R. W. 1945. First record of adult mountain midges from North America (Diptera: Deuterophlebiidae). *Ent. News*, 61: 36.

828. PENNAK, R. W. 1951. Description of the imago of the mountain midge *Deuterophlebia coloradensis* Pennak (Diptera: Deuterophlebiidae). *Amer. Mus. Nov.*, 1534: 1-11.

829. PESTA, O. 1929. Der Hochgebirgssee der Alpen. *Die Binnengewässer*, 8: 1-156.

830. PETERSEN, B. 1954. Some trends of speciation in the cold adapted holarctic fauna. *Zool. Bidr.*, Uppsala, 30: 233-314.

831. PETERSEN, B. 1959. Variation in some northern butterflies and its relation to environment. *Proc. X int. Congr. Ent.*, 1: 687-690.

832. PEYERIMHOFF, P. DE, 1906. Recherches sur la faune cavernicole des Basses-Alpes. *Ann. Soc. ent. France*, 75: 203-222.

833. PEYERIMHOFF, P. DE, 1927. Composition et origine présumée de la faune (Coléoptères) des hautes montagnes de la Berbérie. *C. R. Soc. Biogéogr., Paris*, 4 (32): 95-96.

834. PIC, M. 1927-1929. Entomologische Ergebnisse der schwedischen Kamtchatka-Expedition 1920-1922. Coleoptera. *Ark. Zool.*, (1927) 19 (3): 1-3; 20 (B) (7): 1-4 (1929).

835. PIC, M. 1953. Coléoptères nouveaux du Kilimandjaro. *Ann. Mag. nat. Hist.*, (12) 6: 77-78.

836. PICTET, A. 1915-1917. Influence de la pression barométrique sur le développement des Lepidoptères. *Arch. Sci. Phys. nat.*, (4) 40: 74-77; 44: 413-454 (1917).

837. PITTIONI, B. 1942-1943. Die boreoalpinen Hummeln und Schmarotzerhummeln. *Mitt. Naturw. Inst. Sofia*, 15: 155-218; 16: 1-77.

838. PLETSCH, D. J. 1947. The alpine rock crawler *Grylloblatta campodeiformis campodeiformis* Walker in Montana. *Proc. Montana Acad. Sci.*, 5/6: 17-20.

839. POBOLNY, D. & J. MOUCHA, 1956. On the high mountain Geometridae of the genus *Psodos* Tr. with regard to the question of the origin of species in mountain regions (Lepidoptera: Geometridae) (In: Czech with English summary). *Acta Soc. ent. Csl. Prague*, 52 (1955): 183-188.

840. POULTON, E. B. 1884-1885. The essential nature of the colouring of phytophagous larvae (and their pupae); with an account of some experiment upon the relation between the colour of such larvae and that of their food plants. *Proc. R. Soc. London*, 38: 269-315.

841. POULTON, E. B. 1887. An inquiry into the cause and extent of a special colour relation between certain exposed Lepidopterous pupae and the surfaces, which immediately surround them. *Proc. R. Soc. London*, 42: 94-108 (Abstract); *Phil. Trans. R. Soc. London*, (B) 178: 311-411, pl. xxvi.

842. POULTON, E. B. 1892. Further experiments upon the colour relation between certain Lepidopterous larvae, pupae, cocoon and imagines and their surroundings. *Trans. ent. Soc. London*.

843. POULTON, E. B. 1904. A possible explanation of insect swarms on mountain tops. *Trans. ent. Soc. London*, 1904: xxiii-xxvi.

844. PROCHNOW, O. 1927. Sichtschutztracht oder Wärmeschutztracht? *Biol. Zbl.*, 47: 662-670.

845. PROCHNOW, O. 1929. Die Färbung der Insekten. In: Schröder's Handbuch der Entomologie, 2, Kap. 6, pp. 430-572.

846. PRÜFFER, J. 1923. Studja nad motylami Tatr Polskich. Sprawozdanie Komisji Fiziographicznej obejmyace poglad na czynnósci dokonane w ciagu roku 1922 orez materjay do fiziografji kraji. Polska Akademja Umiejetnosci w Krakowio. pp. 79-170, pl. i-ii.

847. PRYOR, M. E. 1962. Some environmental features of Hallett Station, Antarctica, with special reference to soil Arthropods. *Pacific Insects*, 4 (3): 681-728.

848. PSCHORN-WALCHER, H. 1951. Zur Biologie und Systematik terricoler Milben (i) Die ostalpinen Arten der Gattung *Liacarus* Mich. (Oribatidae). *Bonn. zool. Beitr.*, 2 (1/2): 177-183.

849. PULIKOVSKY, N. 1924. Metamorphosis of *Deuterophlebia* sp. (Diptera: Deuterophlebiidae) *Trans. ent. Soc. London*, pp. 45-62.

850. PULIKOVSKY, N. 1927. Die respiratorischen Anpassungserscheinungen bei den Puppen der Simuliiden und in Bächen lebenden Dipterenpuppen. *Z. Morphol. Ökol. Tiere*, 7.

851. RÄTZER, PF. 1897. Vorkommen von Insekten auf Hochfirn in der Schweiz. *Mitt. schweiz. ent. Ges.*, 10 (1): 3-4.

852. RAFFRAY, A. 1882. Distribution géographique des Coléoptères en Abyssinie et descriptions d'especes nouvelles. *C. R. Acad. Sci. Paris*, 94: 746-748; *Bull. Soc. Ent. Fr.*, (6) 2: v-vi.

853. RAFFRAY, A. 1885. Note sur la dispersion géographique des Coléoptères en Abyssinie et descriptions d'especes nouvelles. *Ann. Soc. Ent. Fr.*, (6) 5: 292-326, pl. vi.

854. RAMME, W. 1929. Die Orthopteren der Elbrus-Expedition Heinrich-Dammholz 1927. *Eos*, 5 (2): 151-152.

855. RAMME, W. 1930. Entomologische Ergebnisse der Deutsch-Russischen Alai-Pamir-Expedition 1928. II. Dermaptera und Orthoptera. *Mitt. zool. Mus. Berlin*, 16: 209-214.

856. RAMME, W. 1941. Die Orthopterenfauna von Kärnten. *Carinthia*, (2) 131: 121-131.

857. RAMME, W. 1948. *Schizonotinus*, a most curious new genus from the Caucasus (Orthoptera, Tettigonidae). *Proc. R. ent. Soc. London*, (B) 17: 133-136, fig. 3.

857a. RAMME, W. 1951. Zur Systematik, Faunistik und Biologie der Orthopteren von Südost-Europa und Vorderasien. *Mitt. zool. Mus. Berlin*, 27: 1-431 (1950).

858. RAND, A. L. 1948. Glaciation. An isolating factor in speciation. *Evolution*, 2: 314-321.

859. RAND, A. L. & L. J. BRASS, 1940. Summary of 1936-1937 New Guinea Expeditions. Results of the Archbold Expeditions. 29. *Bull. Amer. Mus. nat. Hist.*, 77.

860. RAWSON, D. S. 1942. A comparison of some large alpine lakes in Western Colorado. *Ecology*, 23: 143-161.

861. REAKIRT, T. 1866. Coloradian butterflies. *Proc. ent. Soc. Philadelphia*, 6: 122-151.

862. REBEL, H. 1903-1904, 1913. Studien über die Lepidopterenfauna der Balkanländer I. Bulgarien. *Ann. naturhist. Hofmus. Wien*, 18: 123-346 (1903); II. Bosnien und Herzegowina. *ibid.*, 19: 97-377 (1904); III. Montenegro, Albania, Thracien. *ibid.*, 37: 281-334 (1913).

863. REBEL, H. & H. ZERNY, 1931. Die Lepidopterenfauna Albaniens. *Denkschr. Akad. Wiss. Wien*, (math.-natur. Klasse) 103: 37-161.

864. REGEL, A. 1879. Botanicheskaya Exkursiya ot Tashkenta do Kuldzhi. Turkestanskie Verdomosti. pp. 1-13.

865. REGEL, A. 1882. Von Taschkent über durch Naryngebiet. 1880. *Gartenflora*, 31.

866. RÉGIMBART, M. 1906. Voyages de M. Ch. Alluaud dans l'Afrique Orientales: Dytiscidae, Gyrinidae, Hydrophilidae. *Ann. Soc. ent. France*, pp. 235-278.

867. RÉGIMBART, M. 1908. Sjöstedts Kilimandjaro-Meru Expedition. 7. Coleoptera. i. Dytiscidae, Gyrinidae, Hydrophilidae. Uppsala.

868. REHN, J. A. 1945. Man's uninvited fellow-travellers—the cockroach. *Sci. Monthly*, 61: 265-276.

869. REHN, J. A. & H. J. GRANT JR. 1961. A monograph of the Orthoptera of North America (North of Mexico) I. *Monogr. Acad. nat. Sci. Philadelphia*, 12: 1-255, pl. i-viii.

870. REHN, J. A. & M. HEBARD, 1906. A contribution to the knowledge of the Orthoptera of Montana, Yellowstone Park, Utah and Colorado. *Proc. Acad. nat. Sci. Philadelphia*, 58: 358-418.

871. REINIG, W. F. 1929. Untersuchungen zur Ökologie von *Uroprosodes costifera* Kr. *Z. wiss. Insektenbiol.*, 24.

872. REINIG, W. F. 1929. Die zoologischen Arbeiten der deutsch-russischen Pamir-Expedition. *Deutsche Forschung*, 10.

873. REINIG, W. F. 1930. Untersuchungen zur Kenntnis der Hummelfauna des Pamir-Hochlandes. *Z. Morphol. Ökol. Tiere*, 17.

874. REINIG, W. F. 1930. Entomologische Ergebnisse der deutsch-russischen Alai-Pamir-Expedition 1928. II. Coleoptera II. Tenebrionidae. *Mitt. zool. Mus. Berlin*, 16 (6): 867-912.

875. REINIG, W. F. 1930. Entomologische Ergebnisse der deutsch-russischen Alai-Pamir-Expedition 1928. (I): Allgemeines. *Mitt. zool. Mus. Berlin*, 16: 185-208.

876. REINIG, W. F. 1930. Versuch einer zoogeographischen Gliederung des Pamir-Gebietes. *Zool. Jb.*, (Syst.) 59: 449-504.

877. REINIG, W. F. 1931. Entomologische Ergebnisse der deutsch-russischen Alai-Pamir-Expedition 1928, (II): Coleoptera II. Tenebrionidae. *Mitt. münch. ent. Ges.*, 1930: 865-912, fig. 9.

878. REINIG, W. F. 1932. Beiträge zur Faunistik des Pamir Gebietes. *Wiss. Ergeb. Alai-Pamir-Expedition* 1928, I (3). Ökologie und Tiergeographie. pp. 1-195, fig. 29; 2 (3) Syst., pp. 195-312, pl. vi.

879. REINIG, W. F. 1937. Die Holarktis. Ein Beitrag zur diluvialen und alluvialen Geschichte des zirkumpolaren Faunen- und Florengebietes. Jena. pp. 1-124.

880. REIST, D. 1961. Ruwenzori. *The Mountain World*, pp. 157-163, pl. xlii-xlv.

881. RENSCH, B. 1943. Studien über Korrelation und klimatische Parallelität der Rassenmerkmale von *Carabus*-Formen. *Zool. Jb.*, (Syst.) 76.

882. REUTER, O. M. 1908. Monographia generis Heteropterorum *Phimodera* Germ. *Acta Soc. Sci. fenn.*, 33 (8): 1-51, fig. 23, pl. i-ii.

883. RICHARDS, O. W. 1928. On a collection of bumblebees (Hymenoptera: Bombidae) made in Ladak by Col. R. Meinertshagen. *Ann. Mag. nat. Hist.*, 19 (2): 333-336.

884. RICHARDS, O. W. 1954. Two new wingless species of Diptera, Sphaeroceridae (Borboridae) from Ethiopia. *J. Linn. Soc. London*, (Zool.) 42: 387-391.

885. RICHARDS, O. W. 1959. Journey to High Simien (Northern Ethiopia) 1952-53: Diptera, Sphaeroceridae. *J. Linn. Soc. London*, (Zool.) 44 (296): 180-184.

886. RILEY, N. D. 1923. The Rhopalocera of the Mount Everest Expedition. *Trans. ent. Soc. London*, 1922. (iii/iv): 461-483, pl. xxxvi-xxxvii.

887. RILEY, N. D. 1927. The Rhopalocera of the Third Mount Everest Expedition. *Trans. ent. Soc. London*, 75: 119-129.

888. RITTMANN, A. 1936. Vulkane und ihre Tätigkeit. Stuttgart: Ferdinand Enke Verlag, pp. 188.

889. RODENDORF, B. B. 1961. Neue Angaben über das System der Dipteren. *Verh. XI int. Kongr. Ent.*, 1: 151-158.

890. RODENDORF, B. B. 1964. Istoricheskoe razvitie Dvukhrylykh Nacekomik. *Trudi palaeont. Inst. Acad. Nauk SSSR, Moscow*, 100: 1-311.

891. RÖMER, F. 1907. Die Abnahme der Tierarten mit Zunahme der geographischen Breite. Nach einem Vortrage, gehalten in der wissenschaftlichen Sitzung der Senkenbergischen naturforschenden Gesellschaft am 1. Dezember 1906. *Ber. Senkenberg. naturf. Ges., Frankfurt*, pp. 63-112.

892. ROESCHKOE, H. 1889. Carabologische Notizen. IV. *Ent. Nachr.*, 25: 357-358.

893. ROLAND-BROWN, H. 1905. A butterfly hunt in the Pyrenees. *The Entomologist*, pp. 243-249, 273-275.

894. RONDOU, J. P. 1902. Catalogue raisonné des Lépidoptères des Pyrenées. *Acta Soc. Linn. Bordeaux*, 57: 1-179.

895. RONDOU, J. P. 1932. Catalogue des Lépidoptères des Pyrénées. *Ann. Soc. ent. France*, 101: 165-244.

896. ROSEN, K. 1921. Die Rhopaloceren-Ausbeute der Pamir Expedition des deutsch-österreichischen Alpenvereins. *Mitt. münch. ent. Ges.*, 11: 83-100.

897. ROSS, H. H. 1953. On the origin and composition of the Nearctic insect fauna. *Evolution*, 7: 145-158.

898. ROSS, H. H. 1956. Evolution and classification of the mountain caddisflies. Urbana Univ. Illinois Press, pp. vi+213, fig. 370, charts 45.

899. ROSS, H. H. 1958. Affinities and origins of the northern and montane insects of the western North America. *Publ. Amer. Ass. Adv. Sci. Wash.*, 51: 231-252, fig. 15.

900. ROTHSCHILD, W. 1909. Catalogue of the collections of Parnassiinae in the Tring-Museum with systematic notes. *Nov. Zool.*, 16: 1-20.

901. ROTHSCHILD, W. & J. H. DURRANT, 1915. Lepidoptera of the British Ornithologists' Union and Wollaston Expeditions in the Snow Mountains, southern Dutch New Guinea. London 1915. pp. 182, pl. ii. Also *vide*: *Trans. zool. Soc. London*, 20: 497-547, pl. xxxix.

902. ROUBAL, J. 1938. Les Coléoptères boréoalpins et les soi-disant restes de l'époque glaciaire en Tchécoslovaquie. *Sbornik Nár. Mus. Praze, Prag*, pp. 121-141.

903. ROUDIER, A. 1954. Coléoptères de Sierra Nevada: Curculionidae. *Arch. Inst. Aclimatación Almeria*, 2: 123-133.

904. ROUDIER, A. 1957. Zoologisch-systematische Ergebnisse der Studienreise von H. Janetschek und W. Steiner in die spanische Sierra Nevada 1954. VI. Drei neue Curculioniden-Arten (Coleoptera). *Sitzb. österr. Akad. Wiss. Wien*, (math.-naturw. Klasse) (1) 166: 253-263.

905. RÜCKER, F. 1933. Die Farben der Insekten und ihre Bedeutung für den Wärme-haushalt. *Pflüg. Arch. ges. Physiol.*, 231: 729-741.

906. RYLOV, W. M. 1935. Das Zooplankton der Binnengewässer. *Die Binnengewässer*, 15: 1-272.

907. SACHAROV, N. L. 1930. Studies in the cold resistance in insects. *Ecology*, 11: 505-517.

908. SAINTE-CLAIRE DEVILLE, J. 1902. Etude sur divers *Platysma* des Alpes occidentales. *Ann. Soc. Ent. France*, 71: 588-619.

909. SAINTE-CLAIRE DEVILLE, J. 1928. Esquisse du peuplement des Alpes françaises (Coléoptères). *Mem. Soc. Biogéogr.*, 2.

910. SAKHOKIA, M. F. *et al.* 1961. Botanical Excursions over Georgia (Translated from the Russian Text by N.V. Mardjanisvili). *Acad. Sci. Georgian SSR*, Tbilisi, pp. 1-93, pl. lv, map 1.

911. SALMON, J. T. 1954. Melanism and size in relation to climate among New Zealand insects. *Proc. 7th Pacific Sci. Congr.*, 4 (Zool.): 87-90.

912. SALT, G. 1951. The Shira Plateau of Kilimanjaro. *Geogr. J.*, 117: 150-164.

913. SALT, G. 1954. A contribution to the ecology of Upper Kilimanjaro. *J. Ecol.*, 42: 375-423, fig. 3, pl. xvii-xxi.

914. SALT, R. W. 1956. Influence of moisture content and temperature on cold hardiness of hibernating insects. *Canad. J. Zool.*, 34 (4): 283-294.

915. ŠÁMAL, J. 1925. Plecoptera. Wissenschaftlische Ergebnisse der Niederländischen Expedition in den Karakoram. pp. 221-225.

916. ŠÁMAL, J. 1939. Contribution à l'étude de la faune des Plécoptères d'Altai. *Veštnik čs. zoologické Společnosti, Svaz*, vi-vii, Praha, 1938/1939.

917. SANDSTRÖM, J. W. 1913. Eine meteorologische Forschungsreise in den schwedischen Hochgebirgen. *Kungl. Svensk. Vet. Akad. Handl.*, 50 (9).

918. SCHAEFER, L. 1959. Contribution à la connaissance des Coléoptères des Pyrénées orientales. (Troisième note). *Bull. mens. Soc. Linn. Lyon*, 28: 222-235.
919. SCHÄFFER, O. 1900. Collembola. Fauna Arctica, 1 (2).
920. SCHARBAKOV, A. M. 1899. Zur Collembolenfauna Spitzbergens. *Zool. Anz.*, 22: 47.
921. SCHARFF, R. F. 1928. On the terrestrial molluscs of the high Alps and their origin. *Mem. Soc. Biogéogr.*, 2.
922. SCHENKEL, E. 1930. Die Arachniden der schwedischen Kamtchatka-Expedition 1920-1922. *Ark. Zool.*, 21 (A) (15): 1-33.
923. SCHLAGINTWEIT, HERMANN ADOLF & ROBERT, 1858. An account of a journey across the chain of the Kuenluen from Ladak to Khotan. *J. Asiatic Soc. Bengal*, 26: 120-121 (1857).
924. SCHLAGINTWEIT, HERMANN ADOLF & ROBERT, 1862. Bemerkungen über Aufenthalt und Grenzen der Tiere in extremen Höhen und Einfluss der Höhe auf den Menschen. *Arch. Naturg.*, 28 (1): 253-265.
925. SCHMID, F. 1958. Trichoptera du Pakistan. *Tijdschr. Ent.*, 101 (3/4): 181-221, pl. viii-xii.
926. SCHMITZ, H. 1957. Zoologisch-Systematische Ergebnisse der Studienreise von H. Janetschek und W. Steiner in die spanische Sierra Nevada 1954. IV. Phoridae (Diptera). *Sitzb. österr. Akad. Wiss. Wien*, (math.-wissen. Klasse) (1) 166 (5/6): 231-247, pl. i-iii.
927. SCHMOE, F. W. 1925. Our greatest Mountain. pp. 355 (Mt. Rainier).
928. SCHNEIDER, E. C. 1921. Physiological effects of altitude. *Physiol. Rev.*, 1: 631-659.
929. SCHNEIDER, O. 1878. Naturwissenschaftliche Beiträge zur Kenntnis der Kaukasusländer, auf Grund seiner Sammelbeute. Dresden. pp. 160, pl. i-v.
930. SCHNEIDER, O. & H. LEDER, 1878-1879. Beiträge zur Kenntnis der kaukasischen Käferfauna. *Verh. naturf. Ver. Brünn*, 16: 3-258; 17: 3-104.
931. SCHÖNBORN, W. 1961. Untersuchungen über die Sichtung im Hypolithion. *Biol. Zbl.*, 80 (2): 179-197, fig. 5.
932 SCHÖNMANN, R. 1937. Die Artsystematik und geographische Verbreitung der hochalpinen Trechini der Ostalpen. *Zool. Jb.*, 70: 177-226.
933. SCHOSTAKOWITSCH, W. B. 1927. Der ewigerfrorene Boden Sibiriens. *Z. Ges. Erdk.*, Berlin, pp. 394-427.
934. SCHRÖDER, CH. 1927. Wärmeschutztrachttheorie, Anhang zum Abschnitt: Die Färbung der Insekten. Handbuch der Entomologie, Jena: Gustav Fischer Verlag, 2: 573-591.
935. SCHRÖTER, C. 1908. Die Pflanzenleben der Alpen. Zürich: A. Raustem.
936. SCHUBERT, J. 1925. Verdunstungsmessungen an der Küste, im Flach- und Berglande, in Nadel- und Buchenwäldern. *Verh. klimatol. Tagung, Davos*, 39.
937. SCHUBERT, K. 1935. Die Apterygotenfauna des Glatzer Schneeberges. *Beitr. Glaz. Schneeberges*, 1: 89.
938. SCHUBERT, K. 1937. Zur Kenntnis der Apterygotenea des Altvatergebirges. *Z. wiss. Insektenbiol.*, 27: 124-131.
939. SCHULTZ, A. VON, 1916. Landeskundliche Forschungen in Pamir. *Abh. Hamburg Kolonial-Inst.*, 33 (C): 4.
940. SCHULTZ, C. C. 1948. Analiz Noveishei tektoniki i relef Tian-Shan. Zapiski Vasesyuzhnovo geographicheskovo Obsherstva, Novaya Serie 3.
941. SCHUSTER, A. 1926. Bestimmungstabellen der *Laena*-Arten aus dem Himalaya und dem angrenzenden Gebieten mit Beschreibungen neuer Arten. *Koleopt. Rundschau*, 12 (1): 31-48; (2): 49-54.
942. SCHUSTER, A. 1935. Neue *Laena*-Arten aus dem Himalaya (Coleoptera: Tenebrionidae). *Ann. Mag. nat. Hist.*, 16 (10): 448.

943. SCHWARTZ, E. A. 1889. Notes on the comparative vitality of insects in cold water. *Proc. ent. Soc. Wash.*, 1: 208-215.

944. SCHWEIGER, H. 1950. Der Einfluss der Eiszeit auf die Verbreitung ostalpiner *Trechus*-Arten. *Proc. VIII int. Congr. Ent.*, p. 489.

945. SCHWEIGER, H. 1957. Das Phänomen der warmen Hangstufe in den Alpen. *Tagungsb.* 8. *Wandervers. dtsch. Ent. Berlin*, 11: 54-70, fig. 4, map 5.

946. SCHWEIGER, H. 1958. Eine neue Gattung des Tribus Aepini aus den bolivischen Anden (Col. Trechidae). *Opusc. Zool. München*, 1958: 1-3, fig. 2, map 1.

947. SCHWEIGER, H. 1958. Ein neuer mikrophthalmer *Euconus* aus den östlichen Gailtaler Alpen (Col. Scydmaenidae). *Dtsch. ent. Z.*, Berlin, (NF) 5: 382-384, fig. 2.

948. SCHWEIGER, H. 1961. Die Koleopterenfauna der westanatolischen Hochgebirge und ihre Probleme. *Verh. XI int. Kongr. Ent. Wien*, 1960, 1: 496-500, fig. 1.

949. SCHWOERBEL, J. 1962. Subterrane Wassermilben (Hydrachnellae und Trombiidae) aus den Alpen. *Zool. Anz.*, 168 (7/8): 292-300.

950. SCOTT, G. 1952. Journey to the Gughé Highlands (Southern Ethiopia) 1948-1949: Biogeographical Research at high altitudes. *Proc. Linn. Soc. London*, 163: 85-189, fig. 7, pl. iii-xxviii.

951. SCOTT, H. 1958. Biogeographical research in High Simien (Northern Ethiopia) 1953-54. *Proc. Linn. Soc. London*, 170 (1): 1-91, pl. i-xvii.

952. SCUDDER, S. H. 1863. Remarks on some characteristics of the insect-fauna of the White Mountains, New Hampshire. *Boston J. nat. Hist.*, 7: 612-631, pl. i, map 1.

953. SCUDDER, S. H. 1898. The alpine Orthoptera of North America. *Appalachia*, 8: 299-319.

954. SEDYKH, K. F. 1962. Some relict insects from the southern part of the Timan Mountains. *Entomologicheskoe Obozerenie*, 41 (1): 148 (English translation in *Ent. Rev.*, 1: 88)

955. SEITZ, A. 1923. Insektenleben in den Pyrenäen. *Ent. Rundschau*, Stuttgart, 40: 34-50.

956. SEMENOV-TIAN-SHANSKII, A. 1935. Les limites et les subdivisions zoogéographiques de la région paléarctique pour les animaux terrestres basées sur la distribution géographique des insectes Coléoptères. *Trav. Inst. Zool. Acad. Sci. USSR*, 2: 397-410, map 1 (In Russian with summary in French).

957. SEMENOV-TIAN-SHANSKII, A. P. 1946. Memuarii 2. Puteshestvie v Tian Shan v 1856-1857.

958. SHARP, D. 1891. Haliplidae, Dytiscidae, Gyrinidae, Hydrophilidae, Staphylinidae, Scarabaeidae in Scientific Results of the Second Yarkand Mission. Coleoptera. pp. 37-53.

959. SHAW, C. H. 1909. Causes of timber line on mountains. *Plant World*, 12 (8): 169-181.

960. SHAW, C. H. 1909. Vegetation and altitude. *Plant World*, 12 (3): 63-64; *Amer. Nat.*, 43: 420-421.

961. SHELJUZHKO, L. 1914. Neue Rhopaloceren vom Pamir. *Iris*, 28: 18-22.

962. SHERMAN, J. D. 1910. A list of Labrador Coleoptera. *J. N. Y. ent. Soc.*, 18: 173-197.

963. SHEWALL, G. E. 1954. First record of Deuterophlebiidae in Canada. *Canad. Ent.*, 86: 204-206.

964. SHIRÔ, T. & T. SAIGUSA, 1963. Some butterflies from West Pakistan and Iran. In: Masuzo Uéno's Results of the Kyoto University Scientific Expedition to the Karakoram and Hindukush 1955. 4. Insect fauna of Afghanistan and Hindukush. pp. 103-144, fig. 32, pl. i-xvii.

965. SHLAIFER, A. 1938. Studies on mass physiology: Effect of numbers on the oxygen consumption of *Caraussius auratus*. *Physiol. Zool.*, 12: 381-392.

966. SHREVE, F. 1924. Soil temperatures as influenced by altitude and slope exposure. *Ecology*, 5: 128-136.

967. SHTANDEL, A. YE, 1957. Daylight moths of the Altai. *Entomologicheskoe Obozrenie*, 36 (1): 134-141.

968. SHUKLA, S. P. 1960. Entomological Survey of the Himalaya. Part XXX. Chrysomelidae (Coleoptera) from the Northwest Himalaya. *Agra Univ. J. Res.*, (Sci.) 9: 1-24.

969. SINGH, SANTOKH, 1955. Entomological Survey of the Himalaya. Part VII. On a collection of Dermaptera. *Agra Univ. J. Res.*, (Sci.) 4 (1): 179-186.

970. SINGH, SANTOKH, 1957. Third Entomological Expedition to the Northwest Himalaya. *Turtox News*, 35 (8): 170-173.

971. SINGH, SANTOKH, 1958. Entomological Survey of the Himalaya. Part XXVII. Third Entomological Expedition to the Northwest Himalaya. *Proc. nat. Acad. Sci. India*, 28 (B) (1): 1-22.

972. SINGH, SANTOKH, 1958. Entomological Survey of the Himalaya. Part XXI. Notes on some Coleoptera collected by Prof. M. S. Mani's Entomological Expeditions to the Northwest Himalaya. *Agra Univ. J. Res.*, (Sci.) 6 (1): 57-62 (1957).

973. SINGH, SANTOKH, 1958. Entomological Survey of the Himalaya. Part XXIX. On a collection of nival Chironomidae (Diptera) from the Northwest Himalaya. *Proc. nat. Acad. Sci. India*, 28 (B) (4): 308-314.

974. SINGH, SANTOKH, 1961. Entomological Survey of the Himalaya. Part XXXII. A note on the larva of an apparently undescribed species of *Deuterophlebia* Edw. (Deuterophlebiidae: Diptera-Nematocera) from the Northwest Himalaya. *Agra Univ. J. Res.*, (Sci.) 10 (1): 109-114.

975. SINGH, SANTOKH, 1963. Entomological Survey of the Himalaya. Part XXIV. Fourth Annotated check-list of insects from the Northwest Himalaya. *Agra Univ. J. Res.*, (Sci.) 12 (1): 363-394.

976. SINGH, SANTOKH & V. K. GUPTA. 1956. Entomological Survey of the Himalaya. Part XVII. Third annotated check-list of insects from the Northwest Himalaya. *Agra Univ. J. Res.*, (Sci.) 5 (2): 385-442.

977. SINGH, SANTOKH, H. N. BAIJAL, V. K. GUPTA & K. MATHEW. 1955. Entomological Survey of the Himalaya. Part XIV. Notes on some insects collected by the Second Entomological Expedition to the Northwest Himalaya (1955), with descriptions of three new species of Odonata. *Agra Univ. J. Res.*, (Sci.) 4 (Suppl.) 741-766.

978. SINGH, SANTOKH, V. K. GUPTA, K. MATHEW & S. S. KRISHNA. 1955. Entomological Survey of the Himalaya. Part XII. Second annotated check-list of insects from the Northwest Himalaya. *Agra Univ. J. Res.*, (Sci.) 4 (Suppl.): 657-716.

979. SJÖSTEDT, Y. 1906. En Bestigning af Kilimandjaro högsta dalar. *Resebref Ent. Tiskr. Stockholm*, 27: 97-118.

980. SJÖSTEDT, Y. 1910. Wissenschaftliche Ergebnisse der schwedischen zoologischen Expedition nach dem Kilimandjaro, dem Meru und den umgebenden Massaisteppen Deutsch-Ostafrikas 1905-1906. Stockholm. vols. 3.

981. SJÖSTEDT, Y. 1912. Zur Orthopterenfauna des Kamerungebirges. *Ark. Zool.*, 7 (37): 1-30, pl. i-iii.

982. SKINNER, H. 1916. The genus *Parnassius* in America. *Ent. News*, 27: 210-216.

983. SKORIKOV, A. 1900. Zoologische Ergebnisse der russischen Expedition nach Spitzbergen im Jahre 1899. Collembola. *Ann. Mus. Zool. Acad. Sci. Imp. St. Petersburg*, 5: 190-209.

984. SKORIKOV, A. 1931. Die Hummelfauna Turkestans und ihre Beziehungen zur Zentralasiatischen Fauna (Hymenoptera: Bombidae). Abhandlungen der Pamir Expedition 1928. 8. Zoology. pp. 175-247.

985. SLOSSON, A. T. 1893-1906. Lists of insects taken in the alpine region of Mt. Washington. *Ent. News*, 4: 249-252, 287-292 (1893); 5: 271-274, 1-6 (1894); 6: 276-280,

4-7; 6: 316-321 (1895); 7: 262-265 (1896); 8: 237-240 (1897); 9: 251-253 (1898); 11: 319-323 (1900); 13: 4-8, 319-321 (1902); 17: 323-326 (1906).
986. SMITH, A. V. 1958. The resistance of animals to cooling and freezing. *Biol. Rev.*, 33 (2): 197-253 (pp. 201-203 deal with insects).
987. SMITH, F. 1878. Scientific results of the Second Yarkand Mission. Hymenoptera. pp. 1-22.
988. SMITH, R. 1954. The importance of micro-environment in insect ecology. *J. econ. Ent.*, 47: 205-210.
989. SPEYER, O. 1858-1862. Die geographische Verbreitung der Schmetterlinge Deutschlands und der Schweiz. I: 1-478 (1858); II: 1-320 (1862).
990. STACH, J. 1934. Die in den Höhlen Europas vorkommenden Arten der Gattung *Onychiurus* Gervais. *Ann. Mus. Zool. Polon.*, 10.
991. STACH, J. 1946. The species of the genus *Isotomurus* Börn. (Collembola) occurring in European caves. *Prace Mus. Przyrodniczego*, Krakau, 2.
992. STACH, J. 1946. Dziesiec Nowych Gatunków z rzedu Collembola (Skoczgonki) z Alpi i a ich Przedgorza-Ten new species of Collembola from the Alps and the Alpine foreland. *Polsk. Akad. Umiejetnosii Prace Muzeum Przyrodniczego Krakow*, pp. 1-40, pl. i-x.
993. STACH, J. 1959. The Apterygota fauna of the Polish Tatra National Park. *Acta zool. Cracov.*, 4: 1-102, pl. v, map 2 (cavernicole forms).
994. STANDFUSS, M. 1906. Die Resultate 30-jähriger Experimente mit Bezug auf Artbildung und Umgestaltung in der Tierwelt. *Verh. schweiz. naturf. Ges. Luzern.* (Aarau) 88: 263-286.
995. STAUB, R. 1924. Der Bau der Alpen. Bern: Beitr. geolog. Karte der Schweiz. 52.
996. STAUB, R. 1926. Gedanken zur Tektonik Spaniens. *Vierteljahresschr. naturf. Ges. Zürich*, 71.
997. STAUDER, H. 1920. Massenwanderung und Gletschertod von *Pieris rapae* Linn. *Z. wiss. Insektenbiol.*, 95: 263-264.
998. STAUDINGER, O. 1886. Centralasiatische Lepidopteren. *Stett. ent. Z.*, 47: 193-215.
999. STAUDINGER, O. 1892. Lepidopteren des Kentei-Gebirges. *Iris*, 5.
1000. STAUDINGER, O. & A. BANG-HAAS, 1882. Ueber einige neue *Parnassius*- und andere Tagfalter-Arten Central Asiens. *Berl. ent. Z.*, 26: 161-177, pl. i-ii.
1001. STAUDINGER, TH. 1887. Centralasiatische Lepidopteren. *Stett. ent. Z.*, 68: 49-102.
1002. STEGMANN, B. K. 1932. Die Herkunft der paläarktischen Taigafauna. *Arch˙ Naturg.*, 1 (1): 372-397.
1003. STEINBÖCK, O. 1931. Zur Lebensweise einiger Tiere des Ewigschneegebietes. *Z. Morphol. Ökol. Tiere*, 20: 707-718.
1004. STEINBÖCK, O. 1931. Die Tierwelt des Ewigschneegebietes. *Z. dtsch.-österr. Alpenverein*, 70: 29-48.
1005. STEINBÖCK, O. 1934. Die Tierwelt der Gletschergewässer. *Z. dtsch.-österr. Alpenverein*, 45: 263-275.
1006. STEINBÖCK, O. 1939. Der Gletscherfloh. *Z. dtsch.-österr. Alpenverein*, 70: 138-147, pl. xlii.
1007. STEINBÖCK, O. 1939. Die Nunatakfauna der Venter-Berge. *Festchr. Zweig Brandenburg. dtsch. Alpenverein*, München, pp. 64-74, pl. xiv.
1008. STEINHAUSER, F. 1937. Das Klima des Gasteiner Tales. *Beih. Jb. Zentralanstalt Meteorol. Geodyn. Wien*, 4: 25-60.
1009. STEINMANN, P. 1907. Die Tierwelt der Gebirgsbäche: Eine faunistisch biologische Studie. *Ann. Biol. Lac.*, 2: 30-162.
1010. STEMPFER, H. 1962. Notes de voyage en Grèce. *Alexanor*, 11 (7): 263-270; *Rev. Lep. franç.*
1011. STIRLING, J. 1887. Physiography of the Australian Alps. *Rep. Australian Ass. Adv. Sci.*, 1: 359.

1012. STOLICZKA, F. 1878-1891. Scientific results of the Second Yarkand Mission. Calcutta. vols. 2.

1013. STOLJAROV, M. V. 1960. Osobennosti geographicheskovo rasprostraneniya, ecologi i biologii Kuznechikov Abkhazii. *Entomologicheskoe Obozrenie*, 39 (4): 761-774, fig. 4. (The peculiarities of geographical distribution, ecology and biology of Tettigonioidea of Abkhasia).

1014. STOLZ, H. 1915. Über die Käferfauna des Monte Cavallo in den Venetianer Alpen. *Verh. zool.-bot. Ges. Wien*, 45 (4/5): 238-254, fig. 3.

1015. STONE, A. 1919. *Deuterophlebia coloradensis* Pennak in Oregon. *Proc. ent. Soc. Wash.*, 51: 122.

1016. STRENZKE, K. 1950. Systematik, Morphologie und Ökologie der terrestrischen Chironomiden. *Arch. Hydrobiol.*, 18 (Suppl.): 207-414, fig. 100.

1016a. STUARDO, O. C. 1953. Una familia de dipteros nueva para Chile (Diptera: Deuterophlebiidae). *Rev. Chilen. Ent.*, 3: 100.

1017. SUCHLANDT, O. & W. SCHMASSMANN, 1935. Limnologische Beobachtungen an acht Hochgebirgsseen der Landschaft Davos. *Z. Hydrobiol.*, 7: 1-201.

1018. SUSHKIN, P. P. 1925. Outline of the history of the Recent fauna of Palaearctic Asia. *Proc. nat. Acad. Sci. Philadelphia*, 2 (6): 299-302.

1019. SWAINE, J. M., H. C. FALL, C. W. LENG & J. D. SHERMAN JR. 1919. Coleoptera. *Report of the Canadian Arctic Expedition* 1913-1918. Ottawa, 3 (E): 1-27 (1922).

1020. SWAN, L. W. 1952. Some environmental conditions influencing life at high altitudes. *Ecology*, 33 (1): 109-111.

1021. SWAN, L. W. 1961. The ecology of the High Himalayas. *Sci. Amer.*, 205 (4): 68-78.

1022. SWAN, L. W. 1963. Aeolian zone. *Science*, 140 (3562): 77-78, April 5, 1963.

1023. SWAN, L. W. 1963. Ecology of the heights. *Natural History*, pp. 23-29.

1024. SWYNNERTON, G. H. 1949. Report of a safari to explore the Shira Plateau, West Kilimanjaro. *Rep. Game Dept. Tanganyika*, pp. 9-11.

1025. SZILADY, Z. 1918. Über die vertikale Verbreitung der Arthropoden. Mit Beispielen aus der Fauna des Rety-zát. *Z. wiss. Insektenbiol.*, 14: 67-72, 108-117, 172-177, 266-270.

1026. TABUCHI, Y. 1959. The alpine butterflies of Japan, their life-history in Honschu (In Japanese, with English summary). Tokyo: Hobundo Publ. Co. pp. 2-391.

1027. TAMS, W. H. T. 1952. Three new high mountain moths from East Africa. *Ann. Mag. nat. Hist.*, (12) 5: 869-874.

1028. TARBINSKY, S. P. 1926. Some Orthoptera from Pamir and adjacent countries. *Ann. Mag. nat. Hist.*, (9) 17: 83-96.

1029. TARR, R. S. & L. MARTIN, 1913. Alaskan glacier studies. *Nat. geogr. Soc. Wash.*, pp. 448.

1030. TAYLOR, J. W. 1913. Geographical distribution and dominance in relation to evolution and phylogeny. *Trans. II int. Congr. Ent.*, 2: 271-284.

1031. TAYLOR, W. P. 1922. A distributional and ecological study of Mount Rainier, Washington. *Ecology*, 3: 214-236, fig. 4.

1032. TAYLOR, W. P. & W. T. SHAW, 1927. Mammals and birds of Mount Rainier National Park. *U.S. Dept. Inf.*, pp. 1-249, fig. 109.

1033. TCHICATCHEFF, P. A. 1845. Voyage scientifique dans l'Altai Orientale. Paris.

1034. THIELE, H. V. & H. KIRCHNER, 1958. Über die Körpergrösse der Gebirgs- und Flachlandpopulationen einiger Laufkäfer (Carabidae). *Bonn. zool. Beitr.*, 9 (2/4): 294-302.

1035. THIENEMANN, A. 1936. Alpine Chironomiden. *Arch. Hydrobiol.*, 30.

1036. THIENEMANN, A. 1950. Die Verbreitungsgeschichte der Süsswassertierwelt Europas. Stuttgart. pp. 1-809.

1037. THOMPSON, D'ARCY W. 1942. On Growth and Form. Cambridge University Press. pp. 1-1116.
1038. TJEDER, B. 1963. Three subapterous craneflies from Alaska (Dipt. Tipulidae). *Opusc. ent.*, 28: 229-240.
1039. TONNOIR, A. L. 1931. Notes on Indian Blepharocerid larvae and pupae, with remarks on the morphology of Blepharocerid larvae and pupae in general. *Rec. Indian Mus.*, 32 (2): 161-214, (1930).
1040. TORNQUIST, A. 1913. Grundzüge der geologischen Formations- und Gebirgskunde. Berlin: Gebrüder Bornträger Verlag. pp. 296, fig. 127.
1041. TOTTENHAM, C. E. 1953. Staphylinidae from Kilimanjaro, Mt. Kenya and Ruwenzori. *Ann. Mag. nat. Hist.*, (12) 6: 481-512.
1042. TOWER, W. L. 1903. The development of the colours and colour patterns of Coleoptera, with observations upon the development of colour in other orders of insects. *Decennial Publ. Univ. Chicago*, (1) 10: 31-70, pl. iii.
1043. TOXOPEUS, L. J. 1940. Nederlandsch-Indisch Amerikaansche Expeditie naar Nederlandsch Nieuw-Guinea. *Treubia*, 17: 271-279.
1044. TOXOPEUS, L. J. 1950. Geological principles of species evolution in New Guinea. *Proc. VIII int. Congr. Ent.*, pp. 508-522.
1045. TRAVER, J. R. 1939. Himalayan mayflies (Ephemeroptera). *Ann. Mag. nat. Hist.*, (11) 4: 49-65.
1046. TRENTINAGLIA-FELVENBERG, 1876. Ueber die Grenzen der tierischen Organismen in der nivalen und glazialen Region. *Ber. naturw. med. Vereins, Innsbruck*, 7.
1047. TROLL, C. 1959. Die tropischen Gebirge. Ihre dreidimensionale klimatische und pflanzengeographische Zonierung. *Bonn. geogr. Abhandl.*, 25: 1-95.
1048. TROLL, C. 1960. The relation between the climates, ecology and plant geography of the southern cold temperate zone and of the tropical mountains. *Proc. R. Soc. London*, (B) 152: 529-532.
1049. TROLL, C. 1961. Klima und Pflanzenkleid der Erde in dreidimensionaler Sicht. *Die Naturwissenschaften*, 48 (9): 332-348, fig. 21.
1050. TSALKIN, V. I. 1951. Gornii baranii Evropi i Azii. Moscow: Ispiit. Prirodi. Meterl. k pozn. faunii i florii SSSR (Zool.) 27: 1-343.
1051. TSCHETVERRIKOV, S. S. 1911. Cheshekriliya poduostrova Yamala, dobitiya Expediticie B. M. Zhitkova v 1908 godu. Lépidoptères recueilles par l'expédition de Mr B. M. Zhitkov dans la presqu'ile de Yamal en 1908. *Ann. Mus. Zool. Acad. Imp. Sic. St. Petersbourg*, 16: 29-36.
1052. TUTTON, A. E. H. 1927. The Natural History of Ice and Snow. London: Kegan Paul. pp. 319.
1053. TYRRELL, G. W. 1931. Volcanoes. London: Thornton Butterworth Ltd. pp. 252.

1054. UÉNO, M. 1934. Yale North-India Expedition. Report on Amphipod Crustacea of the genus Gammarus. *Mem. Connecticut Acad. Arts & Sci.*, 10 (6): 63-75.
1055. UÉNO, M. 1955. Mayfly nymphs. In: Fauna and Flora of the Nepal Himalaya. Kyoto Univ. pp. 301-316, pl. x.
1056. UMBGROVE, J. H. F. 1930. The amount of maximal lowering of the sea-level in the Pleistocene. *Proc. IV Pacific Sci. Congr.*, (1929) 2: 105-113.
1057. UMNOV, N. N. 1931. Eine Übersicht der Fauna der Heuschrecken (Acridoidea) des Alai-Tales. *Ent. Nachrichtenbl.*, 5: 12-17.
1058. UVAROV, B. P. 1914. Orthoptera recueillis par M. A. N. Avinov dans les plateau de Karakoram. *Rev. russe Ent.*, 14: 231-234.
1059. UVAROV, B. P. 1921. The geographical distribution of Orthopterous insects in the Caucasus and in western Asia. *Proc. zool. Soc. London*, pp. 447-472.
1060. UVAROV, B. P. 1921. Three new alpine Orthoptera from Central Asia. *J. Bombay nat. Hist. Soc.*, 28: 71-75.

1061. UVAROV, B. P. 1921. *Conophyma mitchelli*, sp. n. a new alpine grasshopper from Kashmir. *Ent. mon. Mag.*, 7 (3): 268-270.
1062. UVAROV, B. P. 1922. An interesting new grasshopper from Mt. Everest. *Ann. Mag. nat. Hist.*, (9) 9: 551-553.
1063. UVAROV, B. P. 1925. Some new alpine grasshoppers of the genus *Conophyma* Zub. from Central Asia. *J. Bombay nat. Hist. Soc.*, 30: 551-560.
1064. UVAROV, B. P. 1925. Orthoptera: Acrididae. Guy Babault Mission dans les Provinces Centrales de l'Inde et dans la région occidentale de l'Himalaya 1914. pp. 1-40, pl. i.
1065. UVAROV, B. P. 1925. Grasshoppeis (Acridiidae: Orthoptera) from Mount Everest. *Ann. Mag. nat. Hist.*, (9) 16: 165-173.
1066. UVAROV, B. P. 1927. Die Acridiiden Zentralasiens. Taschkent. pp. 215. (In Russian).
1067. UVAROV, B. P. 1927. A contribution to the zoogeography of the Himalaya. *Ann. Mag. nat. Hist.*, (10) 9: 481-484.
1068. UVAROV, B. P. 1928. Orthoptera of the mountains of Palaearctic Region. *Mem. Soc. Biogéogr.*, 2: 135-141.
1069. UVAROV, B. P. 1929. Composition and origin of the Palaearctic fauna of Orthoptera. *X int. Congr. Zool. Budapest*, 10 (8): 1516-1524.
1070. UVAROV, B. P. 1934. Entomological Expedition to Abyssinia 1926-1927. Orthoptera. *J. Linn. Soc. London* (Zool.), 146 (1933-34): 52-54.
1071. UVAROV, B. P. 1936. Some Orthoptera from Kashmir. *Opusc. ent. Soc. Lund*, 1 (1): 17-20.
1072. UVAROV, B. P. 1938. Ecological and biogeographical relations of Eremian Acrididae. *Mem. Soc. Biogéogr.* 6: 231-273.
1073. UVAROV, B. P. 1939. Some Acrididae from south-eastern Tibet. *J. Linn. Soc. London*, (Zool.) 40: 561.
1074. UVAROV, B. P. 1942. Palaearctic Acrididae new to Indian fauna (Orthoptera). *Eos*, 18: 97-103.
1075. UVAROV, B. P. 1953. Some effects of past climtaic changes on the distribution of African Acrididae. *Trans. IX int. Congr. Ent.*, 2: 157-159.

1076. VALLE, K. 1977. Zur Kenntnis der Odonatenfauna Finnlands III. Ergänzungen und Zusätze. *Acta Soc. Fauna Flora fenn.* 56 (11): 1-36, pl. i, map 1.
1077. VANDEL, A. 1964. Biospéologie: La Biologie des animaux cavernicoles. Paris: Gauthier-Villars. part 6.
1078. VAN DYKE, E. C. 1919. A few observations on the tendency of insects to collect on ridges and mountain snow fields. *Ent. News*, 30: 241-244.
1079. VAN DYKE, E. C. 1929. The influence which geographical distribution has had in the production of the insect fauna of North America. *Trans. IV int. Congr. Ent.*, 1928, 2: 555-566.
1080. VAZIRANI, T. G. 1964. On a new species of aquatic beetle of the genus *Amphizoa* Leconte 1853 (Insecta: Coleoptera: Amphizoidae) from Kashmir, India. *Proc. zool. Soc. Calcutta*, 17: 145-147, fig. 1.
1081. VERHOEFF, K. W. 1929. Studien über Ökologie und Geographie der Diplopoden, hauptsächlich der Ostalpen. *Z. Morphol. Ökol. Tiere*, 15: 35-89.
1082. VERHOEFF, K. W. 1936. Über einen Land-Isopoden aus Ladak. *Mem. Connecticut Acad. Arts & Sci.*, 10 (10): 187-190.
1083. VERSON, E. 1871. Ueber den Einfluss niedriger Temperatur auf die Lebensfähigkeit der Eier des gemeinen Seidenspinners. *Österr. Seidenbau Ztg.*, 3: 57-59.
1084. VOGT, O. 1909. Studien über das Artproblem: Über das Variieren der Hummeln. *Sitzb. Ges. Naturf. Freunde Berlin*, 28.
1085. VOS, A. P. C. 1935. Wasserlebende Insektenlarven. Wissenschaftliche Ergebnisse der Niederländischen Expedition in der Karakoram. pp. 237-240.

1085a. WACE, N. M. 1965. Vascular Plants. In: P. van Oye & J. van Mieghem, Biogeography and Ecology in Antarctica. *Monogr. Biol.* XV, The Hague, Dr. W. Junk N.V.

1086. WADIA, D. N. 1931. Syntaxis of the Northwest Himalaya, its rocks, tectonics and orogeny. *Rec. geol. Surv. India,* 65: 189-220.

1087. WADIA, D. N. 1940. Pleistocene Ice Age deposits of Kashmir. *Proc. nat. Inst. Sci., India,* 7: 49-59.

1088. WADIA, D. N. 1944. Geology of India. London: MacMillan & Co.

1089. WADIA, D. N. 1952. The place of Himalaya in the geography of India. *Himalaya,* 1 (1): 13-15.

1090. WAGNER, E. 1957. Zoologisch-Systematische Ergebnisse der Studienreise von H. Janetschek und W. Steiner in die spanische Sierra Nevada 1954. II. Einige Neue Heteroptera. *Sitzb. österr. Akad. Wiss. Wien,* (math.-naturw. Klasse) (1) 166 (5/6): 219-226.

1091. WAGNER, H. O. 1961. Die Nagetiere einer Gebirgsabdachung in Südmexico und ihre Beziehungen zur Umwelt. *Zool. Jb.,* (Syst.) 89: 177-242.

1092. WALKER, E. M. 1913. The faunal zones of Canada. *Ann. Rep. ent. Soc. Canada,* 43 (1912): 27-33.

1093. WALKER, E. M. 1914. A new species of Orthoptera forming a new family (Grylloblattidae). *Canad. Ent.,* 46: 93.

1094. WALKER, E. M. 1920. Orthoptera. *Report of the Canadian Arctic Expedition* 1913-1918. Ottawa, 3 (J): 1-4 (1922).

1095. WALKER, H. & E. H. PASCOE, 1907. Notes on certain glaciers in Lahaul. *Rec. geol. Survey India,* 25: 139-157, pl. xl-xlvi, lx, lxii.

1096. WALSHINGHAM, Lord, 1885. On some probable causes of a tendency to melanic variation in Lepidoptera of high altitudes. *Trans. Yorks. ent. Union.* pp. 113-140.

1097. WALTER, C. 1919. Hydracarinen aus den peruvianischen Andes und aus Brasilien. *Rev. Suisse Zool.,* 27 (2): 19-59, fig. 45.

1098. WALTER, C. Die Hydracarinen der Alpengewässer. *Denkschr. schweiz. naturf. Ges.,* 58: 54-251.

1099. WARNECKE, G. 1920. Eine lepidopterologische Sammelreise ins Grossglocknergebiet. *Mitt. ent. Ges.,* 10: 43-62.

1100. WARNECKE, G. 1929. Ist *Parnassius apollo* ein Glazialrelikt? *X int. Congr. Zool. Budapest,* 2: 1464-1469.

1101. WARNECKE, G. 1943. Über die Lepidopterologische Literatur des Kaukasus. *Z. wien. ent. Ges.,* 28: 169-176.

1102. WARNECKE, G. 1958. Ursprung und Geschichte der Insektenfauna der nördlichen Paläarktis. *X int. Congr. Ent.,* 1: 719-730.

1103. WARREN, B. C. S. 1936. Monograph of the genus *Erebia.* London: British Museum. pp. 1-407.

1104. WARREN, B. C. S. 1944. A review of the classification of the Argynnidi with a systematic revision of the genus *Boloria* (Lepidoptera: Nymphalidae). *Trans. ent. Soc. London,* 94: 1-101, pl. xlvi.

1105. WATERHOUSE, C. O. 1885. On the insects collected on Kilimanjaro by Mr. H. H. Johnston. *Proc. zool. Soc. London,* pp. 230-235.

1106. WATERHOUSE, C. O. 1895. Insects collected on the summit of Mount Roraima. *Ann. Mag. nat. Hist.,* (6) 15: 494-497.

1107. WATERHOUSE, C. O. 1900. Report on a collection made by Messrs F. V. McConnell and J. J. Quelch at Mt. Roraima in British Guiana. *Trans. Linn. Soc. London,* 8: 74-76.

1108. WELLINGTON, W. G. 1945. Conditions governing the distribution of insects in the free atmosphere. Part I. Atmospheric pressure, temperature and humidity.

Canad. Ent., 77: 7-15; Part II. Surface and upper winds. *ibid.*, 21-28; Part III. Thermal convection. *ibid.*, 44-49; Part IV. Distribution processes of economic importance. *ibid.*, 69-74.

1109. WHALGREN, E. 1900. Collembola, während der schwedischen Grönlands Expedition 1899 auf Jan Mayen und Öst-Grönland eingesammelt. *Öfv. k. Vət. Akad. Forh.*, 57 (3): 353-375.

1110. WHITEHEAD, J. 1893. Exploration of Mount Kina Balu, North Borneo. London. pp. 317.

1111. WHYMPER, E. 1891. Supplementary Appendix to Travels amongst the Great Andes of the Equator. London. p. 147.

1112. WILLIS, R. 1893. The mechanics of Appalachian structure. *U.S. geol. Surv. Ann. Rep.*, 13 (2): 211-281.

1113. WILLMANN, C. 1951. Die hochalpine Milbenfauna der Mittleren Hohen Tauern, insbesondere des Grossglocknergebietes (Acari). *Bonn. zool. Beitr.*, 2 (1/2): 141-174, fig. 1-18.

1114. WILSON, A. T. 1959. Organic nitrogen in New Zealand snows. *Nature, Lond.*, 183: 318-319.

1115. WILSON, A. T. 1959. Surface of the ocean as a source of air-borne nitrogenous material and other plant nutrients. *Nature, Lond.*, 184: 90-101.

1116. WIRTH, W. W. 1951. A new mountain midge from California (Diptera: Deuterophlebiidae). *Pan-Pacific Ent.*, 27: 49-57.

1117. WISLEY, B. 1953. Two wingless stoneflies from Southern New Zealand. *Rec. Cant. Mus.*, New Zealand, 6: 219-231.

1118. WISSMANN, H. VON, 1959. Die heutige Vergletscherung und Schneegrenze in Hochasien mit Hinweisen auf die Vergletcherung der letzten Eiszeit. *Akad. Wiss. Lit. Mainz, Abh. math.-naturw. Klasse.*

1119. WNUKOWSKY, W. 1927. Neue Lepidopterenformen aus Nordost Sibirien und dem mongolischen Altai. *Mitt. münch. ent. Ges.*, 17: 69-72.

1120. WOEIKOFF, A. 1890. Bodentemperatur unter Schnee und ohne Schnee in Katharinenburg am Ural. *Meteorol. Z.*, (Oktober) 7: 381-385.

1121. WÖRNDLE, A. 1950. Die Käfer von Nordtirol. *Schlern-Schriften*, 64: 1-388.

1122. WOJTUSIAK, R. J. & W. NIESIOLOWSKI, 1946-1947. Lepidoptera of the Central Caucasus collected during the Polish Alpine Expedition in 1935 with ecological and zoogeographical remarks. Part I. Macrolepidoptera. *Prace Mus. Przyr.*, 6: 1-74, fig. 3, pl. iii.

1123. WOLF, B. 1934-1938. Animalium cavernarum Catalogus. Dr. W. Junk: Den Haag. Parts 1-14. pp. xxiii+1642.

1124. WOLFF, F. VON, 1913-1914. Der Vulkanismus. 1. Allgemeiner Teil. pp. 711; 2. Spezieller Teil (1923-32). Stuttgart: Ferdinand Enke.

1125. WORKMAN, F. B. & W. H. WORKMAN, 1917. Two summers in the Ice-Wilds of Eastern Karakoram. London.

1126. WRIGHT, W. B. 1937. The Quaternary Ice Age. London: MacMillan & Co. 2nd ed. pp. 478.

1127. XAMBEU, P. 1903. Fauna entomologique des Pyrénées orientales. Beilage zur Zeitschrift l'Exchange. pp. 214.

1128. YABLAKOV-KHNZORYAN, S. M. 1961. Opit vosstanovlieniya genezisa faunii zhestkokriilikh Aremnii. *Akad. Nauk Armyanskoi SSR. Zool. Inst.*, Erivan, pp. 264.

1129. YIE, S. 1933. Observations on a Japanese *Deuterophlebia* (Diptera) (In Japanese with English summary). *Trans. nat. Hist. Soc. Formosa*, 23: 271-296.

1130. YOSHII, R. 1956. Monographie zu den Höhlencollembolen Japans. *Contr. Biol. Lab. Kyoto Univ.*, 3: 1-109, fig. 22, pl. i-l.
1131. YOSHII, R. 1963. On some Collembola of Hindukush, with notes on *Isotoma* Bourlet and its allies. Results of the Kyoto University Sci. Expedition to Karakoram and Hindukush 1955. 4. Insect fauna of Afghanistan and Hindukush. pp. 1-42, fig. 26.
1132. ZELLER, R. 1883. Ueber die Lebensweise einiger Schmetterlinge (Spanner oder Geometriden) der höchsten Alpen. *Kosmos*, 7 (13): 543-546.
1133. ZEUNER, F. 1945. The Pleistocene Period. Ray Society. pp. 130.
1134. ZIMINA, R. P. 1964. Zakonomernosti vertikalnovo rasprostraneniya Mlekopitayushikh. Izdatelstvo Nauka. Moscow. pp. 1-156.
1135. ZNOJKO, D. V. 1930. Carabidae, genus *Dyschirius* Bon. Abhandlungen der Pamir Expedition 1928. II. Zool. pp. 51-54.
1136. Zoological Results of the Ruwenzori Expedition 1905-1906. Insects. *Trans. zool. Soc. London*, 19: 59-232 (1909/1910).
1137. ZSCHOKKE, F. 1894. Fauna hochgelegener Gebirgsseen: Ein Beitrag zur Kenntnis der vertikalen Verbreitung der niederen Tiere. *Verh. naturf. Ges. Basel*, 11.
1138. ZSCHOKKE, F. 1900. Die Tierwelt der Hochgebirgsseen. *Neue Denkschr. allg. schweiz. Ges. gesamt Naturw.*, 37: 1-400, pl. viii.
1139. ZSCHOKKE, F. 1908. Die Beziehungen der mitteleuropäischen Tierwelt zur Eiszeit. *Verh. dtsch. zool. Ges.*, 18. Jahresversammlung Stuttgart, pp. 21-77, pl. i-ii.
1140. ZSCHOKKE, F. 1912. Die tierbiologische Bedeutung der Eiszeit. *Fortschr. Naturwiss. Forsch.*, 4.
1141. ZÜLLICH, R. 1936-1937. Beitrag zur Makrolepidopterenfauna des Rilogebirges in Bulgarien. *Z. österr. ent. Ver.*, 21: 77-79 (1936); 22: 16-19 (1937).

INDEX

MIX
Papier aus verantwortungsvollen Quellen
Paper from responsible sources
FSC® C105338

If you have any concerns about our products,
you can contact us on
ProductSafety@springernature.com

In case Publisher is established outside the EU,
the EU authorized representative is:
Springer Nature Customer Service Center GmbH
Europaplatz 3, 69115 Heidelberg, Germany

Printed by Libri Plureos GmbH
in Hamburg, Germany